Enzyme- and Transporter-Based Drug–Drug Interactions

K. Sandy Pang · A. David Rodrigues · Raimund M. Peter
Editors

Enzyme- and Transporter-Based Drug–Drug Interactions

Progress and Future Challenges

Springer

Editors
K. Sandy Pang
Leslie Dan Faculty of Pharmacy
University of Toronto
144 College Street
Toronto ON M5S 3M2
Canada
ks.pang@utoronto.ca

A. David Rodrigues
Bristol Myers Squibb Co.
P.O.Box 4000
Princeton
NJ 08543
USA
david.rodrigues@bms.com

Raimund M. Peter
AstraZeneca UK
Alderley Park
Macclesfield, Cheshire
United Kingdom SK10 4TF
raimund.peter@astrazeneca.com

ISBN 978-1-4419-0839-1 e-ISBN 978-1-4419-0840-7
DOI 10.1007/978-1-4419-0840-7
Springer New York Dordrecht Heidelberg London

Library of Congress Control Number: 2009938736

Printed on acid-free paper

Springer is part of Springer Science+Business Media (www.springer.com)

Preface

Germination of the thought of "Enzyme- and Transporter-Based Drug–Drug Interactions: Progress and Future Challenges" proceedings came about as part of the annual meeting of The American Association of Pharmaceutical Scientists (AAPS) that was held in San Diego in November 2007. The attendance of the workshop by more than 250 pharmaceutical scientists reflected the increased interest in the area of drug–drug interactions (DDIs), the greater focus of pharmaceutical industry, academia, and regulatory agencies, and the rapid pace of growth in knowledge. The aims of the workshop were to address the progress made in quantitatively predicting enzyme- and transporter-based DDIs as well as highlight areas where such predictions are poor or areas that remain challenging for the future. Because of the serious clinical implications, initiatives have arisen from the FDA (http://www.fda.gov/cber/gdlns/interactstud.htm) to highlight the importance of enzyme- and transporter-based DDIs.

During the past 10–15 years, we have come to realize that transporters, in addition to enzymes, play a vital role in drug elimination. Such insight has been possible because of the continued growth in PK-ADME (pharmacokinetics-absorption-distribution-metabolism-excretion) knowledge, fueled by further advances in molecular biology, greater availability of human tissues, and the development of additional and sophisticated model systems as well as sensitive assay methods for the study of drug metabolism and transport in vitro and in vivo. This has sparked an in-depth probing into mechanisms surrounding DDIs, resulting from ligand-induced changes in nuclear receptors, as well as alterations in transporter and enzyme expression and function. Despite such advances, the in vitro and in vivo study of drug interactions and the integration of various data sets remain challenging. Therefore, it has become apparent that a proceeding that serves to encapsulate current strategies, approaches, methods, and applications is necessary.

As editors, we have assembled a number of opinion leaders and asked them to contribute chapters surrounding these issues. Many of them are the original workshop speakers whereas others had been selected specially to contribute on topics related to basic and applied information that had not been covered in other reference texts on DDI. The resulting volume, entitled *Enzyme- and Transporter-Based Drug–Drug Interactions: Progress and Future Challenges,* comprises of four sections. Twenty-eight chapters dedicated to various topics and perspectives related to the subject of metabolic and transporter-based drug–drug interactions are presented. Section I covers scientific issues and concepts that dwell on the fundamental understanding of transporters and enzymes, their function and regulation by nuclear receptors, and how these work in unison or in competition, in first-pass absorption, transport, and metabolism. Since the first mandate is an understanding of what kinds of transporters, enzymes and eliminating organs are involved in the handling of the drug in terms of deciphering the mechanisms involved in DDI, various organs including the kidney are discussed. Kinetic concepts describing clearance mechanisms and areas under the curve of not only drug but also metabolite have also been introduced. Section II pertains to methodology for the study of DDI. Due to the cost requirement in mounting in vitro vs. in vivo studies, DDI studies are often explored in vitro and the tools, the extrapolation of data in vitro to in vivo from animal to man, together with information retrieval from web data basis for transporters (www.Tp-research.com) and enzymes (www.druginteractioninfo.org) as well as modeling and simulations have been addressed. Section III covers the various topics that impact DDIs and spans competitive to allosteric- and mechanism-based inhibition, inductive, time-dependent alteration in drug elimination rates, inhibition of Phase II pathways, changes in volume and first-pass metabolism, and the final integration of data. Lastly, Section IV describes regulatory aspects and future developments, stressing the use of clearance concepts, PBPK models, and modeling and simulations as well as future challenges that would be faced. It is our hope that the proceedings bring about an improved appreciation of the impact of DDI and a deeper understanding of "where we had been and where we are going."

Toronto, ON — K. Sandy Pang, Ph.D.
Princeton, NJ — A. David Rodrigues, Ph.D.
Macclesfield, UK — Raimund M. Peter, Ph.D.

Contents

Contributors

Guohua An Department of Pharmaceutical Sciences, University at Buffalo, State University of New York, Amherst, NY, USA

Daniel J. Antoine Department of Pharmacology and Therapeutics, University of Liverpool, Liverpool, UK

Praveen Balimane Metabolism and Pharmacokinetics, Department of Pharmaceutical Candidate Optimization, Bristol-Myers Squibb Co., Princeton, NJ, USA

Leslie Z. Benet Department of Biopharmaceutical Sciences, University of California San Francisco, San Francisco, CA, USA

Kelly Bleasby Department of Metabolism and Pharmacokinetics, Merck & Co., In Vitro Technologies, Rahway, NJ, USA

Kim L. R. Brouwer Division of Pharmacotherapy and Experimental Therapeutics, Eshelman School of Pharmacy, The University of North Carolina at Chapel Hill, Chapel Hill, NC, USA

Benjamin J. Burke Department of Worldwide Medicinal Chemistry, Pfizer Inc., Global Research & Development, La Jolla, CA., USA

Edwin C.Y. Chow Leslie Dan Faculty of Pharmacy, University of Toronto, Toronto, ON, Canada

Xiaoyan Chu Department of Metabolism and Pharmacokinetics, Merck & Co., In Vitro Technologies, Rahway, NJ, USA

Christopher J. Endres Department of Pharmaceutics, University of Washington, Seattle, WA, USA

Raymond Evers Department of Drug Metabolism and Pharmacokinetics, Merck & Co., In Vitro Technologies, Rahway, NJ, USA

Odette A. Fahmi Department of Pharmacokinetics, Pharmacodynamics and Drug Metabolism, Pfizer Inc., Global Research & Development, Groton, CT, USA

Aleksandra Galetin School of Pharmacy and Pharmaceutical Sciences, University of Manchester, Manchester, UK

David J. Greenblatt Department of Pharmacology & Experimental Therapeutics, Tufts University School of Medicine, Boston, MA, USA

Anita Grover Department of Biopharmaceutical Sciences, University of California, San Francisco, CA, USA

Houda Hachad Department of Pharmaceutics, University of Washington, Seattle, WA, USA

Yong-hae Han Metabolism and Pharmacokinetics, Department of Pharmaceutical Candidate Optimization, Bristol-Myers Squibb Co., Princeton, NJ, USA

Toshiharu Horie Department of Biopharmaceutics, Graduate School of Pharmaceutical Sciences, Chiba University, Chiba, Japan

J. Brian Houston School of Pharmacy and Pharmaceutical Sciences, University of Manchester, Manchester, UK

Shiew-Mei Huang Office of Clinical Pharmacology, Center for Drug Evaluation and Research, Food and Drug Administration, Silver Spring, MD, USA

Masaki Ishigai Chugai Pharmaceutical Co. Ltd., Gotemba, Shizuoka, Japan

Fred F. Kadlubar Department of Epidemiology, College of Public Health, University of Arkansas for Medical Sciences, Little Rock, AR, USA

Susan Kadlubar Department of Environmental and Occupational Health, College of Public Health, University of Arkansas for Medical Sciences, Little Rock, AR, USA

Motohiro Kato Chugai Pharmaceutical Co. Ltd., Gotemba, Shizuoka, Japan

Reinhold Kerb Dr. Margarete Fischer Bosch Institute of Clinical Pharmacology, University of Tübingen, Stuttgart, Germany

Sean Kim Metabolism and Pharmacokinetics, Department of Pharmaceutical Candidate Optimization, Bristol-Myers Squibb Co., Wallingford, CT, USA

Brian J. Kirby Department of Pharmaceutics, University of Washington, Seattle, WA, USA

Masato Kitajima Fujitsu Kyushu Systems Ltd., Sawara-ku, Fukuoka, Japan

Kathleen M. Knights Clinical Pharmacology, School of Medicine, Flinders University, Adelaide, Australia

René H. Levy Department of Pharmaceutics, University of Washington, Seattle, WA, USA

Peter I. Mackenzie Clinical Pharmacology, School of Medicine, Flinders University, Adelaide, Australia

Soraya Madani Novartis Pharmaceuticals Corp, East Hanover, NJ, USA

Kazuya Maeda Department of Molecular Pharmacokinetics, Graduate School of Pharmaceutical Sciences, The University of Tokyo, Bunkyo-ku, Tokyo, Japan

Punit Marathe Metabolism and Pharmacokinetics, Department of Pharmaceutical Candidate Optimization, Bristol-Myers Squibb Co., Princeton, NJ, USA

John O. Miners Clinical Pharmacology, School of Medicine, Flinders University, Adelaide, Australia

Xin Ming Division of Molecular Pharmaceutics, Eshelman School of Pharmacy, The University of North Carolina at Chapel Hill, Chapel Hill, NC, USA

Marilyn E. Morris Department of Pharmaceutical Sciences, School of Pharmacy and Pharmaceutical Sciences, University at Buffalo, State University of New York, Amherst, NY, USA

R. Scott Obach Department of Pharmacokinetics, Pharmacodynamics and Drug Metabolism, Pfizer Inc., Global Research & Development, Groton, CT, USA

K. Sandy Pang Leslie Dan Faculty of Pharmacy, Department of Pharmaceutical Sciences, University of Toronto, Toronto, ON, Canada

B. Kevin Park Department of Pharmacology and Therapeutics, University of Liverpool, Liverpool, UK

Raimund M. Peter Discovery Drug Metabolism and Pharmacokinetics, CVGI Department, AstraZeneca UK, Macclesfield, Cheshire, UK

Thomas M. Polasek Clinical Pharmacology, School of Medicine, Flinders University, Adelaide, Australia

William R. Proctor Division of Molecular Pharmaceutics, Eshelman School of Pharmacy, The University of North Carolina at Chapel Hill, Chapel Hill, NC, USA

Thomayant Prueksaritanont Department of Drug Metabolism and Pharmacokinetics, Merck Research Laboratories, West Point, PA, USA

Isabelle Ragueneau-Majlessi Department of Pharmaceutics, University of Washington, Seattle, WA, USA

Ahsan Naqi Rizwan Division of Pharmacotherapy and Experimental Therapeutics, Eshelman School of Pharmacy, The University of North Carolina at Chapel Hill, Chapel Hill, NC, USA

A. David Rodrigues Metabolism and Pharmacokinetics, Department of Pharmaceutical Candidate Optimization, Bristol Myers Squibb Co., Princeton, NJ, USA

Amin Rostami-Hodjegan Department of Human Metabolism (M129), The Medical School, University of Sheffield, Sheffield, South Yorkshire, UK

Frans G.M. Russel Department of Pharmacology and Toxicology, Radboud University Nijmegen Medical Centre, Nijmegen Centre for Molecular Life Sciences, Nijmegen, Netherlands

Matthias Schwab Dr Margarete Fischer Bosch Institute of Clinical Pharmacology, University of Tübingen Stuttgart, Germany

Danny D. Shen Department of Pharmacy, University of Washington, Seattle, WA, USA

Yoshihisa Shitara Department of Biopharmaceutics, Graduate School of Pharmaceutical Sciences, Chiba University, Chiba, Japan

Abhishek Srivastava Department of Pharmacology and Therapeutics, University of Liverpool, Liverpool, UK

Chad L. Stoner Department of Pharmacokinetics and Drug Metabolism, Pfizer Inc., Global Research & Development, La Jolla, CA, USA

Murali Subramanian Department of Experimental and Clinical Pharmacology, University of Minnesota, Minneapolis, MN, USA

Yuichi Sugiyama Department of Molecular Pharmacokinetics, Graduate School of Pharmaceutical Sciences, The University of Tokyo, Bunkyo-ku, Tokyo, Japan

Huadong Sun NoAb BioDiscoveries Inc., Mississauga, ON, Canada

Tatsuhiko Tachibana Chugai Pharmaceutical Co. Ltd., Gotemba, Shizuoka, Japan

Dhiren R. Thakker Division of Molecular Pharmaceutics, Eshelman School of Pharmacy, The University of North Carolina at Chapel Hill, Chapel Hill, NC, USA

Kenneth E. Thummel Department of Pharmaceutics, University of Washington, Seattle, WA, USA

Rommel G. Tirona Division of Clinical Pharmacology, Department of Medicine, Department of Physiology and Pharmacology, Schulich School of Medicine and Dentistry, University of Western Ontario, London Health Sciences Centre – University Hospital, London, ON, Canada

Timothy S. Tracy Department of Experimental and Clinical Pharmacology, College of Pharmacy, University of Minnesota, Minneapolis, MN, USA

Jashvant D. Unadkat Department of Pharmaceutics, School of Pharmacy, University of Washington, Seattle, WA, USA

Ragini Vuppugalla Metabolism and Pharmacokinetics, Department of Pharmaceutical Candidate Optimization, Bristol-Myers Squibb Co., Princeton, NJ, USA

Lisa L. von Moltke Millenium Pharmaceutical Institute, Cambridge, MA, USA

Rachel J. Walsh Department of Pharmacology and Therapeutics, University of Liverpool, Liverpool, UK

Robert L. Walsky Department of Pharmacokinetics, Pharmacodynamics and Drug Metabolism, Pfizer Inc., Global Research & Development, Groton, CT, USA

Michael R. Wester Department of Pharmacokinetics and Drug Metabolism, Pfizer Inc., Global Research & Development, La Jolla, CA, USA

Dominic P. Williams Department of Pharmacology and Therapeutics, University of Liverpool, Liverpool, UK

Helen Winter Department of Clinical Pharmacology, Global Alliance for TB Drug Development, New York, NY, USA

Catherine K. Yeung Department of Pharmaceutics, University of Washington, Seattle, WA, USA

Lei Zhang Office of Clinical Pharmacology, Office of Translational Sciences, Center for Drug Evaluation and Research, Food and Drug Administration, Silver Spring, MD, USA

Ping Zhao Office of Clinical Pharmacology, Office of Translational Sciences, Center for Drug Evaluation and Research, Food and Drug Administration, Silver Spring, MD, USA

Joseph K. Zolnerciks Department of Pharmaceutics, University of Washington, Seattle, WA, USA

Tatyana Zvyaga Bristol-Myers Squibb Co., Wallingford, CT, USA

About the Editors

K. Sandy Pang Ph.D. is a professor of Pharmacy and Pharmacology of the Faculties of Pharmacy and Medicine at the University of Toronto. She received her B.S. (Pharmacy) from the University of Toronto, Ph.D. (Pharmaceutical Chemistry) from UCSF, and post-doctoral training with Dr. James R. Gillette as a Fogarty International Fellow at the National Institutes of Health. Dr. Pang's work spans the fields of pharmacokinetics, drug metabolism and transporters, and their regulation, and her works are published in over 200 original articles and chapters. Dr. Pang has served on various committees for NIH: ASPET, AAPS, ISSX, and AAAS, and is the editor-in-chief of *Biopharmaceutics and Drug Disposition* and a member of the editorial review boards of the *American Journal of Physiology*, *Journal of Pharmacology and Experimental Therapeutics*, *Drug Metabolism and Disposition*, *Xenobiotica*, and *AAPS Journal*. She was the recipient of the NIH Research Career Development Award, Faculty Development award from the Medical Research Council of Canada, the McNeil Award from the Faculties of Pharmacies in Canada, and the Research Achievement Award in Pharmacokinetics, Pharmacodynamics and Drug Metabolism from the American Association of Pharmaceutical Scientists (AAPS).

A. David Rodrigues is the Executive Director of the Metabolism and Pharmacokinetics Department, Pharmaceutical Candidate Optimization, Bristol-Myers Squibb, Princeton, New Jersey. The author and co-author of over 90 peer-reviewed journal articles and book chapters, Dr. Rodrigues sits on the Editorial Board of two journals (*Drug Metabolism and Disposition*, and *Xenobiotica*) and is a member of the International Society for the Study of Xenobiotics (ISSX) and the American Association of Pharmaceutical Scientists (AAPS). He serves as a member of the Scientific Affairs Committee (ISSX) and has been elected Fellow (AAPS). He received the B.Sc. degree (1984) in applied science from Kingston-upon-Thames Polytechnic, Surrey, England, and the Ph.D. degree (1988) in biochemistry from the University of Surrey, Guildford, England.

Raimund M. Peter is an associate director of the Drug Metabolism and Pharmacokinetics Section, Cardiovascular and Gastrointestinal Research Department, AstraZeneca, Alderley Park, United Kingdom. The author and co-author of 20 peer-reviewed journal articles, Dr. Peter is the current chairman of the Drug Metabolism Focus Group of AAPS and is a member of the International Society for the Study of Xenobiotics (ISSX), the American Association of Pharmaceutical Scientists (AAPS), and the American Chemical Society. He received the Dipl.-Chem. degree (1986) in chemistry and the Ph.D. degree (1992) in chemistry and biochemical pharmacology from the University of Erlangen-Nürnberg, Germany.

Part I
Determinants of Drug ADME

Chapter 1
Enzymatic Basis of Phase I and Phase II Drug Metabolism

Susan Kadlubar and Fred F. Kadlubar

Abstract Enzyme systems have evolved in humans to maintain cellular homeostasis and to facilitate the disposition of xenobiotics. These enzymes possess complementary, and sometimes overlapping, functions, a feature which can set the stage for drug–drug interactions. In this chapter, we provide an overview of the phase I and phase II families of detoxification enzymes, their potential for induction and inhibition, and the role of genetic variants in the occurrence of drug–drug interactions.

1.1 Introduction to Phase I and Phase II Metabolism

Humans have a myriad of enzymes that function to maintain cellular homeostasis. This includes enzymes that metabolize endogenous lipophilic compounds, such as steroids, and xenobiotics including drugs. Metabolism allows the utilization of nutrients and the subsequent detoxification and excretion of potentially harmful compounds and metabolites. The genes encoding metabolic enzymes are polymorphically expressed in humans; molecular biology and enzymology studies have shown that there are many polymorphisms that have a functional consequence for the expressed protein. This genetic variability can have consequences in the disposition of drugs and the occurrence of drug–drug interactions that in the past have been observed clinically as idiosyncratic drug reactions.

Traditionally, drug metabolism has been classified into two groups based on the types of enzymatic reaction catalyzed by the enzyme family under investigation. Phase I drug metabolism typically involve the functionalization of endo- and xenobiotics to generate more polar derivatives (alcohols, phenols, carboxylic

F.F. Kadlubar (✉)
Department of Epidemiology, College of Public Health, University of Arkansas for Medical Sciences, Little Rock, AR, USA
e-mail: fkadlubar@uams.edu

K.S. Pang et al. (eds.), *Enzyme- and Transporter-Based Drug–Drug Interactions*,
DOI 10.1007/978-1-4419-0840-7_1,

acids). Phase I reactions can also involve reduction reactions, hydrolysis of compounds, and cyclization/decyclization reactions. Phase I oxidative reactions are typically carried out in the liver by mixed function oxidases primarily involving hepatic cytochromes P450 (CYPs) that utilize NADPH and oxygen in their catalytic cycle. However, CYP enzymes are also expressed extrahepatically where they can participate in drug metabolism in target tissues. CYPs expressed in the intestinal tract are particularly important in the first-pass metabolism of many classes of drugs. In some instances, metabolites produced by phase I reactions are of sufficient polarity to undergo excretion, but many others require additional metabolism by phase II detoxification enzymes. Phase II reactions commonly involve conjugation of a substrate with an obligatory cosubstrate specific to each family of phase II enzymes. The most common phase II families include the glutathione *S*-transferases (GSTs), UDP-glucuronosyltransferases (UGTs), sulfotransferases (SULTs), and *N*-acetyltransferases (NATs). It is the hallmark of drug development and discovery to find therapeutic molecular entities that are substrates for more than phase I and phase II enzyme, thereby avoiding drug–drug interactions that can cause deleterious effects.

1.2 Phase I Enzymatic Reactions

1.2.1 Cytochromes P450

The CYP enzyme family constitutes a superfamily of metabolic enzymes that are responsible for as much as 60% of the biotransformation of FDA-approved drugs (Venkatakrishnan et al., 2001). They were originally discovered in microsomal preparations from animal liver (Omura, 1999) but were rapidly identified in all eukaryotic organisms. In recent years, drug–drug interactions involving specific CYP isoforms have been increasingly recognized. For example, CYP3A4, the most abundant hepatic isoform involved in the oxidative metabolism of the majority of drugs, is subject to either induction or inhibition by a variety of therapeutic agents and has been implicated in numerous occurrences of drug–drug interactions (Zhou, 2008). Clearly, understanding the enzymology and factors influencing catalytic activity of this important enzyme superfamily is required for safe use of therapeutics.

1.2.1.1 P450 Catalytic Cycle

P450-catalyzed oxidative metabolism of substrates adheres to a common catalytic cycle, regardless of the isoform involved. The generalized CYP reaction mechanism is depicted as follows:

$$\mathrm{RH} + \mathrm{O_2} + 2\mathrm{H} + 2e^- \xrightarrow{\mathrm{CYP}} \mathrm{ROH} + \mathrm{H_2O}$$

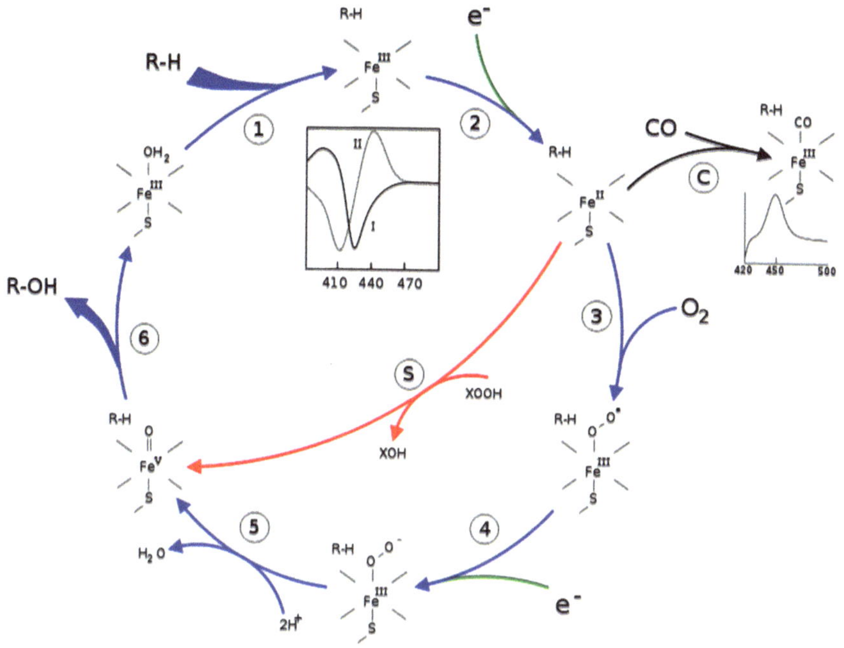

Fig. 1.1 Catalytic cycle of cytochromes P450

Briefly, a substrate (represented by RH), an oxygen molecule, two hydrogens, and two electrons are utilized by a P450 to produce an OH functional group and one molecule of water. The P450 catalytic cycle is shown in Fig. 1.1. Substrate binding causes a conformational change in the active site, thus releasing a water molecule from the heme moiety (Waszkowycz et al., 1994). The result is to modify the spectral properties of the enzyme, increasing UV absorbance at 390 nm with a concomitant decrease in absorbance at 420 nm. This change can be detected spectrophotometrically by "type-I" difference spectrum (see inset graph in Fig. 1.1). Substrates and inhibitors that are capable of binding directly to the heme iron generate a type-II difference spectrum (430/390 nm maximum/minimum – inset graph in Fig. 1.1). Lack of reducing agents will stabilize this complex, thus facilitating the measurement of binding by absorbance differences in *in vitro* assays. Changing the electronic state of the active site allows the transfer of an electron from NAD(P)H. Molecular oxygen can then bind covalently to the distal axial coordination position of the heme iron. A second electron is transferred by the electron-transport system, thus reducing the dioxygen moiety to a negatively charged peroxo intermediate. This intermediate is rapidly protonated twice by local transfer from surrounding amino-acid side chains, releasing one mole of water and forming a highly reactive iron(V)-oxo species (Waszkowycz et al., 1994). The product is released from the active site and the enzyme returns to its original state when a water molecule returns to occupy the distal coordination position of the iron nucleus.

1.2.1.2 Specific P450 Isoforms

CYP1 Subfamily

The CYP1A family consists of three members: CYP1A, CYP1A2, and CYP1B1. Of these, only CYP1A2 is expressed to any appreciable degree in human liver. CYP1A2 comprises approximately 10% of the total hepatic CYP pool (Shimada et al., 1994). CYP1A1 and CYP1A2 share 70% homology and exhibit overlapping substrate specificity. An important feature of the CYP1A family is the inducibility of CYP1A, particularly CYP1A1, by polyaromatic hydrocarbons (PAHs). The discovery of this inducibility stemmed from reports from Conney et al. of PAHs inducing their own metabolism (Conney, 1982). However, as far as the metabolism of therapeutic drugs is concerned, CYP1A2 is the predominant CYP1A isoform involved. CYP1A2 preferentially oxidizes planar aromatic amines and amides, and is subject to induction/inhibition by various xenobiotics. Enzymatic activity of CYP1A2 varies widely in human populations, exhibiting approximately 60-fold differences in activity. Enzymatic activity of CYP1A2 has been assessed using various probe substrates including caffeine (Butler et al., 1992; Fuhr and Rost, 1994), phenacetin (Ching et al., 2001), and theophylline (Bachmann et al., 2003). Other pharmaceuticals metabolized by CYP1A2 include acetaminophen (Snawder et al., 1994; Zaher et al., 1998), tizanidine (Backman et al., 2006), and propranolol (Masubuchi et al., 1994; Johnson et al., 2000). There are some drugs that serve as CYP1A2 inhibitors, including oral contraceptives (Karjalainen et al., 2008), cimetidine (Reilly et al., 1988), ciprofloxacin (Raaska and Neuvonen, 2000), and enoxacin (Kunii et al., 2005). Another area of concern with drug–drug interactions of CYP1A2-metabolized drugs is the fact that CYP1A2 expression and activity can be induced by dietary and environmental exposures. CYP1A2 activity is known to be inducible by cigarette smoke (Mori et al., 1995), PAHs (Shimada et al., 2002), and TCDD (Olson et al., 1994). Dietary constituents such as broccoli and other cruciferous vegetables (Walters et al., 2004) have also been demonstrated to induce the activity of CYP1A2.

CYP1A2 is also genetically polymorphic in humans. According to the National Center for Biotechnology (NCBI) database (http://www.ncbi.nlm.nih.gov/sites/entrez), 158 single nucleotide polymorphisms (SNPs) exist in CYP1A2, and the frequency of these SNPs vary by ethnicity. The large number of SNPs identified is the result of completion of the HapMap project and subsequent resequencing efforts. Many of these SNPs are relatively rare in human populations, and the functional consequences of more common SNPs are yet to be established. However, no genetic variant of CYP1A2 examined thus far for effect on phenotype has shown a marked effect in Caucasian populations. Further functional characterization of identified SNPs is needed.

CYP2B6

CYP2B6 has been shown to comprise approximately 5% of the P450 content in human liver, with at least a 100-fold variation in expression in those livers (Hofmann et al., 2008). While expression of CYP2B6 is low under normal conditions, this enzyme is extremely sensitive to induction by a variety of drugs and chemicals, including rifampin (Rae et al., 2001), phenobarbital (Lee et al., 1998; Faucette et al., 2004), and phenytoin. Induction of CYP2B6 in some, but not all, instances is mediated by the pregnane X receptor (PXR) (Faucette et al., 2004). Induction by phenobarbital-related drugs appears to involve the constitutive androstane receptor (CAR) (Swales et al., 2005). Regulation by PXR and CAR gives rise to cross-regulation of CYP2B6 with CYP3A4, UGT1A1, and numerous drug transporters, increasing the complexity of potential drug–drug interactions. CYP2B6 has also been shown to be highly inducible by organophosphate, organochloride, and pyrethroid pesticides that are persistent in the environment (Lemaire et al., 2004), potentially impacting the metabolism of drugs that are substrates of CYP2B6.

The CYP2B6 primarily oxidizes neutral/weakly basic non-planar molecules having 1–2 hydrogen bond acceptors. Drug substrates of CYP2B6 include bupropion (Zyban) (Faucette et al., 2000), cyclophosphamide (Huang et al., 2000), ketamine (Yanagihara et al., 2001), methadone (Neff and Moody, 2001), and nevirapine (Zanger et al., 2007), and their disposition in humans can be dependent on induction/inhibition of CYP2B6, as well as genetic polymorphisms in the gene. CYP2B6 is one of the most polymorphic of the P450 isoforms, with 451 SNPs deposited in the NCBI SNP database. These variants differ in frequency by ethnicity (reviewed in Wang and Tompkins, 2008), with the most common variant allele *CYP2B6*6* occurring at frequencies between 15 and 60% across populations. Interestingly, the *CYP2B6*6* SNP has been associated with an increased likelihood of relapse in smokers treated with placebo compared to Zyban as part of a smoking cessation program (Lee et al., 2007).

CYP2C Subfamily

There are three major members of the CYP2C family: CYP2C8, CYP2C9, and CYP2C19. A fourth member, CYP2C18, has been identified as an mRNA transcript but for unknown reasons does not get efficiently translated into a protein (Lofgren et al., 2008). Taken together, the CYP2C subfamily constitutes around 30–40% of hepatic P450 content, with CYP2C9 being the most highly expressed, followed by CYP2C8 and CYP2C19. The CYP2C family is strongly homologous, with >82% sequence identity. All three isoforms are controlled transcriptionally by ligands of PXR/CAR and glucocorticoid (GR) nuclear receptors via unique *cis*-acting element in their regulatory regions (Ferguson et al., 2002; Pascussi et al., 2003; Ferguson et al., 2005). Thus, CYP2C isoforms are highly inducible by rifampin, phenobarbital, and dexamethasone (Raucy et al., 2002).

CYP2C8 displays preference for relatively large, weakly acidic substrates. CYP2C8 is notable in its metabolism of the antidiabetic drugs rosiglitazone (Baldwin et al., 1999) and pioglitazone (Yamazaki et al., 2000; Sahi et al., 2003). CYP2C8 is the major P450 catalyzing the 4-hydroxylation of all-*trans* retinoic acid (Nadin and Murray, 1999) and the chemotherapeutic agent taxol (Walle, 1996). CYP2C8 expression/activity is induced by rifampicin. Montelukast, a leukotriene receptor antagonist, is the most potent inhibitor of CYP2C8, raising concerns of clinical drug interactions (Walsky et al., 2005). In another instance, cervistatin was removed from the market because of fatalities resulting from drug interactions with gemfibrozil, which, along with its glucuronide metabolite, were potent inhibitors of CYP2C8 (Wang et al., 2002). Other pharmaceuticals are potent (clotrimazole, felodipine) to moderate (raloxifene, tamoxifen, loratadine) inhibitors of CYP2C8. The NCBI SNP database reports 411 SNPs in CYP2C8; the most common functional alleles identified to date are *CYP2C8*2* and *CYP2C8*3*, both of which are associated with decreased activity and are important in taxol (paclitaxel) pharmacokinetics (Dai et al., 2001).

CYP2C9 metabolizes weakly acidic substrates possessing a hydrogen bond acceptor. Primary substrates for CYP2C9 include *S*-warfarin (Zhang et al., 1997), tolbutamide (Lasker et al., 1998), and several non-steroidal anti-inflammatory drugs such as celecoxib, diclofenac, flurbiprofen, ibuprofen, indomethacin, lornoxicam, meloxicam, naproxen, piroxicam, suprofen, and tenoxicam. Genetic variants of CYP2C9 influence clearance of non-steroidal anti-inflammatory drugs (reviewed in Ali et al., 2009). Like other members of the CYP2C family, it is induced by rifampicin, as well as barbiturates and carbamazepine. Inhibitors include sulfamethoxazole, miconazole, and fluconazole. In fact, the first report of a drug–drug interaction involving CYP2C9 occurred in 1963 and was due to the coadministration of sulfamethoxazole and tolbutamide, resulting in severe hypoglycemia (Christensen et al., 1963). While hundreds of SNPs have been identified, most appear to be rare across populations. The most studied alleles are *CYP2C9*2* and *CYP2C9*3*, both of which appear to decrease enzymatic activity.

CYP2C19 has the lowest hepatic expression levels of this P450 subfamily, but was discovered before any of the others, due to the recognition of *S*-mephenytoin (an anticonvulsant), poor and extensive metabolizer phenotypes (Ibeanu et al., 1998; 1999). It has a marked preference for proton pump inhibitors such as omeprazole and pantoprazole but also plays a role in the metabolism of other classes of drugs such as psychotropics and anticancer agents. CYP2C19 is induced by rifampicin and St. John's wort, and inhibited by chloramphenicol, fluconazole, fluoxetine, and indomethacin. CYP2C19 is also polymorphic, and alone of the CYP2C subfamily, is subject to a gene deletion, resulting in the absence of the enzyme and poor metabolizer phenotype, whereas individuals with at least one functional allele are classified as rapid metabolizers.

CYP2D6

In terms of drug metabolism, CYP2D6 is the most intensely studied CYP because it was the first to be recognized as polymorphic. In the 1970s, volunteers participating in pharmacokinetic studies of debrisoquine, an antihypertensive agent, and sparteine, an antiarrhythmic drug, displayed unexpected adverse reactions. The adverse reactions were found to be due to the individual's inability to oxidize each drug and the defect was under monogenic control (Mahgoub et al., 1977). This defect resulted in categorization of poor metabolizer (PM), extensive metabolizer

(EM), and ultra-rapid metabolizer (UM) phenotypes of patients treated with debrisoquine or sparteine. The specific enzyme responsible for this phenotype was not known at the time, but subsequent studies suggested that a deficiency in a CYP was likely the culprit (Kahn et al., 1982). Gonzalez et al. (1988) isolated a full-length cDNA that was subsequently assigned to the CYP2D family and called CYP2D6. Consequently, CYP2D6 and its variants became one of the most intensively studied drug-metabolizing polymorphisms. Although CYP2D6 accounts for only around 2% of hepatic CYP content, many different classes of drugs are metabolized by it. Substrates include dextromethorphan, metoprolol, propafenone (probe substrates for CYP2D6) (Zanger et al., 2008), haloperidol, risperidone, and imipramine. While there are no known inducers of CYP2D6, it is subject to inhibition by celecoxib, paroxetine, and quinidine.

CYP2D6 has received much attention in the arena of cancer therapy in recent years due to its role in the production of the active metabolites of tamoxifen, 4-hydroxytamoxifen, and endoxifen (Lim et al., 2005; 2006) used in the adjuvant setting to treat estrogen receptor-positive breast cancer. CYP2D6 genotypes are associated with plasma levels of endoxifen and 4-hydroxytamoxifen. Furthermore, patients who were receiving selective serotonin reuptake inhibitors (also CYP2D6 substrates) for tamoxifen treatment-related hot flushes exhibited altered efficacy of therapy that was dependent on CYP2D6 genotype (Jin et al., 2005). CYP2D6 has a host of genetic variants, and these, along with their functional effects, have recently been reviewed in Zanger et al. (2008).

CYP2E1

CYP2E1 makes up around 6% of hepatic P450 content, but it is primarily involved in functions other than therapeutic drug metabolism. Notably, it is induced by ethanol and is known to metabolize acetaminophen to a reactive metabolite that causes liver damage. Several SNPs have been reported, and one allele *CYP2E1*5B* has recently been associated with alcohol-related cirrhosis of the liver (Khan et al., 2009).

CYP3A Subfamily

The CYP3A subfamily is composed of CYP3A4, CYP3A5, CYP3A7, and CYP3A43. These isoforms are the most abundant in human liver, accounting for approximately 30% of total P450 content. They are also highly expressed in the small intestine, where they are positioned to play a role in first-pass metabolism of drugs. Of the top 200 prescribed pharmaceuticals, CYP3A4 (the major adult isoform) metabolizes 37% of them (Zanger et al., 2008). CYP3A5 is expressed in approximately 20% of the population, whereas CYP3A7 is a fetal isoform (Schuetz et al., 1993). The role of CYP3A43 in drug metabolism is uncertain. CYP3A4 prefers lipophilic, large molecules and its substrates include erythromycin, nifedipine, midazolam, testosterone, and verapamil (reviewed in Bu, 2006); inhibitors include ethinylestradiol, antimycotics, and clarithromycin (Racha et al., 2003). CYP3A4 is induced by a host of compounds (Kolars et al., 1992; Roby et al., 2000;

Krusekopf et al., 2003; Matsubara et al., 2007), but much attention has centered on the induction of CYP3A4 by hyperforin, a constituent of the widely used herbal St. John's wort (Madabushi et al., 2006).

1.2.2 Flavin-Containing Monooxygenases

Flavin-containing monooxygenases (FMOs) oxygenate a wide variety of therapeutic agents, pesticides, and dietary constituents (Krueger and Williams, 2005; Cashman and Zhang, 2006), although they have received scant attention compared to the CYPs. Unlike the CYPs, FMOs do not appear to be subject to inhibition or induction (Cashman and Zhang, 2006) and thus are less likely to participate in drug–drug interactions. For this reason, inter-individual variation in FMO activity is predominately due to genetic variation. The general reaction catalyzed by FMO is as follows:

$$RH + O_2 + NADPH + H^+ \xrightarrow{FMO} ROH + H_2O + NADP^+$$

FMOs, in common with the P450s, are expressed in the endoplasmic reticulum of the cell, but they are mechanistically distinct from P450s. FMOs exist in an activated form which then reacts with a nucleophile to complete the catalytic cycle (Fig. 1.2). Requirements for product formation with FMO substrates are simply contact between the substrate and the oxygen of the stable intermediate, 4-hydroperoxy flavin.

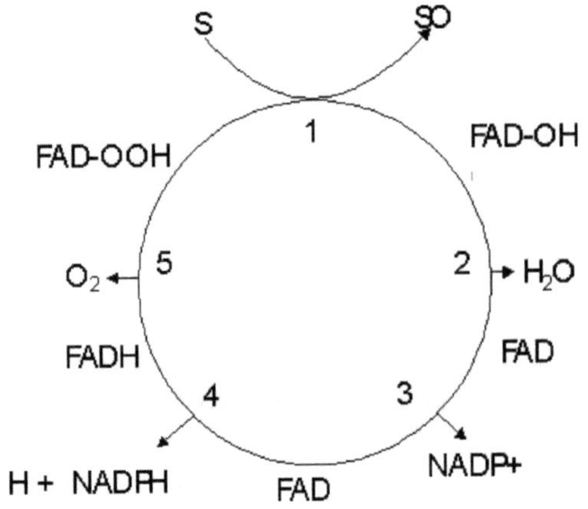

Fig. 1.2 Catalytic cycle of flavin-containing monooxygenases

Humans express five functional FMO isoforms (Mitchell, 2008). Of these, FMOs 1, 2, and 3 appear to be the most important in relation to metabolism of chemicals. FMO1 is primarily expressed in the kidney (Phillips et al., 1995), where the levels of expression exceed that of the P450s, positioning it to play a key role in renal metabolism of drugs. FMO3 is the major isoform expressed in adult human liver. Substrates for FMO3 include amphetamines, tamoxifen, clozapine, deprenyl, sulindac sulfide, methamphetamine, and phenothiothiazine drugs. FMO3 is highly polymorphic and some of these polymorphisms have been associated with altered enzymatic activity. Three SNPs in particular influence activity: 11177C>A results in decreased enzymatic activity in a substrate-dependent manner (Koukouritaki et al., 2007); 21599T>C increases enzymatic activity approximately fivefold (Lattard et al., 2003); 15550C>T shows an intermediate effect on FMO activity but is associated with substrate inhibition of sulindac sulfide *S*-oxygenation (Shimizu et al., 2007). These variants are relatively rare in populations; the NCBI database reports 353 SNPs in the *FMO3* gene, so clearly other studies are needed to examine the functional consequences of genetic variation within this gene.

1.3 Phase II Enzymatic Reactions

1.3.1 Glutathione *S*-Transferases

Glutathione *S*-transferases (GSTs) comprise a family of phase II detoxification enzymes whose function is to defend cellular macromolecules against reactive electrophiles. They accomplish this task by catalyzing the conjugation of reduced glutathione (GSH) to the electrophile, facilitating its removal via the mercapturic acid pathway (Fig. 1.3). GSTs exist as two distinct families: the microsomal GSTs and the cytosolic GSTs. The microsomal GSTs are structurally unrelated to the cytosolic GSTs and are not considered further in this chapter. The cytosolic GSTs consist of at least 16 genes in humans, grouped into eight classes: alpha, kappa, mu, pi, sigma, theta, zeta, and omega.

The GST pi class contains a single gene (*GSTP1*) but the alpha (*GSTA1–4*), mu (GSTM1–5), and theta (*GSTT1*, *GSTT2*) classes consist of clusters of closely related genes coding for closely related GST subunits. Catalytically active GSTs are dimers of subunits within the same class. The GSTs of the alpha, mu, pi, and theta classes are known to accept a wide range of electrophilic substrates, including chemicals, chemotherapeutic drugs, and fatty acid hydroperoxides. Since genotoxic and cytotoxic electrophiles are substrates for these GSTs, they are of particular interest concerning the effectiveness of chemotherapeutic agents.

Tissue-specific expression of GSTs can be strikingly different. For instance, two members of the GST alpha class, *GSTA1* and *GSTA2*, are expressed at high levels in human liver and can represent 3% of total cytosolic protein (van Ommen et al., 1990; Rowe et al., 1997). In contrast, expression of GSTs in colon is relatively low (0.2%) (Peters et al., 1991), with GSTP1 being the predominant GST expressed. Predominant GSTP1 expression has also been demonstrated for other tissues such as lung. GSTM1 is expressed at high levels in a few tissues, including liver, testis, brain, and bladder (Berendsen et al., 1997; Rowe et al., 1997), but in colon and lung, it is expressed at much lower levels than its related isoform, GSTM3 (Anttila et al., 1995; Coles et al., 2000a). These patterns of expression are important in that it may

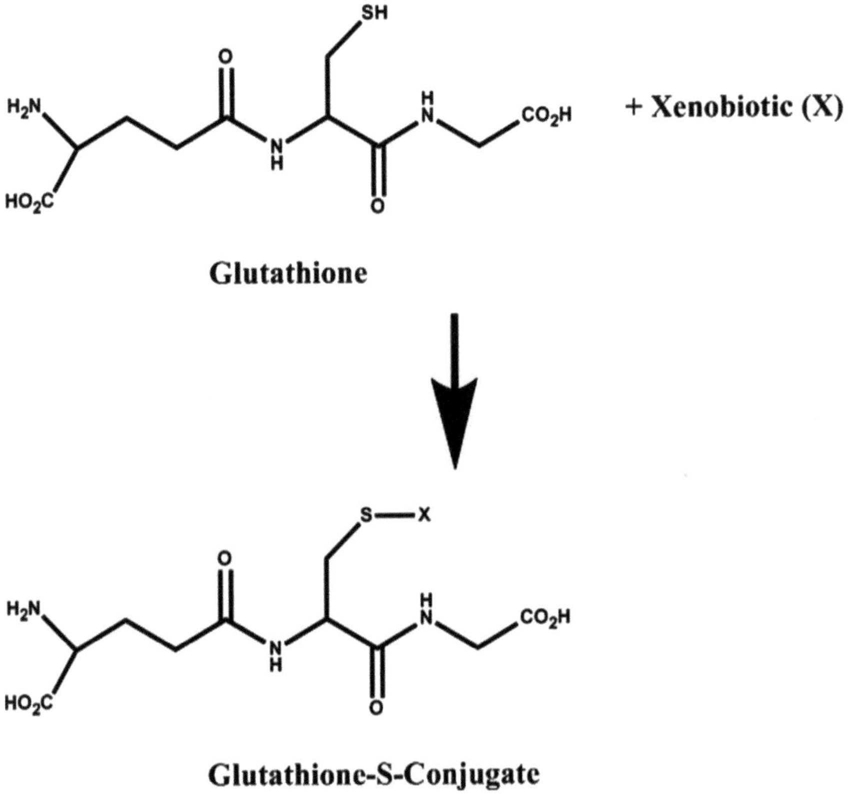

Fig. 1.3 Glutathione conjugation catalyzed by glutathione *S*-transferases

predispose certain organs to genotoxic effects of chemicals and therapeutic agent if they lack a GST with selective activity toward specific electrophiles (Coles and Ketterer, 1990; Hayes and Pulford, 1995).

Drugs that are known to be substrates of GSTs include acetaminophen, valproic acid, and busulfan (Slattery et al., 1987; Kassahun et al., 1991; Poonkuzhali et al., 2001). In addition to these, alkylating chemotherapeutic agents are substrates for GSTs. And while detoxification of reactive electrophiles generated by carcinogen is a desirable function for GSTs, detoxification of alkylating chemotherapy drugs would achieve the opposite result, decreasing therapeutic efficacy of the treatment. GSTA1 catalyzes the detoxification of the therapeutic metabolite of cyclophosphamide, phosphoramide mustard (PM), and GSTP1 catalyzes the detoxification of 4-hydroxy-cyclophosphamide, the precursor of PM (Dirven et al., 1996). Although there have been reports of GST induction in animal models, studies in humans are limited. However, some compounds such as isothiocyanates (ITCs) are chemopreventive agents in animal models, perhaps due in part to their potent effects on inhibition of phase I and induction of phase II enzymes, particularly GSTs (Talalay and Fahey, 2001). Inhibitors of GST activity have also not been well characterized, but recent studies have shown that the natural plant products, ellagic acid and curcumin, can inhibit GSTA1, A2, M1, M2, and P1, while genistein, kaempferol, and quercetin are only capable of inhibiting GSTM1 and M2 (Hayeshi et al., 2007).

Several allelic polymorphisms occur in human GST genes and are known to affect protein expression or to code for proteins with variant catalytic properties. The GSTM1 "null" and GSTT1 "null" polymorphisms represent gene deletions, and individuals who are homozygous null do not express these proteins (Rebbeck, 1997; Hayes and Strange, 2000; Landi, 2000). A polymorphism in intron 6 of GSTM3 shows a minor effect on GSTM3 expression (Coles et al., 2000b). A polymorphism in the promoter of *GSTA1* affects GSTA1 expression in the liver *via* levels of basal expression (Coles et al., 2001).

1.3.2 N-Acetyltransferases

N-Acetyltransferases (NAT) are a family of phase II detoxification enzymes that catalyze several pharmacologically important reactions. In humans, two NAT isoforms (NAT1 and NAT2) are expressed. Of these, NAT2 is expressed in the liver, while NAT1 is primarily extrahepatic. NAT1 and NAT2 have overlapping substrate specificity but also show selectivity for other substrates. For instance, *p*-aminobenzoic acid (PABA) is acetylated by NAT1, while sulfamethazine is selectively acetylated by NAT2 (Grant et al., 1991). The most common NAT-catalyzed reaction is the *N*-acetylation of aromatic amines. Figure 1.4 depicts the acetylation of isoniazid, a substrate that historically brought much attention to this enzyme family. Isoniazid was developed for the treatment of tuberculosis and is the first substrate for which an acetylation polymorphism was clinically apparent. Some patients treated with isoniazid developed peripheral neuropathy that was dependent on the patient's acetylation status (reviewed in Weber and Hein, 1985). Hughes et al. (Hughes, 1953) subsequently demonstrated that other drugs that resembled isoniazid chemically

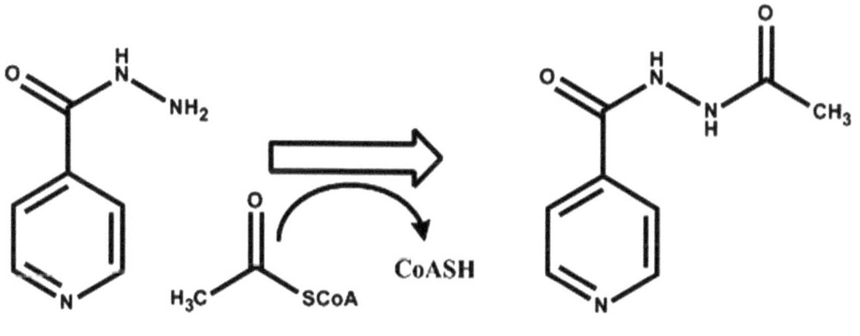

Fig. 1.4 Acetylation of isoniazid by *N*-acetyltransferases

underwent acetylation before excretion and that acetylisoniazid was the main urinary metabolite in humans. Furthermore, patients fell into high and low excretor groups, which led to the discovery of the acetylator polymorphism. Since these early days, the genetic basis of the NAT phenotype has been identified. Both NAT1 and NAT2 are highly polymorphic (reviewed in Sim et al., 2008) and the nomenclature and details of specific SNPs and their functions can be found at the nomenclature website (http://www.louisville.edu/medschool/pharmacology/NAT.html).

NAT activity can be inhibited by nitrosoarene metabolites of various carcinogenic arylamines (Liu et al., 2008), by the chemotherapeutic agent, cisplatin (Ragunathan et al., 2008), and by components of Chinese herbal medicines (Chiu et al., 2004).

1.3.3 UDP-Glucuronosyltransferases

UGTs (EC 2.4.1.17) are a gene superfamily whose principal role is to convert endo- and xenobiotics into water-soluble derivatives. UGTs are divided into two families based on evolutionary divergence. The UGT1A locus is on chromosome 2 and can potentially encode nine functional isoforms and three pseudogenes. The *UGT1A* gene complex is composed of multiple tandem first exons that encode the variable N-terminal part of the enzyme and are linked by differential splicing to common exons that encode the C-terminal region. The first exons have unique TATA elements approximately 30 bp upstream, allowing for independent regulation of the isoforms. Members of the UGT2B family are unique gene products and preferentially glucuronidated steroids and bile acids, in addition to xenobiotics. To date, at least eight 2B isoforms have been identified. UGTs are expressed primarily in the liver but recently findings indicate that extrahepatic glucuronidation contributes significantly to detoxification of endo- and xenobiotics, particularly in the gastrointestinal tract (Guillemette, 2003). UGTs execute their function by transferring the glucuronic acid moiety from the universal donor substrate, uridine diphosphoglucuronic acid, to a host of acceptor molecules (Fig. 1.5). Most UGTs are involved, to a greater or lesser extent, in the metabolism of drugs and other xenobiotics, but the two most notable examples of isoforms relevant to pharmacogenomics are UGT1A1 and UGT2B7.

1.3.3.1 UGT1A1

The primary endogenous substrate of UGT1A1 is bilirubin, and impaired bilirubin glucuronidation by this isoform results in mild (Gilbert's syndrome) to severe (Crigler–Najjar syndrome) metabolic defects. UGT1A1 is also responsible for the metabolism of the chemotherapeutic agent irinotecan (Innocenti et al., 2004). Studies in animal models have shown that phenethyl isothiocyanate, a constituent of cruciferous vegetables, leads to a slight increase in hepatic UGT activity in F344 rats. Induction of UGT activity has been reported in human-derived Hep G2 cells by exposure to extracts of garden cress and white mustard, sprouts which vary in their glucosinolate content (Lhoste et al., 2004). Exposure of Hep G2 cells to the

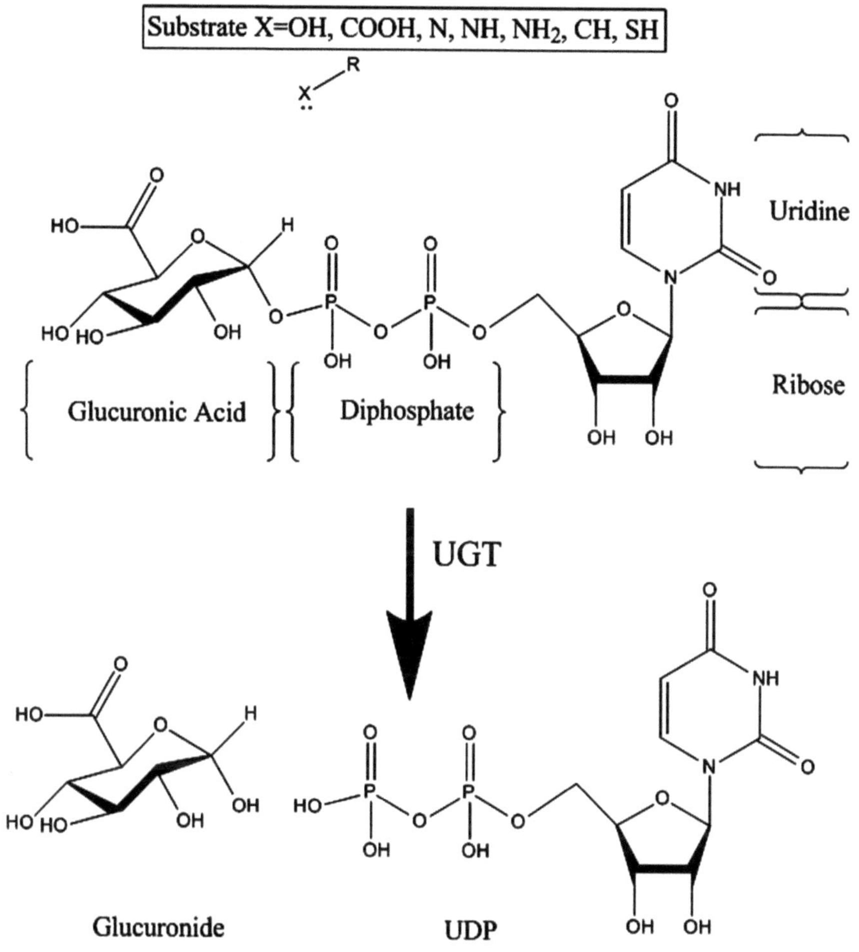

Fig. 1.5 Glucuronidation reaction

extracts resulted in a 1.4- and 1.8-fold induction of UGT activity for garden cress and white mustard, respectively. While these investigations showed induction of UGT activity toward 4-methylumbelliferone, the specific isoform(s) induced were not identified. Studies using the human cell lines Hep G2 and Caco-2 have demonstrated that the flavonoid chrysin is a potent and fairly selective inducer of UGT1A1 (Walle et al., 2000). Treatment of cells with 25 μM chrysin resulted in a 20-fold increase in the activity of UGT1A1 for bilirubin, its primary endogenous substrate. Piperine, which is a major component of black pepper, has been shown to potently inhibit UGT activity in intestinal epithelial cells of guinea pigs (Grancharov et al., 2001a, b). This inhibition was observed to be time- and concentration dependent.

Numerous mutations in UGT1A1 (reviewed in Guillemette, 2003) have been identified but only a few are of sufficient frequency in the general population to be classified as polymorphisms. One particular genetic variation that has been investigated in several studies of cancer risk is a dinucleotide repeat [A(TA)$_n$ TAA] in the atypical TATA box region of the *UGT1A1* promoter. Four variant alleles are the result of variation in the number of dinucleotide repeats. Five repeats generate *UGT1A1*33*; six repeats generate *UGT1A1*1*("wild type"); seven repeats generate *UGT1A1*28*; and eight repeats generate *UGT1A1*34*. Functional studies have shown that increasing numbers of repeats lead to decreased transcription of *UGT1A1*. *UGT1A1*28* is the most common variant allele and has been associated with the occurrence of Gilbert's syndrome, a mild form of unconjugated hyperbilirubinemia (Clarke et al., 1997). The distribution of *UGT1A1* alleles varies by ethnicity, with *UGT1A1*34* much more common in African-Americans than in Caucasians.

1.3.3.2 UGT2B7

UGT2B7 is a member of the steroid UGTs and is responsible for the glucuronidation of a host of endo- and xenobiotics including morphine (Coffman et al., 1997), diclofenac (Daly et al., 2007), lorazepam (Chung et al., 2008), and the active metabolites of tamoxifen (Blevins-Primeau et al., 2009). UGT2B7 has the highest expression levels in liver of the UGTs and has been reported to be upregulated by valproic acid (Valentini et al., 2007). UGT2B7 is inhibited by isolongifolol derivatives (Bichlmaier et al., 2007) and ketoconazole (Takeda et al., 2006). Recent studies have shown that UGT2B7 glucuronidates 4-hydroxytamoxifen and endoxifen (Sun et al., 2007), metabolites responsible for tamoxifen's estrogen receptor antagonist activity. Genetic variants in UGT2B7 include an amino acid change at residue 268 (His>Tyr), and this change is associated with changes in glucuronidation activity toward both 4-hydroxytamoxifen and endoxifen. Therefore, UGT2B7 genotype could influence the efficacy of tamoxifen therapy due to rapid elimination of active metabolites.

1.3.4 Sulfotransferases

SULTs are classified as phase II detoxification enzymes and are responsible for conjugating xeno- and endobiotics with 5′-phosphoadenosine-3′-phosphosulfate, thus rendering the substrate more water soluble. The generalized sulfonation reaction is depicted below:

$$\mathrm{ROH} + \mathrm{PAPS} \xrightarrow{\mathrm{SULT}} \mathrm{PAP} + \mathrm{ROSO_3^-}$$

The pK_a of the sulfate group is around 1.5, so it remains fully ionized at any pH found in biological systems.

In some instances, sulfation of some xenobiotics results in their metabolic activation. In this instance, sulfation is a double-edged sword. Several SULT isoforms are involved in the metabolism of drugs and other chemicals, but SULT1A1 is the most highly expressed hepatic isoform and has the most widespread tissue distribution of all sulfotransferase isoforms. SULT1A1 has preference for phenolic drugs and compounds found in fruits and vegetables and plays a major role in catalyzing the conjugation of dietary phytoestrogens, including daidzein and genistein. SULT1A1 is also potently inhibited by a variety of dietary chemicals (reviewed in Pacifici, 2004). For example, curcumin, which is under investigation as a chemopreventive agent for colorectal cancer, inhibits SULT1A1 activity with an IC_{50} of 12.8 nM, a level easily achievable in vivo (Pacifici, 2004).

SULT1A1 activity, as measured in human platelet preparations, has also been shown to be inhibited by as much as 99% by red wine extracts (Pacifici, 2004). In 1996, Harris and Waring reported the inhibitory effect of more than 30 dietary constituents found in vegetables on SULT1A1 activity (Harris and Waring, 1996). Using vegetable cytosols, they found that, of all cytosols tested, constituents found in radishes, spinach, broccoli, bananas, and leeks were the most potent inhibitors of both SULT1A1 and SULT1A3 activity, although the identity of the inhibitory substance was not identified. Later studies examined the identity of these compounds and found that quercetin, an abundant flavonoid found in fruits and vegetables (and in wine), inhibited SULT1A1 activity with an IC_{50} of 0.1 μM and was also an inhibitor of SULT1A3, SULT1E1, and SULT2A1 (DHEA sulfotransferase) (Pacifici, 2004). This is well within the range of the peak plasma levels of quercetin, which have been reported to be between 0.3 and 0.7 μM. As with quercetin, other dietary compounds inhibit SULT1A1, such as 5-OH-flavone and 3-OH-flavone and both epicatechin gallate and epigallocatechin gallate.

The polymorphic nature of SULTs has long been recognized (Price et al., 1989; Falany, 1997; Weinshilboum et al., 1997; Glatt et al., 2001). These studies have demonstrated a large individual variation (about 50-fold in some studies) in the activity of platelet SULTs in humans. It was reported that a large portion of this variability could be explained by a common polymorphism (a G to A transition) in the coding region (nucleotide 638) of the *SULT1A1* gene (Raftogianis et al., 1997; Ozawa et al., 1998). Another factor that may also affect enzymatic activity is gene deletion and duplication. Recently, Hebbring et al. observed that SULT1A1 enzymatic activity is also correlated with *SULT1A1* gene copy numbers in vitro (Hebbring et al., 2007). The role of gene duplication in metabolism of SULT1A1 substrates remains to be examined.

1.4 Conclusions

Phase I and phase II detoxification enzymes have evolved to protect the organism from toxic insult and from the buildup of endogenous lipophilic molecules. These families show substrate redundancy within a family (i.e., multiple CYPS, UGTs) in most, but not all, cases so that impaired activity in one enzyme will be compensated by another. In other instances, there is overlapping substrate specificity between families. For example, UGTs and SULTs share substrates but differ in their affinity for the substrates. In instances where an exposure is overwhelming, as with pharmaceutical dosing with concomitant medications, impaired functioning of one enzyme family member may be sufficient to produce toxicity due to drug–drug interactions. This impaired functioning can result from inhibition of particular drug-metabolizing enzymes, from induction of enzymes leading to rapid elimination of the drug, or elevated production of toxic metabolites, or from genetic polymorphisms in the enzymes. A thorough understanding of all of these aspects is necessary to predict, and thus prevent, adverse drug reactions.

References

Ali ZK, Kim RJ and Ysla FM (2009) CYP2C9 polymorphisms: considerations in NSAID therapy. *Curr Opin Drug Discov Devel* **12**:108–114.

Anttila S, Luostarinen L, Hirvonen A, Elovaara E, Karjalainen A, Nurminen T, Hayes JD, Vainio H and Ketterer B (1995) Pulmonary expression of glutathione S-transferase M3 in lung cancer patients: association with GSTM1 polymorphism, smoking, and asbestos exposure. *Cancer Res* **55**:3305–3309.

Bachmann K, White D, Jauregui L, Schwartz JI, Agrawal NG, Mazenko R, Larson PJ and Porras AG (2003) An evaluation of the dose-dependent inhibition of CYP1A2 by rofecoxib using theophylline as a CYP1A2 probe. *J Clin Pharmacol* **43**:1082–1090.

Backman JT, Granfors MT and Neuvonen PJ (2006) Rifampicin is only a weak inducer of CYP1A2-mediated presystemic and systemic metabolism: studies with tizanidine and caffeine. *Eur J Clin Pharmacol* **62**:451–461.

Baldwin SJ, Clarke SE and Chenery RJ (1999) Characterization of the cytochrome P450 enzymes involved in the in vitro metabolism of rosiglitazone. *Br J Clin Pharmacol* **48**:424–432.

Berendsen CL, Peters WH, Scheffer PG, Bouman AA, Boven E and Newling DW (1997) Glutathione S-transferase activity and subunit composition in transitional cell cancer and mucosa of the human bladder. *Urology* **49**:644–651.

Bichlmaier I, Kurkela M, Joshi T, Siiskonen A, Ruffer T, Lang H, Suchanova B, Vahermo M, Finel M and Yli-Kauhaluoma J (2007) Isoform-selective inhibition of the human UDP-glucuronosyltransferase 2B7 by isolongifolol derivatives. *J Med Chem* **50**:2655–2664.

Blevins-Primeau AS, Sun D, Chen G, Sharma AK, Gallagher CJ, Amin S and Lazarus P (2009) Functional significance of UDP-glucuronosyltransferase variants in the metabolism of active tamoxifen metabolites. *Cancer Res* **69**:1892–1900.

Bu HZ (2006) A literature review of enzyme kinetic parameters for CYP3A4-mediated metabolic reactions of 113 drugs in human liver microsomes: structure–kinetics relationship assessment. *Curr Drug Metab* **7**:231–249.

Butler MA, Lang NP, Young JF, Caporaso NE, Vineis P, Hayes RB, Teitel CH, Massengill JP, Lawsen MF and Kadlubar FF (1992) Determination of CYP1A2 and NAT2 phenotypes in human populations by analysis of caffeine urinary metabolites. *Pharmacogenetics* **2**:116–127.

Cashman JR and Zhang J (2006) Human flavin-containing monooxygenases. *Annu Rev Pharmacol Toxicol* **46**:65–100.

Ching MS, Blake CL, Malek NA, Angus PW and Ghabrial H (2001) Differential inhibition of human CYP1A1 and CYP1A2 by quinidine and quinine. *Xenobiotica* **31**:757–767.

Chiu TH, Chen JC, Chen LD, Lee JH and Chung JG (2004) Gypenosides inhibited *N*-acetylation of 2-aminofluorene, *N*-acetyltransferase gene expression and DNA adduct formation in human cervix epithelioid carcinoma cells (HeLa). *Res Commun Mol Pathol Pharmacol* **115–116**: 157–174.

Christensen LK, Hansen JM and Kristensen M (1963) Sulphaphenazole-induced hypoglycaemic attacks in tolbutamide-treated diabetics. *Lancet* **2**:1298–1301.

Chung JY, Cho JY, Yu KS, Kim JR, Lim KS, Sohn DR, Shin SG and Jang IJ (2008) Pharmacokinetic and pharmacodynamic interaction of lorazepam and valproic acid in relation to UGT2B7 genetic polymorphism in healthy subjects. *Clin Pharmacol Ther* **83**:595–600.

Clarke DJ, Moghrabi N, Monaghan G, Cassidy A, Boxer M, Hume R and Burchell B (1997) Genetic defects of the UDP-glucuronosyltransferase-1 (UGT1) gene that cause familial non-haemolytic unconjugated hyperbilirubinaemias. *Clin Chim Acta* **266**:63–74.

Coffman BL, Rios GR, King CD and Tephly TR (1997) Human UGT2B7 catalyzes morphine glucuronidation. *Drug Metab Dispos* **25**:1–4.

Coles B and Ketterer B (1990) The role of glutathione and glutathione transferases in chemical carcinogenesis. *Crit Rev Biochem Mol Biol* **25**:47–70.

Coles BF, Anderson KE, Doerge DR, Churchwell MI, Lang NP and Kadlubar FF (2000b) Quantitative analysis of interindividual variation of glutathione S-transferase expression in human pancreas and the ambiguity of correlating genotype with phenotype. *Cancer Res* **60**:573–579.

Coles BF, Morel F, Rauch C, Huber WW, Yang M, Teitel CH, Green B, Lang NP and Kadlubar FF (2001) Effect of polymorphism in the human glutathione S-transferase A1 promoter on hepatic GSTA1 and GSTA2 expression. *Pharmacogenetics* **11**:663–669.

Coles B, Yang M, Lang NP and Kadlubar FF (2000a) Expression of hGSTP1 alleles in human lung and catalytic activity of the native protein variants towards 1-chloro-2,4-dinitrobenzene, 4-vinylpyridine and (+)-anti benzo[a]pyrene-7,8-diol-9,10-oxide. *Cancer Lett* **156**: 167–175.

Conney AH (1982) Induction of microsomal enzymes by foreign chemicals and carcinogenesis by polycyclic aromatic hydrocarbons: G. H. A. Clowes Memorial Lecture. *Cancer Res* **42**: 4875–4917.

Dai D, Zeldin DC, Blaisdell JA, Chanas B, Coulter SJ, Ghanayem BI and Goldstein JA (2001) Polymorphisms in human CYP2C8 decrease metabolism of the anticancer drug paclitaxel and arachidonic acid. *Pharmacogenetics* **11**:597–607.

Daly AK, Aithal GP, Leathart JB, Swainsbury RA, Dang TS and Day CP (2007) Genetic susceptibility to diclofenac-induced hepatotoxicity: contribution of UGT2B7, CYP2C8, and ABCC2 genotypes. *Gastroenterology* **132**:272–281.

Dirven HA, van Ommen B and van Bladeren PJ (1996) Glutathione conjugation of alkylating cytostatic drugs with a nitrogen mustard group and the role of glutathione *S*-transferases. *Chem Res Toxicol* **9**:351–360.

Falany CN (1997) Enzymology of human cytosolic sulfotransferases. *FASEB J* **11**:206–216.

Faucette SR, Hawke RL, Lecluyse EL, Shord SS, Yan B, Laethem RM and Lindley CM (2000) Validation of bupropion hydroxylation as a selective marker of human cytochrome P450 2B6 catalytic activity. *Drug Metab Dispos* **28**:1222–1230.

Faucette SR, Wang H, Hamilton GA, Jolley SL, Gilbert D, Lindley C, Yan B, Negishi M and LeCluyse EL (2004) Regulation of CYP2B6 in primary human hepatocytes by prototypical inducers. *Drug Metab Dispos* **32**:348–358.

Ferguson SS, Chen Y, LeCluyse EL, Negishi M and Goldstein JA (2005) Human CYP2C8 is transcriptionally regulated by the nuclear receptors constitutive androstane receptor, pregnane X receptor, glucocorticoid receptor, and hepatic nuclear factor 4alpha. *Mol Pharmacol* **68**: 747–757.

Ferguson SS, LeCluyse EL, Negishi M and Goldstein JA (2002) Regulation of human CYP2C9 by the constitutive androstane receptor: discovery of a new distal binding site. *Mol Pharmacol* **62**:737–746.

Fuhr U and Rost KL (1994) Simple and reliable CYP1A2 phenotyping by the paraxanthine/caffeine ratio in plasma and in saliva. *Pharmacogenetics* **4**:109–116.

Glatt H, Boeing H, Engelke CE, Ma L, Kuhlow A, Pabel U, Pomplun D, Teubner W and Meinl W (2001) Human cytosolic sulphotransferases: genetics, characteristics, toxicological aspects. *Mutat Res* **482**:27–40.

Gonzalez FJ, Vilbois F, Hardwick JP, McBride OW, Nebert DW, Gelboin HV and Meyer UA (1988) Human debrisoquine 4-hydroxylase (P450IID1): cDNA and deduced amino acid sequence and assignment of the CYP2D locus to chromosome 22. *Genomics* **2**:174–179.

Grancharov K, Engelberg H, Naydenova Z, Muller G, Rettenmeier AW and Golovinsky E (2001a) Inhibition of UDP-glucuronosyltransferases in rat liver microsomes by natural mutagens and carcinogens. *Arch Toxicol* **75**:609–612.

Grancharov K, Naydenova Z, Lozeva S and Golovinsky E (2001b) Natural and synthetic inhibitors of UDP-glucuronosyltransferase. *Pharmacol Ther* **89**:171–186.

Grant DM, Blum M, Beer M and Meyer UA (1991) Monomorphic and polymorphic human arylamine *N*-acetyltransferases: a comparison of liver isozymes and expressed products of two cloned genes. *Mol Pharmacol* **39**:184–191.

Guillemette C (2003) Pharmacogenomics of human UDP-glucuronosyltransferase enzymes. *Pharmacogenomics J* **3**:136–158.

Harris RM and Waring RH (1996) Dietary modulation of human platelet phenolsulphotransferase activity. *Xenobiotica* **26**:1241–1247.

Hayes JD and Pulford DJ (1995) The glutathione S-transferase supergene family: regulation of GST and the contribution of the isoenzymes to cancer chemoprotection and drug resistance. *Crit Rev Biochem Mol Biol* **30**:445–600.

Hayes JD and Strange RC (2000) Glutathione S-transferase polymorphisms and their biological consequences. *Pharmacology* **61**:154–166.

Hayeshi R, Mutingwende I, Mavengere W, Masiyanise V and Mukanganyama S (2007) The inhibition of human glutathione S-transferases activity by plant polyphenolic compounds ellagic acid and curcumin. *Food Chem Toxicol* **45**:286–295.

Hebbring SJ, Adjei AA, Baer JL, Jenkins GD, Zhang J, Cunningham JM, Schaid DJ, Weinshilboum RM and Thibodeau SN (2007) Human SULT1A1 gene: copy number differences and functional implications. *Hum Mol Genet* **16**:463–470.

Hofmann MH, Blievernicht JK, Klein K, Saussele T, Schaeffeler E, Schwab M and Zanger UM (2008) Aberrant splicing caused by single nucleotide polymorphism c.516G>T [Q172H], a marker of CYP2B6*6, is responsible for decreased expression and activity of CYP2B6 in liver. *J Pharmacol Exp Ther* **325**:284–292.

Huang Z, Roy P and Waxman DJ (2000) Role of human liver microsomal CYP3A4 and CYP2B6 in catalyzing *N*-dechloroethylation of cyclophosphamide and ifosfamide. *Biochem Pharmacol* **59**:961–972.

Hughes HB (1953) On the metabolic fate of isoniazid. *J Pharmacol Exp Ther* **109**:444–452.

Ibeanu GC, Blaisdell J, Ferguson RJ, Ghanayem BI, Brosen K, Benhamou S, Bouchardy C, Wilkinson GR, Dayer P and Goldstein JA (1999) A novel transversion in the intron 5 donor splice junction of CYP2C19 and a sequence polymorphism in exon 3 contribute to the poor metabolizer phenotype for the anticonvulsant drug S-mephenytoin. *J Pharmacol Exp Ther* **290**:635–640.

Ibeanu GC, Blaisdell J, Ghanayem BI, Beyeler C, Benhamou S, Bouchardy C, Wilkinson GR, Dayer P, Daly AK and Goldstein JA (1998) An additional defective allele, CYP2C19*5, contributes to the S-mephenytoin poor metabolizer phenotype in Caucasians. *Pharmacogenetics* **8**:129–135.

Innocenti F, Undevia SD, Iyer L, Chen PX, Das S, Kocherginsky M, Karrison T, Janisch L, Ramirez J, Rudin CM, Vokes EE and Ratain MJ (2004) Genetic variants in the UDP-glucuronosyltransferase 1A1 gene predict the risk of severe neutropenia of irinotecan. *J Clin Oncol* **22**:1382–1388.

Jin Y, Desta Z, Stearns V, Ward B, Ho H, Lee KH, Skaar T, Storniolo AM, Li L, Araba A, Blanchard R, Nguyen A, Ullmer L, Hayden J, Lemler S, Weinshilboum RM, Rae JM, Hayes DF and Flockhart DA (2005) CYP2D6 genotype, antidepressant use, and tamoxifen metabolism during adjuvant breast cancer treatment. *J Natl Cancer Inst* **97**:30–39.

Johnson JA, Herring VL, Wolfe MS and Relling MV (2000) CYP1A2 and CYP2D6 4-hydroxylate propranolol and both reactions exhibit racial differences. *J Pharmacol Exp Ther* **294**: 1099–1105.

Kahn GC, Boobis AR, Murray S, Brodie MJ and Davies DS (1982) Assay and characterisation of debrisoquine 4-hydroxylase activity of microsomal fractions of human liver. *Br J Clin Pharmacol* **13**:637–645.

Karjalainen MJ, Neuvonen PJ and Backman JT (2008) In vitro inhibition of CYP1A2 by model inhibitors, anti-inflammatory analgesics and female sex steroids: predictability of in vivo interactions. *Basic Clin Pharmacol Toxicol* **103**:157–165.

Kassahun K, Farrell K and Abbott F (1991) Identification and characterization of the glutathione and *N*-acetylcysteine conjugates of (*E*)-2-propyl-2,4-pentadienoic acid, a toxic metabolite of valproic acid, in rats and humans. *Drug Metab Dispos* **19**:525–535.

Khan AJ, Ruwali M, Choudhuri G, Mathur N, Husain Q and Parmar D (2009) Polymorphism in cytochrome P450 2E1 and interaction with other genetic risk factors and susceptibility to alcoholic liver cirrhosis. *Mutat Res* **664**:55–63.

Kolars JC, Schmiedlin-Ren P, Schuetz JD, Fang C and Watkins PB (1992) Identification of rifampin-inducible P450IIIA4 (CYP3A4) in human small bowel enterocytes. *J Clin Invest* **90**:1871–1878.

Koukouritaki SB, Poch MT, Henderson MC, Siddens LK, Krueger SK, VanDyke JE, Williams DE, Pajewski NM, Wang T and Hines RN (2007) Identification and functional analysis of common human flavin-containing monooxygenase 3 genetic variants. *J Pharmacol Exp Ther* **320**:266–273.

Krueger SK and Williams DE (2005) Mammalian flavin-containing monooxygenases: structure/function, genetic polymorphisms and role in drug metabolism. *Pharmacol Ther* **106**: 357–387.

Krusekopf S, Roots I and Kleeberg U (2003) Differential drug-induced mRNA expression of human CYP3A4 compared to CYP3A5, CYP3A7 and CYP3A43. *Eur J Pharmacol* **466**: 7–12.

Kunii T, Fukasawa T, Yasui-Furukori N, Aoshima T, Suzuki A, Tateishi T, Inoue Y and Otani K (2005) Interaction study between enoxacin and fluvoxamine. *Ther Drug Monit* **27**:349–353.

Landi S (2000) Mammalian class theta GST and differential susceptibility to carcinogens: a review. *Mutat Res* **463**:247–283.

Lasker JM, Wester MR, Aramsombatdee E and Raucy JL (1998) Characterization of CYP2C19 and CYP2C9 from human liver: respective roles in microsomal tolbutamide, S-mephenytoin, and omeprazole hydroxylations. *Arch Biochem Biophys* **353**:16–28.

Lattard V, Zhang J, Tran Q, Furnes B, Schlenk D and Cashman JR (2003) Two new polymorphisms of the FMO3 gene in Caucasian and African-American populations: comparative genetic and functional studies. *Drug Metab Dispos* **31**:854–860.

Lee AM, Jepson C, Hoffmann E, Epstein L, Hawk LW, Lerman C and Tyndale RF (2007) CYP2B6 genotype alters abstinence rates in a bupropion smoking cessation trial. *Biol Psychiatry* **62**: 635–641.

Lee PC, Marquardt M and Lech JJ (1998) Metabolism of nonylphenol by rat and human microsomes. *Toxicol Lett* **99**:117–126.

Lemaire G, de Sousa G and Rahmani R (2004) A PXR reporter gene assay in a stable cell culture system: CYP3A4 and CYP2B6 induction by pesticides. *Biochem Pharmacol* **68**:2347–2358.

Lhoste EF, Gloux K, De Waziers I, Garrido S, Lory S, Philippe C, Rabot S and Knasmuller S (2004) The activities of several detoxication enzymes are differentially induced by juices of garden cress, water cress and mustard in human HepG2 cells. *Chem Biol Interact* **150**:211–219.

Lim YC, Desta Z, Flockhart DA and Skaar TC (2005) Endoxifen (4-hydroxy-*N*-desmethyl-tamoxifen) has anti-estrogenic effects in breast cancer cells with potency similar to 4-hydroxy-tamoxifen. *Cancer Chemother Pharmacol* **55**:471–478.

Lim YC, Li L, Desta Z, Zhao Q, Rae JM, Flockhart DA and Skaar TC (2006) Endoxifen, a secondary metabolite of tamoxifen, and 4-OH-tamoxifen induce similar changes in global gene expression patterns in MCF-7 breast cancer cells. *J Pharmacol Exp Ther* **318**:503–512.

Liu L, Wagner CR and Hanna PE (2008) Human arylamine *N*-acetyltransferase 1: in vitro and intracellular inactivation by nitrosoarene metabolites of toxic and carcinogenic arylamines. *Chem Res Toxicol* **21**:2005–2016.

Lofgren S, Baldwin RM, Hiratsuka M, Lindqvist A, Carlberg A, Sim SC, Schulke M, Snait M, Edenro A, Fransson-Steen R, Terelius Y and Ingelman-Sundberg M (2008) Generation of mice transgenic for human CYP2C18 and CYP2C19: characterization of the sexually dimorphic gene and enzyme expression. *Drug Metab Dispos* **36**:955–962.

Madabushi R, Frank B, Drewelow B, Derendorf H and Butterweck V (2006) Hyperforin in St. John's wort drug interactions. *Eur J Clin Pharmacol* **62**:225–233.

Mahgoub A, Idle JR, Dring LG, Lancaster R and Smith RL (1977) Polymorphic hydroxylation of Debrisoquine in man. *Lancet* **2**:584–586.

Masubuchi Y, Hosokawa S, Horie T, Suzuki T, Ohmori S, Kitada M and Narimatsu S (1994) Cytochrome P450 isozymes involved in propranolol metabolism in human liver microsomes. The role of CYP2D6 as ring-hydroxylase and CYP1A2 as *N*-desisopropylase. *Drug Metab Dispos* **22**:909–915.

Matsubara T, Noracharttiyapot W, Toriyabe T, Yoshinari K, Nagata K and Yamazoe Y (2007) Assessment of human pregnane X receptor involvement in pesticide-mediated activation of CYP3A4 gene. *Drug Metab Dispos* **35**:728–733.

Mitchell SC (2008) Flavin mono-oxygenase (FMO)–the 'other' oxidase. *Curr Drug Metab* **9**: 280–284.

Mori Y, Iimura K, Furukawa F, Nishikawa A, Takahashi M and Konishi Y (1995) Effect of cigarette smoke on the mutagenic activation of various carcinogens in hamster. *Mutat Res* **346**:1–8.

Nadin L and Murray M (1999) Participation of CYP2C8 in retinoic acid 4-hydroxylation in human hepatic microsomes. *Biochem Pharmacol* **58**:1201–1208.

Neff JA and Moody DE (2001) Differential *N*-demethylation of l-alpha-acetylmethadol (LAAM) and norLAAM by cytochrome P450s 2B6, 2C18, and 3A4. *Biochem Biophys Res Commun* **284**:751–756.

Olson JR, McGarrigle BP, Gigliotti PJ, Kumar S and McReynolds JH (1994) Hepatic uptake and metabolism of 2,3,7,8-tetrachlorodibenzo-*p*-dioxin and 2,3,7,8-tetrachlorodibenzofuran. *Fundam Appl Toxicol* **22**:631–640.

Omura T (1999) Forty years of cytochrome P450. *Biochem Biophys Res Commun* **266**:690–698.

Ozawa S, Tang YM, Yamazoe Y, Kato R, Lang NP and Kadlubar FF (1998) Genetic polymorphisms in human liver phenol sulfotransferases involved in the bioactivation of *N*-hydroxy derivatives of carcinogenic arylamines and heterocyclic amines. *Chem Biol Interact* **109**: 237–248.

Pacifici GM (2004) Inhibition of human liver and duodenum sulfotransferases by drugs and dietary chemicals: a review of the literature. *Int J Clin Pharmacol Ther* **42**:488–495.

Pascussi JM, Gerbal-Chaloin S, Drocourt L, Maurel P and Vilarem MJ (2003) The expression of CYP2B6, CYP2C9 and CYP3A4 genes: a tangle of networks of nuclear and steroid receptors. *Biochim Biophys Acta* **1619**:243–253.

Peters WH, Kock L, Nagengast FM and Kremers PG (1991) Biotransformation enzymes in human intestine: critical low levels in the colon? *Gut* **32**:408–412.

Phillips IR, Dolphin CT, Clair P, Hadley MR, Hutt AJ, McCombie RR, Smith RL and Shephard EA (1995) The molecular biology of the flavin-containing monooxygenases of man. *Chem Biol Interact* **96**:17–32.

Poonkuzhali B, Chandy M, Srivastava A, Dennison D and Krishnamoorthy R (2001) Glutathione S-transferase activity influences busulfan pharmacokinetics in patients with beta thalassemia major undergoing bone marrow transplantation. *Drug Metab Dispos* **29**:264–267.

Price RA, Spielman RS, Lucena AL, Van Loon JA, Maidak BL and Weinshilboum RM (1989) Genetic polymorphism for human platelet thermostable phenol sulfotransferase (TS PST) activity. *Genetics* **122**:905–914.

Raaska K and Neuvonen PJ (2000) Ciprofloxacin increases serum clozapine and *N*-desmethylclozapine: a study in patients with schizophrenia. *Eur J Clin Pharmacol* **56**:585–589.

Racha JK, Zhao ZS, Olejnik N, Warner N, Chan R, Moore D and Satoh H (2003) Substrate dependent inhibition profiles of fourteen drugs on CYP3A4 activity measured by a high throughput LCMS/MS method with four probe drugs, midazolam, testosterone, nifedipine and terfenadine. *Drug Metab Pharmacokinet* **18**:128–138.

Rae JM, Johnson MD, Lippman ME and Flockhart DA (2001) Rifampin is a selective, pleiotropic inducer of drug metabolism genes in human hepatocytes: studies with cDNA and oligonucleotide expression arrays. *J Pharmacol Exp Ther* **299**:849–857.

Raftogianis RB, Wood TC, Otterness DM, Van Loon JA and Weinshilboum RM (1997) Phenol sulfotransferase pharmacogenetics in humans: association of common SULT1A1 alleles with TS PST phenotype. *Biochem Biophys Res Commun* **239**:298–304.

Ragunathan N, Dairou J, Pluvinage B, Martins M, Petit E, Janel N, Dupret JM and Rodrigues-Lima F (2008) Identification of the xenobiotic-metabolizing enzyme arylamine *N*-acetyltransferase 1 as a new target of cisplatin in breast cancer cells: molecular and cellular mechanisms of inhibition. *Mol Pharmacol* **73**:1761–1768.

Raucy JL, Mueller L, Duan K, Allen SW, Strom S and Lasker JM (2002) Expression and induction of CYP2C P450 enzymes in primary cultures of human hepatocytes. *J Pharmacol Exp Ther* **302**:475–482.

Rebbeck TR (1997) Molecular epidemiology of the human glutathione S-transferase genotypes GSTM1 and GSTT1 in cancer susceptibility. *Cancer Epidemiol Biomarkers Prev* **6**:733–743.

Reilly PE, Mason SR and Gillam EM (1988) Differential inhibition of human liver phenacetin O-deethylation by histamine and four histamine H2-receptor antagonists. *Xenobiotica* **18**: 381–387.

Roby CA, Anderson GD, Kantor E, Dryer DA and Burstein AH (2000) St John's Wort: effect on CYP3A4 activity. *Clin Pharmacol Ther* **67**:451–457.

Rowe JD, Nieves E and Listowsky I (1997) Subunit diversity and tissue distribution of human glutathione S-transferases: interpretations based on electrospray ionization-MS and peptide sequence-specific antisera. *Biochem J* **325 (Pt 2)**:481–486.

Sahi J, Black CB, Hamilton GA, Zheng X, Jolley S, Rose KA, Gilbert D, LeCluyse EL and Sinz MW (2003) Comparative effects of thiazolidinediones on in vitro P450 enzyme induction and inhibition. *Drug Metab Dispos* **31**:439–446.

Schuetz JD, Kauma S and Guzelian PS (1993) Identification of the fetal liver cytochrome CYP3A7 in human endometrium and placenta. *J Clin Invest* **92**:1018–1024.

Shimada T, Inoue K, Suzuki Y, Kawai T, Azuma E, Nakajima T, Shindo M, Kurose K, Sugie A, Yamagishi Y, Fujii-Kuriyama Y and Hashimoto M (2002) Arylhydrocarbon receptor-dependent induction of liver and lung cytochromes P450 1A1, 1A2, and 1B1 by polycyclic aromatic hydrocarbons and polychlorinated biphenyls in genetically engineered C57BL/6 J mice. *Carcinogenesis* **23**:1199–1207.

Shimada T, Yamazaki H, Mimura M, Inui Y and Guengerich FP (1994) Interindividual variations in human liver cytochrome P-450 enzymes involved in the oxidation of drugs, carcinogens and toxic chemicals: studies with liver microsomes of 30 Japanese and 30 Caucasians. *J Pharmacol Exp Ther* **270**:414–423.

Shimizu M, Yano H, Nagashima S, Murayama N, Zhang J, Cashman JR and Yamazaki H (2007) Effect of genetic variants of the human flavin-containing monooxygenase 3 on *N*- and *S*-oxygenation activities. *Drug Metab Dispos* **35**:328–330.

Sim E, Walters K and Boukouvala S (2008) Arylamine *N*-acetyltransferases: from structure to function. *Drug Metab Rev* **40**:479–510.

Slattery JT, Wilson JM, Kalhorn TF and Nelson SD (1987) Dose-dependent pharmacokinetics of acetaminophen: evidence of glutathione depletion in humans. *Clin Pharmacol Ther* **41**: 413–418.

Snawder JE, Roe AL, Benson RW and Roberts DW (1994) Loss of CYP2E1 and CYP1A2 activity as a function of acetaminophen dose: relation to toxicity. *Biochem Biophys Res Commun* **203**:532–539.

Sun D, Sharma AK, Dellinger RW, Blevins-Primeau AS, Balliet RM, Chen G, Boyiri T, Amin S and Lazarus P (2007) Glucuronidation of active tamoxifen metabolites by the human UDP glucuronosyltransferases. *Drug Metab Dispos* **35**:2006–2014.

Swales K, Kakizaki S, Yamamoto Y, Inoue K, Kobayashi K and Negishi M (2005) Novel CAR-mediated mechanism for synergistic activation of two distinct elements within the human cytochrome P450 2B6 gene in HepG2 cells. *J Biol Chem* **280**:3458–3466.

Takeda S, Kitajima Y, Ishii Y, Nishimura Y, Mackenzie PI, Oguri K and Yamada H (2006) Inhibition of UDP-glucuronosyltransferase 2b7-catalyzed morphine glucuronidation by ketoconazole: dual mechanisms involving a novel noncompetitive mode. *Drug Metab Dispos* **34**:1277–1282.

Talalay P and Fahey JW (2001) Phytochemicals from cruciferous plants protect against cancer by modulating carcinogen metabolism. *J Nutr* **131**:3027S–3033S.

Valentini A, Biancolella M, Amati F, Gravina P, Miano R, Chillemi G, Farcomeni A, Bueno S, Vespasiani G, Desideri A, Federici G, Novelli G and Bernardini S (2007) Valproic acid induces neuroendocrine differentiation and UGT2B7 up-regulation in human prostate carcinoma cell line. *Drug Metab Dispos* **35**:968–972.

van Ommen B, Bogaards JJ, Peters WH, Blaauboer B and van Bladeren PJ (1990) Quantification of human hepatic glutathione S-transferases. *Biochem J* **269**:609–613.

Venkatakrishnan K, Von Moltke LL and Greenblatt DJ (2001) Human drug metabolism and the cytochromes P450: application and relevance of in vitro models. *J Clin Pharmacol* **41**: 1149–1179.

Walle T (1996) Assays of CYP2C8- and CYP3A4-mediated metabolism of taxol in vivo and in vitro. *Methods Enzymol* **272**:145–151.

Walle T, Otake Y, Galijatovic A, Ritter JK and Walle UK (2000) Induction of UDP-glucuronosyltransferase UGT1A1 by the flavonoid chrysin in the human hepatoma cell line hep G2. *Drug Metab Dispos* **28**:1077–1082.

Walsky RL, Gaman EA and Obach RS (2005) Examination of 209 drugs for inhibition of cytochrome P450 2C8. *J Clin Pharmacol* **45**:68–78.

Walters DG, Young PJ, Agus C, Knize MG, Boobis AR, Gooderham NJ and Lake BG (2004) Cruciferous vegetable consumption alters the metabolism of the dietary carcinogen 2-amino-1-methyl-6-phenylimidazo[4,5-b]pyridine (PhIP) in humans. *Carcinogenesis* **25**:1659–1669.

Wang H and Tompkins LM (2008) CYP2B6: new insights into a historically overlooked cytochrome P450 isozyme. *Curr Drug Metab* **9**:598–610.

Wang JS, Neuvonen M, Wen X, Backman JT and Neuvonen PJ (2002) Gemfibrozil inhibits CYP2C8-mediated cerivastatin metabolism in human liver microsomes. *Drug Metab Dispos* **30**:1352–1356.

Waszkowycz B, Clark DE, Frenkel D, Li J, Murray CW, Robson B and Westhead DR (1994) PRO_LIGAND: an approach to de novo molecular design. 2. Design of novel molecules from molecular field analysis (MFA) models and pharmacophores. *J Med Chem* **37**:3994–4002.

Weber WW and Hein DW (1985) *N*-Acetylation pharmacogenetics. *Pharmacol Rev* **37**:25–79.

Weinshilboum RM, Otterness DM, Aksoy IA, Wood TC, Her C and Raftogianis RB (1997) Sulfation and sulfotransferases 1: Sulfotransferase molecular biology: cDNAs and genes. *FASEB J* **11**:3–14.

Yamazaki H, Suzuki M, Tane K, Shimada N, Nakajima M and Yokoi T (2000) In vitro inhibitory effects of troglitazone and its metabolites on drug oxidation activities of human cytochrome P450 enzymes: comparison with pioglitazone and rosiglitazone. *Xenobiotica* **30**:61–70.

Yanagihara Y, Kariya S, Ohtani M, Uchino K, Aoyama T, Yamamura Y and Iga T (2001) Involvement of CYP2B6 in *n*-demethylation of ketamine in human liver microsomes. *Drug Metab Dispos* **29**:887–890.

Zaher H, Buters JT, Ward JM, Bruno MK, Lucas AM, Stern ST, Cohen SD and Gonzalez FJ (1998) Protection against acetaminophen toxicity in CYP1A2 and CYP2E1 double-null mice. *Toxicol Appl Pharmacol* **152**:193–199.

Zanger UM, Klein K, Saussele T, Blievernicht J, Hofmann MH and Schwab M (2007) Polymorphic CYP2B6: molecular mechanisms and emerging clinical significance. *Pharmacogenomics* **8**:743–759.

Zanger UM, Turpeinen M, Klein K and Schwab M (2008) Functional pharmacogenetics/genomics of human cytochromes P450 involved in drug biotransformation. *Anal Bioanal Chem* **392**:1093–1108.

Zhang ZY, Kerr J, Wexler RS, Li HY, Robinson AJ, Harlow PP and Kaminsky LS (1997) Warfarin analog inhibition of human CYP2C9-catalyzed S-warfarin 7-hydroxylation. *Thromb Res* **88**:389–398.

Zhou SF (2008) Drugs behave as substrates, inhibitors and inducers of human cytochrome P450 3A4. *Curr Drug Metab* **9**:310–322.

Chapter 2
Transporters: Importance in Drug Absorption, Distribution, and Removal

Frans G.M. Russel

Abstract There is an increasing appreciation of the role that transport proteins play in the absorption, distribution, and elimination of a wide variety of drugs in clinical use. These transporters can be divided into efflux transporters belonging to the ATP-binding cassette (ABC) family and solute carrier (SLC) family members that mediate the influx or bidirectional movement of drugs across the cell membrane. Their coordinated expression and activities at the basolateral and apical side of transporting epithelia are significant determinants of drug disposition, drug–drug interactions, and variability in drug response and toxicity. This chapter focuses on the major SLC and ABC drug transporters expressed in intestine, liver, and kidney, with special emphasis on their distribution, mode of action, and drug substrate specificity.

2.1 Introduction

During the last 20 years, a large number of membrane transport proteins have been identified. These transporters are important determinants in the absorption, distribution, and elimination of drugs. The involvement of carrier-mediated processes in drug excretion was already appreciated long before the first transporters were cloned, and it has become increasingly apparent that transporters also play a critical role in drug absorption and tissue uptake.

Drug transport proteins can be grouped into two major classes, the solute carriers (SLC) and ATP-binding cassette (ABC) transporters. Over 380 unique SLC sequences have been obtained from the human genome, which can be divided into 48 subfamilies (Fredriksson et al., 2008). The transport activities for xenobiotics for

F.G.M. Russel (✉)
Department Pharmacology and Toxicology, Radboud University Nijmegen Medical Centre, Nijmegen Centre for Molecular Life sciences, Nijmegen, Netherlands
e-mail: f.russel@ncmls.ru.nl

K.S. Pang et al. (eds.), *Enzyme- and Transporter-Based Drug–Drug Interactions*, DOI 10.1007/978-1-4419-0840-7_2,

approximately 19 of these gene families have been described. These transporters include the organic anion transporting polypeptide (*SLCO*), the oligopeptide transporter (*SLC15*), the organic anion/cation/zwitterion transporter (*SLC22*), and the organic cation transporter (*SLC47*) families. Seven subfamilies of ABC transporter genes have been identified, encoding for 49 different proteins (Dean and Allikmets, 2001; Sheps and Ling, 2007). A number of them have specificities for drugs, in particular transporters belonging to the *ABCB*, *ABCC*, and *ABCG* subfamilies (Szakacs et al., 2008).

SLC and ABC transporters share a wide distribution in the body and are involved in the transport of a broad range of substrates. Many of them can potentially contribute to the permeability of drugs into cells and the processes by which drugs gain access to their pharmacological and toxicological targets. There is growing evidence to suggest that it is the rule rather than the exception that a given drug will interact with a set of membrane transporters at some point of its disposition in the body (Dobson and Kell, 2008). Since the intestine, liver, and kidney are the principal organs that determine the absorption, distribution, and elimination of drugs, this chapter focuses on the general characteristics of the major human drug transporters expressed in these organs. The SLC and ABC transporters listed in Tables 2.1 and 2.2 are currently considered to exert the greatest impact on overall drug disposition, pharmacokinetic variability, and drug–drug interactions (Mizuno and Sugiyama, 2002; Tsuji, 2002; Ho and Kim, 2005; Ito et al., 2005; Shitara et al., 2005; Endres et al., 2006; Zair et al., 2008).

Depending on the direction in which carrier proteins translocate the substrate across the cell membrane, they can be categorized as influx or efflux transporters. ABC transporters are by definition efflux transporters because they use energy derived from ATP hydrolysis to mediate the primary active export of drugs from the intracellular to the extracellular milieu, often against a steep diffusion gradient. Many of the SLC family members facilitate the cellular uptake or influx of substrates, either by facilitated diffusion down the electrochemical gradient acting as a channel or uniporter or by secondary active transport against a diffusion gradient coupled to the symport or antiport of inorganic or small organic ions to provide the driving force. Certain SLC transporters exhibit efflux properties or are bidirectional, depending on the concentration gradients of substrate and coupled ion across the membrane.

It is important to understand that the interplay between transporters located on apical and basolateral membranes in epithelial cells is critical in determining the extent and direction of drug movement in organs such as the intestine, liver, and kidney (Figs. 2.1, 2.2, and 2.3). Transport across each of these epithelia may be impeded or facilitated by the asymmetrical membrane distribution of influx and efflux transporters and ultimately contributes to the pharmacokinetic profile of a drug substrate in the body. In this respect, transporters that mediate vectorial drug transfer into the systemic circulation are named as absorptive transporters, regardless of whether they are influx or efflux transporters. In contrast, secretory transporters are involved in the excretion of substrates from the circulation into bile, urine, or gut lumen.

Table 2.1 Major human SLC drug transporters expressed in small intestine, liver, and kidney

Gene	Protein	Mechanism	Tissue distribution	Membrane localization	Examples of drug substrates	References
SLC15 family						
SLC15A1	PEPT1	H^+/peptide symporter	Intestine Kidney	BBM	Ampicillin, amoxicillin, bestatin, cefaclor, cefadroxil, cefixime, enalapril, temocapril, temocaprilat, midodrine, valacyclovir, valganciclovir	Russel et al. (2002), Brandsch et al. (2008), Dobson and Kell (2008), and Rubio-Aliaga and Daniel (2008)
SLC15A2	PEPT2	H^+/peptide symporter	Kidney	BBM	Amoxicillin, bestatin, cefaclor, cefadroxil, valganciclovir	Russel et al. (2002), Brandsch et al. (2008), Dobson and Kell (2008), and Rubio-Aliaga and Daniel (2008)
SLC22 family						
SLC22A1	OCT1	OC uniporter	Intestine Liver	BLM SM	Acyclovir, ganciclovir, metformin, cimetidine, quinine, quinidine, zidovudine	Koepsell et al. (2007), Ciarimboli (2008), and Dobson and Kell (2008)
SLC22A2	OCT2	OC uniporter	Kidney	BLM	Mepiperphenidol, memantine, cimetidine, famotidine, ranitidine, metformin, propranolol, pancuronium, quinine, zidovudine, cisplatin	Koepsell et al. (2007), Ciarimboli (2008), and Dobson and Kell (2008)
SLC22A4	OCTN1	H^+ or OC antiporter	Intestine Kidney	BBM	Mepyramine, quinidine, verapamil, ergothioneine, gabapentin	Koepsell et al. (2007), Dobson and Kell (2008), and Urban et al. (2008)
SLC22A5	OCTN2	OC antiporter Na^+ symporter (carnitine)	Intestine Kidney	BBM	Mepyramine, quinidine, verapamil, valproate, cephaloridine, emetine	Koepsell et al. (2007), Dobson and Kell (2008)

Table 2.1 (continued)

Gene	Protein	Mechanism	Tissue distribution	Membrane localization	Examples of drug substrates	References
SLC22A6	OAT1	DC/OA antiporter	Kidney	BLM	Adefovir, cidofovir, acyclovir, didanosine, stavudine, trifluridine, ganciclovir, PMEG, PMEDAP, tenofovir, zalcitabine, zidovudine, tetracycline, methotrexate, bumetanide, furosemide, ibuprofen, indomethacin, ketoprofen, PAH, cimetidine, ranitidine	Russel et al. (2002), Rizwan and Burckhardt (2007), Dobson and Kell (2008), and Srimaroeng et al. (2008)
SLC22A7	OAT2	OA antiporter	Liver	SM	Erythromycin, cimetidine, ranitidine, zidovudine, 5-fluorouracil, methotrexate, taxol, bumetanide, allopurinol, salicylate, PAH, theophylline	Russel et al. (2002), Rizwan and Burckhardt (2007), Dobson and Kell (2008), and Srimaroeng et al. (2008)
SLC22A8	OAT3	DC/OA antiporter	Kidney	BLM	Benzylpenicillin, tetracycline, valacyclovir, zidovudine, cimetidine, ranitidine, methotrexate, furosemide, ibuprofen, indomethacin, ketoprofen, salicylate, PAH, pravastatin, olmesartan	Russel et al. (2002), Rizwan and Burckhardt (2007), Dobson and Kell (2008), Srimaroeng et al. (2008), and Kusuhara and Sugiyama (2009)
SLC22A9	OAT4	Cl^-/OA antiporter	Kidney	BBM	Tetracycline, zidovudine, methotrexate, bumetanide, ketoprofen, salicylate, PAH	Russel et al. (2002), Rizwan and Burckhardt (2007), Dobson and Kell (2008), and Srimaroeng et al. (2008)
SLC47 family						
SLC47A1	MATE1	H^+/OC antiporter	Liver Kidney (PT, DT)	CM BBM	Cimetidine, procainamide, metformin, cephalexin, cephadrine, fexofenadine	Koepsell et al. (2007), Tanihara et al. (2007), Moriyama et al. (2008), Terada and Inui (2008), and Matsushima et al. (2009)

Table 2.1 (continued)

Gene	Protein	Mechanism	Tissue distribution	Membrane localization	Examples of drug substrates	References
SLC47A2	MATE2-K	H^+/OC antiporter	Kidney	BBM	Cimetidine, procainamide, metformin, fexofenadine, oxaliplatin	Ho and Kim (2005), Koepsell et al. (2007), Moriyama et al. (2008), Terada and Inui (2008), and Matsushima et al. (2009)
SLCO family						
SLCO1A2	OATP1A2	OA antiporter	Intestine Kidney (DT)	BBM BBM	Fexofenadine, indomethacin, ouabain, rocuronium, enalapril, temocaprilat, rosuvastatin, pitavastatin, levofloxacin, methotrexate, imatinib, saquinavir	Dobson and Kell (2008), Hagenbuch and Gui (2008), and Hu et al. (2008)
SLCO1B1	OATP1B1	OA antiporter	Liver	BBM	Benzylpenicillin, rifampicin, atorvastatin, pravastatin, cerivastatin, fluvastatin, pitavastatin, rosuvastatin, simvastatin, valsartan, olmesartan, troglitazone, bosentan, enalapril, caspofungin, fexofenadine, SN-38	Noe et al. (2007), Dobson and Kell (2008), Hagenbuch and Gui (2008), Hu et al. (2008), and Nies et al. (2008)
SLCO1B3	OATP1B3	OA antiporter	Liver (around central vein)	BLM	Digoxin,ouabain, rifampicin, bosentan, enalapril, fluvastatin, pitavastatin, rosuvastatin, telmisartan, valsartan, fexofenadine, methotrexate, paclitaxel	Dobson and Kell (2008), Hagenbuch and Gui (2008), Hu et al. (2008), and Nies et al. (2008)
SLCO2B1	OATP2B1	OA antiporter	Liver Intestine	SM BBM	Benzylpenicillin, bosentan, atorvastatin, pravastatin, pitavastatin, fluvastatin, rosuvastatin, glibenclamide	Dobson and Kell (2008), Hagenbuch and Gui (2008), Hu et al. (2008), and Nies et al. (2008)
SLCO4C1	OATP4C1	ND	Kidney	BLM	Digoxin, ouabain, methotrexate	Dobson and Kell (2008), Hagenbuch and Gui (2008), and Hu et al. (2008)

Abbreviations: PMEG = 9-(2-phosphonylmethoxyethyl)guanine; PMEDAP = 9-(2-phosphonylmethoxyethyl)-2,6-diaminopurine; BBM = brush border membrane; BLM = basolateral membrane; SM = sinusoidal membrane; CM = canalicular membrane; PT = proximal tubule; DT = distal tubule; DC = dicarboxylate; PAH = *p*-aminohippuric acid; SN-38 = active metabolite of irinotecan; ND = not determined

Table 2.2 Major human ABC drug transporters expressed in small intestine, liver, and kidney

Gene	Protein	Mechanism	Tissue distribution	Membrane localization	Examples of drug substrates	References
ABCB family						
ABCB1	MDR1/ P-glycoprotein	Primary active	Intestine Liver Kidney	BBM CM BBM	Vinblastine, vincristine, daunorubicin, doxorubicin, colchicine, docetaxel, paclitaxel, ortataxel, etoposide, imatinib, methotrexate, bisantrene, mitoxantrone, paclitaxel, topotecan, digoxin, digitoxin, celiprolol, talinolol, indinavir, nelfinavir, ritonavir, saquinavir, levofloxacin, grepafloxacin, sparfloxacin, erythromycin, ivermectin, chloroquine, amiodarone, lidocaine, losartan, lovastatin, mibefradil, fexofenadine, terfenadine, carbamazepine, desipramine, loperamide, methadone, morphine, sumatriptan, vecuronium, cyclosporin A, tacrolimus, sirolimus	Russel et al. (2002), Dietrich et al. (2003), Sarkadi et al. (2006), Hu et al. (2008), Murakami and Takano (2008), Zhou (2008), and Oostendorp et al. (2009)
ABCC family						
ABCC2	MRP2	Primary active	Intestine Liver Kidney	BBM CM BBM	Vinblastine, vincristine, doxorubicin, etoposide, cisplatin, methotrexate, indinavir, ritonavir, saquinavir, grepafloxacin, glutathione conjugates, PAH	Russel et al. (2002), van de Water et al. (2005), Nies and Keppler (2007), and Zhou et al. (2008)
ABCC3	MRP3	Primary active	Intestine Liver Kidney	BLM SM BLM (CCD)	Glucuronide conjugates (morphine, acetaminophen, etoposide, ethinylestradiol), methotrexate	Russel et al. (2002), van de Water et al. (2005), Borst et al. (2007), and Zhou et al. (2008)

Table 2.2 (continued)

Gene	Protein	Mechanism	Tissue distribution	Membrane localization	Examples of drug substrates	References
ABCC4	MRP4	Primary active	Intestine Liver Kidney	BBM? SM BBM	Methotrexate, leucovorin, topotecan, 6-mercaptopurine, 6-thioguanine, adefovir, tenofovir, ceftizoxime, cefazolin, cefotaxime, cefmetazole, hydrochlorothiazide, furosemide, olmesartan, edaravone glucuronide, PAH	Russel et al. (2002), van de Water et al. (2005), Nies and Keppler (2007), Russel et al. (2008), and Zhou et al. (2008)
ABCG family						
ABCG2	BCRP	Primary active	Intestine Liver Kidney	BLM CM BBM	Mitoxantrone, flavopiridol, topotecan, SN-38, camptothecin, methotrexate, imatinib, gefitinib, erlotinib, abacavir, lamivudine, zidovudine, nelfinavir, cerivastatin, pitavastatin, rosuvastatin, glibenclamide, olmesartan, dipyridamole, cimetidine, edaravone sulfate, albendazole sulfoxide, oxfendazole, ciprofloxacin, norfloxacin, ofloxacin, sulfasalazine, nitrofurantoin,	Sarkadi et al. (2006), van Herwaarden and Schinkel (2006), Cusatis and Sparreboom (2008), and Robey et al. (2009)

Abbreviations: BBM = brush border membrane; BLM = basolateral membrane; SM = sinusoidal membrane; CM = canalicular membrane; CCD = cortical collecting duct; PAH = *p*-aminohippuric acid; SN-38 = active metabolite of irinotecan

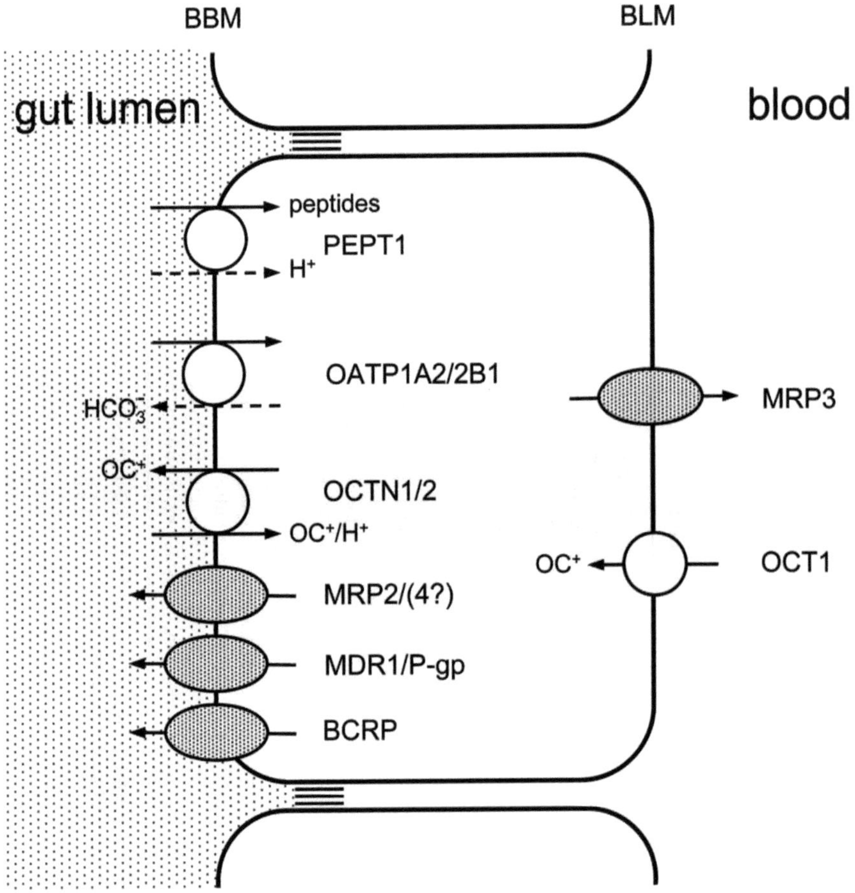

Fig. 2.1 Schematic model of the major drug transporters in enterocytes of human small intestine. SLC transporters are depicted by *open circles* and ABC transporters by *shaded ovals*. *Solid arrows* indicate the direction of drug transport. *Dashed arrows* depict the movement of driving ions. OCT1 is an electrogenic uniporter that transports organic cations (OC^+) from blood into the cell driven by the inside-negative membrane potential. OCTN1 mediates OC^+ uptake from gut lumen as a H^+/OC^+ antiporter or can operate like OCTN2 as a bidirectional cation exchanger, mediating influx or efflux. Peptidomimetic drugs are taken up by a H^+/peptide symporter (PEPT1). Amphipathic drugs are transported into the cell by the organic anion/HCO_3^- antiporters OATP1A2 and OATP2B1 and are extruded as parent compound or metabolites back to the lumen or into blood by the primary active ABC transporters MDR1/P-gp, MRP2, MRP3, MRP4, and BCRP

2.2 SLC Drug Transporters

The SLC subfamilies *SLC15*, *SLC22*, and *SLCO* are considered to have a major role in drug uptake into intestine, liver, and kidney, whereas *SLC47* members mediate drug efflux into bile and urine (Table 2.1, Figs. 2.1, 2.2, and 2.3). Most of these transporters have a similar protein structure in that they consist of 12 putative

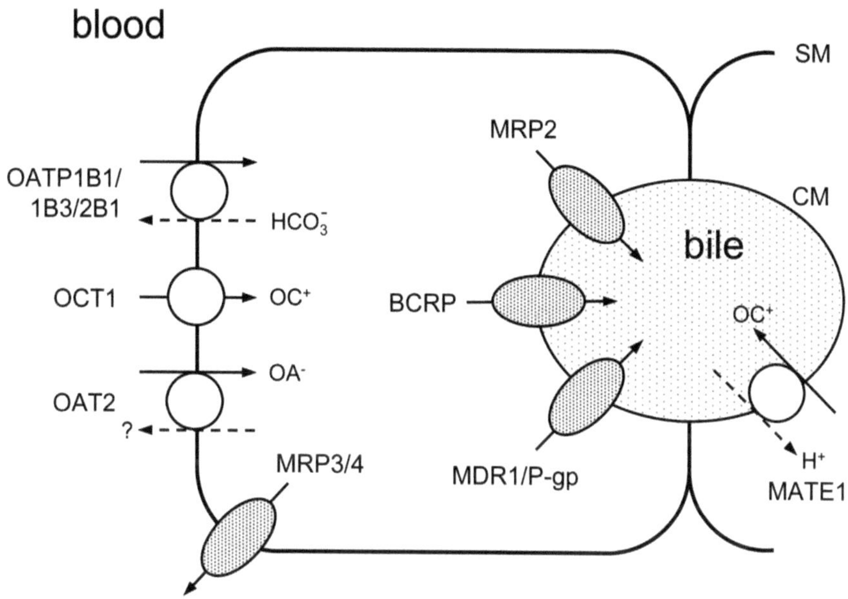

Fig. 2.2 Schematic model of the major drug transporters in human hepatocytes. SLC transporters are depicted by *open circles* and ABC transporters by *shaded ovals*. *Solid arrows* indicate the direction of drug transport. *Dashed arrows* depict the movement of driving ions. OCT1 is an electrogenic uniporter that transports organic cations (OC^+) from blood into the cell driven by the inside negative membrane potential. MATE1 is a biliary OC^+ efflux transporter that operates as a H^+ antiporter. Organic anions (OA^-) are taken up by OAT2, an antiporter for which the driving ion is unknown. Amphipathic drugs are transported into the cell by the organic anion/HCO_3^- antiporters OATP1B1, OATP1B3, and OATP2B1 and are extruded as parent compound or metabolites into bile or blood by the primary active ABC transporters MDR1/P-gp, MRP2, MRP4, and BCRP

membrane-spanning domains and their molecular mass is approximately between 50 and 100 kDa.

Oligopeptide transporters (*SLC15*) operate as H^+-coupled symporters of di- and tripeptides and have the ability to transport peptidomimetics and other drug substrates. PEPT1 is the small intestinal low-affinity, high-capacity peptide transporter, which is also expressed at low levels in the kidney. PEPT2 is the predominant high-affinity, low-capacity renal peptide transporter, although expression has also been observed in other organs (Brandsch et al., 2008; Kamal et al., 2008; Rubio-Aliaga and Daniel, 2008). Typical PEPT1 substrates have amino and carboxylic groups that can be mapped onto a di- or tripeptide skeleton, which allows for rational prodrug design to improve oral drug availability upon increased intestinal absorption (Bailey et al., 2006). PEPT2 reabsorbs renally filtered drugs from tubular fluid into proximal tubules, thereby affecting the systemic pharmacokinetics of some drugs. PEPT1/2 substrates include many important drug classes, including antivirals, β-lactam antibiotics, and angiotensin-converting enzyme (ACE) inhibitors (Table 2.1).

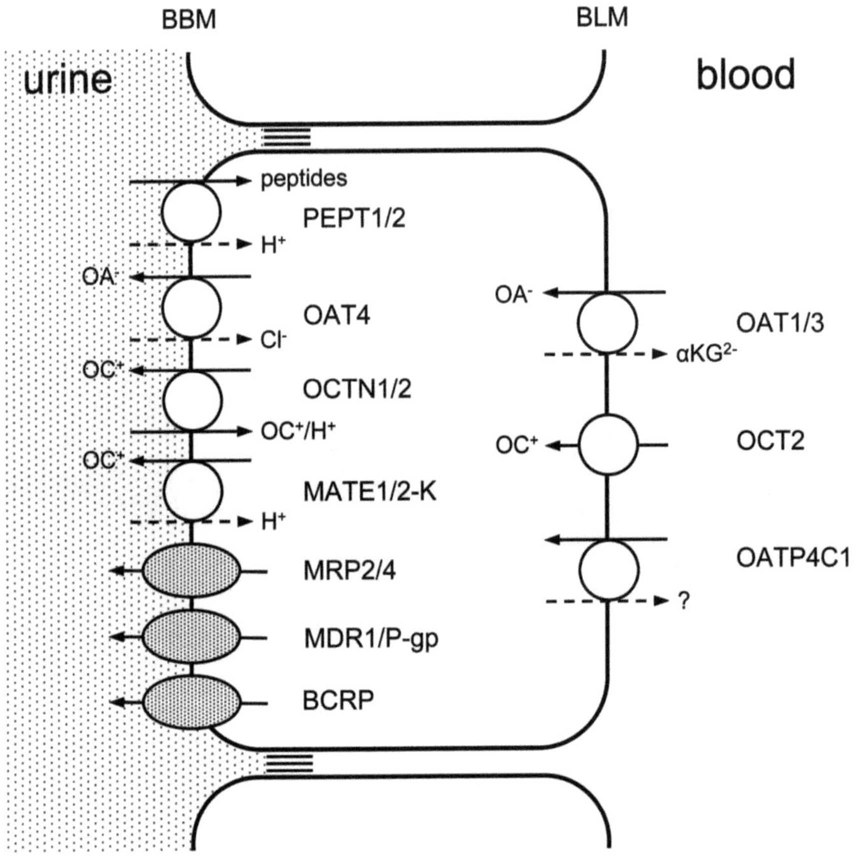

Fig. 2.3 Schematic model of the major drug transporters in human renal proximal tubular cells. SLC transporters are depicted by *open circles* and ABC transporters by *shaded ovals*. *Solid arrows* indicate the direction of drug transport. *Dashed arrows* depict the movement of driving ions. OCT2 is an electrogenic uniporter that transports organic cations (OC^+) from blood into the cell driven by the inside-negative membrane potential. OCTN1 mediates luminal OC^+ uptake as a H^+/OC^+ antiporter or can operate like OCTN2 as a bidirectional cation exchanger, mediating influx or efflux. MATE1 is a urinary OC^+ efflux transporter that operates as a H^+ antiporter. Peptidomimetic drugs are taken up by the H^+/peptide symporters PEPT1 and PEPT2. Organic anions (OA^-) are taken up by the antiporters OAT1 and OAT3, which are driven by the exchange with α-ketoglutarate (α-KG^{2-}), and released by OAT4 in exchange for Cl^-. A few amphipathic drugs are transported into the cell by the organic anion antiporter OATP4C1, for which the driving ion is unknown. The primary active ABC transporters MDR1/P-gp, MRP2, MRP4, and BCRP drive the efflux of a wide variety of amphipathic drugs and metabolites into urine

Organic cation/anion/zwitterion transporters (SLC22) are characterized by a remarkably broad substrate specificity and wide tissue distribution. The major drug transporting members of this family can be found in the liver, kidney, and intestine. The organic anion transporters (OATs) are mainly located in the kidney, but some also occur in the liver, placenta, and brain. They function as antiporters, by coupling the cellular influx of an organic anion to the exchange with dicarboxylates or other organic anions from the cell. The existing inside-out concentration gradient of these anions provides the driving force for the active uptake of anionic drugs against the inside-negative membrane potential. OATs are part of the excretory systems for anionic drug substrates and they typically interact with hydrophilic, small molecular weight (MW < 400–500 Da) substrates, like *p*-aminohippuric acid (PAH), which are categorized as type I organic anions (Table 2.1).

SLC22 subfamily members that specialize in the translocation of cationic drugs consist of the organic cation transporters (OCT1/2) that operate as uniporters, the organic cation and carnitine transporter (OCTN1) that could be a proton antiporter, and the Na^+-carnitine symporter (OCTN2) that also acts as a Na^+-independent organic cation antiporter. The tissue distribution of these transporters differs among species; in human, OCT1 is widely expressed, most strongly in liver, to a lesser extent in intestine and even lower in kidney. OCT2 is present abundantly in the kidney, whereas the expression is low in the small intestine and absent in the liver. OCTN1 is strongly expressed in kidney, skeletal muscle, bone marrow, and trachea, and weaker in intestine and liver. OCTN2 is distributed ubiquitously, with strongest expression in kidney, skeletal muscle, heart, placenta, and liver (Koepsell et al., 2007). Both OCT1 and OCT2 mediate cellular uptake by facilitated diffusion down the inside-negative electrochemical gradient of comparatively small monovalent, type I (MW<400 Da) organic cations (Table 2.1). For OCT2, recent quantitative structure–activity relationship models have emphasized the importance of hydrophobicity, molecular size, and shape as important determinants in defining the binding of drug substrates and may prove useful in the prediction of unwanted drug interactions (Suhre et al., 2005). OCTN1 and OCTN2 exhibit a similar though somewhat less extensive substrate specificity (Koepsell et al., 2007).

The *SLCO* subfamily consists of organic anion transporting polypeptides (OATPs) that are involved in the cellular uptake of more bulky (MW>450 Da) and relatively hydrophobic organic anions that are classified as type II (Table 2.1). In addition, the substrate specificity of the OATPs covers a wide range of amphipathic organic compounds, including bile salts, steroid conjugates, thyroid hormones, and various drugs (Hagenbuch and Gui, 2008). Current evidence suggests that these transporters act as organic anion exchangers, although the precise mechanism remains ambiguous. In rodents, substrate exchange for glutathione was observed but for the human transporters there is evidence to support a role for bicarbonate as a counterion (Li et al., 1998; Mahagita et al., 2007; Leuthold et al., 2009). OATPs seem to operate as pH-dependent bidirectional antiporters, stimulated by an acidic extracellular environment (Hagenbuch and Gui, 2008). The tissue distribution of OATPs differs for each of the isoforms, some are expressed ubiquitously, while for others, the expression is restricted to a single organ. OATP1A2 is expressed in kidney, small intestine, and bile duct, OATP1B1 and OATP1B3 have only been detected in liver, whereas OATP2B1 has a wide tissue distribution, and OATP4C1 is considered to be kidney specific (Hagenbuch and Gui, 2008). The multitude of structurally diverse compounds that are recognized by OATPs strongly suggests the presence of multiple substrate binding sites, and evidence for this is emerging (Hagenbuch and Gui, 2008).

Recently, two human orthologs of the multidrug and toxin extrusion (MATE) family of bacteria have been identified as H^+/organic cation antiport systems and assigned as members of the *SLC47* subfamily (Table 2.1). MATE1 is primarily expressed in the kidney, but also exists in the liver, adrenal gland, testis, and skeletal muscle, whereas MATE2-K has been found exclusively in the kidney (Tanihara et al., 2007; Terada and Inui, 2008) MATE1 and MATE2-K are efflux transporters energized by an inwardly directed proton gradient to overcome the outside positive membrane potential (Moriyama et al., 2008). They typically transport type I organic cations and with a few exceptions their substrate specificity is similar and much like that of the OCT and OCTN transporters (Tanihara et al., 2007; Terada and Inui, 2008).

2.3 ABC Drug Transporters

ABC transporters that are involved in drug transport can be found in the *ABCB*, *ABCC*, and *ABCG* families (Table 2.2). Because of their expression in transporting epithelia, including the intestine, liver, and kidney, the ABC transporters play an important role in the absorption, distribution, and removal of drugs (Figs. 2.1, 2.2, and 2.3). Many of them are also associated with multidrug resistance (MDR) of tumor cells causing treatment failure in cancer. The minimal functional configuration of an ABC transporter has a molecular weight of 150–200 kDa and is made of two transmembrane domains, each consisting of six transmembrane helices and two cytoplasmic ATP-binding domains (Locher, 2009). They bind and hydrolyze ATP to drive primary active drug efflux, which is directly linked to their ATPase activity.

The MDR1/P-glycoprotein (P-gp) is encoded by the *MDR1/ABCB1* gene and has an unusually broad substrate specificity, recognizing hundreds of compounds ranging from small molecules of 350 Da up to polypeptides of 4000 Da. A large number of P-gp substrates fall in the category of bulky, often polyvalent, organic cations (generally >500 Da), which are classified as type II organic cations (Zhou, 2008). Very recently, crystal structures of mammalian P-gp were reported, showing distinct partially overlapping drug-binding sites in the internal cavity of the protein, which provides the first molecular basis for its multispecificity (Aller et al., 2009). P-gp is expressed in many tissues and is located on the apical side of intestine, liver, and kidney epithelia where it reduces systemic drug exposure by limiting oral absorption and promoting urinary and biliary excretion.

A close relative to P-gp is the bile salt export pump, BSEP/ABCB11, which is exclusively expressed in the liver where it is localized at the apical membrane (Stieger et al., 2007). BSEP is predominantly responsible for the excretion of monovalent bile acids into bile and constitutes the major driving force for the generation of bile flow. Its substrate specificity is narrow and largely restricted to bile acids, but a poor rate of transport for taxol, vinblastine, and pravastatin has been reported (Alrefai and Gill, 2007). However, the relevance of BSEP in the overall biliary excretion of these drugs is uncertain. On the other hand, BSEP appears to be a key target of drug-induced cholestasis. Drugs such as glibenclamide, rifampicin, bosentan, and troglitazone can inhibit BSEP, which leads to intracellular accumulation of bile salts, decreased bile flow, and ultimately liver injury (Stieger et al., 2007).

Within the *ABCC* subfamily, nine full multidrug resistance (associated) protein (MRP) members have been identified, among which MRP2/ABCC2, MRP3/ABCC3, and MRP4/ABCC4 are the most important drug transporters (Table 2.2). The MRPs mediate the transport of organic anionic (generally type II) compounds, including glucuronide, glutathione, and sulfate conjugates, but also uncharged amphipathic, and even some cationic substrates in the presence of reduced glutathione (Deeley et al., 2006). MRPs are expressed in intestine, liver, and kidney, and other tissues with a barrier function, such as placenta and brain capillaries (van de Water et al., 2005; Zhou et al., 2008). MRP2 is located at the apical membrane of polarized cells, emphasizing its important function in the terminal excretion of anionic drugs and conjugates (Nies and Keppler, 2007). Because of its basolateral localization, MRP3 mediates the cellular efflux of mainly glucuronidated drug conjugates from the intestine and liver into the blood (Borst et al., 2007). A remarkable feature of MRP4 is its dual membrane localization. In hepatocytes the transporter is localized at the basolateral membrane, whereas it is expressed at the apical membrane of renal proximal tubule cells (van Aubel et al., 2002; Rius et al., 2003). In the intestine, the subcellular distribution of MRP4 has not been established yet, and localization to both the apical and the basolateral membrane was found in a colonic epithelial cell line, with a higher apical abundancy (Li et al., 2007; Russel et al., 2008).

The *ABCG2* gene product, breast cancer resistance protein (BCRP), is the only member of this subfamily that is involved in drug transport. Like other G-subfamily members, BCRP is comprised of one ATP-binding site and one transmembrane region, a structure half the size of a functional ABC transporter. These half transporters are generally thought to homo- or heterodimerize to create the active transporter. BCRP was initially discovered in drug-resistant cancer cell lines and therefore many anticancer drugs are among the first reported substrates. To date, a large number of hydrophobic drug substrates have been described, and although a clear structure–transport relationship has not been identified, it should be noted that many of them are also transported by P-gp (Table 2.2). Similar to P-gp, BCRP is expressed in the apical membrane of intestine, liver, kidney, placenta, and brain capillaries, and often in stem cell populations, where it is thought to play a role in differentiation and protection against xenobiotics (Huls et al., 2009).

2.4 Transporters for Intestinal Drug Absorption

The absorption of drugs from the gastrointestinal tract is a critical factor in determining oral bioavailability. Enterocytes of the small intestine are equipped with an array of influx transporters at the luminal membrane for the absorption of food components and drugs. Although much less is known about the subsequent transport of substrates into the blood stream, ABC transporters in the basolateral membrane could facilitate this step. However, as a first barrier against xenobiotics, the intestine also has a high expression of ABC transporters in the brush border membrane that can effectively pump drugs back into the intestinal lumen, thereby limiting the extent of substrate drug absorption (Fig. 2.1, Tables 2.1 and 2.2).

A number of SLC drug transporting proteins have been described at the brush border membrane of human enterocytes, including PEPT1, OATP1A2, OATP2B1, OCTN1, and OCTN2. Expression levels of some of these transporters appear to vary along the gastrointestinal tract, but results from different studies on mRNA and protein expression do not concord, except for PEPT1, which is predominantly expressed in the small intestine (Glaeser et al., 2007; Hilgendorf et al., 2007; Meier et al., 2007; Oostendorp et al., 2009). PEPT1 recognizes various peptide-like drugs and targeting this transporter has been used to improve the oral bioavailability of poorly absorbed drugs such as nucleoside analogs (Brandsch et al., 2008; Rubio-Aliaga and Daniel, 2008). An example is acyclovir, the bioavailability of which was enhanced by a factor of 2–3 via the oral administration of its valine ester (valacyclovir), which is a PEPT1 substrate. Another promising but yet to be established target for prodrug design is the apical sodium-dependent bile acid transporter (ASBT/SLC10A2) (Balakrishnan and Polli, 2006).

Uptake of cationic drugs from the gut lumen is mediated by OCTN1 and OCTN2, which are energized by electroneutral cation–cation exchange (Koepsell et al., 2007). Mutations in the genes encoding for these transporters have been associated with inflammatory bowel disease and polymorphisms could be of influence on cationic drug absorption (Koepsell et al., 2007; Zair et al., 2008). OATP1A2 and OATP2B1 are responsible for the uptake of a broad range of amphipathic drugs (Hagenbuch and Gui, 2008). While there is quite some overlap in specificity, some substrates are preferentially or exclusively transported by one of them. For example, only OATP1A2 is able to mediate fexofenadine uptake and the likely target of inhibition by grapefruit juice (Glaeser et al., 2007).

The first step in secretion of cationic drugs from blood to gut lumen is mediated by OCT1 in the basolateral membrane, followed by the action of efflux transporters in the brush border membrane (Fig. 2.1). These are the OCTNs that can also operate as secretory transporters by exchanging luminal organic cations against a higher concentration of intracellular cationic drugs. In addition, MDR1/P-gp pumps positively charged hydrophobic drugs back into the lumen, which could have entered the cells by passive diffusion. The ABC transporters P-gp, MRP2, and BCRP are all expressed in the brush border membrane where they have an important role as gatekeeper in the gut, limiting the oral bioavailability of many drug substrates. Modulation of their activity with selective inhibitors could be a useful strategy to increase the oral bioavailability of substrate drugs. Examples of drug inhibitors of P-gp and BCRP are HIV protease inhibitors (ritonavir, indinavir, saquinavir, nelfinavir) and benzimidazole proton pump inhibitors (omeprazole, pantoprazole, lansoprazole, and rabeprazole) (Oostendorp et al., 2009). The exact subcellular distribution of MRP4 needs to be clarified, as this relates to its importance in intestinal drug transport.

The interplay between ABC transporters and drug-metabolizing enzymes makes this barrier even more effective. P-gp and the Phase I enzyme cytochrome P450 (CYP)3A4 appear to be functionally linked as they share the same substrates, while MRP2 and BCRP accept anionic drug conjugates formed by Phase II-metabolizing enzymes, including UDP-glucuronosyltransferases (UGTs), sulfotransferases (SULTs), and glutathione *S*-transferases (GSTs) (Kivisto et al., 2004; Nies et al., 2008). At the basolateral membrane of enterocytes, MRP3 mediates transport into the blood of non-conjugated and mostly conjugated organic anions, with a preference for glucuronidated compounds (Borst et al., 2007). The expression of MRP1 and MRP5 at the basolateral membrane has been described, but their role in facilitating intestinal drug absorption needs further investigation.

The same applies to the recently identified basolateral bile acid carrier, the heteromeric organic solute, and steroid transporter OSTα/OSTβ (Ballatori, 2005). The genes encoding these transporters are unique in the human and mouse genome and do not belong to the SLC or ABC transporter families. OSTα/OSTβ is expressed at relatively high levels in the small intestine and substrate specificity is mainly restricted to steroid-derived molecules, including the cardiac glycoside digoxin (Ballatori et al., 2005).

Expression of MDR1/P-gp varies over the total length of the gastrointestinal tract, gradually increasing from the stomach and duodenum to highest levels in colon. *MRP2* and *BCRP* transcript levels are highest in duodenum, even higher than that of *MDR1*, and they decrease in the direction of colon. *MPR3* expression is higher in duodenum, ileum, and colon as compared to jejunum (Dietrich et al., 2003; Murakami and Takano, 2008; Oostendorp et al., 2009). Regional distribution of these transporters along the gastrointestinal tract could be of influence on the site of drug absorption. In addition, genetic polymorphisms of particularly MDR1/P-gp and BCRP affecting expression level and transporter function have been shown to impact oral drug availability of a number of substrates (Maeda and Sugiyama, 2008; Nakamura et al., 2008).

2.5 Transporters for Hepatic Drug Elimination

The liver has a remarkable ability to efficiently extract drugs with high protein binding from the blood circulation. The hepatic uptake of drugs is frequently followed by Phase I and Phase II biotransformation and efflux of the metabolite(s) into bile and contributes to the hepatic first-pass effect. Influx and efflux transporters expressed at the sinusoidal (basolateral) and canalicular (apical) membrane of hepatocytes have been recognized as critical determinants in drug elimination (Fig. 2.2, Tables 2.1 and 2.2).

Drug influx transporters expressed at the sinusoidal membrane include OATP1B1, OATP1B3, OATP2B1, OAT2, and OCT1. In particular OATP1B1 is recognized as an important uptake transporter for many clinically relevant drugs, such as macrolide antibiotics, statins (HMG-CoA-reductase inhibitors), glitazones (thiazolidinediones), sartans (angiotensin II receptor antagonists), and angiotensin-converting enzyme (ACE) inhibitors (Table 2.1). Clinically relevant drug–drug interactions have been described for OATP1B1-mediated statin transport with the immunosuppressant drug cyclosporine A (Endres et al., 2006). The homolog OATP1B3 has similar substrate specificity, but its expression is more confined to the hepatocytes surrounding the central vein. Because the activity of these transporters is often the rate-limiting step in hepatobiliary elimination, their inhibition and genetic variability are critical factors in the interindividual variation in drug disposition and exposure. A recent genomewide study emphasized the strong association of *SLCO1B1* variants with an increased risk of simvastatin-induced myopathy (Link et al., 2008). These genotypes are known to be associated with higher statin blood concentrations, although surprisingly in vitro studies in cells expressing *SLCO1B1* and its variants were inconsistent in identifying simvastatin as an OATP1B1 substrate (Kameyama et al., 2005; Noe et al., 2007).

OAT2 is moderately expressed and could be involved in the hepatic uptake of type I organic anions, such as salicylate and indomethacin (Rizwan and Burckhardt, 2007). Although OAT2 likely functions as an antiporter, its transport mode and particularly the identity of the intracellular counterions have not yet been resolved. OCT1 mediates the influx of type I organic cations into hepatocytes; however, as a bidirectional electrogenic uniporter it can also facilitate the efflux of cationic drugs back into the blood, depending on the electrochemical gradient (Koepsell et al., 2007). A clinically important substrate is metformin, which is among the most widely prescribed drugs for the treatment of type 2 diabetes. Its antidiabetic action is dependent on uptake into hepatocytes and certain loss of function variants of the transporter could be associated with reduced therapeutic efficacy, although recent clinical studies have not confirmed this supposition (Shikata et al., 2007; Shu et al., 2007; Zhou et al., 2009).

Efflux transporters expressed in the canalicular membrane represent the final step in the vectorial transport of drugs and drug metabolites from blood into bile. Excretion of type I and II cationic drugs across the canalicular membrane is mediated by MATE1 and MDR1/P-gp, respectively (Tables 2.1 and 2.2). Metformin is a good substrate of MATE1, although the drug is mainly excreted into urine by active tubular secretion. MRP2 and BCRP are primarily responsible for the canalicular efflux of unconjugated and conjugated anionic drugs, including glucuronide-, sulfo-, and glutathione conjugates (Table 2.2) (Nies and Keppler, 2007; Robey et al., 2009). Not all Phase II drug metabolites formed in the hepatocyte are transferred to bile. The localization of MRP3 and MRP4 at the sinusoidal membrane indicates that these conjugates are also transported back into the circulation so that they can undergo renal elimination. Interindividual variation in MRP3 protein levels is about 80-fold, whereas MRP4 abundance is very low under normal conditions (Lang et al., 2004; Nies et al., 2008). Under cholestatic conditions, protein levels of both transporters are increased and are able to mediate the efflux of bile salts since these are substrates of both MRP3 and MRP4 (Nies et al., 2008). Whether MRP3 and MRP4 have an impact on the overall hepatic clearance depends on the kinetic properties of the drug. Their contribution will be negligible if uptake is the rate-limiting step in elimination (Kusuhara and Sugiyama, 2009).

2.6 Transporters for Renal Drug Elimination

The renal handling of drugs involves passive processes, including glomerular filtration and back diffusion along the nephron, and carrier-mediated secretion and reabsorption that are mainly located in the proximal tubule. For most drugs that undergo carrier-mediated transport in the kidney, renal secretion can be considered as a vectorial process involving the uptake of substances from the blood across the basolateral membrane of proximal tubular cells, followed by their efflux across the brush border membrane into urine. At the basolateral membrane, separate influx transporters exist for the uptake of mainly type I organic anions and cations, which are notable for their high clearance capacity, wide variety of substrates accepted, and involvement in drug–drug interactions (Masereeuw and Russel, 2001). Because of efficient uptake, many drugs tend to accumulate in the cell sometimes causing nephrotoxicity. The large number of efflux transporters expressed at the brush border membrane emphasizes the importance to ensure rapid efflux of potentially toxic compounds into urine (Fig. 2.3, Table 2.1 and 2.2).

The uptake of anionic drugs at the basolateral membrane of renal proximal tubule is regulated by OAT1 and OAT3. Both transporters have overlapping substrate specificities and share the same mode of transport driven by the exchange of organic anions with dicarboxylates (Rizwan and Burckhardt, 2007). OAT1 has a higher affinity for hydrophilic organic anions with small molecular weights (type I), like PAH, adefovir, cidofovir, and tenofovir (Table 2.1). OAT3 also transports some amphipathic organic anions (type II) that are liver OATP substrates, including benzylpenicillin, pravastatin, and olmesartan, and even some cationic drugs, such as cimetidine and ranitidine (Table 2.1). The broader specificity, as well as the relatively higher renal expression levels of OAT3 compared to OAT1, suggests a more pronounced role of OAT3 in human renal organic anion transport (Masereeuw and Russel, 2001; El-Sheikh et al., 2008). Severe drug–drug interactions have been reported between methotrexate and NSAIDs due to competition for OAT1- and OAT3-mediated uptake, although the interaction at the level of the apical efflux transporters MRP2 and MRP4 probably also contributes to this mechanism (El-Sheikh et al., 2007).

The first step in tubular secretion of cationic drugs is mediated by OCT2, the predominant organic cation uniporter in the basolateral membrane. A splice variant of OCT2 (OCT2A), which shares 81% identity, exhibits some transport activity for cationic drugs with different affinity (Urakami et al., 2002). Metformin is also transported by OCT2, even with a higher affinity than by OCT1. Several studies have identified race-specific OCT2 variants, but little is known regarding their effects on pharmacokinetic variability (Zair et al., 2008). Coadministration of cimetidine with metformin has been shown to reduce the renal clearance of metformin, leading to a clinically relevant increase in plasma concentration (Wang et al., 2008).

In human kidney, only OATP4C1 is expressed in the basolateral membrane of proximal tubular cells, and a remarkable species difference exist with rodents with regard to the renal expression of OATPs. Except for the ortholog Oatp4c1, at least three Oatps, absent in humans, are located in the brush border membrane of the rodent kidney (Sekine et al., 2006). The substrate specificity of OATP4C1 is restricted to only a few drugs that are mainly excreted by the kidney. Important examples are methotrexate, the cardiac glycosides, digoxin, and ouabain, as well as thyroid hormones (Table 2.1). The transport mechanism of OATP4C1 and the counter ion it exchanges its substrates for are not yet identified.

At the proximal tubular brush border membrane, a team of four ABC transporters mediate the primary active efflux of drugs, viz., MDR1/P-gp, MRP2, MRP4, and BCRP. P-gp likely provides the efflux pathway for digoxin and a number of hydrophobic cationic drugs (Masereeuw and Russel, 2001; Zhou, 2008). MRP2 and MRP4 are involved in the efflux of anionic drugs and drug conjugates that have been either formed in the proximal tubular cell or released from the liver and taken up from the circulation (van de Water et al., 2005). MRP4 appears to have a higher affinity for type I organic anions and its protein expression is approximately fivefold higher (Smeets et al., 2004; Russel et al., 2008). BCRP has recently been localized to the proximal tubule brush border membrane, suggesting its potential involvement in renal drug excretion (Huls et al., 2008).

The SLC organic cation transporters MATE1, MATE2-K, OCTN1, and OCTN2 mediate the secondary active efflux of cationic drugs across the brush border membrane. The H^+/organic cation or organic cation/organic cation antiport used by these transporters helps to overcome the outside positive membrane potential. Because of their bidirectionality, OCTNs could be also involved in organic cation reabsorption. The outside-in H^+ gradient across the brush border membrane provides the driving force for the MATE transporters. Several genetic variants of MATE1 and MATE2-K with decreased activity have been recently described that could contribute to the variability in renal handling of various cationic drugs such as metformin and lead to accumulation of oxaliplatin, causing drug-induced nephrotoxicity (Kajiwara et al., 2009).

The H^+/peptide symporters, PEPT1 and PEPT2, are both expressed in a sequential order along the renal proximal tubule (Brandsch et al., 2008). However, PEPT2, the high affinity, low capacity transporter appears to be the major player in the renal reabsorption of peptide-like drugs (Kamal et al., 2008).

OAT4 exhibits characteristics of an asymmetric antiporter mediating efflux of anionic drugs into urine, perhaps in exchange of Cl^-, and reabsorption of endogenous substrates like urate and estrone sulfate into the proximal tubular cell. The substrate specificity of OAT4 has not been fully explored, but the number of drugs accepted seems smaller than for OAT1 and OAT3 (Rizwan and Burckhardt, 2007).

2.7 Conclusions

SLC and ABC transporters expressed in the intestine, liver, and kidney are increasingly being recognized as significant determinants of drug disposition and drug–drug interactions. Many examples have emerged in the literature describing the impact of transport proteins on the pharmacokinetics of established drugs and new chemical entities and in recent years pharmaceutical companies and drug regulatory authorities have realized the need for including transporter studies in the early stages of drug development, in particular, for predicting the impact of transporter-based drug interactions and genetic polymorphisms. Since there exists a considerable functional redundancy in transporters, especially for apical efflux at the intestine, liver, and kidney membranes, clinically relevant drug–drug interactions and genetic variability are difficult to predict from in vitro experiments. An important step is to elucidate the relative contribution of the target transporter to the overall membrane transport of a specific drug (Endres et al., 2006; Kusuhara and Sugiyama, 2009). Though not addressed in this chapter, the creation of a variety of transporter knockout mouse models has contributed greatly to our understanding of the pharmacological and toxicological roles of transporter proteins, despite their species differences.

A wealth of information has been accumulated about the structure and function of transporter proteins, but much remains unresolved regarding their molecular mechanisms, structure–function relationships, and regulation. Many of the necessary tools are now available to gain a greater insight into each of these fundamental questions. A next challenge is to study the coordinated action of influx and efflux transporters, drug-metabolizing enzymes, and their coregulation by nuclear receptors as an integrated system (see Chapter 17). Such a system that includes quantitative information on distribution, kinetics, genetic variation, and abundance of transport proteins throughout the body, as already available for many metabolizing enzymes, will eventually lead to an in silico human model which has the ability to predict and simulate effectively the entire disposition and exposure of drugs in terms of pharmacological networks. As genetic variability in pharmacokinetics is concerned, it seems an oversimplification to suggest that one gene or single nucleotide polymorphism really matters. Rather, it is more likely that genes combine to create interindividual variability (Maeda and Sugiyama, 2008; Robey et al., 2009). One example discussed in this chapter is the interaction of metformin with different organic cation transporters. To reach the goal of true individualized therapy, a systems pharmacology approach would likely prove to be essential. A similar case can be made for studying drug–drug and drug–nutrient interactions in a comprehensive model.

This chapter aims to give a concise overview of the importance of major human transporters involved in the handling of clinically relevant drugs by intestine, liver, and kidney and therefore has only briefly touched upon the manifold aspects regarding their significance for drug–drug interactions, pharmacokinetic variability, and drug-induced toxicity. Many of them will be discussed in more depth in the chapters to follow.

References

Aller SG, Yu J, Ward A, Weng Y, Chittaboina S, Zhuo R, Harrell PM, Trinh YT, Zhang Q, Urbatsch IL and Chang G (2009) Structure of P-glycoprotein reveals a molecular basis for poly-specific drug binding. *Science* **323**:1718–1722.

Alrefai WA and Gill RK (2007) Bile acid transporters: structure, function, regulation and pathophysiological implications. *Pharm Res* **24**:1803–1823.

Bailey PD, Boyd CA, Collier ID, George JP, Kellett GL, Meredith D, Morgan KM, Pettecrew R and Price RA (2006) Affinity prediction for substrates of the peptide transporter PepT1. *Chem Commun (Camb)*3:323–325.

Balakrishnan A and Polli JE (2006) Apical sodium dependent bile acid transporter (ASBT, SLC10A2): a potential prodrug target. *Mol Pharm* **3**:223–230.

Ballatori N (2005) Biology of a novel organic solute and steroid transporter, OSTalpha-OSTbeta. *Exp Biol Med (Maywood)* **230**:689–698.

Ballatori N, Christian WV, Lee JY, Dawson PA, Soroka CJ, Boyer JL, Madejczyk MS and Li N (2005) OSTalpha-OSTbeta: a major basolateral bile acid and steroid transporter in human intestinal, renal, and biliary epithelia. *Hepatology* **42**:1270–1279.

Borst P, de Wolf C and van de Wetering K (2007) Multidrug resistance-associated proteins 3, 4, and 5. *Pflugers Arch* **453**:661–673.

Brandsch M, Knutter I and Bosse-Doenecke E (2008) Pharmaceutical and pharmacological importance of peptide transporters. *J Pharm Pharmacol* **60**:543–585.

Ciarimboli G (2008) Organic cation transporters. *Xenobiotica* **38**:936–971.

Cusatis G and Sparreboom A (2008) Pharmacogenomic importance of ABCG2. *Pharmacogenomics* **9**:1005–1009.

Dean M and Allikmets R (2001) Complete characterization of the human ABC gene family. *J Bioenerg Biomembr* **33**:475–479.

Deeley RG, Westlake C and Cole SP (2006) Transmembrane transport of endo- and xenobiotics by mammalian ATP-binding cassette multidrug resistance proteins. *Physiol Rev* **86**:849–899.

Dietrich CG, Geier A and Oude Elferink RP (2003) ABC of oral bioavailability: transporters as gatekeepers in the gut. *Gut* **52**:1788–1795.

Dobson PD and Kell DB (2008) Carrier-mediated cellular uptake of pharmaceutical drugs: an exception or the rule?. *Nat Rev Drug Discov* **7**:205–220.

El-Sheikh AA, Masereeuw R and Russel FG (2008) Mechanisms of renal anionic drug transport. *Eur J Pharmacol* **585**:245–255.

El-Sheikh AA, van den Heuvel JJ, Koenderink JB and Russel FG (2007) Interaction of non-steroidal anti-inflammatory drugs with multidrug resistance protein (MRP) 2/ABCC2- and MRP4/ABCC4-mediated methotrexate transport. *J Pharmacol Exp Ther* **320**:229–235.

Endres CJ, Hsiao P, Chung FS and Unadkat JD (2006) The role of transporters in drug interactions. *Eur J Pharm Sci* **27**:501–517.

Fredriksson R, Nordstrom KJ, Stephansson O, Hagglund MG and Schioth HB (2008) The solute carrier (SLC) complement of the human genome: phylogenetic classification reveals four major families. *FEBS Lett* **582**:3811–3816.

Glaeser H, Bailey DG, Dresser GK, Gregor JC, Schwarz UI, McGrath JS, Jolicoeur E, Lee W, Leake BF, Tirona RG and Kim RB (2007) Intestinal drug transporter expression and the impact of grapefruit juice in humans. *Clin Pharmacol Ther* **81**:362–370.

Hagenbuch B and Gui C (2008) Xenobiotic transporters of the human organic anion transporting polypeptides (OATP) family. *Xenobiotica* **38**:778–801.

Hilgendorf C, Ahlin G, Seithel A, Artursson P, Ungell AL and Karlsson J (2007) Expression of thirty-six drug transporter genes in human intestine, liver, kidney, and organotypic cell lines. *Drug Metab Dispos* **35**:1333–1340.

Ho RH and Kim RB (2005) Transporters and drug therapy: implications for drug disposition and disease. *Clin Pharmacol Ther* **78**:260–277.

Hu S, Franke RM, Filipski KK, Hu C, Orwick SJ, de Bruijn EA, Burger H, Baker SD and Sparreboom A (2008) Interaction of imatinib with human organic ion carriers. *Clin Cancer Res* **14**:3141–3148.

Huls M, Brown CD, Windass AS, Sayer R, van den Heuvel JJ, Heemskerk S, Russel FG and Masereeuw R (2008) The breast cancer resistance protein transporter ABCG2 is expressed in the human kidney proximal tubule apical membrane. *Kidney Int* **73**: 220–225.

Huls M, Russel FG and Masereeuw R (2009) The role of ATP binding cassette transporters in tissue defense and organ regeneration. *J Pharmacol Exp Ther* **328**:3–9.

Ito K, Suzuki H, Horie T and Sugiyama Y (2005) Apical/basolateral surface expression of drug transporters and its role in vectorial drug transport. *Pharm Res* **22**:1559–1577.

Kajiwara M, Terada T, Ogasawara K, Iwano J, Katsura T, Fukatsu A, Doi T and Inui K (2009) Identification of multidrug and toxin extrusion (MATE1 and MATE2-K) variants with complete loss of transport activity. *J Hum Genet* **54**:40–46.

Kamal MA, Keep RF and Smith DE (2008) Role and relevance of PEPT2 in drug disposition, dynamics, and toxicity. *Drug Metab Pharmacokinet* **23**:236–242.

Kameyama Y, Yamashita K, Kobayashi K, Hosokawa M and Chiba K (2005) Functional characterization of SLCO1B1 (OATP-C) variants, SLCO1B1∗5, SLCO1B1∗15 and SLCO1B1∗15+C1007G, by using transient expression systems of HeLa and HEK293 cells. *Pharmacogenet Genomics* **15**:513–522.

Kivisto KT, Niemi M and Fromm MF (2004) Functional interaction of intestinal CYP3A4 and P-glycoprotein. *Fundam Clin Pharmacol* **18**:621–626.

Koepsell H, Lips K and Volk C (2007) Polyspecific organic cation transporters: structure, function, physiological roles, and biopharmaceutical implications. *Pharm Res* **24**:1227–1251.

Kusuhara H and Sugiyama Y (2009) In vitro-in vivo extrapolation of transporter-mediated clearance in the liver and kidney. *Drug Metab Pharmacokinet* **24**:37–52.

Lang T, Hitzl M, Burk O, Mornhinweg E, Keil A, Kerb R, Klein K, Zanger UM, Eichelbaum M and Fromm MF (2004) Genetic polymorphisms in the multidrug resistance-associated protein 3 (ABCC3, MRP3) gene and relationship to its mRNA and protein expression in human liver. *Pharmacogenetics* **14**:155–164.

Leuthold S, Hagenbuch B, Mohebbi N, Wagner CA, Meier PJ and Stieger B (2009) Mechanisms of pH-gradient driven transport mediated by organic anion polypeptide transporters. *Am J Physiol Cell Physiol* **296**:C570–C582.

Li C, Krishnamurthy PC, Penmatsa H, Marrs KL, Wang XQ, Zaccolo M, Jalink K, Li M, Nelson DJ, Schuetz JD and Naren AP (2007) Spatiotemporal coupling of cAMP transporter to CFTR chloride channel function in the gut epithelia. *Cell* **131**:940–951.

Li L, Lee TK, Meier PJ and Ballatori N (1998) Identification of glutathione as a driving force and leukotriene C4 as a substrate for oatp1, the hepatic sinusoidal organic solute transporter. *J Biol Chem* **273**:16184–16191.

Link E, Parish S, Armitage J, Bowman L, Heath S, Matsuda F, Gut I, Lathrop M and Collins R (2008) SLCO1B1 variants and statin-induced myopathy – a genomewide study. *N Engl J Med* **359**:789–799.

Locher KP (2009) Review. Structure and mechanism of ATP-binding cassette transporters. *Philos Trans R Soc Lond B Biol Sci* **364**:239–245.

Maeda K and Sugiyama Y (2008) Impact of genetic polymorphisms of transporters on the pharmacokinetic, pharmacodynamic and toxicological properties of anionic drugs. *Drug Metab Pharmacokinet* **23**:223–235.

Mahagita C, Grassl SM, Piyachaturawat P and Ballatori N (2007) Human organic anion transporter 1B1 and 1B3 function as bidirectional carriers and do not mediate GSH-bile acid cotransport. *Am J Physiol Gastrointest Liver Physiol* **293**:G271–G278.

Masereeuw R and Russel FG (2001) Mechanisms and clinical implications of renal drug excretion. *Drug Metab Rev* **33**:299–351.

Matsushima S, Maeda K, Inoue K, Ohta KY, Yuasa H, Kondo T, Nakayama H, Horita S, Kusuhara H and Sugiyama Y (2009) The inhibition of human multidrug and toxin extrusion 1 is involved in the drug–drug interaction caused by cimetidine. *Drug Metab Dispos* **37**:555–559.
Meier Y, Eloranta JJ, Darimont J, Ismair MG, Hiller C, Fried M, Kullak-Ublick GA and Vavricka SR (2007) Regional distribution of solute carrier mRNA expression along the human intestinal tract. *Drug Metab Dispos* **35**:590–594.
Mizuno N and Sugiyama Y (2002) Drug transporters: their role and importance in the selection and development of new drugs. *Drug Metab Pharmacokinet* **17**:93–108.
Moriyama Y, Hiasa M, Matsumoto T and Omote H (2008) Multidrug and toxic compound extrusion (MATE)-type proteins as anchor transporters for the excretion of metabolic waste products and xenobiotics. *Xenobiotica* **38**:1107–1118.
Murakami T and Takano M (2008) Intestinal efflux transporters and drug absorption. *Expert Opin Drug Metab Toxicol* **4**:923–939.
Nakamura T, Yamamori M and Sakaeda T (2008) Pharmacogenetics of intestinal absorption. *Curr Drug Deliv* **5**:153–169.
Nies AT and Keppler D (2007) The apical conjugate efflux pump ABCC2 (MRP2). *Pflugers Arch* **453**:643–659.
Nies AT, Schwab M and Keppler D (2008) Interplay of conjugating enzymes with OATP uptake transporters and ABCC/MRP efflux pumps in the elimination of drugs. *Expert Opin Drug Metab Toxicol* **4**:545–568.
Noe J, Portmann R, Brun ME and Funk C (2007) Substrate-dependent drug–drug interactions between gemfibrozil, fluvastatin and other organic anion-transporting peptide (OATP) substrates on OATP1B1, OATP2B1, and OATP1B3. *Drug Metab Dispos* **35**:1308–1314.
Oostendorp RL, Beijnen JH and Schellens JH (2009) The biological and clinical role of drug transporters at the intestinal barrier. *Cancer Treat Rev* **35**:137–147.
Rius M, Nies AT, Hummel-Eisenbeiss J, Jedlitschky G and Keppler D (2003) Cotransport of reduced glutathione with bile salts by MRP4 (ABCC4) localized to the basolateral hepatocyte membrane. *Hepatology* **38**:374–384.
Rizwan AN and Burckhardt G (2007) Organic anion transporters of the SLC22 family: biopharmaceutical, physiological, and pathological roles. *Pharm Res* **24**:450–470.
Robey RW, To KK, Polgar O, Dohse M, Fetsch P, Dean M and Bates SE (2009) ABCG2: a perspective. *Adv Drug Deliv Rev* **61**:3–13.
Rubio-Aliaga I and Daniel H (2008) Peptide transporters and their roles in physiological processes and drug disposition. *Xenobiotica* **38**:1022–1042.
Russel FG, Koenderink JB and Masereeuw R (2008) Multidrug resistance protein 4 (MRP4/ABCC4): a versatile efflux transporter for drugs and signalling molecules. *Trends Pharmacol Sci* **29**:200–207.
Russel FG, Masereeuw R and van Aubel RA (2002) Molecular aspects of renal anionic drug transport. *Annu Rev Physiol* **64**:563–594.
Sarkadi B, Homolya L, Szakacs G and Varadi A (2006) Human multidrug resistance ABCB and ABCG transporters: participation in a chemoimmunity defense system. *Physiol Rev* **86**: 1179–1236.
Sekine T, Miyazaki H and Endou H (2006) Molecular physiology of renal organic anion transporters. *Am J Physiol Renal Physiol* **290**:F251–F261.
Sheps JA and Ling V (2007) Preface: the concept and consequences of multidrug resistance. *Pflugers Arch* **453**:545–553.
Shikata E, Yamamoto R, Takane H, Shigemasa C, Ikeda T, Otsubo K and Ieiri I (2007) Human organic cation transporter (OCT1 and OCT2) gene polymorphisms and therapeutic effects of metformin. *J Hum Genet* **52**:117–122.
Shitara Y, Sato H and Sugiyama Y (2005) Evaluation of drug–drug interaction in the hepatobiliary and renal transport of drugs. *Annu Rev Pharmacol Toxicol* **45**:689–723.
Shu Y, Sheardown SA, Brown C, Owen RP, Zhang S, Castro RA, Ianculescu AG, Yue L, Lo JC, Burchard EG, Brett CM and Giacomini KM (2007) Effect of genetic variation in the organic cation transporter 1 (OCT1) on metformin action. *J Clin Invest* **117**:1422–1431.

Smeets PH, van Aubel RA, Wouterse AC, van den Heuvel JJ and Russel FG (2004) Contribution of multidrug resistance protein 2 (MRP2/ABCC2) to the renal excretion of p-aminohippurate (PAH) and identification of MRP4 (ABCC4) as a novel PAH transporter. *J Am Soc Nephrol* **15**:2828–2835.

Srimaroeng C, Perry JL and Pritchard JB (2008) Physiology, structure, and regulation of the cloned organic anion transporters. *Xenobiotica* **38**:889–935.

Stieger B, Meier Y and Meier PJ (2007) The bile salt export pump. *Pflugers Arch* **453**:611–620.

Suhre WM, Ekins S, Chang C, Swaan PW and Wright SH (2005) Molecular determinants of substrate/inhibitor binding to the human and rabbit renal organic cation transporters hOCT2 and rbOCT2. *Mol Pharmacol* **67**:1067–1077.

Szakacs G, Varadi A, Ozvegy-Laczka C and Sarkadi B (2008) The role of ABC transporters in drug absorption, distribution, metabolism, excretion and toxicity (ADME-Tox). *Drug Discov Today* **13**:379–393.

Tanihara Y, Masuda S, Sato T, Katsura T, Ogawa O and Inui K (2007) Substrate specificity of MATE1 and MATE2-K, human multidrug and toxin extrusions/H(+)-organic cation antiporters. *Biochem Pharmacol* **74**:359–371.

Terada T and Inui K (2008) Physiological and pharmacokinetic roles of H^+/organic cation antiporters (MATE/SLC47A). *Biochem Pharmacol* **75**:1689–1696.

Tsuji A (2002) Transporter-mediated drug interactions. *Drug Metab Pharmacokinet* **17**:253–274.

Urakami Y, Akazawa M, Saito H, Okuda M and Inui K (2002) cDNA cloning, functional characterization, and tissue distribution of an alternatively spliced variant of organic cation transporter hOCT2 predominantly expressed in the human kidney. *J Am Soc Nephrol* **13**:1703–1710.

Urban TJ, Brown C, Castro RA, Shah N, Mercer R, Huang Y, Brett CM, Burchard EG and Giacomini KM (2008) Effects of genetic variation in the novel organic cation transporter, OCTN1, on the renal clearance of gabapentin. *Clin Pharmacol Ther* **83**:416–421.

van Aubel RA, Smeets PH, Peters JG, Bindels RJ and Russel FG (2002) The MRP4/ABCC4 gene encodes a novel apical organic anion transporter in human kidney proximal tubules: putative efflux pump for urinary cAMP and cGMP. *J Am Soc Nephrol* **13**:595–603.

van de Water FM, Masereeuw R and Russel FG (2005) Function and regulation of multidrug resistance proteins (MRPs) in the renal elimination of organic anions. *Drug Metab Rev* **37**: 443–471.

van Herwaarden AE and Schinkel AH (2006) The function of breast cancer resistance protein in epithelial barriers, stem cells and milk secretion of drugs and xenotoxins. *Trends Pharmacol Sci* **27**:10–16.

Wang ZJ, Yin OQ, Tomlinson B and Chow MS (2008) OCT2 polymorphisms and in-vivo renal functional consequence: studies with metformin and cimetidine. *Pharmacogenet Genomics* **18**:637–645.

Zair ZM, Eloranta JJ, Stieger B and Kullak-Ublick GA (2008) Pharmacogenetics of OATP (SLC21/SLCO), OAT and OCT (SLC22) and PEPT (SLC15) transporters in the intestine, liver and kidney. *Pharmacogenomics* **9**:597–624.

Zhou SF (2008) Structure, function and regulation of P-glycoprotein and its clinical relevance in drug disposition. *Xenobiotica* **38**:802–832.

Zhou K, Donnelly LA, Kimber CH, Donnan PT, Doney AS, Leese G, Hattersley AT, McCarthy MI, Morris AD, Palmer CN and Pearson ER (2009) Reduced function SLC22A1 polymorphisms encoding organic cation transporter 1 (OCT1) and glycaemic response to metformin: a Go-DARTS study. *Diabetes*: Mar 31. [Epub ahead of print].

Zhou SF, Wang LL, Di YM, Xue CC, Duan W, Li CG and Li Y (2008) Substrates and inhibitors of human multidrug resistance associated proteins and the implications in drug development. *Curr Med Chem* **15**:1981–2039.

Chapter 3
ADME Pharmacogenetics and Its Impact on Drug–Drug Interactions

Reinhold Kerb and Matthias Schwab

Abstract The extent of drug metabolism or drug transport-based pharmacokinetic drug–drug interactions is highly variable between individuals. CYP enzymes such as CYP2B6, CYP2C9, CYP2C19, and CYP2D6 and drug efflux and uptake transporters such as ABCB1 and OAT1B1 display genetic polymorphisms (presence of genetic variants in at least 1% of a population) that may result in altered drug metabolism or transport capacities, respectively. These polymorphisms explain the interindividual variable magnitude of a drug–drug interaction to a significant extent by determining either the substrate susceptibility for interactions or the interaction potential of an inducer or inhibitor. Knowledge of the activity of the enzyme that is responsible for the metabolism of the affected drug or inhibitor can offer vital information when assessing drug interactions. Drug efflux and uptake are increasingly recognized to supplement this information. Nowadays, metabolic and transport statuses are easily accessible by genotyping and are an important prerequisite to fully judge the potential of a drug for drug interactions.

3.1 Introduction

Pharmacogenetics and pharmacogenomics study the variation in single and multiple genes in relation to interindividual variation in drug response – differences that vary from potentially life-threatening adverse drug reactions (ADR) to a lack of desired therapeutic effect. Pharmacokinetic pathways largely dictate drug response, and thus variations in genes involved in absorption, distribution, metabolism, and excretion (ADME) are major determinants of their interindividual variability. Functional variants of drug-metabolizing enzymes determine to which extent

M. Schwab (✉)
Dr Margarete Fischer-Bosch Institute of Clinical Pharmacology, University of Tübingen, Stuttgart, Germany
e-mail: matthias.schwab@ikp-stuttgart.de

K.S. Pang et al. (eds.), *Enzyme- and Transporter-Based Drug–Drug Interactions*,
DOI 10.1007/978-1-4419-0840-7_3,

a specific agent is metabolized in an individual patient, thereby affecting drug exposure and subsequently efficacy and/or susceptibility to ADR. While the field of pharmacogenetics began with a focus on drug-metabolizing enzymes (Meyer, 2004), it has been extended to membrane transporters (uptake and efflux) that influence drug absorption, distribution, and excretion (Kerb, 2006).

Drug–Drug interactions (DDIs), especially through inhibition of P450-mediated drug metabolism or P-glycoprotein (P-gp)-mediated transport by a coadministered drug, are one of the primary causes of serious adverse events occurring in clinical practice (Kongkaew et al., 2008). Considering the prominent role of drug-metabolizing enzymes and transporters as source for DDIs, it is obvious that the presence of polymorphic variants may also affect the interindividual variable occurrence of DDIs and complicate their prediction. Genetic variants not only determine to which extent a drug is metabolized, taken up or excreted by ADME genes, but they also influence the various mechanisms of how two or more drugs interact with the resulting protein. In the simplest case of two drugs competing for the same metabolizing enzyme or transporter, the extent of interaction depends directly on the amount of protein available. Genetic variants, which cause more or fewer protein being expressed, increase or lower the likelihood for the competition to become clinically relevant. On the other hand, if the protein is absent or truncated and inactive as it is the case with, e.g. partial or total gene deletions, the DDI can naturally not occur. More difficult to predict is the impact of genetic variation which alter the amino acid sequence of the protein. They may affect protein function (catalytic efficiency) not for all drugs to the same extent. As a consequence, ADME genetic variation may also change the DDI profile of drugs.

Furthermore, certain genotypes may directly affect the induction potential of a drug. In general, the lower the baseline enzymatic activity, the higher the induction achieved with inducers (Vesell and Page, 1969; McCune et al., 2000) while absent or low baseline activity caused by null alleles or other genetic factors causing complete absence or inactivity of the enzyme is naturally not inducible and no conclusive data are available, whether, e.g., genetic variants in regulatory regions may change the induction potential. However, the relationship between genetic variations and DDI can be much more complex. Genetic variation in drug metabolism or transport or in the receptors mediating the induction (e.g., PXR, CAR) may modify the systemic availability of an inducer and subsequently the risk for induction-related failure of coadministered drugs. Pharmacogenetics tools hold the potential to identify patients who will respond favorably to a particular drug and hence avoid ADRs and diminish treatment failure. They may also define the individual susceptibility for or the degree of clinically relevant drug–drug interactions.

While the effects of genetic polymorphisms in drug-metabolizing enzymes and transporters on drug exposure have been extensively studied, only little attention has been given until very recently to whether these polymorphisms alter the interindividual susceptibility to or the degree of induction or inhibition. In this review we present an overview of studies investigating how polymorphisms in ADME-related genes affect drug–drug interactions. This pharmacogenetic review is limited to those genetically differential metabolic enzymes and drug transporters that have been investigated in relation to drug interactions. In addition, other polymorphisms in ADME genes exist, whose clinical consequences have not yet been addressed and which are therefore not considered here.

3.1.1 CYP450s

The cytochrome P450 isoenzymes (CYPs) are responsible for about 75% of phase I-dependent drug metabolism. About 40% of CYPs are genetically polymorphic: *CYP1A2, CYP2A6, CYP2B6, CYP2C8, CYP2C9, CYP2C19, CYP2D6, CYP3A5* (Ingelman-Sundberg et al., 2007; Zanger et al., 2008). Genetic variation in these enzymes can alter drug metabolism and may result in patients that are either extensive metabolizers (EMs), poor metabolizers (PMs), or intermediate metabolizers (IMs) of drugs. For some isozymes an ultrarapid metabolizer phenotype (UM) has been described in part based on gene amplifications (e.g., *CYP2D6*). Hence, the same dose of drug could yield poor or no effect in EMs or a toxic response in PMs, whereas PMs may suffer from therapy failure when given prodrugs that require bioactivation by a polymorphic enzyme (e.g., activation of codein to morphine by CYP2D6). ADRs due to defective alleles have been frequently reported (Eichelbaum et al., 2006). In the following, a brief overview about the most relevant metabolic polymorphisms is given, which is limited to common polymorphisms, which occur with frequencies of >5% in at least one ethnicity. The clinically most relevant CYP450 polymorphisms including their frequencies in African, Asian, and Caucasian populations are summarized in Table 3.1. For a comprehensive overview on P450 pharmacogenomics the reader is referred to a recent review (Ingelman-Sundberg et al., 2007; Zanger et al., 2008).

3.1.2 CYP450 Polymorphism-Related Drug–Drug Interactions

3.1.2.1 CYP2B6

The *CYP2B6* gene is highly polymorphic (Lang et al., 2001) and is subject to high interindividual variability in expression and activity (Zanger et al., 2007).The most common variant allele, *CYP2B6*6*, consisting of two single nucleotide polymorphisms (SNPs), *c.516G>T* (Q172H) combined with *c.785A>G* (K262R), occurs with frequencies between 12 and 49% across different populations and has been associated with elevated plasma concentrations of efavirenz and nevirapine- and efavirenz-related neurotoxicity in HIV-infected individuals. This is in agreement with the lower activity of *CYP2B6*6*, which however may be substrate dependent. In contrast, isolated occurrence of *c.785A>G* (*CYP2B6*4*) was detected with enhanced activity in Caucasian (4%) and Asian (7%) populations, while a null allele lacking any activity, *CYP2B6*18*, is present (2–8%) in African populations (Zanger et al., 2008). The possibility of diminished genotype-dependent drug interaction resulting

Table 3.1 Common alleles in the functionally most important metabolizing enzymes and their frequency distribution in Africans, Asians, and Caucasians

			Frequency [%]		
Gene	Allele[a]	Phenotype	African	Asian	Caucasian
CYP2B6	Total frequency of PM phenotypes:				
	*4 (Lys262**Arg**)	PM	0	7	4
	*6 (Gln172**His**/Lys262**Arg**)	PM	25–49	12–19	15–25
	*18 (Ile328**Thr**)	PM	2–8	0	0
CYP2C9	Total frequency of PM phenotypes:		1–6	<1	2–6
	*2	PM	2–4	0	10–15
	*3	PM	1–2	1–4	5–10
	*5	PM	1–2	0	0
	*6	PM	0.6–3	0	0
	*8	PM/UM[b]	3–9	0	0
	*11	PM	2–5	0	<1
CYP2C19	Total frequency of PM phenotypes:		2–7	10–30	2–5
	*2	PM	10–25	20–50	13–19
	*3	PM	0–2	5–13	<1
	*17	UM	18	18	4
CYP2D6	Total frequency of PM phenotypes:		1–3	1–2	2–10
	*3	PM	<1	0.8–1	1–3
	*4	PM	1–8	0.5–1	12–23
	*5	PM	1–7	5–6	2–7
	*6	PM	<1	–	0.7–1.4
	*10	IM	3–9	39–70	2–8
	*17	IM	9–34	–	<1
	*29	IM	5–7 (20)[c]	–	<1
	*1/2×N	UM	2–5 (29)[d]	0.5–1	1–2
CYP3A5					
	*3	PM[e]	12–40[f]	60–75	85–95
	*6	PM[e]	22	0	0
	*7	PM[e]	10	0	0

PM, poor metabolizer; UM, ultrarapid metabolizer; IM, intermediate metabolizer
[a]Variant nucleotide or amino acid in bold
[b]Substrate-dependent phenotype
[c]In Tanzanians
[d]In Ethiopians
[e]Since CYP3A4 and CYP3A5 have largely overlapping substrate specificities, lacking CYP3A5 activity may be compensated by CYP3A4 for many drugs.
[f]12–15% in African, 30–40% in African-American populations

from CYP2B6 inhibition has been suggested after demonstrating lower sertraline and clopidogrel inhibitor potency for CYP2B6-catalyzed efavirenz hydroxylation by K262R (**4*) alone and particularly in combination with Q172H (**6*) in recombinant expression systems (Talakad et al., 2009). Importantly, this genotype effect may be inhibitor-specific, because previous studies have demonstrated that **4*, in contrast to **6*, is refractory to mechanism-based inactivation by 17alpha-ethynylestradiol or efavirenz, while susceptibility of **4* to inactivation is preserved with bergamottin, *N,N′,N′*-triethylenethiophosphoramide, and 8-hydroxy-efavirenz (Bumpus et al., 2005; 2006).

Whether coadministration of an inducer such as rifampin has differential effects on *CYP2B6* genotypes has not yet systematically been studied. *CYP2B6*6* polymorphism results in three distinctive phenotypes of efavirenz metabolism: PMs, IMs, and EMs in homozygous *CYP2B6*6* carriers, heterozygous *CYP2B6* 1/6* genotypes, and homozygous *CYP2B6* 1/1* genotypes, respectively (Zanger et al., 2007). After coadministration of rifampin in HIV/TB-coinfected patients efavirenz clearance did not any longer differ between *CYP2B6* 1/6* and *-* 1/1* genotypes suggesting a differential genotype-dependent induction. On the other hand, efavirenz clearance was significantly lower in homozygous carriers of the **6* allele, indicating that rifampin induction cannot reverse the PM phenotype (Kwara et al., 2008). This is in agreement with recent in vitro work showing that the *CYP2B6*6*-causing *c.516G>T* polymorphism results in aberrant mRNA splicing with deletion of critical protein domains, and subsequently nonfunctional protein (Hofmann et al., 2008). Thus in *CYP2B6*6* PMs rifampin enhances only aberrantly spliced mRNA, which will not produce active enzyme. However, firm conclusions from this study are limited by the lack of an untreated control group (Kwara et al., 2008). Similarly, the lower hydroxybupropion-to-bupropion AUC ratio in *CYP2B6*6/6* compared to ** 1/1* genotypes became more pronounced after induction with the herbal medicine baicalin in a small clinical crossover trial (Fan et al., 2008a).

3.1.2.2 CYP2C9

To date, 30 *CYP2C9* allelic variants located within the coding region have been reported (Ingelman-Sundberg et al., 2008). The **2* and **3* alleles are present in approximately 25% of Caucasian individuals, but are much less prevalent in black and Asian populations (García-Martín et al., 2006). In vitro data have consistently demonstrated that the *CYP2C9*2* and **3* alleles are associated with significant but highly variable reductions in intrinsic clearance, depending on the particular substrate (Lee et al., 2002). The **3* allele is more strongly affected than **2*, with a reduction in activity of up to 90% for some substrates. Numerous clinical studies have demonstrated the clinical significance of the *CYP2C9*2* and **3* polymorphisms for the oral clearance of substrate drugs and the occurrence of adverse drug reactions like hypoglycemia from oral antidiabetic drugs, gastrointestinal bleeding from NSAIDs, and serious bleeding from warfarin treatment (Zanger et al., 2008).

In vitro studies reported a genotype-dependent inhibitory profile of CYP2C9. Kumar et al. compared the degree of inhibition of flurbiprofen oral clearance by the prototypical inhibitor fluconazole between healthy volunteers of various *CYP2C9* 1/1*, ** 1/3*, and **3/3* genotypes (Kumar et al., 2006; 2008). Lowering of flurbiprofen apparent oral clearance after coadministration of fluconazole was gene dose dependent. The degree of DDI observed decreased with the metabolic capacity of CYP2C9, as determined by the numerical presence of *CYP2C *3* alleles, with virtually no change occurring in **3/3* subjects (Kumar et al., 2008). These results suggested that the contribution of the <20–30% residual activity of the **3* allele to overall flurbiprofen clearance is to minor to be inhibited by fluconazole.

3.1.2.3 CYP2C19

The CYP2C19 PM phenotype results from two null alleles, leading to the absence of functional CYP2C19 protein. About 2–7% of white and black populations but up to 30% of Asians are CYP2C19 PMs. The two most common null alleles are *CYP2C19*2*, which occurs almost exclusively in Caucasians, and *CYP2C19*3*, which occurs primarily in Asians (Shimizu et al., 2003). A promoter gene variant (*CYP2C19*17*) has recently been identified which appears to be related to increased substrate turnover (UM) by an as yet unexplained mechanism (Sim et al., 2006). The clinical impact of *CYP2C19* gene polymorphism is well established. CYP2C19 PMs benefit from higher omeprazole plasma levels resulting in accelerated ulcer healing (Schwab et al., 2004) but suffer from lower antiplatelet effect due to reduced bioactivation of the prodrug clopidogrel (Geisler et al., 2008).

Most thoroughly investigated has been the genetically differential modulation of CYP2C19 using proton pump inhibitors such as omeprazole as both substrate and inhibitor. Omeprazole significantly inhibited (up to 2-fold) moclobemide metabolism in EMs (*CYP2C19*1/1*) but not in PMs (*CYP2C19*2/3* and *CYP2C19*2/2*) (Yu et al., 2001). After steady-state administration of omeprazole (40 mg), the mean AUC of moclobemide more than doubled in the EMs, while the AUC ratio of the major metabolite to parent compound was significantly reduced indicating reduced activity of the CYP2C19 enzyme (Yu et al., 2001). Interestingly, in a subsequent study in the same subjects, the same genotype-dependent interaction was demonstrated for moclobemide on omeprazole reflecting the competitive mechanism of inhibition. A single 300 mg moclobemide dose inhibited CYP2C19 enzyme activity causing an approximately 2-fold increase in AUC and maximum plasma concentration ($C_{\max}$) and of the mean AUC ratios of omeprazole to 5-OH-omeprazole, whereas only insignificant changes were observed in PMs (Cho et al., 2002). The same differential inhibitory effects of *CYP2C19* genotype on omeprazole metabolism were also seen for the CYP2C19 inhibitor fluvoxamine (Yasui-Furukori et al., 2004b), which were confirmed with the use of lansoprazole, another proton pump inhibitor and CYP2C19 substrate (Yasui-Furukori et al., 2004a). Accordingly, the omeprazole–R-warfarin (Uno et al., 2008) or allicin (Yang et al., 2009) interaction was only detectable in CYP2C19 EMs. This is consistent with the expectation of insignificant changes in pharmacokinetic parameters of a CYP2C19 substrate among CYP2C19 PMs because of the PMs' decreased baseline enzyme activity.

The magnitude of interaction is also strongly affected by the plasma concentration of inhibitors or inducers. Therefore, if an inhibitor or inducer is metabolized by a genetically differential enzyme, genotype becomes the determinant of its interaction potential. Clinically relevant DDIs have been observed between tacrolimus and proton pump inhibitors in those *CYP2C19* gene variants coding for PMs and IMs, who exhibited higher proton pump inhibitor plasma concentrations, compared to EMs (Itagaki et al., 2004; Miura et al., 2007; Hosohata et al., 2009). Although the triazole antifungal voriconazole is metabolized by several CYPs (CYP3A4, CYP2C19, and CYP2C9), the presence of deficient *CYP2C19*2* alleles resulted in >3 times higher AUC and lower apparent oral clearance compared with the wild type (Mikus et al., 2006). Surprisingly, inhibition of CYP3A4 by single dose coadministration of ritonavir was most prominent in CYP2C19 PMs. With a reduction of 86% voriconazole oral clearance was almost abolished in PMs compared to a change of only 34% in EMs (Mikus et al., 2006). Thus, having one metabolic pathway less, in CYP2C19 PMs the remaining CYP3A4 was more susceptible for inhibition by ritonavir.

In agreement with the findings in other CYP450 the induction of CYP2C19-mediated metabolism of mephenytoin by St. John's Wort was limited to CYP2C19 EMs (Wang et al., 2004). More complex was the voriconazole interaction with the CYP3A inducer St. John's Wort, which after an initial significant but small increase in voriconazole exposure limited to the absorption phase, resulted in an extensive 2 times reduction of exposure after 15-day chronic application. Interestingly, the increase in the CL/F and drop in exposure were smaller in the presence of the *CYP2C19*2* allele compared with the wild-type group, which was explained by the lack of co-induction of CYP2C19 metabolism in the PMs.

Another example of the complex interplay between genetic makeup in one CYP pathway and induction potential for another CYP is the induction of CYP1A2 by standard dose of omeprazole (40 mg), which is only seen in CYP2C19 PMs of omeprazole (Ma and Lu, 2007). A 3 times higher dose of omeprazole is required to achieve sufficiently high plasma concentrations of omeprazole needed to induce CYP1A2 in EMs (Ma and Lu, 2007).

3.1.2.4 CYP3A4/5

The cytochrome P450 3A (CYP3A) subfamily members are the most important drug-metabolizing enzymes in humans and participate in the metabolism of 40% of the mostly prescribed drugs (Zanger et al., 2008). CYP3A expression and function are highly variable. There is little evidence for a relevant contribution of *CYP3A4* gene polymorphisms in determining CYP3A4 activity, although in very rare individual cases, defective *CYP3A4* mutants may provide an explanation for toxicity (Westlind-Johnsson et al., 2006). In contrast, genetic determinants have been identified for much of the variable expression of CYP3A5. It has been shown that the higher incidence of the inactive *CYP3A5*3* allele in Caucasians (85–95%) vs. African Americans (30–50%) causes the lower expression levels of CYP3A5 protein in Caucasians over African Americans (<30 vs. ~50%) (Wojnowski, 2004). Other common genetic *CYP3A5* polymorphisms are *CYP3A5*6* and **7*; both are without any functional activity and are found in subjects of African origin. Apart from a clear effect of CYP3A5 expression status for therapies with the immunosuppressant tacrolimus, the contribution of polymorphic CYP3A5 on CYP3A-mediated metabolism remains controversial (Zanger et al., 2008). It is difficult to delineate the relative contribution of CYP3A4 and CYP3A5 as their protein structure, function, and substrates are so similar and functionally one may compensate for the lack of the other.

The prediction of DDIs and the evaluation of the contribution of genetic variants gets further complicated. From in vitro investigations it is known that the extent of inhibition of CYP3A activity varies with substrate (Kenworthy et al., 1999; Stresser et al., 2000; Wang et al., 2000). Several studies suggested that CYP3A5 expression can influence the extent of hepatic CYP3A-mediated inhibition (Isoherranen et al., 2008), and inhibitors and inducers may have different affinity for CYP3A4 and -A5 (Gibbs et al., 1999). As a consequence, the genotype effect on the extent of inhibitory or inducing CYP3A interactions may also vary depending on the CYP3A substrate and the interacting drug used.

In several studies the extent of hepatic CYP3A inhibition in vitro and in vivo varies with substrate and inheritance of CYP3A5. Surprisingly, the CYP3A5*1 allele was significantly associated with reduced inhibitory effects of itraconazole and fluconazole on midazolam-mediated CYP3A activity (Yu et al., 2004; Isoherranen et al., 2008), an effect which was not detectable using erythromycin as CYP3A substrate (Isoherranen et al., 2008). At first view these findings are contradictory to those in other CYP450s that inhibitory potency increases with metabolic activity. However, considering that ketoconazole and fluconazole inhibit CYP3A5 4-fold and 10-fold weaker than CYP3A4, respectively (Gibbs et al., 1999) and midazolam, in contrast to macrolide antibiotics, is effectively cleared by CYP3A5 (Williams et al., 2002). Therefore, one can speculate that in carriers of the *CYP3A5*1* allele remaining CYP3A5 metabolizes midazolam, while CYP3A4 is effectively suppressed.

Similarly, *CYP3A5* polymorphism appeared also to influence susceptibility to induction-type drug interactions only for some inducers. While *CYP3A5* polymorphism had no effects on the ability of the potent CYP3A4 inducer rifampin to induce CYP3A activity when assessed by systemic clearance of midazolam or the erythromycin breath test (Floyd et al., 2003; Yu et al., 2004; Roberts et al., 2008), the inductive effects of dexamethasone on the erythromycin breath test were only detected in the absence of the *CYP3A5*1* allele as a result of lower basal CYP3A activity (Roberts et al., 2008).

3.1.2.5 CYP2D6

Although CYP2D6 expression constitutes only 2–5% of the total CYP content, approximately 15% of the top 200 prescription drugs are metabolized by CYP2D6 (Zanger et al., 2008). In addition to the wild-type allele (*CYP2D6*1A*), more than 80 variant alleles have been identified. *CYP2D6*1,*2, *33*, and **35* have normal enzyme activity, whereas *CYP2D6*3, *4, *6, *7*, and **8* have null enzyme activity. *CYP2D6*5* represents deletion of the enzyme, and *CYP2D6*9, *10, *17*, and **41* code for decreased enzyme activity (Ingelman-Sundberg et al., 2008).

Studies evaluating drug interactions caused by CYP2D6 inhibition have consistently shown a larger magnitude of inhibition in CYP2D6 EMs compared with CYP2D6 PMs (Abdel-Rahman et al., 1999; Hamelin et al., 2000b; Llerena et al., 2001; Bondolfi et al., 2002). Particularly relevant are CYP2D6-based DDIs in the treatment of psychiatric disorders because many antidepressants and neuroleptics are substrates and/or inhibitors of CYP2D6 (Flockhart, 2008). The genetic differences in the extent of DDIs have been demonstrated using the CYP2D6-metabolized selective serotonin reuptake inhibitor (SSRI) venlafaxine. Coadministration of diphenhydramine in EMs more than doubled venlafaxine AUC and decreased the metabolic clearance to O-demethylated metabolites by approximately 3 times, while the potent CYP2D6 inhibitor quinidine increased in EMs the AUC of (*S*)- and (*R*)-venlafaxine by 4- and 12-fold, respectively, and reduced the conversion to O-demethylated metabolites by 7- and 113-fold, respectively. In contrast, no significant changes in the pharmacokinetic parameters occurred in *CYP2D6**4/4 -*3/4, *7/7* PMs and *CYP2D6*3/4, *4/4* PMs after coadministration of diphenhydramine or quinidine, respectively (Lessard et al., 2001; Eap et al., 2003). These results further imply that concurrent administration of a CYP2D6 inhibitor with venlafaxine to CYP2D6 EMs may require a dose reduction of venlafaxine, whereas a dose reduction in CYP2D6 PMs could result in therapeutic failure. As indicated above, SSRIs are also often potent CYP450 inhibitors (Flockhart, 2008). For paroxetine and fluoxetine, plasma concentrations and dosage strongly influence the magnitude of enzyme inhibition. Moreover, the potential of paroxetine (a CYP2D6 substrate) as an inhibitor was affected by *CYP2D6* genotype and metabolic capacity (Lam et al., 2002).

Less is known of how remaining metabolic activity in CYP2D6 IMs affects inhibitory drug interactions. This question is of particular clinical importance for Asian populations, because the population shift in Asians toward lower metabolic rates is caused by a high frequency (<40%) of the reduced function allele, *CYP2D6*10* (Bradford, 2002). Subjects with the *CYP2D6*10/10* IM geno-/phenotype had lower flecainide basal clearance, which appeared unaffected by coadministration of paroxetine, while the higher flecainide clearance of carriers of at least one allele with normal CYP2D6 activity (EMs) was clearly inhibited by paroxetine down to the IM levels (Lim et al., 2008). Therefore with respect to the absent interaction potential IMs seem comparable to PMs.

3.2 Drug Transporters

Membrane-bound proteins involved in the in- or outward transport of compounds, broadly referred to as transporters, play critical roles in delivery and overall disposition of numerous clinically used drugs (Nies, 2007; Nies et al., 2008). Multiple genetic polymorphisms and haplotypes have been described in efflux and uptake transporters and have been investigated with respect to potential implications for drug disposition and to the apparent intersubject variability in drug response (Kerb, 2006). The following section considers only transporter polymorphisms, which have been evaluated in relation to DDI and which are more frequent than 5% in at least one ethnicity. The pharmacologically most relevant polymorphisms of membrane transport proteins including their frequencies in African, Asian, and Caucasian populations are summarized in Table 3.2. For a comprehensive overview on transporter pharmacogenomics the reader is referred to recent reviews (Kerb, 2006; Seithel et al., 2008; Zaïr et al., 2008).

3.2.1 ABCB1 (P-gp) Polymorphisms and Drug Interactions

To date, more than 100 variants have been identified in the *ABCB1* (*MDR1*) gene (Kerb, 2006). A synonymous SNP in exon 26 (*c.3435C>T*) attracted particular attention because it was found to be associated with altered P-glycoprotein (P-gp) activity and function (Hoffmeyer et al., 2000). It has been demonstrated that codon usage played a role in the effect of this silent polymorphisms on P-gp function by altering the structure of substrate and inhibitor interaction sites (Kimchi-Sarfaty et al., 2007). *ABCB1 c.3435C>T* is part of a common haplotype with *c.1236C>T* and *c.2677G>T* (Kim et al., 2001). Although challenged by discordant results, numerous studies reported *c.3435C>T* (alone or in combination) to be associated with mRNA expression or ex vivo rhodamine 123 efflux in lymphocytes, pharmacokinetics of typical P-gp drugs such as digoxin, fexofenadine, nelfinavir, and cyclosporine in clinical studies, as well as with drug responsiveness in HIV or epilepsy patients (reviewed in Kerb, 2006; Chinn and Kroetz, 2007).

Only very few studies have addressed the *ABCB1* genotype-based interaction with P-gp inhibitors or inducers. This is partly based on the fact that applicable in vitro assays are experimentally and technically elaborative and of limited availability. Furthermore, many P-gp substrates and inhibitors are also substrates and inhibitors of CYP3A hampering to differentiate inhibition of P-gp transport from inhibition of CYP3A-mediated metabolism in clinical studies. As a consequence, clinical studies on P-gp-related DDIs have employed probe drugs with negligible metabolism such as digoxin, fexofenadine, or talinolol. By use of a vaccine virus expression system, the specificity of P-gp for model substrates was not substantially affected by any of the tested polymorphisms, while the *c.1236T—c.2677T—c.3435T* haplotype exhibited a functional change in the presence of P-gp inhibitors verapamil and cyclosporine suggesting an altered interaction potential (Kimchi-Sarfaty et al., 2007). In vitro testing of rhodamine 123 and calcein-AM transport in a stable recombinant P-gp expression system demonstrated that inhibition activities varied with *ABCB1* gene polymorphisms and with the inhibitor. Inhibitory activities with decreasing potency were cyclosporine A > verapamil > phenytoin > carbamazepine > lamotrigine > phenobarbital > valproic acid = levetiracetam = gabapentin. The *c.1236T—c.2677A/T—c.3435T* haplotype resulted in profoundly less inhibitory activities against rhodamine and calcein-AM with significantly lower intracellular substrate concentration (Hung et al., 2008). Kurata et al.

Table 3.2 Clinically relevant polymorphisms of drug uptake and efflux transporters

Drug transporter	Localization	Polymorphism[a]	Phenotype	Frequency [%] African	Frequency [%] Asian	Frequency [%] Caucasian
Efflux transporters						
ABCB1 (MDR1)	Liver, intestine	c.1236C>**T**	Reduced function/expression	21	69	46
	Kidney, brain	c.2677G>**T/A** (Ala893**Ser**)	Higher digoxin levels	<1–3[b]	45/7	46/2
		c.3435C>**T**		10–16[b]	40	56
Uptake transporters						
SLCO1B1 (OATP1B1)	Liver	c.521T>**C** (Val174**Ala**), *5/*15	Reduced function, higher prava-/atorva-/rosuvastatin levels	1–2	11	12–15
SLC22A1 (OCT1)	Liver, (kidney)	c.1201G>A (Gly401Ser)	Reduced function, higher metformin levels, reduced effect	1	0	1–3
		c.1258–1260delATG (Met420del)		3	0	16–19
		c.1393G>A (Gly465Arg)		2	0	<1
SLC22A2 (OCT2)	Kidney	c.808G>T (Ala270Ser)	Reduced function, reduced renal metformin clearance	11	9	16

[a] Variant nucleotide or amino acid in bold
[b] Increasing prevalence from south to north

reported a higher digoxin bioavailability and reduced digoxin renal clearance in subjects with the *ABCB1 c.3435T* variant. Coadministration of the potent inhibitor of P-gp drug efflux activity, clarithromycin, however, significantly increased digoxin bioavailability only in *c.2677C—c.3435C* subjects causing the genetic differences on digoxin bioavailability to disappear, although the effect of *ABCB1* variants on renal and secretory clearance was still observed suggesting that the digoxin–clarithromycin interaction is attributable entirely to processes occurring in the gut (Kurata et al., 2002). Thus, although gene expression or function is determined by genotype to the same extent in all tissues, the genetic influence on the extent of DDIs may differ in different tissues depending on the inhibitor levels to which various tissues and organs are exposed. Furthermore, membrane transport proteins are not only involved in the systemic availability of drugs but they also control the access of drugs to their target site of action. In these sanctuaries the local situation with respect to fraction transported or the presence of competing transporters may be completely different. As an example, loperamide, a potent opiate, which normally lacks CNS effects, because P-gp efflux at the blood–brain barrier effectively prevents uptake into the brain, produced pronounced miosis (opiate CNS effect) in *ABCB1 3435T* carriers, when P-gp was additionally inhibited by quinidine, an effect which was not reflected in the loperamide plasma levels (Skarke et al., 2003).

Dipyridamole is an in vitro and in vivo P-gp inhibitor that increases digoxin plasma concentrations by enhancement of its intestinal absorption. Although subjects harboring the *ABCB1 c.3435TT* genotype had significantly higher digoxin AUCs than carriers of *ABCB1 3435CC*, the modulation of digoxin pharmacokinetics by dipyridamole was similar in both genotypes (Verstuyft et al., 2003). Even contrary results were reported by Shon et al., who found that the effect of *ABCB1 c.2677T—c.3435T* haplotype on fexofenadine disposition was amplified in the presence of the potent P-gp inhibitor itraconazole (Shon et al., 2005). In fact, itraconazole pretreatment increased fexofenadine pharmacokinetic parameters particularly in carriers of *c.2677T—c.3435T* and the moderate difference in the pharmacokinetic parameters between the two *ABCB1* haplotypes became more prominent to reach statistical significance. However, the inhibitory effect of itraconazole appeared strong enough to influence peripheral antihistamine effects of fexofenadine, as determined by the suppression of histamine-induced wheal and flare reactions, whereas the effect of the *ABCB1* haplotype remained clinically insignificant (Shon et al., 2005).

The influence of *ABCB1* genotype on P-gp induction was addressed in a clinical study investigating the inductive effects of St. John's Wort. Subjects harboring the combined *c.1236T—c.2677T—c.3435T* polymorphism had lower basal intestinal ABCB1 mRNA level and displayed an attenuated inductive response to St. John's Wort resulting in a 25% lower talinolol bioavailability, while talinolol disposition remained unchanged in carriers of the wild-type allele (*c.1236C—c.2677C—c.3435C*) (Schwarz et al., 2007). Under the conditions of P-gp induction, the genetic differences in talinolol disposition minimized, whereas the genetic differences in ABCB1 mRNA expression became apparent.

Because many P-gp substrates are also substrates of CYP3A, both are regulated via pregnane X receptor (PXR) (Geick et al., 2001), P-gp may also affect drug metabolism and, indeed, in vitro experiments have demonstrated that intra-hepatocyte concentrations and metabolic rates of drugs are influenced by P-gp (Lan et al., 2000). However, the dynamic interplay between CYP3A and P-gp may be even more complex. *ABCB1* SNPs may also contribute to human variation in basal CYP3A activity as well as to variation in CYP3A induction by rifampin. The magnitude of CYP3A induction was inversely related to CYP3A constitutive expression in liver and intestine. Individuals harboring the *ABCB1 c.3435T* or the linked *c.2677T* allele had higher basal hepatic CYP3A4 expression but a lower magnitude of induction as assessed by testosterone 6β-hydroxylation in primary human hepatocytes (Lamba et al., 2006). Consistent with these results a clinical study showed that in persons induced with steroids, carriers of the *ABCB1 c.3435T* allele had higher induction of etoposide clearance than those with the *c.3435C* allele (Kishi et al., 2004). As a consequence, *ABCB1* genetic variation may predict the extent of CYP3A-mediated DDIs.

3.3 Drug Uptake Transporters

3.3.1 OATP Polymorphisms and Drug Interactions

OATP1B1 mediates the uptake on the sinusoidal membrane of human liver. Of more than 40 sequence variations, 17 nonsynonymous (change of an amino acid) have been identified in the *SLCO1B1* gene encoding OATP1B1 (reviewed in Kerb, 2006). Some variants have been associated with altered activity of OATP1B1. In particular, the nonsynonymous *c.521T>C* (Val174Ala) SNP has consistently been associated with reduced transport activity both in vitro (Tirona et al., 2001; Iwai et al., 2004; Ho et al., 2006) and in vivo (Nishizato et al., 2003; Niemi et al., 2004; Kalliokoski et al., 2008b). Particularly important are the 174Ala-encoding **5*, **15*, and **17* alleles for the hepatic uptake and disposition of water-soluble statins (pravastatin, pitavastatin, rosuvastatin, atorvastatin, and simvastatin acid) but not for simvastatin lactone and fluvastatin (Niemi, 2007). The use of statins in humans is influenced by the potential for pharmacokinetic DDIs. Gemfibrozil and cyclosporine are known to increase the systemic exposure of statins increasing the risk for myopathy and rhabdomyolysis; inhibition of OATP1B1 has been suggested to play a role in these DDIs (Neuvonen et al., 2006). Most interestingly, a genome-wide scan of approximately 300,000 markers in 85 myopathy patients yielded a single strong association of myopathy with the *c.521T>C* SNP (Link et al., 2008).

Analogous to the negligible effects of PMs in CYP2C19- and CYP2D6-mediated interactions, cyclosporine increased the AUC of another OATP1B1 substrate, repaglinide, 42% less in subjects harboring the *SLCO1B1 c.521TC* genotype compared to the *c.521TT* genotype (Kajosaari et al., 2005). However, another study using gemfibrozil as inhibitor produced a larger increase of repaglinide AUC in carriers of the *c.521C* allele (Kalliokoski et al., 2008a) and in both studies, the relative increases of AUC did not differ significantly between genotype groups (Kajosaari et al., 2005; Kalliokoski et al., 2008a). While *SLCO1B1* genotype was, concordant to earlier studies (Kalliokoski et al., 2008b), a major determinant of repaglinide pharmacokinetics, these data suggest that, in contrast to the inhibition of CYP2C8- (gemfibrozil) or CYP3A4-mediated metabolism (cyclosporine), inhibition of OATP1B1 plays only a limited role in the interaction between gemfibrozil or cyclosporine and repaglinide.

Because *c.521C* reduced the in vitro uptake of rifampin, a reduced PXR activation and subsequently reduced induction of drug metabolism and transport were suggested (Tirona et al., 2003). Although men carrying the *SLCO1B1 c.521C* SNP were more likely to have low CYP3A4 levels in primary human hepatocytes, no association of *SLCO1B1* genotype with CYP3A4 induction by rifampin was detected (Lamba et al., 2006). Moreover, a clinical study in 38 subjects including five carriers of the *c.521*TT* genotype did not detect an effect of *SLCO1B1* genotype on the extent of induction of hepatic CYP3A4 as assessed by 4β-hydroxycholesterol plasma concentration (Niemi et al., 2006). In contrast, an interaction was shown for other OATP1B1 inductors such as the herbal medicine baicalin, which increased rosuvastatin plasma concentrations in **15* in a gene dose-dependent manner (Fan et al., 2008b).

3.3.2 OCT Polymorphisms and Drug Interactions

Currently, three OCTs are considered to be relevant for the transport of drugs. While a number of polymorphisms in the *SLC22A1* gene have been shown to affect in vitro transport function of OCT1 (Kerb et al., 2002) and in vivo exposure and glucose-lowering effects of metformin (Shu et al., 2007; 2008), their influence on drug interactions has not yet been addressed. More than 40 genetic variations in the OCT2 encoding *SLC22A2* gene have been reported (Kerb, 2006), including mostly rare variants (population frequency <1%), which exhibited altered transport function in vitro (Leabman et al., 2002). However, the only common nonsynonymous SNP, *c.808G>T* (Ala270Ser) occurring with a frequency of 9–16% in various populations showed a decreased inhibitory sensitivity in recombinant oocytes (Leabman et al., 2002). Recently, in Chinese healthy volunteers the inhibitory effect of cimetidine on the OCT2-mediated renal clearance of metformin appeared to be dependent on this polymorphism and was only detectable in the *c.808*TT* genotype (Wang et al., 2008). While the *c.808G>T* polymorphism was associated with a reduced metformin renal tubular clearance yet without inhibition, the genotype effect on AUC was only detected in the presence of cimetidine (Wang et al., 2008).

3.4 Clinical Relevance of Pharmacogenetics for DDI

While the effects of genetic polymorphisms in drug-metabolizing enzymes and transporters on drug exposure have been extensively studied, only little attention has been given until very recently to whether these polymorphisms alter the interindividual susceptibility to or the degree of induction or inhibition.

With a few exceptions (e.g., refractoriness of *CYP2B6*4* to mechanism-based inhibition), a greater magnitude of inhibitory drug interactions can be expected in EMs than in PMs and the extent of inhibition can be predicted by genotyping of the baseline metabolic activity of the responsible CYP enzyme. These consistent findings are not surprising, given that the underlying genetic polymorphisms result in the expression of inactive or absent proteins, such that no residual activity is present to be inhibited. A similar concept applies to inducing drug interactions; the lower the baseline enzymatic activity, the higher the induction achieved with inducers (Vesell and Page, 1969; McCune et al., 2000) unless the low baseline activity is caused by null alleles or other genetic factors causing complete absence or inactivity of the enzyme.

Generally, the same concept should also apply to transport proteins. However, being expressed at various tissues and organs throughout the whole body, membrane transporters not only contribute to systemic exposure of drugs but also determine drug access to pharmacological sanctuaries requiring a more sophisticated approach. For instance, local concentrations of the interacting drug or the presence of redundant transporters resulting in distinct fraction of the transported drug need to be considered.

A number of factors may neutralize or abolish the clinical relevance of a genetic polymorphism. Genetic polymorphisms are unlikely to have a major clinical impact on inhibitory drug interaction unless the following applies for the drug being affected:

- The drug has a low therapeutic index. If a drug is safe over a wide range of concentrations that encompass the variation caused by genetics, then drug interactions are most likely clinically not relevant as it is not important to consider genetic polymorphisms in the therapeutic context.
- There is a clear association of plasma concentration with clinical endpoints, either with desired effects or with concentration-related adverse effects. If pharmacokinetis effects does not translate into clinical effects a pharmacokinetic DDI is unlikely to be of clinical relevance unless a transporter influences clinical effects by mediating drug entry to the target tissue (e.g., brain).
- Metabolism or transport is mainly via a single pathway mediated by the polymorphism. With multiple pathways the fraction of the drug metabolized or transported by a single of these pathways is so minor that inhibition either has little effect or may be compensated by the other pathways. However, if one of several metabolic pathways is omitted by PM geno-/phenotype, the interaction susceptibility of the remaining pathway can increase.

3.4.1 Implications for Drug Treatment

A greater degree of drug interaction may lead to a greater likelihood of toxicity or therapeutic failure. The above-summarized findings can be further interpreted as EMs being at greater risk for clinically significant drug interactions when an inhibitory drug of the same CYP enzyme is introduced. Although changes in pharmacokinetics cannot always be translated into expected changes in pharmacodynamics, altered pharmacodynamics caused by increased drug exposure have been shown with some commonly used drugs. For instance, coadministration of diphenhydramine and metoprolol to EMs not only significantly altered pharmacokinetics but also prolonged negative chronotropic and inotropic effects (Hamelin et al., 2000a). Such changes were not seen in the PMs. When an inhibitor impairs the metabolism of a drug with a narrow therapeutic range dosed on a long-term basis in an EM, there is a considerable chance of reaching toxic drug exposures with a subsequent occurrence of ADRs. This may be further interpreted in the clinical setting to indicate that dose adjustment may be necessary only in EMs who are taking a genetically differentially metabolized substrate when an inhibitor of this enzyme is initiated, but not in PMs.

Importantly, antipsychotics and antidepressants are often prescribed concurrently to treat comorbid symptoms. Antidepressants such as paroxetine and fluoxetine are potent inhibitors of CYP2D6, the enzyme responsible for metabolizing many antipsychotic medications, most of which have a narrow therapeutic range (Flockhart, 2008). Studies have reported that a significant rise in the concentrations of the antipsychotic drug, resulting from initiation of paroxetine or fluoxetine, can lead to an increased frequency or worsening of extrapyramidal symptoms (Spina et al., 2008). Such adverse drug interactions are more likely to be seen in EMs because the extent of inhibition is more significant in EMs than PMs.

Fewer information is available on the interaction potential of reduced function proteins such as the *CYP2D6*10* IM or the residual activity *CYP2C9*3* allele. Studies on CYP2C9-mediated flurbiprofen–fluconazole interaction and on CYP2D6-mediated flecainide–paroxetine interaction indicated that these reduced function alleles behaved, as if they were inactive (Kumar et al., 2008; Lim et al., 2008), although results await confirmation with other inhibitor–drug combinations.

In this context, it is worthwhile to note that enzyme activity also varies broadly among EMs (Ma and Lu, 2007). Consequently, the extent of inhibitory drug interactions may also vary among EMs (Yasui-Furukori et al., 2004a; b). Whether heterozygous EMs have the same change in drug exposure after the addition of an inhibitory agent like homozygous EMs has not yet been elaborated. Therefore data from homozygous EMs should be applied with caution to heterozygous EMs.

This concept further suggests that individuals with duplicated or amplified genes of the CYP2D6 enzyme, also known as ultrarapid metabolizers, would have the greatest extent of inhibitory effect as a result of elevated baseline enzyme activity and an increased baseline drug dose.

While investigating DDIs between substrates and inhibitors of the same CYP enzyme yielded consistent results, using a substrate that is metabolized through several CYP pathways in EMs and PMs of one CYP enzyme and adding an inhibitor of another CYP enzyme that is also responsible for metabolizing the substrate may result in contradictory data as exemplified in the inhibitory effect of ritonavir on voriconazole metabolism in CYP2C19 EMs vs. PMs (Mikus et al., 2006).

For many medications not only several metabolic pathways but also drug transport is involved. As shown for rifampin, *SLCO1B1* genotype may affect hepatic uptake and subsequently could contribute to variable ABCB1 transport or CYP3A metabolism, advising the conclusion that transporters, metabolizing enzymes, and regulatory factors should be viewed and evaluated as an integrative system rather than single components.

3.4.2 Implications for Drug Development

A greater magnitude of inhibitory drug interactions occurs in EMs than in PMs and the extent of inhibition can be predicted by the genotyping of the baseline metabolic activity of the responsible CYP enzyme. Although it is logical that the extent of drug interactions should be greater in EMs, only a limited number of drug interaction studies are currently performed that distinguish genotyped EMs and PMs during the drug development process. When metabolizer status is not determined, the role of genetic polymorphisms is ignored, and the potential for, and range of, variation in drug interactions cannot be evaluated in relation to genotype. For example, a drug interaction study that primarily includes PMs might miss inhibitory effects leading to the false conclusion that there is no drug interaction. However, very different results can be expected if only EMs are enrolled in the study. In addition, because a different extent of drug interactions exists between EMs and PMs, a study with an anonymous number or combination of EMs and null alleles will likely show a large range in the magnitude of a drug interaction. This is a frequent outcome of drug interaction studies. For these reasons, drug interaction information on the product label or common medical references can be misleading if presented as a mean or a range without identification of the subjects' metabolizer status.

Thus, a properly designed DDI study involving a drug that is metabolized by a polymorphic enzyme should preselect volunteers based on their genotype. While PMs can be excluded as they are inappropriate to investigate DDIs, inclusion of various EM genotypes may be necessary to consider that genetic variants can confer variable susceptibility to inactivation, particularly if in vitro experiments suggest mechanism-based inhibition (Bumpus et al., 2005; 2006). The latter example further emphasizes that one should be careful to uncritically generalize results from DDI studies, which typically involve only certain model substrates, inhibitors, or inducers (FDA/CBER, 2006). The same principles and considerations also apply to drug transporters which have been meanwhile adopted in regulatory guidance that gives detailed recommendations on study design, substrates, dosing and data analysis, and labeling to assess metabolic- and transporter-based drug–drug interactions (FDA/CBER, 2006; Huang et al., 2007; 2008). However, if the investigational drug in question is metabolized or transported by a genetically polymorphic protein, to a clinically relevant extent (> 25%), genotyping of this pathway should be incorporated into the drug development and decision-making process and the extent of inhibitory drug interactions should be reported according to, e.g., metabolizer or transporter status.

3.5 Conclusions

Understanding of genetic influences on metabolism, transport, and drug–drug interactions is critical to the benefit/risk assessment of a drug. With improved understanding of the molecular bases of drug–drug interactions, and the interplay of various genetic and non-genetic factors affecting these interactions, risks associated with DDIs can be assessed and managed to minimize unwanted effects. Although it is difficult to untangle genetic factors from non-genetic factors when evaluating and predicting DDIs the influence of genetic factors on interindividual variability can be reduced by genotyping. While this may challenge the drug development process and the clinical use of drugs, it is essential for providing more accurate information (package insert, drug labeling) and allows clinicians to better predict potential DDIs and recommend appropriate dose adjustments when necessary. Clinicians should be cautious when using commonly available drug information resources to determine drug dosage adjustments with interacting drugs, as these are mostly mean data not considering the large ranges of the interactions in the population, and this variability may be a result of genetic differential metabolism or transport. Thus, pharmacogenetic tools not only hold the potential to identify patients who will respond favorably to a particular drug and hence avoid ADRs and diminish treatment failure but may also define the individual susceptibility for or the degree of clinically relevant DDIs.

3.5.1 WEB-Based Scientific Resources

Pharmacogenetics and genomics:

- PharmGkb. The Pharmacogenetics and Pharmacogenomics Knowledge Base. Available from: http://www.pharmgkb.org/

Nomenclature of drug-metabolizing enzymes including a detailed description of polymorphic alleles with their functional consequences:

- Sim SC. Home Page of the Human Cytochrome P450 (CYP) Allele Nomenclature Committee. Last update Sep 9, 2009. Available from http://www.cypalleles.ki.se/ [Accessed Jan 19, 2009]
- Nelson D. Cytochrome P450 Homepage. 2007 Available from http://drnelson.utmem.edu/CytochromeP450.html [Accessed Nov 22, 2007]
- Flockhart DA. Drug Interactions: Cytochrome P450 Drug Interaction Table. 2008 Version 4.0 released on Aug 20, 2007. Available from http://medicine.iupui.edu/flockhart/table.htm [Accessed Apr 28, 2008]

References

Abdel-Rahman SM, Gotschall RR, Kauffman RE, Leeder JS and Kearns GL (1999) Investigation of terbinafine as a CYP2D6 inhibitor in vivo. *Clin Pharmacol Ther* **65**: 465–472.

Bondolfi G, Eap CB, Bertschy G, Zullino D, Vermeulen A and Baumann P (2002) The effect of fluoxetine on the pharmacokinetics and safety of risperidone in psychotic patients. *Pharmacopsychiatry* **35**:50–56.

Bradford LD (2002) CYP2D6 allele frequency in European Caucasians, Asians, Africans and their descendants. *Pharmacogenomics* **3**:229–243.

Bumpus NN, Kent UM and Hollenberg PF (2006) Metabolism of efavirenz and 8-hydroxyefavirenz by P450 2B6 leads to inactivation by two distinct mechanisms. *J Pharmacol Exp Ther* **318**: 345–351.

Bumpus NN, Sridar C, Kent UM and Hollenberg PF (2005) The naturally occurring cytochrome P450 (P450) 2B6 K262R mutant of P450 2B6 exhibits alterations in substrate metabolism and inactivation. *Drug Metab Dispos* **33**:795–802.

Chinn LW and Kroetz DL (2007) ABCB1 pharmacogenetics: progress, pitfalls, and promise. *Clin Pharmacol Ther* **81**:265–269.

Cho JY, Yu KS, Jang IJ, Yang BH, Shin SG and Yim DS (2002) Omeprazole hydroxylation is inhibited by a single dose of moclobemide in homozygotic EM genotype for CYP2C19. *Br J Clin Pharmacol* **53**:393–397.

Eap CB, Lessard E, Baumann P, Brawand-Amey M, Yessine M-A, O'Hara G and Turgeon J (2003) Role of CYP2D6 in the stereoselective disposition of venlafaxine in humans. *Pharmacogenetics* **13**:39–47.

Eichelbaum M, Ingelman-Sundberg M and Evans WE (2006) Pharmacogenomics and individualized drug therapy. *Annu Rev Med* **57**:119–137.

Fan L, Wang JC, Jiang F, Tan ZR, Chen Y, Li Q, Zhang W, Wang G, Lei HP, Hu DL, Wang D and Zhou HH (2008a) Induction of cytochrome P450 2B6 activity by the herbal medicine baicalin as measured by bupropion hydroxylation. *Eur J Clin Pharmacol* **65**:403–409.

Fan L, Zhang W, Guo D, Tan ZR, Xu P, Li Q, Liu YZ, Zhang L, He TY, Hu DL, Wang D and Zhou HH (2008b) The effect of herbal medicine baicalin on pharmacokinetics of rosuvastatin, substrate of organic anion-transporting polypeptide 1B1. *Clin Pharmacol Ther* **83**: 471–476.

FDA/CBER (2006) Draft Guidance for Industry: Drug Interaction Studies – Study Design, Data Analysis, and Implications for Dosing and Labeling, in, U.S. Food and Drug Administration.

Flockhart DA (2008) Drug Interactions: Cytochrome P450 Drug Interaction Table, Available at: http://medicine.iupui.edu/flockhart/table.htm.

Floyd MD, Gervasini G, Masica AL, Mayo G, George AL, Bhat K, Kim RB and Wilkinson GR (2003) Genotype-phenotype associations for common CYP3A4 and CYP3A5 variants in the basal and induced metabolism of midazolam in European- and African-American men and women. *Pharmacogenetics* **13**:595–606.

García-Martín E, Martínez C, Ladero JM and Agúndez JAG (2006) Interethnic and intraethnic variability of CYP2C8 and CYP2C9 polymorphisms in healthy individuals. *Mol Diagn Ther* **10**:29–40.

Geick A, Eichelbaum M and Burk O (2001) Nuclear receptor response elements mediate induction of intestinal MDR1 by rifampin. *J Biol Chem* **276**:14581–14587.

Geisler T, Schaeffeler E, Dippon J, Winter S, Buse V, Bischofs C, Zuern C, Moerike K, Gawaz M and Schwab M (2008) CYP2C19 and nongenetic factors predict poor responsiveness to clopidogrel loading dose after coronary stent implantation. *Pharmacogenomics* **9**: 1251–1259.

Gibbs MA, Thummel KE, Shen DD and Kunze KL (1999) Inhibition of cytochrome P-450 3A (CYP3A) in human intestinal and liver microsomes: comparison of Ki values and impact of CYP3A5 expression. *Drug Metab Dispos* **27**:180–187.

Hamelin BA, Bouayad A, Methot J, Jobin J, Desgagnes P, Poirier P, Allaire J, Dumesnil J and Turgeon J (2000a) Significant interaction between the nonprescription antihistamine diphenhydramine and the CYP2D6 substrate metoprolol in healthy men with high or low CYP2D6 activity. *Clin Pharmacol Ther* **67**:466–477.

Hamelin BA, Bouayad A, Méthot J, Jobin J, Desgagnés P, Poirier P, Allaire J, Dumesnil J and Turgeon J (2000b) Significant interaction between the nonprescription antihistamine diphenhydramine and the CYP2D6 substrate metoprolol in healthy men with high or low CYP2D6 activity. *Clin Pharmacol Ther* **67**:466–477.

Ho RH, Tirona RG, Leake BF, Glaeser H, Lee W, Lemke CJ, Wang Y and Kim RB (2006) Drug and bile acid transporters in rosuvastatin hepatic uptake: function, expression, and pharmacogenetics. *Gastroenterology* **130**:1793–1806.

Hoffmeyer S, Burk O, von Richter O, Arnold HP, Brockmoller J, Johne A, Cascorbi I, Gerloff T, Roots I, Eichelbaum M and Brinkmann U (2000) Functional polymorphisms of the human multidrug-resistance gene: multiple sequence variations and correlation of one allele with P-glycoprotein expression and activity in vivo. *Proc Natl Acad Sci U S A* **97**:3473–3478.

Hofmann MH, Blievernicht JK, Klein K, Saussele T, Schaeffeler E, Schwab M and Zanger UM (2008) Aberrant splicing caused by single nucleotide polymorphism c.516G>T [Q172H], a marker of CYP2B6*6, is responsible for decreased expression and activity of CYP2B6 in liver. *J Pharmacol Exp Ther* **325**:284–292.

Hosohata K, Masuda S, Katsura T, Takada Y, Kaido T, Ogura Y, Oike F, Egawa H, Uemoto S and Inui KI (2009) Impact of intestinal CYP2C19 genotypes on the interaction between tacrolimus and omeprazole, but not lansoprazole, in adult living-donor liver transplant patients. *Drug Metab Dispos* **37**:821–826.

Huang S-M, Strong JM, Zhang L, Reynolds KS, Nallani S, Temple R, Abraham S, Habet SA, Baweja RK, Burckart GJ, Chung S, Colangelo P, Frucht D, Green MD, Hepp P, Karnaukhova E, Ko H-S, Lee J-I, Marroum PJ, Norden JM, Qiu W, Rahman A, Sobel S, Stifano T, Thummel K, Wei X-X, Yasuda S, Zheng JH, Zhao H and Lesko LJ (2008) New era in drug interaction evaluation: US Food and Drug Administration update on CYP enzymes, transporters, and the guidance process. *J Clin Pharmacol* **48**:662–670.

Huang SM, Temple R, Throckmorton DC and Lesko LJ (2007) Drug interaction studies: study design, data analysis, and implications for dosing and labeling. *Clin Pharmacol Ther* **81**: 298–304.

Hung CC, Chen CC, Lin CJ and Liou HH (2008) Functional evaluation of polymorphisms in the human ABCB1 gene and the impact on clinical responses of antiepileptic drugs. *Pharmacogenet Genomics* **18**:390–402.

Ingelman-Sundberg M, Daly AK and Nebert DW (2008) Home Page of the Human Cytochrome P450 (CYP) Allele Nomenclature Committee, Available at: http://www.cypalleles.ki.se/.

Ingelman-Sundberg M, Sim SC, Gomez A and Rodriguez-Antona C (2007) Influence of cytochrome P450 polymorphisms on drug therapies: pharmacogenetic, pharmacoepigenetic and clinical aspects. *Pharmacol Ther* **116**:496–526.

Isoherranen N, Ludington SR, Givens RC, Lamba JK, Pusek SN, Dees EC, Blough DK, Iwanaga K, Hawke RL, Schuetz EG, Watkins PB, Thummel KE and Paine MF (2008) The influence of CYP3A5 expression on the extent of hepatic CYP3A inhibition is substrate-dependent: an in vitro-in vivo evaluation. *Drug Metab Dispos* **36**:146–154.

Itagaki F, Homma M, Yuzawa K, Nishimura M, Naito S, Ueda N, Ohkohchi N and Kohda Y (2004) Effect of lansoprazole and rabeprazole on tacrolimus pharmacokinetics in healthy volunteers with CYP2C19 mutations. *J Pharm Pharmacol* **56**:1055–1059.

Iwai M, Suzuki H, Ieiri I, Otsubo K and Sugiyama Y (2004) Functional analysis of single nucleotide polymorphisms of hepatic organic anion transporter OATP1B1 (OATP-C). *Pharmacogenetics* **14**:749–757.

Kajosaari LI, Niemi M, Neuvonen M, Laitila J, Neuvonen PJ and Backman JT (2005) Cyclosporine markedly raises the plasma concentrations of repaglinide. *Clin Pharmacol Ther* **78**:388–399.

Kalliokoski A, Backman JT, Kurkinen KJ, Neuvonen PJ and Niemi M (2008a) Effects of gemfibrozil and atorvastatin on the pharmacokinetics of repaglinide in relation to SLCO1B1 polymorphism. *Clin Pharmacol Ther* **84**:488–496.

Kalliokoski A, Backman JT, Neuvonen PJ and Niemi M (2008b) Effects of the SLCO1B1*1B haplotype on the pharmacokinetics and pharmacodynamics of repaglinide and nateglinide. *Pharmacogenetics Genom* **18**:937–942.

Kenworthy KE, Bloomer JC, Clarke SE and Houston JB (1999) CYP3A4 drug interactions: correlation of 10 in vitro probe substrates. *Br J Clin Pharmacol* **48**:716–727.

Kerb R (2006) Implications of genetic polymorphisms in drug transporters for pharmacotherapy. *Cancer Lett* **234**:4–33.

Kerb R, Brinkmann U, Chatskaia N, Gorbunov D, Gorboulev V, Mornhinweg E, Keil A, Eichelbaum M and Koepsell H (2002) Identification of genetic variations of the human organic cation transporter hOCT1 and their functional consequences. *Pharmacogenetics* **12**:591–595.

Kim RB, Leake BF, Choo EF, Dresser GK, Kubba SV, Schwarz UI, Taylor A, Xie HG, McKinsey J, Zhou S, Lan LB, Schuetz JD, Schuetz EG and Wilkinson GR (2001) Identification of functionally variant MDR1 alleles among European Americans and African Americans. *Clin Pharmacol Ther* **70**:189–199.

Kimchi-Sarfaty C, Oh JM, Kim I-W, Sauna ZE, Calcagno AM, Ambudkar SV and Gottesman MM (2007) A "silent" polymorphism in the MDR1 gene changes substrate specificity. *Science (New York, NY)* **315**:525–528.

Kishi S, Yang W, Boureau B, Morand S, Das S, Chen P, Cook EH, Rosner GL, Schuetz E, Pui C-H and Relling MV (2004) Effects of prednisone and genetic polymorphisms on etoposide disposition in children with acute lymphoblastic leukemia. *Blood* **103**:67–72.

Kongkaew C, Noyce PR and Ashcroft DM (2008) Hospital admissions associated with adverse drug reactions: a systematic review of prospective observational studies. *Ann Pharmacother* **42**:1017–1025.

Kumar V, Brundage RC, Oetting WS, Leppik IE and Tracy TS (2008) Differential genotype dependent inhibition of CYP2C9 in humans. *Drug Metab Dispos* **36**:1242–1248.

Kumar V, Wahlstrom JL, Rock DA, Warren CJ, Gorman LA and Tracy TS (2006) CYP2C9 inhibition: impact of probe selection and pharmacogenetics on in vitro inhibition profiles. *Drug Metab Dispos* **34**:1966–1975.

Kurata Y, Ieiri I, Kimura M, Morita T, Irie S, Urae A, Ohdo S, Ohtani H, Sawada Y, Higuchi S and Otsubo K (2002) Role of human MDR1 gene polymorphism in bioavailability and interaction of digoxin, a substrate of P-glycoprotein. *Clin Pharmacol Ther* **72**:209–219.

Kwara A, Lartey M, Sagoe KW, Xexemeku F, Kenu E, Oliver-Commey J, Boima V, Sagoe A, Boamah I, Greenblatt DJ and Court MH (2008) Pharmacokinetics of efavirenz when co-administered with rifampin in TB/HIV co-infected patients: pharmacogenetic effect of CYP2B6 variation. *J Clin Pharmacol* **48**:1032–1040.

Lam YWF, Gaedigk A, Ereshefsky L, Alfaro CL and Simpson J (2002) CYP2D6 inhibition by selective serotonin reuptake inhibitors: analysis of achievable steady-state plasma concentrations and the effect of ultrarapid metabolism at CYP2D6. *Pharmacotherapy* **22**:1001–1006.

Lamba J, Strom S, Venkataramanan R, Thummel KE, Lin YS, Liu W, Cheng C, Lamba V, Watkins PB and Schuetz E (2006) MDR1 genotype is associated with hepatic cytochrome P450 3A4 basal and induction phenotype. *Clin Pharmacol Ther* **79**:325–338.

Lan LB, Dalton JT and Schuetz EG (2000) Mdr1 limits CYP3A metabolism in vivo. *Mol Pharmacol* **58**:863–869.

Lang T, Klein K, Fischer J, Nüssler AK, Neuhaus P, Hofmann U, Eichelbaum M, Schwab M and Zanger UM (2001) Extensive genetic polymorphism in the human CYP2B6 gene with impact on expression and function in human liver. *Pharmacogenetics* **11**:399–415.

Leabman MK, Huang CC, Kawamoto M, Johns SJ, Stryke D, Ferrin TE, DeYoung J, Taylor T, Clark AG, Herskowitz I and Giacomini KM (2002) Polymorphisms in a human kidney xenobiotic transporter, OCT2, exhibit altered function. *Pharmacogenetics* **12**:395–405.

Lee CR, Goldstein JA and Pieper JA (2002) Cytochrome P450 2C9 polymorphisms: a comprehensive review of the in-vitro and human data. *Pharmacogenetics* **12**:251–263.

Lessard E, Yessine MA, Hamelin BA, Gauvin C, Labbé L, O'Hara G, LeBlanc J and Turgeon J (2001) Diphenhydramine alters the disposition of venlafaxine through inhibition of CYP2D6 activity in humans. *J Clin Psychopharmacol* **21**:175–184.

Lim KS, Cho JY, Jang IJ, Kim BH, Kim J, Jeon JY, Tae YM, Yi S, Eum S, Shin SG and Yu KS (2008) Pharmacokinetic interaction of flecainide and paroxetine in relation to the CYP2D6*10 allele in healthy Korean subjects. *Br J Clin Pharmacol* **66**:660–666.

Link E, Parish S, Armitage J, Bowman L, Heath S, Matsuda F, Gut I, Lathrop M and Collins R (2008) SLCO1B1 variants and statin-induced myopathy – a genomewide study. *N Engl J Med* **359**:789–799.

Llerena A, Berecz R, de la Rubia A, Fernández-Salguero P and Dorado P (2001) Effect of thioridazine dosage on the debrisoquine hydroxylation phenotype in psychiatric patients with different CYP2D6 genotypes. *Ther Drug Monit* **23**:616–620.

Ma Q and Lu AYH (2007) CYP1A induction and human risk assessment: an evolving tale of in vitro and in vivo studies. *Drug Metab Dispos* **35**:1009–1016.

McCune JS, Hawke RL, LeCluyse EL, Gillenwater HH, Hamilton G, Ritchie J and Lindley C (2000) In vivo and in vitro induction of human cytochrome P4503A4 by dexamethasone. *Clin Pharmacol Ther* **68**:356–366.

Meyer UA (2004) Pharmacogenetics – five decades of therapeutic lessons from genetic diversity. *Nat Rev Genet* **5**:669–676.

Mikus G, Schowel V, Drzewinska M, Rengelshausen J, Ding R, Riedel KD, Burhenne J, Weiss J, Thomsen T and Haefeli WE (2006) Potent cytochrome P450 2C19 genotype-related interaction between voriconazole and the cytochrome P450 3A4 inhibitor ritonavir. *Clin Pharmacol Ther* **80**:126–135.

Miura M, Inoue K, Kagaya H, Satoh S, Tada H, Sagae Y, Habuchi T and Suzuki T (2007) Influence of rabeprazole and lansoprazole on the pharmacokinetics of tacrolimus in relation to CYP2C19, CYP3A5 and MDR1 polymorphisms in renal transplant recipients. *Biopharm Drug Dispos* **28**:167–175.

Neuvonen PJ, Niemi M and Backman JT (2006) Drug interactions with lipid-lowering drugs: mechanisms and clinical relevance. *Clin Pharmacol Ther* **80**:565–581.

Niemi M (2007) Role of OATP transporters in the disposition of drugs. *Pharmacogenomics* **8**: 787–802.

Niemi M, Kivisto KT, Diczfalusy U, Bodin K, Bertilsson L, Fromm MF and Eichelbaum M (2006) Effect of SLCO1B1 polymorphism on induction of CYP3A4 by rifampicin. *Pharmacogenet Genomics* **16**:565–568.

Niemi M, Schaeffeler E, Lang T, Fromm MF, Neuvonen M, Kyrklund C, Backman JT, Kerb R, Schwab M, Neuvonen PJ, Eichelbaum M and Kivistö KT (2004) High plasma pravastatin concentrations are associated with single nucleotide polymorphisms and haplotypes of organic anion transporting polypeptide-C (OATP-C, SLCO1B1). *Pharmacogenetics* **14**: 429–440.

Nies AT (2007) The role of membrane transporters in drug delivery to brain tumors. *Cancer Lett* **254**:11–29.

Nies AT, Schwab M and Keppler D (2008) Interplay of conjugating enzymes with OATP uptake transporters and ABCC/MRP efflux pumps in the elimination of drugs. *Expert Opin Drug Metab Toxicol* **4**:545–568.

Nishizato Y, Ieiri I, Suzuki H, Kimura M, Kawabata K, Hirota T, Takane H, Irie S, Kusuhara H, Urasaki Y, Urae A, Higuchi S, Otsubo K and Sugiyama Y (2003) Polymorphisms of OATP-C (SLC21A6) and OAT3 (SLC22A8) genes: consequences for pravastatin pharmacokinetics. *Clin Pharmacol Ther* **73**:554–565.

Roberts PJ, Rollins KD, Kashuba AD, Paine MF, Nelsen AC, Williams EE, Moran C, Lamba JK, Schuetz EG and Hawke RL (2008) The influence of CYP3A5 genotype on dexamethasone induction of CYP3A activity in African Americans. *Drug Metab Dispos* **36**:1465–1469.

Schwab M, Schaeffeler E, Klotz U and Treiber G (2004) CYP2C19 polymorphism is a major predictor of treatment failure in white patients by use of lansoprazole-based quadruple therapy for eradication of Helicobacter pylori. *Clin Pharmacol Ther* **76**:201–209.

Schwarz UI, Hanso H, Oertel R, Miehlke S, Kuhlisch E, Glaeser H, Hitzl M, Dresser GK, Kim RB and Kirch W (2007) Induction of intestinal P-glycoprotein by St John's wort reduces the oral bioavailability of talinolol. *Clin Pharmacol Ther* **81**:669–678.

Seithel A, Glaeser H, Fromm MF and König J (2008) The functional consequences of genetic variations in transporter genes encoding human organic anion-transporting polypeptide family members. *Expert Opin Drug Metab Toxicol* **4**:51–64.

Shimizu T, Ochiai H, Asell F, Shimizu H, Saitoh R, Hama Y, Katada J, Hashimoto M, Matsui H, Taki K, Kaminuma T, Yamamoto M, Aida Y, Ohashi A and Ozawa N (2003) Bioinformatics research on inter-racial difference in drug metabolism I. Analysis on frequencies of mutant alleles and poor metabolizers on CYP2D6 and CYP2C19. *Drug Metab Pharmacokinet* **18**: 48–70.

Shon JH, Yoon YR, Hong WS, Nguyen PM, Lee SS, Choi YG, Cha IJ and Shin JG (2005) Effect of itraconazole on the pharmacokinetics and pharmacodynamics of fexofenadine in relation to the MDR1 genetic polymorphism. *Clin Pharmacol Ther* **78**:191–201.

Shu Y, Brown C, Castro RA, Shi RJ, Lin ET, Owen RP, Sheardown SA, Yue L, Burchard EG, Brett CM and Giacomini KM (2008) Effect of genetic variation in the organic cation transporter 1, OCT1, on metformin pharmacokinetics. *Clin Pharmacol Ther* **83**:273–280.

Shu Y, Sheardown SA, Brown C, Owen RP, Zhang S, Castro RA, Ianculescu AG, Yue L, Lo JC, Burchard EG, Brett CM and Giacomini KM (2007) Effect of genetic variation in the organic cation transporter 1 (OCT1) on metformin action. *J Clin Invest* **117**:1422–1431.

Sim SC, Risinger C, Dahl M-L, Aklillu E, Christensen M, Bertilsson L and Ingelman-Sundberg M (2006) A common novel CYP2C19 gene variant causes ultrarapid drug metabolism relevant for the drug response to proton pump inhibitors and antidepressants. *Clin Pharmacol Ther* **79**:103–113.

Skarke C, Jarrar M, Schmidt H, Kauert G, Langer M, Geisslinger G and Lotsch J (2003) Effects of ABCB1 (multidrug resistance transporter) gene mutations on disposition and central nervous effects of loperamide in healthy volunteers. *Pharmacogenetics* **13**:651–660.

Spina E, Santoro V and D'Arrigo C (2008) Clinically relevant pharmacokinetic drug interactions with second-generation antidepressants: an update. *Clin Ther* **30**:1206–1227.

Stresser DM, Blanchard AP, Turner SD, Erve JC, Dandeneau AA, Miller VP and Crespi CL (2000) Substrate-dependent modulation of CYP3A4 catalytic activity: analysis of 27 test compounds with four fluorometric substrates. *Drug Metab Dispos* **28**:1440–1448.

Talakad JC, Kumar S and Halpert JR (2009) Decreased susceptibility of the cytochrome P450 2B6 variant K262R to inhibition by several clinically important drugs. *Drug Metab Dispos* **37**:644–650.

Tirona RG, Leake BF, Merino G and Kim RB (2001) Polymorphisms in OATP-C: identification of multiple allelic variants associated with altered transport activity among European- and African-Americans. *J Biol Chem* **276**:35669–35675.

Tirona RG, Leake BF, Wolkoff AW and Kim RB (2003) Human organic anion transporting polypeptide-C (SLC21A6) is a major determinant of rifampin-mediated pregnane X receptor activation. *J Pharmacol Exp Ther* **304**:223–228.

Uno T, Sugimoto K, Sugawara K and Tateishi T (2008) The role of cytochrome P2C19 in R-warfarin pharmacokinetics and its interaction with omeprazole. *Ther Drug Monit* **30**:276–281.

Verstuyft C, Strabach S, El-Morabet H, Kerb R, Brinkmann U, Dubert L, Jaillon P, Funck-Brentano C, Trugnan G and Becquemont L (2003) Dipyridamole enhances digoxin bioavailability via P-glycoprotein inhibition. *Clin Pharmacol Ther* **73**:51–60.

Vesell ES and Page JG (1969) Genetic control of the phenobarbital-induced shortening of plasma antipyrine half-lives in man. *J Clin Invest* **48**:2202–2209.

Wang RW, Newton DJ, Liu N, Atkins WM and Lu AY (2000) Human cytochrome P-450 3A4: in vitro drug–drug interaction patterns are substrate-dependent. *Drug Metab Dispos* **28**:360–366.

Wang ZJ, Yin OQ, Tomlinson B and Chow MS (2008) OCT2 polymorphisms and in-vivo renal functional consequence: studies with metformin and cimetidine. *Pharmacogenet Genomics* **18**:637–645.

Wang LS, Zhu B, Abd El-Aty AM, Zhou G, Li Z, Wu J, Chen GL, Liu J, Tang ZR, An W, Li Q, Wang D and Zhou HH (2004) The influence of St John's Wort on CYP2C19 activity with respect to genotype. *J Clin Pharmacol* **44**:577–581.

Westlind-Johnsson A, Hermann R, Huennemeyer A, Hauns B, Lahu G, Nassr N, Zech K, Ingelman-Sundberg M and von Richter O (2006) Identification and characterization of CYP3A4*20, a novel rare CYP3A4 allele without functional activity. *Clin Pharmacol Ther* **79**:339–349.

Williams JA, Ring BJ, Cantrell VE, Jones DR, Eckstein J, Ruterbories K, Hamman MA, Hall SD and Wrighton SA (2002) Comparative metabolic capabilities of CYP3A4, CYP3A5, and CYP3A7. *Drug Metab Dispos* **30**:883–891.

Wojnowski L (2004) Genetics of the variable expression of CYP3A in humans. *Ther Drug Monit* **26**:192–199.

Yang LJ, Fan L, Liu ZQ, Mao YM, Guo D, Liu LH, Tan ZR, Peng L, Han CT, Hu DL, Wang D and Zhou HH (2009) Effects of allicin on CYP2C19 and CYP3A4 activity in healthy volunteers with different CYP2C19 genotypes. *Eur J Clin Pharmacol* **65**:601–608.

Yasui-Furukori N, Saito M, Uno T, Takahata T, Sugawara K and Tateishi T (2004a) Effects of fluvoxamine on lansoprazole pharmacokinetics in relation to CYP2C19 genotypes. *J Clin Pharmacol* **44**:1223–1229.

Yasui-Furukori N, Takahata T, Nakagami T, Yoshiya G, Inoue Y, Kaneko S and Tateishi T (2004b) Different inhibitory effect of fluvoxamine on omeprazole metabolism between CYP2C19 genotypes. *Br J Clin Pharmacol* **57**:487–494.

Yu K-S, Cho J-Y, Jang I-J, Hong K-S, Chung J-Y, Kim J-R, Lim H-S, Oh D-S, Yi S-Y, Liu K-H, Shin J-G and Shin S-G (2004) Effect of the CYP3A5 genotype on the pharmacokinetics of intravenous midazolam during inhibited and induced metabolic states. *Clin Pharmacol Ther* **76**:104–112.

Yu KS, Yim DS, Cho JY, Park SS, Park JY, Lee KH, Jang IJ, Yi SY, Bae KS and Shin SG (2001) Effect of omeprazole on the pharmacokinetics of moclobemide according to the genetic polymorphism of CYP2C19. *Clin Pharmacol Ther* **69**:266–273.

Zaïr ZM, Eloranta JJ, Stieger B and Kullak-Ublick GA (2008) Pharmacogenetics of OATP (SLC21/SLCO), OAT and OCT (SLC22) and PEPT (SLC15) transporters in the intestine, liver and kidney. *Pharmacogenomics* **9**:597–624.

Zanger UM, Klein K, Saussele T, Blievernicht J, Hofmann MH and Schwab M (2007) Polymorphic CYP2B6: molecular mechanisms and emerging clinical significance. *Pharmacogenomics* **8**:743–759.

Zanger U, Turpeinen M, Klein K and Schwab M (2008) Functional pharmacogenetics/genomics of human cytochromes P450 involved in drug biotransformation. *Anal Bioanal Chem* **392**:1093–1108.

Chapter 4
Impact of Nuclear Receptors CAR, PXR, FXR, and VDR, and Their Ligands On Enzymes and Transporters

Rommel G. Tirona

Abstract Nuclear receptors play a central role on the mechanism of induction-type drug–drug interactions by acting as xenosensing transcription factors. Together, PXR, CAR, FXR, and VDR form a core group of nuclear receptors that regulate the expression of a number of important drug-metabolizing enzymes and drug transporters. In this chapter, the molecular determinants of adaptive response to drug exposure are detailed in the context of clinical relevance and drug development.

4.1 Introduction

It is estimated that 20% of all adverse drug effects are the result of drug–drug interactions (Kohler et al., 2000). When one considers statistics from the Institute of Medicine which indicates that 2 million adverse drug reactions occur yearly in the United States, it is not difficult to believe that drug reactions are the fourth leading cause of death among hospitalized patients (Lazarou et al., 1998). Therefore, a better mechanistic understanding, improved reporting, and heightened awareness of drug–drug interactions will likely have a positive impact on the health of patients. Among the mechanisms of drug–drug interactions, inhibition of drug metabolism or drug transport is considered to be most important because this often leads to increased systemic drug levels, resulting in overt toxicities. Perhaps less appreciated are drug–drug interactions that involve changes (increase or decrease) in the expression of metabolic and transport proteins. Commonly, this is in the form of an induction-type interaction that results in reduction of plasma levels of drugs that are concomitantly administered with the so-called offending drug. Hence, induction-type drug–drug interactions often manifest as a loss of pharmacological response

R.G. Tirona (✉)
Division of Clinical Pharmacology, Department of Medicine, Department of Physiology & Pharmacology, Schulich School of Medicine and Dentisity, University of Western Ontario, London Health Sciences Centre - University Hospital, London, Ontario, Canada
e-mail: rommel.tirona@schulich.uwo.ca

K.S. Pang et al. (eds.), *Enzyme- and Transporter-Based Drug–Drug Interactions*,
DOI 10.1007/978-1-4419-0840-7_4,

or withdrawal syndrome which, for some types of drug action, are subtle and can be overlooked clinically. If one considers that only 10% of adverse drug effects are reported and that for the most part, these comprise those that involve noticeable reactions from inhibition of metabolic or transport pathways, one would expect that the prevalence of induction-type drug–drug interactions is probably underappreciated.

There has been recognition over at least the last 50 years that drug or chemical exposure can cause an adaptive biological response aimed at minimizing the potential toxic effects of xenobiotics that one may have been in contact. Indeed, studies in the 1960s had shown that pretreatment with the anticonvulsant drug phenobarbital increased the hepatic biotransformation and shortened the sedative effects of hexobarbital when administered to rats (Conney et al., 1960). Shortly thereafter, it was observed that phenobarbital decreased the plasma concentration of the anti-seizure medication phenytoin and coumarin anticoagulants in humans (Cucinell et al., 1963). By the 1970s, it was determined that the hepatic cytochrome P450 (CYP) content was elevated in patients treated with phenytoin, phenobarbital, and rifampin (Schoene et al., 1972; Remmer et al., 1973). Later in the 1980s, it became apparent that induction of hepatic drug metabolism by exposure to certain drugs was due to a transcriptional mechanism (Adesnik et al., 1981). The precise molecular mechanisms for the majority of xenobiotic sensing and inductive response were defined in the late 1990s with the discovery of the ligand-activated nuclear receptors pregnane X receptor (PXR) (Bertilsson et al., 1998; Blumberg et al., 1998; Kliewer et al., 1998; Lehmann et al., 1998; Goodwin et al., 1999) and the constitutive androstane receptor (CAR) (Baes et al., 1994; Forman et al., 1998; Honkakoski et al., 1998; Sueyoshi et al., 1999). Concurrently, the bile acid-sensing nuclear receptor, farnesoid X receptor (FXR), was cloned (Makishima et al., 1999; Parks et al., 1999; Wang et al., 1999) and has been recognized lately as an important regulator of drug metabolism and transporter genes. Moreover, the receptor for the hormone 1α,25-dihydroxyvitamin D_3 [$1,25(OH)_2D_3$], vitamin D receptor (VDR), was identified (Baker et al., 1988) and later shown important in regulating intestinal drug metabolism by acting as a bile acid sensor (Makishima et al., 2002). On a parallel path, the signaling pathways involved in the adaptive metabolic response to exposure to polyaromatic hydrocarbons such as tetrachlorodibenzo-*p*-dioxin were clarified in the 1970s and the 1980s with the isolation and cloning of the aryl hydrocarbon (Ah) receptor (AhR) (Poland et al., 1976; Hoffman et al., 1991; Burbach et al., 1992). Furthermore, within the last decade, there has been significant progress in understanding the role of oxidative stress in adaptive response to xenobiotic exposure with the discovery of the nuclear factor-E2-p45-related factor 2 (Nrf2) signaling pathway that regulates the transcription of select drug detoxication genes (Venugopal and Jaiswal, 1996; Nguyen et al., 2000; Hayashi et al., 2003; Kwak et al., 2003; Kang et al., 2004; Wakabayashi et al., 2004).

This chapter will review drug–drug interactions with respect to the pharmacology of induction, clinically relevant induction-type drug interactions, and therapeutic aspects of nuclear receptor-mediated regulation of drug-metabolizing enzyme and transporter gene expression. Focus will be given on pathways involving the nuclear receptors PXR, CAR, FXR, and VDR since these represent key transcriptional regulators for the most important drug-metabolizing enzymes and transporters. Although as clinically relevant, this chapter will not discuss the roles of AhR and Nrf2 in drug–drug interactions and the reader is directed to recent reviews (Beischlag et al., 2008; Pickett et al., 2009).

4.2 Pharmacology of Induction

4.2.1 Nuclear Receptor-Mediated Regulation of Drug Disposition Gene Expression

A simple and elegant system of xenosensing and hormonal regulation of gene expression is mediated through the actions of the superfamily of nuclear receptor genes (Mangelsdorf and Evans, 1995; Chawla et al., 2001). Typically, these nuclear receptors are resident in the cellular cytoplasm and upon binding to ligands, a common signaling mechanism is elicited whereby the bound receptor translocates to the nucleus, heterodimerizes with the 9-*cis* retinoic acid receptor (RXR) and binding of the complex to regulatory regions of target genes. Upon release of co-repressor proteins and the recruitment of co-activators, there is a stimulation of the general transcriptional machinery. Ligand binding to each nuclear receptor within the heterodimer causes differential allosteric communication that results in permissive, conditional, or non-permissive transcriptional activation (Shulman et al., 2004).

Arguably, the most important nuclear receptor involved in induction-type drug–drug interactions is PXR. The title rests upon this protein in part because PXR is a promiscuous receptor that is bound and activated by chemically diverse compounds owing to a large and flexible ligand-binding cavity (Watkins et al., 2001; 2003a; 2003b). Secondly, PXR regulates the expression of the phase I enzyme CYP3A4 and the ATP-binding cassette (ABC) efflux transporter P-glycoprotein (P-gp) encoded by the *MDR1* gene (Table 4.1). Both of these proteins are responsible for the metabolism and transport of 50% of all drugs on the market. Wide interindividual variation exists in the tissue activities of CYP3A4 and P-gp and this is not the result of common non-synonymous polymorphisms in these genes (Hoffmeyer et al., 2000; Kim et al., 2001; Lamba et al., 2002) but probably because of differences in gene expression which are controlled by environmental chemical exposure and PXR signaling. It should be noted that there are clear differences in the activation of PXR orthologs between species that account for variability in CYP3A inducibility by various drugs (Kocarek et al., 1995). For example, pregnenolone carbonitrile (PCN) induces the expression of rodent *Cyp3a* genes more powerfully than human CYP3A4, while rifampin upregulates human CYP3A4 much more strongly than Cyp3a in rodents. Studies using humanized PXR transgenic mice have elegantly demonstrated that the species difference in CYP3A inducibility relates to differential activation of PXR orthologs by xenobiotics (Xie et al., 2000). Part of this species difference in drug-dependent activation is due to a single amino acid

Table 4.1 Human drug disposition genes regulated by PXR and/or CAR

Gene	References
Oxidative metabolism (phase I)	
CYP2A6	Itoh et al. (2006)
CYP2B6	Goodwin et al. (2001) and Wang et al. (2003)
CYP2C9	Ferguson et al. (2002), Gerbal-Chaloin et al. (2002) and Chen et al. (2004)
CYP2C19	Gerbal-Chaloin et al. (2001) and Chen et al. (2003)
CYP3A4	Goodwin et al. (1999)
CYP3A7	Pascussi et al. (1999), Bertilsson et al. (2001)
HCE2	Yang and Yan (2007)
Conjugation (phase II)	
UGT1A1	Sugatani et al. (2001, 2004)
UGT1A3	Gardner-Stephen et al. (2004)
UGT1A4	Gardner-Stephen et al. (2004)
UGT2B7	Gallicano et al. (1999)
SULT2A1	Echchgadda et al. (2007) and Fang et al. (2007)
Transport (phase III)	
MDR1	Geick et al. (2001)
MRP2	Kast et al. (2002)
MRP4	Assem et al. (2004)
OATP1A2	Miki et al. (2006) and Meyer zu Schwabedissen et al. (2008)

difference between rodent and human PXR that lies in a location which lines the pore to the ligand-binding cavity (Tirona et al., 2004).

The nuclear receptor CAR is activated by fewer compounds than PXR but is responsible for regulating similar target genes such as *CYP3A4* and *MDR1* (Table 4.1) and hence is a significant contributor to drug–drug interactions. Interestingly, phenobarbital causes inductive responses through CAR signaling despite that it is not a direct ligand of the receptor (Moore et al., 2000b). Recent reports have demonstrated an important role of kinase activation in the non-ligand activation of CAR (Hosseinpour et al., 2007; Inoue and Negishi, 2008; Sueyoshi et al., 2008). Moreover, studies on CAR structure have shed insight into the molecular mechanisms that determine the constitutive activity of this receptor (Suino et al., 2004; Xu et al., 2004).

FXR activation has become a relevant nuclear signaling pathway regulating drug metabolism and transport. Recent studies have identified functional FXR response elements in genes such as *CYP3A4* (Gnerre et al., 2004) and *ABCC2* (MRP2) (Kast et al., 2002) suggesting an important role in determining the pharmacokinetics of numerous drugs (Table 4.2). While bile acids are endogenous ligands for FXR, there have been no reports to date that indicate that prescription drugs are functional FXR agonists. However, a number of FXR agonists are under drug development (Dussault et al., 2003a; Hartman et al., 2009) or have been isolated from natural products such as coffee (e.g. cafestol) (Ricketts et al., 2007). Moreover, several FXR antagonists have been identified including the gugulipid constituent guggulsterone

Table 4.2 Human drug disposition genes regulated by FXR

Gene	References
Oxidative metabolism (phase I)	
CYP3A4	Gnerre et al. (2004)
Conjugation (phase II)	
SULT2A1	Miyata et al. (2006)
UGT2B4	Barbier et al. (2003)
UGT2B15	Kaeding et al. (2008)
UGT2B17	Kaeding et al. (2008)
Transport (phase III)	
BSEP	Ananthanarayanan et al. (2001), Schuetz et al. (2001), and Plass et al. (2002)
MRP2	Kast et al. (2002)
NTCP	Denson et al. (2001)
OATP1B3	Jung et al. (2002) and Ohtsuka et al. (2006)
OSTα/β	Boyer et al. (2006) and Landrier et al. (2006)

(Urizar et al., 2002) and the soya lipid, stigmasterol (Carter et al., 2007). Hence, there exists the possibility that food–drug interactions may occur due to interactions at the level of FXR.

The role of vitamin D as a modulating factor in drug metabolism came to the attention of scientists after the observation that the hormone was capable of inducing CYP3A4 in the intestinal Caco-2 cell line (Schmiedlin-Ren et al., 1997). It was later appreciated that VDR regulated intestinal CYP3A4 expression and potentially first-pass intestinal drug metabolism (Thummel et al., 2001).

4.2.2 Drug-Metabolizing Enzymes Regulated by Nuclear Receptors

The drug-metabolizing enzymes regulated by PXR, CAR, FXR, and VDR are summarized in Tables 4.1, 4.2, and 4.3, respectively. Among those genes regulated by PXR signaling, CYP3A4 expression is the most upregulated of all enzymes when examined by microarray analysis of inducer-treated cultured human hepatocytes (Healan-Greenberg et al., 2008). On the other hand, in human hepatocytes, CYP2B6 is most sensitive to the inductive effects of CAR activation (Finkelstein et al., 2006). In precision-cut human liver slices and in the human hepatocellular carcinoma cell model HepG2, treatment with the FXR agonist, chenodeoxycholate (CDCA), produced modest induction of CYP3A4 (Gnerre et al., 2004; Jung et al., 2007). Recent studies have demonstrated a role for PXR in the induction of carboxylesterase isoform 2 by 8-methoxypsoralen exposure (Yang and Yan, 2007).

Table 4.3 Human drug disposition genes regulated by VDR

Gene	References
Oxidative metabolism (phase I)	
CYP2B6	Drocourt et al. (2002)
CYP2C9	Drocourt et al. (2002)
CYP3A4	Thummel et al. (2001) and Makishima et al. (2002)
Conjugation (phase II)	
SULT2A1	Echchgadda et al. (2004)
Transport (phase III)	
MDR1	Saeki et al. (2008)
MRP3	McCarthy et al. (2005)

4.2.3 Drug Transporters Regulated by Nuclear Receptors

Drug transporters whose expression is modulated by nuclear receptors are listed in Tables 4.1, 4.2, and 4.3. With respect to PXR activation, P-gp expression appears to be strongly regulated (Geick et al., 2001). Indeed, rifampin-treated subjects had significant upregulation of P-gp and multidrug resistance-associated protein 2 (MRP2) in enterocytes (Greiner et al., 1999; Fromm et al., 2000). FXR activation is an important component of homeostatic control of enterohepatic bile acid recirculation. Hence, CDCA is known to induce the canalicular bile salt export pump (BSEP) (Ananthanarayanan et al., 2001; Schuetz et al., 2001; Plass et al., 2002), the basolateral organic anion transporting polypetide 1B3 (Jung et al., 2002; Briz et al., 2006; Jung et al., 2007) in liver, and the apical organic solute transporter α/β in enterocytes (Boyer et al., 2006; Landrier et al., 2006), while reducing the expression of the hepatic basolateral transporter, sodium-dependent taurocholate cotransporting polypeptide (NTCP) (Denson et al., 2001).

Table 4.4 Selected modulators of CAR

Compound	References
Activators	
Artemisinin	Burk et al. (2005)
Phenobarbital	Kawamoto et al. (1999)
CITCO	Maglich et al. (2003)
Oltipraz	Merrell et al. (2008)
Di-(2-ethylhexyl) phthalate	Dekeyser et al. (2009)
Phenytoin	Wang et al. (2004a)
Cerivastatin	Kobayashi et al. (2005)
Fluvastatin	Kobayashi et al. (2005)
Simvastatin	Kobayashi et al. (2005)
Atorvastatin	Kobayashi et al. (2005)
Inhibitors	
PK11195	Li et al. (2008)

4.2.4 Nuclear Receptor Regulatory Networks in Drug Disposition

There is much complexity in the regulatory networks for drug disposition genes due to variety of mechanisms. For example, any given drug can activate a number of different nuclear receptors (lack of specificity). Such is the case for phenytoin, a drug that will activate PXR and CAR (Wang et al., 2004a) (see Tables 4.4 and 4.5). Each nuclear receptor can be activated or deactivated by many drugs, which may or may not be co-adminstered in a patient (promiscuity). For instance, PXR binds to a large variety of chemically diverse compounds (Table 4.5) where binding for a drug can occur in different orientations and multiple ligands are thought to be able to occupy the ligand-binding cavity (Watkins et al., 2001). Furthermore, each nuclear receptor can activate a large gene battery which may

Table 4.5 Selected activators and inhibitors of PXR

Compound	References
Activators	
Rifampin	Luo et al. (2002)
Hyperforin	Moore et al. (2000a)
Clotrimazole	Luo et al. (2002)
Phenytoin	Luo et al. (2002)
Phenobarbital	Luo et al. (2002)
Dexamethasone	Pascussi et al. (2001)
Zearalenone	Ding et al. (2006)
Flucloxacillin	Huwyler et al. (2006)
Artemisinin	Huang et al. (2004) and Burk et al. (2005)
Nicotine	Lamba et al. (2004)
Nifedipine	Drocourt et al. (2001)
Troleandomycin	Trubetskoy et al. (2005)
Omeprazole	Drocourt et al. (2001) and Raucy et al. (2002)
Paclitaxel	Mani et al. (2005)
Forskolin	Ding and Staudinger (2005a)
Carbamazepine	Luo et al. (2002)
Topotecan	Schuetz et al. (2002)
Etoposide	Schuetz et al. (2002)
Mevastatin	Raucy et al. (2002)
Sulfinpyrazone	Luo et al. (2002)
Ritonavir	Luo et al. (2002)
Pyrethroids	Yang et al. (2009)
Genistein	Li et al. (2009b)
Nafcillin	Yasuda et al. (2008)
Colupulone	Teotico et al. (2008)
Inhibitors	
Ketoconazole	Huang et al. (2007)
Trabectedin (ET-742)	Synold et al. (2001)
Sulforaphane	Zhou et al. (2007)
Stigmasterol	Carter et al. (2007)
Coumestrol	Wang et al. (2008a)
A-792611	Healan-Greenberg et al. (2008)

overlap with those regulated by other xenosensors (broad target selectivity). Such is the case for PXR and CAR whose target genes largely overlap, but the degree of regulation by each receptor may differ (Maglich et al., 2002). Lastly, each drug disposition gene can be regulated by a number of nuclear receptors (redundancy) which may compete or synergize with respect to target gene transcription. Such is the case for the sulfotransferase enzyme SULT2A1 which is regulated by PXR (Sonoda et al., 2002), CAR (Assem et al., 2004), VDR (Echchgadda et al., 2004), and FXR (Miyata et al., 2006). To further add to the complexity of drug disposition gene regulation, the nuclear receptors themselves regulate each other. PXR is regulated by FXR (Jung et al., 2006), CAR (Maglich et al., 2002), AhR (Maglich et al., 2002), and itself (Maglich et al., 2002).

4.2.5 Prediction of Induction-Type Drug Interactions

Prediction of in vivo induction-type drug–drug interactions from in vitro systems is of certain interest to those involved in the discovery and development of drugs. Hence, a number of experimental systems have been employed including cultured primary human hepatocytes and liver slices, humanized mouse models, transformed hepatocytes or cell lines, reporter gene assays, co-activator recruitment assays, and receptor binding assays. Each system differs in the degree of biological complexity and experimental ease. Examination of gene expression changes in cultures of primary human hepatocytes after drug challenge is considered the "gold standard" approach (http://www.fda.gov/CDER/guidance/6695dft.htm). A number of "humanized" mouse models have been recently described, each with differing variations in nuclear receptor composition (PXR/CAR) (Xie et al., 2000; Huang et al., 2004; Scheer et al., 2008) or those also harbouring human drug-metabolizing enzyme genes such as CYP3A4 (Ma et al., 2008). Given the cost and limited availability of primary human hepatocytes, various transformed cell systems have been developed to mimic the adaptive response in liver. These include the Fa2N-4 immortalized human hepatocyte clone (Mills et al., 2004; Ripp et al., 2006; Hariparsad et al., 2008) as well as the HepaRG human hepatoma cell line (Aninat et al., 2006; Kanebratt and Andersson, 2008; Lambert et al., 2009a, b; McGinnity et al., 2009). Of popular use are reporter gene assays whereby cell lines are stably or transiently transfected with luciferase reporters and nuclear receptors (Goodwin et al., 1999). The advantages of this system are its technical simplicity and predictability of in vivo effects.

We have shown that the PXR reporter gene assay predicts well the magnitude of CYP3A4 induction in vivo as assessed in high-quality clinical drug–drug interaction studies (Tirona and Kim, 2009). In that analysis, data derived from reporter gene assays were compiled with information pertaining to the potency (EC_{50}) and efficacy (E_{max}) of PXR activation for a number of drugs. Similar to other pharmacological responses, the magnitude of inductive response is related to the target potency and the concentration of the active compound at the receptor site. When considering adaptive responses in liver, the maximum concentration of the drug in plasma (C_{max}) after (oral) dosing is thought to reflect the intracellular concentration of drug in hepatocytes. Therefore, the ratio of C_{max} to EC_{50} is a useful metric to predict the propensity of a compound to elicit an inductive response on a PXR target gene such as CYP3A4. To test this metric, data from drug–drug interaction studies that examined the effects of chronically administered compounds on the pharmacokinetics of CYP3A4 probe drugs such as midazolam and simvastatin were used. The results of this exercise demonstrated that when C_{max}/EC_{50} was less than 0.1, such as in the case of statin drugs and calcium channel blockers, no in vivo CYP3A4 induction was evident. For drugs whose C_{max}/EC_{50} were between values of 0.1 and 1, for example, bosentan, rifabutin, and efavirenz, moderate induction of CYP3A4 has been documented as demonstrated by a reduction of probe drug exposure (area under the plasma concentration–time curve; AUC) of 15–55%. Now for drugs with C_{max}/EC_{50} values greater than 1, marked induction of CYP3A4 activity was evident. Indeed, the AUC of CYP3A4 probe drugs decreased 61–94%, after exposure when PXR activators in this group such as carbamazepine, hyperforin, rifampin, and phenobarbital. Although this analysis was performed to specifically understand CYP3A4 inducibility by PXR activation, it is likely that a similar evaluation using in vitro data for other nuclear receptor (CAR or FXR) responses by drugs and their in vivo target gene (enzymes and transporters) induction will also yield comparable results.

For some instances, however, the plasma concentrations of drugs (C_{max}) do not serve as good surrogate measures of the intracellular levels at the site of nuclear receptors. One reason is because intracellular drug accumulation is modulated by the actions of drug uptake and efflux transporters. One example of such an effect is given by OATP1B1, the transporter for rifampin uptake, whose overexpression caused an enhancement of PXR activation in vitro (Tirona et al., 2003). In another scenario, induction of Cyp3a in the livers of mice that were P-gp deficient was greater than that in wild-type mice when both were treated with rifampin (Schuetz et al., 1996). This was consistent with elevated levels of rifampin in livers of P-gp knockout compared to wild-type mice, presumably because P-gp transports rifampin.

4.3 Drugs and Herbal Medicines Involved in Clinically Relevant Induction-Type Drug–Drug Interactions

While the list of compounds that activate the nuclear receptors PXR, CAR, and FXR has rapidly grown over the last decade, it is interesting to note that from a therapeutic perspective relatively few prescription drugs are known to cause clinically relevant induction-type drug interactions. The number of such clinical inducers amounts to roughly 20 drugs and these are found in only five major therapeutic categories: anticonvulsants, antibiotics, human immunodeficiency virus (HIV) protease inhibitors, non-nucleoside reverse transcriptase inhibitors, and miscellaneous (Table 4.6). It can be appreciated that within each category, not all drugs are

Table 4.6 Prescription medicines known to cause clinically relevant induction-type drug interactions

Drug	References
Anticonvulsants	
Phenobarbital	Conney et al. (1965) and Kawamoto et al. (1999)
Carbamazepine	Luo et al. (2002)
Oxcarbazepine	Lloyd et al. (1994)
Phenytoin	Conney et al. (1965)
Valproic acid	DeVane (2003)
Lamotrigine	Benedetti (2000)
Topiramate	Benedetti (2000)
Felbamate	Benedetti (2000)
Antibiotics	
Nafcillin	Qureshi et al. (1984)
Rifampin	Acocella (1978)
Rifabutin	Finch et al. (2002)
HIV protease inhibitors	
Ritonavir	Product monograph
Nelfinavir	Product monograph
Lopinavir	Product monograph
Tipranavir	King and Acosta (2006)
Amprenavir	Justesen et al. (2003)
Atazanavir	Perloff et al. (2005)
Non-nucleoside reverse Transcriptase inhibitors	
Efavirenz	Mouly et al. (2002)
Nevirapine	Murphy et al. (1999)
Other	
Bosentan	van Giersbergen et al. (2002)

inducers and therefore a “class effect” for inductive drug interactions does not exist. This is highlighted in the case of HIV protease inhibitors. While many of the newer compounds such as tipranavir and atazanavir are CYP inducers (PXR activators), a handful of the older HIV protease inhibitors including indinavir and saquinavir are devoid of this propensity. The inductive potential of many HIV protease inhibitors is however negated somewhat by the fact that they are often co-administered with the potent CYP3A4 inhibitor ritonavir in “boosted” therapy. The result of the ritonavir and other HIV protease inhibitor co-treatment is net inhibition of CYP3A4 activity. However, since ritonavir does not inhibit other PXR target genes such as CYP2B6, several inductive drug interactions have been documented involving the CYP2B6 substrate methadone during boosted HIV protease inhibitor therapy (McCance-Katz et al., 2003; Hendrix et al., 2004).

Within the rifamycin class of antibiotics, there appears to be differences in the magnitude of inductive drug interactions when you compare the older agent rifampin with that of the relatively newer rifabutin. Here, the reduction in antiretroviral drug plasma exposure is less marked with rifabutin than rifampin (Baciewicz et al., 2008) despite that both drugs are equipotent with regard to PXR activation

(EC_{50} values ~1.5–1.9 μM). An explanation for this differential inductive effect is that rifabutin concentrations in plasma are low (C_{max} ~0.5 μM) while those for rifampin are high enough (C_{max} ~16.5 μM) to elicit an effect on PXR signaling. One interesting drug in the rifamycin class is rifaximin. Rifaximin is used in the treatment of irritable bowel syndrome and is structurally similar to other rifamycins. Not surprisingly, rifaximin is a PXR activator in vitro (Ma et al., 2007) but uncharacteristically, it does not cause inductive drug interactions as was demonstrated in the absence of effect on either ethinylestradiol or midazolam pharmacokinetics (Pentikis et al., 2007; Trapnell et al., 2007). Here, the lack of in vivo inductive effect is related to the fact that rifaximin is not absorbed to any great extent into the systemic circulation to cause an elevation of hepatic drug elimination (Ma et al., 2007).

Herbal medicine use has become popular with an estimated 38 million Americans (19% of the population) having used this treatment modality within 12 months of a 2002 survey (Kennedy et al., 2008). This statistic may be somewhat alarming if you consider that only one-third of herbal medicine users inform their physician (Kennedy et al., 2008) and that approximately 43% of herbal users also take prescription medications (Peng et al., 2004). Moreover, these numbers are unsettling because by 1999, it became clear that herbal medicine use was not innocuous as generally believed, particularly because of the possibility of significant drug–herb

Table 4.7 Induction-type herb–drug interactions

Herb	Drugs affected	References
Hypericum perforatum (St John's Wort)	Cyclosporine	Ruschitzka et al. (2000)
	Indinavir	Piscitelli et al. (2000)
	Omeprazole	Wang et al. (2004b)
	Digoxin	Johne et al. (1999) and Durr et al. (2000)
	Imatinib	Frye et al. (2004)
	Oral contraceptives	Schwarz et al. (2003)
	Tacrolimus	Mai et al. (2003)
	Warfarin	Yue et al. (2000)
	Theophylline	Nebel et al. (1999)
	Midazolam	Dresser et al. (2003)
	Fexofenadine	Dresser et al. (2003)
	Alprazolam	Markowitz et al. (2003)
	Quazepam	Kawaguchi et al. (2004)
	Talinolol	Schwarz et al. (2007)
	Ivabradine	Portoles et al. (2006)
	Irinotecan	Mathijssen et al. (2002)
Echinacea purpurea or augustifolia	Midazolam	Gorski et al. (2004)
Allium sativum (garlic)	Saquinavir	Piscitelli et al. (2002)
Ginkgo biloba	Omeprazole	Yin et al. (2004)
	Midazolam	Robertson et al. (2008)
Panax quinquefolius (American ginseng)	Warfarin	Yuan et al. (2004)

interactions. It was around this time that a number of reports documenting that treatment with preparations of St. John's Wort, commonly used for mild depression, caused reduction in the plasma concentration of co-administered prescription medications, including theophylline (Nebel et al., 1999), digoxin (Johne et al., 1999), cyclosporine (Ruschitzka et al., 2000), and indinavir (Piscitelli et al., 2000). In the cases with cyclosporine, the results of the herb–drug interaction were episodes of heart transplant rejection stemming from inadequate immunosuppressant plasma concentrations. Later, the active inducing component in St. John's Wort, hyperforin, was isolated and interestingly, it is a compound with the greatest potency of all activators of PXR ($EC_{50} = 23$ nM) (Moore et al., 2000a). In addition to St. John's Wort, Ginko biloba extract appears to also contain PXR and CAR activating constituents (Yeung et al., 2008; Li et al., 2009a) and that formal drug interaction studies with midazolam demonstrate that this herbal induces CYP3A4 activity in vivo (Robertson et al., 2008). Currently, there are less compelling data for other herbal medicines and their capacity for causing inductive herb–drug interactions (Table 4.7)

4.4 Therapeutic Aspects of Induction-Type Drug Interactions

4.4.1 Time Course of Nuclear Receptor Responses

Understanding the time course of nuclear receptor responses in vivo has practical implications. For instance, clinicians need to know how long induction and de-induction phases persist before new steady-state drug responses are reached for patients initiating or discontinuing medications that are known to have induction properties. This would allow one to titrate co-medications when necessary at appropriate time frames during drug therapy. Moreover, from a drug development perspective, the design of induction-type drug–drug interaction studies requires knowledge of the inductive response in order to plan the appropriate duration for the intervention.

While increased transcription rate of target genes occurs within hours of xenobiotic exposure and nuclear receptor activation, there is a lag in the time for protein levels to change. A number of studies have mapped the time course of this induction process by following the plasma levels of drugs and/or metabolites that are susceptible to changes in metabolic activity. One such study involved monitoring of carbamazepine levels in children during treatment initiation as the auto-induction process occurred. Within the first few doses, there was a decrease in carbamazepine levels which reached steady state only by 1 month of therapy (Bertilsson et al., 1980). In another study, the quantity of the 6β-hydroxylated metabolite of cortisol in urine was measured as a marker of CYP3A4 activity in subjects administered rifampin (Tran et al., 1999). Here, CYP3A4 induction was obvious within a day of rifampin initiation, then enzyme activity peaked and reached a plateau by 5 days. Upon discontinuation of rifampin, CYP3A4 activity declined after 2 days then returned to baseline levels by 7 days. Hence rifampin caused both relatively rapid CYP3A4 induction and de-induction owing to the fact that it has a short half-life of 3 h and that transcriptional enhancement is fast and CYP3A4 protein turnover is somewhat high. These data reflect the time course of inductive response for CYP3A4 via PXR signaling. More studies are required to understand the time course of adaptive response for other drug-metabolizing enzymes and transporters.

4.4.2 Inhibition of Nuclear Receptors

There has been recent interest in identifying inhibitors of nuclear receptors for use as therapeutic agents or as chemical tools to explore gene expression pathways (Tables 4.4 and 4.5). PXR antagonists such as ketonconazole are being considered as adjunct treatment to counteract cancer chemotherapeutic drug resistance by acting as agents to suppress the tumoral expression of drug-metabolizing enzymes and transporters such as P-gp (Huang et al., 2007; Wang et al., 2007). In breast cancer, PXR is highly expressed and regulates the expression of the estrogen uptake transporter OATP1A2 (Miki et al., 2006; Meyer zu Schwabedissen et al., 2008). Blockade of PXR activity by treatment with the HIV protease inhibitor A-792611 (Healan-Greenberg et al., 2008) inhibits estrogen-mediated breast cancer cell proliferation in vitro (Meyer zu Schwabedissen et al., 2008). The antineoplastic agent trabectedin (ET-743), which is derived from the sea squirt, is among the most potent PXR antagonists (IC_{50} = 50 nM). Other PXR antagonists include the broccoli constituent sulforaphane (Zhou et al., 2007), the soya lipid stigmasterol (Carter et al., 2007), and the phytoestrogen coumestrol (Wang et al., 2008a). While a number of PXR antagonists have been identified, there has not been clinical demonstration that when co-administered with PXR agonists such as rifampin, these compounds would attenuate the induction of target genes such as CYP3A4.

4.4.3 Nuclear Receptors and Drug Side Effects

In the treatment of tuberculosis, HIV infection, or epilepsy, chronic or lifelong treatment with drugs occurs where nuclear receptors, as an off-target, are continually activated. Such sustained nuclear receptor activation has been associated with a number of drug side effects. For instance, anticonvulsants (Kodama et al., 1989; Eiris-Punal et al., 1999; Castro-Gago et al., 2007) and rifampin (Kodama et al., 1989; Kim et al., 2007) cause hypothyroidism because of increased turnover of thyroid hormone in liver resulting from induction of glucuronidation and sulfation (Qatanani et al., 2005) or biliary excretion (Curran and DeGroot, 1991; Wong et al., 2005). Interruption of vitamin D signaling is thought to cause rifampin-induced bone loss or osteomalacia (Shah et al., 1981). By acting on PXR, rifampin upregulates the vitamin D-metabolizing enzyme, CYP24, hence promoting degradation of active hormone (Pascussi et al., 2005; Zhou et al., 2006a). In addition, pseudo-Cushing's syndrome (hypercortisolism) is caused by PXR activation by rifampin and subsequent disruption of glucocorticoid homeostasis (Terzolo et al., 1995; Zhai et al., 2007). Lastly, chronic PXR and CAR activation causes hepatic steatosis (Zhou et al., 2008a; Finn et al., 2009; Hoekstra et al., 2009), in part due to upregulation of the fatty acid transporter cd36 (Zhou et al., 2008a). Hence, treatment with HIV protease inhibitors is associated with the development of fatty liver (Riddle et al., 2001; Sulkowski et al., 2005; Moreno-Torres et al., 2007).

4.4.4 Interindividual Variation in Drug Response – Pharmacogenetics of Nuclear Receptors

Polymorphisms in the *NR1I2* (PXR) gene have been extensively studied (Zhang et al., 2001; Koyano et al., 2002). Few, rare non-synonymous polymorphisms in PXR are demonstrated to have reduced transcriptional activity (Hustert et al., 2001; Lim et al., 2005). Other single-nucleotide polymorphisms (SNP) in PXR are located in regulatory regions or introns of the gene and are linked with variation in hepatic CYP3A4 content (Lamba et al., 2008). In a formal study of the PXR genotype and phenotype, it was observed that subjects with the H1/H1 PXR haplotype possessed lower basal CYP3A4 activity but increased CYP3A4 inducibility upon treatment with St. John's Wort, when nifedipine was used as a drug probe (Wang et al., 2009). An influence of PXR genetics was also observed for doxorubicin pharmacokinetics in breast cancer patients, whereby the PXR∗1b haplotype was associated with decreased drug clearance (Sandanaraj et al., 2008). Moreover, a PXR SNP is associated with reduced and subtherapeutic levels of the HIV protease inhibitor atazanavir (Siccardi et al., 2008). Genetic polymorphisms in PXR have also been linked to increased risk for inflammatory bowel diseases such as Crohn's disease and ulcerative colitis (Langmann et al., 2004; Dring et al., 2006). In support of these findings, mice with genetic deficiency in pxr show signs of intestinal inflammation (Shah et al., 2007).The mechanism for control of gut inflammation appears to result from mutual inhibition of PXR and NF-κB signaling (Zhou et al., 2006b; Shah et al., 2007).

Interestingly, FXR *(NR1I4)* gene polymorphisms have become associated with a number of phenotypes. The relatively common polymorphism within the Kozak sequence (FXR∗1b) shows reduced function in vitro and is linked with decreased hepatic target gene expression (e.g. OATP1B3) (Marzolini et al., 2007). At present, there is no information to indicate that FXR polymorphisms relate to interindividual differences in drug clearance. However, the FXR∗1b genotype is associated with an increased risk for intrahepatic cholestasis of pregnancy (Van Mil et al., 2007) and for cholesterol gallstone disease (Kovacs et al., 2008).

4.4.5 Xenobiotic Receptors as Drug Targets

The nuclear receptors that regulate drug disposition genes have been studied not only for their relevance to drug–drug interactions but also for their potential as therapeutic drug targets. This is not a novel concept as activating CAR by treatment with phenobarbital has long been used in the treatment of neonatal hyperbilirubinemia to upregulate the expression of bilirubin glucuronidation (UGT1A1) and biliary conjugate excretion (MRP2) (Sugatani et al., 2001; Huang et al., 2003; 2004). CAR may also be a druggable target because of its role in energy homeostasis. For example, CAR-deficient mice do not exhibit hypertriglyceridemia when fed a high-fat diet (Maglich et al., 2009). There is also reason to believe that CAR may be critical in the development of non-alcoholic steatohepatitis (Yamazaki et al., 2007). Given that a specific CAR antagonist (1-(2-chlorophenyl-methylpropyl)-3-isoquinoline-carboxamide) has already been discovered (Li et al., 2008), it follows that drugs that inhibit CAR may find benefit in the treatment of lipid disorders.

The use of PXR antagonists to circumvent drug resistance in various cancers has already been discussed previously as a way to suppress drug metabolism and transport in tumors. However, recent findings also indicate another mechanism for modulating PXR activity in cancer. It appears that PXR has antiapoptotic effect in colon cancer unrelated to regulation of drug metabolism but due to a combination of upregulation of antiapoptotic genes and downregulation of proapoptotic genes (Zhou et al., 2008b). Similar to CAR and FXR, PXR is involved in lipid and cholesterol homeostasis and therefore targeting of this receptor may have utility in the prevention or treatment of cardiovascular disease. Indeed, PXR activation has complex effects on serum lipoproteins (Masson et al., 2005; Sporstol et al., 2005; Ricketts et al., 2007; de Haan et al., 2009; Hoekstra et al., 2009). The potential for therapeutic modulation of PXR activity for other diseases of lipid excess has also been suggested for cerebrotendinous xanthomatosis (Dussault et al., 2003b) and Niemann–Pick type C1 disease (Langmade et al., 2006).

FXR agonists have received attention recently as a number of compounds under drug development are beginning to show promise. For example, studies with the FXR agonists INT-747 (Mencarelli et al., 2009) and WAY-362450 (Evans et al., 2009; Flatt et al., 2009; Hartman et al., 2009) have compelling data to demonstrate antiatherosclerotic activities in preclinical models. Moreover, FXR activation results in hepatic and vascular anti-inflammatory responses (Li et al., 2007; Wang et al., 2008b; Zhang et al., 2009) which may augment other cardiovascular benefits. FXR agonism may also be a novel strategy in the treatment or prevention of cholesterol gallstone disease by increasing phospholipid and bile acid concentrations in the maintenance of cholesterol solubility in bile (Moschetta et al., 2004). Furthermore, there is experimental evidence that activation of FXR may be of benefit in the treatment of non-alcoholic fatty liver disease (Figge et al., 2004; Kong et al., 2009).

4.5 Perspectives

There has been much progress in understanding the role of nuclear receptors as important determinants of drug–drug interactions and contributors to drug response and toxicity. However, there are areas in nuclear receptor research that require further study and are the topics of recent studies. In this regard, there remains more to clarify with respect to mechanisms of epigenetic control of nuclear receptor expression and function. Indeed, studies have shown that PXR expression is regulated by micro-RNAs which subsequently influence CYP3A4 levels (Takagi et al., 2008). Furthermore, studies are beginning to clarify the roles of histone methyltransferases in regulating PXR (Xie et al., 2009) and FXR (Rizzo et al., 2005) transcriptional responses. Likewise, as what has become important for CAR signaling, nuclear receptor phosphorylation of PXR (Ding and Staudinger, 2005b; Pondugula et al., 2009) and FXR (Frankenberg et al., 2008; Gineste et al., 2008) determines their activities.

Another new area of drug disposition research is the role of nuclear receptors in the maintenance of the blood–brain barrier. It appears that PXR is expressed in brain capillaries to regulate the expression of drug transporters (Bauer et al., 2004; 2006; Bauer et al., 2008; Narang et al., 2008; Zastre et al., 2009). How regulation of drug transport in the brain influences pharmacologic and toxic drug effects in humans remains to be seen.

Lastly, there is compelling evidence for a clinical impact of nuclear receptor-mediated control of CYP enzymes, UGTs, and MDR1. However, the in vitro and animal models indicate that other drug-metabolizing enzymes and transporters are induced by nuclear receptor activation. It will be important in the next few years to demonstrate functionally in vivo that such similar regulation occurs and that there may be an impact on drug response.

It can be expected that, as we gain further insight from molecular mechanisms to clinical consequences, studies in the nuclear receptor field will provide knowledge and means to develop safer and more effective drugs as well as better strategies for the optimal use of medicines in current practice.

References

Acocella G (1978) Clinical pharmacokinetics of rifampicin. *Clin Pharmacokinet* **3**:108–127.

Adesnik M, Bar-Nun S, Maschio F, Zunich M, Lippman A and Bard E (1981) Mechanism of induction of cytochrome P-450 by phenobarbital. *J Biol Chem* **256**:10340–10345.

Ananthanarayanan M, Balasubramanian N, Makishima M, Mangelsdorf DJ and Suchy FJ (2001) Human bile salt export pump promoter is transactivated by the farnesoid X receptor/bile acid receptor. *J Biol Chem* **276**:28857–28865.

Aninat C, Piton A, Glaise D, Le Charpentier T, Langouet S, Morel F, Guguen-Guillouzo C and Guillouzo A (2006) Expression of cytochromes P450, conjugating enzymes and nuclear receptors in human hepatoma HepaRG cells. *Drug Metab Dispos* **34**:75–83.

Assem M, Schuetz EG, Leggas M, Sun D, Yasuda K, Reid G, Zelcer N, Adachi M, Strom S, Evans RM, Moore DD, Borst P and Schuetz JD (2004) Interactions between hepatic Mrp4 and Sult2a as revealed by the constitutive androstane receptor and Mrp4 knockout mice. *J Biol Chem* **279**:22250–22257.

Baciewicz AM, Chrisman CR, Finch CK and Self TH (2008) Update on rifampin and rifabutin drug interactions. *Am J Med Sci* **335**:126–136.

Baes M, Gulick T, Choi HS, Martinoli MG, Simha D and Moore DD (1994) A new orphan member of the nuclear hormone receptor superfamily that interacts with a subset of retinoic acid response elements. *Mol Cell Biol* **14**:1544–1551.

Baker AR, McDonnell DP, Hughes M, Crisp TM, Mangelsdorf DJ, Haussler MR, Pike JW, Shine J and O'Malley BW (1988) Cloning and expression of full-length cDNA encoding human vitamin D receptor. *Proc Natl Acad Sci U S A* **85**:3294–3298.

Barbier O, Torra IP, Sirvent A, Claudel T, Blanquart C, Duran-Sandoval D, Kuipers F, Kosykh V, Fruchart JC and Staels B (2003) FXR induces the UGT2B4 enzyme in hepatocytes: a potential mechanism of negative feedback control of FXR activity. *Gastroenterology* **124**:1926–1940.

Bauer B, Hartz AM, Fricker G and Miller DS (2004) Pregnane X receptor up-regulation of P-glycoprotein expression and transport function at the blood-brain barrier. *Mol Pharmacol* **66**:413–419.

Bauer B, Hartz AM, Lucking JR, Yang X, Pollack GM and Miller DS (2008) Coordinated nuclear receptor regulation of the efflux transporter, Mrp2, and the phase-II metabolizing enzyme, GSTpi, at the blood-brain barrier. *J Cereb Blood Flow Metab* **28**:1222–1234.

Bauer B, Yang X, Hartz AM, Olson ER, Zhao R, Kalvass JC, Pollack GM and Miller DS (2006) In vivo activation of human pregnane X receptor tightens the blood-brain barrier to methadone through P-glycoprotein up-regulation. *Mol Pharmacol* **70**:1212–1219.

Beischlag TV, Luis Morales J, Hollingshead BD and Perdew GH (2008) The aryl hydrocarbon receptor complex and the control of gene expression. *Crit Rev Eukaryot Gene Expr* **18**:207–250.

Benedetti MS (2000) Enzyme induction and inhibition by new antiepileptic drugs: a review of human studies. *Fundam Clin Pharmacol* **14**:301–319.

Bertilsson G, Berkenstam A and Blomquist P (2001) Functionally conserved xenobiotic responsive enhancer in cytochrome P450 3A7. *Biochem Biophys Res Commun* **280**:139–144.

Bertilsson G, Heidrich J, Svensson K, Asman M, Jendeberg L, Sydow-Backman M, Ohlsson R, Postlind H, Blomquist P and Berkenstam A (1998) Identification of a human nuclear receptor defines a new signaling pathway for CYP3A induction. *Proc Natl Acad Sci U S A* **95**: 12208–12213.

Bertilsson L, Hojer B, Tybring G, Osterloh J and Rane A (1980) Autoinduction of carbamazepine metabolism in children examined by a stable isotope technique. *Clin Pharmacol Ther* **27**: 83–88.

Blumberg B, Sabbagh W, Jr., Juguilon H, Bolado J, Jr., van Meter CM, Ong ES and Evans RM (1998) SXR, a novel steroid and xenobiotic-sensing nuclear receptor. *Genes Dev* **12**: 3195–3205.

Boyer JL, Trauner M, Mennone A, Soroka CJ, Cai SY, Moustafa T, Zollner G, Lee JY and Ballatori N (2006) Upregulation of a basolateral FXR-dependent bile acid efflux transporter OSTalpha-OSTbeta in cholestasis in humans and rodents. *Am J Physiol* **290**:G1124–G1130.

Briz O, Romero MR, Martinez-Becerra P, Macias RI, Perez MJ, Jimenez F, San Martin FG and Marin JJ (2006) Oatp8/1B3-mediated cotransport of bile acids and glutathione. An export pathway for organic anions from hepatocytes?. *J Biol Chem* **281**:30326–30335.

Burbach KM, Poland A and Bradfield CA (1992) Cloning of the Ah-receptor cDNA reveals a distinctive ligand-activated transcription factor. *Proc Natl Acad Sci USA* **89**:8185–8189.

Burk O, Arnold KA, Nussler AK, Schaeffeler E, Efimova E, Avery BA, Avery MA, Fromm MF and Eichelbaum M (2005) Antimalarial artemisinin drugs induce cytochrome P450 and MDR1 expression by activation of xenosensors pregnane X receptor and constitutive androstane receptor. *Mol Pharmacol* **67**:1954–1965.

Carter BA, Taylor OA, Prendergast DR, Zimmerman TL, Von Furstenberg R, Moore DD and Karpen SJ (2007) Stigmasterol, a soy lipid-derived phytosterol, is an antagonist of the bile acid nuclear receptor FXR. *Pediatr Res* **62**:301–306.

Castro-Gago M, Novo-Rodriguez MI, Gomez-Lado C, Rodriguez-Garcia J, Rodriguez-Segade S and Eiris-Punal J (2007) Evolution of subclinical hypothyroidism in children treated with antiepileptic drugs. *Pediatr Neurol* **37**:426–430.

Chawla A, Repa JJ, Evans RM and Mangelsdorf DJ (2001) Nuclear receptors and lipid physiology: opening the X-files. *Science* **294**:1866–1870.

Chen Y, Ferguson SS, Negishi M and Goldstein JA (2003) Identification of constitutive androstane receptor and glucocorticoid receptor binding sites in the CYP2C19 promoter. *Mol Pharmacol* **64**:316–324.

Chen Y, Ferguson SS, Negishi M and Goldstein JA (2004) Induction of human CYP2C9 by rifampicin, hyperforin, and phenobarbital is mediated by the pregnane X receptor. *J Pharmacol Exp Ther* **308**:495–501.

Conney AH, Davison C, Gastel R and Burns JJ (1960) Adaptive increases in drug-metabolizing enzymes induced by phenobarbital and other drugs. *J Pharmacol Exp Ther* **130**:1–8.

Conney AH, Jacobson M, Schneidman K and Kuntzman R (1965) Induction of liver microsomal cortisol 6beta-hydroxylase by diphenylhydantoin or phenobarbital: an explanation for the increased excretion of 6-hydroxycortisol in humans treated with these drugs. *Life Sci* **4**:1091–1098.

Cucinell SA, Koster R, Conney AH and Burns JJ (1963) Stimulatory Effect Of Phenobarbital On The Metabolism Of Diphenylhydantoin. *J Pharmacol Exp Ther* **141**:157–160.

Curran PG and DeGroot LJ (1991) The effect of hepatic enzyme-inducing drugs on thyroid hormones and the thyroid gland. *Endocr Rev* **12**:135–150.

de Haan W, de Vries-van der Weij J, Mol IM, Hoekstra M, Romijn JA, Jukema JW, Havekes LM, Princen HM and Rensen PC (2009) PXR agonism decreases plasma HDL levels in ApoE3-Leiden.CETP mice. *Biochim Biophys Acta* **1791**:191–197.

Dekeyser JG, Stagliano MC, Auerbach SS, Prabu KS, Jones AD and Omiecinski CJ (2009) Di(2-ethylhexyl) phthalate is a highly potent agonist for the human constitutive androstane receptor splice variant, CAR2. *Mol Pharmacol* **75**:1003–1013.

Denson LA, Sturm E, Echevarria W, Zimmerman TL, Makishima M, Mangelsdorf DJ and Karpen SJ (2001) The orphan nuclear receptor, shp, mediates bile acid-induced inhibition of the rat bile acid transporter, ntcp. *Gastroenterology* **121**:140–147.

DeVane CL (2003) Pharmacokinetics, drug interactions, and tolerability of valproate. *Psychopharmacol Bull* **37 Suppl 2**:25–42.

Ding X, Lichti K and Staudinger JL (2006) The mycoestrogen zearalenone induces CYP3A through activation of the pregnane X receptor. *Toxicol Sci* **91**:448–455.

Ding X and Staudinger JL (2005a) Induction of drug metabolism by forskolin: the role of the pregnane X receptor and the protein kinase a signal transduction pathway. *J Pharmacol Exp Ther* **312**:849–856.

Ding X and Staudinger JL (2005b) Repression of PXR-mediated induction of hepatic CYP3A gene expression by protein kinase C. *Biochem Pharmacol* **69**:867–873.

Dresser GK, Schwarz UI, Wilkinson GR and Kim RB (2003) Coordinate induction of both cytochrome P4503A and MDR1 by St John's wort in healthy subjects. *Clin Pharmacol Ther* **73**:41–50.

Dring MM, Goulding CA, Trimble VI, Keegan D, Ryan AW, Brophy KM, Smyth CM, Keeling PW, O'Donoghue D, O'Sullivan M, O'Morain C, Mahmud N, Wikstrom AC, Kelleher D and McManus R (2006) The pregnane X receptor locus is associated with susceptibility to inflammatory bowel disease. *Gastroenterology* **130**:341–348 quiz 592.

Drocourt L, Ourlin JC, Pascussi JM, Maurel P and Vilarem MJ (2002) Expression of CYP3A4, CYP2B6, and CYP2C9 is regulated by the vitamin D receptor pathway in primary human hepatocytes. *J Biol Chem* **277**:25125–25132.

Drocourt L, Pascussi JM, Assenat E, Fabre JM, Maurel P and Vilarem MJ (2001) Calcium channel modulators of the dihydropyridine family are human pregnane X receptor activators and inducers of CYP3A, CYP2B, and CYP2C in human hepatocytes. *Drug Metab Dispos* **29**:1325–1331.

Durr D, Stieger B, Kullak-Ublick GA, Rentsch KM, Steinert HC, Meier PJ and Fattinger K (2000) St John's Wort induces intestinal P-glycoprotein/MDR1 and intestinal and hepatic CYP3A4. *Clin Pharmacol Ther* **68**:598–604.

Dussault I, Beard R, Lin M, Hollister K, Chen J, Xiao JH, Chandraratna R and Forman BM (2003a) Identification of gene-selective modulators of the bile acid receptor FXR. *J Biol Chem* **278**:7027–7033.

Dussault I, Yoo HD, Lin M, Wang E, Fan M, Batta AK, Salen G, Erickson SK and Forman BM (2003b) Identification of an endogenous ligand that activates pregnane X receptor-mediated sterol clearance. *Proc Natl Acad Sci USA* **100**:833–838.

Echchgadda I, Song CS, Oh T, Ahmed M, De La Cruz IJ and Chatterjee B (2007) The xenobiotic-sensing nuclear receptors pregnane X receptor, constitutive androstane receptor, and orphan nuclear receptor hepatocyte nuclear factor 4 in the regulation of human steroid-/bile acid-sulfotransferase. *Mol Endocrinol* **21**:2099–2111.

Echchgadda I, Song CS, Roy AK and Chatterjee B (2004) Dehydroepiandrosterone sulfotransferase is a target for transcriptional induction by the vitamin D receptor. *Mol Pharmacol* **65**:720–729.

Eiris-Punal J, Del Rio-Garma M, Del Rio-Garma MC, Lojo-Rocamonde S, Novo-Rodriguez I and Castro-Gago M (1999) Long-term treatment of children with epilepsy with valproate or carbamazepine may cause subclinical hypothyroidism. *Epilepsia* **40**:1761–1766.

Evans MJ, Mahaney PE, Borges-Marcucci L, Lai K, Wang S, Krueger JA, Gardell SJ, Huard C, Martinez R, Vlasuk GP and Harnish DC (2009) A synthetic farnesoid X receptor (FXR) agonist promotes cholesterol lowering in models of dyslipidemia. *Am J Physiol* **296**:G543–G552.

Fang HL, Strom SC, Ellis E, Duanmu Z, Fu J, Duniec-Dmuchowski Z, Falany CN, Falany JL, Kocarek TA and Runge-Morris M (2007) Positive and Negative Regulation of Human Hepatic Hydroxysteroid Sulfotransferase (SULT2A1) Gene Transcription by Rifampicin: Roles of Hepatocyte Nuclear Factor 4? and Pregnane X Receptor. *J Pharmacol Exp Ther* **323**: 586–590.

Ferguson SS, LeCluyse EL, Negishi M and Goldstein JA (2002) Regulation of human CYP2C9 by the constitutive androstane receptor: discovery of a new distal binding site. *Mol Pharmacol* **62**:737–746.

Figge A, Lammert F, Paigen B, Henkel A, Matern S, Korstanje R, Shneider BL, Chen F, Stoltenberg E, Spatz K, Hoda F, Cohen DE and Green RM (2004) Hepatic overexpression of murine Abcb11 increases hepatobiliary lipid secretion and reduces hepatic steatosis. *J Biol Chem* **279**:2790–2799.

Finch CK, Chrisman CR, Baciewicz AM and Self TH (2002) Rifampin and rifabutin drug interactions: an update. *Arch Intern Med* **162**:985–992.

Finkelstein D, Lamba V, Assem M, Rengelshausen J, Yasuda K, Strom S and Schuetz E (2006) ADME transcriptome in Hispanic versus White donor livers: evidence of a globally enhanced NR1I3 (CAR, constitutive androstane receptor) gene signature in Hispanics. *Xenobiotica* **36**:989–1012.

Finn RD, Henderson CJ, Scott CL and Wolf CR (2009) Unsaturated fatty acid regulation of cytochrome P450 expression via a CAR-dependent pathway. *Biochem J* **417**:43–54.

Flatt B, Martin R, Wang TL, Mahaney P, Murphy B, Gu XH, Foster P, Li J, Pircher P, Petrowski M, Schulman I, Westin S, Wrobel J, Yan G, Bischoff E, Daige C and Mohan R (2009) Discovery of XL335 (WAY-362450), a highly potent, selective, and orally active agonist of the farnesoid X receptor (FXR). *J Med Chem* Jan 21. [Epub ahead of print].

Forman BM, Tzameli I, Choi HS, Chen J, Simha D, Seol W, Evans RM and Moore DD (1998) Androstane metabolites bind to and deactivate the nuclear receptor CAR-beta. *Nature* **395**: 612–615.

Frankenberg T, Miloh T, Chen FY, Ananthanarayanan M, Sun AQ, Balasubramaniyan N, Arias I, Setchell KD, Suchy FJ and Shneider BL (2008) The membrane protein ATPase class I type 8B member 1 signals through protein kinase C zeta to activate the farnesoid X receptor. *Hepatology* **48**:1896–1905.

Fromm MF, Kauffmann HM, Fritz P, Burk O, Kroemer HK, Warzok RW, Eichelbaum M, Siegmund W and Schrenk D (2000) The effect of rifampin treatment on intestinal expression of human MRP transporters. *Am J Pathol* **157**:1575–1580.

Frye RF, Fitzgerald SM, Lagattuta TF, Hruska MW and Egorin MJ (2004) Effect of St John's wort on imatinib mesylate pharmacokinetics. *Clin Pharmacol Ther* **76**:323–329.

Gallicano KD, Sahai J, Shukla VK, Seguin I, Pakuts A, Kwok D, Foster BC and Cameron DW (1999) Induction of zidovudine glucuronidation and amination pathways by rifampicin in HIV-infected patients. *Br J Clin Pharmacol* **48**:168–179.

Gardner-Stephen D, Heydel JM, Goyal A, Lu Y, Xie W, Lindblom T, Mackenzie P and Radominska-Pandya A (2004) Human PXR variants and their differential effects on the regulation of human UDP-glucuronosyltransferase gene expression. *Drug Metab Dispos* **32**:340–347.

Geick A, Eichelbaum M and Burk O (2001) Nuclear receptor response elements mediate induction of intestinal MDR1 by rifampin. *J Biol Chem* **276**:14581–14587.

Gerbal-Chaloin S, Daujat M, Pascussi JM, Pichard-Garcia L, Vilarem MJ and Maurel P (2002) Transcriptional regulation of CYP2C9 gene. Role of glucocorticoid receptor and constitutive androstane receptor. *J Biol Chem* **277**:209–217.

Gerbal-Chaloin S, Pascussi JM, Pichard-Garcia L, Daujat M, Waechter F, Fabre JM, Carrere N and Maurel P (2001) Induction of CYP2C genes in human hepatocytes in primary culture. *Drug Metab Dispos* **29**:242–251.

Gineste R, Sirvent A, Paumelle R, Helleboid S, Aquilina A, Darteil R, Hum DW, Fruchart JC and Staels B (2008) Phosphorylation of farnesoid X receptor by protein kinase C promotes its transcriptional activity. *Mol Endocrinol* **22**:2433–2447.

Gnerre C, Blattler S, Kaufmann MR, Looser R and Meyer UA (2004) Regulation of CYP3A4 by the bile acid receptor FXR: evidence for functional binding sites in the CYP3A4 gene. *Pharmacogenetics* **14**:635–645.

Goodwin B, Hodgson E and Liddle C (1999) The orphan human pregnane X receptor mediates the transcriptional activation of CYP3A4 by rifampicin through a distal enhancer module. *Mol Pharmacol* **56**:1329–1339.

Goodwin B, Moore LB, Stoltz CM, McKee DD and Kliewer SA (2001) Regulation of the human CYP2B6 gene by the nuclear pregnane X receptor. *Mol Pharmacol* **60**:427–431.

Gorski JC, Huang SM, Pinto A, Hamman MA, Hilligoss JK, Zaheer NA, Desai M, Miller M and Hall SD (2004) The effect of echinacea (Echinacea purpurea root) on cytochrome P450 activity in vivo. *Clin Pharmacol Ther* **75**:89–100.

Greiner B, Eichelbaum M, Fritz P, Kreichgauer HP, von Richter O, Zundler J and Kroemer HK (1999) The role of intestinal P-glycoprotein in the interaction of digoxin and rifampin. *J Clin Invest* **104**:147–153.

Hariparsad N, Carr BA, Evers R and Chu X (2008) Comparison of immortalized Fa2N-4 cells and human hepatocytes as in vitro models for cytochrome P450 induction. *Drug Metab Dispos* **36**:1046–1055.

Hartman HB, Gardell SJ, Petucci CJ, Wang S, Krueger JA and Evans MJ (2009) Activation of farnesoid X receptor prevents atherosclerotic lesion formation in LDLR-/- and apoE-/- mice. *J Lipid Res* Jan 27. [Epub ahead of print].

Hayashi A, Suzuki H, Itoh K, Yamamoto M and Sugiyama Y (2003) Transcription factor Nrf2 is required for the constitutive and inducible expression of multidrug resistance-associated protein 1 in mouse embryo fibroblasts. *Biochem Biophys Res Commun* **310**:824–829.

Healan-Greenberg C, Waring JF, Kempf DJ, Blomme EA, Tirona RG and Kim RB (2008) A human immunodeficiency virus protease inhibitor is a novel functional inhibitor of human pregnane X receptor. *Drug Metab Dispos* **36**:500–507.

Hendrix CW, Wakeford J, Wire MB, Lou Y, Bigelow GE, Martinez E, Christopher J, Fuchs EJ and Snidow JW (2004) Pharmacokinetics and pharmacodynamics of methadone enantiomers after coadministration with amprenavir in opioid-dependent subjects. *Pharmacotherapy* **24**: 1110–1121.

Hoekstra M, Lammers B, Out R, Li Z, Van Eck M and Van Berkel TJ (2009) Activation of the nuclear receptor PXR decreases plasma LDL-cholesterol levels and induces hepatic steatosis in LDL receptor knockout mice. *Mol Pharm* **6**:182–189.

Hoffman EC, Reyes H, Chu FF, Sander F, Conley LH, Brooks BA and Hankinson O (1991) Cloning of a factor required for activity of the Ah (dioxin) receptor. *Science* **252**:954–958.

Hoffmeyer S, Burk O, von Richter O, Arnold HP, Brockmoller J, Johne A, Cascorbi I, Gerloff T, Roots I, Eichelbaum M and Brinkmann U (2000) Functional polymorphisms of the human multidrug-resistance gene: multiple sequence variations and correlation of one allele with P-glycoprotein expression and activity in vivo. *Proc Natl Acad Sci USA* **97**: 3473–3478.

Honkakoski P, Zelko I, Sueyoshi T and Negishi M (1998) The nuclear orphan receptor CAR-retinoid X receptor heterodimer activates the phenobarbital-responsive enhancer module of the CYP2B gene. *Mol Cell Biol* **18**:5652–5658.

Hosseinpour F, Timsit Y, Koike C, Matsui K, Yamamoto Y, Moore R and Negishi M (2007) Overexpression of the Rho-guanine nucleotide exchange factor ECT2 inhibits nuclear translocation of nuclear receptor CAR in the mouse liver. *FEBS Lett* **581**:4937–4942.

Huang H, Wang H, Sinz M, Zoeckler M, Staudinger J, Redinbo MR, Teotico DG, Locker J, Kalpana GV and Mani S (2007) Inhibition of drug metabolism by blocking the activation of nuclear receptors by ketoconazole. *Oncogene* **26**:258–268.

Huang W, Zhang J, Chua SS, Qatanani M, Han Y, Granata R and Moore DD (2003) Induction of bilirubin clearance by the constitutive androstane receptor (CAR). *Proc Natl Acad Sci USA* **100**:4156–4161.

Huang W, Zhang J and Moore DD (2004) A traditional herbal medicine enhances bilirubin clearance by activating the nuclear receptor CAR. *J Clin Invest* **113**:137–143.

Hustert E, Zibat A, Presecan-Siedel E, Eiselt R, Mueller R, Fuss C, Brehm I, Brinkmann U, Eichelbaum M, Wojnowski L and Burk O (2001) Natural protein variants of pregnane X receptor with altered transactivation activity toward CYP3A4. *Drug Metab Dispos* **29**:1454–1459.

Huwyler J, Wright MB, Gutmann H and Drewe J (2006) Induction of cytochrome P450 3A4 and P-glycoprotein by the isoxazolyl-penicillin antibiotic flucloxacillin. *Curr Drug Metab* **7**: 119–126.

Inoue K and Negishi M (2008) Nuclear receptor CAR requires early growth response 1 to activate the human cytochrome P450 2B6 gene. *J Biol Chem* **283**:10425–10432.

Itoh M, Nakajima M, Higashi E, Yoshida R, Nagata K, Yamazoe Y and Yokoi T (2006) Induction of human CYP2A6 is mediated by the pregnane X receptor with peroxisome proliferator-activated receptor-gamma coactivator 1alpha. *J Pharmacol Exp Ther* **319**:693–702.

Johne A, Brockmoller J, Bauer S, Maurer A, Langheinrich M and Roots I (1999) Pharmacokinetic interaction of digoxin with an herbal extract from St John's wort (Hypericum perforatum). *Clin Pharmacol Ther* **66**:338–345.

Jung D, Elferink MG, Stellaard F and Groothuis GM (2007) Analysis of bile acid-induced regulation of FXR target genes in human liver slices. *Liver Int* **27**:137–144.

Jung D, Mangelsdorf DJ and Meyer UA (2006) Pregnane X receptor is a target of farnesoid X receptor. *J Biol Chem* **281**:19081–19091.

Jung D, Podvinec M, Meyer UA, Mangelsdorf DJ, Fried M, Meier PJ and Kullak-Ublick GA (2002) Human organic anion transporting polypeptide 8 promoter is transactivated by the farnesoid X receptor/bile acid receptor. *Gastroenterology* **122**:1954–1966.

Justesen US, Klitgaard NA, Brosen K and Pedersen C (2003) Pharmacokinetic interaction between amprenavir and delavirdine after multiple-dose administration in healthy volunteers. *Br J Clin Pharmacol* **55**:100–106.

Kaeding J, Bouchaert E, Belanger J, Caron P, Chouinard S, Verreault M, Larouche O, Pelletier G, Staels B, Belanger A and Barbier O (2008) Activators of the farnesoid X receptor negatively regulate androgen glucuronidation in human prostate cancer LNCAP cells. *Biochem J* **410**: 245–253.

Kanebratt KP and Andersson TB (2008) HepaRG cells as an in vitro model for evaluation of cytochrome P450 induction in humans. *Drug Metab Dispos* **36**:137–145.

Kang MI, Kobayashi A, Wakabayashi N, Kim SG and Yamamoto M (2004) Scaffolding of Keap1 to the actin cytoskeleton controls the function of Nrf2 as key regulator of cytoprotective phase 2 genes. *Proc Natl Acad Sci USA* **101**:2046–2051.

Kast HR, Goodwin B, Tarr PT, Jones SA, Anisfeld AM, Stoltz CM, Tontonoz P, Kliewer S, Willson TM and Edwards PA (2002) Regulation of multidrug resistance-associated protein 2 (ABCC2) by the nuclear receptors pregnane X receptor, farnesoid X-activated receptor, and constitutive androstane receptor. *J Biol Chem* **277**:2908–2915.

Kawaguchi A, Ohmori M, Tsuruoka S, Nishiki K, Harada K, Miyamori I, Yano R, Nakamura T, Masada M and Fujimura A (2004) Drug interaction between St John's Wort and quazepam. *Br J Clin Pharmacol* **58**:403–410.

Kawamoto T, Sueyoshi T, Zelko I, Moore R, Washburn K and Negishi M (1999) Phenobarbital-responsive nuclear translocation of the receptor CAR in induction of the CYP2B gene. *Mol Cell Biol* **19**:6318–6322.

Kennedy J, Wang CC and Wu CH (2008) Patient disclosure about herb and supplement use among adults in the US. *Evid Based Complement Alternat Med* **5**:451–456.

Kim RB, Leake BF, Choo EF, Dresser GK, Kubba SV, Schwarz UI, Taylor A, Xie HG, McKinsey J, Zhou S, Lan LB, Schuetz JD, Schuetz EG and Wilkinson GR (2001) Identification of functionally variant MDR1 alleles among European Americans and African Americans. *Clin Pharmacol Ther* **70**:189–199.

Kim DL, Song KH, Lee JH, Lee KY and Kim SK (2007) Rifampin-induced hypothyroidism without underlying thyroid disease. *Thyroid* **17**:793–795.

King JR and Acosta EP (2006) Tipranavir: a novel nonpeptidic protease inhibitor of HIV. *Clin Pharmacokinet* **45**:665–682.

Kliewer SA, Moore JT, Wade L, Staudinger JL, Watson MA, Jones SA, McKee DD, Oliver BB, Willson TM, Zetterstrom RH, Perlmann T and Lehmann JM (1998) An orphan nuclear receptor activated by pregnanes defines a novel steroid signaling pathway. *Cell* **92**:73–82.

Kobayashi K, Yamanaka Y, Iwazaki N, Nakajo I, Hosokawa M, Negishi M and Chiba K (2005) Identification of HMG-CoA reductase inhibitors as activators for human, mouse and rat constitutive androstane receptor. *Drug Metab Dispos* **33**:924–929.

Kocarek TA, Schuetz EG, Strom SC, Fisher RA and Guzelian PS (1995) Comparative analysis of cytochrome P4503A induction in primary cultures of rat, rabbit, and human hepatocytes. *Drug Metab Dispos* **23**:415–421.

Kodama S, Tanaka K, Konishi H, Momota K, Nakasako H, Nakayama S, Yagi J and Koderazawa K (1989) Supplementary thyroxine therapy in patients with hypothyroidism induced by long-term anticonvulsant therapy. *Acta Paediatr Jpn* **31**:555–562.

Kohler GI, Bode-Boger SM, Busse R, Hoopmann M, Welte T and Boger RH (2000) Drug–drug interactions in medical patients: effects of in-hospital treatment and relation to multiple drug use. *Int J Clin Pharmacol Ther* **38**:504–513.

Kong B, Luyendyk JP, Tawfik O and Guo GL (2009) Farnesoid X receptor deficiency induces nonalcoholic steatohepatitis in low-density lipoprotein receptor-knockout mice fed a high-fat diet. *J Pharmacol Exp Ther* **328**:116–122.

Kovacs P, Kress R, Rocha J, Kurtz U, Miquel JF, Nervi F, Mendez-Sanchez N, Uribe M, Bock HH, Schirin-Sokhan R, Stumvoll M, Mossner J, Lammert F and Wittenburg H (2008) Variation of the gene encoding the nuclear bile salt receptor FXR and gallstone susceptibility in mice and humans. *J Hepatol* **48**:116–124.

Koyano S, Kurose K, Ozawa S, Saeki M, Nakajima Y, Hasegawa R, Komamura K, Ueno K, Kamakura S, Nakajima T, Saito H, Kimura H, Goto Y, Saitoh O, Katoh M, Ohnuma T, Kawai M, Sugai K, Ohtsuki T, Suzuki C, Minami N, Saito Y and Sawada J (2002) Eleven novel single nucleotide polymorphisms in the NR1I2 (PXR) gene, four of which induce non-synonymous amino acid alterations. *Drug Metab Pharmacokinet* **17**:561–565.

Kwak MK, Wakabayashi N, Itoh K, Motohashi H, Yamamoto M and Kensler TW (2003) Modulation of gene expression by cancer chemopreventive dithiolethiones through the Keap1-Nrf2 pathway. Identification of novel gene clusters for cell survival. *J Biol Chem* **278**:8135–8145.

Lamba J, Lamba V, Strom S, Venkataramanan R and Schuetz E (2008) Novel single nucleotide polymorphisms in the promoter and intron 1 of human pregnane X receptor/NR1I2 and their association with CYP3A4 expression. *Drug Metab Dispos* **36**:169–181.

Lamba JK, Lin YS, Schuetz EG and Thummel KE (2002) Genetic contribution to variable human CYP3A-mediated metabolism. *Adv Drug Deliv Rev* **54**:1271–1294.

Lamba V, Yasuda K, Lamba JK, Assem M, Davila J, Strom S and Schuetz EG (2004) PXR (NR1I2): splice variants in human tissues, including brain, and identification of neurosteroids and nicotine as PXR activators. *Toxicol Appl Pharmacol* **199**:251–265.

Lambert CB, Spire C, Claude N and Guillouzo A (2009a) Dose- and time-dependent effects of phenobarbital on gene expression profiling in human hepatoma HepaRG cells. *Toxicol Appl Pharmacol* **234**:345–360.

Lambert CB, Spire C, Renaud MP, Claude N and Guillouzo A (2009b) Reproducible chemical-induced changes in gene expression profiles in human hepatoma HepaRG cells under various experimental conditions. *Toxicol In Vitro* **23**:466–475.

Landrier JF, Eloranta JJ, Vavricka SR and Kullak-Ublick GA (2006) The nuclear receptor for bile acids, FXR, transactivates human organic solute transporter-alpha and -beta genes. *Am J Physiol Gastrointest Liver Physiol* **290**:G476–G485.

Langmade SJ, Gale SE, Frolov A, Mohri I, Suzuki K, Mellon SH, Walkley SU, Covey DF, Schaffer JE and Ory DS (2006) Pregnane X receptor (PXR) activation: a mechanism for neuroprotection in a mouse model of Niemann-Pick C disease. *Proc Natl Acad Sci USA* **103**:13807–13812.

Langmann T, Moehle C, Mauerer R, Scharl M, Liebisch G, Zahn A, Stremmel W and Schmitz G (2004) Loss of detoxification in inflammatory bowel disease: dysregulation of pregnane X receptor target genes. *Gastroenterology* **127**:26–40.

Lazarou J, Pomeranz BH and Corey PN (1998) Incidence of adverse drug reactions in hospitalized patients: a meta-analysis of prospective studies. *J Am Med Assoc* **279**:1200–1205.

Lehmann JM, McKee DD, Watson MA, Willson TM, Moore JT and Kliewer SA (1998) The human orphan nuclear receptor PXR is activated by compounds that regulate CYP3A4 gene expression and cause drug interactions. *J Clin Invest* **102**:1016–1023.

Li L, Chen T, Stanton JD, Sueyoshi T, Negishi M and Wang H (2008) The peripheral benzodiazepine receptor ligand 1-(2-chlorophenyl-methylpropyl)-3-isoquinoline-carboxamide is a novel antagonist of human constitutive androstane receptor. *Mol Pharmacol* **74**:443–453.

Li Y, Ross-Viola JS, Shay NF, Moore DD and Ricketts ML (2009b) Human CYP3A4 and murine Cyp3A11 are regulated by equol and genistein via the pregnane X receptor in a species-specific manner. *J Nutr* **139:**:898–904.

Li L, Stanton JD, Tolson AH, Luo Y and Wang H (2009a) Bioactive terpenoids and flavonoids from Ginkgo biloba extract induce the expression of hepatic drug-metabolizing enzymes through pregnane X receptor, constitutive androstane receptor, and aryl hydrocarbon receptor-mediated pathways. *Pharm Res* **26**:872–882.

Li YT, Swales KE, Thomas GJ, Warner TD and Bishop-Bailey D (2007) Farnesoid X receptor ligands inhibit vascular smooth muscle cell inflammation and migration. *Arterioscler Thromb Vasc Biol* **27**:2606–2611.

Lim YP, Liu CH, Shyu LJ and Huang JD (2005) Functional characterization of a novel polymorphism of pregnane X receptor, Q158K, in Chinese subjects. *Pharmacogenet Genomics* **15**:337–341.

Lloyd P, Flesch G and Dieterle W (1994) Clinical pharmacology and pharmacokinetics of oxcarbazepine. *Epilepsia* **35 Suppl 3**:S10–S13.

Luo G, Cunningham M, Kim S, Burn T, Lin J, Sinz M, Hamilton G, Rizzo C, Jolley S, Gilbert D, Downey A, Mudra D, Graham R, Carroll K, Xie J, Madan A, Parkinson A, Christ D, Selling B,

LeCluyse E and Gan LS (2002) CYP3A4 induction by drugs: correlation between a pregnane X receptor reporter gene assay and CYP3A4 expression in human hepatocytes. *Drug Metab Dispos* **30**:795–804.

Ma X, Cheung C, Krausz KW, Shah YM, Wang T, Idle JR and Gonzalez FJ (2008) A double transgenic mouse model expressing human pregnane X receptor and cytochrome P450 3A4. *Drug Metab Dispos* **36**:2506–2512.

Ma X, Shah YM, Guo GL, Wang T, Krausz KW, Idle JR and Gonzalez FJ (2007) Rifaximin is a gut-specific human pregnane X receptor activator. *J Pharmacol Exp Ther* **322**:391–398.

Maglich JM, Lobe DC and Moore JT (2009) The nuclear receptor CAR (NR1I3) regulates serum triglyceride levels under conditions of metabolic stress. *J Lipid Res* **50**:439–445.

Maglich JM, Parks DJ, Moore LB, Collins JL, Goodwin B, Billin AN, Stoltz CA, Kliewer SA, Lambert MH, Willson TM and Moore JT (2003) Identification of a novel human constitutive androstane receptor (CAR) agonist and its use in the identification of CAR target genes. *J Biol Chem* **278**:17277–17283.

Maglich JM, Stoltz CM, Goodwin B, Hawkins-Brown D, Moore JT and Kliewer SA (2002) Nuclear pregnane x receptor and constitutive androstane receptor regulate overlapping but distinct sets of genes involved in xenobiotic detoxification. *Mol Pharmacol* **62**: 638–646.

Mai I, Stormer E, Bauer S, Kruger H, Budde K and Roots I (2003) Impact of St John's wort treatment on the pharmacokinetics of tacrolimus and mycophenolic acid in renal transplant patients. *Nephrol Dial Transplant* **18**:819–822.

Makishima M, Lu TT, Xie W, Whitfield GK, Domoto H, Evans RM, Haussler MR and Mangelsdorf DJ (2002) Vitamin D receptor as an intestinal bile acid sensor. *Science* **296**:1313–1316.

Makishima M, Okamoto AY, Repa JJ, Tu H, Learned RM, Luk A, Hull MV, Lustig KD, Mangelsdorf DJ and Shan B (1999) Identification of a nuclear receptor for bile acids. *Science* **284**:1362–1365.

Mangelsdorf DJ and Evans RM (1995) The RXR heterodimers and orphan receptors. *Cell* **83**: 841–850.

Mani S, Huang H, Sundarababu S, Liu W, Kalpana G, Smith AB and Horwitz SB (2005) Activation of the steroid and xenobiotic receptor (human pregnane X receptor) by nontaxane microtubule-stabilizing agents. *Clin Cancer Res* **11**:6359–6369.

Markowitz JS, Donovan JL, DeVane CL, Taylor RM, Ruan Y, Wang JS and Chavin KD (2003) Effect of St John's wort on drug metabolism by induction of cytochrome P450 3A4 enzyme. *J Am Med Assoc* **290**:1500–1504.

Marzolini C, Tirona RG, Gervasini G, Poonkuzhali B, Assem M, Lee W, Leake BF, Schuetz JD, Schuetz EG and Kim RB (2007) A common polymorphism in the bile acid receptor farnesoid X receptor is associated with decreased hepatic target gene expression. *Mol Endocrinol* **21**: 1769–1780.

Masson D, Lagrost L, Athias A, Gambert P, Brimer-Cline C, Lan L, Schuetz JD, Schuetz EG and Assem M (2005) Expression of the pregnane X receptor in mice antagonizes the cholic acid-mediated changes in plasma lipoprotein profile. *Arterioscler Thromb Vasc Biol* **25**:2164–2169.

Mathijssen RH, Verweij J, de Bruijn P, Loos WJ and Sparreboom A (2002) Effects of St. John's wort on irinotecan metabolism. *J Natl Cancer Inst* **94**:1247–1249.

McCance-Katz EF, Rainey PM, Friedland G and Jatlow P (2003) The protease inhibitor lopinavir-ritonavir may produce opiate withdrawal in methadone-maintained patients. *Clin Infect Dis* **37**:476–482.

McCarthy TC, Li X and Sinal CJ (2005) Vitamin D receptor-dependent regulation of colon multidrug resistance-associated protein 3 gene expression by bile acids. *J Biol Chem* **280**:23232–23242.

McGinnity DF, Zhang G, Kenny J, Hamilton G, Otmani S, Stams K, Haney S, Brassil P, Stresser DM and Riley RJ (2009) Evaluation of multiple in vitro systems for assessment of CYP3A4 induction in drug discovery: human hepatocytes, PXR reporter gene, Fa2N-4 and HepaRG cells. *Drug Metab Dispos* Mar 23. [Epub ahead of print].

Mencarelli A, Renga B, Distrutti E and Fiorucci S (2009) Antiatherosclerotic effect of farnesoid X receptor. *Am J Physiol Heart Circ Physiol* **296**:H272–H281.

Merrell MD, Jackson JP, Augustine LM, Fisher CD, Slitt AL, Maher JM, Huang W, Moore DD, Zhang Y, Klaassen CD and Cherrington NJ (2008) The Nrf2 activator oltipraz also activates the constitutive androstane receptor. *Drug Metab Dispos* **36**:1716–1721.

Meyer zu Schwabedissen HE, Tirona RG, Yip CS, Ho RH and Kim RB (2008) Interplay between the nuclear receptor pregnane X receptor and the uptake transporter organic anion transporter polypeptide 1A2 selectively enhances estrogen effects in breast cancer. *Cancer Res* **68**: 9338–9347.

Miki Y, Suzuki T, Kitada K, Yabuki N, Shibuya R, Moriya T, Ishida T, Ohuchi N, Blumberg B and Sasano H (2006) Expression of the steroid and xenobiotic receptor and its possible target gene, organic anion transporting polypeptide-A, in human breast carcinoma. *Cancer Res* **66**:535–542.

Mills JB, Rose KA, Sadagopan N, Sahi J and de Morais SM (2004) Induction of drug metabolism enzymes and MDR1 using a novel human hepatocyte cell line. *J Pharmacol Exp Ther* **309**: 303–309.

Miyata M, Matsuda Y, Tsuchiya H, Kitada H, Akase T, Shimada M, Nagata K, Gonzalez FJ and Yamazoe Y (2006) Chenodeoxycholic acid-mediated activation of the farnesoid X receptor negatively regulates hydroxysteroid sulfotransferase. *Drug Metab Pharmacokinet* **21**:315–323.

Moore LB, Goodwin B, Jones SA, Wisely GB, Serabjit-Singh CJ, Willson TM, Collins JL and Kliewer SA (2000a) St. John's wort induces hepatic drug metabolism through activation of the pregnane X receptor. *Proc Natl Acad Sci USA* **97**:7500–7502.

Moore LB, Parks DJ, Jones SA, Bledsoe RK, Consler TG, Stimmel JB, Goodwin B, Liddle C, Blanchard SG, Willson TM, Collins JL and Kliewer SA (2000b) Orphan nuclear receptors constitutive androstane receptor and pregnane X receptor share xenobiotic and steroid ligands. *J Biol Chem* **275**:15122–15127.

Moreno-Torres A, Domingo P, Pujol J, Blanco-Vaca F, Arroyo JA and Sambeat MA (2007) Liver triglyceride content in HIV-1-infected patients on combination antiretroviral therapy studied with 1H-MR spectroscopy. *Antivir Ther* **12**:195–203.

Moschetta A, Bookout AL and Mangelsdorf DJ (2004) Prevention of cholesterol gallstone disease by FXR agonists in a mouse model. *Nat Med* **10**:1352–1358.

Mouly S, Lown KS, Kornhauser D, Joseph JL, Fiske WD, Benedek IH and Watkins PB (2002) Hepatic but not intestinal CYP3A4 displays dose-dependent induction by efavirenz in humans. *Clin Pharmacol Ther* **72**:1–9.

Murphy RL, Sommadossi JP, Lamson M, Hall DB, Myers M and Dusek A (1999) Antiviral effect and pharmacokinetic interaction between nevirapine and indinavir in persons infected with human immunodeficiency virus type 1. *J Infect Dis* **179**:1116–1123.

Narang VS, Fraga C, Kumar N, Shen J, Throm S, Stewart CF and Waters CM (2008) Dexamethasone increases expression and activity of multidrug resistance transporters at the rat blood-brain barrier. *Am J Physiol Cell Physiol* **295**:C440–C450.

Nebel A, Schneider BJ, Baker RK and Kroll DJ (1999) Potential metabolic interaction between St. John's wort and theophylline. *Ann Pharmacother* **33**:502.

Nguyen T, Huang HC and Pickett CB (2000) Transcriptional regulation of the antioxidant response element. Activation by Nrf2 and repression by MafK. *J Biol Chem* **275**:15466–15473.

Ohtsuka H, Abe T, Onogawa T, Kondo N, Sato T, Oshio H, Mizutamari H, Mikkaichi T, Oikawa M, Rikiyama T, Katayose Y and Unno M (2006) Farnesoid X receptor, hepatocyte nuclear factors 1alpha and 3beta are essential for transcriptional activation of the liver-specific organic anion transporter-2 gene. *J Gastroenterol* **41**:369–377.

Parks DJ, Blanchard SG, Bledsoe RK, Chandra G, Consler TG, Kliewer SA, Stimmel JB, Willson TM, Zavacki AM, Moore DD and Lehmann JM (1999) Bile acids: natural ligands for an orphan nuclear receptor. *Science* **284**:1365–1368.

Pascussi JM, Drocourt L, Gerbal-Chaloin S, Fabre JM, Maurel P and Vilarem MJ (2001) Dual effect of dexamethasone on CYP3A4 gene expression in human hepatocytes. Sequential role of glucocorticoid receptor and pregnane X receptor. *Eur J Biochem* **268**:6346–6358.

Pascussi JM, Jounaidi Y, Drocourt L, Domergue J, Balabaud C, Maurel P and Vilarem MJ (1999) Evidence for the presence of a functional pregnane X receptor response element in the CYP3A7 promoter gene. *Biochem Biophys Res Commun* **260**:377–381.

Pascussi JM, Robert A, Nguyen M, Walrant-Debray O, Garabedian M, Martin P, Pineau T, Saric J, Navarro F, Maurel P and Vilarem MJ (2005) Possible involvement of pregnane X receptor-enhanced CYP24 expression in drug-induced osteomalacia. *J Clin Invest* **115**:177–186.

Peng CC, Glassman PA, Trilli LE, Hayes-Hunter J and Good CB (2004) Incidence and severity of potential drug-dietary supplement interactions in primary care patients: an exploratory study of 2 outpatient practices. *Arch Intern Med* **164**:630–636.

Pentikis HS, Connolly M, Trapnell CB, Forbes WP and Bettenhausen DK (2007) The effect of multiple-dose, oral rifaximin on the pharmacokinetics of intravenous and oral midazolam in healthy volunteers. *Pharmacotherapy* **27**:1361–1369.

Perloff ES, Duan SX, Skolnik PR, Greenblatt DJ and von Moltke LL (2005) Atazanavir: effects on P-glycoprotein transport and CYP3A metabolism in vitro. *Drug Metab Dispos* **33**: 764–770.

Pickett CBPD, Nguyen T and Nioi P (2009) The Nrf2-ARE signaling pathway and its activation by oxidative stress. *J Biol Chem* Jan 30. [Epub ahead of print].

Piscitelli SC, Burstein AH, Chaitt D, Alfaro RM and Falloon J (2000) Indinavir concentrations and St John's wort. *Lancet* **355**:547–548.

Piscitelli SC, Burstein AH, Welden N, Gallicano KD and Falloon J (2002) The effect of garlic supplements on the pharmacokinetics of saquinavir. *Clin Infect Dis* **34**:234–238.

Plass JR, Mol O, Heegsma J, Geuken M, Faber KN, Jansen PL and Muller M (2002) Farnesoid X receptor and bile salts are involved in transcriptional regulation of the gene encoding the human bile salt export pump. *Hepatology* **35**:589–596.

Poland A, Glover E and Kende AS (1976) Stereospecific, high affinity binding of 2,3,7,8-tetrachlorodibenzo-p-dioxin by hepatic cytosol. Evidence that the binding species is receptor for induction of aryl hydrocarbon hydroxylase. *J Biol Chem* **251**:4936–4946.

Pondugula SR, Brimer-Cline C, Wu J, Schuetz EG, Tyagi RK and Chen T (2009) A phosphomimetic mutation at threonine-57 abolishes transactivation activity and alters nuclear localization pattern of human pregnane x receptor. *Drug Metab Dispos* **37**: 719–730.

Portoles A, Terleira A, Calvo A, Martinez I and Resplandy G (2006) Effects of Hypericum perforatum on ivabradine pharmacokinetics in healthy volunteers: an open-label, pharmacokinetic interaction clinical trial. *J Clin Pharmacol* **46**:1188–1194.

Qatanani M, Zhang J and Moore DD (2005) Role of the constitutive androstane receptor in xenobiotic-induced thyroid hormone metabolism. *Endocrinology* **146**:995–1002.

Qureshi GD, Reinders TP, Somori GJ and Evans HJ (1984) Warfarin resistance with nafcillin therapy. *Ann Intern Med* **100**:527–529.

Raucy J, Warfe L, Yueh MF and Allen SW (2002) A cell-based reporter gene assay for determining induction of CYP3A4 in a high-volume system. *J Pharmacol Exp Ther* **303**:412–423.

Remmer H, Schoene B and Fleischmann RA (1973) Induction of the unspecific microsomal hydroxylase in the human liver. *Drug Metab Dispos* **1**:224–230.

Ricketts ML, Boekschoten MV, Kreeft AJ, Hooiveld GJ, Moen CJ, Muller M, Frants RR, Kasanmoentalib S, Post SM, Princen HM, Porter JG, Katan MB, Hofker MH and Moore DD (2007) The cholesterol-raising factor from coffee beans, cafestol, as an agonist ligand for the farnesoid and pregnane X receptors. *Mol Endocrinol* **21**:1603–1616.

Riddle TM, Kuhel DG, Woollett LA, Fichtenbaum CJ and Hui DY (2001) HIV protease inhibitor induces fatty acid and sterol biosynthesis in liver and adipose tissues due to the accumulation of activated sterol regulatory element-binding proteins in the nucleus. *J Biol Chem* **276**: 37514–37519.

Ripp SL, Mills JB, Fahmi OA, Trevena KA, Liras JL, Maurer TS and de Morais SM (2006) Use of immortalized human hepatocytes to predict the magnitude of clinical drug–drug interactions caused by CYP3A4 induction. *Drug Metab Dispos* **34**:1742–1748.

Rizzo G, Renga B, Antonelli E, Passeri D, Pellicciari R and Fiorucci S (2005) The methyl transferase PRMT1 functions as co-activator of farnesoid X receptor (FXR)/9-cis retinoid X receptor and regulates transcription of FXR responsive genes. *Mol Pharmacol* **68**:551–558.

Robertson SM, Davey RT, Voell J, Formentini E, Alfaro RM and Penzak SR (2008) Effect of Ginkgo biloba extract on lopinavir, midazolam and fexofenadine pharmacokinetics in healthy subjects. *Curr Med Res Opin* **24**:591–599.

Ruschitzka F, Meier PJ, Turina M, Luscher TF and Noll G (2000) Acute heart transplant rejection due to Saint John's wort. *Lancet* **355**:548–549.

Saeki M, Kurose K, Tohkin M and Hasegawa R (2008) Identification of the functional vitamin D response elements in the human MDR1 gene. *Biochem Pharmacol* **76**:531–542.

Sandanaraj E, Lal S, Selvarajan V, Ooi LL, Wong ZW, Wong NS, Ang PC, Lee EJ and Chowbay B (2008) PXR pharmacogenetics: association of haplotypes with hepatic CYP3A4 and ABCB1 messenger RNA expression and doxorubicin clearance in Asian breast cancer patients. *Clin Cancer Res* **14**:7116–7126.

Scheer N, Ross J, Rode A, Zevnik B, Niehaves S, Faust N and Wolf CR (2008) A novel panel of mouse models to evaluate the role of human pregnane X receptor and constitutive androstane receptor in drug response. *J Clin Invest* **118**:3228–3239.

Schmiedlin-Ren P, Thummel KE, Fisher JM, Paine MF, Lown KS and Watkins PB (1997) Expression of enzymatically active CYP3A4 by Caco-2 cells grown on extracellular matrix-coated permeable supports in the presence of 1?,25-dihydroxyvitamin D_3. *Mol Pharmacol* **51**:741–754.

Schoene B, Fleischmann RA, Remmer H and von Oldershausen HF (1972) Determination of drug metabolizing enzymes in needle biopsies of human liver. *Eur J Clin Pharmacol* **4**:65–73.

Schuetz E, Lan L, Yasuda K, Kim R, Kocarek TA, Schuetz J and Strom S (2002) development of a real-time in vivo transcription assay: application reveals pregnane X receptor-mediated induction of CYP3A4 by cancer chemotherapeutic agents. *Mol Pharmacol* **62**: 439–445.

Schuetz EG, Schinkel AH, Relling MV and Schuetz JD (1996) P-glycoprotein: a major determinant of rifampicin-inducible expression of cytochrome P4503A in mice and humans. *Proc Natl Acad Sci USA* **93**:4001–4005.

Schuetz EG, Strom S, Yasuda K, Lecureur V, Assem M, Brimer C, Lamba J, Kim RB, Ramachandran V, Komoroski BJ, Venkataramanan R, Cai H, Sinal CJ, Gonzalez FJ and Schuetz JD (2001) Disrupted bile acid homeostasis reveals an unexpected interaction among nuclear hormone receptors, transporters, and cytochrome P450. *J Biol Chem* **276**:39411–39418.

Schwarz UI, Buschel B and Kirch W (2003) Unwanted pregnancy on self-medication with St John's wort despite hormonal contraception. *Br J Clin Pharmacol* **55**:112–113.

Schwarz UI, Hanso H, Oertel R, Miehlke S, Kuhlisch E, Glaeser H, Hitzl M, Dresser GK, Kim RB and Kirch W (2007) Induction of intestinal P-glycoprotein by St John's wort reduces the oral bioavailability of talinolol. *Clin Pharmacol Ther* **81**:669–678.

Shah YM, Ma X, Morimura K, Kim I and Gonzalez FJ (2007) Pregnane X receptor activation ameliorates DSS-induced inflammatory bowel disease via inhibition of NF-kappaB target gene expression. *Am J Physiol* **292**:G1114–G1122.

Shah SC, Sharma RK, Hemangini and Chitle AR (1981) Rifampicin induced osteomalacia. *Tubercle* **62**:207–209.

Shulman AI, Larson C, Mangelsdorf DJ and Ranganathan R (2004) Structural determinants of allosteric ligand activation in RXR heterodimers. *Cell* **116**:417–429.

Siccardi M, D'Avolio A, Baietto L, Gibbons S, Sciandra M, Colucci D, Bonora S, Khoo S, Back DJ, Di Perri G and Owen A (2008) Association of a single-nucleotide polymorphism in the pregnane X receptor (PXR 63396—>T) with reduced concentrations of unboosted atazanavir. *Clin Infect Dis* **47**:1222–1225.

Sonoda J, Xie W, Rosenfeld JM, Barwick JL, Guzelian PS and Evans RM (2002) Regulation of a xenobiotic sulfonation cascade by nuclear pregnane X receptor (PXR). *Proc Natl Acad Sci USA* **99**:13801–13806.

Sporstol M, Tapia G, Malerod L, Mousavi SA and Berg T (2005) Pregnane X receptor-agonists down-regulate hepatic ATP-binding cassette transporter A1 and scavenger receptor class B type I. *Biochem Biophys Res Commun* **331**:1533–1541.

Sueyoshi T, Kawamoto T, Zelko I, Honkakoski P and Negishi M (1999) The repressed nuclear receptor CAR responds to phenobarbital in activating the human CYP2B6 gene. *J Biol Chem* **274**:6043–6046.

Sueyoshi T, Moore R, Sugatani J, Matsumura Y and Negishi M (2008) PPP1R16A, the membrane subunit of protein phosphatase 1beta, signals nuclear translocation of the nuclear receptor constitutive active/androstane receptor. *Mol Pharmacol* **73**:1113–1121.

Sugatani J, Kojima H, Ueda A, Kakizaki S, Yoshinari K, Gong QH, Owens IS, Negishi M and Sueyoshi T (2001) The phenobarbital response enhancer module in the human bilirubin UDP-glucuronosyltransferase UGT1A1 gene and regulation by the nuclear receptor CAR. *Hepatology* **33**:1232–1238.

Sugatani J, Yamakawa K, Tonda E, Nishitani S, Yoshinari K, Degawa M, Abe I, Noguchi H and Miwa M (2004) The induction of human UDP-glucuronosyltransferase 1A1 mediated through a distal enhancer module by flavonoids and xenobiotics. *Biochem Pharmacol* **67**:989–1000.

Suino K, Peng L, Reynolds R, Li Y, Cha JY, Repa JJ, Kliewer SA and Xu HE (2004) The nuclear xenobiotic receptor CAR: structural determinants of constitutive activation and heterodimerization. *Mol Cell* **16**:893–905.

Sulkowski MS, Mehta SH, Torbenson M, Afdhal NH, Mirel L, Moore RD and Thomas DL (2005) Hepatic steatosis and antiretroviral drug use among adults coinfected with HIV and hepatitis C virus. *Aids* **19**:585–592.

Synold TW, Dussault I and Forman BM (2001) The orphan nuclear receptor SXR coordinately regulates drug metabolism and efflux. *Nat Med* **7**:584–590.

Takagi S, Nakajima M, Mohri T and Yokoi T (2008) Post-transcriptional regulation of human pregnane X receptor by micro-RNA affects the expression of cytochrome P450 3A4. *J Biol Chem* **283**:9674–9680.

Teotico DG, Bischof JJ, Peng L, Kliewer SA and Redinbo MR (2008) Structural basis of human pregnane X receptor activation by the hops constituent colupulone. *Mol Pharmacol* **74**: 1512–1520.

Terzolo M, Borretta G, Ali A, Cesario F, Magro G, Boccuzzi A, Reimondo G and Angeli A (1995) Misdiagnosis of Cushing's syndrome in a patient receiving rifampicin therapy for tuberculosis. *Horm Metab Res* **27**:148–150.

Thummel KE, Brimer C, Yasuda K, Thottassery J, Senn T, Lin Y, Ishizuka H, Kharasch E, Schuetz J and Schuetz E (2001) Transcriptional control of intestinal cytochrome P-4503A by 1α,25-dihydroxyvitamin D_3. *Mol Pharmacol* **60**:1399–1406.

Tirona RG and Kim RB (2009) Nuclear receptors and drug–drug interactions with prescription and herbal medicines, in *Nuclear Receptors in Drug Metabolism* (Xie W ed) pp. 211–239, Wiley, Hoboken, NJ.

Tirona RG, Leake BF, Podust LM and Kim RB (2004) Identification of amino acids in rat pregnane X receptor that determine species-specific activation. *Mol Pharmacol* **65**:36–44.

Tirona RG, Leake BF, Wolkoff AW and Kim RB (2003) Human organic anion transporting polypeptide-C (SLC21A6) is a major determinant of rifampin-mediated pregnane X receptor activation. *J Pharmacol Exp Ther* **304**:223–228.

Tran JQ, Kovacs SJ, McIntosh TS, Davis HM and Martin DE (1999) Morning spot and 24-hour urinary 6 beta-hydroxycortisol to cortisol ratios: intraindividual variability and correlation under basal conditions and conditions of CYP 3A4 induction. *J Clin Pharmacol* **39**: 487–494.

Trapnell CB, Connolly M, Pentikis H, Forbes WP and Bettenhausen DK (2007) Absence of effect of oral rifaximin on the pharmacokinetics of ethinyl estradiol/norgestimate in healthy females. *Ann Pharmacother* **41**:222–228.

Trubetskoy O, Marks B, Zielinski T, Yueh MF and Raucy J (2005) A simultaneous assessment of CYP3A4 metabolism and induction in the DPX-2 cell line. *Aaps J* **7**:E6–E13.

Urizar NL, Liverman AB, Dodds DT, Silva FV, Ordentlich P, Yan Y, Gonzalez FJ, Heyman RA, Mangelsdorf DJ and Moore DD (2002) A natural product that lowers cholesterol as an antagonist ligand for FXR. *Science* **296**:1703–1706.

van Giersbergen PL, Gnerre C, Treiber A, Dingemanse J and Meyer UA (2002) Bosentan, a dual endothelin receptor antagonist, activates the pregnane X nuclear receptor. *Eur J Pharmacol* **450**:115–121.

Van Mil SW, Milona A, Dixon PH, Mullenbach R, Geenes VL, Chambers J, Shevchuk V, Moore GE, Lammert F, Glantz AG, Mattsson LA, Whittaker J, Parker MG, White R and Williamson C (2007) Functional variants of the central bile acid sensor FXR identified in intrahepatic cholestasis of pregnancy. *Gastroenterology* **133**:507–516.

Venugopal R and Jaiswal AK (1996) Nrf1 and Nrf2 positively and c-Fos and Fra1 negatively regulate the human antioxidant response element-mediated expression of NAD(P)H:quinone oxidoreductase1 gene. *Proc Natl Acad Sci U S A* **93**:14960–14965.

Wakabayashi N, Dinkova-Kostova AT, Holtzclaw WD, Kang MI, Kobayashi A, Yamamoto M, Kensler TW and Talalay P (2004) Protection against electrophile and oxidant stress by induction of the phase 2 response: fate of cysteines of the Keap1 sensor modified by inducers. *Proc Natl Acad Sci USA* **101**:2040–2045.

Wang H, Chen J, Hollister K, Sowers LC and Forman BM (1999) Endogenous bile acids are ligands for the nuclear receptor FXR/BAR. *Mol Cell* **3**:543–553.

Wang YD, Chen WD, Wang M, Yu D, Forman BM and Huang W (2008b) Farnesoid X receptor antagonizes nuclear factor kappaB in hepatic inflammatory response. *Hepatology* **48**: 1632–1643.

Wang H, Faucette SR, Moore R, Sueyoshi T, Negishi M and LeCluyse EL (2004a) Human constitutive androstane receptor mediates induction of CYP2B6 gene expression by phenytoin. *J Biol Chem* **279**:29295–29301.

Wang H, Faucette S, Sueyoshi T, Moore R, Ferguson S, Negishi M and LeCluyse EL (2003) A novel distal enhancer module regulated by pregnane X receptor/constitutive androstane receptor is essential for the maximal induction of CYP2B6 gene expression. *J Biol Chem* **278**: 14146–14152.

Wang H, Huang H, Li H, Teotico DG, Sinz M, Baker SD, Staudinger J, Kalpana G, Redinbo MR and Mani S (2007) Activated pregnenolone X-receptor is a target for ketoconazole and its analogs. *Clin Cancer Res* **13**:2488–2495.

Wang H, Li H, Moore LB, Johnson MD, Maglich JM, Goodwin B, Ittoop OR, Wisely B, Creech K, Parks DJ, Collins JL, Willson TM, Kalpana GV, Venkatesh M, Xie W, Cho SY, Roboz J, Redinbo M, Moore JT and Mani S (2008a) The phytoestrogen coumestrol is a naturally occurring antagonist of the human pregnane X receptor. *Mol Endocrinol* **22**:838–857.

Wang XD, Li JL, Su QB, Guan S, Chen J, Du J, He YW, Zeng J, Zhang JX, Chen X, Huang M and Zhou SF (2009) Impact of the haplotypes of the human pregnane X receptor gene on the basal and St John's wort-induced activity of cytochrome P450 3A4 enzyme. *Br J Clin Pharmacol* **67**:255–261.

Wang LS, Zhou G, Zhu B, Wu J, Wang JG, Abd El-Aty AM, Li T, Liu J, Yang TL, Wang D, Zhong XY and Zhou HH (2004b) St John's wort induces both cytochrome P450 3A4-catalyzed sulfoxidation and 2C19-dependent hydroxylation of omeprazole. *Clin Pharmacol Ther* **75**: 191–197.

Watkins RE, Davis-Searles PR, Lambert MH and Redinbo MR (2003a) Coactivator binding promotes the specific interaction between ligand and the pregnane X receptor. *J Mol Biol* **331**:815–828.

Watkins RE, Maglich JM, Moore LB, Wisely GB, Noble SM, Davis-Searles PR, Lambert MH, Kliewer SA and Redinbo MR (2003b) 2.1 A crystal structure of human PXR in complex with the St. John's wort compound hyperforin. *Biochemistry* **42**:1430–1438.

Watkins RE, Wisely GB, Moore LB, Collins JL, Lambert MH, Williams SP, Willson TM, Kliewer SA and Redinbo MR (2001) The human nuclear xenobiotic receptor PXR: structural determinants of directed promiscuity. *Science* **292**:2329–2333.

Wong H, Lehman-McKeeman LD, Grubb MF, Grossman SJ, Bhaskaran VM, Solon EG, Shen HS, Gerson RJ, Car BD, Zhao B and Gemzik B (2005) Increased hepatobiliary clearance of unconjugated thyroxine determines DMP 904-induced alterations in thyroid hormone homeostasis in rats. *Toxicol Sci* **84**:232–242.

Xie W, Barwick JL, Downes M, Blumberg B, Simon CM, Nelson MC, Neuschwander-Tetri BA, Brunt EM, Guzelian PS and Evans RM (2000) Humanized xenobiotic response in mice expressing nuclear receptor SXR. *Nature* **406**:435–439.

Xie Y, Ke S, Ouyang N, He J, Xie W, Bedford MT and Tian Y (2009) Epigenetic regulation of transcriptional activity of pregnane X receptor by protein arginine methyltransferase 1. *J Biol Chem* **284**:9199–9205.

Xu RX, Lambert MH, Wisely BB, Warren EN, Weinert EE, Waitt GM, Williams JD, Collins JL, Moore LB, Willson TM and Moore JT (2004) A structural basis for constitutive activity in the human CAR/RXRalpha heterodimer. *Mol Cell* **16**:919–928.

Yamazaki Y, Kakizaki S, Horiguchi N, Sohara N, Sato K, Takagi H, Mori M and Negishi M (2007) The role of the nuclear receptor constitutive androstane receptor in the pathogenesis of non-alcoholic steatohepatitis. *Gut* **56**:565–574.

Yang D, Wang X, Chen YT, Deng R and Yan B (2009) Pyrethroid insecticides: Isoform-dependent hydrolysis, induction of cytochrome P450 3A4 and evidence on the involvement of the pregnane X receptor. *Toxicol Appl Pharmacol* Feb 26. [Epub ahead of print].

Yang J and Yan B (2007) Photochemotherapeutic agent 8-methoxypsoralen induces cytochrome P450 3A4 and carboxylesterase HCE2: evidence on an involvement of the pregnane X receptor. *Toxicol Sci* **95**:13–22.

Yasuda K, Ranade A, Venkataramanan R, Strom S, Chupka J, Ekins S, Schuetz E and Bachmann K (2008) A comprehensive in vitro and *in silico* analysis of antibiotics that activate pregnane X receptor and induce CYP3A4 in liver and intestine. *Drug Metab Dispos* **36**:1689–1697.

Yeung EY, Sueyoshi T, Negishi M and Chang TK (2008) Identification of Ginkgo biloba as a novel activator of pregnane X receptor. *Drug Metab Dispos* **36**:2270–2276.

Yin OQ, Tomlinson B, Waye MM, Chow AH and Chow MS (2004) Pharmacogenetics and herb–drug interactions: experience with Ginkgo biloba and omeprazole. *Pharmacogenetics* **14**:841–850.

Yuan CS, Wei G, Dey L, Karrison T, Nahlik L, Maleckar S, Kasza K, Ang-Lee M and Moss J (2004) Brief communication: American ginseng reduces warfarin's effect in healthy patients: a randomized, controlled trial. *Ann Intern Med* **141**:23–27.

Yue QY, Bergquist C and Gerden B (2000) Safety of St John's wort (Hypericum perforatum). *Lancet* **355**:576–577.

Zastre JA, Chan GN, Ronaldson PT, Ramaswamy M, Couraud PO, Romero IA, Weksler B, Bendayan M and Bendayan R (2009) Up-regulation of P-glycoprotein by HIV protease inhibitors in a human brain microvessel endothelial cell line. *J Neurosci Res* **87**:1023–1036.

Zhai Y, Pai HV, Zhou J, Amico JA, Vollmer RR and Xie W (2007) Activation of pregnane X receptor disrupts glucocorticoid and mineralocorticoid homeostasis. *Mol Endocrinol* **21**: 138–147.

Zhang J, Kuehl P, Green ED, Touchman JW, Watkins PB, Daly A, Hall SD, Maurel P, Relling M, Brimer C, Yasuda K, Wrighton SA, Hancock M, Kim RB, Strom S, Thummel K, Russell CG, Hudson JR, Jr., Schuetz EG and Boguski MS (2001) The human pregnane X receptor: genomic structure and identification and functional characterization of natural allelic variants. *Pharmacogenetics* **11**:555–572.

Zhang S, Liu Q, Wang J and Harnish DC (2009) Suppression of interleukin-6-induced C-reactive protein expression by FXR agonists. *Biochem Biophys Res Commun* **379**:476–479.

Zhou C, Assem M, Tay JC, Watkins PB, Blumberg B, Schuetz EG and Thummel KE (2006a) Steroid and xenobiotic receptor and vitamin D receptor crosstalk mediates CYP24 expression and drug-induced osteomalacia. *J Clin Invest* **116**:1703–1712.

Zhou J, Febbraio M, Wada T, Zhai Y, Kuruba R, He J, Lee JH, Khadem S, Ren S, Li S, Silverstein RL and Xie W (2008a) Hepatic fatty acid transporter Cd36 is a common target of LXR, PXR, and PPARgamma in promoting steatosis. *Gastroenterology* **134**:556–567.

Zhou J, Liu M, Zhai Y and Xie W (2008b) The antiapoptotic role of pregnane X receptor in human colon cancer cells. *Mol Endocrinol* **22**:868–880.

Zhou C, Poulton EJ, Grun F, Bammler TK, Blumberg B, Thummel KE and Eaton DL (2007) The dietary isothiocyanate sulforaphane is an antagonist of the human steroid and xenobiotic nuclear receptor. *Mol Pharmacol* **71**:220–229.

Zhou C, Tabb MM, Nelson EL, Grun F, Verma S, Sadatrafiei A, Lin M, Mallick S, Forman BM, Thummel KE and Blumberg B (2006b) Mutual repression between steroid and xenobiotic receptor and NF-kappaB signaling pathways links xenobiotic metabolism and inflammation. *J Clin Invest* **116**:2280–2289.

Chapter 5
Impact of Physiological Determinants: Flow, Binding, Transporters and Enzymes on Organ and Total Body Clearances

K. Sandy Pang, Huadong Sun, and Edwin C.Y. Chow

Abstract Physiologically based pharmacokinetic (PBPK) models of the intestine, liver, and kidney were developed to examine the influence of transporters as well as enzymes on the area under the curve and clearances of drugs and metabolites. Whole body PBPK models were then developed, with the kidney and intestine or the kidney and liver as the organs for excretion and metabolism. From these PBPK models, the influence of flow, binding, transporters, and enzymes and the presence of competing pathways and competing organs on the areas of the drug and metabolite and total body clearance were defined. These relationships on the drug and metabolite AUC data are extremely useful for understanding the DDI mechanism, transport or metabolism, or both.

5.1 Introduction

The absorption, distribution, metabolism, and excretion (ADME) of a drug determine the concentration–time profile of the drug in the body. These topics on drug absorption, distribution, and the processing of drug or metabolite within eliminating organs by enzymes and transmembrane transporters on ADME have been covered in some detail in Chapters 1 and 2 of Part I. The composite effects of drug binding, enzymes, and transporters on drug entry and excretion, together with blood flow to the organ, may be collectively integrated as organ clearance. Addis (1917) and Möller and coworkers (1928) were among the few who first commented on the renal clearance of urea. They defined the renal clearance of urea as the volume of biological fluid that is cleared of the contained urea per unit time. Needless to say, the volume is not a real volume since, as blood flows through the kidneys, no single milliliter has all of its urea removed in any transit. Rather, a portion of the urea is removed from each of the many milliliter of blood that is perfusing the kidneys. Clearance (CL) is obtained upon summation of the urea removed, and by expression

K.S. Pang (✉)
Leslie Dan Faculty of Pharmacy, University of Toronto, Toronto, Ontario, Canada
e-mail: ks.pang@utoronto.ca

K.S. Pang et al. (eds.), *Enzyme- and Transporter-Based Drug–Drug Interactions*,
DOI 10.1007/978-1-4419-0840-7_5,

of this as if elimination arose from the blood clearing all of the contained urea in a much small volume.

The concept of clearance applies not only to the kidneys but also to other eliminating organs. Hepatic drug clearance is used to describe removal of drugs by the liver. The concept was first introduced by Lewis (1949) who related hepatic clearance to the ability of the liver to remove drugs. Clearance concepts surrounding the liver has evolved, through the efforts of Rowland (1972) and coworkers (1973; 1995; Pang and Rowland, 1977) and Wilkinson and colleagues (Wilkinson and Shand 1975; Wilkinson, 1987). The field has evolved for the consideration of intestinal clearance (Klippert and Noordhoek, 1985; Yu and Amidon, 1998; Ito et al., 1999; Cong et al., 2000; Pang, 2003, Yang et al., 2007).

Clearance relates to the rate of removal and determines the steady-state concentration C_{ss} in infusion situations as infusion rate/CL. More importantly, clearance relates to the efficiency of an organ to remove a drug (Rowland et al., 1973). Clearance, together with the volume of distribution (V), serves as determinants of the half-life ($t_{1/2}$), in the relation $t_{1/2} = 0.693V/\mathrm{CL}$ (Rowland and Tozer, 1995). The clearance of a prototypical substrate such as indocyanine green (ICG) (Faybik and Hetz, 2006; Sakka, 2007), sorbitol (Molino et al., 1998; Li et al., 2003), bromosulfophthalein (Häcki et al., 1976; Vaubourdolle et al., 1991), or lidocaine (Ercolani et al., 2000; Kaneko et al., 2001) has been used to assess liver function, and that for *p*-aminohippurate, inulin, and creatinine, to appraise renal function (Brewer et al., 1990; Toto, 1995). Clearance concepts play an important role in pharmacokinetics and biopharmaceutics. Constancy in clearance is the basis of bioavailability (F) estimates for the comparison of the dose-corrected area under the curve (AUC) of the oral vs. intravenous dose (Wagner, 1972; Greenblatt et al., 1973; Koch-Weser, 1974a, b). In this context, CL is a sensitive marker reflective of changes in the handling of a drug and is a determinant of the drug exposure (area under the curve) and drug toxicity.

5.2 Common Determinants of Clearance in Eliminating Organs

For most organs and tissues, the rate of removal, v, may be estimated as the product of flow (Q) and the difference in input (C_{In}) and output (C_{Out}) concentrations (Fig. 5.1a). Under steady-state conditions, the rate of loss (v_{ss}) is attributed to elimination only, since the distribution and binding of the drug to tissue proteins are complete. Upon normalization of v_{ss} to the input concentration, one obtains clearance, whereas when v_{ss} is normalized to the rate of drug presentation (QC_{In}), one obtains the extraction ratio (E) (Fig. 5.1b). It is readily recognized that clearance at the steady-state (CL_{ss}) of an organ is the product, $Q\mathrm{E}_{ss}$ (Rowland et al., 1973; Wilkinson and Shand, 1975; Pang and Rowland, 1977), though in this simple relationship, the identity of the physiological variables that govern CL is not revealed. Excretion, metabolism, and transport mechanisms need to be considered, with concomitant monitor of circulatory metabolites and unchanged drug in excretory fluids.

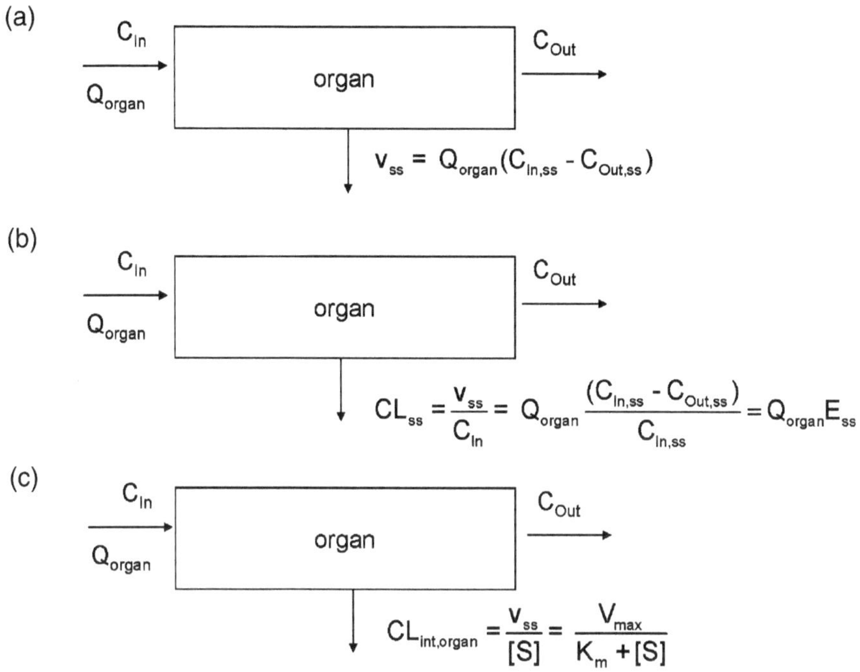

Fig. 5.1 Schematic depiction of drug removal across an eliminating organ: the steady-state elimination rate (v_{ss}) according to Fick's law, and how v_{ss} relates to the steady-state clearance, CL_{ss}, extraction ratio, E_{ss}, and the intrinsic clearance, CL_{int}

Removal pathways are mostly saturable processes that are characterized by the maximum velocity, V_{max}, and the Michaelis–Menten constant, K_m. In the absence of a transmembrane barrier (flow-limited distribution), the rate of removal for a substrate equals $(V_{max}\ [S])/(K_m + [S])$, where $[S]$, the unbound substrate concentration within the liver, equals the unbound drug concentration in liver blood that is exiting the organ. The term, intrinsic clearance (CL_{int}), is developed to relate to the rate of removal, v_{ss}, which equals $CL_{int}[S]$ (Fig. 5.1c) (Gillette, 1971). The intrinsic clearance may be used to define the activity pertaining to a transporter or enzyme; multiplication of the intrinsic clearance to the (unbound) substrate concentration $[S]$ provides the rate, and hence the intrinsic clearance is the volume of cellular water that is cleared of the unbound drug per unit time. Under first-order conditions, CL_{int} equals V_{max}/K_m (Gillette, 1971; Wilkinson and Shand, 1975).

5.2.1 Protein and Red Blood Cell Binding

Usually, the unbound drug is the species that enters the organ and becomes eliminated. Binding to vascular components, including protein and red blood cell (rbc), is regarded as an inhibitory factor for drug clearance. Plasma protein binding is a key factor in the clearance of poorly extracted drugs (Yacobi et al., 1977; 1979;

Bekersky et al., 1984; Schary and Rowland, 1983; Rodriguez and Smith, 1991; Chiba and Pang, 1993; Boffito et al., 2003). The elimination rate may be rate-limited by binding (expressed as the unbound or free fraction of the drug in the plasma) and less on the intrinsic clearance when the CL_{int} is high. Drug binding to proteins is usually a reversible process, denoted by the on- and off-rate constants; the ratio of these is the binding association constant, K_A. The unbound concentration of drug, $C_{P,u}$, that is bound to a single class of binding site of n sites and protein concentration, P_t, is given by

$$C_{P,u} = \frac{-(1 + nK_AP_t - K_AC_P) + \sqrt{(1 + nK_AP_t - K_AC_P)^2 + 4K_AC_P}}{2K_A} \quad (5.1)$$
$$\Rightarrow C_{P,u} = C_P - C_{P,b}$$

and is the difference between the total (C_P) and bound ($C_{P,b}$) plasma concentrations. A binding association constant (K_A) of $< 10^4$ M^{-1} suggests poor binding, whereas one $> 10^4$ M^{-1} suggests tighter binding. The higher the K_A, the greater will be the likelihood of fluctuation due to saturation of binding sites (Coffey et al., 1971). It is often assumed that only the unbound drug could enter the cells by passive diffusion or active transport. The uptake of drug may be dependent on the dissociation of the drug–protein bound complex when the dissociation rate constant is very slow (Weisiger, 1985). The unbound fraction in plasma (f_P), given by the ratio of the unbound plasma concentration, $C_{P,u}$, and the total plasma concentration, C_P, is related to the binding dissociation constant (K_D) or $1/K_A$, the binding capacity, nP_t, and the unbound plasma concentration.

$$f_P = \frac{C_{P,u}}{C_P} = \frac{K_D + C_{P,u}}{nP_t + K_D + C_{P,u}} \quad (5.2)$$

In similar fashions, appreciable distribution of a drug into red blood cells, accompanied by slow efflux from red blood cells (rbc), constitutes an impediment to drug removal within an eliminating organ (Goresky et al., 1975; Hinderling, 1984; Goresky et al., 1988; Pang et al., 1995). It is therefore important to investigate the effect of red cell distribution in drug disposition. The partitioning of drug from plasma in the red blood cell is given by the following mass balance equation in relation to the blood and plasma concentrations, C_B and C_P, respectively,

$$C_{rbc} = \frac{C_B - C_P(1 - \mathrm{Hct})}{\mathrm{Hct}} \quad (5.3)$$

showing the influence of the hematocrit, Hct. The equilibration of the unbound drug in plasma and that within the red cells is shown in Fig. 5.2a. The equilibrative rate constants for the exchange of unbound drug between that plasma and rbc are expressed as k_{pr} and k_{rp}, respectively. Due to the difficulty in the determination of red cell binding, k'_{rp} or k_{rp}/f_{rbc} (where f_{rbc} is the unbound fraction in rbc) is used in lieu of k_{rp} to relate to the movement of total and not the unbound concentration from the red blood cell to plasma. The drug in the rbc is not readily available for

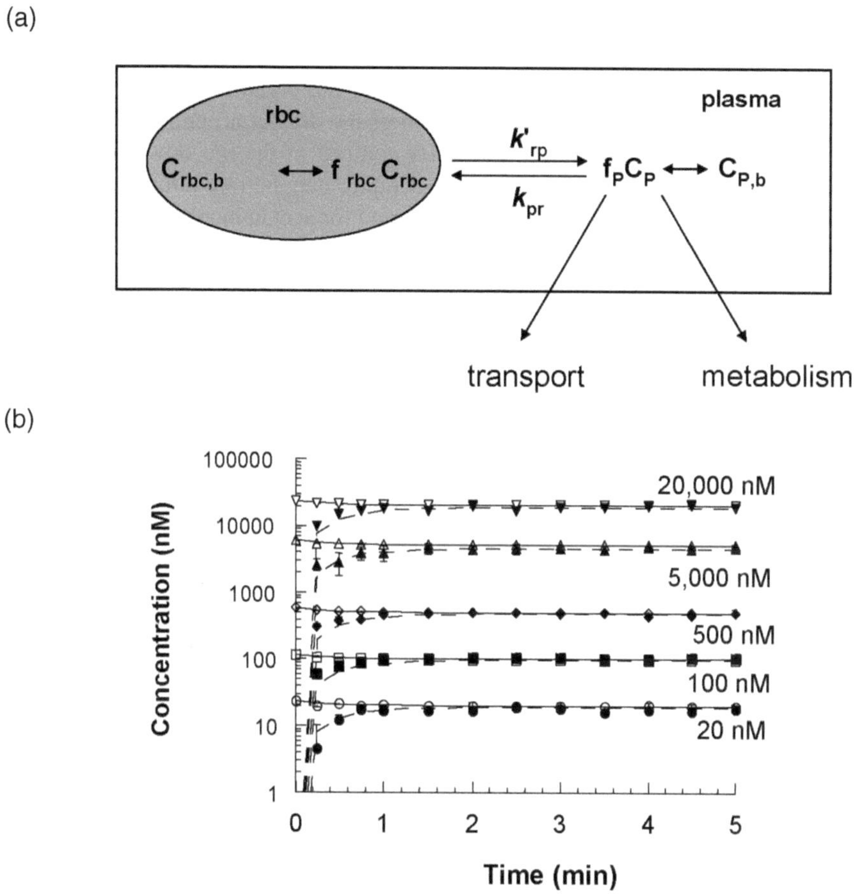

Fig. 5.2 (**a**) Drug binding to plasma protein and red cell distribution and (**b**) the slow equilibration between digoxin in red blood cell and plasma (from Liu et al., 2005, with permission)

transport or metabolism when the release or off-rate constant, k'_{rp}, is slow (Goresky et al., 1975; Pang et al., 1995; Liu et al., 2005).

Experimentally, these distribution rate constants may be ascertained by mixing the drug-containing plasma with blank whole blood. The rates of change of drug concentrations in rbc (C_{rbc}) and plasma (C_p) are

$$\frac{dC_P}{dt} = k'_{rp}\, C_{rbc}\frac{\text{Hct}}{(1-\text{Hct})} - k_{pr}f_P C_P \tag{5.4}$$

$$\frac{dC_{rbc}}{dt} = k_{pr}f_P C_P\frac{(1-\text{Hct})}{\text{Hct}} - k'_{rp}\, C_{rbc} \tag{5.5}$$

Immediately after admixture, there is movement of drug from the plasma into rbc until equilibrium is reached (Fig. 5.2b). This method revealed a slow influx of digoxin from plasma to rbc (k_{pr} of 0.468 min^{-1}) and a slightly faster efflux from rbc to plasma (k'_{rp} of 1.81 min^{-1}) (Fig. 5.2b). If movement of drug from the plasma into rbc is slow and efflux is fast in relation to the organ transit time, there will be some but not an extensive accumulation of the drug concentration in rbc, as is the case for digoxin (Liu et al., 2005). By contrast, if the rbc slowly releases the drug content into plasma, drug removal, if rapid, may be rate-limited by the slow release of bound drug from the rbc, as observed for acetaminophen sulfation (Pang et al., 1995). When the equilibration time of drugs is fast, a red cell capacity effect exists, whereas when the equilibration time is slow, a red cell carriage effect exists (Goresky et al., 1975; 1988; Pang et al., 1995).

For rapidly equilibrative red cell binding, the unbound fraction in blood (f_u) is related to the unbound fractions in rbc (f_{rbc}) and f_P, as shown in Equation (5.6) below (Pang and Rolwand, 1977), with the assumption that there is no concentrative transport of drug into the rbc and the unbound concentrations are the same in blood, red cell, and plasma.

$$f_{rbc} = \frac{\text{Hct}}{\dfrac{1}{f_u} - \dfrac{(1-\text{Hct})}{f_P}} \tag{5.6}$$

It is recognized that drugs compete for similar binding sites and cause displacement of other drugs from their binding sites on plasma or the rbc. Significant drug–drug interactions have been observed for displacement reactions in plasma (McElnay and D'Arcy, 1983; Toon et al., 1986; Goulden et al., 1987; McNamara et al., 1990; Yu et al., 1990; Orlando et al., 2009). Displacement of drug by its metabolite for rbc (carbonic anhydrase) binding sites has been observed (Wong et al., 1996), resulting in higher unbound fractions. Under these cases, the clearance and the volume of distribution would increase.

5.2.2 Blood Flow

Blood flow is an important determinant of clearance, especially for drugs that are removed with high intrinsic clearances and enter the organ rapidly, in a flow-limited sense. The pattern of blood flowing into an organ is very important since it influences the pattern of mixing. The flow pattern: bulk flow, plug flow, or dispersive flow, affects the degree of mixing within the liver and the manner in which the drug recruits enzymatic and transporter activities (Perl and Chinard, 1968; Winkler et al., 1973; Pang and Rowland, 1977; Roberts and Rowland, 1985; Pang, 1995). Moreover, not all the total flow rate reaches the metabolic zones of the kidney or the intestine. In the kidney, part of the plasma is filtered due to glomerular filtration (Hekman and van Ginnekan, 1983), and in the intestine, only a small part of the blood reaches the enterocyte region where absorptive and secretory apical transporters and enzymes are present (Cong et al., 2000; Pang, 2003).

5.2.3 Enzymatic Activity

The metabolism (biotransformation) of drugs has been traditionally regarded as an important factor in drug clearance, especially for the liver. The reactions (and responsible enzymes) in phase I metabolism include oxidation (cytochrome P450 and NADPH–cytochrome P450 reductase, flavin monooxygenases, monoamine oxidase, and alcohol dehydrogenase), reduction (cytochrome P450 and NADPH–cytochrome P450 reductase, carbonyl reductase, sulfatase, and glucuronidase), and hydrolysis (carboxylesterases, peptidases, and epoxide hydrolase). The enzymes in phase II metabolism include glucuronidation (UGT), sulfation (SULT), glutathione conjugation (GST), methylation (methyltransferase), acetylation (*N*-acetyltransferase), and amino acid conjugation (amino acid *N*-acetyltransferase) (see Chapter 1). Phase II reactions involve the addition of a specific cosubstrate to form conjugates, which include uridine-5′-diphosphoglucuronic acid (UDPGA), 3′-phosphoadenosine-5′-phosphosulfate (PAPS), glutathione (GSH), *S*-adenosylmethionine (SAM), and acetyl coenzyme A in glucuronidation, sulfation, glutathione conjugation, methylation, and acetylation, respectively. The cosubstrate could become a rate-limiting factor in phase II metabolism when it becomes depleted (Gregus et al., 1992; Kim et al., 1995; Tirona et al., 1999).

5.2.4 Excretory Activity

Transporters present at the apical membrane mediate the excretion of both the drug and metabolite species. For example, canalicular transporters of the liver: P-glycoprotein (P-gp), the multidrug resistance-associated protein 2 (MRP2), bile salt export pump (BSEP), and breast cancer resistance protein (BCRP) belong to the ATP-binding cassette (ABC) gene superfamily and directly utilize ATP to excrete drugs and metabolites into the bile (Trauner and Boyer, 2003; van Montfoort et al., 2003; Eloranta and Kullak-Bulick, 2005; Shitara et al., 2006; Robey et al., 2007) (see Chapter 2). P-gp is the most widely studied excretory transporter and primarily mediates the transport of simple ions, complex lipids, hydrophobic cations, and xenobiotics (Thiebaut et al., 1987; Ambudkar et al., 2008). MRP2 usually excretes phase II products such as glucuronides, sulfates, and glutathione conjugates (Zamek-Gliszczynski et al., 2006). Being a half transporter, BCRP has been found to involve in the excretion of sulfate conjugates of steroids and xenobiotics (Mizuno et al., 2004; Krishnamurthy and Schuetz, 2006; Vlaming et al., 2009) and plays a role in drug resistance to anticancer agents (Sugimoto et al., 2005).

5.2.5 Basolateral Transporters

Uptake transporters in the liver, kidney, and intestine have been thoroughly reviewed, and these mediate the entry of substrate into the organ (for reviews, see Trauner and Boyer, 2003; van Montfoort et al., 2003; Eloranta and Kullak-Ublick, 2005; Shitara et al., 2006). Many uptake transporters on the basolateral membrane belong to the gene superfamily of solute carriers (*SLC*) that have been covered in Chapter 2. The basolateral transporters include the organic anion transporters (OAT), the organic cation transporters (OCT), the monocarboxylic acid transporter (MCT), the bile acid-specific sodium-dependent taurocholate cotransporting polypeptide (NTCP), and organic anion transporting polypeptides (OATPs) that transport anionic, cationic, and neutral molecules into the liver. Intestinal and renal absorptive transporters at the apical membrane, such as the oligopeptide transporter 1 (PEPT1) and apical sodium-dependent bile acid transporter (ASBT), are uniquely present to increase the cellular accumulation of di- and tri-peptides and bile acids, respectively.

Efflux from the basolateral membrane of the organ back into the circulation may occur via passive diffusion and/or transporters. These transporters, which belong to the ABC superfamily, include MRP1 and MRP3–MRP8, members of the MRP subfamily (Chandra and Brouwer, 2004; Kruh et al., 2007). Similar to the uptake (influx) transporters, the efflux transporters could modulate drug clearance by readily altering the drug concentration in the cell.

5.3 Physiologically-Based Pharmacokinetic (PBPK) Models

The roles of flow rate, red cell and plasma protein binding, and enzymes and transporters on organ clearances are integratively summarized in physiologically-based pharmacokinetic (PBPK) models. PBPK modeling relates organ or tissue structures to the physiology of the organ or tissue, and is based on the concept that compartments are homogeneous and well-stirred. PBPK models describe compartments of discrete volumes being perfused uniformly by blood flow. The organ–tissues are interconnected by the circulation according to their anatomical arrangements. Venous equilibration is assumed to occur, and the venous drug concentration is in equilibrium with that within the blood of the organ.

Rate equations may be constructed to describe processes of binding and debinding, flow, entry, efflux, metabolism, and removal. These models encompass transporters and enzymes and their associated intrinsic clearances (CL_{in} and CL_{ef} for influx and efflux at the basolateral membrane, respectively, and $CL_{int,sec}$ and $CL_{int,met}$ for excretion at the apical membrane and metabolism within the cell, respectively). These PBPK models have been developed to examine hepatic (de Lannoy and Pang, 1993; Tirona et al., 1999; Liu et al., 2005), renal (de Lannoy et al., 1990; de Lannoy et al., 1993), and intestinal (Cong et al., 2000) clearances and will be covered in detail in the ensuing sections. One of the greatest advantages of PBPK models is that mathematical expressions that relate to flow, binding, and transporter and enzymatic activities may be solved by matrix inversion for the area under the curve under linear conditions. This method had been applied to single, eliminating organs: the intestine, liver, and kidney (Pang et al., 2008; Sun and Pang, 2009a).

These may be extended to whole body PBPK models that include, in addition, the renal excretion of the intact drug and its metabolites (Sun and Pang, 2009b).

5.4 Rate-Limiting Step in Clearance

After attaining an understanding of the various determinants of CL, another important concept is the identification of the rate-limiting step of clearance. As shown in the ensuing section, clearance is influenced by flow, binding, the influx and efflux clearances at the basolateral membrane, and the metabolic and excretory intrinsic clearances. For a drug that is extremely highly cleared, meaning that the intrinsic clearance for elimination is extremely large, clearance approaches the value of blood flow and the drug enters the organ readily (flow-limited distribution). In contrast, a drug will be extremely poorly cleared, either because it cannot enter the organ due to poor permeation properties or that the intrinsic clearance for elimination is extremely low. Wilkinson and coworkers described these two conditions as "non-restrictive" and "restrictive" clearances, respectively (Wilkinson and Shand, 1975; Wilkinson, 1987). In the first instance, clearance is limited by blood flow (Benowitz et al., 1974; Xu and Pang, 1989; Sun et al., 2006), and in the second instance, clearance is limited by the unbound fraction and/or the intrinsic clearance for elimination. For instances where uptake is rapid, metabolism and/or excretion is rate-limiting, as exemplified by the glutathione conjugation of bromosulfophthalein in the dog liver (Goresky, 1964), hydrolysis of enalapril (Abu-Zahra et al., 2000; Abu-Zahra and Pang, 2000) and sulfation of estrone (Tan and Pang, 2001) in the perfused rat liver preparation. In other cases, the membrane can pose as a barrier, barring the entry of drugs, hence drug clearance is diffusion rate limited (de Lannoy and Pang, 1987). On occasion, transporters that mediate the uptake of drugs could also be the rate-limiting step, when influx (uptake clearance) is much slower than the metabolic and/or secretory intrinsic clearance. Poor transmembrane entry had been found to rate-limit the metabolism or excretion of ethacrynic acid (Tirona et al., 1999) and enalaprilat (Pang et al., 1985; de Lannoy and Pang, 1987; Schwab et al., 1990; 1992). Lastly, the binding of drugs to plasma proteins is known to delimit the entry of drugs and therefore removal of drugs. A good example is the highly fluorescent substrate, bromosulfophthalein or BSP, which binds strongly to albumin. In the presence of albumin, the intrahepatic gradient of BSP concentration was very shallow. However, in absence of albumin, BSP became highly cleared and exhibited a steep concentration grade across the liver lobule (Gumucio et al., 1981; 1984).

5.5 Models for Hepatic Drug Clearance

The anatomical position of the liver renders it as a gatekeeper, acting as a major first-pass organ for orally administered drugs, between the intestine and the systemic circulation. About 1.5 liter of blood flow or 25% of the cardiac output enters the human liver in each minute, rendering it one of the most highly perfused organs in the body. The venous blood from the entire gastrointestinal tract is enriched with nutrients that are absorbed from the intestine and brought to the liver by the portal vein (PV) that constitutes about 75% of the liver blood flow. The remaining 25% is oxygenated blood that is supplied by the hepatic artery (HA). Within the liver, branches of PV and HA continue in parallel and converge at the sinusoid, ending eventually at the terminal hepatic venule before drainage into the hepatic vein (HV). The microcirculatory unit of the circulation in the liver is known as the acinus that comprises of three zonal regions, 1, 2, and 3, regions (Rappaport, 1958) that are analogous to the periportal, midzonal, and pericentral metabolic zones of the liver.

The concept of hepatic drug clearance has been firmly established in the early 1970s. Models of hepatic drug clearance that embellish physiological variables such as blood flow, binding, and the enzymatic and excretory activities surfaced with respect to their ability to predict drug removal by the liver. Among these models, the "well-stirred" model is the simplest since it assumes venous equilibration, with drug emerging from the outflow being in equilibrium with the drug within the liver; the concentration within the liver is the same throughout (Rowland et al., 1973; Pang and Rowland, 1977). The "parallel tube" (Winkler et al., 1973) and dispersion (Perl and Chianrd, 1968; Roberts and Rowland, 1985) models, and the barrier-limited, variable transit-time model of Goresky and coworkers (Goresky, 1964; Goresky et al., 1973; Schwab et al., 1990; 1992) have been used to account for the observed sinusoidal concentration gradient from the inlet and outlet and the binding of drugs to vascular components (Pang et al., 1995). Other hepatic clearance models include the series-compartment model of Gray and Tam (1987). A departure from these models also exists, since, in reality, heterogeneity in flow, enzymes, and transporters is not an unusual occurrence. For the improved description of metabolite formation and sequential metabolism, the zonal-liver or enzyme-distributed model adds heterogeneity in enzymatic and transporter distribution within zonal regions to account for drug metabolism and transport (Pang and Stillwell, 1983; Xu and Pang, 1989; Xu et al., 1990; Kwon and Morris, 1997; Abu-Zahra and Pang, 2000).

The flow pattern and the degree of mixing constitute different inferences on what the substrate concentration should be (Pang, 1995). For the well-stirred model, the venous unbound concentration reflects the substrate concentration within the liver. Substitution of this relation into the rate equation, with the assumption that the unbound concentration leaving the organ, $f_u C_{Out}$, given by the unbound fraction in blood, f_u, multiplied to the outflow concentration C_{Out}, equals that in the organ (venous equilibration). However, these assumptions usually view the lack of the transmembrane barrier for the drug to enter the liver cell. The relationship will not hold when a membrane barrier exists.

5.5.1 The Well-Stirred Model for Hepatic Drug Clearance

The well-stirred model, the most popular among the models of hepatic drug clearance, is conceptually congruent with compartmental analysis for homogeneous or well-mixed compartments. To date, the liver blood flow, vascular (protein and cellular) and tissue binding, transporters, and metabolic enzymes are regarded as determinants that affect hepatic drug clearance according to the well-stirred model (Pang and Rowland, 1977; Pang et al., 2008). A simple PBPK model for the liver is based on the organ being the only eliminating organ, in a situation akin to the recirculating perfused rat liver preparation (Fig. 5.3) (Liu and Pang, 2006; Pang et al., 2008). The unbound drug enters the cell by passive diffusion or relies on transporters for entry. The liver is comprised of three subcompartments: liver tissue (L), liver blood (LB), and the bile compartmnt (bile). The reservoir and liver compartments of volumes V_R and V_L are interconnected by the blood flow, Q_H. V_{LB} and V_{Bile} represent the volumes of the liver blood and bile, respectively. The unbound fractions in liver blood and liver are denoted by f_u and f_L, respectively. The recognition that the membrane of the organ contains transporters or exhibits membrane-limited transport has led to concepts of influx (CL_{in}^{H}) and efflux (CL_{ef}^{H}) clearances, or sums of passive diffusive clearance and the carrier-mediated transport clearance that facilitate drug entry and efflux at the basolateral membrane. The drug within the tissue is metabolized by enzymes of intrinsic metabolic clearance, $CL_{int,met1,H}$ (V_{max}/K_m for first-order conditions), to form the primary metabolite, Mi_L, and $CL_{int,met2,H}$ to form other primary metabolites; biliary excretion of the drug is a function of the

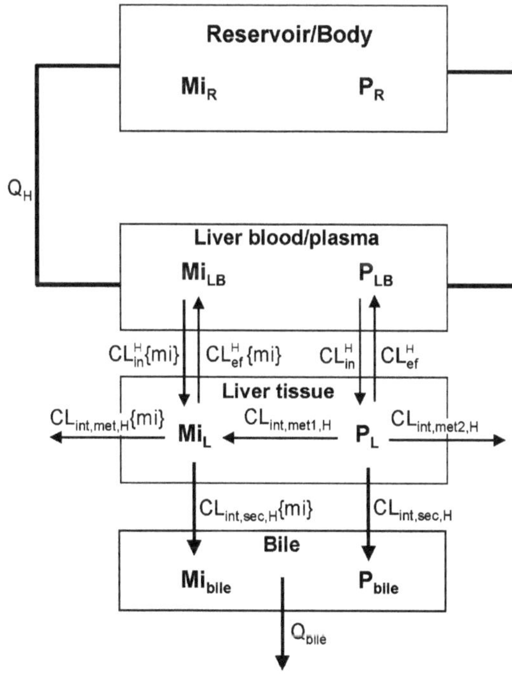

Fig. 5.3 Physiologically-based pharmacokinetic (PBPK) liver model, with liver as the only elimination organ, as in the perfused rat liver preparation. See text for description (modified from Pang et al., 2008)

biliary intrinsic clearance, $CL_{int,sec,H}$. A similar scenario may be described for the metabolite, Mi_L, that is formed within the liver. The formed metabolite may be further metabolized to other species with the intrinsic metabolic clearance of the metabolite, $CL_{int,met,H}\{mi\}$, and/or secreted into bile with the secretory intrinsic clearance of the metabolite, $CL_{int,sec,H}\{mi\}$. The metabolite enters and leaves the cell at the basolateral membrane with transport clearances, $CL_{in}^{H}\{mi\}$ and $CL_{ef}^{H}\{mi\}$, respectively.

5.5.2 *Solutions for AUC and CL*

The method of matrix inversion of the coefficients derived from rate equations for drug and metabolite in the various compartments was used (Pang et al., 2008; Sun and Pang, 2009a) to yield the area under the curve (AUC) from time 0 to ∞ for the drug and the metabolite. The AUC for the drug after iv administration was first solved for the liver, viewed as the only eliminating organ, and later with other eliminating organs.

For the liver, the AUC_{iv} is

$$AUC_{iv} = \frac{Dose_{iv}\left[Q_H\left(CL_{ef}^{H} + CL_{int,H}\right) + f_u CL_{in}^{H} CL_{int,H}\right]}{f_u CL_{in}^{H} Q_H CL_{int,H}} \tag{5.7}$$

and the corresponding area under the curve for the formed metabolite, $AUC_{iv}\{mi,P\}$, is

$$AUC_{iv}\{mi,P\} = \frac{Dose_{iv} CL_{int,met1,H} CL_{ef}^{H}\{mi\}}{\left(CL_{int,met1,H} + CL_{int,met2,H} + CL_{int,sec,H}\right) f_u\{mi\} CL_{in}^{H}\{mi\} CL_{int,H}\{mi\}} \tag{5.8}$$

Note that the $AUC_{iv}\{mi,P\}$ of the metabolite is dependent on its formation intrinsic clearance, $CL_{int,met1,H}$ from drug, and individual components of the total $CL_{int,H}$ or sum of $CL_{int,met1,H}$, $CL_{int,met2H}$, and $CL_{int,sec,H}$, as well as binding and the transport clearances of metabolite ($CL_{in}^{H}\{mi\}$ and $CL_{ef}^{H}\{mi\}$) and the metabolite's total hepatic intrinsic clearance, $CL_{int,H}\{mi\}$.

The AUC_{iv} in turn allows the total body clearance, CL, to be estimated, and, since the liver is the only organ for removal, CL equals CL_H, the hepatic clearance.

$$CL = CL_H = \frac{Dose_{iv}}{AUC_{iv}} = Q_H \frac{f_u CL_{int,H} CL_{in}^{H}}{Q_H CL_{ef}^{H} + CL_{int,H}\left(Q_H + f_u CL_{in}^{H}\right)} \tag{5.9}$$

When CL_{in}^{H} and $CL_{ef}^{H} >> Q_H$, Equations (5.7) and (5.9) collapse to yield Equations (5.10) and (5.11), as shown by Rowland et al. (1973) and Pang and Rowland (1977) for drugs that exhibit flow-limited distribution.

$$AUC_{iv} = \frac{Dose_{iv}(Q_H + f_u CL_{int,H})}{Q_H f_u CL_{int,H}} \tag{5.10}$$

$$CL_H = \frac{Q_H f_u CL_{int,H}}{(Q_H + f_u CL_{int,H})} \tag{5.11}$$

Note that CL_H is related to the total hepatic blood flow (Q_H), the unbound fraction in blood (f_u), and the total hepatic intrinsic clearance $CL_{int,H}$, sum of the metabolic ($CL_{int,met,H}$) and excretory ($CL_{int,sec,H}$) intrinsic clearances.

Equation (5.11) relates to the special case of absence of a transport barrier (flow-limited case) (Rowland et al., 1973). Under this condition, the drug permeability is large and venous equilibration exists, with the unbound liver concentration being equal to $f_u C_{Out}$. For this special condition where drug removal is solely by hepatic metabolism, the ratio $v/(f_u C_{Out})$ estimates the hepatic intrinsic clearance or V_{max}/K_m under first-order conditions and the hepatic intrinsic clearance may be estimated as dose/(unbound AUC_{po}) when the orally administered dose is completely absorbed. Otherwise, the transport clearance terms in Equation (5.8) must be included in the definition of clearance (the longer formats), and $v/(f_u C_{Out})$ no longer yields the intrinsic clearance. The total hepatic clearance (CL_H) is the sum of the metabolic ($CL_{H,met}$) and biliary ($CL_{H,ex}$) clearances, and it may be deduced that the ratio of the metabolic to biliary clearance is the ratio of (total) $CL_{int,met,H}$ to $CL_{int,sec,H}$ under linear kinetic conditions (Liu and Pang, 2006; Pang et al., 2008).

$$\frac{CL_{H,met}}{CL_{H,ex}} = \frac{CL_{int,met,H}}{CL_{int,sec,H}} \tag{5.12}$$

Equations (5.7)–(5.12) are useful for enabling inferences to be made on the mechanism of drug–drug interactions within the liver. A change in one eliminatory mechanism, secretion or alternate metabolism, will alter $CL_{int,H}$ in Equations (5.7) and (5.8) and would result in an increase or a decrease in AUC of the drug and metabolite. For example, a decrease in the metabolite formation intrinsic clearance, $CL_{int,met1,H}$, would increase AUC of drug but decrease AUC{mi,P}. The AUC_{iv} of the drug would be increased upon inhibition of the secretory pathway, the alternate pathway (Equation (5.7)). The corresponding area of the formed metabolite, AUC_{iv}\{mi,P\}, would be decreased as a result of inhibition of the formation pathway ($CL_{int,met1,H}$) but would increase upon inhibition of the competing pathways, secretion or alternate metabolism ($CL_{int,sec,H}$ and $CL_{int,met2,H}$ in Equation (5.7)) (Sirianni and Pang, 1997). Generally speaking, an increase in basolateral efflux increases the AUC, whereas an increase in basolateral influx decreases the AUC (Liu and Pang, 2006). Concomitantly, the metabolite AUC{mi,P}s and the amounts secreted would show no change. Solutions for the oral case are similar, with the exception that F_{abs} term is present (Sun and Pang, 2009b).

5.5.3 Examples

The effects of physiological variables on clearance and AUC are illustrated by many examples in rat liver perfusion studies. Digoxin is a poorly cleared lipophilic drug that is mostly excreted by P-gp and to a lesser degree, metabolized. In a simulation based on protein and red cell binding and removal characteristics of digoxin (Liu et al., 2005), it was shown that changing the flow rate from 10 to 40 ml/min failed to alter the AUC or clearance of this poorly extracted compound (Fig. 5.4a), whereas decreasing the binding to red cells and albumin enhanced clearance (Fig. 5.4b). Moreover, decreasing CL_{in}^{H} and increasing CL_{ef}^{H} would increase the AUC, whereas decreasing both $CL_{int,sec,H}$ and $CL_{int,met,H}$ would increase the AUC (Fig. 5.4c–f).

Contrasting to this example is the fate of estradiol 17β-glucuronide (E_217G), a compound that is extremely highly cleared such that the hepatic extraction ratio, E_H, approaches 1 and CL_H approaches Q_H. E_217G enters the liver rapidly by the rat Oatp family of transporters and is eliminated almost equally via sulfation by the estrogen sulfotransferase, Sult1e1, and excreted via Mrp2 in the rat liver. A 50% reduction in the E_217G influx in tumor-bearing livers failed to alter levels of E_217G and its CL (Fig. 5.5a). The reduction in sinusoidal activity that commensurates with tumor development proved to be unimportant since basolateral entry of E_217G was extremely rapid (50 × flow rate), and normally, CL_{in}^{H} was extremely fast and not the rate-limiting step. CL would remain relatively constant unless CL_{in}^{H} was significantly reduced by > 60% (Fig. 5.5b). With tumor metastasis, however, induction of Sult1e1 increased formation of the sulfate metabolite, E_23S17G, and evoked a compensatory decrease in the biliary excretion of E_217G (Fig. 5.5c), even though the activities of Mrp2 that was responsible for the excretion of both E_217G and E_23S17G were unimpaired. An increase in one pathway (formation of the 3-sulfate metabolite) evoked a compensatory decrease in the biliary excretion of E_217G. The observations illustrate nicely the interplay between the enzyme and the canalicular transporter.

5.6 PBPK Modeling of Renal Drug Clearance

The kidney is an organ endowed with excretory mechanisms as well as enzymes capable of drug biotransformation. The inadequate excretion of drugs in the kidney is evidenced by the serious complications of aminoglycosides such as gentamicin and amikacin on nephrotoxicity and ototoxicity since a large proportion of the intravenously administered dose accumulates in the kidney (about 10% of dose), whereas little distribution of aminoglycosides to other tissues is observed (Nagai and Takano, 2004). Aminoglycosides are eliminated by glomerular filtration (Martínez-Salgado et al., 2007), and a fraction is reabsorbed in the proximal tubule. Aminoglycosides are taken up in epithelial cells of the renal proximal tubule and stay inside there for a long time, resulting in nephrotoxicity (Nagai and Takano, 2004). Polycationic

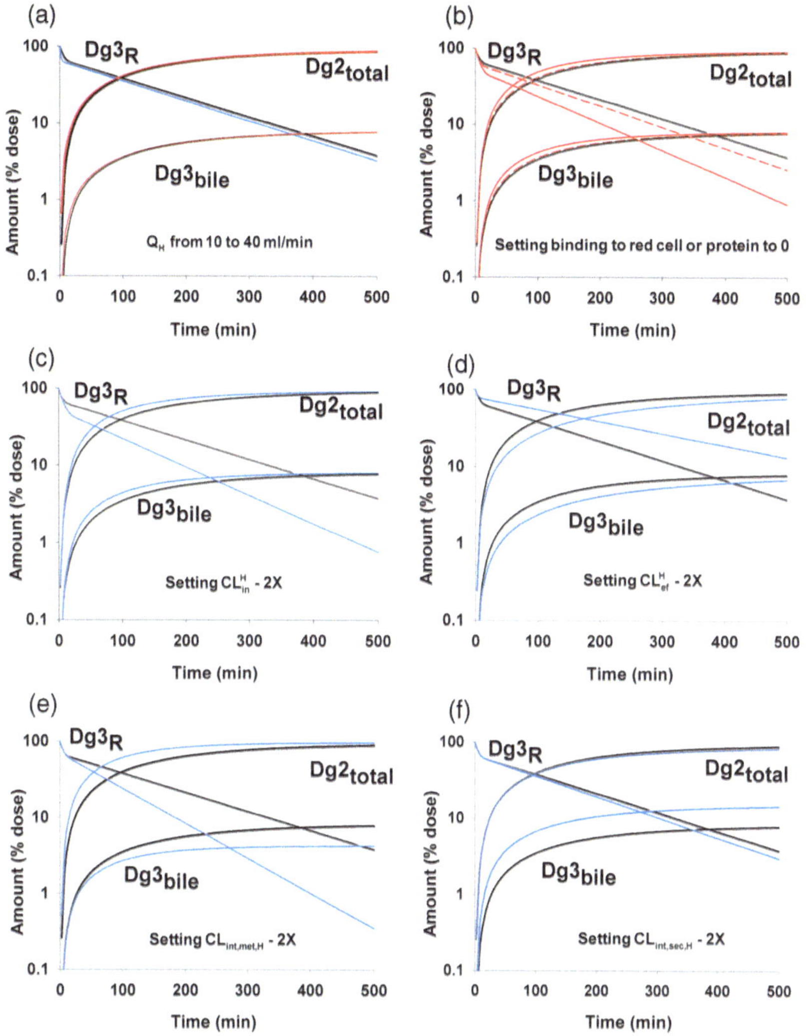

Fig. 5.4 Simulations of the effect of flow rate (**a**), red cell and albumin binding (**b**), CL_{in} or influx clearance (**c**), CL_{ef} or efflux clearance (**d**), $CL_{int,met,H}$ (**e**), and $CL_{int,sec,H}$ (**d**) on changes in Dg3 (digoxin) reservoir perfusate ($Dg3_R$), bile ($Dg3_{bile}$), and total metabolite formation ($Dg2_{total}$) with the PBPK model for the rbc–albumin-perfused rat liver preparation (Liu et al., 2005). The simulation was based on physiological parameters obtained from the perfused rat liver preparation on digoxin disposition. (**a**) The flow rate was changed from 10 ml/min (*solid black line*) to 40 ml/min (*blue and red lines*); (**b**) when binding to red blood cell (*red line*, - - - -) or to serum bovine albumin (*red solid line*) was set as zero. Also, parameter values for CL_{in}^{H} (**c**), CL_{ef}^{H} (**d**), $CL_{int,met,H}$ (**e**), and $CL_{int,sec,H}$ (**f**) were doubled those of the rbc-perfused livers. The original, controlled condition was denoted as the *black solid line*; changes were shown in *blue and red* (from Liu et al., 2005 and Liu and Pang, 2006, with permission)

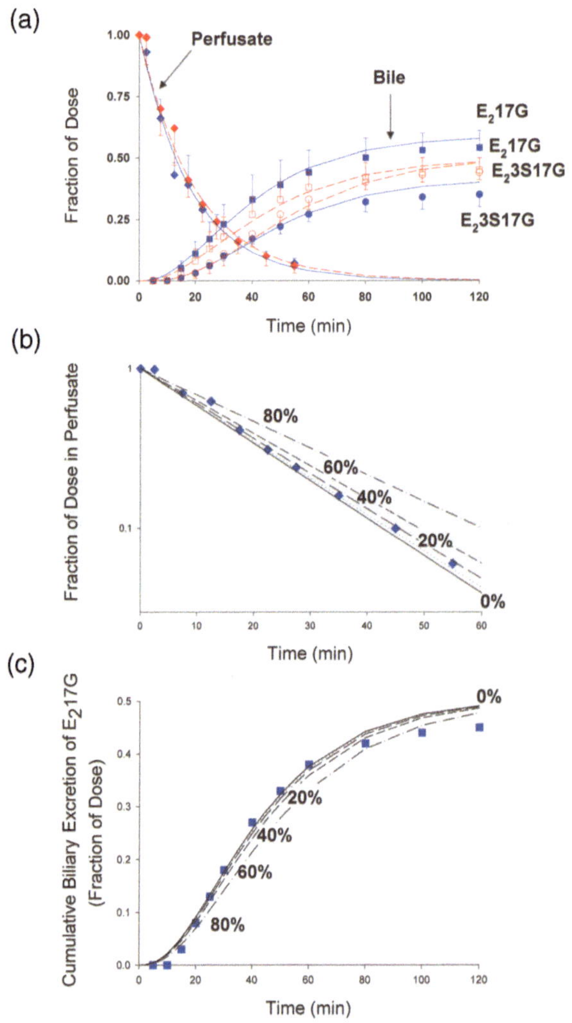

Fig. 5.5 The fates of estradiol-17β-glucuronide (E_217G) and its 3-sulfated metabolite, E_23S17G; both underwent biliary excretion in the perfused rat liver preparations of the sham-operated control and Wag–Rij tumor rats that were inoculated with vehicle or CC-531 cells (4 weeks) (from Sun et al., 2006, with permission). The CL_H was extremely high (approaching flow rate). (**a**) According to PBPK liver modeling, the optimized fit of the data on E_217G in perfusate (diamond symbols) showed rapid decaying and unchanged profiles, whereas the biliary excretion of E_23S17G (●) and E_217G (■) of the sham-operated liver (blue solid symbols) was increased and decreased, respectively in tumor-bearing livers (*red open symbols*). The change corresponded to the increase in protein expression of Sult1e1 as detected by immunoblotting. This resulted in higher E_23S17G formation and excretion into bile, causing a compensatory decrease in the excretion of E_217G in the metastatic tumor group. The 50% reduction in uptake sinusoidal clearance (from 469 to 230 ml/min or > 20× flow rate of 10 ml/min) as a result of tumor failed to affect the CL_H of E_217G, whose entry remained flow-limited. Simulations with the PBPK model predicted that (**b**) perfusate decay and (**c**) biliary excretion of E_217G were slowed down only when CL_{in}^{H} was reduced by >60% (numbers on graphs denote % reduction of CL_{in}^{H}). The data inferred that the sinusoidal entry of E_217G was not the rate-limiting step. Changes from control (*solid line*) were observed upon reduction of CL_{in}^{H} by > 60% (— — —) or 80% (—•—•—) but not for 20% (••••••) or 40% (- - -) reduction in CL_{in}^{H}

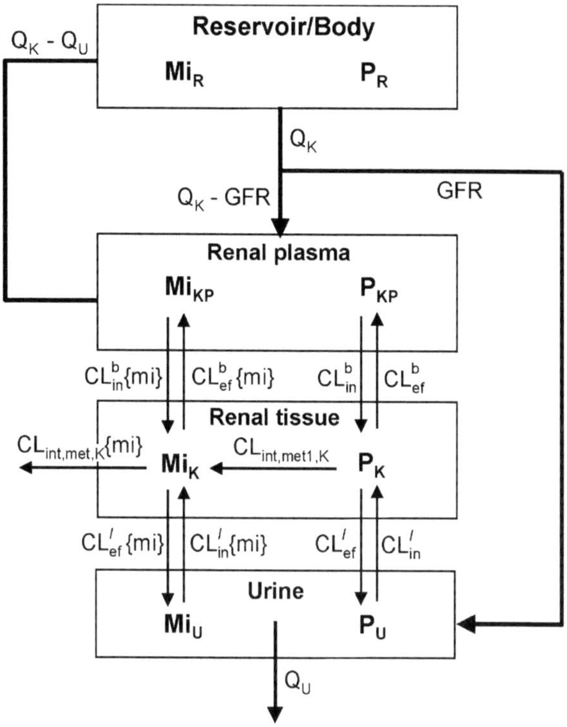

Fig. 5.6 Physiologically-based pharmacokinetic (PBPK) kidney model, with kidney as the only elimination organ. The administered species is filtered and appear in the urine; the plasma flow returning to the venous blood is (Q_K–Q_U). Transport, secretion, metabolism, and reabsorption of drugs and metabolite occur (from Pang et al., 2008, with modifications). See text for description

aminoglycosides bind to anionic, brush border, phospholipid membranes and are transported intracellularly (Swan, 1997).

PBPK modeling of the kidney was developed and described in its simplistic format, as in the isolated perfused kidney preparation (Fig. 5.6). Drug influx and efflux at the basolateral membrane are denoted as CL^b_{in} and CL^b_{ef}, respectively. The renal tubular cells are perfused by plasma that is partially removed due to glomerular filtration (rate is GFR), and the remaining flow perfuses the renal tissue, where metabolism and excretion at the luminal membrane, denoted by $CL_{int,met,K}$ and CL^l_{ef}, respectively, take place. Solutes that are prone to reabsorption are defined by the reabsorptive clearance, CL^l_{in}. The returning flow is (Q_K – Q_U) or the renal plasma flow minus the loss due to urinary flow. The well-known phenomenon of change in pH in the renal tubule that can affect drug reabsorption has also been considered in PBPK modeling (Boom et al., 1994). These PBPK models enabled an accurate description and analysis of the measured drug plasma levels and renal excretion (Russel et al., 1987), saturable secretion and reabsorption (Boom et al., 1994; Masereeuw et al., 1996), and effects of tubular (Russel et al., 1989) or basolateral (Boom et al., 1998) inhibitors such as probenecid that interferes with secretion. The model also views renal metabolism/excretion of a drug in the formation of its metabolite (de Lannoy et al., 1989).

5.6.1 Solutions for AUC and CL_r

The equations for the AUC and renal clearance (CL_r) for the kidney are highly dependent on whether alternate pathways exist for the drug and the metabolite. For a drug that is only renally excreted, the AUC is given by Equation (5.13).

$$AUC_{iv} = \frac{Dose_{iv}\left\{(Q_K - Q_U)[CL_{in}^{l}CL_{ef}^{b} + Q_U(CL_{ef}^{b} + CL_{ef}^{l})] + Q_U f_u CL_{in}^{b}CL_{ef}^{l}\right\}}{Q_U f_u\left[Q_K CL_{in}^{b}CL_{ef}^{l} + GFR(Q_K - Q_U)(CL_{ef}^{b}+CL_{ef}^{l})\right]} \tag{5.13}$$

Note that all of the transport clearances at the basolateral ($CL_{in}^{b}$and CL_{ef}^{b}) and luminal (CL_{in}^{l} and CL_{ef}^{l}) membranes, GFR, and the flow terms, Q_K and Q_U, are present in the solution of AUC, and renal clearance, CL_r, is

$$CL_r = \frac{Q_U f_u\left[Q_K CL_{in}^{b}CL_{ef}^{l} + GFR(Q_K - Q_U)(CL_{ef}^{b} + CL_{ef}^{l})\right]}{(Q_K - Q_U)[CL_{in}^{l}CL_{ef}^{b} + Q_U(CL_{ef}^{b} + CL_{ef}^{l})] + Q_U f_u CL_{in}^{b}CL_{ef}^{l}} \tag{5.14}$$

and the ratio of the CL_r/filtration clearance or $CL_r/(f_p GFR)$ is the fractional excretion (FE) or excretion ratio (ER).

When the drug further suffers metabolism, with the metabolic intrinsic clearance $CL_{int,met,K}$, the AUC_{iv} and CL_r are

$$AUC_{iv} = \frac{Dose_{iv}\left\{(Q_K - Q_U)\left[CL_{in}^{l}(CL_{ef}^{b} + CL_{int,met,K}) + Q_U(CL_{ef}^{b} + CL_{ef}^{l} + CL_{int,met,K})\right] + f_u CL_{in}^{b}\left[Q_U CL_{ef}^{l} + CL_{int,met,K}(CL_{in}^{l} + Q_U)\right]\right\}}{GFR f_u(Q_K - Q_U)\left[CL_{int,met,K}(CL_{in}^{l} + Q_U) + Q_u(CL_{ef}^{b} + CL_{ef}^{I})\right] + Q_K f_u CL_{in}^{b}\left[Q_U CL_{ef}^{l} + CL_{int,met,K}(CL_{in}^{l} + Q_U)\right]} \tag{5.15}$$

$$CL_r = \frac{GFR f_u(Q_K - Q_U)\left[CL_{int,met,K}(CL_{in}^{l} + Q_U) + Q_u(CL_{ef}^{b} + CL_{ef}^{l})\right] + Q_K f_u CL_{in}^{b}\left[Q_U CL_{ef}^{l} + CL_{int,met,K}(CL_{in}^{l} + Q_U)\right]}{(Q_K - Q_U)\left[CL_{in}^{l}(CL_{ef}^{b} + CL_{int,met,K}) + Q_U(CL_{ef}^{b} + CL_{ef}^{l} + CL_{int,met,K})\right] + f_u CL_{in}^{b}\left[Q_U CL_{ef}^{l} + CL_{int,met,K}(CL_{in}^{l} + Q_U)\right]} \tag{5.16}$$

Under this instance, the total renal clearance exceeds the urinary clearance and is the sum of the renal metabolic and excretory clearances. Values of FE would be underestimated in the presence of renal metabolism (Sirianni and Pang, 1997).

Again, this PBPK modeling approach accounts for sequential metabolism and excretion readily despite that both models are well-stirred and assume venous equilibration, namely, the concentration of drug leaving the organ is the same as that in the outflow blood. Solutions for AUC{mi,P} showed that flow, binding, and transport and intrinsic clearances of both the metabolite and the drug are determinants of clearance of the renally formed metabolite (Pang et al., 2008). However, the solutions were too big to be presented, regardless of whether the formed metabolite is further metabolized by the kidney or not (Pang et al., 2008).

5.6.2 Examples

PBPK modeling of secretory and metabolic events in the kidney has generated concepts on transporters for uptake and reabsorption, GFR, as well as metabolism. Upon simultaneous administration of tracer [^{14}C]benzoate and preformed [^{3}H]hippurate to the isolated perfused rat kidney, venous outflow, and urinary concentrations of [^{14}C]benzoate, formed [^{14}C]hippurate, and [^{3}H]hippurate from the single-pass preparation were well predicted by the physiological model of the kidney (Geng and Pang, 1999). The extraction of the administered metabolite, E{hippurate} = (0.24), was lower than that of the hippurate formed from benzoate (E{formed hippurate} or 0.39). The discrepancy was explained by the PBPK model, that, due to facile formation of hippurate from benzoate and a faster luminal efflux of hippurate, E{formed hippurate} > E{hippurate}.

The renal PBPK model was able to describe the metabolism and excretion of the ACE inhibitor, enalapril, and explained the observed discrepancies between [^{14}C]-generated and preformed [^{3}H]enalaprilat (metabolite) elimination in the constant flow, single-pass and recirculating isolated perfused rat kidney preparations (IPKs). The difference: a high extraction of the formed [^{14}C]enalaprilat vs. a low extraction for preformed [^{3}H]enalaprilat was the result of the different origins of the metabolites – formed within the kidney or entering into the kidney from blood. The finding suggests the presence of a barrier for the entry of preformed enalaprilat into the kidney, as also found for the liver (de Lannoy and Pang, 1987; 1993; de Lannoy et al., 1989; 1993; Schwab et al., 1990; 1992). PBPK modeling further divulged an interesting phenomenon for the excretion of a metabolite within its formation organ: for the preformed metabolite, there was a relatively constant renal clearance of preformed [^{3}H]enalaprilat but there existed a greater, time-dependent renal clearance for formed [^{14}C]enalaprilat (de Lannoy et al., 1989; 1993). The time-dependent renal clearance of formed [^{14}C]enalaprilat was due to two components contributing to the renal clearance: one from the [^{14}C]enalaprilat in circulation and the other from [^{14}C]enalaprilat formed in situ the kidney that was immediately excreted (de Lannoy et al., 1989; 1990; 1993). The conventional method of estimating urinary clearance for the metabolite [(total) excretion rate/midpoint plasma metabolite concentration] results in a greater metabolite clearance than that expected from the administration of preformed metabolite (de Lannoy et al., 1989).

Concepts on metabolism and excretion in the kidney have unraveled the concept that the presence of competing pathways, enzymatic or secretory, tends to decrease substrate concentration in the organ and reduces the apparent clearance of the pathway (Sirianni and Pang, 1997). Metabolism and reabsorptive transporters decrease the observed renal clearance or FE. A decrease in the enalapril metabolism by paraoxon, an inhibitor of ester hydrolysis, in the single-pass isolated perfused kidney resulted in doubling of the apparent urinary clearance and FE. The observation was due to a greater concentration of substrate that becomes available for more excretion (Sirianni and Pang, 1999). When facile metabolism ($CL_{int,met,K}$) for benzoate (Geng and Pang, 1999) was set as zero, the value of FE predicted by the PBPK model was found to dramatically increase from 0.27 to 4.

5.7 Models for Intestinal Drug Clearance

Intestinal absorption is affected by both drug characteristics and physiology of the gastrointestinal tract (GIT). The pKa of the weak acid or conjugate acid of the weak base determines the extent of ionization at various pH values (pH 1.3 for stomach and 6 for intestine) and the extent of passive drug absorption. It is generally considered that lipophilicity, often defined by the organic:water partition coefficient, is a major determinant of extent of transmembrane permeation when aqueous solubility is adequate and when the unstirred water layer is not an imposing barrier (Ungell et al., 1998). Drugs that are unionized or undergo hydrogen bonding exhibit a much greater lipophilicity toward membrane permeation than the ionic counterparts. Agents that exhibit extremely high lipophilicity do not traverse the unstirred water layer well, whereas hydrophilic agents experience the lipoidal membrane as an austere barrier. Drugs possessing good hydrophilic and lipophilic balance penetrate the unstirred water layer and membrane equally well; in this case, the blood flow rate poses as the rate-limiting step for transmembrane transport.

Drug absorption occurs mostly in the small intestine because of the villi and microvilli that greatly magnify the surface area many-fold, especially for the duodenum and jejunum whose surface area is larger than the ileum (Magee and Dalley, 1986). As in the liver and kidney, intestinal enzymes and transporters are abundant within enterocytes and these would reduce the extent of absorption. The effects of drug secretion may be negated if rapid reabsorption takes place (Lin et al., 1999; Sun and Pang, 2009a). However, a faster GI transit reduces the extent of reabsorption and promotes irreversible drug loss, whereas an increased intestinal residence time allows time for the drug to dwell longer in the intestinal lumen for absorption.

A simple PBPK model for the intestine is based on the tissue being the only eliminating organ, in a situation akin to the recirculating perfused intestinal preparation (Fig. 5.7) (Cong et al., 2000). The unbound drug is absorbed by passive diffusion or relies on transporters for entry; the summative process is described by the rate constant of absorption, k_a. The intestine is comprised of three subcompartments:

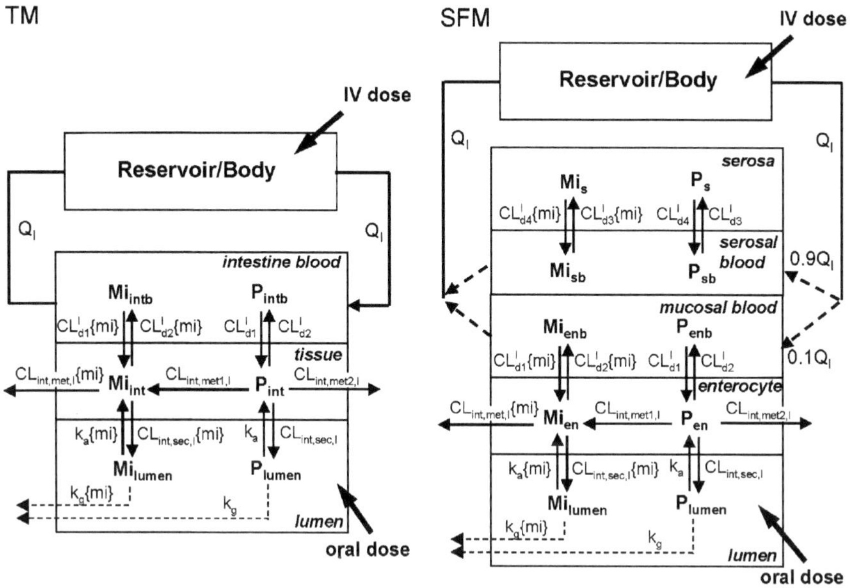

Fig. 5.7 Physiologically based pharmacokinetic (PBPK) intestine models, the traditional model (TM), and the segregated flow model (SFM), with the intestine as the only elimination tissue. For the SFM, the intestine tissue is divided into the enterocyte (en) region and the serosal (s) region. See text for description (from Cong et al., 2000, with modifications)

with intestinal tissue, intestinal blood, and the intestinal lumen. The body or reservoir (R, or blood compartment) and intestine compartments are interconnected by the intestinal or portal venous blood flow Q_I or Q_{PV}, and the volumes are V_R and V_{int}, respectively. V_{intb}, V_{int}, and V_{lumen} represent volumes of the various intestine subcompartments, intestine blood, intestine, and lumen, respectively. The unbound fractions in plasma and intestine are denoted by subscripts "p" and "int," respectively. Recognition that the apical membrane of the intestine contains transporters for absorption (denoted by k_a) or secretion ($CL_{int,sec,I}$) or exhibit membrane-limited transport has led to concepts of influx (CL^I_{d1}) and efflux (CL^I_{d2}) clearances, sums of passive diffusive clearance, and the carrier-mediated transport clearance that facilitate drug entry and efflux at the basolateral membrane. The rate constant, k_g, represents the net loss in lumen, either due to degradation or ineffective absorption; F_{abs} is the net fraction of dose absorbed into the superior mesenteric artery and is given by the ratio, $k_a/(k_a+k_g)$. These various apical transporters for organic anions and cations have been reviewed (see Chapter 2). Within the intestine, the drug is metabolized by enzymes of intrinsic clearance, $CL_{int,met1,I}$ (V_{max}/K_m for first-order conditions) that forms the primary metabolite, Mi_{int}, and $CL_{int,met2,I}$ that forms other primary metabolites; luminal secretion of the drug is a function of the secretory intrinsic clearance, $CL_{int,sec,I}$. The intestinally formed metabolite, Mi_{int}, may be further metabolized to other species with the metabolic intrinsic clearance of the metabolite, $CL_{int,met,I}\{mi\}$, and/or secreted into lumen with the secretory intrinsic clearance, $CL_{int,sec,I}\{mi\}$. The metabolite enters and leaves the cell to enter blood at the basolateral membrane with transport clearances, $CL_{d1}^{I}\{mi\}$ and $CL_{d2}^{I}\{mi\}$.

The simple PBPK model, also known as the traditional model (TM), was first introduced to describe the metabolism of morphine to morphine 6-glucuronide in the perfused small intestine preparation (Cong et al., 2000). The phenomenon, known as route-dependent intestine metabolism, reveals a greater extent of intestinal metabolism for drugs given orally than those given systemically (Cong et al., 2000; Pang, 2003). The TM was modified to yield the segregated flow model (SFM) (Fig. 5.7b) that consists of a split flow to the enterocytic region (Q_{en}, varying from 5 to 30% of total intestinal flow) and the remaining flow to the serosal region (Q_s). The corresponding influx (CL_{d3}^{I}) and efflux (CL_{d4}^{I}) clearances facilitate drug entry and efflux at the serosal membrane (Cong et al., 2000).

5.7.1 Solutions for AUC and CL_I

The inception of the TM and SFM had led to the development of a more flexible model that incorporates alternate metabolism of the drug to form other metabolites with $CL_{int,met2,I}$ and sequential metabolism of the metabolite with $CL_{int,met,I}\{mi\}$. Corresponding AUCs for the drug and metabolite were obtained (Sun and Pang, 2009a; 2009b). The area under the curve for drug after po administration is the same for both the TM and SFM.

$$AUC_{po}^{TM\&SFM} = \frac{F_{abs} dose_{po} CL_{d2}^{I}}{CL_{d1}^{I}\left[(1 - F_{abs})\, CL_{int,sec,I} + CL_{int,met1,I} + CL_{int,met2,I}\right]} \tag{5.17}$$

whereas the equations for the AUC_{iv} for TM and SFM differed only for the flow terms, Q_{PV} or Q_I for the TM and Q_{en} for the SFM.

$$AUC_{iv}^{TM} = \frac{Dose_{iv}\left[(1 - F_{abs})CL_{int,sec,I}(Q_I + CL_{d1}^{I}) + Q_I(CL_{d1}^{I} + CL_{int,met,I}) + CL_{d1}^{I}CL_{int,met,I}\right]}{Q_I CL_{d1}^{I}\left[(1 - F_{abs})CL_{int,sec,I} + CL_{int,met,I} + CL_{int,met2,I}\right]} \tag{5.18}$$

$$AUC_{iv}^{SFM} = \frac{Dose_{iv}\left[(1 - F_{abs})CL_{int,sec,I}(Q_{en} + CL_{d1}^{I}) + Q_{en}(CL_{d1}^{I} + CL_{int,met,I}) + CL_{d1}^{I}CL_{int,met,I}\right]}{Q_{en} CL_{d1}^{I}\left[(1 - F_{abs})CL_{int,sec,I} + CL_{int,met,I} + CL_{int,met2,I}\right]} \tag{5.19}$$

Several notable observations may be made. First, the transport clearances ($CL_{d3}^{I}$and CL_{d4}^{I}) for the serosal compartment, a non-eliminating compartment, are irrelevant and absent in the equations (Equations 5.17 and 5.19). Second, the intrinsic secretory clearance of drug was modified by ($1-F_{abs}$), suggesting that if k_a is high and k_g is low, secretion would become nil as F_{abs} approaches 1 [since F_{abs} is $k_a/(k_a+k_g)$].

For the formed metabolite, the $AUC_{iv}\{mi,P\}$ are identical for both the TM and SFM. The equation for $AUC_{po}\{mi,P\}$ is similar, except that F_{abs} is present. The area under the curve for the metabolite is influenced by the effective secretory clearance and metabolic intrinsic clearances of drug and the metabolite, as well as the formation intrinsic clearance, $CL_{int,met1,I}$, and the transport clearance of the metabolite, $CL^{I}_{d1}\{mi\}$ and $CL^{I}_{d2}\{mi\}$.

$$AUC_{po}\{mi,P\}^{TM\&SFM} = \frac{Dose_{po}F_{abs}CL_{int,met1,I}CL^{I}_{d2}\{mi\}}{\left[(1-F_{abs})CL_{int,sec,I}+CL_{int,met1,I}+CL_{int,met2,I}\right]CL^{I}_{d1}\{mi\}\left[(1-F_{abs}\{mi\})CL_{int,sec,I}\{mi\}+CL_{int,met,I}\{mi\}\right]} \tag{5.20}$$

$$AUC_{iv}\{mi,P\}^{TM\&SFM} = \frac{Dose_{iv}CL_{int,met1,I}CL^{I}_{d2}\{mi\}}{\left[(1-F_{abs})CL_{int,sec,I}+CL_{int,met1,I}+CL_{int,met2,I}\right]CL^{I}_{d1}\{mi\}\left[(1-F_{abs}\{mi\})CL_{int,sec,I}\{mi\}+CL_{int,met,I}\{mi\}\right]} \tag{5.21}$$

5.7.2 Examples

The SFM is needed to describe the greater extent of intestinal metabolism observed for drugs given orally over intravenously (Pang, 2003). The SFM model was found to be superior over the TM in explaining the absence of morphine glucuronidation when the compound given systemically to the perfused rat intestine preparation, whereas marked formation of morphine glucuronide occurred when given orally (Fig. 5.8). The concept is being adopted by various simulation packages, including Simcyp (Yang et al., 2007), in which a reduced intestinal flow rate was utilized to predict intestinal clearance. Due to enzyme and transporter heterogeneity, the TM and SFM have been expanded to include the three segmental regions – the duodenum, jejunum, and ileum in the segmental segregated flow model (SSFM) and the segmental, traditional model (STM) (Tam et al., 2003, Liu et al., 2006). This type of modeling has been adopted by Badhan et al. (2009) to simulate intestinal absorption in man.

5.8 Whole Body PBPK Modeling

One of the limitations of organ PBPK models is that the body consists of multiple organs, and it would be prudent to examine PBPK modeling of enzymes and transporters in the whole body. The renal clearances for the parent drug (CL_r) and the

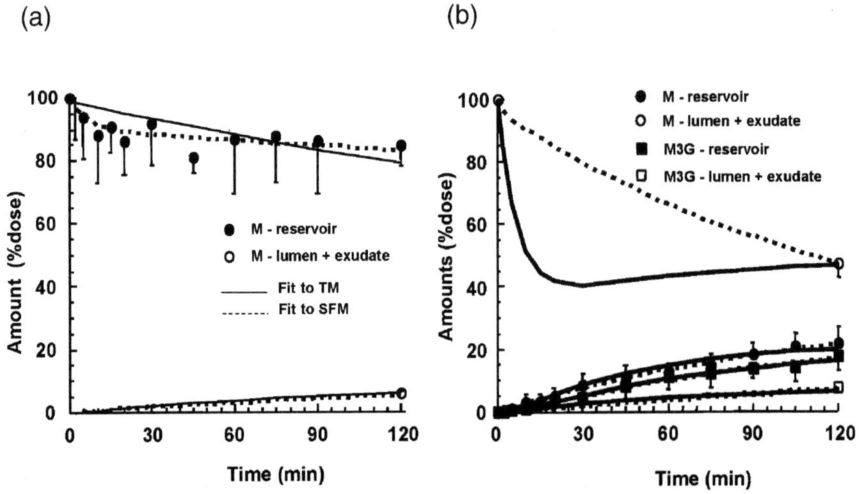

Fig. 5.8 Fit of the data of morphine (**M**) and its metabolite, morphine 6-glucuronide (M6G) following intravenous (a) and oral (b) administration of morphine in the recirculating perfused rat intestine preparation to the TM and the SFM. Glucuronidation of morphine was absent with systemic dosing (**a**), whereas the metabolite, morphine 6-glucuronide, was detected in both lumen and perfusate after oral dosing (**b**). The SFM fitted the data better than the TM. See text for description (from Cong et al., 2000, with modifications)

metabolite ($CL_r\{mi\}$) are placed in the central or blood compartment for simplification; the influence of renal transporters or enzymes was not considered in these examples. Due to the complexity in the solutions for AUC, PBPK models for renally excreted drugs that also undergo intestine (case I) or liver (case II) metabolism only were considered (Sun and Pang, 2009b). Otherwise, even when solutions are found, these would be too lengthy to be expressed concisely. More general PBPK intestinal and kidney (Fig. 5.9a) and liver and kidney (Fig. 5.9b) models that consider transporters and enzymes for the handling of a drug and its metabolite with competing pathways present for the drug and metabolite are presented. The drug forms more than one metabolite that may undergo sequential metabolism.

Solutions for the AUCs of the drug and metabolite for case I and case II are summarized in Tables 5.1 and 5.2 respectively. For case I, renal excretion and intestinal metabolism exist. It is conceivable that examples may be found for the scenario of intestinal metabolism only. The AUC_{iv} and AUC_{po} are found influenced by the intestinal transport clearances for drug CL^{I}_{d1} and CL^{I}_{d2}, the intestinal drug metabolic ($CL_{int,met1,I}$ and $CL_{int,met2,I}$) and secretory clearance ($CL_{int,sec,I}$), and the intestinal flow (Q_{pv}) and the renal clearance (CL_r). $CL_{int,sec,I}$ was reduced by reabsorption, denoted by the fraction, $1–F_{abs}$; and the AUC_{po} was further affected by F_{abs}. The area under the curve of the intestinally formed metabolite (AUC{mi,P}) following po and iv drug dosing is influenced by the parent drug transport, metabolic, and excretory parameters, plus those of the metabolite: the (basolateral) transport clearances of metabolite $CL^{I}_{d1}\{mi\}$ and $CL^{I}_{d2}\{mi\}$, the metabolic ($CL_{int,met,I}\{mi\}$)

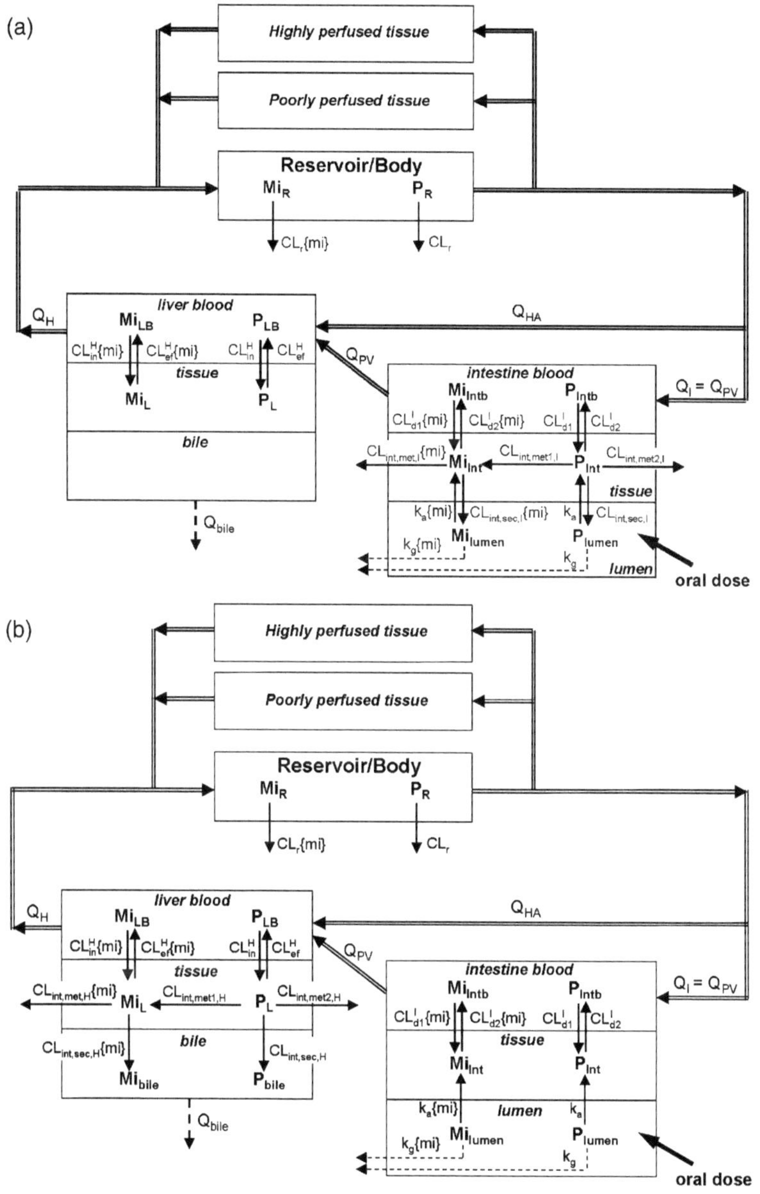

Fig. 5.9 Whole body, physiologically based pharmacokinetic (PBPK) model, with drug and metabolite being excreted by the kidneys with CL_r and $CL_r\{mi\}$, and metabolized only by the intestine (**a**) or liver (**b**). Sequential metabolism of the formed metabolite occurs with the organ of formation, and the drug may be further metabolized to other metabolites. The intestinal traditional model (TM) was presented here pictorially. See text for description (from Sun and Pang, 2009b, with modifications)

Table 5.1 Solutions for AUCs for whole body PBPK models, case I: intestinal metabolism only (Fig. 5.9a). Equations for AUC_{iv} differed only for the flow terms, Q_{PV} or Q_I for the TM and Q_{en} for the SFM (taken from Sun and Pang, 2009b)

PBPK modeling, case IA CL_r and $CL_r\{mi\} > 0$; F_{abs} is the ratio of $k_a/(k_a+k_g)$
$AUC_{po} = \dfrac{F_{abs} Dose_{po} Q_{PV} CL^{I}_{d2}}{CL_r Q_{PV} CL^{I}_{d2} + \left[CL_r Q_{PV} + CL^{I}_{d1}(CL_r + Q_{PV})\right]\left[CL_{int,met1,I} + CL_{int,met2,I} + (1 - F_{abs}) CL_{int,sec,I}\right]}$
$AUC_{iv} = Dose_{iv} \dfrac{Q_{PV} CL^{I}_{d2} + (Q_{PV} + CL^{I}_{d1})\left[CL_{int,met1,I} + CL_{int,met2,I} + (1 - F_{abs}) CL_{int,sec,I}\right]}{CL_r Q_{PV} CL^{I}_{d2} + \left[CL_r Q_{PV} + CL^{I}_{d1}(CL_r + Q_{PV})\right]\left[CL_{int,met1,I} + CL_{int,met2,I} + (1 - F_{abs}) CL_{int,sec,I}\right]}$
$AUC_{po}\{mi,P\} = \dfrac{F_{abs} Dose_{po} CL_{int,met1,I}\left[CL^{I}_{d1}(CL_r + Q_{PV}) + CL_r Q_{PV}\right]}{CL_r Q_{PV} CL^{I}_{d2} + \left[CL_r Q_{PV} + CL^{I}_{d1}(CL_r + Q_{PV})\right]\left[CL_{int,met1,I} + CL_{int,met2,I} + (1 - F_{abs}) CL_{int,sec,I}\right]} \times$
$\dfrac{Q_{PV} CL^{I}_{d2}\{mi\}}{CL_r\{mi\} Q_{PV} CL^{I}_{d2}\{mi\} + \left[CL_r\{mi\} Q_{PV} + CL^{I}_{d1}\{mi\}(CL_r\{mi\} + Q_{PV})\right]\left[CL_{int,met,I}\{mi\} + (1 - F_{abs}\{mi\}) CL_{int,sec,I}\{mi\}\right]}$
$AUC_{iv}\{mi,P\} = \dfrac{Dose_{iv} CL_{int,met1,I} Q_{PV} CL^{I}_{d1}}{CL_r Q_{PV} CL^{I}_{d2} + \left[CL_r Q_{PV} + CL^{I}_{d1}(CL_r + Q_{PV})\right]\left[CL_{int,met1,I} + CL_{int,met2,I} + (1 - F_{abs}) CL_{int,sec,I}\right]} \times$
$\dfrac{Q_{PV} CL^{I}_{d2}\{mi\}}{CL_r\{mi\} Q_{PV} CL^{I}_{d2}\{mi\} + \left[CL_r\{mi\} Q_{PV} + CL^{I}_{d1}\{mi\}(CL_r\{mi\} + Q_{PV})\right]\left[CL_{int,met,I}\{mi\} + (1 - F_{abs}\{mi\}) CL_{int,sec,I}\{mi\}\right]}$

Table 5.1 (continued)

PBPK modeling, case IB – CL_r and $CL_r\{mi\} = 0$

$$AUC_{po} = \frac{F_{abs} Dose_{po} CL^{I}_{d2}}{CL^{I}_{d1}\left[CL_{int,met1,I} + CL_{int,met2,I} + (1 - F_{abs}) CL_{int,sec,I}\right]}$$

$$AUC_{iv} = Dose_{iv} \frac{Q_{PV} CL^{I}_{d2} + (CL^{I}_{d1} + Q_{PV})\left[CL_{int,met1,I} + CL_{int,met2,I} + (1 - F_{abs}) CL_{int,sec,I}\right]}{Q_{PV} CL^{I}_{d1}\left[CL_{int,met1,I} + CL_{int,met2,I} + (1 - F_{abs}) CL_{int,sec,I}\right]}$$

$$AUC_{po}\{mi,P\} = \frac{F_{abs} Dose_{po} CL_{int,met1,I} CL^{I}_{d2}\{mi\}}{\left[CL_{int,met1,I} + CL_{int,met2,I} + (1 - F_{abs}) CL_{int,sec,I}\right]\left[CL_{int,met,I}\{mi\} + (1 - F_{abs}) CL_{int,sec,I}\{mi\}\right] CL^{I}_{d1}\{mi\}}$$

$$AUC_{iv}\{mi,P\} = \frac{Dose_{iv} CL_{int,met1,I} CL^{I}_{d2}\{mi\}}{\left[CL_{int,met,I}\{mi\} + (1 - F_{abs}) CL_{int,sec,I}\{mi\}\right]\left[CL_{int,met1,I} + CL_{int,met2,I} + (1 - F_{abs}) CL_{int,sec,I}\right] CL^{I}_{d1}\{mi\}}$$

Table 5.2 Solutions for AUCs for whole body PBPK models, case II: liver metabolism only (Fig. 5.9b) (taken from Sun and Pang, 2009b)

$$Q_{\mathrm{H}} = Q_{\mathrm{HA}} + Q_{\mathrm{PV}};\ \mathrm{CL_{int,H}} = \mathrm{CL_{int,met1,H}} + \mathrm{CL_{int,met2,H}} + \mathrm{CL_{int,sec,H}}$$

PBPK modeling, case IIA – $\mathrm{CL_r}$ and $\mathrm{CL_r\{mi\}} > 0$

$$\mathrm{AUC_{po}} = \frac{\mathrm{F_{abs}Dose_{po}}Q_{\mathrm{H}}(\mathrm{CL_{ef}^{H}} + \mathrm{CL_{int,H}})}{\mathrm{CL_r}Q_{\mathrm{H}}(\mathrm{CL_{ef}^{H}} + \mathrm{CL_{int,H}}) + \mathrm{CL_{in}^{H}CL_{int,H}}(\mathrm{CL_r} + Q_{\mathrm{H}})}$$

$$\mathrm{AUC_{iv}} = \frac{\mathrm{Dose_{iv}}\left[Q_{\mathrm{H}}(\mathrm{CL_{ef}^{H}} + \mathrm{CL_{int,H}}) + \mathrm{CL_{in}^{H}CL_{int,H}}\right]}{\mathrm{CL_r}Q_{\mathrm{H}}(\mathrm{CL_{ef}^{H}} + \mathrm{CL_{int,H}}) + \mathrm{CL_{in}^{H}CL_{int,H}}(\mathrm{CL_r} + Q_{\mathrm{H}})}$$

$$\mathrm{AUC_{po}\{mi,P\}} = \frac{F_{\mathrm{abs}}\mathrm{Dose_{po}}Q_{\mathrm{H}}\mathrm{CL_{in}^{H}CL_{int,met1,H}}(\mathrm{CL_r} + Q_{\mathrm{H}})}{\mathrm{CL_r}Q_{\mathrm{H}}(\mathrm{CL_{ef}^{H}} + \mathrm{CL_{int,H}}) + \mathrm{CL_{in}^{H}CL_{int,H}}(\mathrm{CL_r} + Q_{\mathrm{H}})} \times$$

$$\frac{\mathrm{CL_{ef}^{H}\{mi\}}}{\mathrm{CL_r\{mi\}}Q_{\mathrm{H}}(\mathrm{CL_{d2}^{H}\{mi\}} + \mathrm{CL_{int,H}\{mi\}}) + \mathrm{CL_{in}^{H}\{mi\}CL_{int,H}\{mi\}}(\mathrm{CL_r\{mi\}} + Q_{\mathrm{H}})}$$

$$\mathrm{AUC_{iv}\{mi,P\}} = \frac{\mathrm{Dose_{iv}}Q_{\mathrm{H}}\mathrm{CL_{in}^{H}CL_{int,met1,H}}}{\mathrm{CL_r}Q_{\mathrm{H}}(\mathrm{CL_{ef}^{H}} + \mathrm{CL_{int,,H}}) + \mathrm{CL_{in}^{H}CL_{int,H}}(\mathrm{CL_r} + Q_{\mathrm{H}})} \times$$

$$\frac{Q_{\mathrm{H}}\mathrm{CL_{ef}^{H}\{mi\}}}{\mathrm{CL_r\{mi\}}Q_{\mathrm{H}}(\mathrm{CL_{ef}^{H}\{mi\}} + \mathrm{CL_{int,H}\{mi\}}) + \mathrm{CL_{in}^{H}\{mi\}CL_{int,H}\{mi\}}(\mathrm{CL_r\{mi\}} + Q_{\mathrm{H}})}$$

PBPK modeling, case IIB – $\mathrm{CL_r}$ and $\mathrm{CL_r\{mi\}} = 0$

$$\mathrm{AUC_{po}} = \frac{F_{\mathrm{abs}}\mathrm{Dose_{po}}(\mathrm{CL_{ef}^{H}} + \mathrm{CL_{int,H}})}{\mathrm{CL_{in}^{H}CL_{int,H}}}$$

$$\mathrm{AUC_{iv}} = \frac{\mathrm{Dose_{iv}}\left[Q_{\mathrm{H}}(\mathrm{CL_{ef}^{H}} + \mathrm{CL_{int,H}}) + \mathrm{CL_{in}^{H}CL_{int,H}}\right]}{Q_{\mathrm{H}}\mathrm{CL_{in}^{H}CL_{int,H}}}$$

$$\mathrm{AUC_{po}\{mi,P\}} = \frac{F_{\mathrm{abs}}\mathrm{Dose_{po}CL_{int,met1,H}CL_{ef}^{H}\{mi\}}}{\mathrm{CL_{int,H}CL_{in}^{H}\{mi\}CL_{int,H}\{mi\}}}$$

$$\mathrm{AUC_{iv}\{mi,P\}} = \frac{\mathrm{Dose_{iv}CL_{int,met1,H}CL_{ef}^{H}\{mi\}}}{\mathrm{CL_{int,H}CL_{in}^{H}\{mi\}CL_{int,H}\{mi\}}}$$

and the secretory ($\mathrm{CL_{int,sec,I}\{mi\}}$) intrinsic clearances of the metabolite; it is further noted that $[1-F_{\mathrm{abs}}\{\mathrm{mi}\}]$ or $[k_{\mathrm{g}}\{\mathrm{mi}\}/((k_{\mathrm{a}}\{\mathrm{mi}\}+k_{\mathrm{g}}\{\mathrm{mi}\})]$, appeared as a product with $\mathrm{CL_{int,sec,I}\{mi\}}$ to denote the net secretion intrinsic clearance of *M*i. The solutions become much simplified when there is no excretion of the drug or the metabolite ($\mathrm{CL_r}$ and $\mathrm{CL_r\{mi\}} = 0$; case IB, Table 5.1). Although Table 5.1 expresses only the condition for the TM, solutions for the SFM are similar. However, Q_{en} should be substituted for Q_{PV} in the equations for the SFM.

The second case of metabolism in the liver only with excretion in the kidneys is a more common occurrence for many drugs (case II). For a drug that undergoes liver metabolism and renal excretion, solutions for AUC_{po} and AUC_{iv} and $AUC_{po}\{mi\}$ and $AUC_{iv}\{mi\}$ were found (Sun and Pang, 2009b). The total liver blood flow, Q_H, the transport clearances, CL_{d1}^{H} and CL_{d2}^{H}, and the components of hepatic intrinsic clearance, $CL_{int,met1,H}$, $CL_{int,met2,H}$, and $CL_{int,sec,H}$, and CL_r all affect AUC_{iv} and AUC_{po} (Table 5.2; case II). $AUC_{po}\{mi\}$ and $AUC_{iv}\{mi\}$ were additionally affected by parameters pertaining to the metabolite and $F_{abs}\{mi\}$ (Table 5.2). For orally absorbed drugs, the effective oral dose was reduced as $F_{abs}Dose_{po}$, and the solutions were similar.

5.9 Transporter–Enzyme, Transporter–Transporter, and Enzyme–Enzyme Interplay

The equations presented in this chapter and those in Tables 5.1 and 5.2 are useful for visualization of how changes of transporters or enzymes or reduction of their intrinsic clearances affect the area under the curve of drugs and metabolites and in turn, clearance. A clear view of these events allows one to make inferences on the mechanism of drug–drug interactions. A change in one eliminatory mechanism, secretion, or metabolism in the liver alters $CL_{int,H}$, whereas a decrease in $CL_{int,met1,H}$ due to DDI would result in an increase in AUC of the drug, but a decrease in the AUC of the metabolite, AUC{mi,P}, as defined in Equations (5.7) and (5.8) and equations in Tables 5.1 and 5.2. Upon inhibition of the secretory and/or alternate metabolic intrinsic clearance, $CL_{int,met2,H}$, both AUC_{iv} and $AUC_{iv}\{mi,P\}$ are increased (Equation (5.7)) (Liu and Pang, 2006). These observations are predicted outcomes of compensatory pathways (Morris and Pang, 1987; Sirianni and Pang, 1997) and reveal a see-saw phenomenon (Liu and Pang, 2006). Generally speaking, the presence of two enzymes or two pathways for drug metabolism/elimination within an organ compete for the same substrate, and each would mutually decrease the reaction rate of the other, leading one to assume incorrectly that the enzymatic activities are lower. The same holds for competition between an enzyme and apical transporters that compete for the substrate for elimination (Sirianni and Pang, 1997). Inhibition of one pathway would apparently result in higher reaction rates of alternate pathways although no induction exists due to a higher availability of substrate (Morris and Pang, 1987). In addition, an increase in efflux clearance increases the AUC, whereas an increase in the influx clearance would decrease the AUC (Liu and Pang, 2006); concomitantly, metabolite AUC{mi,P}s and the amounts secreted would exhibit opposite trends. Similar solutions and trends are found for the oral case (Sun and Pang, 2009b).

An examination of the changes associated with AUC and AUC{mi,P} allows one to decipher various DDI mechanisms, whether inhibition of enzymes or transporters. This may be illustrated upon examination of the inter-relationship between the hepatic formation intrinsic clearance, $CL_{int,met1,H}$, the secretory intrinsic clearance, $CL_{int,sec,H}$, and the renal clearance, CL_r, for drugs that are metabolized by the liver and excreted by the kidney (Fig. 5.10, Table 5.2). Inhibition of the metabolite formation intrinsic clearance increased the AUC of drug but decreased AUC{mi,P} whereas a reduction in $CL_{intsec,H}$ would bring about higher AUC and AUC{mi,P} (Fig. 5.10a). A decrease in renal clearance of the drug, CL_r, would also result in increased AUC of drug and AUC{mi,P} (Figs. 5.10b and 5.10d).

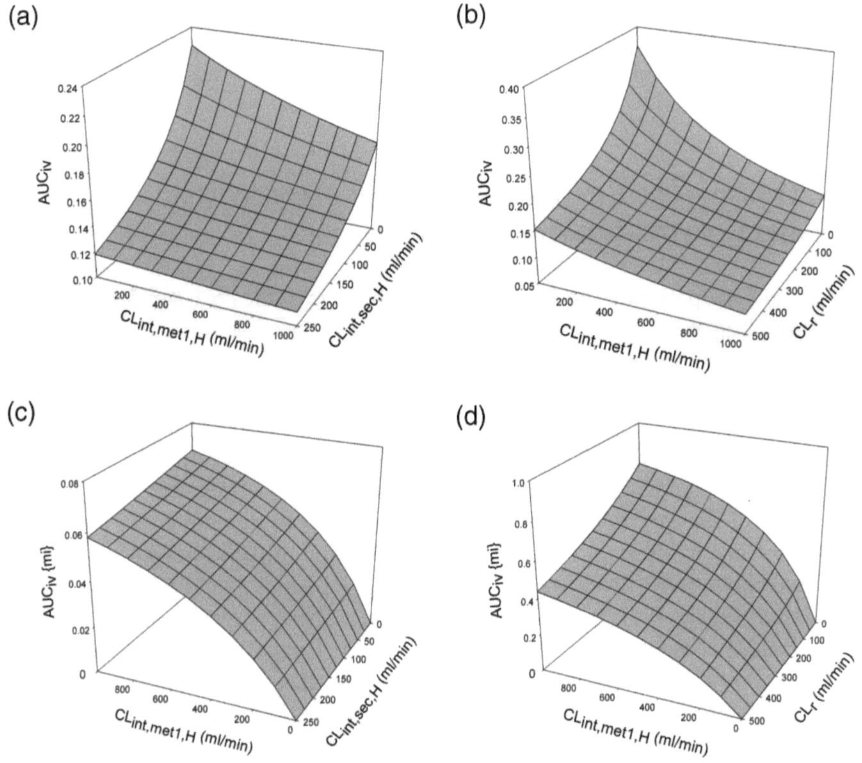

Fig. 5.10 Simulations of the influence of $CL_{int,met1,H}$, $CL_{int,sec,H}$ (**a**) and CL_r (**b**) on the area under the curve for drug (AUC) and AUC{mi,P} (**c**) and (**d**). These simulations were preformed by setting Q_H as 1500 ml/min, $CL_{int,met2,H}$ as 300 ml/min, CL_{in}^{H} and CL_{ef}^{H} as 9000 ml/min, $CL_r\{mi\} = 500$ ml/min, and $CL_{in}^{H}\{mi\}$ and $CL_{ef}^{H}\{mi\}$ as 1000 ml/min as constants for the dose of 100 units given intravenously (with equations in Table 5.2, case IIA). For (**a**) and (**c**), the value of CL_r was set as 200 ml/min; For (**b**) and (**d**), $CL_{int,sec,H}$ was set as 50 ml/min

Examples on changes in the AUC of drug and AUC{mi,P} for hepatically formed metabolites due to DDI are found in Table 5.3. These changes encompass an increase in drug AUC but an increase or decrease in AUC{mi,P}, pending on the mechanism of inhibition. With knowledge of how the drug is being handled and taking advantage of the AUC{mi,P} as a metric to the understanding of DDI, the application of the conceptual frameworks outlined in this chapter allows the DDI mechanisms to be deciphered. The success is high when the metabolite formed in question is a major part of the total drug clearance.

Table 5.3 Possible mechanisms due to inhibition of secretion or metabolism

Drug/Metabolite	Inhibitor	Route and Dose	AUC Before	AUC After	AUC{mi,P} Before	AUC{mi,P} After	Suggested inhibitory mechanism	References
Venlafaxine	Terbinafine	Oral (75 mg)	883	2888			CYP2D6	Hynninen et al. (2008)
O-Desmethylvenlafaxane	Voriconazole				2483	1226	CYP3A4, CYP2C9 or CYP2C19	
Cerivastatin	Gemfibrozil	Oral (0.3 mg)	20.9	91.1			CYP2C8 for M-23 formation	Backman et al. (2002)
Cerivastatin lactone					1.9	6.5		
M1 acid					1.3	4.6		
M-23 acid					5.8	1		
Repaglinide	Gemfibrozil	Oral (0.25 mg)	4.9	34.2			CYP2C8 for M4 formation*	Tornio et al. (2008)
M1 (metabolite)					9.8	43.8		
M2 (metabolite)					17.7	30.8		
M4 (metabolite)					12	2.5		
Repaglinide	Telithromycin	Oral (0.25 mg)	3.9	6.9			CYP3A4 for M1 formation*	Kajosaari et al. (2006)
M1 (metabolite)					2.7	1.5		
M2 (metabolite)					5.3	11.4		
M4 (metabolite)					1.3	3.5		
Pioglitazone	Gemfibrozil	Oral (15 mg)	5.25	16.9			CYP2C8	Jaakkola et al. (2005)
M3 (metabolite)					14.5	0		
M5 (metabolite)					24.8	0		
Midazolam	Voriconazole	iv (0.05 mg/kg)	151	534			(Liver) enzyme for alternate metabolism	Saari et al. (2006)
1′α-Hydroxy metabolite					16	27		
Midazolam	Voriconazole	Oral (7.5 mg)	91	855			Intestinal CYPs; enzyme for alternate metabolism	Saari et al. (2006)
1′α-Hydroxy metabolite					24.4	60.8		

Table 5.3 (continued)

Midazolam	Saquinavir	Oral (7.5 mg)	108	559			Intestinal CYP3A4	Palkama et al. (1999)
1′α-Hydroxy metabolite					0.42	0.08		
		Route	AUC		AUC{mi,P}		Suggested inhibitory	
Drug/Metabolite	Inhibitor	and Dose	Before	After	Before	After	mechanism	References
Midazolam	Saquinavir	iv (0.05 mg/kg)	119	296			Liver CYP3A4	Palkama et al. (1999)
1′α-Hydroxy metabolite					0.19	0.08		
Morphine	Quinidine	Oral (30 mg)	40.8	65.1			Intestinal/liver P-gp (and not UGT)	Kharasch et al. (2003)
Morphine-3-glucuronide					14.4	9.0		
Morphine-6-glucuronide					134	80		
Oxycodone	Voriconazole	Oral (10 mg)	6.1	21.8			CYP3A	Hagelberg et al. (2009)
Noroxycodone					6.5	2.1		
Oxymorphone					0.15	1.05		
Noroxymorphone					1.7	0.86		
Alfentanil	Ritonavir	Oral (4.3 and 4.3 mg/kg)	78	182			P-gp and CYP3A4	Kharasch et al. (2008)
Fexofenadine					1010	2780		
Risperidone	Verapamil	Oral (1 mg)	39.4	62.8			mostly P-gp and less CYP3A4	Nakagami et al. (2005)
9-Hydroxyrisperidone					105.9	136.8		
Diltiazem[a]	Resveratrol	Oral (15 mg/kg)	342	547			P-gp and CYP3A4	Hong et al. (2008)
Desacetyldiltiazem					283	439		

*Assuming no change in elimination of formed metabolites
[a]Rat study

These concepts may be applied a priori to clinical observations on DDI, in addition to the seminal equation of Rowland and Matin (1973) on the AUC ratio of drug (inhibited vs. control) being equal to (1 + [I]/K_i), where [I] is the inhibitor concentration and K_i is the inhibition constant. Among this and other reports that proposed further refinement of the equation (Ito et al., 1998; 2005; Wang et al., 2004; Obach et al., 2006; Brown et al., 2005; 2006), there is the uncertainty of what should be selected as the surrogate inhibitor concentration to reflect [I] (Tucker et al., 2001; Ito et al., 2002). The most appropriate, of course, is that concentration at the enzyme site. Moreover, most of these emphasize DDI due to enzymes, and less exists to examine transporter inhibition. The present development of AUC solutions from PBPK models that contain transporters and enzymes should fill some of the void. The emphasis is to consider not only changes in AUC of drug but also those of metabolites.

5.10 Conclusions and Perspectives

This chapter highlighted the usefulness of PBPK models to deduce the influence of transporters, enzymes, flow, and binding on organ drug clearances and metabolite formation. The PBPK models may also be used to fit or simulate data on DDI. For example, Lin et al. (1982; 1984) had successfully examined the handling of ethoxybenzamide in the rabbit and later defined the inhibition mechanism as product inhibition with PBPK modeling. The understanding of the absorption, distribution, and clearance by generic PBPK models by combining compound-specific, physicochemical, and pharmacokinetic properties with physiological processes allows, via fitting and simulations, predictions of concentration–time profiles of plasma and target tissues (Poulin and Theil, 2002; Lüpfert and Reichel, 2005; Peters, 2008a; 2008b). The utility of in silico- and in vitro-based prediction tools for absorption, distribution, metabolism, and excretion (ADME) in these generic PBPK models to rational drug development allows the prediction of DDI (Theil et al., 2003). Many of these in vitro techniques and their applications are covered in Part II. Generic PBPK models may be used to integrate data not only during drug discovery and in vivo animal studies to obtain tissue:plasma partition coefficients, plasma protein binding as sole input parameters, and intrinsic clearance (CL_{int}) for scale up to whole body and man. The prevalent use of PBPK modeling in risk assessment is a clear indication of the advantages (Andersen et al., 1987; 1997; Sarangapani et al., 2002; Clewell et al., 2001; Corley and McMartin, 2005; Dobrev et al., 2002; Pang, 2009). In view of the MIST document, FDA guidance for Industry. Safety Testing of Drug Metabolites. February 2008 Pharmcology and Toxicology. http://www.fda.gov/cder/guidance/, the need intensifies to improve PBPK modeling of drug and metabolite kinetics for the prediction of DDI. It is the hope that more PBPK models involving transporters and metabolism other than the liver are included for DDI considerations, and that a dynamic and not a static concentration of the inhibitor is used. A recent development has shed some light that this type of modeling is able to describe DDIs between the mechanism-based inhibitor diltiazem and its *N*-desmethylated metabolite and the CYP3A4 substrate, midazolam, and the time- and concentration-dependent clearance of the inhibitor (Zhang et al., 2009a). Complex models of simultaneous reversible and irreversible inhibition would need to be included (Zhang et al., 2009b). Metabolite monitoring is paramount, and the contribution of metabolites as inhibitors of transport and metabolism should be included (Shitara et al., 2004).

Acknowledgments The work is supported by the Canadian Institute for Health Research, CIHR (MOP89850).

References

Abu-Zahra TN and Pang KS (2000) Effect of zonal transport and metabolism on hepatic removal: enalapril hydrolysis in zonal, isolated rat hepatocytes in vitro and correlation with perfusion data. *Drug Metab Dispos* **28**:807–813.

Abu-Zahra TN, Wolkoff AW, Kim RB, and Pang KS (2000) Uptake of enalapril and expression of organic anion transporting polypeptide 1 in zonal, isolated rat hepatocytes. *Drug Metab Dispos* **28**:801–806.

Addis T (1917) Ratio between the urea content of the urine and blood after administration of large quantities of urea. *J Urol* **1**:263–287.

Ambudkar SV, Kim W-W, Booth-Genthe C (2008) Relationship between drugs and function activity of various mammalian P-glycoproteins (ABCB1). *Mini-Rev Med Chem* **8**:193–200.

Andersen ME, Brinbaum IS, Barton HA, and Eklund CR (1997) Regional hepatic CYP1A1 and CYP1A2 induction with 2,3,7,8-tetrachlorodibenzo-p-dioxin evaluated with a multicompartment geometric model of hepatic zonation. *Toxicol Appl Pharmacol* **144**:145–155.

Andersen ME, Clewell 3rd HJ, Gargas ML, Smith FA, and Reitz RH (1987) Physiologically based pharmacokinetics and the risk assessment process for methylene chloride. *Toxicol Appl Pharmacol* **87**:185–205.

Badhan R, Penny J, Galetin A, and Houston JB (2009) Methodology for development of a physiological model incorporating CYP3A and P-glycoprotein for the prediction of intestinal drug absorption. *J Pharm Sci* **98**:2180–2197.

Backman JT, Kyrklund C, Neuvonen M, and Neuvonen PJ (2002) Gemfibrozil greatly increases plasma concentrations of cerivastatin. *Clin Pharmacol Ther* **7**:685–691.

Bekersky I, Popick AC, and Colburn WA (1984) Influence of protein binding and metabolic interconversion on the disposition of sulfisoxazole and its N_4-acetyl metabolite by the isolated perfused rat kidney. *Drug Metab Dispos* **12**:607–613.

Benowitz N Forsyth FP, Melmon KL, and Rowland M (1974) Lidocaine disposition kinetics in monkey and man. I. Prediction by a perfusion model. *Clin Pharmacol Ther* **16**:87–98.

Boffito M, Back DJ, Blaschke TF, Rowland M, Bertz RJ, Gerber JG, and Miller V (2003) Protein binding in antiretroviral therapies. *AIDS Res Hum Retroviruses* **19**:825–835.

Boom SP, Meyer I, Wouterse A, and Russel FG (1998) A physiologically based kidney model for the renal clearance of ranitidine and the interaction with cimetidine and probenecid in the dog. *Biopharm Drug Dispos* **19**:199–208.

Boom SP, Moons MM, and Russel FG (1994) Renal tubular transport of cimetidine in the isolated perfused kidney of the rat. *Drug Metab Dispos* **22**:148–153.

Brewer BD, Clement SF, Lotz WS, Gronwall R (1990) A comparison of inulin, para-aminohippuric acid, and endogenous creatinine clearances as measures as renal function of neonatal foals. *J Vet Intern Med* **4**:301–305.

Brown HS, Galetin A, Halifax D, and Houston JB (2006) Prediction of in vivo drug-drug interactions from in vitro data factors affecting prototypic drug–drug interactions involving CYP2C9, CYP2D6 and CYP3A4. *Clin Pharmacokinet* **45**:1035–1050.
Brown HS, Ito K, Galetin A and Houston JB (2005) Prediction of in vivo drug–drug interactions from in vitro data: impact of incorporating parallel pathways of drug elimination and inhibitor absorption rate constant. *Br J Clin Pharmacol* **60**:508–518.
Chandra P and Brouwer KL (2004) The complexities of hepatic drug transport: current knowledge and emerging concepts. *Pharm Res* **21**:719–735.
Chiba M and Pang KS (1993) Effect of protein binding of 4-methylumbelliferyl sulfate desulfation kinetics in perfused rat liver. *J Pharmacol Exp Ther* **266**:492–499.
Clewell HJ, Gentry PR, Gearhart JM, Allen BC, Andersen ME (2001) Comparison of cancer risk estimates for vinyl chloride using animal and human data with a PBPK model. *Sci Total Environ* **274**:37–66.
Coffey JJ, Bullock FJ, and Schoenemann PT (1971) Numerical solution of nonlinear pharmacokinetic equations: effects of plasma protein binding on drug distribution and elimination. *J Pharm Sci* **60**:1623–1628.
Cong D, Doherty M, and Pang KS (2000) A new physiologically-based segregated flow model to explain route-intestinal metabolism. *Drug Metab Dispos* **28**:224–235.
Corley RA and McMartin KE (2005) Incorporation of therapeutic interventions in physiologically based pharmacokinetic modeling of human clinical case reports of accidental or intentional overdosing with ethylene glycol. *Toxicol Sci* **85**:491–501.
de Lannoy IAM and Pang KS (1987) Effect of diffusional barriers on drug and metabolite kinetics. *Drug Metab Dispos* **15**:51–58.
de Lannoy IAM and Pang KS (1993) Combined recirculation of the rat liver and kidney: studies with enalapril and enalaprilat. *J Pharmacokinet Biopharm* **21**:423–456.
de Lannoy IAM, Barker BF 3rd and Pang KS (1993) Combined recirculation of the rat liver and kidney: studies with enalapril and enalaprilat. *J Pharmacokinet Biopharm* **21**:423–456.
de Lannoy IAM, Hirayama H, and Pang KS (1990) A physiological model for renal drug metabolism: enalapril esterolysis to enalaprilat in the isolated perfused rat kidney. *J Pharmacokinet Biopharm* **18**:561–588.
de Lannoy IAM, Nespeca R, and Pang KS (1989) Renal handling of enalapril and its metabolite, enalaprilat, in the isolated red blood cell-perfused rat kidney. *J Pharmacol Exp Ther* **251**: 1211–1222.
Dobrev ID, Anderson ME, and Yang YSH (2002) *In silico* toxicology: simulating interaction thresholds for human exposure to mixtures of trichloroethylene, tetrachloroethylene, and 1,1,1-trichloroethane. *Environ Health Perspect* **110**:1031–1039.
Eloranta JJ and Kullak-Ubllick GA (2005) Coordinate transcriptional regulation of bile acid homeostasis and drug metabolism. *Arch Biochim Biophys* **433**:397–412.
Ercolani G, Grazi GL, Callivà R, Pierangeli F, Cescon M, Cavallari A, and Mazziotti A (2000) The lidocaine (MEGX) test as an index of hepatic function: its clinical usefulness in liver surgery. *Surgery* **127**:464–471.
Faybik P and Hetz H (2006) Plasma disappearance of indocyanine green. *Transplant Proc* **38**: 801–802.
Geng W and Pang KS (1999) Differences in excretion of hippurate, as a metabolite of benzoate and as an administered species in the single pass isolated perfused rat kidney explained. *J Pharmacol Exp Ther* **288**:597–606.
Gillette JR (1971) Factors affecting drug metabolism. *Ann NY Acad Sci* **179**:43–66.
Goresky CA (1964) Initial distribution and rate of uptake of sulfobromophthalein in the liver. *Am J Physiol* **207**:13–26.
Goresky CA, Bach GC, and Nadeau BE (1973) On the uptake of materials by the intact liver. The transport and net removal of galactose. *J Clin Invest* **52**:991–1009.
Goresky CA, Bach GG, and Nadeau BE (1975) Red cell carriage of label. Its limiting effect on the exchange of materials in the liver. *Circ Res* **36**:328–351.

Goresky CA, Schwab AJ, and Rose CJ (1988) Xenon handling by the liver: red cell capacity effect. *Circ Res* **63**:767–778.

Goulden KJ, Dooley JM, Camfield PR, and Fraser AD (1987) Clinical valproate toxicity induced by acetylsalicylic acid. *Neurology* **37**:1392–1394.

Gray MR and Tam YK (1987) The series-compartment model for hepatic elimination. *Drug Metab Dispos* **15**:27–31.

Greenblatt DJ, Duhme DW, Koch-Weser J, and Smith TW (1973) Evaluation of digoxin bioavailability in single-dose studies. *N Engl J Med* **289**:651–654.

Gregus Z, Fekete T, Varga F, and Klaassen CD (1992) Availability of glycine and Coenzyme A limits glycine conjugation in vivo. *Drug Metab Dispos* **20**:234–240.

Gumucio DL, Gumucio JJ, Wilson JAP, Cutter C, Krauss M, Caldwell R, and Chen E (1984) Albumin influences sulfobromophthalein transport by hepatocytes of each acinar zone. *Am J Physiol* **246**:G86–G95.

Gumucio JJ, Miller DL, Krauss MD, and Cutter Zanollli C (1981) Transport of fluorescent compounds into hepatocytes and the resultant zonal labelling of the hepatic acinus in the rat. *Gastroenterology* **80**:639–646.

Hagelberg NM, Nieminen TH, Saari TI, Neuvonen M, Neuvonen PJ, Laine K, and Olkkola KT (2009) Voriconazole drastically increases exposure to oral oxycodone. *Eur J Clin Pharmacol* **65**:263–271.

Hekman P and van Ginneken CA (1983) Simultaneous kinetic modelling of plasma levels and uriary excretion of salicyluric acid, and the influence of probenecid. *Eur J Drug Metab Pharmacokinet* **8**:239–249.

Hinderling PH (1984) Kinetics of partitioning and binding of digoxin and its analogues in the subcompartments of blood. *J Pharm Sci* **73**:1042–1053.

Hong SP, Choi DH, and Choi JS (2008) Effects of resveratrol on the pharmacokinetics of diltiazem and its major metabolite, desacetyldiltiazem, in rats. *Cardiovasc Ther* **26**:269–275.

Hynninen VV, Olkkola KT, Bertilsson L, Kurkinen K, Neuvonen PJ and Laine K (2008) Effect of terbinafine and voriconazole on the pharmacokinetics of the antidepressant venlafaxine. *Clin Pharmacol Ther* **83**:342–348.

Häcki W, Bircher J, and Presig R (1976) A new look at the plasma disappearance of sulfobromphthalein (BSP): correlation with the BSP transport maximum and the hepatic plasma flow in man. *J Lab Clin Med* **88**:1019–1031.

Ito K, Chiba K, Horikawa M, Ishigami M, Mizuno N, Aoki J, Gotoh Y, Iwatsubo T, Kanamitsu S, Kato M, et al. (2002) Which concentration of the inhibitor should be used to predict in vivo drug interactions from in vitro data? *AAPS PharmSci* **4**:E25.

Ito K, Hallifax D, Obach RS, and Houston JB (2005) Impact of parallel pathways of drug elimination and multiple cytochrome P450 involvement on drug-drug interactions: CYP2D6 paradigm. *Drug Metab Dispos* **33**:837–844.

Ito K, Iwatsubo T, Kanamitsu S, Ueda K, Suzuki H, and Sugiyama Y (1998) Prediction of pharmacokinetic alterations caused by drug-drug interactions: metabolic interaction in the liver. *Pharmacol Rev* **50**:387–411.

Ito K, Kusuhara H, and Sugiyama Y (1999) Effects of intestinal CYP3A4 and P-glycoprotein on oral drug absorption – theoretical approach. *Pharm Res* **16**:225–231.

Jaakkola T, Backman JT, Neuvonen M and Neuvonen PJ (2005) Effects of gemfibrozil, itraconazole, and their combination on the pharmacokinetics of pioglitazone. *Clin Pharmacol Ther* **77**:404–414.

Kajosaari LI, Niemi M, Backman JT and Neuvonen PJ (2006) Telithromycin, but not montelukast, increases the plasma concentrations and effects of the cytochrome P450 3A4 and 2C8 substrate repaglinide. *Clin Pharmacol Ther* **79**:231–242.

Kaneko H, Otsuka Y, Katagiri M, Maeda T, Tsuchiya M, Tamura M, Tamura A, Ishii T, Takagi S, and Shiba T (2001) Reassessment of monoethylglycinexylidide as preoperative liver function test in a rat model of liver cirrhosis and man. *Clin Exp Med* **1**:19–26.

Kharasch ED, Bedynek PS, Walker A, Whittington D and Hoffer C (2008) Mechanism of ritonavir changes in methadone pharmacokinetics and pharmacodynamics: II. Ritonavir effects on CYP3A and P-glycoprotein activities. *Clin Pharmacol Ther* **84**:506–512.

Kharasch ED, Hoffer C, Whittington D, and Sheffels P (2003) Role of P-glycoprotein in the intestinal absorption and clinical effects of morphine. *Clin Pharmacol Ther* **74**:543–554.

Kim HJ, Cho JH, and Klaassen CD (1995) Depletion of hepatic 3′-phosphoadenosine 5′-phosphosulfate (PAPS) and sulfate in rats by xenobiotics that are sulfated. *J Pharmacol Exp Ther* **275**:654–658.

Klippert PJ and Noordhoek J (1985) The area under the curve of metabolites for drugs and metabolites cleared by the liver and extrahepatic organs. Its dependence on the administration route of precursor drug. *Drug Metab Dispos* **13**:97–101.

Koch-Weser J (1974a) Bioavailability of drugs (first of two parts). *N Engl J Med* **291**:233–237.

Koch-Weser J (1974b) Bioavailability of drugs (second of two parts). *N Engl J Med* **291**: 503–506.

Krishnamurthy P and Schuetz JD (2006) Role of ABCG2/BCRP in biology and medicine. *Annu Rev Pharmacol Toxicol* **46**:381–410.

Kruh GD, Belinksky MG, Gallo JM, and Lee K (2007) Physiological and pharmacological functions of Mrp2, Mrp3 and Mrp4 as determined from recent studies on gene-disrupted mice. *Cancer Mestastis Rev* **26**:5–14.

Kwon Y and Morris ME (1997) Membrane transport in hepatic clearance of drugs. II. Zonal distribution patterns of concentration-dependent transport and elimination process. *Pharm Res* **14:**:780–785.

Lewis AE (1949) The concept of hepatic clearance. *Am J Clin Pathol* **18**:789–795.

Li YM, Lv F, Xu X, Ji H, Gao WT, Lei TJ, Ren GB, Bai ZL, and Li Q (2003) Evaluation of liver functional reserve by combining D-sorbitol clearance rate and CT-measured liver volume. *World J Gastroenterol* **9**:2092–2095.

Lin JH, Chiba M, and Baillie TA (1999) Is the role of the small intestine in first-pass metabolism overemphasized? *Pharmacol Rev* **51**:135–158.

Lin JH, Sugiyama Y, Awazu S, and Hanano M (1982) Physiological pharmacokinetics of ethoxybenzamide based on biochemical data obtained in vitro as well as on physiological data. *J Pharmacokinet Biopharm* **10**:649–661.

Lin JH, Sugiyama Y, Hanano M, and Awazu S (1984) Effect of product inhibition on elimination kinetics of ethoxybenzamide in rabbits. Analysis by physiological pharmacokinetic model. *Drug Metab Dispos* **12**:253–256.

Liu L, Mak E, Tirona RG, Tan E, Novikoff PM, Wang P, Wolkoff AW, and Pang KS (2005) Vascular binding, blood flow, transporter and enzyme interactions on the processing of digoxin in rat liver. *J Pharmacol Exp Ther* **315**:433–448.

Liu L and Pang KS (2006) An integrated approach to model hepatic drug clearance. *Eur J Pharm Sci* **29**:215–230.

Liu S, Tam D, Chen X, and Pang KS (2006) An unstirred water layer and P-glycoprotein barring digoxin absorption by the perfused rat small intestine preparation: induction studies with and without pregnenolone 16α-carbonitrile (PCN) induction. *Drug Metab Dispos* **34**: 1468–1479.

Lüpfert C and Reichel A (2005) Development and application of physiologically based pharmacokinetic-modeling tools to support drug discovery. *Chem Biodivers* **2**:1462–1486.

Magee DF and Dalley AF II (1986) *Digestion and The Structure and Function of The Gut*, (*Karger Continuing Education Series*, vol. 8). Karger, Basel.

Martínez-Salgado C, López-Hernández FJ, and López-Novoa JM (2007) Glomerular nephrotoxicity of aminoglycosides. *Toxicol Appl Pharmacol* **15**(223):86–98.

Masereeuw R, Moons MM, Smits P, and Russel FG (1996) Glomerular filtration and saturable absorption of iohexol in the rat isolated perfused kidney. *Br J Pharmacol* **119**:57–64.

McElnay JC and D'Arcy PF (1983) Protein binding displacement interactions and their clinical importance. *Drugs* **25**:495–513.

McNamara PJ, Trueb V, and Stoeckel K (1990) Ceftriaxone binding to human serum albumin. Indirect displacement by probenecid and diazepam. *Biochem Pharmacol* **40**:1247–1253.

Mizuno N, Suzuki M, Kusuhara H, Suzuki H, Takeuchi K, Niwa T, Jonker JW, and Sugiyama Y (2004) Impaired renal excretion of 6-hydroxy-5,7-dimethyl-2-methylamino-4-(3-pyridylmethyl) benzothiazole (E3040) sulfate in breast cancer resistance protein (BCRP1/ABCG2) knockout mice. *Drug Metab Dispos* **32**:898–901.

Molino G, Avagnina P, Belforte G, and Bircher J (1998) Assessment of the hepatic circulation in humans: new concepts based on evidence derived from a D-sorbitol clearance method. *J Lab Clin Med* **131**:393–405.

Morris ME and Pang KS (1987) Competition between two enzymes for substrate removal in liver: modulating effects of competitive pathways. *J Pharmacokinet Biopharm* **15**:473–496.

Möller E, McIntosh JR, and van Slyke DD (1928) Studies of urea excretion II: relationship between urine volume and the rate of urea excretion by normal adults. *J Clin Invest* **6**:427–465.

Nagai J and Tokano M (2004) Molecular aspects of renal handling of aminoglycosides and strategies for preventing the nephrotoxicity. *Drug Metab Pharmacokinet* **19**:159–170.

Nakagami T, Yasui-Furukori N, Saito M, Tateishi T, and Kaneo S (2005) Effect of verapamil on pharmacokinetics and pharmacodynamics of risperidone: in vivo evidence of involvement of P-glycoprotein in risperidone disposition. *Clin Pharmacol Ther* **78**:43–51.

Obach RS, Walsky RL, Venkatakrishnan K, Emily A Gaman EA, Houston JB and Tremaine LM (2006) The utility of in vitro cytochrome P450 inhibition data in the prediction of drug-drug interactions. *J Pharmacol Exp Ther* **316**:336–348.

Orlando R, De Martin S, Pegoraro P, Quintieri L, and Palatini P (2009) Irreversible CYP3A inhibition accompanied by plasma protein-binding displacement: a comparative analysis in subjects with normal and impaired liver function. *Clin Pharmacol Ther* **85**:319–326.

Palkama VJ, Ahonen J, Neuvonen PJ, and Olkkola KT (1999) Effect of saquinavir on the pharmacokinetics and pharmacodynamics of oral and intravenous midazolam. *Clin Pharmacol Ther* **66**:33–39.

Pang KS (1995) Modeling of metabolite disposition. In, *Advanced Methods of Pharmacokinetic and Pharmacodynamic System Analysis Volume II*, (DZ D'Argenio ed), Plenum Press, New York, pp 3–26.

Pang KS (2003) Modeling of intestinal drug absorption: roles of transporters and metabolic enzymes (For the Gillette Review Series). *Drug Metab Dispos* **31**:1507–1519.

Pang KS (2009) Safety testing of metabolites: expectations and outcomes. *ChemicoBiol Interac* **179**:45–59.

Pang KS, Barker III F, Simard A, Schwab AJ, and Goresky CA (1995) Sulfation of acetaminophen by the perfused rat liver: Effect of red blood cell carriage. *Hepatology* **22**:267–282.

Pang KS, Morris ME, and Sun H (2008) Formed and preformed metabolite: Facts and comparisons. *J Pharm Pharmacol* **60**:1247–1275.

Pang KS and Rowland M (1977) Hepatic clearance of drugs. I. Theoretical considerations of a "well-stirred" model and a "parallel tube" model. Influence of hepatic blood flow, plasma and blood cell binding, and the hepatocellular enzymatic activity on hepatic drug clearance. *J Pharmacokinet Biopharm* **5**:625–653.

Pang KS and Stillwell RN (1983) An understanding of the role of enzymic localization of the liver on metabolite kinetics: a computer simulation. *J Pharmacokinet Biopharm* **11**:451–468.

Pang KS, Cherry WF, and Ulm EH (1985) Disposition of enalapril in the perfused rat intestine-liver preparation: Absorption, metabolism, and first-pass effects. *J Pharmacol Exp Ther* **233**: 788–795.

Perl W and Chinard FP (1968) A convection-diffusion model of indicator transport through an organ. *Circ Res* **22**:273–298.

Peters SA (2008a) Identification of intestinal loss of a drug through physiologically based pharmacokinetic simulation of plasma concentration-time profiles. *Clin Pharmacokinet* **47**:245–259.

Peters SA (2008b) Evaluation of a generic physiologically based pharmacokinetic model for lineshape analysis. *Clin Pharmacokinet* **47**:261–275.

Poulin P and Theil FP (2002) Prediction of pharmacokinetics prior to in vivo studies. II. Generic physiologically based pharmacokinetic models of drug disposition. *J Pharm Sci* **91**: 1358–1370.

Rappaport AM (1958) The structural and functional unit in the human liver (liver acinus). *Anat Rec* **130**:673–689.

Roberts MS and Rowland M (1985) Hepatic elimination – dispersion model. *J Pharm Sci* **74**: 585–587.

Robey RW, Polgar O, Deeken J, To KW, and Bates SE (2007) ABCG2: determining its relevance in clinical drug resistance. *Cancer Metastasi Rev* **26**:39–57.

Rodriguez CA and Smith DE (1991) Influence of unbound concentration of cefonicid on its renal elimination in isolated perfused rat kidney. *Antimicrob Agents Chemother* **35**:2395–2400.

Rowland M (1972) Influence of route of administration on drug availability. *J Pharm Sci* **61**: 70–74.

Rowland M, Benet LZ, and Graham GG (1973) Clearance concepts in pharmacokinetics. *J Pharmacokinet Biopharm* **1**:123–136.

Rowland M and Matin SB (1973) Kinetics of drug–drug interactions. *J Pharmacokinet Biopharm* **1**::553–567.

Rowland M and Tozer TN (1995) *Clinical Pharmacokinetics: Concepts and Applications* 3rd edition, Williams & Wilkins, Philadelphia, PA

Russel FG, Wouterse AC, and van Ginnekan CA (1987) Physiologically based pharmacokinetic model for the renal clearance of phenolsulfonphthalein and the interaction with probenecid and salicyluric acid in the dog. *J Pharmacokinet Biopharm* **15**:349–368.

Russel FG, Wouterse AC, and van Ginnekan CA (1989) Physiologically based pharmacokinetic model for the renal clearance of iodopyracet and the interaction with probenecid in the dog. *Biopharm Drug Dispos* **10**:137–152.

Saari TI, Laine K, Leino K, Valtonen M, Neuvonen PJ and Olkkola KT (2006) Effect of voriconazole on the pharmacokinetics and pharmacodynamics of intravenous and oral midazolam. *Clin Pharmacol Ther* **79**:362–370.

Sakka SG (2007) Assessing liver function. *Curr Opin Crit Care* **13**:207–214.

Sarangapani R, Teeguarden JG, Cruzan G, Clewell HJ, and Andersen ME (2002) Physiologically based pharmacokinetic modeling of styrene and styrene oxide respiratory-tract dosimetry in rodents and humans. *Inhal Toxicol* **12**:789–834.

Schary WL and Rowland M (1983) Protein binding and hepatic clearance: studies with tolbutamde, a drug of low intrinsic clearance, in the isolated perfused rat liver preparation. *J Pharmacokinet Biopharm* **11**:225–243.

Schwab AJ, Barker III F, Goresky CA, and Pang KS (1990) Transfer of enalaprilat across rat liver cell membranes is barrier-limited. *Am J Physiol* **258**:G461–G475.

Schwab AJ, de Lannoy IAM, Poon K, Goresky CA, and Pang KS (1992) Enalaprilat handling by the rat kidney: barrier-limited cell entry. *Am J Physiol* **263**:F858–F869.

Shitara Y, Hirano M, Sata H, and Sugiyama Y (2004) Gemfibrozil and its glucuronide inhibit the organic anion transporting polypeptide 2 (OATP2/OATP1B1:SLC21A6)-mediated hepatic uptake and CYP2C8-mediated metabolism of cerivastatin: analysis of the mechanism of the clinically relevant drug-drug interaction between cerivastatin and gemfibrozil. *J Pharmacol Exp Ther* **311**:228–236.

Shitara Y, Horie T, and Sugiyama Y (2006) Transporters as a determinant of drug clearance and tissue distribution. *Eur J Pharm Sci* **27**:425–446.

Sirianni GL and Pang KS (1997) Organ clearance concepts: new perspectives on old principles. *J Pharmacokinet Biopharm* **25**:449–470.

Sirianni GL and Pang KS (1999) Inhibition of esterolysis of enalapril by paraoxon increases the urinary clearance in isolated perfused rat kidney. *Drug Metab Dispos* **27**:931–936.

Sugimoto Y, Tsukahara S, Ishikawa E, and Mitsuhashi J (2005) Breast cancer resistance protein: molecular target for anticancer drug resistance and pharmacokinetics/pharmacodynamics. *Cancer Sci* **96**:457–465.

Sun H and Pang KS (2009a) Disparity in intestine disposition between formed and preformed metabolites and implications: a theoretical study. *Drug Metab Dispos* **37**:187–202.

Sun and Pang (2009b) Physiological modeling to understand the impact of enzymes and transporters on drug and metabolite data and bioavailability estimates. *Pharm Res*, Under revision.

Sun H, Liu L, and Pang KS (2006) Increased estrogen sulfation of estradiol 17β-D glucuronide in rat metastasis tumor livers. *J Pharmacol Exp Ther* **319**:818–831.

Swan SK (1997) Aminoglycoside toxicity. *Semin Nephrol* **17**:27–33.

Tam D, Tirona RG, and Pang KS (2003) Segmental intestinal transporters and metabolic enzymes on intestinal drug absorption. *Drug Metab Dispos* **33**:373–383.

Tan E and Pang KS (2001) Sulfation is rate limiting in the futile cycling between estrone and estrone sulfate in enriched periportal and perivenous rat hepatocytes. *Drug Metab Dispos* **29**:335–346.

Theil FP, Guentert TW, Haddad S, and Poullin P (2003) Utility of physiologically based pharmacokinetic models to drug development and rational drug discovery candidate selection. *Toxicol Lett* **138**:29–49.

Thiebaut F, Tsuruo T, Hamada H, Gottesman MM, Pastan I, and Willingham MC (1987) Cellular localization of the multidrug-resistance gene product P-glycoprotein in normal human tissues. *Proc Natl Acad Sci USA* **84**:7735–7738.

Tirona RG, Tan E, Meier G, and Pang KS (1999) Uptake and glutathione conjugation kinetics of ethacrynic acid in rat liver: in vitro and perfusion studies. *J Pharmacol Exp Ther* **291**: 1210–1219.

Toon S, Low LK, Gibaldi M, Trager WF, O'Reilly RA, Motley CH, and Goulart DA (1986) The warfarin-sulfinpyrazone interaction: stereochemical considerations. *Clin Pharmacol Ther* **39**:15–24.

Tornio A, Niemi M, Neuvonen M, Laitila J, Kalliokoski A, Neuvonen PJ and Backman JT (2008) The effect of gemfibrozil on repaglinide pharmacokinetics persists for at least 12 h after the dose: evidence for mechanism-based inhibition of CYP2C8 in vivo. *Clin Pharmacol Ther* **84**:403–411.

Toto RD (1995) Conventional measurement of renal function using serum creatinine, creatinine clearance, inulin and para-aminohippuric acid clearance. *Curr Opin Nephrol Hypertens* **4**: 505–509.

Trauner M and Boyer JL (2003) Bile salt transporters: molecular characterization, function and regulation. *Physiol Rev* **83**:633–671.

Tucker GT, Houston JB, and Huang S-M (2001) Optimising drug development: strategies to assess drug metabolism/transporter interaction potential – toward a consensus. *Clin Pharmacol Ther* **70**:103–114.

Ungell AL, Nylander S, Bergstrand S, Sjöberg A, and Lennernäs H (1998) Membrane transport of drugs in different regions of the intestinal tract of the rat. *J Pharm Sci* **87**:360–366.

van Montfoort JE, Hagenbuch B, Groothuis GM, Koepsell H, Meier PJ, and Meijer DK (2003) Drug uptake systems in liver and kidney. *Curr Drug Metab* **4**:185–211.

Vaubourdolle M, Guffet V, Chazouillères O, Biboudeau J, and Poupon R (1991) Indocyanine green-sulfobromophthalein pharmacokinetics for diagnosing primary biliary cirrhosis and assessing histological severity. *Clin Chem* **37**:1688–1690.

Vlaming ML, Lagas JS, and Schinkel AH (2009) Physiological and pharmacological roles of ABCG2 (BCRP): recent findings in Abcg2 knockout mice. *Adv Drug Deliv Rev* **62**: 14–25.

Wagner JG (1972) An overview of the analysis and interpretation of bioavailability studies in man. *Pharmacology* **8**:102–117.

Wang Y-H, Jones DR, and Hall SD (2004) Prediction of cytochrome P450 3A inhibition by verapamil enantiomers and their metabolites. *Drug Metab Dispos* **32**:259–266.

Weisiger RA (1985) Dissociation from albumin: a potentially rate-limiting step in the clearance of substances by the liver. *Proc Natl Acad Sci USA* **82**:1563–1567.

Wilkinson GR (1987) Clearance approaches in pharmacology. *Pharmacol Rev* **39**:1–47.

Wilkinson GR and Shand DG (1975) Commentary: a physiological approach to hepatic drug clearance. *Clin Pharmacol Ther* **18**:377–390.

Winkler K, Keiding S, and Tygstrup N (1973) Clearance as a quantitative measure of liver function. In, *The Liver: Quantitative Aspects of Structure and Functions* (P Paumgartner, and R Presig, eds), Karger, Basel, pp 144–155.

Wong BK, Bruhin PJ, Barrish A, and Lin JH (1996) Nonlinear dorzolamide pharmacokinetics in rats: concentration-dependent erythrocyte distribution and drug-metabolite displacement interaction. *Drug Metab Dispos* **24**:659–663.

Xu X and Pang KS (1989) Hepatic modeling of metabolite kinetics in sequential and parallel pathways: salicylamide and gentisamide metabolism in perfused rat liver. *J Pharmacokinet Biopharm* **17**:645–671.

Xu X, Tang BK, and Pang KS (1990) Sequential metabolism of salicylamide exclusively to gentisamide-5-glucuronide and not gentisamide sulfate conjugates in single pass in situ perfused rat liver. *J Pharmacol Exp Ther* **253**:965–973.

Yacobi A, Øie S and Levy G (1977) Relationship between protein binding of bilirubin, salicylic acid and sulfisoxazole in serum of unmediated and phenobarbital-treated rats. Frequency distribution of bilirubin intrinsic clearance in adult male Sprague-Dawley rats. *J Pharm Sci* **66**:1025–1027.

Yacobi A, Øie S and Levy G (1979) Effect of serum protein binding on sulfisoxazole distribution, metabolism, and excretin in rats. *J Pharm Sci* **68**:742–746.

Yang J, Jamei M, Yeo KR, Tucker GT and Rostami-Hodjegan A (2007) Prediction of intestinal first-pass drug metabolism. *Curr Drug Metab* **8**:676–684.

Yu LX and Amidon GL (1998) Saturable small intestinal drug absorption in humans: modeling and interpretation of cetrizine data. *Eur J Pharmaceut Biopharm* **45**:199–203.

Yu HY, Shen YZ, Sugiyama Y, and Hanano M (1990) Drug interaction. Effects of salicylate on pharmacokinetics of valproic acid in rats. *Drug Metab Dispos* **18**:121–126.

Zamek-Gliszczynski MJ, Hoffmaster KA, Nezasa K, Tallamn MN, and Brouwer KL (2006) Integration of hepatic drug transporters and phase II metabolizing enzymes: mechanisms of hepatic excretion of sulfate, glucuronide, and glutathione metabolites. *Eur J Pharm Sci* **27**:447–486.

Zhang X, Jones DR, and Hall SD (2009b) Prediction of the effect of erythromycin, diltiazem, and their metabolites, alone and in combination, on CYP3A4 inhibition. *Drug Metab Dispos* **37**:150–160.

Zhang X, Quinney SK, Gorski JC, Jones DR, and Hall SD (2009a) Semi-physiologically-based pharmacokinetic models for the inhibition of midazolam clearance by diltiazem and its major metabolite. *Drug Metab Dispos* **37**:1587–1597.

Part II
Methods for the Study of Drug–Drug Interactions

Chapter 6
In Silico Approaches to Predict DDIs

Chad L. Stoner, Michael R. Wester, and Benjamin J. Burke

Abstract This chapter will briefly describe in silico methodologies for the prediction of drug–drug interactions (DDIs) and highlight the broad application of computational tools to study DDIs. This chapter outlines the main methodologies currently applied including QSAR modeling, pharmacophore modeling, docking, and the combination of in silico and experimental approaches. There is an emphasis on cytochrome P450 and how in silico models are used in current drug discovery efforts to reduce the risk of DDIs. The discussion of the limitations associated with the various approaches as well as future aspects of DDI modeling and simulation can give researchers helpful guidance to this useful and growing area.

6.1 Introduction

This chapter will introduce in silico methodologies for the prediction of drug–drug interactions (DDIs) and describe the broad application of computational methodologies to study DDIs. These in silico or computer-based methodologies for predicting DDIs have been developed for a large number of the experimental endpoints described in the previous chapters. In silico models of DDIs have for the most part focused on the cytochrome P450 family of enzymes. A good introduction to 3D ligand-based methods and structure-based approaches for predicting DDIs can be found in the following references (de Groot, 2006; Refsgaard et al., 2006).

C.L. Stoner (✉)
Department of Pharmacokinetics and Drug Metabolism, Pfizer Inc., Global Research & Development, La Jolla, CA, USA
e-mail: chad.stoner@pfizer.com

K.S. Pang et al. (eds.), *Enzyme- and Transporter-Based Drug–Drug Interactions*,
DOI 10.1007/978-1-4419-0840-7_6,

6.2 QSAR Modeling

In general, most of the models applied in early drug discovery are quantitative structure–activity relationship (QSAR) models. A large number of QSAR models for predicting DDIs have been developed based on chemical descriptors. Chemical descriptors are simple properties or structural fragments (sometimes called fingerprints) that can be calculated based solely on a compound structure (Refsgaard et al., 2006). A variety of statistical and machine-learning computational algorithms have been used to derive predictive models of DDI endpoints. These models tend to be the least computationally intensive and thousands of compounds can be processed in seconds. Data sets are generally split into two groups, a validation set (which is excluded during the building of the model) and a training set (which is used to build the model). The outputs from these models range from a simple Yes or No for a specific DDI, to a percent inhibition, to a predicted IC50 against a particular P450 isoform (Ekins et al., 2000b, c). They also vary in the type of confidence metric which is reported to aid in interpretation of the result. For example, some models may report the number of nearest neighbors; that is, the number of compounds in the training set which are structurally similar to the compound being predicted. Other models might report the structural fragment that is likely causing the DDI. Interpretability of the model, which comes from understanding the relationships between chemical descriptors and DDI, facilitates future compound designs and testing decisions.

Another subset of QSAR models for predicting DDIs have been developed based on 3D descriptors using molecular fields (see Table 1 in de Graaf et al., 2005). These models capture the important interactions between a set of ligands and the P450 and can reflect the SAR differences between P450 isoforms (e.g., in terms of size, lipophilicity, planarity, or basicity) (Refsgaard et al., 2006). Therefore, these models were particularly important for developing an understanding of the 3D features of CYPs prior to P450 crystal structures. Certain 3D-QSAR models for P450 inhibition require neither a postulated bioactive conformation nor an alignment rule (Afzelius et al., 2004), but many 3D-QSAR models essentially resemble pharmacophore models.

6.3 Pharmacophore Modeling

The second most commonly applied approach to studying DDIs in silico is a pharmacophore approach. Pharmacophore models have been built for P450 inhibition, P450 heteroactivation, P450 induction as well as transporter function and induction. As is the case with a QSAR model, a pharmacophore model requires no structural information regarding the protein, because the model takes into account only the user-specified physical properties of the ligand. These properties usually include geometry, charge, and hydrophilic/hydrophobic information. What is returned is a 3D arrangement of pharmacophoric features common to the training molecules that describes the elements necessary for recognition in the binding site.

Pharmacophore approaches typically require fewer training set compounds than statistical or machine-learning models and perform best when the training set contains active, rigid ligands.

Early application of pharmacophores to pharmaceutically relevant ligands of CYP1A2 (Fuhr et al., 1993), CYP2C9, (Ekins et al., 2000a) and CYP3A4 (Ekins et al., 1999; 2003) has been described. Some more recent work by Jones et al. highlights a pharmacophore model that aids in the prediction of reactive metabolic intermediates implicated in time-dependent inhibition of P450 3A4 (Jones et al., 2007). Two other groups used pharmacophores to design potent inhibitors of CYP19 or aromatase (Neves et al., 2007; Schuster et al., 2006). Finally, one must remember that often the best approach integrates several sources of information, for example, approaches combining pharmacophore and protein modeling have been used in efforts to understand P450 inhibitors (Masimirembwa et al., 2002; de Groot et al., 2002).

6.4 Docking

The third approach to modeling of DDIs does require structural information of the protein. The structures of P450s have yielded detailed molecular interactions occurring between the P450s active site and inhibitors (Fig. 6.1). As more P450 crystal structures are published in the public domain, the popularity of docking and structure-based approaches will continue to expand. Some limitations of structure-based approaches have been the accuracy of force fields and properly characterizing the water and protein movements inherent in this class of enzymes. Advances in computing power could allow for a more routine use of rigorous techniques like QM/MM scoring (Bathelt et al., 2008) , MD sampling (Hritz et al., 2008), including waters in scoring (de Graaf et al., 2006) or entropic calculations (Chang et al., 2007). Scoring in these structure-based models is more complex than that of standard QSAR approaches and requires an expert end user.

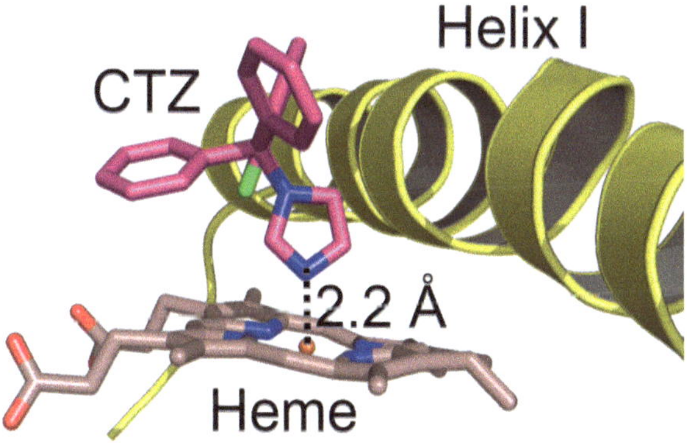

Fig. 6.1 Crystal structure of the inhibitor clotrimazole in complex with P450 3A4

6.5 In Combo Modeling

A fourth approach, which we refer to as in combo is the use of in silico predictions in combination with wet experimental data (Smith et al., 2007). We believe that the increased pressures of drug discovery will drive scientists to maximize the value derived from every experiment and result in increased use of this approach. A simple in combo approach could be, for example the combination of an in silico prediction of P450 inhibition with a spectral shift assay to elucidate the mechanism of inhibition. One could then model the ligand in the active site of the P450 to identify the likely coordinating moieties. A more complex in combo approach is the PBPK approach described in the subsequent chapters. SIMCYP® is one commercially available PBPK modeling package that allows for studying the effects of a DDI in a patient population. This approach can provide robust estimates of a pharmacokinetic changes caused by the DDI, including predicting the changes in clinical exposure reported as AUC (area under the curve) relative to an AUCi (area under the curve with inhibitor coadministered) driven by various concomitantly administered drugs (Youdim et al., 2008).

Recent advances in the screening of compounds for time-dependent inhibition (TDI) has also resulted in models that can predict compounds likely to be mechanism-based inhibitors (MBIs) (Ekins and Swaan, 2004; Jones et al., 2007) as well as those compounds that may form reactive metabolites (RMs) that could produce a DDI. It should be noted however that not all RMs cause a TDI.

6.6 Industrial Application of DDI Modeling

The application of in silico methodologies to study DDIs has been on the rise across the pharmaceutical industry. In silico, methodologies complements the majority of the experimental endpoints and in some cases are used in advance of the actual "wet" experiments. As drug discovery cost continues to increase, the pressure to accelerate and improve design and maximize the value derived from experimentation will increase.

In silico models can generally be categorized as a "global model," a "local model", or a "compound-specific" model. Global models tend to be the most general models and tend to have the largest and most diverse chemical matter in the training sets. These models can be used to predict endpoints for a very diverse set of new chemical entities (NCEs); however the accuracy of these predictions can be highly variable depending on whether or not the NCEs are well described by the training set used to build the model. Local models tend to be built for one chemical series and are generally more predictive than a global model, when used to predict NCEs from the same series. The downside to local models is the cost and time required to generate a large enough training set to build this type of model. A compound-specific model would generally refer to how a specific compound is likely to dock into a specific P450 through a docking experiment (Ahlstrom et al., 2007).

It is fairly common across the industry for a high-throughput screen (HTS) to look at millions of potential chemical leads for a new therapeutic target. As the "hits" from an HTS can result in thousands of leads, the pressure and cost to do follow-up work to determine the drug-like qualities of these potential new lead series have resulted in the incorporation of a large number of in silico filters being applied to immediately remove those compounds and series that have the greatest risks for ADME and safety liabilities. Along these same lines, global models for DDI are used to filter virtual libraries to eliminate NCE designs that might produce a DDI. The ability to assess a risk before a compound is even synthesized has a significant cost and time savings potential.

Once a series has moved into lead development, a much more robust assessment of DDI liability is likely. Just as the experimental scrutiny in this phase of drug discovery increases from, for example, a cocktail DDI assay to a single probe substrate with a specific clinically relevant probe, the in silico approach can also be optimized. Series-specific models (local models) generally provide a significant increase in predictive confidence and can be assessed for isosteric (chemical) substitutions to help with optimization.

6.7 The Current State of Enzymatic Predictions for DDI

In this section, we will provide an overview of the current statistical approaches used to predict DDIs and provide some examples of how these models would be used in contemporary drug discovery. We will also cover some recent advances in P450 crystallography and the development of in silico docking approaches to aid in drug design.

6.7.1 Overview of Statistical Approches

As mentioned above, the basis of statistical learning or QSAR models involves establishing a relationship between a set of descriptors which characterize molecular features to an "activity" (e.g., assayed inhibition of a P450). The data from a single source eliminate lab-to-lab variation. For P450s, there is an important dependence upon the molecular probe used to characterize the P450 inhibition (Bell et al., 2008). Researchers often have their favorite sets of descriptors and mathematical methods which in combination have varying degrees of interpretability. Two-dimensional descriptors try to describe molecular features of shape, size, electrostatics, polarizability, lipophilicity, and hydrogen bonding properties typically using graph theoretic, topological, electronic, and fragment-based descriptions of molecules (Helguera et al., 2008). Fingerprints, E-state and MOE descriptors (Labute, 2000) would fall into this category. Three-dimensional descriptors arose from the concept of a pharmacophore where certain features in a particular orientation elicit a molecular response (Kier, 1971). Three-dimensional descriptors are generally categorized as alignment dependent where molecular overlays are a prerequisite to comparing the features between compounds, or alignment independent where the feature description is based on autocorrelations, internal distances, principal moments, or spectral comparisons (Doweyko et al., 2007). Pharmacophores and comparative molecular field analysis (CoMFA) fall into the first category, while GRIND (Pastor et al., 2000), WHIM (Bravi et al., 1997; Bravi and Wikel, 2000) and EVA (Ferguson et al., 1997) fall into the second category. Quantum mechanical descriptors related to the ability to donate or accept electrons can come into play in some in silico models (Kriegl et al., 2005b).

Popular machine-learning techniques such as support vector machines (SVM), decision trees (DT), and k-nearest neighbor (kNN) methods were recently reviewed with an eye toward applications in ADMET (de Groot et al., 2007) (Li et al., 2007). These methods, along with Bayesian methods (O′Brien and De Groot, 2005), have found utility in building P450 inhibition classification models (Chohan et al., 2005). Regression methods such as multiple linear regression (MLR) are useful for simpler QSARs and, while PLS and its variants are the standards for regressions with large systematic blocks of data, like those found in molecular field applications like CoMFA or GRIND.

6.7.2 Examples of P450 Inhibition

Four reviews form a very strong foundation for understanding in silico approaches to P450 inhibition (Refsgaard et al., 2006; de Graaf et al., 2005; de Groot, 2006; de Groot et al., 2007). More recent updates to the progress in modeling P450 inhibition can be found in two more recent reports (Fox and Kriegl, 2007; Stjernschantz et al., 2008). Recent accomplishments include models that apply to broad chemical space, provide measures of confidence in the model, emphasize QSAR interpretation, suggest important protein–ligand interactions, or that use consensus as a way to derive higher quality composite models.

The current state of the P450 inhibition statistical models range from the very simple (multiple linear regression of single physicochemical properties like logD) to complex (nonlinear mathematical relationships between several hundred descriptors). One of the simplest relationships is the correlation between logD and P450 inhibition (Lewis et al., 2006a, b, 2007). Several models of P450 inhibition include logP or logD terms, and some researchers have combined this feature with the presence of unhindered N-containing heterocycles to characterize DDI potential (Fox and Kriegl, 2007). In general, this sort of simple correlation should be a minimum benchmark to which more complicated models are compared.

Global DDI models can be used to triage HTS series or filter virtual compounds to synthesize only those with desired properties (Byvatov et al., 2007). For example, Aureus-Pharma developed a three-level (high, medium, low) inhibition classification model for P450 1A2 and 2D6 compounds with classification accuracy between 78 and 96% (Burton et al., 2006). Novo Nordisk developed a classification model for P450 2D6 and 3A4 inhibition with classification accuracy between 82 and 88% (Jensen et al., 2007). For global models, the confidence in the prediction often varies with the similarity in structure to the compounds in the model and/or the number of nearest neighbors which reflects whether or not the model covers the chemical space for the test compound. Chemical space coverage was an important characteristic for successful prediction of an external set of compounds for two SVM models built using several hundreds to several thousands of compounds (Arimoto et al., 2005; Zhou et al., 2007).

"Local" models using a few compounds can be developed when there is not enough resolution in a global model to redesign compounds within a series. For example, Wang built a Kernel-PLS model for a series of 55 nicotine analogs with a Q^2 of 0.70 that highlighted the relationship between E-state descriptors demonstrating a relationship between hydrogen bonding and aromatic carbon descriptors and P450 2A6 inhibition (Wang et al., 2007).

The level of interpretation depends upon the descriptors and the mathematical relationships used in the in silico models. Astra Zeneca undertook an effort to develop models against five P450s (1A2, 2C9, 2C19, 2D6, and 3A4) using descriptors that were "familiar to chemists" to help interpretation (Gleeson et al., 2007). The descriptors focused on chemical fragments represented as SMART strings and bulk properties (Daylight Chemical Information Systems, Inc. www.daylight.com).

Consensus modeling is an approach that is gaining popularity because multiple models can be used to derive a better overall result than any single model. Hudelson and coworkers used the consensus of three models to predict P450 2C9 inhibition (Hudelson et al., 2008). When the models agreed, the predictive rate was 90%. Boehringer Ingelheim used several P450 3A4 inhibition models in parallel to reduce false positives (Kriegl et al., 2005a). A combination of three models boosted the overall prediction rate for P450 2D6 inhibitors to 99% (O′Brien and De Groot, 2005).

A grand, unified model of the interaction between several P450s and a variety of ligands was developed to yield "a single general model that represents all studied entities" (Kontijevskis et al., 2008). This "proteochemometric" model used *z*-scale descriptors to describe the protein active sites and GRIND descriptors to describe ligand characteristics, and included a full set of interaction cross-terms.

6.7.3 P450 Crystallization

Modern structure-based (e.g., docking) approaches toward understanding P450-mediated DDI are rooted in the relatively recent advances in the crystallography of human drug-metabolizing cytochrome P450 enzymes. We believe that the application of such in silico approaches is immature, but worthy of discussion since these efforts appear to be useful to understand the ligand-binding events that give rise to the DDI. This information can aid in the generation of new design ideas. That redesign is aimed at minimizing the DDI risk by reducing affinity to a particular P450 while maintaining affinity against the therapeutic target. With an eye toward understanding P450-mediated DDI and docking, this section will include a high-level discussion of structural models available, the data needed to properly design docking experiments, common computational methods employed, and some limitations of SBDD approaches.

The docking approaches used to investigate ligand-binding modes in cytochrome P450s are similar, whether the investigator is seeking to understand metabolism or inhibition giving rise to DDI. Investigations into DDI might be classified into four general areas: perpetrators, victims, mechanism-based inhibitors (MBI), and heteroactivators. Perpetrator docking studies are used to identify the active pose for redesign to reduce the affinity to the P450 of interest. Victim docking experiments are used to increase the diversity of metabolic liability (metabolic switching) by altering compounds such that they can be metabolized by more than one P450. MBI docking studies are used to identify the binding mode giving rise to the metabolic activation as opposed to the inhibition event. Heteroactivators docking studies have focused on docking in the presence of another ligand or docking to peripheral binding sites.

Clearly the quality of the starting protein structure plays a significant role in the likelihood of successful docking. To date, several structures of human drug-metabolizing P450s have been published, including the structures of the P450s 3A4, 2C9, 2C8, 2E1, 2A6, and 2D6 (Williams et al., 2003, 2004; Wester et al., 2004; Yano et al., 2004, 2006; Ekroos and Sjogren, 2006; Rowland et al., 2006; Sansen et al., 2007; Porubsky et al., 2008). The ideal starting point for a docking experiment would be a ligand-bound crystal structure, in the P450 isozyme of interest, where the ligand is derived from the same series as the proposed compound to be docked. However, due to the limited number of human P450 crystal structures solved and the relatively complex process to obtain new structures, this is usually not attainable. Prior to the recent breakthroughs in P450 crystallization, and in instances where the crystal structure has yet to be solved for the specific isozyme of interest, homology models built upon structures of soluble bacterial P450 enzymes or recently solved mammalian P450 enzymes have been used in docking experiments with some success. An example of the later would be the design of P450 1A1 inhibitors using a homology model based on the recently solved 1A2 structure (Sangamwar et al., 2008). A review of many homology models is available (de Graaf et al., 2005).

6.7.4 P450 Docking

In order to initiate a docking experiment that will offer results that are useful in the redesign of a compound (or series) facing DDI liabilities, several key pieces of information are needed. While these requirements may seem obvious to some, the details can be lost in a fast-paced drug discovery team setting. The first piece of information is an understanding of the mechanism of DDI. We find that in vitro inhibition experiments using recombinant P450 enzymes and known probe substrates are straightforward to run and interpret. In addition, if the experiment is designed properly, time-dependent or mechanism-based inhibition can also be assessed (Hyland et al., 2009). In the instance of competitive inhibition, a researcher would routinely seek to understand the binding in the active site of the P450 enzyme being inhibited, while for heteroactivators, one may want to know whether a compound binds to a particular part of the active site (Locuson et al., 2007; Kang et al., 2008). Generally, the activating enzyme is the one that is also inhibited, but this is not always the case. The second, related piece of information is the identification of the P450 isoform implicated in the DDI. In addition, if the experiment is designed properly, time-dependent or mechanism-based inhibition can also be assessed (Hyland et al., 2009). Ideally, a software package similar to isoCYP (Terfloth et al., 2007) aimed at inhibitors instead of substrates will be developed to predict isoform specificity of P450 inhibition. Finally, any information regarding binding modes of the inhibitors that can be attained through experiments such as spectral binding assays will aid in the interpretation of docking results (Jefcoate, 1978). For example, if one knows that a nitrogen-containing compound is giving rise to DDI through competitive inhibition, and that it displays a type II binding spectrum when incubated with the P450 isoform that is inhibited, one can pay particular attention to those low energy binding poses that orient the nitrogen in proximity of the heme iron (Peng et al., 2008). Ultimately, regardless of the structure-based method used, the final poses must be assessed to determine if results are compatible with what is known about the biochemistry of the DDI being studied and fit with the overall knowledge of the metabolic cascade.

Some key tools to characterize P450 inhibitor binding in a structure-based format are docking and molecular dynamics simulations. Docking comes in several varieties and can use simple rigid protein representations. Some docking protocols, like the docking package AUTODOCK, employ side chain flexibility that may be able to effectively sample some of the various side chain conformations possible (Goodsell et al., 1996; Morris et al., 2008). It may be important to include solvent molecules in the active site. The poses will heavily depend on the force field parameters (Kirton et al., 2005). Computationally more expensive approaches like molecular dynamics simulations additionally allow a dynamically modeled active site cavity with solvent that responds to the ligand. A listing of specific docking and molecular dynamics packages is outside of the scope of this review; however a 2005 review article highlights many of the commercially available and public domain software packages available for homology modeling, molecular dynamics simulations, and automated docking (de Graaf et al., 2005).

The largest challenge facing computational ADME scientist attempting to model P450 ligand-binding interactions stems from the conformational flexibility of the protein. Large conformational changes have been observed upon ligand binding for several of the mammalian drug-metabolizing P450s. The most striking example is observed in the crystal structures of 3A4 published by Ekroos and Sjorgren in 2006 where large shifts in the entire F–G helix region are observed (Ekroos and Sjogren, 2006). However, even in the instance where major tertiary structure changes are not observed, minor side-chain movements can result in rather striking differences in active site geometry. In the case of P450 3A4 independently bound to either bromocriptine or clotrimazole, side chain conformation changes

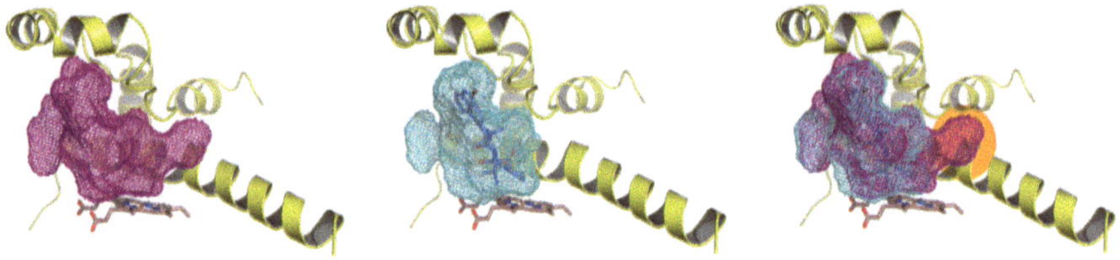

Fig. 6.2 Three panel view of P450 3A4 crystal structure, showing active site volumes in (**left**) ligand free, (**middle**) bromocriptine, (**right**) overlay of A and B, differences in *orange*

results in greatly modified active site space (Fig. 6.2 unpublished data, manuscript in preparation). Ways to effectively handle this problem include docking into an ensemble of P450 structures and using scoring functions to sort the resulting poses based upon predicted energies.

6.7.5 Mechanism Based Inhibitors

The development of in silico models to predict mechanism-based inhibitors (MBIs) of cytochrome P450 3A4 has been reported by a number of groups (Jones et al., 2007; Lightning et al.,2000). It is important for computational ADME scientist trying to mitigate this problem to keep in mind a number of critical features. First, when designing compounds with reduced MBI, any changes in the structure of an NCE may shift the metabolism to another P450 or phase 2 mechanism. Second, to potentially form an MBI, the NCE must be electronically appropriate to form a reactive intermediate. Third, the compound must have the correct features to be able to approach the heme for metabolic activation. Finally, the reactive moiety must either coordinate with the heme or apoprotein in a destructive fashion. The existing models tend to focus on electronics or identifying well-described reactive moieties. One Pfizer in silico approach predicts the likelihood of a TDI based on a Bayesian/fingerprint model. This model identifies the specific fragments that may lead to the TDI (Zientek in preparation) and can aid in prioritizing which molecules to test (high risk) as well as prioritize synthetic targets. Other groups have been working on a subset of MBIs that lead to a time-dependent inhibition (TDI) of the major P450s (Grime et al., 2009). For any of these reactions to occur in vivo, one must consider the "free drug" available, the amount of turnover in the P450 system, and the projected dose (concentration) of the NCE. One commercial product from Molecular Discovery is a model that flags compounds likely to be mechanism-based inhibitors in the Metasite^{TM} v. 3.0. Metasite^{TM} predictions for MBI are based on whether the compound is a good P450 substrate, contains potentially reactive moieties, and these moieties are likely to be exposed to the heme. This is a good multistep approach, which when combined with biotransformation experimentation can help aid project teams.

6.8 Current State of Transporter-Based DDI Models

Just as is the case for P450s, DDIs for transporters can occur due to competition from a more potent substrate or via direct inhibition. Most in silico models that have been published currently focus on substrates for a specific transporter (Ekins et al., 2007; Winiwarter and Hilgendorf, 2008; Zhang et al., 2002). However, there have been a number of direct inhibition models developed. Many central targets in drug discovery have transporter inhibition as the goal in therapeutic intervention. One such report covers the QSAR relationship of a local series inhibiting dopamine transporter (DAT), but there are many for Sert, Nat, Gaba, and system-L (Srivastava et al., 2008). These QSAR approaches have added significantly to the scholarship of transporter science over the last 30 years.

One of the more confusing aspects of transporter research has been the constant shifting between common name, alias, and gene name and of course species variations making it more difficult to determine if an in silico model exists for the correct endpoint of interest. The following reviews (Hagenbuch and Meier, 2003; Gottesman, 2002) provide some guidance to help understand the nomenclature, and we suggest the reader also refer to the previous chapters.

One of the most broadly studied transporters for drug distribution is the human MDR1 protein, commonly called P-glycoprotein (P-gp). P-gp has been among the most studied transporters for DDI. Since many P-gp substrates are commonly prescribed medications, in silico predictions of a significant DDI via P-gp may warrant follow-up in vitro or in vivo studies (Wang et al., 2005). However, Fenner (Fenner et al., 2009) has suggested that drug interactions solely related to P-gp may not be generally clinically significant.

Pregnane X receptor (PXR) regulates the expression of P-gp, multidrug resistance proteins, and cytochrome P450 enzymes. In silico models of PXR induction were built for PXR systems from human and other orthologs (Ung et al., 2007; Khandelwal et al., 2008). Pharmacophore modeling has identified the essential features for binding, and both pharmacophore and docking models have been used to identify additional agonists and antagonists (Lemaire et al., 2007; Ekins et al., 2007). The ability to identify antibiotics that may cause drug–drug interactions using pharmacophores was investigated (Yasuda et al., 2008).

Another transporter that has been linked with adverse events due to inhibition is the bile salt export pump (BSEP). This transporter is also called ATP-binding cassette transporter (ABCB11) and has become associated with hepatotoxicity. A number of in silico methodologies to identify inhibitors are becoming more common as experimental screens are being incorporated into discovery funnels at many pharmaceutical companies (Lin, 2007).

A growing number of academic groups are also actively contributing to the generation of in silico methods to identify inhibitors. One recent human organic transporter (OCT1) inhibition model was published applying PCA analysis and pharmacophore approaches looking for shared inhibitor structural fragments (Ahlin et al., 2008).

While direct inhibition of transporters is clearly a concern, a number of in silico methods also exist to identify those compounds that are good substrates of a particular transporter and could inhibit the absorption or distribution of an NCE competitively. Current example of this are OATP substrate models that are aimed at identifying compounds likely to inhibit the hepatic distribution of statins (Lau et al., 2007; Lin, 2007). Researchers at Pfizer have developed an in silico model of OATP uptake that can identify compounds that may cause competitive inhibition with other substrates of this transporter. As we continue to gain more knowledge about transporter involvement in distribution, more in silico methodologies will be applied. These models should aid in optimizing distribution across the blood–brain barrier, distribution into the liver, and eventually aid in designing compounds that inhibit efflux transporters in tumors.

6.9 Limitations in Current In Silico Methodologies

There have also been significant advances made in the predictive power of in silico models and how predictions fit into drug design. Most major pharmaceutical companies have spent significant resources to make in silico tools for DDI prediction available to the teams designing new chemical entities (NCEs). It should be noted however that these models can have major limitations and end users need to have adequate training to know how to best apply these approaches.

One potential pitfall that in silico methods can have is called "model overtraining." This can be a significant problem, especially when dealing with small data sets. If the number of descriptors used is too large when compared to the training set, the model can have substantial biases (Li et al., 2008).

Current statistical models are powerful for screening and provide rapid results; however, this simplicity can be their Achilles heel. Current structure-based models can be improved in the areas of protein and ligand sampling, solvation descriptions, and energetics of complex formation.

Another potential source of error is the experimental data set used to generate the model. Errors can arise due to variability between laboratories, enzyme batches, expression levels, NCE physicochemical properties (e.g., solubility), and even experimentalists (Li et al., 2008). Most groups are striving to reduce microsomal and recombinant batch-to-batch variability by using positive controls to maintain data quality and consistency, but there is still need for improvement. As an example of the impact NCE physicochemical properties can play: If in an automated screen, a large series of insoluble 3A4 inhibitors is tested at an experimental concentration that yields erroneous data and these data are mistakenly incorporated into the training set of the model, then the model will likely misclassify similar inhibitors.

6.10 Future Perspectives, Emerging Science, Conclusions

Data integration, visualization, and translation into new decisions and designs remain as one of the greatest hurdles faced by drug discovery scientist (Stoner et al., 2004; Wishart, 2007). For decisions about testing, synthesis, and further compound progression to be made in a rapid manner, more integrated tools will need to be developed. Many of these tools will link in silico predictions with in vitro, in vivo, and biomarker data. Analyzing this vast amount of data will require the development of significantly more complex data decision tools.

Future models of DDI will need to become much more mechanistic and take into account free drug theory. As microsomal binding and protein binding assays continue to become standard approaches, in silico methods for free fraction will become as common for DDI predictions as they are currently for free clearance predictions. In addition, as one P450 is inhibited, in silico models will eventually be able to predict the fraction metabolized (FM) shift to another clearance mechanism, and the predicted impact on exposure in terms of AUC, $C_{\max}$, and associated pharmacokinetic parameters. Systems biology and the link between transporters throughout the body and their impact on exposure related distribution can be modeled. Some in silico approaches have attempted to link P450 3A4 and P-gp, but there are a significant number of transporter enzyme systems that have yet to be understood at a fundamental level.

Further revelations of the crystal structures of P450s and transporters and a greater understanding of the importance of isoforms such as P450s 3A5 and 3A7 will continue to drive the evolution of DDI science. A greater understanding of inhibitors of the phase 2 metabolic pathways such as aldehyde oxidase or sulfonyl transferases is an area that is also poised for further modeling. While some basic understanding of induction and inactivation exists experimentally, this knowledge has yet to be fully translated into in silico approaches that have the ability to take into account the effects of time and concentration on DDI. Eventually "grand consensus" models will simulate the effects of DDI, taking into account induction, inactivation, inhibition, and concomitant medication effects on fraction metabolized and exposure.

References

Afzelius L, Zamora I, Masimirembwa CM, Karlen A, Andersson TB, Mecucci S, Baroni M and Cruciani G (2004) Conformer- and alignment-independent model for predicting structurally diverse competitive CYP2C9 inhibitors. *J Med Chem* **47**:907–914.

Ahlin G, Karlsson J, Pedersen JM, Gustavsson L, Larsson R, Matsson P, Norinder U, Bergstroem CAS and Artursson P (2008) Structural requirements for drug inhibition of the liver specific human organic cation transport protein. *J Med Chem* **51**:5932–5942.

Ahlstrom MM, Ridderstrom M and Zamora I (2007) CYP2C9 structure-metabolism relationships: substrates, inhibitors, and metabolites. *J Med Chem* **50**:5382–5391.

Arimoto R, Prasad MA and Gifford EM (2005) Development of CYP3A4 inhibition models: comparisons of machine-learning techniques and molecular descriptors. *J Biomol Screening* **10**:197–205.

Bathelt CM, Mulholland AJ and Harvey JN (2008) QM/MM modeling of benzene hydroxylation in human cytochrome P450 2C9. *J Phys Chem A* **112**:13149–13156.

Bell L, Bickford S, Nguyen PH, Wang J, He T, Zhang B, Friche Y, Zimmerlin A, Urban L and Bojanic D (2008) Evaluation of fluorescence- and mass spectrometry-based CYP inhibition assays for use in drug discovery. *J Biomol Screening* **13**:343–353.

Bravi G, Gancia E, Mascagni P, Pegna M, Todeschini R and Zaliani A (1997) MS-WHIM, new 3D theoretical descriptors derived from molecular surface properties: a comparative 3D QSAR study in a series of steroids. *J Comput Aided Mol Des* **11**:79–92.

Bravi G and Wikel JH (2000) Application of MS-WHIM descriptors: 1. Introduction of new molecular surface properties and 2. Prediction of binding affinity data. *Quant Struct Act Relat* **19**:29–38.

Burton J, Ijjaali I, Barberan O, Petitet F, Vercauteren DP and Michel A (2006) Recursive partitioning for the prediction of cytochromes P450 2D6 and 1A2 inhibition: importance of the quality of the dataset. *J Med Chem* **49**:6231–6240.

Byvatov E, Baringhaus K-H, Schneider G and Matter H (2007) A virtual screening filter for identification of cytochrome P450 2C9 (CYP2C9) inhibitors. *QSAR Comb Sci* **26**:618–628.

Chang C-eA, Chen W and Gilson MK (2007) Ligand configurational entropy and protein binding. *Proc Natl Acad Sci USA* **104**:1534–1539.

Chohan KK, Paine SW, Mistry J, Barton P and Davis AM (2005) A rapid computational filter for cytochrome P450 1A2 inhibition potential of compound libraries. *J Med Chem* **48**:5154–5161.

de Graaf C, Oostenbrink C, Keizers PHJ, van der Wijst T, Jongejan A and Vermeulen NPE (2006) Catalytic site prediction and virtual screening of cytochrome P450 2D6 substrates by consideration of water and rescoring in automated docking. *J Med Chem* **49**:2417–2430.

de Graaf C, Vermeulen NPE and Feenstra KA (2005) Cytochrome P450 *in silico*: an integrative modeling approach. *J Med Chem* **48**:2725–2755.

de Groot M (2006) Designing better drugs: predicting cytochrome P450 metabolism. *Drug Discov Today* **11**:601–606.

de Groot MJ, Alex AA and Jones BC (2002) Development of a combined protein and pharmacophore model for cytochrome P450 2C9. *J Med Chem* **45**:1983–1993.

de Groot M, Lewis DFV and Modi S (2007) Molecular modeling and quantitative structure-activity relationship of substrates and inhibitors of drug metabolism enzymes, in pp 809–825, Elsevier.

Doweyko AM, John BT and David JT (2007) Three-dimensional quantitative structure-activity relationship: the state of the art, in *Comprehensive Medicinal Chemistry II* pp 575–595, Elsevier, Oxford.

Ekins S, Bravi G, Binkley S, Gillespie JS, Ring BJ, Wikel JH and Wrighton SA (1999) Three and four dimensional-quantitative structure activity relationship analyses of CYP3A4 inhibitors. *J Pharm Exp Ther* **290**:429–438.

Ekins S, Bravi G, Binkley S, Gillespie JS, Ring BJ, Wikel JH and Wrighton SA (2000a) Three- and four-dimensional-quantitative structure activity relationship (3D/4D-QSAR) analyses of CYP2C9 inhibitors. *Drug Metab Dispos* **28**:994–1002.

Ekins S, Mestres J and Testa B (2007) *In silico* pharmacology for drug discovery: applications to targets and beyond. *Br J Pharmacol* **152**:21–37.

Ekins S, Ring BJ, Bravi G, Wikel JH and Wrighton SA (2000b) Predicting drug-drug interactions *in silico* using pharmacophores: a paradigm for the next millennium, in *Pharmacophore perception, development, and use in drug design* (Guner OF ed) pp 269–299, IUL, San Diego.

Ekins S, Ring BJ, Grace J, McRobie-Belle DJ and Wrighton SA (2000c) Present and future in vitro approaches for drug metabolism. *J Pharm Tox Methods* **44**:313–324.

Ekins S, Stresser DM and Williams JA (2003) In vitro and pharmacophore insights into CYP3A enzymes. *Trends Pharmacol Sci* **24**:191–196.

Ekins S and Swaan PW (2004) Development of computational models for enzymes, transporters, channels and receptors relevant to ADME/TOX. *Rev Comp Chem* **20**:333–415.

Ekroos M and Sjogren T (2006) Structural basis for ligand promiscuity in cytochrome P450 3A4. *Proc Natl Acad Sci USA* **103**:13682–13687.

Fenner KS, Troutman MD, Kempshall S, Cook JA, Ware JA, Smith DA and Lee CA (2009) Drug-drug interactions mediated through p-glycoprotein: clinical relevance and in vitro-in vivo correlation using digoxin as a probe drug. *Clin Pharmacol Ther (NY, NY, US)* **85**: 173–181.

Ferguson AM, Heritage T, Jonathon P, Pack SE, Phillips L, Rogan J and Snaith PJ (1997) EVA: a new theoretically based molecular descriptor for use in QSAR/QSPR analysis. *J Comput Aided Mol Des* **11**:143–152.

Fox T and Kriegl JM (2007) Linear quantitative structure-activity relationships for the interaction of small molecules with human cytochrome P450 isoenzymes. *Annu Rep Comput Chem* **3**: 63–81.

Fuhr U, Strobl G, Manaut F, Anders EM, Soergel F, Lopez-de-Binas E, Chu DTW, Pernet AG, Mahr G and et al. (1993) Quinolone antibacterial agents: relationship between structure and in vitro inhibition of the human cytochrome P450 isoform CYP1A2. *Mol Pharmacol* **43**: 191–199.

Gleeson MP, Davis AM, Chohan KK, Paine SW, Boyer S, Gavaghan CL, Arnby CH, Kankkonen C and Albertson N (2007) Generation of in-silico cytochrome P450 1A2, 2C9, 2C19, 2D6, and 3A4 inhibition QSAR models. *J Comput Aided Mol Des* **21**:559–573.

Goodsell DS, Morris GM and Olson AJ (1996) Automated docking of flexible ligands: applications of AutoDock. *J Mol Recognit* **9**:1–5.

Gottesman M (2002) Multidrug resistance in cancer: role of ATP-dependent transporters. *Nat Rev* **2**:48–58.

Grime KH, Bird J, Ferguson D and Riley RJ (2009) Mechanism-based inhibition of cytochrome P450 enzymes: an evaluation of early decision making in vitro approaches and drug-drug interaction prediction methods. *Eur J Pharm Sci* **36**:175–191.

Hagenbuch B and Meier PJ (2003) The superfamily of organic anion transporting polypeptides. *Biochim Biophys Acta Biomembr* **1609**:1–18.

Helguera AM, Combes RD, Gonzalez MP and Cordeiro MNDS (2008) Applications of 2D descriptors in drug design: a DRAGON tale. *Curr Top Med Chem (Sharjah, UAE)* **8**:1628–1655.

Hritz J, de Ruiter A and Oostenbrink C (2008) Impact of plasticity and flexibility on docking results for cytochrome P450 2D6: a combined approach of molecular dynamics and ligand docking. *J Med Chem* **51**:7469–7477.

Hudelson MG, Ketkar NS, Holder LB, Carlson TJ, Peng C-C, Waldher BJ and Jones JP (2008) High confidence predictions of drug-drug interactions: predicting affinities for cytochrome P450 2C9 with multiple computational methods. *J Med Chem* **51**:648–654.

Hyland R OR, Stoner C, West M, Wester MR, Youdim K, Zientek M (2009) Drug-drug interactions: Screening for liability and assessment of risk, in *Hit and Lead Profiling* (Urban L, Faller B. eds), Wiley-VCH, Weinheim.

Jefcoate CR (1978) Measurement of substrate and inhibitor binding to microsomal cytochrome P-450 by optical-difference spectroscopy. *Methods Enzymol* **52**:258–279.

Jensen BF, Vind C, Padkjr SB, Brockhoff PB and Refsgaard HHF (2007) *In Silico* prediction of cytochrome P450 2D6 and 3A4 inhibition using Gaussian kernel weighted k-nearest neighbor and extended connectivity fingerprints, including structural fragment analysis of inhibitors versus noninhibitors. *J Med Chem* **50**:501–511.

Jones DR, Ekins S, Li L and Hall SD (2007) Computational approaches that predict metabolic intermediate complex formation with CYP3A4 (+b5). *Drug Metab Dispos* **35**: 1466–1475.

Kang P, Liao M, Wester MR, Leeder JS, Pearce RE and Correia MA (2008) CYP3A4-mediated carbamazepine (CBZ) metabolism: formation of a covalent CBZ-CYP3A4 adduct and alteration of the enzyme kinetic profile. *Drug Metab Dispos* **36**:490–499.

Khandelwal A, Krasowski MD, Reschly EJ, Sinz MW, Swaan PW and Ekins S (2008) Machine learning methods and docking for predicting human pregnane X receptor activation. *Chem Res Toxicol* **21**:1457–1467.

Kier LB (1971) *Molecular Orbital Theory in Drug Research. (Medicinal Chemistry, Vol. 10)*, Academic Press, New York.

Kirton SB, Murray CW, Verdonk ML and Taylor RD (2005) Prediction of binding modes for ligands in the cytochromes P450 and other heme-containing proteins. *Proteins: Struct Funct Bioinf* **58**:836–844.

Kontijevskis A, Komorowski J and Wikberg JES (2008) Generalized proteochemometric model of multiple cytochrome P450 enzymes and their inhibitors. *J Chem Inf Model* **48**:1840–1850.

Kriegl JM, Arnhold T, Beck B and Fox T (2005a) A support vector machine approach to classify human cytochrome P450 3A4 inhibitors. *J Comput Aided Mol Des* **19**:189–201.

Kriegl JM, Eriksson L, Arnhold T, Beck B, Johansson E and Fox T (2005b) Multivariate modeling of cytochrome P450 3A4 inhibition. *Eur J Pharm Sci* **24**:451–463.

Labute P (2000) A widely applicable set of descriptors. *J Mol Graphics Modell* **18**:464–477.

Lau YY, Huang Y, Frassetto L and Benet LZ (2007) Effect of OATP1B transporter inhibition on the pharmacokinetics of atorvastatin in healthy volunteers. *Clin Pharmacol Ther (NY, NY, US)* **81**:194–204.

Lemaire G, Benod C, Nahoum V, Pillon A, Boussioux A-M, Guichou J-F, Subra G, Pascussi J-M, Bourguet W, Chavanieu A and Balaguer P (2007) Discovery of a highly active ligand of human pregnane X receptor: a case study from pharmacophore modeling and virtual screening to "in vivo" biological activity. *Mol Pharmacol* **72**:572–581.

Lewis DFV, Lake BG and Dickins M (2006a) Quantitative structure-activity relationships (QSARs) in CYP3A4 inhibitors: the importance of lipophilic character and hydrogen bonding. *J Enzyme Inhib Med Chem* **21**:127–132.

Lewis DFV, Lake BG and Dickins M (2007) Quantitative structure-activity relationships (QSARs) in inhibitors of various cytochromes P450: the importance of compound lipophilicity. *J Enzyme Inhib Med Chem* **22**:1–6.

Lewis DFV, Lake BG, Ito Y and Anzenbacher P (2006b) Quantitative structure-activity relationships (QSARs) within cytochromes P450 2B (CYP2B) subfamily enzymes: the importance of lipophilicity for binding and metabolism. *Drug Metab Drug Interact* **21**:213–231.

Li H, Sun J, Fan X, Sui X, Zhang L, Wang Y and He Z (2008) Considerations and recent advances in QSAR models for cytochrome P450-mediated drug metabolism prediction. *J Comput Aided Mol Des* **22**:843–855.

Li H, Yap CW, Ung CY, Xue Y, Li ZR, Han LY, Lin HH and Chen YZ (2007) Machine learning approaches for predicting compounds that interact with therapeutic and ADMET related proteins. *J Pharm Sci* **96**:2838–2860.

Lightning LK, Jones JP, Friedberg T, Pritchard MP, Shou M, Rushmore TH, Trager WF (2000) Mechanism-based inactivation of cytochrome P450 3A4 by L-754,394. *Biochemistry* **39**: 4276–4287.

Lin JH (2007) Transporter-mediated drug interactions: clinical implications and in vitro assessment. *Expert Opin Drug Metab Toxicol* **3**:81–92.

Locuson CW, Gannett PM, Ayscue R and Tracy TS (2007) Use of simple docking methods to screen a virtual library for heteroactivators of cytochrome P450 2C9. *J Med Chem* **50**: 1158–1165.

Masimirembwa CM, Ridderstrom M, Zamora I and Andersson TB (2002) Combining pharmacophore and protein modeling to predict CYP450 inhibitors and substrates. *Methods Enzymol* **357**:133–144.

Morris GM, Huey R and Olson AJ (2008) Using AutoDock for ligand-receptor docking. *Curr Protoc Bioinformatics*, **Chapter 8**:Unit 8.14.

Neves MA, Dinis TC, Colombo G and Sa e Melo ML (2007) Combining computational and biochemical studies for a rationale on the anti-aromatase activity of natural polyphenols. *Chem Med Chem* **2**:1750–1762.

O'Brien SE and De Groot MJ (2005) Greater than the sum of its parts: combining models for useful ADMET prediction. *J Med Chem* **48**:1287–1291.

Pastor M, Cruciani G, McLay I, Pickett S and Clementi S (2000) GRid-INdependent Descriptors (GRIND): a novel class of alignment-independent three-dimensional molecular descriptors. *J Med Chem* **43**:3233–3243.

Peng C-C, Cape JL, Rushmore T, Crouch GJ and Jones JP (2008) Cytochrome P450 2C9 type II binding studies on quinoline-4-carboxamide analogues. *J Med Chem* **51**:8000–8011.

Porubsky PR, Meneely KM and Scott EE (2008) Structures of human cytochrome P-450 2E1. Insights into the binding of inhibitors and both small molecular weight and fatty acid substrates. *J Biol Chem* **283**:33698–33707.

Refsgaard HHF, Jensen BF, Christensen IT, Hagen N and Brockhoff PB (2006) *In silico* prediction of cytochrome P450 inhibitors. *Drug Dev Res* **67**:417–429.

Rowland P, Blaney FE, Smyth MG, Jones JJ, Leydon VR, Oxbrow AK, Lewis CJ, Tennant MG, Modi S, Eggleston DS, Chenery RJ and Bridges AM (2006) Crystal structure of human cytochrome P450 2D6. *J Biol Chem* **281**:7614–7622.

Sangamwar AT, Labhsetwar LB and Kuberkar SV (2008) Exploring CYP1A1 as anticancer target: homology modeling and *in silico* inhibitor design. *J Mol Model* **14**:1101–1109.

Sansen S, Yano JK, Reynald RL, Schoch GA, Griffin KJ, Stout CD and Johnson EF (2007) Adaptations for the oxidation of polycyclic aromatic hydrocarbons exhibited by the structure of human P450 1A2. *J Biol Chem* **282**:14348–14355.

Schuster D, Laggner C, Steindl TM, Palusczak A, Hartmann RW and Langer T (2006) Pharmacophore modeling and *in silico* screening for new P450 19 (aromatase) inhibitors. *J Chem Inf Model* **46**:1301–1311.

Smith DA, Cucurull-Sanchez L, John BT and David JT (2007) The adaptive in combo strategy, in *Comprehensive Medicinal Chemistry II* pp 957–969, Elsevier, Oxford.

Srivastava AK, Jaiswal M, Archana and Chaurasia S (2008) QSAR of substituted N-benzyl piperidines in the GBR series. *J Indian Chem Soc* **85**:842–848.

Stjernschantz E, Vermeulen NPE and Oostenbrink C (2008) Computational prediction of drug binding and rationalisation of selectivity towards cytochromes P450. *Expert Opin Drug Metab Toxicol* **4**:513–527.

Stoner CL, Gifford E, Stankovic C, Lepsy CS, Brodfuehrer J, Prasad JVNV and Surendran N (2004) Implementation of an ADME enabling selection and visualization tool for drug discovery. *J Pharm Sci* **93**:1131–1141.

Terfloth L, Bienfait B and Gasteiger J (2007) Ligand-based models for the isoform specificity of cytochrome P450 3A4, 2D6, and 2C9 substrates. *J Chem Inf Model* **47**:1688–1701.

Ung CY, Li H, Yap CW and Chen YZ (2007) *In silico* prediction of pregnane X receptor activators by machine learning approaches. *Mol Pharmacol* **71**:158–168.

Wang Y, Li Y and Wang B (2007) An *in silico* method for screening nicotine derivatives as cytochrome P450 2A6 selective inhibitors based on kernel partial least squares. *Int J Mol Sci* **8**:166–179.

Wang Y-H, Li Y, Yang S-L and Yang L (2005) An *in silico* approach for screening flavonoids as P-glycoprotein inhibitors based on a Bayesian-regularized neural network. *J Comput Aided Mol Des* **19**:137–147.

Wester MR, Yano JK, Schoch GA, Yang C, Griffin KJ, Stout CD and Johnson EF (2004) The structure of human cytochrome P450 2C9 complexed with flurbiprofen at 2.0-A resolution. *J Biol Chem* **279**:35630–35637.

Williams PA, Cosme J, Vinkovic DM, Ward A, Angove HC, Day PJ, Vonrhein C, Tickle IJ and Jhoti H (2004) Crystal structures of human cytochrome P450 3A4 bound to metyrapone and progesterone. *Science* **305**:683–686.

Williams PA, Cosme J, Ward A, Angove HC, Matak Vinkovic D and Jhoti H (2003) Crystal structure of human cytochrome P450 2C9 with bound warfarin. *Nature* **424**: 464–468.

Winiwarter S and Hilgendorf C (2008) Modeling of drug-transporter interactions using structural information. *Curr Opin Drug Discov Devel* **11**:95–103.

Wishart DS (2007) Improving early drug discovery through ADME modelling: an overview. *Drugs R&D* **8**:349–362.

Yano JK, Denton TT, Cerny MA, Zhang X, Johnson EF and Cashman JR (2006) Synthetic inhibitors of cytochrome P-450 2A6: inhibitory activity, difference spectra, mechanism of inhibition, and protein cocrystallization. *J Med Chem* **49**:6987–7001.

Yano JK, Wester MR, Schoch GA, Griffin KJ, Stout CD and Johnson EF (2004) The structure of human microsomal cytochrome P450 3A4 determined by X-ray crystallography to 2.05-A resolution. *J Biol Chem* **279**:38091–38094.

Yasuda K, Ranade A, Venkataramanan R, Strom S, Chupka J, Ekins S, Schuetz E and Bachmann K (2008) A comprehensive in vitro and *in silico* analysis of antibiotics that activate pregnane X receptor and induce CYP3A4 in liver and intestine. *Drug Metab Dispos* **36**:1689–1697.

Youdim KA, Zayed A, Dickins M, Phipps A, Griffiths M, Darekar A, Hyland R, Fahmi O, Hurst S, Plowchalk DR, Cook J, Guo F and Obach RS (2008) Application of CYP3A4 in vitro data to predict clinical drug–drug interactions; predictions of compounds as objects of interaction. *Br J Clin Pharmacol* **65**:680–692.

Zhang EY, Phelps MA, Cheng C, Ekins S and Swaan PW (2002) Modeling of active transport systems. *Adv Drug Delivery Rev* **54**:329–354.

Zhou D, Liu R, Otmani SA, Grimm SW, Zauhar RJ and Zamora I (2007) Rapid classification of CYP3A4 inhibition potential using support vector machine approach. *Lett Drug Des Discov* **4**:192–200.

Chapter 7
In Vitro Techniques to Study Drug–Drug Interactions of Drug Metabolism: Cytochrome P450

J. Brian Houston and Aleksandra Galetin

Abstract Approaches are discussed for the generation and use of in vitro data on metabolic reversible inhibition, time-dependent inhibition and induction to make predictions of in vivo drug–drug interactions. The in vitro experimental conduct, choice of probe substrates appropriate for individual P450 enzymes and the importance of nonspecific binding within in vitro systems are discussed. In addition to the inhibitor/inducer properties, the importance of enzyme turnover and the victim drug properties (e.g. parallel elimination pathways and gut enzymes) for quantitative prediction via either of three interaction mechanisms are discussed. Contribution of multiple inhibitors (or metabolites) and/or consequences of a multiple inhibition mechanisms are addressed, as well as mechanisms by which false negatives and false positives may result. Finally, perspectives on future application and improvements of these prediction strategies are outlined.

7.1 Introduction

Changes in cytochrome P450 activity represent an important mechanism for drug–drug interactions (DDIs) and are of major interest to drug metabolism scientists both in academia and within the pharmaceutical industry. The use of in vitro data to delineate drug-related changes in P450-mediated drug metabolism and hence identify potential serious DDIs in humans is widespread. Over recent years, substantial technological advances have been made in the conduct of in vitro studies in terms of automation and generic protocols. However, there remains a lack of confidence in the interpretation of the outcome, in particular with regard to false-negative and

J.B. Houston (✉)
School of Pharmacy and Pharmaceutical Sciences, University of Manchester, Manchester M13 9PT, UK
e-mail: brian.houston@manchester.ac.uk

K.S. Pang et al. (eds.), *Enzyme- and Transporter-Based Drug–Drug Interactions*,
DOI 10.1007/978-1-4419-0840-7_7,

false-positive predictions. This may reflect a lack of understanding of the importance of integrating in vitro data with other pharmacokinetic phenomena to achieve a holistic view of potential DDIs.

Mechanistically, a number of scenarios may result. First, there may be competitive inhibition between two or more drug substrates for metabolism by a particular cytochrome P450 enzyme. Second, there may be an alteration in the regulation of these particular enzymes. This may take the form of enzyme inactivation as a result of irreversible inhibition (time-dependent inhibition) or up-regulation of the synthesis of particular P450 enzymes – induction. All three mechanisms will be considered in this chapter. Emphasis will be on the utility of different in vitro approaches with human systems in terms of the choices available, the caveats which must be considered and also the particular needs and inherent limitations.

A major aim of in vitro studies is to generate kinetic parameters that can be used within prediction strategies to assess possible in vivo consequences. The most common in vivo metric used to assess DDIs is the change in area under the plasma concentration–time curve (AUC) of the victim drug following multiple dosing of a second interacting drug relative to the control state (Ito et al., 1998; Tucker et al., 2001; Yao and Levy, 2002). The AUC is the most reliable parameter as it inversely reflects drug clearance. It is assumed that the second drug (putative modifier or perpetrator) has reached steady state by the second phase of the study. Often the change in the clearance is related to the average steady-state concentration of a modifier within the dosing interval. In contrast to a time-averaged value, certain simulation programmes (for example, Simcyp®) (Jamei et al., 2009b) can incorporate the time course of the modifier concentration and hence generate a temporal profile of the process. This is a particularly valuable option when dealing with reversible inhibition mechanisms where marked changes may occur within a relatively short period of time but is less critical for time-dependent inhibition and induction where the altered state of activity is relatively stable, at least during the dosing interval period of the study.

The link between AUC and the in vitro assessment of hepatocellular enzyme-mediated clearance (CL_{int}) can be seen through consideration of the well-stirred liver model (Pang and Rowland, 1977) shown in Equation (7.1):

$$\text{AUC} = \frac{F_A \cdot F_G \cdot \text{dose}}{fu_b \cdot CL_{int}} \tag{7.1}$$

where F_A and F_G are the fractions of administered drug dose absorbed from the intestine and escaping metabolism by the intestine, respectively, and fu_b is the fraction of unbound drug in the blood.

Thus on the assumption that the change in P450-mediated clearance is not accompanied by any effect on intestinal absorption or plasma protein binding, comparison of AUC under control conditions and in the presence of the modifier would depend on both hepatic and intestinal events, as illustrated in Equation (7.2):

$$\frac{\mathrm{AUC}'}{\mathrm{AUC}} = \frac{F_{\mathrm{G}}'}{F_{\mathrm{G}}} \cdot \frac{\mathrm{CL_{int}}}{\mathrm{CL_{int}}'} \tag{7.2}$$

where the prime superscript indicates the parameter in the presence of the modifier.

It is valuable to expand this relationship to indicate the specific role of individual P450 enzymes in clearance (Rowland and Matin, 1973; Ito et al., 2005) and to consider a change in AUC for a particular drug resulting in the generic equation shown below.

$$\frac{\mathrm{AUC}'}{\mathrm{AUC}} = \frac{F_{\mathrm{G}}'}{F_{\mathrm{G}}} \cdot \frac{1}{\sum_i^n \frac{fm_{\mathrm{CYP}i}}{\mathrm{CL_{int}}/\mathrm{CL_{int}}'} + \left(1 - \sum_i^n fm_{\mathrm{CYP}i}\right)} \tag{7.3}$$

where $fm_{\mathrm{CYP}i}$ denotes the fraction of the victim drug clearance accountable to the particular P450 subject to the inhibition/induction effect; the term i indicates the existence of multiple enzymes (n). The overall consequence being dependent on the sum of n P450s involved in the victim drug clearance and their respective change in activity. The $(1-\Sigma fm_{\mathrm{CYP}})$ term in the denominator accounts for other clearance pathways (other metabolic enzymes, biliary or renal excretion) for the victim drug that are not altered by the modifier under investigation. While the second term in Equation (7.3) relates to hepatic changes, the first comprises of the F_{G} ratio and allows for intestinal changes. For the latter the predominant expression of CYP3A in the intestine (Paine et al., 2006) means that an fm_{CYP} term is not necessary. Interaction with intestinal CYP3A enzymes is accommodated in an analogous way to the liver and the F_{G} ratio can be expanded as indicated by Equation (7.4).

$$\frac{F_{\mathrm{G}}'}{F_{\mathrm{G}}} = \frac{1}{F_{\mathrm{G}} + (1 - F_{\mathrm{G}})\,\frac{\mathrm{CL_{intG}}'}{\mathrm{CL_{intG}}}} \tag{7.4}$$

It is clear from the generic equations that a number of other parameters are involved in addition to the in vitro parameters that describe the change in $\mathrm{CL_{int}}$. These particular terms (fm_{CYP} and F_{G}) are victim drug characteristics, rather than inherent perpetrator properties. Thus, the success of a prediction relies as much on how well the pharmacokinetic properties of the victim drug have been characterized as much as how accurate the in vitro kinetics of inhibition/induction have been determined.

Each of the three mechanisms described above can be defined by a specific mechanistic equation for the $\mathrm{CL_{int}}$ ratio which can be substituted within the generic equations (7.3 and 7.4) for the hepatic and intestinal components (see Table 7.1). In these equations the interactant [I] (inhibitor/inducer) concentration at the enzyme/effector site cannot be measured. Therefore, several surrogate concentration terms have been proposed – both total and unbound plasma concentration in either systemic or hepatic portal vein (for the liver) and intestinal luminal concentration (for the intestine) (Kanamitsu et al., 2000b; Ito et al., 2004; Obach et al., 2006; Galetin et al., 2007). The hepatic input concentration combines the circulating

Table 7.1 Mechanistic equations describing changes in hepatic and intestinal CL_{int}

Mechanism	Mechanistic equation for CL_{int} ratio: Hepatic $\frac{CL_{int}}{CL_{int}'}$	Intestinal $\frac{CL_{int\,G}}{CL_{int\,G}'}$
Reversible inhibition[a]	$1+\sum_{j}^{m}[\mathrm{I}]_j/K_{ij}$	$1+\sum_{j}^{m}[\mathrm{I_G}]_j/K_{ij}$
Irreversible inhibition[b]	$1+\sum_{j}^{m}\frac{k_{\mathrm{inact}\,j}\,[\mathrm{I}]}{k_{\mathrm{deg}}\left(K_{Ij}+[\mathrm{I}]\right)}$	$1+\sum_{j}^{m}\frac{k_{\mathrm{inact}\,j}\,[\mathrm{I_G}]}{k_{\mathrm{deg}}\left(K_{Ij}+[\mathrm{I_G}]\right)}$
Induction[c]	$\frac{1}{1+\sum_{j}^{m}\frac{E_{\mathrm{max}\,j}\,[\mathrm{I}]}{[\mathrm{I}]+EC_{50j}}}$	$\frac{1}{1+\sum_{j}^{m}\frac{E_{\mathrm{max}\,j}\,[\mathrm{I_G}]}{[\mathrm{I_G}]+EC_{50j}}}$

[a] where [I] is the inhibitor concentration available to the enzyme and K_i is the inhibition constant, $[\mathrm{I_G}]$ refers to concentration in the intestine during absorption phase.
[b] where K_{inact} is the maximal inactivation rate constant, K_I the inhibitor concentration at 50% of K_{inact} and K_{deg} is the endogenous degradation rate constant of the enzyme.
[c] where E_{max} is the maximal induction effect and EC_{50} is the concentration at 50% of E_{max}. In each case the possibility of in inhibitory species is accounted for with the subscript j.

systemic plasma concentration and the concentration of an inhibitor occurring during the absorption phase.

Multiple inhibitory species are accounted for in the mechanistic equations in Table 7.1. The generic equations are also flexible in that they can be adapted to accommodate simultaneously two or even all three mechanisms for one particular DDI (Equation 7.5) (Rostami-Hodjegan and Tucker, 2004; Fahmi et al., 2008b; Hinton et al., 2008). This is far from an academic requirement as often perpetrators cause induction as well as inhibition and inhibition may occur by both of the mechanisms outlined above.

$$\frac{\mathrm{AUC}'}{\mathrm{AUC}}=\frac{1}{\frac{fm_{\mathrm{CYP}}}{\prod_{k}^{p}\mathrm{CL_{int}}/\mathrm{CL_{int}}_k'}+(1-fm_{\mathrm{CYP}})} \quad (7.5)$$

where k indicates the number of mechanisms p involved.

7.2 General Approach and Practices

The prediction of DDIs can be envisaged in a number of stages, as summarized in the four-step scheme shown in Fig. 7.1 Although a general strategy, it is instructive to consider first the specific case of reversible inhibition as this is the most developed prediction scenario. As described by Equation (7.6) the CL_{int} of a substrate is reduced by a factor that relates inhibitor concentration [I] to its potency (K_i). The distinction between competitive and non-competitive inhibition mechanisms is not relevant when the substrate concentration is much lower than the K_m value, which is a commonly encountered in vivo situation.

$$CL_{int}' = \frac{CL_{int}}{1 + [I]/K_i} \tag{7.6}$$

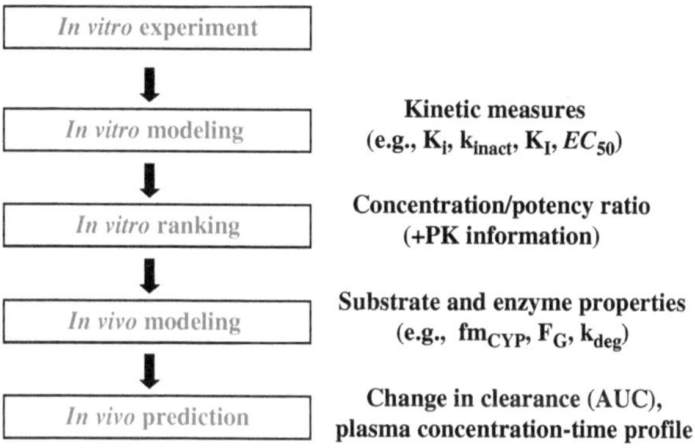

Fig. 7.1 General strategies for prediction of metabolic drug–drug interactions from in vitro kinetic data

1. Data from the in vitro experiment are first modelled using traditional biochemical models to determine K_i. The origins of the in vitro procedures used to obtain the parameter values in the mechanistic equations are from standard enzymology and are well documented in biochemistry textbooks; however, their application to a multi-enzyme family such as the cytochrome P450 enzymes has always been a subject for debate. There are several caveats associated with the application of Michaelis–Menten kinetics to cytochrome P450 enzyme kinetics (see later).
2. An in vitro ranking can be made, provided that information from pharmacokinetic studies on the estimated in vivo concentration of inhibitor is available, to place the in vitro data in context. The $[I]/K_i$ rank order for different P450 enzymes helps in prioritizing the in vivo events, starting with the P450 with largest $[I]/K_i$ and therefore strongest inhibition potential (Obach et al., 2005; Huang et al., 2007). At this level only hepatic interactions are accounted for in the analysis. However, this can be valuable to qualitatively zone DDIs based on inhibition/induction in vitro information (see later).
3. In the subsequent steps the data are modelled to predict an interaction for a particular victim drug. At this stage, input of the specific victim drug-related parameters (e.g. fm_{CYP}, F_G) and the inhibited enzyme (e.g. k_{deg}) is required to allow the final prediction. Following the full strategy allows quantitative prediction that is comprehensive with consideration of the existence of more than one metabolic/elimination pathway, impact of multisite kinetics for CYP3A enzyme, contribution of the intestinal inhibition, impact of enzyme properties and the role of multiple inhibitors and mechanisms.
4. As part of the implementation of a comprehensive prediction exercise the choice of [I] used may be explored. Although pharmacokinetic theory would favor the use of unbound plasma concentration, in practice total concentration is often as accurate.

This basic relationship between the AUC ratio and $[I]/K_i$ allows predictions of inhibition DDIs to be categorized into four zones: true positives (AUC ratio>2, $[I]/K_i$ >1), true negatives (AUC ratio<2, $[I]/K_i$ <1), false positives (AUC ratio<2, $[I]/K_i$ >1) or false negatives (AUC ratio>2, $[I]/K_i$ <1) – see Fig. 7.2. Analogous rationales can be envisaged for irreversible inhibition and induction. FDA has suggested a lower cut-off ($[I]/K_i$ of 0.1) as a decision-making criteria for whether a clinical trial should be initiated (Huang et al., 2007; 2008). Examples of drugs as inhibitors and inducers of CYP3A4 classified in accord with the AUC criteria are shown in Table 7.2.

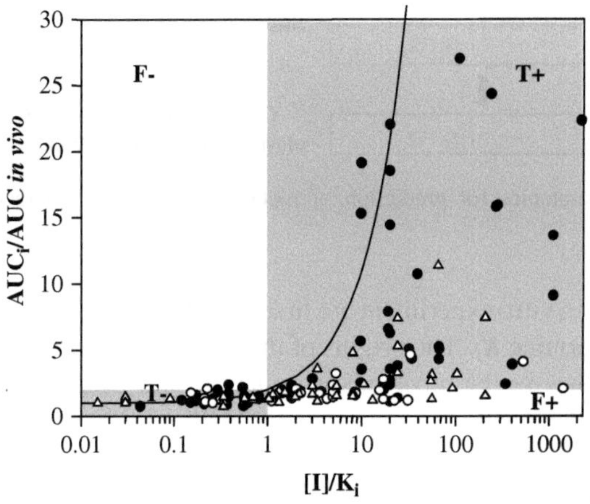

Fig. 7.2 Qualitative zoning for the prediction of drug–drug interactions involving P450 inhibition. The curve represents the theoretical curve based on the relationship AUC ratio = 1+ $[I]/K_{i,}$. F– represents false-negative, T–, true-negative, F+, false-positive and T+, true-positive DDI prediction. ● represents CYP3A4, △, CYP2D6 and ○ CYP2C9 DDIs. Reproduced from Ito et al. (2004) with the permission from Wiley-Blackwell

Table 7.2 Classification of CYP3A inhibitors and inducers[a] adapted from FDA

	Strong CYP3A inhibitors/inducers	Moderate CYP3A inhibitors/inducers	Weak CYP3A inhibitors/inducers
Change in AUC	≥5-fold change in AUC	≥2-fold but <5-fold change in AUC	≥1.25-fold but <2-fold change in AUC
Inhibitors	Ketoconazole, itraconazole, clarithromycin, indinavir, nelfinavir, ritonavir, saquinavir	Amprenavir, aprepitant, diltiazem, erythromycin, fluconazole, fosamprenavir, grapefruit juice, verapamil	Cimetidine, azithromycin
Inducers	Rifampicin, troglitazone	Phenobarbital, phenytoin, efavirenz	Eletriptan, gemfibrozil

[a] The dominant mechanism is shown

Our experience with in vitro aproaches for each of the three mechanisms differs substantially and hence each is currently at a different level of development. There is general confidence in the experimental generation of in vitro parameters and the steps 1 and 2, and hence qualitative prediction/zoning. Further advances to steps 3 and 4 to a quantitative level tend to be made more cautiously. The most explored mechanism is reversible inhibition with the in vitro derived input parameter K_i. Time-dependent inhibition requires more extensive in vitro inputs in terms of both the inhibitor properties (k_{inact} and K_I) and also the enzyme properties (k_{deg}). The third mechanism, induction, is the least studied in terms of both optimizing in vitro experimentation and extrapolating the data obtained in order to gain insight into in vivo effects. However, the latter is an area of considerable activity at the moment and it is likely that many advances will be made in the near future.

7.2.1 Experimental Conduct

Michaelis–Menten principles are commonly assumed for the determination of kinetic constants from in vitro data on P450 metabolism using the steady-state and rapid equilibrium approach. First it is assumed that there is only one substrate, although strictly speaking this is not the case; P450-mediated reactions also have NADPH and oxygen as substrates in addition to the test drug. NADPH is generally ignored as it is added to the reaction at a saturating concentration (e.g. concentrations of NADPH of 0.5–1 mM are typically used). Oxygen concentration, while difficult to manipulate, is assumed to be constant. A second assumption is that the enzyme is present at low concentrations relative to the substrate; thus at steady state the formation of the enzyme–substrate complex is rapid. This ratio should be maintained for the duration of the experiment (typically minutes) and there should be no significant impact on the substrate concentration as a consequence of enzyme binding or metabolism of the substrate. Under pre-steady-state kinetics or at high enzyme concentrations the mathematics is far more complex, hence the steady-state approach is used (Segel, 1993). The ionic strength, pH and temperature should be physiologically relevant (e.g. the use of phosphate or tris buffers at concentrations of 50–100 mM at pH 7.4 and 37°C) and should remain constant throughout the reaction.

The amount of product formed should be negligible compared to the substrate concentration. Up to 10% conversion of substrate to product over the duration of the assay is generally considered acceptable, i.e. only one-tenth of the substrate is exhausted. If too much substrate is exhausted the rate of reaction could decline due to either the reduction in substrate concentration or the significant buildup of product causing product inhibition. In order to select a suitable enzyme concentration the linearity of the reaction with respect to the enzyme/protein concentration should be determined. If the reaction is linear at a low substrate concentration it is likely that it will also be suitable at higher substrate concentrations. The incubation time should also be tailored such that the amount of product formed is negligible compared to the substrate concentration. Very short incubation times are in reality impractical and may introduce further errors. Long incubation times unnecessarily prolong the experiment and may result in degradation of the enzyme and reagents/substrates used. Linear conditions should be established using a higher enzyme concentration than required, which would then also be suitable for a lower enzyme concentration if used. In practice, for many drug metabolizing enzymes a protein concentration of <0.1 mg/mL and an incubation time of 5–10 min will produce a linear reaction with readily detectable rates of reaction.

7.2.2 Probe Substrates

The first step in the general strategy is to measure the change in enzyme activity, whether this is due to competitive inhibition, irreversible inhibition or enzyme induction. This is best assessed by using a selective probe for a particular P450 enzyme; the list of recommended probes by FDA is shown in Table 7.3 (Bjornsson et al., 2003; Huang et al., 2007). There is much discussion over the appropriate choice of probes and the table gives a selection of these. They are all well-studied compounds and appropriate conditions in terms of linearity and analytical detail are fully documented in the scientific literature. It is important to emphazise that these probes have gained acceptance because of their selectivity and analytical benefits and not because of their therapeutic importance. As a consequence, the change in enzyme activity can be kinetically fully characterized and then extrapolated to a variety of victim drugs at the further stages in the prediction strategy. Also these probe compounds can be readily used in the high-throughput mode and hence are used extensively to screen a number of possible perpetrators. Since these probes are extensively studied any unusual kinetic aspects in terms of multisite enzyme kinetics for CYP3A4 (Galetin et al., 2003) can be readily accommodated and in fact the choice can be varied in order to minimize such complexities (Galetin et al., 2005).

7.2.3 Correction for Nonspecific Binding In Vitro

The incorporation of the fraction unbound in microsomes (fu_{mic}) has been recognized as an important parameter in the in vitro–in vivo extrapolation strategies (Obach, 1999; Margolis and Obach, 2003; Galetin et al., 2005; Ito and Houston, 2005; Rostami-Hodjegan and Tucker, 2007; Houston and Galetin, 2008). Significant nonspecific binding to microsomal phospholipids and proteins may lead to a reduced free inhibitor concentration and biased estimation of inactivator potency (Brown et al., 2006; Grime and Riley, 2006; Houston and Galetin, 2008; Walsky and Boldt, 2008). Although generally accepted to improve the accuracy of clearance and DDI predictions, the assessment of microsomal binding is still challenging, especially for highly lipophilic drugs (Gertz et al., 2008c). Use of recombinant enzymes in inhibition screening reduces the opportunity of nonspecific binding to protein/lipid, as

Table 7.3 Examples of in vivo substrates, inhibitors (reversible and irreversible) and inducers for specific P450 enzymes adapted from FDA recommendations (Huang et al., 2007)

P450	Substrate	Inhibitor	Inducer
1A2	Theophylline, caffeine	Fluvoxamine	Smokers vs. non-smokers
2B6	Efavirenz	Ticlopidine	Rifampicin
2C8	Repaglinide, rosiglitazone	Gemfibrozil and its glucuronide, phenelzine	Rifampicin
2C9	S-Warfarin, tolbutamide	Fluconazole, tienilic acid, amiodarone (use of PM vs. EM subjects)	
2C19	Omeprazole, esoprazole, lansoprazole, pantoprazole	Omeprazole, fluvoxamine, moclobemide (use of PM vs. EM subjects)	Rifampicin
2D6	Desipramine, dextromethorphan, atomoxetine	Paroxetine, quinidine, fluoxetine (use of PM vs. EM subjects)	None identified
2E1	Chlorzoxazone	Disulfiram	Ethanol
3A4/ 3A5	Midazolam, buspirone, felodipine, sildenafil, simvastatin, triazolam	Clarithromycin, indinavir, mifepristone, ritonavir, itraconazole, ketoconazole, nefazodone, nelfinavir, saquinavir, telithromycin	Rifampicin, carbamazepine

high protein concentrations can be avoided due to the high enzyme activity (Obach et al., 2006). However, certain experimental investigations require relatively high microsomal protein concentrations; for example, most in vitro assessments of the time-dependent inhibition potential use the two-step experimental procedure with a high protein concentration (1–2 mg/mL) in the pre-incubation in order to allow adequate dilution in the second step (Ghanbari et al., 2006; Venkatakrishnan and Obach, 2007); thus, if binding is not accounted for bias in K_I estimates may be introduced (Venkatakrishnan and Obach, 2005; Houston and Galetin, 2008; Grime et al., 2009).

Recently, two in silico approaches have been described to estimate the unbound fraction in a microsomal preparation (Austin et al., 2002; Hallifax and Houston, 2006); both predictive tools are based on the lipophilicity of the compounds (log P or log $D_{7.4}$ for bases and acids/neutrals, respectively) and the microsomal protein concentration. The limitations of these empirical predictive tools and their utility for fu_{mic} predictions over a range of lipophilicity have been addressed recently (Gertz et al., 2008c). A significant difference in the fu_{mic} estimates was seen in the area of intermediate lipophilicity (log P/D = 2.5–5) due to the nature of the prediction equations and their sensitivity on the variability in the log P estimates. The extent of nonspecific binding for highly lipophilic drugs (log P/D ≥ 5) was poorly predicted by both equations, suggesting that the fraction unbound should be determined experimentally for these drugs; this cutoff should be lower (log P/D $\geq$ 3) if a microsomal protein concentration >0.1 mg/mL is used. As overall prediction accuracy was the highest at low microsomal protein concentration, it is prudent to perform kinetic and inhibition studies at the lowest microsomal protein concentration possible.

Analogous to applying a correction for microsomal drug binding, the fraction unbound in hepatocyte incubations (fu_{hep}) is also required for prediction of clearance and DDIs (McGinnity et al., 2006; Brown et al., 2007a; Stringer et al., 2008). Different experimental methods are employed to determine the drug binding in hepatocyte incubations, namely oil centrifugation, dialysis and ultrafiltration using either dead (freeze-thawed) or "live" hepatocytes (Austin et al., 2005; Henshall et al., 2008). An issue with the freeze-thawed approach is that cells break open during the process, which in turn can affect the binding of drugs to the hepatocytes resulting in poor estimate of the fu_{hep}. In contrast, use of "live" cells requires the presence of inhibitors to prevent any potential metabolism. In addition to the experimental methods, the relationship between the extent of binding in microsomal and hepatocyte incubations has been used as an empirical tool for the prediction of fu_{hep} either directly from the drug lipophilicity metric or from a previously obtained fu_{mic} value (Kilford et al., 2008). Analogous to microsomes, prediction of hepatocyte binding must be undertaken with caution for drugs with log P/D $\geq$ 3 due to the impact of inaccuracies in log P/D estimates and the general inaccuracy of the predictive tools in this lipophilicity area. In addition, for drugs for which hepatocellular uptake is not limited to simple partitioning (e.g. lysosomal uptake for bases at low drug concentrations (Hallifax and Houston, 2007) or active uptake in hepatocytes (Hirano et al., 2006; Ho et al., 2006; Yamada et al., 2007)), the prediction of fu_{hep} will be further complicated. However, the use of these calculated unbound fractions, while imprecise in many cases, is far preferable to ignoring this important phenomenon.

7.3 Enzyme and Victim Drug Properties

As indicated earlier, it is evident from the equations that knowledge of the enzyme and victim drug properties is as important as the inhibitory/induction properties of the interacting drug in governing the magnitude of a DDI. This amounts to the need for accurate estimation of the enzyme turnover (k_{deg}), the relative importance of various clearance processes (fm_{CYP}) and the role of the intestine as an eliminating organ (F_G) and will be addressed in the sections below.

7.3.1 Enzyme Turnover

At steady-state, the rate of change of active enzyme concentration ($d[E]/dt$) is determined by the equilibrium between the rates of de novo synthesis (R_o) and degradation of the enzyme, as illustrated in the following equations (Equations 7.7 and 7.8) (Wang et al., 2004; Yang et al., 2008):

$$\frac{d[E]}{dt} = R_o - k_{deg}[E] \quad (7.7)$$

$$[E]_{ss} = \frac{R_o}{k_{deg}} \quad (7.8)$$

where R_o and k_{deg} are the rate of enzyme synthesis and degradation rate constant, respectively, $[E]_{ss}$ is the enzyme content at steady state. When the balance between synthesis and degradation is disturbed (as in the case of either induction or irreversible inhibition), the abundance of particular P450 enzyme will change to reach a new steady-state; the E_{ss} and the time to reach steady state are determined by the enzyme k_{deg}. Therefore, an accurate estimate of enzyme turnover has implications for the successful prediction of both induction and time-dependent DDIs. Intrinsic clearance of a substrate of the enzyme can be described by Equation (7.9):

$$CL_{int} = \frac{[E]_{ss} \cdot k_{cat}}{K_m} \quad (7.9)$$

The assumption is that at the enzyme active site $[S] << K_m$, where k_{cat} represents the first-order rate constant that relates V_{max} to E_{ss} and K_m is the Michaelis–Menten constant. The relationship between CL_{int} and E_{ss} shown in Equation (7.9) is applicable both in the absence and in the presence of a modifier (either inducer- or mechanism-based inhibitor), but the CL_{int} and E_{ss} will differ.

In the presence of an irreversible inhibitor, the enzyme level will decrease as described by Equation (7.10), reaching the new steady-state enzyme level and intrinsic clearance, defined by Equations (7.11) and (7.12), respectively.

$$\frac{d[E]}{dt} = R_o - k_{deg}E - k_{obs}E \quad (7.10)$$

$$[E]_{ss} = \frac{R_o}{k_{deg} + k_{obs}} \quad (7.11)$$

$$CL_{int}' = \frac{[E]_{ss}' \cdot k_{cat}}{K_m} \quad (7.12)$$

where k_{obs} is the rate constant for inhibition observed in vitro over a range of inhibitor concentrations and is dependent on k_{inact}, K_I and inhibitor concentration according to the following relationship:

$$k_{obs} = \frac{k_{inact} I}{K_I + I} \quad (7.13)$$

The ratio of CL_{int} in the absence and presence of an irreversible inhibitor is expressed by Equation (7.14):

$$\frac{\mathrm{CL_{int}}}{\mathrm{CL_{int}}'} = \frac{[E]_{\mathrm{ss}}}{[E]_{\mathrm{ss}}'} = 1 + \frac{k_{\mathrm{inact}}I}{k_{\mathrm{deg}}(\mathrm{K_I} + \mathrm{I})} \tag{7.14}$$

In the presence of an inducer, the enzyme level will increase as defined in Equation (7.15). At the new steady state the enzyme level and intrinsic clearance are defined by Equations (7.16) and (7.12), respectively.

$$\begin{aligned}\frac{\mathrm{d}\,[E]'}{\mathrm{d}t} &= R_{\mathrm{o}} + k_{\mathrm{ind}} \cdot [E]' - k_{\mathrm{deg}} \cdot [E]' \\ &= R_{\mathrm{o}} + (R_{\max} - R_{\mathrm{o}}) \cdot \frac{[\mathrm{I}]}{EC_{50} + [\mathrm{I}]} - k_{\mathrm{deg}} \cdot [E]'\end{aligned} \tag{7.15}$$

$$[E]_{\mathrm{ss}}' = \frac{R_{\mathrm{o}} + (R_{\max} - R_{\mathrm{o}})}{K_{\mathrm{deg}}} \bullet \frac{[\mathrm{I}]}{EC_{50} + [\mathrm{I}]} \tag{7.16}$$

where k_{ind} is the rate constant for enzyme induction, $R_{\max}$ the maximum rate of enzyme synthesis, $\mathrm{EC_{50}}$ the concentration of inducer at 50% of maximal induction and $[E]'_{\mathrm{ss}}$ the enzyme content at steady state in the presence of an inducer.

The ratio of $\mathrm{CL_{int}}$ in the absence and presence of an inducer is expressed by Equation (7.17), where $E_{\max}$ is the maximum induction of enzyme (net fold increase in enzyme) or ($R_{\max}$–R_{o}):

$$\frac{\mathrm{CL_{int}}}{\mathrm{CL_{int}}'} = \frac{[E]_{\mathrm{ss}}}{[E]_{\mathrm{ss}}'} = \frac{1}{1 + \dfrac{E_{\max} \cdot [\mathrm{I}]}{EC_{50} + [\mathrm{I}]}} \tag{7.17}$$

Equations (7.13) and (7.17) are also shown in more general forms in Table 7.1.

7.3.2 Impact of Parallel Elimination Pathways

The fraction of the metabolic clearance of the victim drug that is catalyzed by the enzyme subject to inhibition (fm_{CYP}) depends on the fraction of the dose metabolized and the contribution of each individual P450 to the total metabolic clearance. Due to high sensitivity of the prediction models to this victim drug parameter (Fig. 7.3), in particular for victim drugs where the enzyme affected contributes >80% to the overall elimination, its accuracy has a significant impact on the success of DDI predictions (Brown et al., 2005; Ito et al., 2005; Chien et al., 2006; Galetin et al., 2006; Hinton et al., 2008). The knowledge of the major routes of clearance is also beneficial in order to avoid significant inter-individual differences due to metabolism via polymorphic enzyme (Collins et al., 2006; Zhang et al., 2007).

A number of in vivo approaches can be employed to estimate fm_{CYP} value, as summarized recently (Houston and Galetin, 2008). The most unequivocal method is based on the comparison of phenotyping data in extensive and poor metabolizers, as illustrated in the case of CYP2D6 (Ito et al., 2005). A good alternative to this approach is "pheno-copying" based on the difference between the urinary recovery

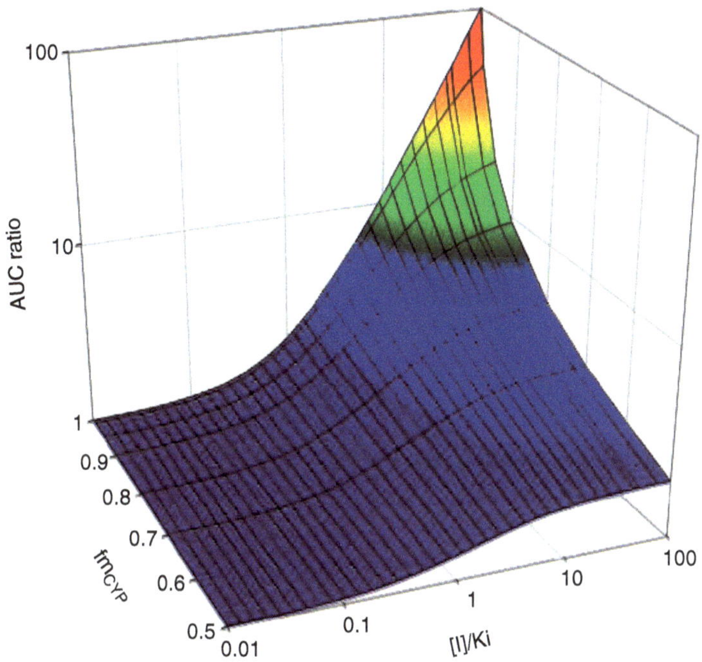

Fig. 7.3 Effect of fm_{CYP} on the prediction of drug–drug interactions. The sensitivity of the predicted AUC ratios on the fraction of the drug metabolized by the inhibited pathway via a particular P450 enzyme (fm_{CYP}) over a range of inhibitor potency defined by $[I]/K_i$

of metabolites in the presence and absence of a selective inhibitor (e.g. tolbutamide in the presence of sulfaphenazole as CYP2C9 inhibitor) (Brown et al., 2005; Kusama et al., 2009). In the case of CYP3A4, the most problematic P450 enzyme, combined information on the urinary recovery of metabolites, biliary excretion and the recovery of unchanged drug has been used (Brown et al., 2005; Galetin et al., 2005; 2006). In addition to excretion data, the extent of increase in the victim drug AUC in the presence of a selective CYP3A4 inhibitor (e.g. ketoconazole, itraconazole) has been suggested to estimate fm_{CYP3A4} (Shou et al., 2008). A limitation of this approach is that it is based solely on the effect of the inhibitor on the hepatic metabolism, even though a number of clinical studies have indicated a more pronounced inhibition (and induction) of intestinal CYP3A4 in comparison to the liver (Gomez et al., 1995; Palkama et al., 1999; Tsunoda et al., 1999; Saari et al., 2006; Galetin et al., 2007).

Reaction phenotyping is commonly used in pharmaceutical industry to determine the relative contribution of various enzymes to the metabolism of a new chemical entity (Zhang et al., 2007). A number of methods are used including the use of recombinant proteins, enzyme-selective chemical inhibitors and antibodies (Lu et al., 2003; Shou and Lu, 2009). As all these methodologies are readily available for human P450, in vitro reaction-phenotyping data are now routinely included in most regulatory documents (Williams et al., 2003; Zhang et al., 2007). Recent studies have shown the use of inhibition and reaction-phenotyping recombinant P450 data in the prediction of the magnitude of in vivo DDIs (McGinnity et al., 2008; Youdim et al., 2008) using the modeling and simulation software Simcyp® (Jamei et al., 2009b). In addition to fm_{CYP}, recent studies have reported the use of in vitro data to estimate the extent of glucuronidation (fm_{UGT}) (Cubitt et al., 2009; Kilford et al., 2009). The contribution of glucuronidation can be determined in both liver and intestine using separate incubations with either NADPH (P450) or UDPGA (UGT) cofactors. Considering the increasing importance of glucuronidation elimination, this approach should be incorporated in the prediction of clearance or DDIs. However, without accurate information on the role of renal elimination, an overestimation of either fm_{CYP} or fm_{UGT} can result; hence, cautious interpretation of the in vitro data is necessary.

7.3.3 Prediction of F_G from In Vitro Data

The contribution of an intestinal interaction can be incorporated into the prediction equation using hepatic enzyme interaction data in the form of the ratio of the intestinal wall availability in the presence and absence of the modifier ($F_G{}'$ and F_G, respectively); this approach is applicable for both inhibition and induction interactions (Table 7.1) (Rostami-Hodjegan and Tucker, 2004; Wang et al., 2004; Galetin et al., 2006; Obach et al., 2006; Fahmi et al., 2008b; Galetin et al., 2008). Recent analysis has highlighted the sensitivity of DDI prediction models to the accuracy of the estimated basal F_G value, in particular for drugs with a high intestinal first-pass extraction (>75%), either as a result of extensive intestinal metabolism (e.g. buspirone) or proposed interplay with efflux transporter (e.g. tacrolimus) (Galetin et al., 2007).

Direct measurement of F_G is rarely performed in humans due to practical and ethical reasons, with the exception of studies in anhepatic patients during liver transplant operations or patients whose portal blood circulation bypassed the liver (Paine et al., 1996; Thummel et al., 1996). One approach to estimate F_G is to use plasma concentration–time profiles after oral and i.v. administration, based on the assumptions that there is negligible metabolism in enterocytes after i.v. dose and that the systemic clearance of a drug after i.v. administration reflects only hepatic elimination (Hall et al., 1999; Galetin et al., 2008). Alternatively, F_G can be estimated from interaction studies with grapefruit juice (Bailey et al., 1998; Saito et al., 2005), assuming complete irreversible inhibition of CYP3A-mediated metabolism in the intestine but no effect on hepatic enzymes or the fraction absorbed of the investigated drug. Recent analysis (Gertz et al., 2008a) has addressed advantages and limitations of both of these methods, highlighting the limitation of the grapefruit juice approach for the estimation of F_G for drugs whose disposition is co-dependent on efflux/uptake transporters and metabolic enzymes.

In silico approaches to estimate F_G using physiologically based pharmacokinetic models require the incorporation of the drug absorption process together with zonal and cellular heterogeneous distribution of metabolic enzyme and efflux/uptake transporters along the length of the intestine and enterocytic, rather than total intestinal, blood flow. A number of models have been reported with different levels of complexity and integration of passive permeability with the activity of metabolic enzymes and transporters (Ito et al., 1999; Pang, 2003; Tam et al., 2003; Dokoumetzidis et al., 2007; Badhan et al., 2009). Adaptations of the original compartmental absorption transit (CAT) (Yu and Amidon, 1999) and advanced CAT model (Agoram et al., 2001; Tubic et al., 2006) are incorporated in the commercially available softwares (GastroPlus and Simcyp®). A major disadvantage of some of these models is their complexity and limited availability of the parameters required for the prediction of the F_G. Therefore, addressing all the complexities of the intestinal first-pass metabolism adequately and yet retaining a relatively practical model remains a challenge. The quality and use of standardized experimental conditions for in vitro data generation are essential in addition to the knowledgeable integration of the data from different in vitro systems.

In contrast to complex physiologically based models, Yang et al. (2007a) proposed a "minimal" Q_{gut} model. This model overcomes the inadequacy of the well-stirred approach adapted from the commonly used liver model (Rowland et al., 1973; Wilkinson and Shand, 1975) to estimate F_G (Lin et al., 1997; Thummel et al., 1997). A mechanistic model to predict F_G values from in vitro metabolism and transporter data is shown in the Equations (7.18) and (7.19) (Yang et al., 2007a):

$$F_G = \frac{Q_{\text{gut}}}{Q_{\text{gut}} + fu_G \cdot \text{CL}_{\text{intG}}} \tag{7.18}$$

$$Q_{\text{gut}} = \frac{\text{CL}_{\text{perm}} \cdot Q_{\text{ent}}}{\text{CL}_{\text{perm}} + Q_{\text{ent}}} \tag{7.19}$$

where fu_G represents the fraction unbound in the enterocytes, CLu_{intG} is the intestinal clearance and Q_{gut} is "the nominal blood flow" in the small intestine. Q_{gut} is a hybrid function of the permeability clearance (CL_{perm}) and enterocytic blood flow (Q_{ent}), as illustrated in the Equation (7.19). Permeability input in the Q_{gut} model can be incorporated either in the form of apparent permeability data (P_{app} $(A\text{–}B)$) determined in either MDCK-MDR1 or Caco-2 cells or by predicting effective permeability (P_{eff}) from the P_{app} values or in silico parameters (e.g. hydrogen bond donors or polar surface area) (Winiwarter et al., 1998; Lennernas, 2007; Yang et al., 2007a).

Sensitivity analysis of this model has indicated that intestinal extraction of greater than 50% ($F_G < 0.5$) is observed for drugs with approximate $\text{CLu}_{\text{intG}} \geq 1$ μL/min/pmolCYP3A) regardless of Q_{gut} if fu_G is 1 (Galetin et al., 2008). Prediction of F_G values from in vitro permeability and clearance data performed for a large range of CYP3A4 substrates (Gertz et al., 2008b) indicated that the Q_{gut} model is less sensitive to the drug permeability in comparison to drug clearance. Therefore, the incorporation of a permeability parameter is only significant for drugs where $P_{\text{app}} < 100$ nm/s, whereas for other drugs the well-stirred gut model represents an equally valid model to predict F_G (Galetin et al., 2008; Gertz et al., 2008b). Different surrogates can be used for the effective free fraction in the gut, namely unity or plasma or blood binding; however, an accurate estimate of this parameter is difficult. It is clear however that the use of plasma or blood binding to obtain fu_G results in a general loss of prediction accuracy in comparison to assumption of no binding (fu_G=1) (Yang et al., 2007a; Gertz et al., 2008b).

7.4 Reversible Enzyme Inhibition

Evaluation of enzyme inhibition in vitro has become a routine method in drug metabolism for the prediction and evaluation of DDIs, particularly for P450 (Ito et al., 2004; Galetin et al., 2005; Brown et al., 2006; Obach et al., 2006; Einolf, 2007; Houston and Galetin, 2008; Walsky and Boldt, 2008; Youdim et al., 2008). The rapid and simple experimental procedures developed have allowed substantial technological advances in the conduct of in vitro studies and such information is now expected from regulatory authorities (Tucker et al., 2001; Bjornsson et al., 2003; Huang et al., 2007). The FDA-recommended inhibitors are listed in Tables 7.2 and 7.3, including ketoconazole and itraconazole (CYP3A4), quinidine (CYP2D6), gemfibrozil (CYP2C8) and fluvoxamine (CYP1A2).

The main types of reversible inhibition are as follows:

(i) Competitive inhibition: Competitive inhibitor combines with an enzyme in such a manner that it prevents substrate binding. It may bind to exactly the same site, or partly share or mask the substrate-binding site. In competitive inhibition substrate and inhibitor binding is mutually exclusive and there is reciprocity of inhibition.
(ii) Non-competitive inhibition: A simple non-competitive inhibitor has no effect on substrate binding and vice versa. S and I bind reversibly and independently, but bound inhibitor does not inactivate the enzyme, but has an effect of apparently decreasing the amount of enzyme present.

In order to assess reversible inhibition potential, IC_{50} values are obtained across the range of P450 enzymes. IC_{50} represents the inhibitor concentration that reduces the control rate by 50% and is frequently used, as these experiments are quicker and easier to conduct than full K_i determinations. However, no information is obtained on the inhibition mechanism and the IC_{50} can be substantially different from the K_i depending on the substrate concentration used. The in vitro inhibition studies are designed based on steady-state kinetic principles, i.e. that the inhibitor concentration should be much higher than the enzyme concentration, such that inhibitor binding to the enzyme does not alter the free concentration of the inhibitor. In the case of IC_{50} determinations the experiment should be designed to cover a range from virtually no inhibition to virtually complete inhibition in concentrations evenly spread on a log scale. The substrate concentration range should be in a similar range as that for determination of V_{max} and K_m. The highly lipophilic nature of many developing compounds may result in solubility problems and different strategies to address this issue have been discussed (Fowler and Zhang, 2008). IC_{50} plots are conventionally plotted as rate vs. log [I], and the curve can be described by the following equation:

$$v = \frac{v_0}{1 + \left(\frac{[I]}{IC_{50}}\right)^s} + b \tag{7.20}$$

where v_0 is the control rate, b is the background or non-inhibitable activity and s is the slope factor. For selective assays there should be no background activity and the slope should equal 1 irrespective of the mechanism of inhibition. The IC_{50} should always be calculated by nonlinear regression of the untransformed data. Incorporation of the pre-incubation step may be used as an indicator of potential irreversible inhibition and this is addressed in Section 7.5.

An indication of the inhibition potency of a potential drug candidate can be obtained from IC_{50} data from Michaelis–Menten principles; the K_i values represent $IC_{50}/2$ in the case of competitive inhibition and IC_{50} for non-competitive inhibitors at substrate concentration equal to K_m of the probe used. Determination of the K_i is necessary when complete in vitro characterization of an inhibitor is required and for the prediction of potential DDI in vivo (Table 7.1). The value of K_i parameter is independent of substrate concentration used, unlike IC_{50}. In an ideal in vitro experimental design the inhibitor concentrations should span from 0.5 to 5 expected K_i. In order to determine the mechanism of inhibition involved, preliminary analysis of the direct plots of reaction rate vs. substrate concentration should be performed as an indicator of changes in kinetic parameters in the presence of an inhibitor (for example, increase in K_m with no effect on V_{max} would suggest competitive inhibition). Data may also be linearized in the form of an Eadie–Hofstee plot (v vs. v/[S]) or as a Dixon plot (1/v vs. [I]). The K_i is not readily determined from the Eadie–Hofstee plot, but this may be a useful way of presenting the data and identifying the inhibition mechanism; in the Dixon plot the K_i is the point where all the lines intercept (Houston et al., 2003). K_i should always be determined by simultaneous nonlinear regression of the entire untransformed data set using a suitable software package (Houston and Galetin, 2003; Fowler and Zhang, 2008; Walsky and Boldt, 2008).

7.4.1 In Vitro Systems Used for the Assessment of Reversible Inhibition

The in vitro screens used in drug discovery focus on the six major human drug metabolising P450s, CYP1A2, CYP2C8, CYP2C9, CYP2C19, CYP2D6 and CYP3A4 (Walsky and Obach, 2004; Obach ct al., 2005; Huang et al., 2007; Youdim et al., 2008), although some other human P450 enzymes, such as CYP2B6, have recently been gaining increased attention (Walsky et al., 2006). In terms of in vitro systems, assessment of reversible inhibition potential mainly relies on the use of human liver microsomes and recombinant enzymes rather than hepatocytes. Human liver microsomes remain the most used system, as their practical convenience overrides the common knowledge that hepatic microsomes do not contain transporter proteins and all the drug metabolizing enzymes. The dilemma of selection of appropriate donor tissue that arises due to the large inter-individual expression of the various P450s (Parkinson et al., 2004; Rowland-Yeo et al., 2004; Paine et al., 2006) is often overcome by the use of pools of numerous donors. However, the use of hepatic microsomal inhibition data may significantly underestimate in vivo potency in the case of accumulation of the inhibitor, as the occurrence of the active processes will lead to high cellular concentrations of inhibitor (e.g. HIV protease inhibitors, fluoroquinolones).

Inhibition data obtained in hepatocytes should be more representative of the in vivo situation, due to the presence of intracellular proteins, organelles and intact plasma membrane containing transporter proteins. Recent comparison of unbound K_i values in rat microsomes and freshly isolated hepatocytes (Brown et al., 2007a) of a series of P450 inhibitors illustrated very good agreement in the in vitro inhibitory potency between the systems despite high cell-to-medium ratio observed for some inhibitors (e.g. fluoxetine, ketoconazole). In these cases, although significant intracellular binding of inhibitor to cellular proteins occurs, there is no significant difference between the drug concentration in the hepatic cytosol and the incubation medium and hence no difference in unbound K_i values between systems. In contrast, for inhibitors that undergo active transport processes more potent hepatocyte K_i values are expected in comparison to microsomal values, as the unbound intracellular and extracellular inhibitor concentrations are not equivalent (e.g. gemfibrozil and its glucuronide, enoxacin, clarithromycin, unpublished data). Comparison of IC_{50} or K_i values obtained in microsomes and hepatocytes represents a useful initial tool for assessing potential hepatic accumulation of inhibitors prior to transporter-specific inhibition studies and uptake studies in isolated hepatocytes.

7.4.2 Inhibition at Multiple Sites

The potential of substrates and modifiers of CYP3A4 to show homo- and heterotropic cooperative effects (Kenworthy et al., 2001; Shou et al., 2001; Galetin et al., 2003; Henshall et al., 2008) confounds any straightforward in vitro–in vivo correlation. Atypical kinetic phenomena attributed to the existence of multiple binding sites have been associated predominantly with CYP3A4 (Galetin et al., 2002; Yano et al., 2004); however, recent studies indicate similar behavior for other P450s and UDP glucuronosyltransferase enzymes (Egnell et al., 2003; Hutzler et al., 2003; Uchaipichat et al., 2004; 2008).

Interactions with multiple binding sites may result in inhibition at only one of the binding sites or a differential effect at each site (Galetin et al., 2003; Houston et al., 2003). In order to explore the range of possible consequences of heterotropic interactions the use of multiple substrates in vitro at various substrate concentrations has been recommended (Kenworthy et al., 1999). Modeling of inhibition at multiple binding sites, alterations in binding affinity or in rate of metabolite formation and the lack of reciprocity between competing binders have been addressed (Galetin et al., 2005; Houston and Galetin, 2005). In order to assess the importance of cooperativity and predict changes in the in vivo plasma concentration–time profile from CYP3A4 in vitro data, an equation was derived based on the same rapid equilibrium /steady-state assumptions as the single-site model inhibition models (Galetin et al., 2005). In addition to $[I]/K_i$ ratio, this two-site model equation also incorporates the cooperative changes in the catalytic efficacy and binding affinity in the presence of the inhibitor. The complexity of multisite kinetic analysis observed with certain CYP3A4 substrates (e.g. testosterone) (Kenworthy et al., 2001; Galetin et al., 2003) indicated the need for an assessment of alternative and/or more pragmatic approaches and substrates for the prediction of potential DDI involving this enzyme. Midazolam and quinidine were suggested as a more pragmatic choice of CYP3A4 substrates in comparison to nifedipine and testosterone, due to their apparent hyperbolic kinetics and good DDI prediction accuracy and precision observed with azole inhibitors (Galetin et al., 2005).

7.5 Time-Dependent Inhibition

Mechanism-based or time-dependent inhibition interactions (TDI) involve the metabolism of an inhibitor (i.e. require NADPH for P450 reactions) to reactive species which inactivate the catalyzing enzyme by different mechanisms in a concentration- and time-dependent manner. In the case of quasi-irreversible inhibitors, these reactive intermediates coordinate with the heme prosthetic group leading to the formation of a catalytically inactive metabolite–inhibitor complex of the P450 enzyme. In contrast, reactive intermediates generated from the metabolism of irreversible inactivators covalently react with an active site within the apoprotein (Silverman, 1995; Kent et al., 2001; Zhou et al., 2005; Kalgutkar et al., 2007). Due to the requirement for metabolic activation and therefore interaction with the enzyme active site, time-dependent inhibitors will also inhibit the same enzyme in a competitive reversible manner to some extent. The design and conduct of in vitro irreversible inactivation studies are more complex compared to reversible inhibition due to the time-dependent nature of the inhibition. Two main parameters are important for prediction of TDI – the capacity parameter (k_{inact}) defining maximal inactivation rate constant and the potency term (K_{I}) representing the inhibitor concentration leading to 50% of k_{inact} (Silverman, 1995; Zhou et al., 2005; Venkatakrishnan and Obach, 2007); both parameters are derived from inactivation data using multiple inhibitor concentrations after multiple pre-incubation time points. Potent time-dependent inhibitors include mibefradil, ritonavir and clarithromycin (CYP3A4), tienilic acid (CYP2C9), phenelzine (CYP2C8), furafylline (CYP1A2) and paroxetine (CYP2D6) (Polasek et al., 2004; Ernest et al., 2005; Pinto et al., 2005; Venkatakrishnan and Obach, 2005; Obach et al., 2007).

The most commonly applied method to asses this type of inhibition interaction in vitro uses varying concentrations of inhibitor preincubated with the enzyme source (human liver microsomes or recombinant P450 preparation) and NADPH for various times prior to dilution into a secondary incubation (activity assay) containing NADPH and a probe substrate of the P450 of interest. This method is often referred to as the two-step dilution assay and a number of recent studies have addressed the critical aspects of its experimental design, all of which may affect the estimates of inhibitor potency (Ito et al., 2003; Ernest et al., 2005; Pinto et al., 2005; Galetin et al., 2006; Ghanbari et al., 2006; Polasek and Miners, 2007; Riley et al., 2007; Fowler and Zhang, 2008). This approach assumes that the inhibitor concentration in the final incubation is negligible and therefore the rate of metabolism of a probe substrate reflects the remaining enzyme activity; this may not be correct depending on the conditions mentioned below. Ideally, pre-incubation time should exceed the incubation, i.e. short secondary incubation time (5–10 min depending on the enzyme) compared to pre-incubation in order to minimize further time-dependent inhibition. In addition, careful selection of preincubation time points is crucial to define the slope of the inactivation rate curve adequately for both slow and rapid inhibitors while avoiding an excessive number of incubations (Yang et al., 2005; Ghanbari et al., 2006).

After different pre-incubation times, aliquots of the enzyme–inhibitor mix are diluted into the substrate solution and the remaining enzyme activity is determined. A minimal 1:10 dilution factor should be applied to reduce the occurrence of competitive inhibition and to allow enzyme inactivation to be distinguished from inhibitory metabolite generation. Contrary to earlier studies where great inconsistency in the experimental design was reported (Ghanbari et al., 2006), there is now an increasing awareness of the importance of appropriate enzyme dilution to minimize the residual effects of the inhibitors. Current improved analytical sensitivity of the LC-MS/MS methods allows the use of high dilution factors (20 or even 50) (Obach et al., 2007; Walsky and Boldt, 2008) and much shorter incubation times reducing the potential bias in in vitro parameter estimates. In addition, the use of high substrate concentrations (at concentrations in excess of its K_m, equivalent to V_{max}) and nonlinear regression analysis are recommended when obtaining the k_{inact} and K_I parameters (Van et al., 2006; Yang et al., 2007b). The use of saturating substrate concentrations in the second metabolic incubation contributes to reducing the effect of any potential reversible inhibition which may otherwise confound estimation of the TDI kinetic parameters.

7.5.1 In Vitro Systems Used for the Assessment of Irreversible Inhibition

Analogous to reversible inhibition, assessment of time-dependent inhibition potential is predominantly performed in human liver microsomes and recombinant enzymes (McConn et al., 2004; Ernest et al., 2005; Obach et al., 2007; Venkatakrishnan and Obach, 2007; Fowler and Zhang, 2008; Zhang et al., 2009). Recombinant P450 enzymes offer many attractions as they avoid issues associated with the involvement of multiple enzymes, allowing the assessment of inhibition on a specific enzyme in a membrane environment. However, the choice of recombinant system (with generally higher P450 content per mg of protein than human liver microsomes), the concentration of accessory proteins (e.g. NADPH P450-oxidoreductase and cytochrome b_5) and their relative ratio to the P450 protein may result in differential catalysis rates and degrees of uncoupling in comparison to liver microsomes (Voice et al., 1999; Venkatakrishnan et al., 2000; Jushchyshyn et al., 2005). Figure 7.4 shows a comparison of k_{inact}/K_I estimates obtained in recombinant CYP3A4 and human liver microsomes from six different studies (Kanamitsu et al., 2000a; McConn et al., 2004; Wang et al., 2004; Ernest et al., 2005; McGinnity et al., 2006; Polasek and Miners, 2006). The k_{inact} and K_I values for 16 CYP3A4 inhibitors covered a 100- and 3000-fold range across studies, respectively. Comparison indicated on average 4-fold higher k_{inact}/K_I in recombinant enzymes in comparison to the liver microsomes.

The use of intact hepatocytes (in suspension or culture) in the assessment of time-dependent inhibition is less common (McGinnity et al., 2006; Van et al., 2007) in comparison to the assessment of clearance and induction studies (Brown et al., 2007b; Hewitt et al., 2007b; Soars et al., 2007; Hallifax et al., 2008; Shou et al., 2008). The studies have indicated a clear trend of lower k_{inact}/K_I in hepatocytes compared to human liver microsomes and recombinant P450, mainly driven by a significantly lower k_{inact} estimates in hepatocyte systems, as observed for the time-dependent inhibitors fluoxetine (CYP3A4 and CYP2C19) (McGinnity et al., 2006) and MDMA (CYP2D6) (Van et al., 2007).

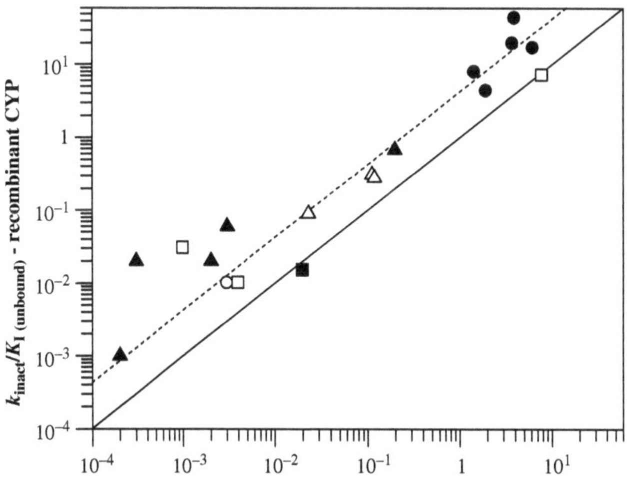

Fig. 7.4 Comparison of the TDI metric k_{inact}/K_I obtained in recombinant enzymes and human liver microsomes. Inhibitor data taken from (○) (Kanamitsu et al., 2000a), (□) (McConn et al., 2004), (△) (Wang et al., 2004), (●) (Ernest et al., 2005), (■) (McGinnity et al., 2006) and (▲) (Polasek and Miners, 2006). The *solid line* represents unity and the *dotted line* represents the average 4.3-fold bias

The extrapolation of the in vitro k_{inact}/K_I to the in vivo situation depends on the relative magnitude of this parameter to enzyme degradation; in the case when the inactivation index – $k_{inact} \cdot I/k_{deg} \cdot (K_I + I)$ is << k_{deg} (<10%, e.g. azithromycin) a weak inhibitory effect can be expected, with an AUC ratio ≤2. Early predictions of TDI (Ito et al., 2003; Wang et al., 2004) used CYP3A4 degradation half-life estimates ($t_{1/2\,deg}$) obtained in either rat or Caco-2 cells (Correia, 1991), resulting in $t_{1/2\,deg}$ of 14–35 h. Recently, the estimates of human CYP3A4 k_{deg} from both induction studies and in vitro investigations in liver slices have been collated (Galetin et al., 2006; Ghanbari et al., 2006; Yang et al., 2008). The estimated decay of CYP3A4 activity is 2-fold longer in comparison to some other cytochrome P450 enzymes (Ghanbari et al., 2006) but comparable to CYP2D6 (Renwick et al., 2000; Venkatakrishnan and Obach, 2005). The differential degradation half-lives reported for CYP3A4 and CYP3A5 in vitro (79 vs. 35 h, respectively) (Renwick et al., 2000) and the lesser susceptibility to inhibition observed for CYP3A5 (McConn et al., 2004; Pinto et al., 2005; Wang et al., 2005) may contribute to the extent of inter-individual variability observed in the magnitude of interactions.

The importance of accurate enzyme degradation estimates has been shown for the prediction of CYP2D6 inactivation by paroxetine (Venkatakrishnan and Obach, 2005); a recent comparable study was reported for CYP3A4 (Galetin et al., 2006). The impact of inter-individual variability of human CYP3A4 $t_{1/2\,deg}$ (20–146 h) (k_{deg} of 0.8–5 × 10^{-4} min^{-1}) on the assessment of TDI potential and DDI prediction accuracy was investigated. The sensitivity of the predicted magnitude of a DDI to a CYP3A4 degradation rate was dependent on fm_{CYP3A4}. The prediction accuracy was very sensitive to CYP3A4 degradation rate for victim drugs mainly eliminated by this enzyme ($fm_{CYP3A4} \geq 0.8$); minimal effects were observed when CYP3A4 contributed less than 50% to the overall elimination in cases when the parallel elimination pathway was not subject to inhibition. The study also recommended the use of hepatic CYP3A4 $t_{1/2\,deg}$ of 3 days (i.e. k_{deg} of 1.6 × 10^{-4} min^{-1}) in the assessment of time-dependent interaction potential, as this approach yielded an 89% success rate (defined by 2-fold of observed in vivo values criteria).

CYP3A4 intestinal k_{deg} can be estimated from the reported changes in the AUC of the victim drug at different times after ingestion of grapefruit juice. This approach allows the estimation of intestinal CYP3A4 $t_{1/2\,deg}$ due to the selective and irreversible inhibition of intestinal CYP3A4 by components of grapefruit juice with no effect on hepatic enzymes (Bailey et al., 1998; Saito et al., 2005). CYP3A4 recovery $t_{1/2}$ of 13–33 h was obtained (mean value of 23±10 h) resulting in the corresponding intestinal k_{deg} of 4.8 × 10^{-4} min^{-1} (Obach et al., 2007; Gertz et al., 2008a; Yang et al., 2008). The estimated intestinal CYP3A4 $t_{1/2\,deg}$ is shorter in comparison to hepatic estimates, probably reflecting the turnover of intestinal mucosal cells, and will impact on the accurate assessment of the contribution of the intestinal interaction to the overall magnitude of time-dependent inhibition or induction.

7.5.2 Alternative Approaches for the Assessment of Time-Dependent Drug–Drug Interactions

Recently, the use of IC_{50} shifts following various pre-incubation times has received attention as a potential alternative approach to the resource intensive two-step dilution assay (Obach et al., 2007; Walsky and Boldt, 2008; Grime et al., 2009; Perloff et al., 2009). In this method, the IC_{50} experiment is performed in the same manner as in the case of reversible inhibition, but with the additional pre-incubation of the inhibitor (usually 20–30 min) with the enzyme source (liver microsomes or recombinant enzymes) and NADPH. Following the pre-incubation, probe substrate is added at approximately K_m concentration and incubated further (shorter incubation, 2–10 min). In order to assess the impact of pre-incubation, an identical experiment is performed without this step or with a pre-incubation lacking NADPH; the shift to lower IC_{50} following pre-incubation compared to an IC_{50} obtained without pre-incubation indicates time-dependent inhibition (von Moltke et al., 2000; Yeo and Yeo, 2001; Yamamoto et al., 2004; Obach et al., 2007; Perloff et al., 2009). Typically, a single pre-incubation time point is used which allows sufficient time for time-dependent inhibition to be manifested (usually 30 min). However, a single time-point approach means that the kinetics of IC_{50} reduction during the pre-incubation is uncertain and a set pre-incubation of 30 min may not be suitable for all inhibitors. Perloff et al. (2009) have investigated the utility of two pre-incubation time points (10 and 30 min) for the assessment of time-dependent inhibitors of a range of P450 enzymes. In the case of rapid inactivators (e.g. azamulin and CYP3A4) the use of long pre-incubation times had no beneficial impact on the IC_{50} reduction compared to shorter pre-incubation times, opposite to the effect seen with slow-acting inhibitors (e.g. verapamil and CYP3A4). The use of both short and long pre-incubation time points should facilitate the subsequent selection of the optimal conditions for the k_{inact}/K_I estimation. Therefore, short pre-incubation times with closer time points would be more appropriate for the k_{inact}/K_I assays for rapid inactivators (e.g. mibefradil/CYP3A4 and tienilic acid/CYP2C9), whereas longer pre-incubation and more wider spread of time points are more suitable for slower inhibitors (e.g. diltiazem/CYP3A4).

Significant depletion of an inhibitor over the pre-incubation time (e.g. mibefradil, saquinavir) may result in a lack of any significant decrease in IC_{50} and a false-negative result for a potent time-dependent inhibitor. In order to overcome this and allow detection of inhibition despite inhibitor depletion, recent studies have introduced the dilution step in the IC_{50} method, analogous to the k_{inact}/K_I assay (Obach et al., 2007; Perloff et al., 2009). Although a dilution step has proved to be useful in the assessment of saquinavir, it has also resulted in the artefactual shift in ketoconazole IC_{50} due to enzyme/protein dilution (Fowler and Zhang, 2008). Inclusion of a 10-fold dilution step resulted in a marginal IC_{50} shift of ritonavir, a potent time-dependent and reversible CYP3A4 inhibitor (1.6- and 2.1-fold depending on probe substrate) (Obach et al., 2007), suggesting that a greater dilution step may be required to overcome the confounding issue of a potent reversible inhibition.

Recently, attempts have been made to correlate IC_{50} data, obtained following pre-incubation, with k_{inact} and K_I (Atkinson et al., 2005; Obach et al., 2007; Grime et al., 2009); a mechanistic relationship between these parameters was proposed by Maurer et al. (2000). Recent studies (Obach et al., 2007; Grime et al., 2009) have shown good agreement between $1/IC_{50}$ values (after pre-incubation) and k_{inact}/K_I data obtained from more detailed experiments for a range of inhibitors and over a wide range of potency; the rank order in inhibitor potency was also comparable between two approaches. These preliminary estimates of k_{inact} and K_I obtained from the IC_{50} shift data should reduce the necessity for more detailed k_{inact}/K_I assays depending on the potential risk of TDI and assist experimental design of those studies.

A third possible approach involves the use of metabolite progress curves to monitor time-dependent inhibition (Fairman et al., 2007). This is the least explored method, yet potentially it may offer solutions to some of the limitations outlined above. It allows measurement of a probe substrate metabolism independently of the metabolism of an inhibitor; good reproducibility of the in vitro parameters obtained by this method in comparison to the two-step dilution assay has been reported (Wimalasena and Haines, 1996); further evaluation of the predictive utility of the parameters obtained by this method is required.

7.6 Induction

Methods for in vitro assessment of P450 induction are less well characterized and consequently the relationship between in vitro and in vivo is less clear in comparison to the inhibition studies. However, in recent years considerable advances have been made with a range of in vitro systems for P450 induction. In developing prediction strategies, advantage has been taken of experiences with inhibition, and modeling approaches that integrate both induction and inhibition are available.

The nature of the induction response, which results in an increase in the basal level of P450 enzymes following exposure to certain drugs, makes the development of a robust, sensitive in vitro system capable of generating parameter values for in vivo modeling challenging. In comparison to inhibition, enzyme induction is a slow process resulting from the change in the rate of enzyme synthesis and enzyme degradation. An increase in the steady-state enzyme concentration can result from increased synthesis rate or decreased degradation rate. Induction effects become evident on multiple dosing when the steady-state concentration of inducing drug is maintained long enough to result in an elevated enzyme level. This may be manifest in the form of auto-induction where the clearance of the inducer itself decreases with time or hetero-induction where a DDI results.

Although the concept of induction of drug metabolizing enzymes has been known for several decades an understanding of the mechanisms of P450 induction has developed slowly. CYP3A4 has received most attention due to its high inducibility and dominant role in the metabolism of many drugs. Also, of interest are the highly inducible CYP1A2 and the modest responses evident with CYP2C9, CYP2C8, CYP2A6 and CYP2B6. Induction of CYP3A4 has been well characterized for an increasing number of drugs, in particular for the very potent inducer rifampicin ($EC_{50} < 1\ \mu M$). Other classic inducers such as phenobarbital, phenytoin and carbamazepine are 100 times less potent but are associated with relatively high therapeutic plasma concentrations. More recently developed drugs of interest for their inducing properties include troglitazone, bosentan and efavirenz (Tables 7.2 and 7.3).

The major mechanism for CYP3A4 induction has been shown to occur via activation of a human orphan nuclear receptor known as pregnane X-receptor (PXR) (Sinz et al., 2006; McGinnity et al., 2009). The inducer binds to PXR in the cytosol and is translocated into the nucleus where it heterodimerizes with retinoid X-receptor which in turn binds to a DNA responsive element upstream from the *CYP3A4* gene and activates its promoter leading to transcription. PXR also upregulates the transcription of a number of other P450s albeit less potent, CYP2B6 and CYP2C9. Other nuclear hormone receptors that play a role in the induction response for certain drugs include the constitutive androstane receptor (CAR) and aryl hydrocarbon receptor (AHR – responsible for CYP1A2 induction).

Induction data are usually reported as a comparison of the test compound with a negative and/or a positive control (rifampicin often being used). Hence the fold induction can be expressed and this provides a crude measure of defining the capability of induction. When induction experiments are carried out over a wide enough concentration range it is possible to characterize a response in terms of the classic E_{max} model (Lin, 2006) with an EC_{50} and E_{max} as in vitro parameters which can be used to make an in vivo prediction.

7.6.1 In Vitro Systems Used for the Assessment of Induction

Several in vitro tools for induction assessment are available. Primary human hepatocytes are considered the gold standard for in vitro induction assessment and are the FDA recommendation (Huang et al., 2008). Hepatocytes are typically treated with a test compound for 2–3 days; then P450 levels are compared with those in a vehicle treated cells. P450 expression can be evaluated by measuring enzymatic activity with a selective probe substrate or by measuring protein or mRNA. While primary hepatocytes (fresh or cryopreserved) are considered the most reliable way to assess induction, quantity, quality and availability of human hepatocytes remains limited. Significant inter-donor variability in the induction response in primary hepatocytes and the question of what is an average liver response are of major concern. The option of generating pools of numerous donors is not a practical one in the case of human hepatocytes; some preparations show very little response while others show a high degree (Parkinson et al., 2004; Tang et al., 2005).

The balance of data would suggest that variability between donors predominantly comes from E_{max} rather than EC_{50} (Silva et al., 1998; Meunier et al.,

2000; LeCluyse, 2001a; Madan et al., 2003; McGinnity et al., 2009). Significant inter-individual difference is observed in basal levels of P450 activity (and mRNA expression) and this can result in a significant change in the fold induction observed between donors as there appears to be a ceiling for the maximum inducible activity (LeCluyse, 2001a). A number of laboratories include rifampicin as a positive control and express their data relative to the effects seen with this particular inducer. While the reasons for the choice of rifampicin are understandable it has a number of other complications due to its reliance on uptake transporters to access the cells (Niemi, 2007; Zheng et al., 2009) and certain alternatives to primary human hepatocytes (e.g. Fa2N-4 cells) do not express these particular proteins.

The use of cryopreserved hepatocytes for induction is increasing with the commercial availability of donor cells with good adhesion properties and hence are suitable for culture (Lin, 2006; Hewitt et al., 2007a; b; Fahmi et al., 2008b). Other investigators favor the use of alternative systems to human primary hepatocytes and a number of comparative evaluation studies have been reported (Hariparsad et al., 2008; Kanebratt and Andersson, 2008; Kenny et al., 2008). Because species differences in induction are known to be substantial in terms of both the spectrum of enzymes affected and the extent of induction, the use of animal tissue sources is inappropriate for assessing human induction (LeCluyse, 2001b).

The immortalized hepatocytes are considered to offer advantages as they are easily maintained, propagated in culture and amenable to higher throughput applications and at the same time showing less intrinsic variability (Mills et al., 2004; Hariparsad et al., 2008). An example being the Fa2N-4-cell line, which is derived from immortalization of primary human hepatocytes and maintains expression and inducibility of various drug metabolizing enzymes as well as maintaining the morphological characteristics of primary hepatocytes. Fa2N-4 cells are responsive to the prototypic inducers of CYP3A4 and 1A2 (Mills et al., 2004; Hariparsad et al., 2008; Kenny et al., 2008) and appear to be a good surrogate for primary human hepatocytes when assessing AHR- and PXR-mediated induction but not CAR-mediated (CYP2B6) induction (Kenny et al., 2008). Transfected HepG2 cells (HepaRG) have also been used for this purpose and have additional advantages as they express several P450s and the full range of known receptors plus conjugating enzymes and membrane transporters. Currently HepaRG cell line appears to be the best alternative to primary human hepatocytes due to its comprehensive expression of inducible proteins and good sensitivity and discriminating nature between various inducers (Guillouzo et al., 2007; Kanebratt and Andersson, 2008; McGinnity et al., 2009).

Cellular systems offer the option to measure mRNA and protein (immunoblotting) in addition to P450 functional activity using probe substrates. The preferred option is to monitor a probe activity, for example, formation of 6β-hydroxytestosterone or 1′-hydroxymidazolam in the case of CYP3A4. The availability of RT-PCR kits means that mRNA can be very conveniently assayed and this has become a very attractive option. However, demonstrable correlations between enzyme activity and mRNA transcription levels remain elusive.

Another endpoint that has the advantages of relatively high throughput and practicality for screening large number of compounds involves the use of reporter gene assays (El-Sankary et al., 2001; Luo et al., 2002; Lin, 2006). Induction potential is measured by the ability of the test compound to activate transcription of a reporter gene. PXR-based reporter assays involve cultured cell lines that have been transfected with an expression vector for PXR and a reported gene construct containing a responsive element for PXR. While these systems have a wide range of response and are generally good indicators they do have a number of drawbacks. As they are often run in tumor lines with limited or no metabolic capability, there is a concern over whether the co-regulatory factors required for full transcription activation are present. Also as they do not measure native gene expression or native enzyme activity there is no base line. Consequently Kozawa et al. (2009) normalized data on various inducers to that from rifampicin, rather than reporting an induction ratio. Recent studies (Sinz et al., 2006; McGinnity et al., 2009) are very supportive of the use of the PXR receptor assay. In contrast, Luo et al. (2002) reported very poor correlations relative to human hepatocytes; thus, there are opposing views on the value of this system and its correlation with primary human hepatocytes.

McGinnity and coworkers (McGinnity et al., 2009) have carried out an extensive study involving 24 prototypic CYP3A4 inducers which were tested with the PXR reporter gene assay, Fa2N-4 cells, HepaRG cells and primary human hepatocytes measuring mRNA and enzyme activity as endpoints. Overall mRNA endpoints showed a greater response than enzyme activity, in some cases this was associated with time-dependent inhibition as well as induction properties (e.g. ritonavir and verapamil). Marked difference between these two endpoints has also been noted for rifampicin by a number of investigators (Mills et al., 2004; Hariparsad et al., 2008; Kanebratt and Andersson, 2008; Shou et al., 2008) and in this instance TDI was not a contributing factor. Overall, there was a good correlation between EC_{50} for two human donors and the reporter gene assay and the two cell lines (McGinnity et al., 2009). However, a similar relationship was not established for E_{max} values, again this is consistent with the reports from previous studies. This is particularly evident with rifampicin where the E_{max} response can vary > 30-fold based on the enzyme activity (Mills et al., 2004; Hariparsad et al., 2008; Kanebratt and Andersson, 2008; Shou et al., 2008). In conclusion, these authors propose that the HepaRG cells are a valuable addition to the armory of in vitro tools for CYP3A4 induction and appear to be an excellent surrogate for primary human hepatocytes. However, this conclusion must be tempered by the limited amount of information which has been reported with the so-called gold standard human primary hepatocytes. Until this database grows and some consistency is achieved, any evaluation is going to be somewhat subjective.

In some in vitro studies it has become clear that it is important to include a Hill function in the E_{max} model (Shou et al., 2008); in other cases the classic hyperbolic curve is not achieved and a bell-shaped concentration response is observed (Ripp et al., 2006; Hariparsad et al., 2008). Clearly it is important to investigate whether cytotoxicity is the reason for the falling off of response. The Fa2N-4 cell line is particularly sensitive to chemical insult relative to primary hepatocytes; however, HepaRG cells also show this phenomenon.

7.7 Predictive Utility of In Vitro Inhibition and Induction Parameters

In recent years numerous efforts have been made to predict the magnitude of in vivo DDI from in vitro data with varying degrees of success (Ito et al., 2004; Galetin et al., 2005; Brown et al., 2006; Obach et al., 2006; Einolf, 2007; Obach et al., 2007; Fahmi et al., 2008b; Shou et al., 2008; Youdim et al., 2008). The accurate determination of perpetrator concentration in vivo is problematic as direct measurement is not possible and there is no generally accepted approach for the extrapolation of inhibitor/inducer concentration in the plasma to that at the enzyme site. A number of predictions of the drug–drug interactions have been attempted with differential of success using a range of [I] values as a surrogate, including the average plasma total or unbound concentration or hepatic input concentration (Kanamitsu et al., 2000b; Ito et al., 2004).

7.7.1 Qualitative Zoning and Ranking Approach of Drug–Drug Interactions

Comprehensive analysis of 193 DDIs involving inhibition of CYP2C9, CYP2D6 and CYP3A4 (Ito et al., 2004) has shown that the use of hepatic input (hepatic portal vein) concentration was the most successful for categorizing P450 inhibitors and identifying true-negative interactions. This approach identified true-positive and true-negative predictions, but resulted in a significant number of false positives and marked over-prediction of most true positives (Fig. 7.2). The $[I]/K_i$ ratio represents a useful tool in qualitative zoning and ranking of putative metabolic inhibitors (Ito et al., 2004; Obach et al., 2006; Fowler and Zhang, 2008; Walsky and Boldt, 2008). Current FDA guidelines (Huang et al., 2007; 2008) propose that for cases where $[I]/K_i$ ratio for a particular P450 (based on maximum plasma concentration at steady state after the highest proposed clinical dose of an inhibitor) is <0.1, no in vivo inhibition study is required for victim drugs metabolized by that enzyme.

In an analogous way the relative induction score has been proposed for the qualitative assessment of potential inducers, as shown in Equation (7.21) (Ripp et al., 2006; Fahmi et al., 2008a):

$$\text{Relative induction score} = \frac{E_{\text{max}}[\text{I}]}{[\text{I}] + EC_{50}} \tag{7.21}$$

The values of the relative induction scores obtained are correlated to the magnitude of clinical DDI for midazolam or ethinyl estradiol. The relationship established for a range of inducers and in a particular in vitro system allows predictions of potential induction DDIs for any new developing compound. In cases when it is not possible to achieve sufficiently high inducer concentrations to characterize E_{max} and EC_{50}, the slope of the response concentration curve can be used (Shou et al., 2008). However, the use of the slope is acceptable providing the projected in vivo concentration of inducer is likely to be below its EC_{50}. In the case of induction, the FDA recommends that in vitro fold increase in enzyme activity <40% of the positive control (e.g. rifampicin for CYP3A) indicates no need for a subsequent in vivo induction study.

A qualitative approach and decision tree based on $[I]/K_i$ ratio is currently also applied for the assessment of potential transporter-mediated DDIs (Zhang et al., 2008). Analogous to the metabolic models, in the case of transporters, the assumption is that transport occurs exclusively via the particular transport protein subject to inhibition and that no passive uptake occurs.

These pragmatic approaches ignore the specific properties of the victim drugs; therefore, significant number of over-predictions of true-positive metabolic and transporter-mediated DDIs occurring may not be surprising (Ito et al., 2004; Hinton et al., 2008). These qualitative approaches should be considered as initial discriminating screens due to their empirical nature; however, subsequent mechanistic studies are required for comprehensive evaluation of a positive result (Houston and Galetin, 2008).

7.7.2 Quantitative Prediction of Inhibition

In order to progress towards a quantitative prediction, use of appropriate models and consideration of a number of enzyme-, inhibitor- and substrate-related parameters are required, including existence of more than one metabolic/elimination pathway (Rostami-Hodjegan and Tucker, 2004; Brown et al., 2005; Ito and Houston, 2005; Galetin et al., 2006; Obach et al., 2006), contribution of the intestinal interaction (Wang et al., 2004; Einolf, 2007; Galetin et al., 2007; Obach et al., 2007; Fahmi et al., 2008b) and the role of multiple inhibitors and mechanisms (Hinton et al., 2008; Templeton et al., 2008; Zhang et al., 2009). The impact of these factors on the magnitude of the predicted DDI is discussed in the following sections.

Application of an adequate prediction model is essential for quantitative prediction, in particular in the case of more complex time-dependent and induction DDIs. For example, the prediction of TDI using the $[I]/K_i$ approach with the assumption of reversible inhibition generally results in under-prediction of the interaction with a significant number of predictions being classified as false negatives (Ito et al., 2004; Rostami-Hodjegan and Tucker, 2004; Galetin et al., 2006). An assessment performed on 37 CYP3A4 time-dependent DDIs (Galetin et al., 2006) resulted in a marked reduction in the incidence of false negatives and in a significant increase in the number of studies predicted within 2-fold (89%) when time-dependent prediction model was applied. In the evaluation of prediction success it is also important to cover a wide range of in vivo effects, i.e. to incorporate a representative number of weak, medium and potent DDIs (Bjornsson et al., 2003; Huang et al., 2007) in the validation data sets as illustrated in a number of studies (Galetin et al., 2005; Brown et al., 2006; Einolf, 2007; Fahmi et al., 2008b).

There is inconsistency within the literature in the choice of surrogate inhibitor/inducer concentration at the enzyme (or transporter) active site (either hepatic input or average plasma concentration) and the use of either single point or full inhibitor time profile (Brown et al., 2006; Galetin et al., 2006; Obach et al., 2006; Einolf, 2007; Fahmi et al., 2008b; Youdim et al., 2008) in prediction models. Although the use of plasma inhibitor concentration may be adequate for certain perpetrators, it will underestimate the liver concentration in the case of hepatic uptake into the cell resulting in false-negative predictions for those inhibitors. Recent studies (Brown et al., 2005) have shown the impact of accuracy of absorption rate constants (*k*a) on the estimated inhibitor concentration for 10 inhibitors. Refinement of this parameter from the widely assumed maximum value (0.1 min^{-1} assuming the gastric emptying is the rate-limiting step for absorption) (Ito et al., 2004) resulted in 2- to 14-fold lower *k*a, affecting significantly the relative ratio of the absorption to systemic component. A number of recent studies have applied a standard value of 0.03 min^{-1} (Obach et al., 2007; Fahmi et al., 2008a) due to lack of any prior knowledge on the absorption process which may not be correct for some inhibitors and may also vary with the dose and intake of food (Daneshmend et al., 1981). Considering the multifactorial nature of DDI models, the use of various inhibitor concentrations will also affect the evaluation of the role of other parameters in the prediction model. For example, the use of unrealistically high hepatic input concentrations may compensate for ignoring the input of the intestinal interaction, whereas incorporation of the same inhibitor concentration corrected for the plasma binding in conjunction with the intestinal inhibition may result in a comparable prediction.

The databases used so far to assess the impact of individual victim drug properties (e.g. impact of intestinal inhibition or fm_{CYP}) often differ in the parameter estimates used in the algorithms and this contributes significantly to the DDI prediction success (Ito et al., 2004; Wang et al., 2004; Galetin et al., 2005, 2006; Brown et al., 2006; Obach et al., 2006; Einolf, 2007; Fahmi et al., 2008b; Shou et al., 2008). Figure 7.3 illustrates the sensitivity of the predicted change in the victim drug AUC to the fm_{CYP} term in the case of reversible inhibition; even minor changes in this parameter (e.g. from 1 to 0.95) alter prediction accuracy significantly (Brown et al., 2005; Ito et al., 2005; Galetin et al., 2006). The change in the AUC decreases progressively with the decreasing contribution of the enzyme of interest to the overall elimination of a substrate; this is equally applicable for induction and inhibition DDIs. Therefore, it can be concluded that for victim drugs, where the enzyme of interest contributes less than 50% to the overall elimination, the fold change in the AUC will not exceed 2 in the case of inhibition/induction of that particular pathway. In general, the incorporation of fm_{CYP} in the prediction models reduces significantly the extent of over-predictions of true positives and corrects several false-positive predictions. The importance of this parameter in the DDI prediction model is illustrated in the extensive analysis of 115 in vivo inhibition DDI studies for CYP2C9, CYP2D6 and CYP3A4 (Brown et al., 2006), where the incorporation of parallel pathways of substrate elimination with the in vitro

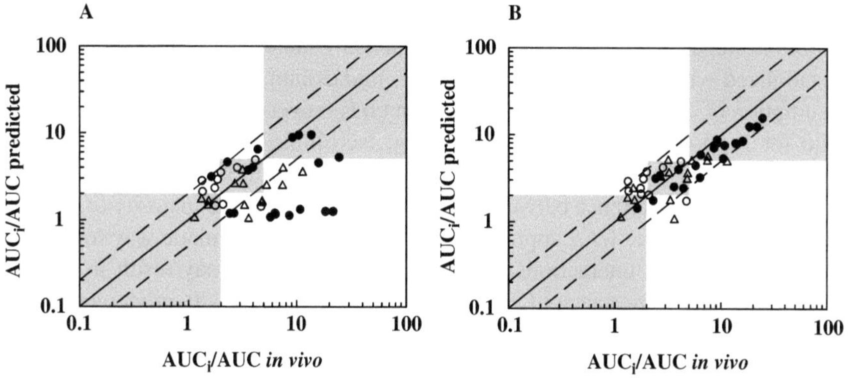

Fig. 7.5 Relationship between predicted and observed AUC ratios for reversible drug–drug interactions showing the improvement obtained by incorporating fm_{CYP}. Predicted AUC ratios obtained using the average systemic total drug plasma concentration, standardized in vitro data and incorporating fm_{CYP} for CYP2C9 (○), CYP2D6 (△) and CYP3A4 (●) in the drug–drug interaction database from Ito et al. (2004) (**A**), and the corresponding studies in the database from Brown et al. (2006) (**B**). The *grey boxes* shown relate to potent, moderate and weak DDI using the AUC ratio criteria (Bjornsson et al., 2003). Reproduced from Ito et al. (2004) and Brown et al. (2006) with the permission from Wiley-Blackwell and Adis, respectively

inhibition data obtained under optimal standardized conditions substantially improved the prediction accuracy (Fig. 7.5).

There is an increasing interest in a potential contribution of metabolic intestinal interactions due to high concentrations of the perpetrator drug achieved during the absorption phase (in some cases >100 μM, e.g. fluconazole and gemfibrozil) (Galetin et al., 2007). Relative ratio of intestinal concentration of the inhibitor to its estimated potency (I_G/K_i) represents a useful initial indicator of a potential for intestinal interaction to occur. However, the properties of the victim drug (as defined by F_G) also need to be considered, as minimal interaction is expected if the intestinal extraction is low (F_G >0.5), irrespective of the potency of the inhibitor and its inhibition mechanism (Galetin et al., 2007). Physiological variability in enterocytic blood flow (0.1–0.5 L/min, 2–10% cardiac output), inhibitor/inducer dose and intake of food may all affect the concentration in the gut wall and therefore the magnitude of potential interaction (Galetin et al., 2008).

A comprehensive assessment of the role of F_G ratio has been performed recently (Galetin et al., 2007). The analysis was based on the extensive database of 93 reversible and time-dependent inhibition CYP3A4 DDIs with 11 victim drugs with differential importance of intestinal first-pass metabolism. More than half of the interaction studies involved victim drugs with an intestinal extraction >50% (e.g. felodipine, buspirone) and were classified as either moderate or potent (increase in AUC of >2- or >5-fold, respectively). The analysis has shown a very good agreement between maximal F_G ratios (i.e. $1/F_G$ when $F_G'=1$ due to complete intestinal inhibition) and observed in vivo values for interactions involving potent inhibitors (regardless of the inhibition mechanism) and victim drugs predominantly metabolized with no transporter issues (e.g. midazolam, nifedipine). However, the assumption of complete intestinal inhibition ($1/F_G$) over-predicted the observed F_G ratio for P-gp-influenced victim drugs (e.g. tacrolimus). Comparison of the predicted (Equation 7.4) and maximal F_G ratios for 36 inhibitors in the data set resulted in no significant difference between the two approaches for the majority of the studies (91%). Therefore, $1/F_G$ approach was proposed as a pragmatic way to estimate intestinal inhibition interactions; however, this assumption may result in an over-prediction of the importance of intestinal inhibition in DDIs with moderate to weak inhibitors (<2-fold change in AUC of a victim drug). No comprehensive assessment has been performed for intestinal induction interactions.

The incorporation of the intestinal interaction into the DDI prediction strategy in the form of F_G'/F_G ratio has shown variable improvements (Rostami-Hodjegan and Tucker, 2004; Wang et al., 2004; Obach et al., 2006, 2007; Einolf, 2007; Galetin et al., 2007). Differences in the success cannot be associated solely with the incorporation of the intestine, as other parameters were not constant between the data sets (e.g. limited number of victim drugs investigated, value of fm_{CYP3A4}, use of different inhibitors concentration in the prediction model). Galetin et al. (2006) have investigated the impact of the maximal intestinal inhibition on the prediction of 28 DDIs with nine victim drugs, covering a wide range of F_G, from 0.21 to 0.98 in case of buspirone and alprazolam, respectively. The use of the $1/F_G$ approach minimized the number of false-negative predictions; however, the number of predictions within 2-fold of in vivo value was also reduced by 20%. The prediction of interactions with midazolam, triazolam and nifedipine as victim drugs was generally improved by the incorporation of the intestinal interaction, whereas pronounced over-predictions were observed for the interactions with cyclosporine and buspirone, as illustrated in Fig. 7.6. The over-prediction trend was in agreement with overestimation of atorvastatin and tacrolimus interaction observed when intestinal interaction was incorporated into the prediction strategy (Brown et al., 2006). As highlighted above, the potential intestinal contribution needs to be investigated in conjunction with other perpetrator- and victim drug-related properties considering the multifactorial nature of the DDI prediction models.

7.7.3 Importance of Multiple Inhibitors and Multiple Inhibition Mechanisms

The contribution of multiple inhibitors (or metabolites) and/or the consequence of multiple inhibition mechanisms is becoming increasingly important in the DDI assessment (Isoherranen et al., 2004; Rostami-Hodjegan and Tucker, 2004; Houston and Galetin, 2008; Zhang et al., 2009). Recent studies have reported appreciable in vivo concentrations of itraconazole metabolites (hydroxy, keto and *N*-desalkyl-itraconazole) (Templeton et al., 2008), in particular hydroxy-itraconazole which

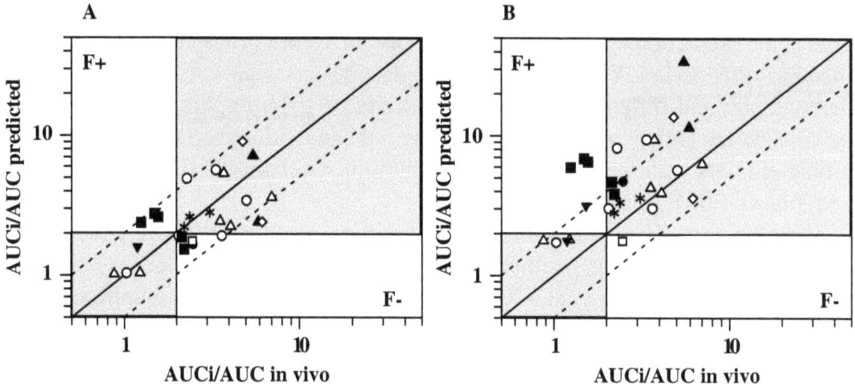

Fig. 7.6 Relationship between predicted and observed AUC ratios for time-dependent interactions showing the consequences of incorporating maximal intestinal inhibition. Predictions of 28 TDI using irreversible inhibition equation (Table 7.1) applying the average unbound plasma concentration of the inhibitor, corresponding fm_{CYP3A4} and CYP3A4 $t_{1/2\,deg}$ of 3 days. (**A**) No intestinal interaction and (**B**) the effect of incorporating the maximal intestinal inhibition (F_G'=1). The *solid line* represents line of unity and *dashed lines* represent the 2-fold limit in prediction accuracy. The *shaded* areas correspond to the true-negative and true-positive time-dependent interactions defined by the 2-fold increase in the AUC; F+ and F– represent false-positive and false-negative predictions, respectively. Interactions identified according to the substrates: △ represents midazolam, ○ triazolam, □ alprazolam, ▲ buspirone, ▼ quinidine, ◇ simvastatin, ■ cyclosporine, ● felodipine and ∗ nifedipine. Reproduced from Galetin et al. (2006) with permission from ASPET

circulates in plasma at equal or higher concentrations than itraconazole. Based on unbound plasma concentrations and IC_{50} data, all the metabolites of itraconazole were predicted to contribute to the magnitude of CYP3A inhibition observed. The time course of CYP3A4 inhibition was linked to the relative contribution of inhibitory metabolites, as their relative importance varied with time after dosing of the parent drug; for example, *N*-desalkyl itraconazole was predicted to contribute 30–40% to overall CYP3A4 inhibition 24 h after last itraconazole dose. In addition, comparable in vitro potency towards CYP3A4 inhibition between ketoconazole and its *N*-desacetyl metabolite has been reported (Davis et al., 2008); the implications of these in vitro findings in the clinical settings need to be investigated further.

In the case of inhibitors that act via inhibition of different metabolic pathways the overall inhibitory effect is additive. However, when two inhibitors act on the same enzyme and via the same mechanism, the impact of the least potent inhibitor is negligible, in particular if the relative ratio of $[I]/K_i$ of two inhibitors is > 100-fold (Hinton et al., 2008). Therefore, in such cases it would be sufficient to include only the more potent inhibitor in the prediction model, as the magnitude of DDI is driven by more potent inhibitor. When inhibitors act via different inhibition mechanisms (e.g. reversible and irreversible inhibitions of CYP2C8 in the case of gemfibrozil and its acyl glucuronide, respectively) the overall effect is not additive (Equation 7.5) and will depend on the potency of inhibitors and the individual contribution of the inhibited enzymes to the overall elimination. Hinton et al. (2008) have shown that incorporation of the time-dependent inhibition of CYP2C8 into prediction model affects the prediction outcome only in the case of victim drugs eliminated predominantly (>80%) via CYP2C8. In addition, disease- and age-specific differences in demography and differential metabolic activity of polymorphic enzymes need to be considered as they will contribute to the inter-individual variability in the magnitude of DDIs with multiple inhibitors/inhibition mechanisms (Rostami-Hodjegan and Tucker, 2004; Jamei et al., 2009a).

Experimental design to assess the inhibitory potential of acyl glucuronides on P450 enzymes using gemfibrozil glucuronide as an example has been proposed recently (Ogilvie et al., 2006). The study has shown that gemfibrozil acyl glucuronide is hydroxylated at a distal site to the glucuronide moiety and the formation of this hydroxyglucuronide results in the inhibition of CYP2C8 in a time-dependent, rather than a reversible manner. Experimental method proposed to assess inhibitory potential of glucuronides consisted of two activation steps: preliminary pre-incubation in alamethicin-activated microsomes and in the presence of UGT cofactors (allows formation of the acyl glucuronide), followed by NADPH pre-incubation before measuring P450 activity using specific probe, as described in sections above. These examples illustrate the need to consider the contribution of inhibitory metabolites in the DDI prediction for novel compounds.

In addition to metabolic DDIs, a number of recent studies have emphasized the contribution of the hepatic uptake transporter OATP1B1 to the disposition of a wide range of therapeutically used drugs and also the magnitude of observed DDI associated with these drugs (Niemi et al., 2005; Maeda et al., 2006; Lau et al., 2007; Zheng et al., 2009). Current FDA methods for ranking of potential transporter-mediated DDIs are based on the qualitative $[I]/K_i$ approach (Huang et al., 2008; Zhang et al., 2008). Hinton et al. (2008) proposed a refinement of this basic transporter model with the incorporation of a substrate-specific property, defined as fraction of drug transported by a particular transporter protein (*ft*). This parameter is analogous to fm_{CYP} in the metabolic prediction model and allows differential contribution of the transporter of interest to the overall uptake. Prediction of transporter-mediated DDIs will be addressed in more detail in other chapters.

7.7.4 Quantitative Prediction of Induction

In contrast to inhibition, the number of studies investigating prediction of induction DDIs has been small. The use of the relative induction score or other induction ratios (Kato et al., 2005; Fahmi et al., 2008a) is promising in illustrating a good rank order between in vitro and in vivo and if it is assumed that fm_{CYP} is 1, then certain conclusions can be drawn regarding the relative potency. Using data on both mRNA and activity from hepatocytes and Fa2N-4 cells, strong correlations with in vivo studies for two victim drugs (midazolam and ethinyl estradiol) with six inducers have been demonstrated (Ripp et al., 2006; Fahmi et al., 2008b).

Recently, a comprehensive set of 103 DDI studies involving six inducers of CYP3A4 and 38 victim drugs (fm_{CYP3A4} ranging from 0.05 to 0.94 for ropivacaine and midazolam, respectively) was reported (Shou et al., 2008). Their database is illustrated in Fig. 7.7 and is based on the generic DDI prediction model with the induction mechanistic term (Table 7.1). In vitro data originated from human hepatocytes and from a single donor, but similar success was seen with a second donor. The predictions shown are based on the in vitro data obtained using testosterone activity as an endpoint; when mRNA was used, a marked reduction in precision was noted. Success was also evident in earlier studies (Kato et al., 2005) which used in vitro data for eight inducers from a variety of laboratory sources with 12 victim drugs. Also of interest is the study by Onho et al. (2008) which examined seven inducers and 22 victim drugs (total 37 DDIs) from a predictive stance. While not using in vitro data they were able to illustrate the value of the generic equation approach stressing the applicability to both induction and inhibition through the use of two elements (enzyme response and victim drug PK characteristics) (Ohno et al., 2008).

Two other recent studies have investigated in vivo predictions of induction from in vitro data. Fahmi et al. (2008b) predicted 28 induction studies as a part of comprehensive investigation of a mixed induction and inhibition DDI mechanism using mRNA determination in cryopreserved human hepatocytes. Recently, reporter gene assay data on five inducers from various laboratories were analyzed in order to predict DDIs from 24 clinical studies (Kozawa et al., 2009). Both reports found that

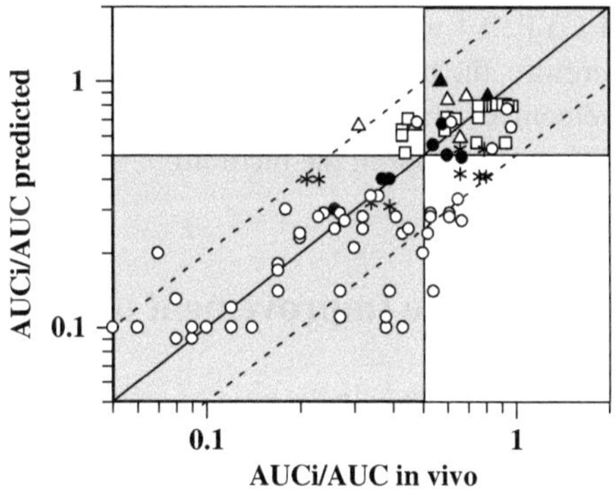

Fig. 7.7 Relationship between predicted and observed AUC ratios for induction drug–drug interactions. Predictions of 103 DDIs reported by Shou et al. (2008) using induction equation (Table 7.1). The *solid line* represents line of unity and *dashed lines* represent the 2-fold limit in prediction accuracy. The *shaded areas* correspond to the true-negative and true-positive induction interactions defined by the 2-fold change in the AUC; F+ and F– represent false positive and false-negative predictions, respectively. Interactions identified according to the inducers: △ represents bosentan, ▲ dexamethasone, □ efavirenz, ∗ phenobarbitone, ● carbamazepine and ○ rifampicin

with incorporation of empirical calibration factors (that are donor specific) excellent predictions were possible for the majority of studies.

7.7.5 Mechanisms for False-Negative and False-Positive DDI Predictions

The potential for falsely categorizing a potential DDI based on in vitro data should not be ignored and several mechanisms can be identified based on the above considerations.

False negatives. This may result from unrecognized TDI inhibition mechanism as well as an unrecognized contribution from inhibitory metabolites. Another potential source of false negatives would be the existence of transporter involvement which may lead to either an incorrect value for K_i or the inhibitor concentration available to the enzyme. Alternatively, if induction occurs in addition to inhibitory effects, the balance may differ in vitro from in vivo.

False positives. The most common mechanism for this is likely to be an incorrect assignment of the fm_{CYP} value. Considering the difficulties in estimating this parameter (Section 7.3.2) this is undoubtedly a very common shortcoming. This situation also provides a rationale for some of the inconsistencies in different analyses in the literature of the same in vivo data. A particularly striking example being alprazolam DDIs; certain investigations use fm_{CYP} value of 1 despite the literature documenting renal clearance accounting for 20% of total elimination.

False negatives and false positives. In addition there are some mechanisms, which may result in either a false-positive or false-negative conclusion. This would include inhibition of CYP3A4 when the K_i values are substantially influenced by multisite binding; therefore, the possibility of too high or too low K_i value can arise. Similarly the involvement of transporters may create an inhibitor concentration within the cell that differs from that in the medium.

7.8 Perspectives for Future Improvement of Prediction Strategies

Substantial advances have been made in the ease of generation of in vitro parameters to assess DDI potential, fuelled by the need for high-throughput procedures and the general acceptance of generic protocols. The availability of simulation programmes (e.g. Simcyp®) is assisting the better use of in vitro data for prediction purposes by allowing the incorporation of inter-individual variability and the easy exploration of various model options. Also, the realization that in vitro data per se are only part of the requirements for a quantitative prediction is becoming more widely appreciated.

In the literature there are several independent demonstrations of good correlations for prediction of DDI potential from in vitro data. This is particularly true for inhibition and the numbers of reports for induction are rapidly growing. However, the prediction algorithms required are multifactorial and involve various parameters that often have uncertain precision. In the literature it is clear that there is inconsistency over the values of certain key parameters that are used in conjunction with the in vitro data (e.g. appropriate [I]). Unfortunately there is a large subjective element in the selection of these additional parameters which is not always supported by adequate documentation. Table 7.4 lists the values for enzyme and victim drug properties that have in our personal opinion the strongest body of evidence to support their use in prediction strategies. A consensus on these values would prevent further confusing analyses and allow a clearer view of the outstanding issues.

Table 7.4 Recommended parameter values for enzyme (k_{deg}) and victim drug properties (fm_{CYP} and F_G). Victim drug listed are preferred by FDA for the assessment of potential drug–drug interactions (Huang et al., 2007)

P450 (Recommended k_{deg})	Victim drug	Recommended fm_{CYP}	Recommended F_G	References
CYP3A4[a] (1.6 and 4.8 × 10^{-4} min^{-1})	Buspirone	0.99	0.21	Galetin et al. (2006, 2008)
	Felodipine	0.81	0.45	Brown et al. (2005) and Galetin et al. (2008)
	Midazolam	0.94	0.51[d]	Brown et al. (2005) and Galetin et al. (2008)
	Nifedipine	0.71	0.78	Brown et al. (2005) and Galetin et al. (2008)
	Sildenafil	0.9	0.54	Galetin et al. (2008) and Houston and Galetin (2008)
	Simvastatin	0.99	0.14[e]	Gertz et al. (2008a) and Houston and Galetin (2008)
	Triazolam	0.92	0.75	Brown et al. (2005) and Galetin et al. (2008)
CYP2C9[b] (1.1 × 10^{-4} min^{-1})	Tolbutamide	0.8	1	Brown et al. (2005)
	S-warfarin	0.87	1	Brown et al. (2005)
CYP2D6[c] (2.3 × 10^{-4} min^{-1})	Desipramine	0.88	1	Ito et al. (2005)

[a] Hepatic and intestinal value, respectively (Galetin et al., 2006; Gertz et al., 2008a; Yang et al., 2008).
[b] Ghanbari et al. (2006) and Yang et al. (2008).
[c] Venkatakrishnan and Obach (2005) and Obach et al. (2007).
[d] Estimated from i.v./oral data from 315 individuals, F_G ranging from 0.40 to 0.79 (Galetin et al., 2008).
[e] Estimated from grapefruit juice interaction data in 30 individuals, F_G ranging from 0.06 to 0.28 (Gertz et al., 2008a).

Current trends in the type of new chemical entities under development as drugs have necessitated a broadening of the scope of in vitro systems, and hence prediction strategies. In addition to liver, many proteins responsible for clearance processes are also expressed extrahepatically. Hence, prediction strategies need to be expanded to include metabolic process in the gut and kidneys and to incorporate conjugation enzymes in addition to the P450s. In the commonly used liver models the term intrinsic clearance relies on the fact that clearance occurs via a number of parallel (usually metabolic) processes hence individual clearance terms are additive. When transporters operate for a drug the hepatic clearance may involve sequential processes rather than parallel ones and this presents challenges for the future for both in vitro and modeling methodologies. Combined physiologically based models incorporating both uptake transporters and metabolic pathways in a sequential manner to provide a comprehensive prediction tool are needed to explore and accommodate the range of possible outcomes and the added complexity that may result.

References

Agoram B, Woltosz WS and Bolger MB (2001) Predicting the impact of physiological and biochemical processes on oral drug bioavailability. *Adv Drug Deliv Rev* **50 Suppl 1**:S41–S67.

Atkinson A, Kenny JR and Grime K (2005) Automated assessment of time-dependent inhibition of human cytochrome P450 enzymes using liquid chromatography-tandem mass spectrometry analysis. *Drug Metab Dispos* **33**:1637–1647.

Austin RP, Barton P, Cockroft SL, Wenlock MC and Riley RJ (2002) The influence of nonspecific microsomal binding on apparent intrinsic clearance, and its prediction from physicochemical properties. *Drug Metab Dispos* **30**:1497–1503.

Austin RP, Barton P, Mohmed S and Riley RJ (2005) The binding of drugs to hepatocytes and its relationship to physicochemical properties. *Drug Metab Dispos* **33**:419–425.

Badhan R, Penny J, Galetin A and Houston JB (2009) Methodology for development of a physiological model incorporating CYP3A and P-glycoprotein for the prediction of intestinal drug absorption. *J Pharm Sci* **98**:2180–2197.

Bailey DG, Malcolm J, Arnold O and Spence JD (1998) Grapefruit juice-drug interactions. *Br J Clin Pharmacol* **46**:101–110.

Bjornsson TD, Callaghan JT, Einolf HJ, Fischer V, Gan L, Grimm S, Kao J, King SP, Miwa G, Ni L, Kumar G, McLeod J, Obach RS, Roberts S, Roe A, Shah A, Snikeris F, Sullivan JT, Tweedie D, Vega JM, Walsh J and Wrighton SA (2003) The conduct of in vitro and in vivo drug–drug interaction studies: a Pharmaceutical Research and Manufacturers of America (PhRMA) perspective. *Drug Metab Dispos* **31**:815–832.

Brown HS, Chadwick A and Houston JB (2007a) Use of isolated hepatocyte preparations for cytochrome P450 inhibition studies: comparison with microsomes for Ki determination. *Drug Metab Dispos* **35**:2119–2126.

Brown HS, Galetin A, Hallifax D and Houston JB (2006) Prediction of in vivo drug–drug interactions from in vitro data: factors affecting prototypic drug–drug interactions involving CYP2C9, CYP2D6 and CYP3A4. *Clin Pharmacokinet* **45**:1035–1050.

Brown HS, Griffin M and Houston JB (2007b) Evaluation of cryopreserved human hepatocytes as an alternative in vitro system to microsomes for the prediction of metabolic clearance. *Drug Metab Dispos* **35**:293–301.

Brown HS, Ito K, Galetin A and Houston JB (2005) Prediction of in vivo drug–drug interactions from in vitro data: impact of incorporating parallel pathways of drug elimination and inhibitor absorption rate constant. *Br J Clin Pharmacol* **60**:508–518.

Chien JY, Lucksiri A, Ernest CS, 2nd, Gorski JC, Wrighton SA and Hall SD (2006) Stochastic prediction of CYP3A-mediated inhibition of midazolam clearance by ketoconazole. *Drug Metab Dispos* **34**:1208–1219.

Collins C, Levy R, Ragueneau-Majlessi I and Hachad H (2006) Prediction of maximum exposure in poor metabolizers following inhibition of nonpolymorphic pathways. *Curr Drug Metab* **7**:295–299.

Correia MA (1991) Cytochrome P450 turnover. *Methods Enzymol* **206**:315–325.

Cubitt HE, Houston JB and Galetin A (2009) Relative importance of intestinal and hepatic glucuronidation-impact on the prediction of drug clearance. *Pharm Res* **26**:1073–1083.

Daneshmend TK, Warnock DW, Turner A and Roberts CJ (1981) Pharmacokinetics of ketoconazole in normal subjects. *J Antimicrob Chemother* **8**:299–304.

Davis J, Galetin A and Jones B (2008) Ketoconazole based drug–drug interactions – an update. *Clin Pharmacol Ther* **83 (Suppl. 1)**:33.

Dokoumetzidis A, Kalantzi L and Fotaki N (2007) Predictive models for oral drug absorption: from *in silico* methods to integrated dynamical models. *Expert Opin Drug Metab Toxicol* **3**:491–505.

Egnell AC, Eriksson C, Albertson N, Houston B and Boyer S (2003) Generation and evaluation of a CYP2C9 heteroactivation pharmacophore. *J Pharmacol Exp Ther* **307**:878–887.

Einolf HJ (2007) Comparison of different approaches to predict metabolic drug–drug interactions. *Xenobiotica* **37**:1257–1294.

El-Sankary W, Gibson GG, Ayrton A and Plant N (2001) Use of a reporter gene assay to predict and rank the potency and efficacy of CYP3A4 inducers. *Drug Metab Dispos* **29**:1499–1504.

Ernest CS, 2nd, Hall SD and Jones DR (2005) Mechanism-based inactivation of CYP3A by HIV protease inhibitors. *J Pharmacol Exp Ther* **312**:583–591.

Fahmi OA, Boldt S, Kish M, Obach RS and Tremaine LM (2008a) Prediction of drug–drug interactions from in vitro induction data: application of the relative induction score approach using cryopreserved human hepatocytes. *Drug Metab Dispos* **36**:1971–1974.

Fahmi OA, Maurer TS, Kish M, Cardenas E, Boldt S and Nettleton D (2008b) A combined model for predicting CYP3A4 clinical net drug–drug interaction based on CYP3A4 inhibition, inactivation, and induction determined in vitro. *Drug Metab Dispos* **36**:1698–1708.

Fairman DA, Collins C and Chapple S (2007) Progress curve analysis of CYP1A2 inhibition: a more informative approach to the assessment of mechanism-based inactivation?. *Drug Metab Dispos* **35**:2159–2165.

Fowler S and Zhang H (2008) In vitro evaluation of reversible and irreversible cytochrome P450 inhibition: current status on methodologies and their utility for predicting drug–drug interactions. *Aaps J* **10**:410–424.

Galetin A, Burt H, Gibbons L and Houston JB (2006) Prediction of time-dependent CYP3A4 drug–drug interactions: impact of enzyme degradation, parallel elimination pathways, and intestinal inhibition. *Drug Metab Dispos* **34**:166–175.

Galetin A, Clarke SE and Houston JB (2002) Quinidine and haloperidol as modifiers of CYP3A4 activity: multisite kinetic model approach. *Drug Metab Dispos* **30**:1512–1522.

Galetin A, Clarke SE and Houston JB (2003) Multisite kinetic analysis of interactions between prototypical CYP3A4 subgroup substrates: midazolam, testosterone, and nifedipine. *Drug Metab Dispos* **31**:1108–1116.

Galetin A, Gertz M and Houston JB (2008) Potential role of intestinal first-pass metabolism in the prediction of drug–drug interactions. *Expert Opin Drug Metab Toxicol* **4**:909–922.

Galetin A, Hinton LK, Burt H, Obach RS and Houston JB (2007) Maximal inhibition of intestinal first-pass metabolism as a pragmatic indicator of intestinal contribution to the drug–drug interactions for CYP3A4 cleared drugs. *Curr Drug Metab* **8**:685–693.

Galetin A, Ito K, Hallifax D and Houston JB (2005) CYP3A4 substrate selection and substitution in the prediction of potential drug–drug interactions. *J Pharmacol Exp Ther* **314**:180–190.

Gertz M, Davis JD, Harrison A, Houston JB and Galetin A (2008a) Grapefruit juice-drug interaction studies as a method to assess the extent of intestinal availability: utility and limitations. *Curr Drug Metab* **9**:785–795.

Gertz M, Fenner KS, Harrison A, Davis J, Houston JB and Galetin A (2008b) Impact of permeability on in vitro predictions of intestinal availability in the QGut model. *Drug Metab Rev* **40 (Suppl. 1)**:92.

Gertz M, Kilford PJ, Houston JB and Galetin A (2008c) Drug lipophilicity and microsomal protein concentration as determinants in the prediction of the fraction unbound in microsomal incubations. *Drug Metab Dispos* **36**:535–542.

Ghanbari F, Rowland-Yeo K, Bloomer JC, Clarke SE, Lennard MS, Tucker GT and Rostami-Hodjegan A (2006) A critical evaluation of the experimental design of studies of mechanism based enzyme inhibition, with implications for in vitro-in vivo extrapolation. *Curr Drug Metab* **7**:315–334.

Gomez DY, Wacher VJ, Tomlanovich SJ, Hebert MF and Benet LZ (1995) The effects of ketoconazole on the intestinal metabolism and bioavailability of cyclosporine. *Clin Pharmacol Ther* **58**:15–19.

Grime KH, Bird J, Ferguson D and Riley RJ (2009) Mechanism-based inhibition of cytochrome P450 enzymes: an evaluation of early decision making in vitro approaches and drug–drug interaction prediction methods. *Eur J Pharm Sci* **36**:175–191.

Grime K and Riley RJ (2006) The impact of in vitro binding on in vitro-in vivo extrapolations, projections of metabolic clearance and clinical drug–drug interactions. *Curr Drug Metab* **7**:251–264.

Guillouzo A, Corlu A, Aninat C, Glaise D, Morel F and Guguen-Guillouzo C (2007) The human hepatoma HepaRG cells: a highly differentiated model for studies of liver metabolism and toxicity of xenobiotics. *Chem Biol Interact* **168**:66–73.

Hall SD, Thummel KE, Watkins PB, Lown KS, Benet LZ, Paine MF, Mayo RR, Turgeon DK, Bailey DG, Fontana RJ and Wrighton SA (1999) Molecular and physical mechanisms of first-pass extraction. *Drug Metab Dispos* **27**:161–166.

Hallifax D, Galetin A and Houston JB (2008) Prediction of metabolic clearance using fresh human hepatocytes: Comparison with cryopreserved hepatocytes and hepatic microsomes for five benzodiazepines. *Xenobiotica* **38**:353–367.

Hallifax D and Houston JB (2006) Binding of drugs to hepatic microsomes: comment and assessment of current prediction methodology with recommendation for improvement. *Drug Metab Dispos* **34**:724–726.

Hallifax D and Houston JB (2007) Saturable uptake of lipophilic amine drugs into isolated hepatocytes: mechanisms and consequences for quantitative clearance prediction. *Drug Metab Dispos* **35**:1325–1332.

Hariparsad N, Carr BA, Evers R and Chu X (2008) Comparison of immortalized Fa2N-4 cells and human hepatocytes as in vitro models for cytochrome P450 induction. *Drug Metab Dispos* **36**:1046–1055.

Henshall J, Galetin A, Harrison A and Houston JB (2008) Comparative analysis of CYP3A heteroactivation by steroid hormones and flavonoids in different in vitro systems and potential in vivo implications. *Drug Metab Dispos* **36**:1332–1340.

Hewitt NJ, Lechón MJ, Houston JB, Hallifax D, Brown HS, Maurel P, Kenna JG, Gustavsson L, Lohmann C, Skonberg C, Guillouzo A, Tuschl G, Li AP, LeCluyse E, Groothuis GMM and Hengstler JG (2007b) Primary hepatocytes: current understanding of the regulation of metabolic enzymes and transporter proteins, and pharmaceutical practice for the use of hepatocytes in metabolism, enzyme induction, transporter, clearance, and hepatotoxicity studies. *Drug Metab Rev* **39**:159–234.

Hewitt NJ, de Kanter R and LeCluyse E (2007a) Induction of drug metabolizing enzymes: a survey of in vitro methodologies and interpretations used in the pharmaceutical industry – do they comply with FDA recommendations?. *Chem Biol Interact* **168**:51–65.

Hinton LK, Galetin A and Houston JB (2008) Multiple inhibition mechanisms and prediction of drug–drug interactions: status of metabolism and transporter models as exemplified by gemfibrozil-drug interactions. *Pharm Res* **25**:1063–1074.

Hirano M, Maeda K, Shitara Y and Sugiyama Y (2006) Drug–drug interaction between pitavastatin and various drugs via OATP1B1. *Drug Metab Dispos* **34**:1229–1236.

Ho RH, Tirona RG, Leake BF, Glaeser H, Lee W, Lemke CJ, Wang Y and Kim RB (2006) Drug and bile acid transporters in rosuvastatin hepatic uptake: function, expression, and pharmacogenetics. *Gastroenterology* **130**:1793–1806.

Houston JB and Galetin A (2003) Progress towards prediction of human pharmacokinetic parameters from in vitro technologies. *Drug Metab Rev* **35**:393–415.

Houston JB and Galetin A (2005) Modelling atypical CYP3A4 kinetics: principles and pragmatism. *Arch Biochem Biophys* **433**:351–360.

Houston JB and Galetin A (2008) Methods for predicting in vivo pharmacokinetics using data from in vitro assays. *Curr Drug Metab* **9**:940–951.

Houston JB, Kenworthy KE and Galetin A (2003) Typical and atypical enzyme kinetics, in *Drug Metabolizing Enzymes: Cytochrome P450 and Other Enzymes in Drug Discovery and Development* (Fisher M, Lee J, Obach S, eds) 211–254, Fontis Media. Lausanne.

Huang SM, Strong JM, Zhang L, Reynolds KS, Nallani S, Temple R, Abraham S, Habet SA, Baweja RK, Burckart GJ, Chung S, Colangelo P, Frucht D, Green MD, Hepp P, Karnaukhova E, Ko HS, Lee JI, Marroum PJ, Norden JM, Qiu W, Rahman A, Sobel S, Stifano T, Thummel K, Wei XX, Yasuda S, Zheng JH, Zhao H and Lesko LJ (2008) New era in drug interaction evaluation: US Food and Drug Administration update on CYP enzymes, transporters, and the guidance process. *J Clin Pharmacol* **48**:662–670.

Huang SM, Temple R, Throckmorton DC and Lesko LJ (2007) Drug interaction studies: study design, data analysis, and implications for dosing and labeling. *Clin Pharmacol Ther* **81**:298–304.

Hutzler JM, Wienkers LC, Wahlstrom JL, Carlson TJ and Tracy TS (2003) Activation of cytochrome P450 2C9-mediated metabolism: mechanistic evidence in support of kinetic observations. *Arch Biochem Biophys* **410**:16–24.

Isoherranen N, Kunze KL, Allen KE, Nelson WL and Thummel KE (2004) Role of itraconazole metabolites in CYP3A4 inhibition. *Drug Metab Dispos* **32**: 1121–1131.

Ito K, Brown HS and Houston JB (2004) Database analyses for the prediction of in vivo drug–drug interactions from in vitro data. *Br J Clin Pharmacol* **57**:473–486.

Ito K, Hallifax D, Obach RS and Houston JB (2005) Impact of parallel pathways of drug elimination and multiple cytochrome P450 involvement on drug–drug interactions: CYP2D6 paradigm. *Drug Metab Dispos* **33**:837–844.

Ito K and Houston JB (2005) Prediction of human drug clearance from in vitro and preclinical data using physiologically based and empirical approaches. *Pharm Res* **22**:103–112.

Ito K, Iwatsubo T, Kanamitsu S, Ueda K, Suzuki H and Sugiyama Y (1998) Prediction of pharmacokinetic alterations caused by drug–drug interactions: metabolic interaction in the liver. *Pharmacol Rev* **50**:387–412.

Ito K, Kusuhara H and Sugiyama Y (1999) Effects of intestinal CYP3A4 and P-glycoprotein on oral drug absorption – theoretical approach. *Pharm Res* **16**:225–231.

Ito K, Ogihara K, Kanamitsu S and Itoh T (2003) Prediction of the in vivo interaction between midazolam and macrolides based on in vitro studies using human liver microsomes. *Drug Metab Dispos* **31**:945–954.

Jamei M, Dickinson GL and Rostami-Hodjegan A (2009a) A framework for assessing inter-individual variability in pharmacokinetics using virtual human populations and integrating general knowledge of physical chemistry, biology, anatomy, physiology and genetics: a tale of 'bottom-up' vs 'top-down' recognition of covariates. *Drug Metab Pharmacokinet* **24**:53–75.

Jamei M, Marciniak S, Feng K, Barnett A, Tucker G and Rostami-Hodjegan A (2009b) The Simcyp((R)) population-based ADME simulator. *Expert Opin Drug Metab Toxicol.*

Jushchyshyn MI, Hutzler JM, Schrag ML and Wienkers LC (2005) Catalytic turnover of pyrene by CYP3A4: evidence that cytochrome b5 directly induces positive cooperativity. *Arch Biochem Biophys* **438**:21–28.

Kalgutkar AS, Obach RS and Maurer TS (2007) Mechanism-based inactivation of cytochrome P450 enzymes: chemical mechanisms, structure-activity relationships and relationship to clinical drug–drug interactions and idiosyncratic adverse drug reactions. *Curr Drug Metab* **8**:407–447.

Kanamitsu S, Ito K, Green CE, Tyson CA, Shimada N and Sugiyama Y (2000a) Prediction of in vivo interaction between triazolam and erythromycin based on in vitro studies using human liver microsomes and recombinant human CYP3A4. *Pharm Res* **17**:419–426.

Kanamitsu S, Ito K and Sugiyama Y (2000b) Quantitative prediction of in vivo drug–drug interactions from in vitro data based on physiological pharmacokinetics: use of maximum unbound concentration of inhibitor at the inlet to the liver. *Pharm Res* **17**:336–343.

Kanebratt KP and Andersson TB (2008) HepaRG cells as an in vitro model for evaluation of cytochrome P450 induction in humans. *Drug Metab Dispos* **36**:137–145.

Kato M, Chiba K, Horikawa M and Sugiyama Y (2005) The quantitative prediction of in vivo enzyme-induction caused by drug exposure from in vitro information on human hepatocytes. *Drug Metab Pharmacokinet* **20**:236–243.

Kenny JR, Chen L, McGinnity DF, Grime K, Shakesheff KM, Thomson B and Riley R (2008) Efficient assessment of the utility of immortalized Fa2N-4 cells for cytochrome P450 (CYP) induction studies using multiplex quantitative reverse transcriptase-polymerase chain reaction (qRT-PCR) and substrate cassette methodologies. *Xenobiotica* **38**:1500–1517.

Kent UM, Juschyshyn MI and Hollenberg PF (2001) Mechanism-based inactivators as probes of cytochrome P450 structure and function. *Curr Drug Metab* **2**:215–243.

Kenworthy KE, Bloomer JC, Clarke SE and Houston JB (1999) CYP3A4 drug interactions: correlation of 10 in vitro probe substrates. *Br J Clin Pharmacol* **48**:716–727.

Kenworthy KE, Clarke SE, Andrews J and Houston JB (2001) Multisite kinetic models for CYP3A4: simultaneous activation and inhibition of diazepam and testosterone metabolism. *Drug Metab Dispos* **29**:1644–1651.

Kilford PJ, Gertz M, Houston JB and Galetin A (2008) Hepatocellular binding of drugs: correction for unbound fraction in hepatocyte incubations using microsomal binding or drug lipophilicity data. *Drug Metab Dispos* **36**:1194–1197.

Kilford PJ, Stringer R, Sohal B, Houston JB and Galetin A (2009) Prediction of drug clearance by glucuronidation from in vitro data: use of combined cytochrome P450 and UDP-glucuronosyltransferase cofactors in alamethicin-activated human liver microsomes. *Drug Metab Dispos* **37**:82–89.

Kozawa M, Honma M and Suzuki H (2009) Quantitative prediction of in vivo profiles of CYP3A4 induction in humans from in vitro results with reporter gene assay. *Drug Metab Dispos* **37**:1234–1241.

Kusama M, Maeda K, Chiba K, Aoyama A and Sugiyama Y (2009) Prediction of the effects of genetic polymorphism on the pharmacokinetics of CYP2C9 substrates from in vitro data. *Pharm Res* **26**:822–835.

Lau YY, Huang Y, Frassetto L and Benet LZ (2007) Effect of OATP1B transporter inhibition on the pharmacokinetics of atorvastatin in healthy volunteers. *Clin Pharmacol Ther* **81**:194–204.

LeCluyse EL (2001a) Human hepatocyte culture systems for the in vitro evaluation of cytochrome P450 expression and regulation. *Eur J Pharm Sci* **13**:343–368.

LeCluyse EL (2001b) Pregnane X receptor: molecular basis for species differences in CYP3A induction by xenobiotics. *Chem Biol Interact* **134**:283–289.

Lennernas H (2007) Modeling gastrointestinal drug absorption requires more in vivo biopharmaceutical data: experience from in vivo dissolution and permeability studies in humans. *Curr Drug Metab* **8**:645–657.

Lin JH (2006) CYP induction-mediated drug interactions: in vitro assessment and clinical implications. *Pharm Res* **23**:1089–1116.

Lin JH, Chiba M and Baillie TA (1997) In vivo assessment of intestinal drug metabolism. *Drug Metab Dispos* **25**:1107–1109.

Lu AY, Wang RW and Lin JH (2003) Cytochrome P450 in vitro reaction phenotyping: a re-evaluation of approaches used for P450 isoform identification. *Drug Metab Dispos* **31**:345–350.

Luo G, Cunningham M, Kim S, Burn T, Lin J, Sinz M, Hamilton G, Rizzo C, Jolley S, Gilbert D, Downey A, Mudra D, Graham R, Carroll K, Xie J, Madan A, Parkinson A, Christ D, Selling B,

LeCluyse E and Gan LS (2002) CYP3A4 induction by drugs: correlation between a pregnane X receptor reporter gene assay and CYP3A4 expression in human hepatocytes. *Drug Metab Dispos* **30**:795–804.

Madan A, Graham RA, Carroll KM, Mudra DR, Burton LA, Krueger LA, Downey AD, Czerwinski M, Forster J, Ribadeneira MD, Gan LS, LeCluyse EL, Zech K, Robertson P, Jr., Koch P, Antonian L, Wagner G, Yu L and Parkinson A (2003) Effects of prototypical microsomal enzyme inducers on cytochrome P450 expression in cultured human hepatocytes. *Drug Metab Dispos* **31**:421–431.

Maeda K, Ieiri I, Yasuda K, Fujino A, Fujiwara H, Otsubo K, Hirano M, Watanabe T, Kitamura Y, Kusuhara H and Sugiyama Y (2006) Effects of organic anion transporting polypeptide 1B1 haplotype on pharmacokinetics of pravastatin, valsartan, and temocapril. *Clin Pharmacol Ther* **79**:427–439.

Margolis JM and Obach RS (2003) Impact of nonspecific binding to microsomes and phospholipid on the inhibition of cytochrome P4502D6: implications for relating in vitro inhibition data to in vivo drug interactions. *Drug Metab Dispos* **31**:606–611.

Maurer TS, Tabrizi-Fard MA and Fung HL (2000) Impact of mechanism-based enzyme inactivation on inhibitor potency: implications for rational drug discovery. *J Pharm Sci* **89**:1404–1414.

McConn DJ, 2nd, Lin YS, Allen K, Kunze KL and Thummel KE (2004) Differences in the inhibition of cytochromes P450 3A4 and 3A5 by metabolite-inhibitor complex-forming drugs. *Drug Metab Dispos* **32**:1083–1091.

McGinnity DF, Berry AJ, Kenny JR, Grime K and Riley RJ (2006) Evaluation of time-dependent cytochrome P450 inhibition using cultured human hepatocytes. *Drug Metab Dispos* **34**: 1291–1300.

McGinnity DF, Waters NJ, Tucker J and Riley RJ (2008) Integrated in vitro analysis for the in vivo prediction of cytochrome P450-mediated drug–drug interactions. *Drug Metab Dispos* **36**: 1126–1134.

McGinnity DF, Zhang G, Kenny J, Hamilton G, Otmani S, Stams K, Haney S, Brassil P, Stresser DM and Riley RJ (2009) Evaluation of multiple in vitro systems for assessment of CYP3A4 Induction in Drug Discovery: Human hepatocytes, PXR reporter gene, Fa2N-4 and HepaRG cells. *Drug Metab Dispos* **37**:1259–1268.

Meunier V, Bourrie M, Julian B, Marti E, Guillou F, Berger Y and Fabre G (2000) Expression and induction of CYP1A1/1A2, CYP2A6 and CYP3A4 in primary cultures of human hepatocytes: a 10-year follow-up. *Xenobiotica* **30**:589–607.

Mills JB, Rose KA, Sadagopan N, Sahi J and de Morais SM (2004) Induction of drug metabolism enzymes and MDR1 using a novel human hepatocyte cell line. *J Pharmacol Exp Ther* **309**: 303–309.

Niemi M (2007) Role of OATP transporters in the disposition of drugs. *Pharmacogenomics* **8**: 787–802.

Niemi M, Backman JT, Kajosaari LI, Leathart JB, Neuvonen M, Daly AK, Eichelbaum M, Kivisto KT and Neuvonen PJ (2005) Polymorphic organic anion transporting polypeptide 1B1 is a major determinant of repaglinide pharmacokinetics. *Clin Pharmacol Ther* **77**: 468–478.

Obach RS (1999) Prediction of human clearance of twenty-nine drugs from hepatic microsomal intrinsic clearance data: An examination of in vitro half-life approach and nonspecific binding to microsomes. *Drug Metab Dispos* **27**:1350–1359.

Obach RS, Walsky RL and Venkatakrishnan K (2007) Mechanism-based inactivation of human cytochrome p450 enzymes and the prediction of drug–drug interactions. *Drug Metab Dispos* **35**:246–255.

Obach RS, Walsky RL, Venkatakrishnan K, Gaman EA, Houston JB and Tremaine LM (2006) The utility of in vitro cytochrome P450 inhibition data in the prediction of drug–drug interactions. *J Pharmacol Exp Ther* **316**:336–348.

Obach RS, Walsky RL, Venkatakrishnan K, Houston JB and Tremaine LM (2005) In vitro cytochrome P450 inhibition data and the prediction of drug–drug interactions: qualitative relationships, quantitative predictions, and the rank-order approach. *Clin Pharmacol Ther* **78**:582–592.

Ogilvie BW, Zhang D, Li W, Rodrigues AD, Gipson AE, Holsapple J, Toren P and Parkinson A (2006) Glucuronidation converts gemfibrozil to a potent, metabolism-dependent inhibitor of CYP2C8: implications for drug–drug interactions. *Drug Metab Dispos* **34**:191–197.

Ohno Y, Hisaka A, Ueno M and Suzuki H (2008) General framework for the prediction of oral drug interactions caused by CYP3A4 induction from in vivo information. *Clin Pharmacokinet* **47**:669–680.

Paine MF, Hart HL, Ludington SS, Haining RL, Rettie AE and Zeldin DC (2006) The human intestinal cytochrome P450 "pie". *Drug Metab Dispos* **34**:880–886.

Paine MF, Shen DD, Kunze KL, Perkins JD, Marsh CL, McVicar JP, Barr DM, Gillies BS and Thummel KE (1996) First-pass metabolism of midazolam by the human intestine. *Clin Pharmacol Ther* **60**:14–24.

Palkama VJ, Ahonen J, Neuvonen PJ and Olkkola KT (1999) Effect of saquinavir on the pharmacokinetics and pharmacodynamics of oral and intravenous midazolam. *Clin Pharmacol Ther* **66**:33–39.

Pang KS (2003) Modeling of intestinal drug absorption: roles of transporters and metabolic enzymes (for the Gillette Review Series). *Drug Metab Dispos* **31**:1507–1519.

Pang KS and Rowland M (1977) Hepatic clearance of drugs. I. Theoretical considerations of a "well-stirred" model and a "parallel tube" model. Influence of hepatic blood flow, plasma and blood cell binding, and the hepatocellular enzymatic activity on hepatic drug clearance. *J Pharmacokinet Biopharm* **5**:625–653.

Parkinson A, Mudra DR, Johnson C, Dwyer A and Carroll KM (2004) The effects of gender, age, ethnicity, and liver cirrhosis on cytochrome P450 enzyme activity in human liver microsomes and inducibility in cultured human hepatocytes. *Toxicol Appl Pharmacol* **199**:193–209.

Perloff ES, Mason AK, Dehal SS, Blanchard AP, Morgan L, Ho T, Dandeneau A, Crocker RM, Chandler CM, Boily N, Crespi CL and Stresser DM (2009) Validation of cytochrome P450 time-dependent inhibition assays: a two-time point IC50 shift approach facilitates kinact assay design. *Xenobiotica* **39**:99–112.

Pinto AG, Wang YH, Chalasani N, Skaar T, Kolwankar D, Gorski JC, Liangpunsakul S, Hamman MA, Arefayene M and Hall SD (2005) Inhibition of human intestinal wall metabolism by macrolide antibiotics: effect of clarithromycin on cytochrome P450 3A4/5 activity and expression. *Clin Pharmacol Ther* **77**:178–188.

Polasek TM, Elliot DJ, Lewis BC and Miners JO (2004) Mechanism-based inactivation of human cytochrome P4502C8 by drugs in vitro. *J Pharmacol Exp Ther* **311**:996–1007.

Polasek TM and Miners JO (2006) Quantitative prediction of macrolide drug–drug interaction potential from in vitro studies using testosterone as the human cytochrome P4503A substrate. *Eur J Clin Pharmacol* **62**:203–208.

Polasek TM and Miners JO (2007) In vitro approaches to investigate mechanism-based inactivation of CYP enzymes. *Expert Opin Drug Metab Toxicol* **3**:321–329.

Renwick AB, Watts PS, Edwards RJ, Barton PT, Guyonnet I, Price RJ, Tredger JM, Pelkonen O, Boobis AR and Lake BG (2000) Differential maintenance of cytochrome P450 enzymes in cultured precision-cut human liver slices. *Drug Metab Dispos* **28**: 1202–1209.

Riley RJ, Grime K and Weaver R (2007) Time-dependent CYP inhibition. *Expert Opin Drug Metab Toxicol* **3**:51–66.

Ripp SL, Mills JB, Fahmi OA, Trevena KA, Liras JL, Maurer TS and de Morais SM (2006) Use of immortalized human hepatocytes to predict the magnitude of clinical drug–drug interactions caused by CYP3A4 induction. *Drug Metab Dispos* **34**:1742–1748.

Rostami-Hodjegan A and Tucker G (2004) `*In silico*′ simulations to assess the `in vivo′ consequences of `in vitro′ metabolic drug–drug interactions. *Drug Discov Today Tech* **1**:441.

Rostami-Hodjegan A and Tucker GT (2007) Simulation and prediction of in vivo drug metabolism in human populations from in vitro data. *Nat Rev Drug Discov* **6**:140.

Rowland M, Benet LZ and Graham GG (1973) Clearance concepts in pharmacokinetics. *J Pharmacokinet Biopharm* **1**:123–136.

Rowland M and Matin SB (1973) Kinetics of drug–drug interactions. *J Pharmacokinet Biopharm* **1**:553–567.

Rowland-Yeo K, Rostami-Hodjegan A and Tucker GT (2004) Abundance of cytochrome P450 in human liver: a meta-analysis. *Br J Clin Pharmacol* **57**:687–688.

Saari TI, Laine K, Leino K, Valtonen M, Neuvonen PJ and Olkkola KT (2006) Effect of voriconazole on the pharmacokinetics and pharmacodynamics of intravenous and oral midazolam. *Clin Pharmacol Ther* **79**:362–370.

Saito M, Hirata-Koizumi M, Matsumoto M, Urano T and Hasegawa R (2005) Undesirable effects of citrus juice on the pharmacokinetics of drugs: focus on recent studies. *Drug Saf* **28**:677–694.

Segel I (1993) *Enzyme Kinetics: Behavior and Analysis of Rapid Equilibrium and Steady-State Enzyme Systems*, Wiley & Sons Inc, New York.

Shou M, Hayashi M, Pan Y, Xu Y, Morrissey K, Xu L and Skiles GL (2008) Modeling, prediction, and in vitro in vivo correlation of CYP3A4 induction. *Drug Metab Dispos* **36**:2355–2370.

Shou M, Lin Y, Lu P, Tang C, Mei Q, Cui D, Tang W, Ngui JS, Lin CC, Singh R, Wong BK, Yergey JA, Lin JH, Pearson PG, Baillie TA, Rodrigues AD and Rushmore TH (2001) Enzyme kinetics of cytochrome P450-mediated reactions. *Curr Drug Metab* **2**:17–36.

Shou M and Lu AY (2009) Antibodies as a probe in cytochrome P450 research. *Drug Metab Dispos* **37**:925–931.

Silva JM, Morin PE, Day SH, Kennedy BP, Payette P, Rushmore T, Yergey JA and Nicoll-Griffith DA (1998) Refinement of an in vitro cell model for cytochrome P450 induction. *Drug Metab Dispos* **26**:490–496.

Silverman RB (1995) Mechanism-based enzyme inactivators. *Methods Enzymol* **249**:240–283.

Sinz M, Kim S, Zhu Z, Chen T, Anthony M, Dickinson K and Rodrigues AD (2006) Evaluation of 170 xenobiotics as transactivators of human pregnane X receptor (hPXR) and correlation to known CYP3A4 drug interactions. *Curr Drug Metab* **7**:375–388.

Soars MG, Grime K, Sproston JL, Webborn PJ and Riley RJ (2007) Use of hepatocytes to assess the contribution of hepatic uptake to clearance in vivo. *Drug Metab Dispos* **35**: 859–865.

Stringer R, Nicklin PL and Houston JB (2008) Reliability of human cryopreserved hepatocytes and liver microsomes as in vitro systems to predict metabolic clearance. *Xenobiotica* **38**: 1313–1329.

Tam D, Tirona RG and Pang KS (2003) Segmental intestinal transporters and metabolic enzymes on intestinal drug absorption. *Drug Metab Dispos* **31**:373–383.

Tang C, Lin JH and Lu AY (2005) Metabolism-based drug–drug interactions: what determines individual variability in cytochrome P450 induction?. *Drug Metab Dispos* **33**:603–613.

Templeton IE, Thummel KE, Kharasch ED, Kunze KL, Hoffer C, Nelson WL and Isoherranen N (2008) Contribution of itraconazole metabolites to inhibition of CYP3A4 in vivo. *Clin Pharmacol Ther* **83**:77–85.

Thummel KE, Kunze KL and Shen DD (1997) Enzyme-catalyzed processes of first-pass hepatic and intestinal drug extraction. *Adv Drug Deliv Rev* **27**:99–127.

Thummel KE, O'Shea D, Paine MF, Shen DD, Kunze KL, Perkins JD and Wilkinson GR (1996) Oral first-pass elimination of midazolam involves both gastrointestinal and hepatic CYP3A-mediated metabolism. *Clin Pharmacol Ther* **59**:491–502.

Tsunoda SM, Velez RL, von Moltke LL and Greenblatt DJ (1999) Differentiation of intestinal and hepatic cytochrome P450 3A activity with use of midazolam as an in vivo probe: effect of ketoconazole. *Clin Pharmacol Ther* **66**:461–471.

Tubic M, Wagner D, Spahn-Langguth H, Bolger MB and Langguth P (2006) *In silico* modeling of non-linear drug absorption for the P-gp substrate talinolol and of consequences for the resulting pharmacodynamic effect. *Pharm Res* **23**:1712–1720.

Tucker GT, Houston JB and Huang SM (2001) Optimizing drug development: strategies to assess drug metabolism/transporter interaction potential-toward a consensus. *Clin Pharmacol Ther* **70**:103–114.

Uchaipichat V, Galetin A, Houston JB, Mackenzie PI, Williams JA and Miners JO (2008) Kinetic modeling of the interactions between 4-methylumbelliferone, 1-naphthol, and zidovudine glucuronidation by UDP-glucuronosyltransferase 2B7 (UGT2B7) provides evidence for multiple substrate binding and effector sites. *Mol Pharmacol* **74**:1152–1162.

Uchaipichat V, Mackenzie PI, Guo XH, Gardner-Stephen D, Galetin A, Houston JB and Miners JO (2004) Human udp-glucuronosyltransferases: isoform selectivity and kinetics of 4-methylumbelliferone and 1-naphthol glucuronidation, effects of organic solvents, and inhibition by diclofenac and probenecid. *Drug Metab Dispos* **32**:413–423.

Van LM, Heydari A, Yang J, Hargreaves J, Rowland-Yeo K, Lennard MS, Tucker GT and Rostami-Hodjegan A (2006) The impact of experimental design on assessing mechanism-based inactivation of CYP2D6 by MDMA (Ecstasy). *J Psychopharmacol* **20**:834–841.

Van LM, Swales J, Hammond C, Wilson C, Hargreaves JA and Rostami-Hodjegan A (2007) Kinetics of the time-dependent inactivation of CYP2D6 in cryopreserved human hepatocytes by methylenedioxymethamphetamine (MDMA). *Eur J Pharm Sci* **31**:53–61.

Venkatakrishnan K and Obach RS (2005) In vitro-in vivo extrapolation of CYP2D6 inactivation by paroxetine: prediction of nonstationary pharmacokinetics and drug interaction magnitude. *Drug Metab Dispos* **33**:845–852.

Venkatakrishnan K and Obach RS (2007) Drug–drug interactions via mechanism-based cytochrome P450 inactivation: points to consider for risk assessment from in vitro data and clinical pharmacologic evaluation. *Curr Drug Metab* **8**:449–462.

Venkatakrishnan K, von Moltke LL, Court MH, Harmatz JS, Crespi CL and Greenblatt DJ (2000) Comparison between cytochrome P450 (CYP) content and relative activity approaches to scaling from cDNA-expressed CYPs to human liver microsomes: ratios of accessory proteins as sources of discrepancies between the approaches. *Drug Metab Dispos* **28**:1493–1504.

Voice MW, Zhang Y, Wolf CR, Burchell B and Friedberg T (1999) Effects of human cytochrome b5 on CYP3A4 activity and stability in vivo. *Arch Biochem Biophys* **366**:116–124.

von Moltke LL, Durol AL, Duan SX and Greenblatt DJ (2000) Potent mechanism-based inhibition of human CYP3A in vitro by amprenavir and ritonavir: comparison with ketoconazole. *Eur J Clin Pharmacol* **56**:259–261.

Walsky RL, Astuccio AV and Obach RS (2006) Evaluation of 227 drugs for in vitro inhibition of cytochrome P450 2B6. *J Clin Pharmacol* **46**:1426–1438.

Walsky RL and Boldt SE (2008) In vitro cytochrome P450 inhibition and induction. *Curr Drug Metab* **9**:928–939.

Walsky RL and Obach RS (2004) Validated assays for human cytochrome P450 activities. *Drug Metab Dispos* **32**:647–660.

Wang YH, Jones DR and Hall SD (2004) Prediction of cytochrome P450 3A inhibition by verapamil enantiomers and their metabolites. *Drug Metab Dispos* **32**:259–266.

Wang YH, Jones DR and Hall SD (2005) Differential mechanism-based inhibition of CYP3A4 and CYP3A5 by verapamil. *Drug Metab Dispos* **33**:664–671.

Wilkinson GR and Shand DG (1975) Commentary: a physiological approach to hepatic drug clearance. *Clin Pharmacol Ther* **18**:377–390.

Williams JA, Hurst SI, Bauman J, Jones BC, Hyland R, Gibbs JP, Obach RS and Ball SE (2003) Reaction phenotyping in drug discovery: moving forward with confidence?. *Curr Drug Metab* **4**:527–534.

Wimalasena K and Haines DC (1996) A general progress curve method for the kinetic analysis of suicide enzyme inhibitors. *Anal Biochem* **234**:175–182.

Winiwarter S, Bonham NM, Ax F, Hallberg A, Lennernas H and Karlen A (1998) Correlation of human jejunal permeability (in vivo) of drugs with experimentally and theoretically derived parameters. A multivariate data analysis approach. *J Med Chem* **41**:4939–4949.

Yamada A, Maeda K, Kamiyama E, Sugiyama D, Kondo T, Shiroyanagi Y, Nakazawa H, Okano T, Adachi M, Schuetz JD, Adachi Y, Hu Z, Kusuhara H and Sugiyama Y (2007) Multiple human isoforms of drug transporters contribute to the hepatic and renal transport of olmesartan, a selective antagonist of the angiotensin II AT1-receptor. *Drug Metab Dispos* **35**:2166–2176.

Yamamoto T, Suzuki A and Kohno Y (2004) High-throughput screening for the assessment of time-dependent inhibitions of new drug candidates on recombinant CYP2D6 and CYP3A4 using a single concentration method. *Xenobiotica* **34**:87–101.

Yang J, Jamei M, Yeo KR, Tucker GT and Rostami-Hodjegan A (2005) Kinetic values for mechanism-based enzyme inhibition: assessing the bias introduced by the conventional experimental protocol. *Eur J Pharm Sci* **26**:334–340.

Yang J, Jamei M, Yeo KR, Tucker GT and Rostami-Hodjegan A (2007a) Prediction of intestinal first-pass drug metabolism. *Curr Drug Metab* **8**:676–684.

Yang J, Jamei M, Yeo KR, Tucker GT and Rostami-Hodjegan A (2007b) Theoretical assessment of a new experimental protocol for determining kinetic values describing mechanism (time)-based enzyme inhibition. *Eur J Pharm Sci* **31**:232–241.

Yang J, Liao M, Shou M, Jamei M, Yeo KR, Tucker GT and Rostami-Hodjegan A (2008) Cytochrome p450 turnover: regulation of synthesis and degradation, methods for determining rates, and implications for the prediction of drug interactions. *Curr Drug Metab* **9**:384–394.

Yano JK, Wester MR, Schoch GA, Griffin KJ, Stout CD and Johnson EF (2004) The structure of human microsomal cytochrome P450 3A4 determined by X-ray crystallography to 2.05-A resolution. *J Biol Chem* **279**:38091–38094.

Yao C and Levy RH (2002) Inhibition-based metabolic drug–drug interactions: predictions from in vitro data. *J Pharm Sci* **91**:1923–1935.

Yeo KR and Yeo WW (2001) Inhibitory effects of verapamil and diltiazem on simvastatin metabolism in human liver microsomes. *Br J Clin Pharmacol* **51**:461–470.

Youdim KA, Zayed A, Dickins M, Phipps A, Griffiths M, Darekar A, Hyland R, Fahmi O, Hurst S, Plowchalk DR, Cook J, Guo F and Obach RS (2008) Application of CYP3A4 in vitro data to predict clinical drug–drug interactions; predictions of compounds as objects of interaction. *Br J Clin Pharmacol* **65**:680–692.

Yu LX and Amidon GL (1999) A compartmental absorption and transit model for estimating oral drug absorption. *Int J Pharm* **186**:119–125.

Zhang H, Davis CD, Sinz MW and Rodrigues AD (2007) Cytochrome P450 reaction-phenotyping: an industrial perspective. *Expert Opin Drug Metab Toxicol* **3**:667–687.

Zhang X, Jones DR and Hall SD (2009) Prediction of the effect of erythromycin, diltiazem, and their metabolites, alone and in combination, on CYP3A4 inhibition. *Drug Metab Dispos* **37**:150–160.

Zhang L, Zhang YD, Strong JM, Reynolds KS and Huang SM (2008) A regulatory viewpoint on transporter-based drug interactions. *Xenobiotica* **38**:709–724.

Zheng HX, Huang Y, Frassetto LA and Benet LZ (2009) Elucidating rifampin's inducing and inhibiting effects on glyburide pharmacokinetics and blood glucose in healthy volunteers: unmasking the differential effects of enzyme induction and transporter inhibition for a drug and its primary metabolite. *Clin Pharmacol Ther* **85**:78–85.

Zhou S, Yung Chan S, Cher Goh B, Chan E, Duan W, Huang M and McLeod HL (2005) Mechanism-based inhibition of cytochrome P450 3A4 by therapeutic drugs. *Clin Pharmacokinet* **44**:279–304.

Chapter 8
The In Vitro Characterization of Inhibitory Drug–Drug Interactions Involving UDP-Glucuronosyltransferase

John O. Miners, Thomas M. Polasek, Peter I. Mackenzie, and Kathleen M. Knights

Abstract Inhibition of UDP-glucuronosyltransferase (UGT) activity gives rise to both drug–drug (DDIs) and drug–endobiotic interactions in vivo. Furthermore, several glucuronides have been shown to reduce the metabolic clearances of cytochrome P4502C8 substrates. Experimental paradigms, based on the use of human liver microsomes (HLM), hepatocytes, and recombinant UGTs as the enzyme sources, are now available for the investigation of drug glucuronidation in vitro including the prediction and characterization of DDIs. The reaction phenotyping of drug glucuronidation is becoming increasingly feasible with the availability of 'batteries' of recombinant enzymes along with substrate and inhibitor probes for the major hepatic drug metabolizing UGTs. However, the occurrence of homo- and heterotropic activation potentially complicates screening for DDIs in vitro. Recent advances in knowledge of factors that influence UGT activity in vitro have also led to the development of experimental approaches that accurately predict the magnitude of known DDIs involving glucuronidated drugs, but further work in this area is required to demonstrate the generalizability of these models.

8.1 Inhibitory Drug–Drug Interactions Involving Conjugation Enzymes

Glucuronidation is the conjugation pathway of greatest significance in the clearance of currently marketed therapeutic drugs. Indeed, drugs from almost all therapeutic classes (e.g., analgesics, anticonvulsants, antidepressants, antihypertensives, antipsychotics, antiviral agents, cancer chemotherapeutics, general anesthetics, hormones and hormone antagonists, hypnosedative anxiolytics, hypolipidemics, and nonsteroidal anti-inflammatory drugs) are metabolized to a significant extent via

J.O. Miners (✉)
Clinical Pharmacology, School of Medicine, Flinders University, Adelaide, Australia
e-mail: john.miners@flinders.edu.au

K.S. Pang et al. (eds.), *Enzyme- and Transporter-Based Drug–Drug Interactions*,
DOI 10.1007/978-1-4419-0840-7_8,

glucuronidation (Miners and Mackenzie, 1991; Kiang et al., 2005). Drug–drug interactions (DDIs) involving inhibition of metabolic clearance by glucuronidation have been reported, many of which involve valproic acid (lamotrigine, lorazepam, and zidovudine) and probenecid (e.g., acetaminophen, clofibric acid, lorazepam, and zidovudine) (Miners and Mackenzie, 1991; Kiang et al., 2005). Other examples include the zidovudine – fluconazole (Sahai et al., 1994) and lopinavir/ritonavir – SN38 (the active metabolite of irinotecan) (Corona et al., 2008) interactions. Drugs may also inhibit the glucuronidation of endogenous compounds, notably the hyperbilirubinemia arising from inhibition of UGT1A1 activity by indinavir (Boyd et al., 2006).

After glucuronidation, sulfation and N-acetylation are generally considered to be the conjugation pathways of most importance in drug clearance. However, in comparison to glucuronidation and cytochrome P450 catalyzed biotransformation, N-acetylation and sulfation represent the primary clearance mechanism for relatively few drugs and documented interactions involving inhibition of sulfotransferase and *N*-acetyltransferase are uncommon. Indeed, there appear to be no reports of clinically significant interactions involving inhibition of N-acetylation. Acetaminophen (paracetamol) is the agent most commonly implicated in the inhibition of drug sulfation. Coadministration of acetaminophen is known to impair the sulfation of ethinylestradiol, fenoldopam, minoxidil, and salicylamide, but none of these interactions are of therapeutic significance.Decreased sulfation in vivo is thought to involve both enzyme inhibition and cofactor (3′-phosphoadenosine-5′-phosphosulfate) depletion.

Other known conjugation pathways in humans include glucosidation, methylation, and amino acid and glutathione conjugation. Although only a limited number of drugs are cleared by these mechanisms, DDIs may occur. For example, olsalazine has been reported to inhibit thiopurine methyltransferase and exacerbate the toxicity of coadministered 6-mercaptopurine (Lewis et al., 1997).

8.2 The Glucuronidation Reaction and UDP-Glucuronosyltransferase (UGT)

Glucuronidation reactions involve the covalent linkage (or conjugation) of glucuronic acid, derived from the cofactor UDP-glucuronic acid (UDPGA), to a typically lipophilic substrate bearing a nucleophilic acceptor functional group, namely alcohol (aliphatic and phenolic), carboxylic acid, amine, acidic carbon atom, or thiol. The reaction is catalyzed by UDP-glucuronosyltransferase (UGT) and proceeds according to a second-order nucleophilic substitution mechanism. Most DDIs involving glucuronidated substrates arise from inhibition of UGT activity, but other mechanisms are possible. Glucuronides may inhibit cytochrome P450 (CYP) activity (see Section 8.7) and cofactor availability is potentially rate limiting in the glucuronidation reaction. UDPGA is synthesized from UDP-glucose by the cytosolic enzyme UDP-glucose dehydrogenase and then transported across the membrane of the endoplasmic reticulum (ER) to the luminally oriented UGT protein (Fig. 8.1). Although it is generally believed that constitutive concentrations of UDPGA in hepatocytes and other tissues are sufficient to support drug glucuronidation at normal exposures, mean cofactor concentrations in liver homogenate are within the range of UDPGA K_m values of UGT enzymes. While altered UDPGA availability at the intra-luminal UGT active site may potentially affect drug clearance via glucuronidation, there is no direct evidence at this stage for depletion of cytosolic UDPGA or inhibition of cofactor synthesis or membrane transport as mechanisms of DDIs in humans.

UGT comprises a superfamily of enzymes. Nineteen human UGT proteins have been identified to date and these have been classified in families and subfamilies (viz. 1A, 2A, and 2B) based on amino acid identity (Fig. 8.2) (Mackenzie et al., 2005). The individual UGT enzymes exhibit distinct, but overlapping, substrate and inhibitor selectivities (Section 8.4). With some exceptions, most UGT enzymes are expressed in liver, kidney, and the gastrointestinal tract. However, UGT1A1 is expressed in liver but not kidney while UGT1A7, 1A8, and 1A10 are expressed primarily in the gastrointestinal tract and not liver (Tukey and Strassburg, 2000). Apart from tissue-specific expression, numerous other factors affect UGT activity in vivo; age (especially the neonatal period), diet, disease states, DDIs (induction and inhibition), ethnicity, genetic polymorphism, and hormonal factors (Miners and Mackenzie 1991). While most of these influences would be expected to differ from enzyme to enzyme, the development of generalizable conclusions is limited by the previous use of nonselective substrates in both in vitro and in vivo studies. The exception in this regard is knowledge of genetic polymorphism. As might be expected, the individual UGTs exhibit differences in genetic polymorphism, in terms of occurrence, frequency, and functional significance (Miners et al., 2002; Guillemette, 2003).

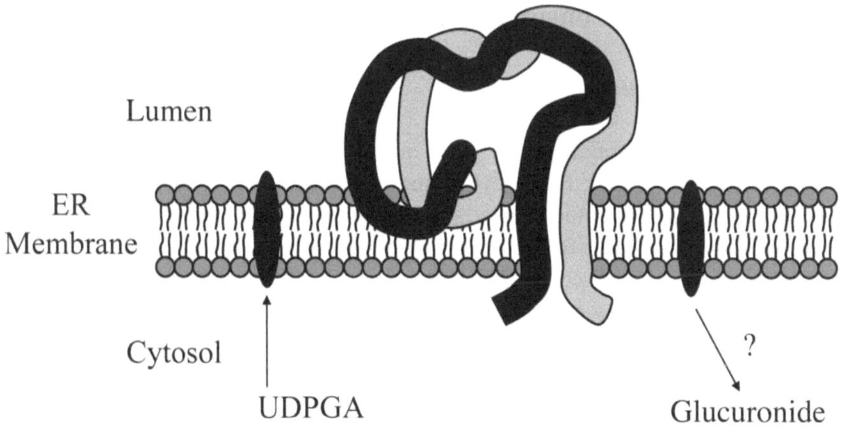

Fig. 8.1 UDP-glucuronosyltransferase topology model showing membrane localization

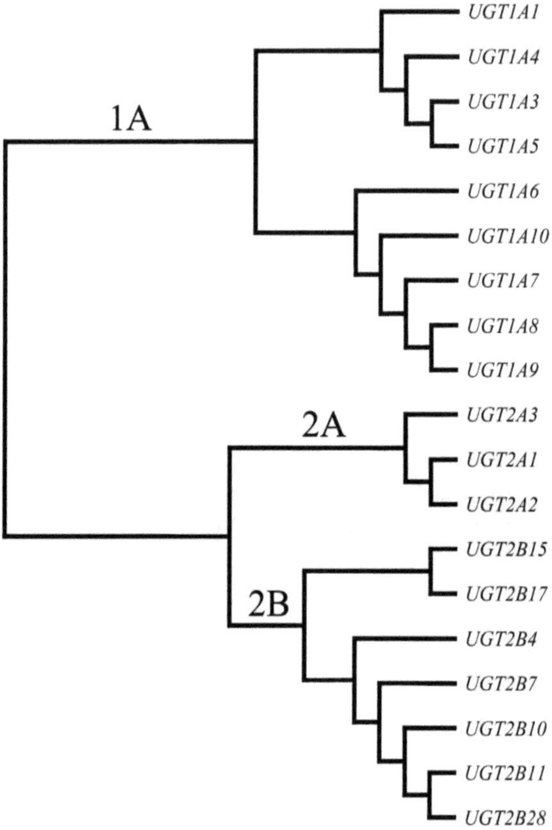

Fig. 8.2 Human UDP-glucuronosyltransferase proteins

8.3 The Characterization of Drug Glucuronidation In Vitro: Experimental Considerations

8.3.1 Enzyme Sources

Studies of drug glucuronidation in vitro, including DDIs, variably employ human liver microsomes (HLM), hepatocytes, and/or recombinant UGT enzymes as the enzyme sources. Several studies have shown that prediction bias of in vivo intrinsic clearance (CL_{int}) is lower for kinetic constants generated with human hepatocytes than with HLM (Brown et al., 2007; Riley et al., 2005), and hepatocytes have been recommended as a superior enzyme source for in vitro–in vivo extrapolation (IV-IVE) (Engtrakul et al., 2005). Apart from metabolic enzymes, hepatocytes express uptake and efflux transporters and thus provide a theoretical advantage for in vitro kinetic studies with compounds requiring transporter-mediated membrane translocation. However, activities of low hepatic clearance glucuronidated drugs may be difficult to measure with human hepatocytes (Soars et al., 2002). HLM offer benefits compared to hepatocytes, particularly cost and ease of preparation and handling. Indeed, under appropriate experimental conditions (Section 8.6), values of the Michaelis-Menten (K_m) and inhibitor (K_i) constants obtained with HLM are essentially identical to those reported for cryopreserved hepatocytes (Uchaipichat et al., 2006a; Rowland et al., 2007, 2008a, b).

Like HLM, recombinant human UGTs are a readily available enzyme source for IV-IVE and, under optimized experimental conditions, can provide a reliable estimate of K_i (Uchaipichat et al., 2006a). UGTs are commonly stably expressed in human embryonic kidney (HEK) 293 or insect cells (baculoviral-mediated expression). Most commercial sources market recombinant UGTs in the form of insect cell microsomes. Irrespective of the enzyme source, experimental systems for the characterization of drug glucuronidation must be optimized for the various factors that influence UGT enzyme activity in vitro (see following sections).

8.3.2 Latency of Microsomal UGT

As noted previously, glucuronidation is a bisubstrate reaction that requires both the substrate (aglycone) and cofactor. Membrane translocation of UDPGA is a bidirectional carrier-mediated ATP-independent process in intact microsomal vesicles and the intact ER of permeabilized rat hepatocytes. The antiport mechanism in both systems appears to primarily exchange UDPGA with UDP-*N*-acetylglucosamine. Fully functional UDPGA transporters are assumed to exist in hepatocytes, but optimal microsomal glucuronidation activity requires an activation step which disrupts the ER membrane thereby permitting access of cofactor to the luminally oriented UGT protein (Fig. 8.1). In the absence of activation, microsomal UGT remains in a "latent" state. The glucuronidation activities of fully activated rat and human liver microsomes are 1.5- to 10-fold higher than those of "native" microsomes depending on the quality of the microsomal preparation and the UGT enzyme activity being measured (Miners et al., 1990; Court et al., 2001).

Several treatments activate the glucuronidation activity of HLM particularly detergents, physical disruption, sonication, and the pore-forming agent alamethicin. The latter is the preferred agent. Pre-incubation of HLM with alamethicin (50 μg/mg microsomal protein) fully activates UGT activities without affecting cytochromes P450 (Boase and Miners, 2002; Fisher et al., 2000). While the use of a nonionic detergent such as Brij58 normally results in similar activation to alamethicin (Boase and Miners, 2002; Soars et al., 2003), particular care is required for optimization of the rate of glucuronidation. Activation by detergents exhibits a "bell-shaped" relationship that may vary from substrate to substrate (Miners et al., 1990). Although alamethicin activation is generally employed with recombinant UGTs expressed in insect cells, conditions appear not to have been optimized. Activation is not required for HEK293 cell expression when cells are lysed by sonication. However, care is required with sonication since UGT enzyme activities are heat labile to a variable extent (Uchaipichat et al., 2004). The glucuronidation activities of HLM and recombinant UGTs are also stimulated by Mg^{2+} and hence incubations are routinely supplemented with $MgCl_2$ (usually 4 mM) (Boase and Miners, 2002).

8.3.3 Dependence of UGT Activity on pH, Buffer Type, Ionic Strength, and Organic Solvents

The glucuronidation activities of both recombinant UGTs and HLM vary with pH. For example, human liver microsomal zidovudine glucuronidation is maximal at pH 8 (Boase and Miners, 2002). The mechanism is unclear but probably reflects the relative charge states of the substrate and active site amino acids. Similarly, enzyme activity may change with buffer type and ionic strength (Boase and Miners, 2002; Soars et al., 2003; Engtrakul et al., 2005). Interestingly, highest human liver microsomal zidovudine glucuronidation activity was observed for incubations in carbonate buffer, apparently due to a reduction in K_m (Engtrakul et al., 2005). Most studies of UGT activity tend to use phosphate buffer, 100 mM at pH 7.4 but there is still no consensus in this regard.

Since most UGT substrates and inhibitors are moderately lipophilic, organic solvents are often required for solubilization. However, like cytochromes P450, activities of the individual UGT enzymes are variably affected by organic solvents (Uchaipichat et al., 2004). For example, 1% (v/v) methanol or ethanol reduces UGT1A6 activity by 25–35% while DMSO has the greatest inhibitory effect on UGT1A9. Thus, solvent selection should take into account the UGT activity under investigation.

8.3.4 Glucuronide Stability

The glucuronide conjugates of alcohols and amines are stable at pH 7.4. However, acyl glucuronides hydrolyze at pH values > 7. Where the rate of acyl glucuronide hydrolysis is significant, incubations conducted at pH 6.8 may provide a more accurate estimate of glucuronide formation rate (Miners et al., 1997). As most glucuronides are substrates of β-glucuronidase, saccharic acid lactone, an inhibitor of this enzyme, is often added to incubations of HLM or recombinant UGTs. However, the β-glucuronidase activity of human liver, kidney, intestinal and lung microsomes, and recombinant UGTs expressed in HEK293 and insect cells is low or negligible (Oleson and Court, 2008). In addition, saccharic acid lactone tends to lower the pH of incubations, which potentially alters UGT activity. Thus, there appears to be no advantage from the inclusion of saccharic acid lactone in incubations, unless there is evidence that the enzyme source possesses β-glucuronidase activity that is likely to result in product degradation.

8.3.5 Cofactor Concentration

As noted in Section 8.2, glucuronidation is a bisubstrate reaction and hence rates of glucuronide formation are potentially rate limited by UDPGA concentration. It is generally believed that constitutive concentrations of UDPGA in hepatocytes and other tissues are sufficient to support drug and chemical glucuronidation. Interestingly, however, the UDPGA K_m values for UGT enzymes (ca. 50–500 μM) span the mean cofactor concentration of human liver (280 μM). In order to simplify the analysis of kinetic data generated using human liver microsomal and recombinant UGTs as the enzyme sources and ensure adequate cofactor supply, the UDPGA concentration of incubations is typically in the range 2–5 mM. Moreover, HLM should be activated to ensure that transporter-mediated delivery of UDPGA to the enzyme active site does not limit glucuronidation activity in vitro (Section 8.3.2).

8.3.6 Nonspecific Binding of Substrate and Inhibitor

The nonspecific binding of substrates and inhibitors to membranes of enzyme sources (particularly HLM) must be taken into account in the calculation of K_m, IC_{50}, and inhibitor constant (K_i). Where binding of substrate or inhibitor to the incubation milieu is significant, concentrations at the enzyme active site are lower than the added concentration leading to underestimation of K_m, CL_{int} (as V_{max}/K_m), IC_{50}, and K_i (McLure et al., 2000; Austin et al., 2002). Consequently, nonspecific binding must be characterized and accounted for in the calculation of kinetic constants. Several empirically derived models have been generated for the estimation of the unbound fraction in incubations (Austin et al., 2002; Sykes et al., 2006; Gertz et al., 2008) but the generalizability of most models is yet to be demonstrated. Nonspecific microsomal binding is normally highest for organic bases, but binding of acidic and neutral compounds cannot be discounted. Binding may occur to both HLM and recombinant enzyme preparations. For example, Uchaipichat et al. (2006b) reported extensive nonspecific binding of the lipophilic glucuronidated base trifluoperazine to HLM and HEK293 cell lysate such that the K_m values were overestimated approximately 10-fold in the absence of correction for binding.

8.4 Reaction Phenotyping and the Qualitative Prediction of DDIs

Reaction phenotyping refers to identification of the enzyme(s) responsible for the metabolism of any given compound. If factors that influence the activity of that enzyme(s) in vivo are known, then perturbation of the hepatic clearance and extraction ratio of the compound in defined patient groups (e.g., DDIs due to inhibition or induction, genetic polymorphism) may be predicted, at least at the qualitative level. Reaction phenotyping is most commonly undertaken with HLM or hepatocytes as the enzyme source, since they express the full complement of hepatic UGTs.

Several approaches may be adopted for reaction phenotyping in vitro (Miners et al., 2004 and 2006): (i) reduction in the glucuronidation of the test compound by UGT enzyme-selective chemical and antibody inhibitors; (ii) competitive inhibition of the glucuronidation of a UGT enzyme-selective substrate by the test compound, with K_i matching its known K_m; (iii) a significant correlation between rates of glucuronidation of the test compound and immunoreactive UGT enzyme content or activity in microsomes or hepatocytes from a panel of human livers; and (iv) screening for metabolism by a battery of recombinant UGTs along with comparison of K_m values for glucuronidation by the individual recombinant enzymes and HLM/hepatocytes.

The use of UGT enzyme-selective inhibitors is considered the most definitive approach for reaction phenotyping, at least when experimental conditions for inhibitor selectivity (particularly concentration present in incubations) are well established. For example, if a selective UGT inhibitor abolishes the glucuronidation of the test compound by HLM and/or hepatocytes, then sole involvement of that UGT enzyme can generally be assumed. However, the number of UGT enzyme-selective inhibitors is currently limited. Hecogenin (10 μM) is a highly selective inhibitor of UGT1A4 (Uchaipichat et al., 2006b) and inhibition by this compound is generally taken as diagnostic of UGT1A4 involvement in a metabolic pathway. Fluconazole, at an added concentration of 2.5 mM, is a moderately selective of UGT2B7 (Uchaipichat et al., 2006a). Niflumic acid, 2 μM, selectively inhibits UGT1A9, but inhibition of other UGT1A enzymes occurs at higher concentrations (JO Miners, unpublished data). Some studies have utilized UGT enzyme-selective substrates as inhibitors, but substrate and inhibitor selectivities are not necessarily identical. For example, bilirubin, which is a specific substrate of UGT1A1, has been reported to inhibit UGT1A4 activity. Kinetic considerations, particularly the occurrence of homo- and heterotropic activation, must also be taken into account in the interpretation of data from glucuronidation inhibition experiments (see Section 8.5).

Whereas the availability of selective inhibitors is limited, selective substrates are available for the major hepatically expressed drug metabolizing UGT enzymes (Table 8.1). Special care is required in kinetic studies with several of these compounds. In particular, bilirubin and trifluoperazine exhibit extensive nonselective binding to HLM and HEK293 cell lysate (Uchaipichat et al., 2006b; Udomuksorn et al., 2007). In addition, kinetic parameters (K_m, V_{max}, CL_{int}) may differ markedly between selective substrates for the same enzyme, which may influence the choice of probe substrate. It should also be noted that some of these compounds may also be metabolized by extra-hepatic enzymes. For example, propofol is glucuronidated by UGT1A8 and 1A9, but the former is expressed exclusively in the gastrointestinal tract. Thus, careful selection of probe substrate is required for studies comparing metabolism in different tissues. As noted previously, glucuronidation activities may vary with pH, buffer type and strength, cofactor concentration, and the presence of activators. It is therefore essential that the glucuronidation rates of the test compound and a UGT enzyme-selective substrate be compared under identical experimental conditions.

Table 8.1 Selective substrates of the major human hepatic UDP-glucuronosyltransferases[a]

Enzyme	Selective substrate(s)
UGT1A1	bilirubin, estradiol (3-glucuronidation)[b], etoposide
UGT1A3	hexafluoro-1α,25-dihydroxyvitamin D_3
UGT1A4	trifluoperazine
UGT1A6	deferiprone
UGT1A9	propofol, sulfinpyrazone
UGT2B7	morphine (3-glucuronidation), 6α-hydroxyprogesterone, zidovudine
UGT2B15	*S*-oxazepam

[a]References: see Table 8.1, Miners et al. (2006) and Benoit-Biancamono et al. (2009); Bowalgaha et al. (2007); Court (2005); Kerdpin et al. (2006); Wen et al. (2007).
[b]Partially selective (also metabolized by UGT1A3).

Comparative kinetic studies with HLM, hepatocytes, and recombinant UGT enzymes must also be interpreted with caution. The respective K_m and K_i values of UGT substrates and inhibitors can differ from one enzyme source to another due to the effects of endogenous inhibitors. For the same reasons, comparisons of the rates of glucuronide formation of a test substrate by different UGT enzymes may not reflect the relative contribution of the individual enzymes to metabolic clearance via glucuronidation. At least with UGT1A9 and UGT2B7, however, these differences can be eliminated by the addition of bovine serum albumin to incubations (Section 8.6).

8.5 Screening for Inhibition of Drug Glucuronidation In Vitro – Kinetic and Pharmacogenetic Considerations

8.5.1 Atypical Glucuronidation Kinetics

The predictability of in vitro–in vivo extrapolation (IV-IVE) of kinetic data for glucuronidated drugs is dependent on how closely key kinetic parameters, including K_i for assessment of DDI potential, reflect altered enzyme activity in vivo. Apart from the experimental considerations outlined previously, appropriate kinetic analysis is essential. Impairment of drug glucuronidation is known to occur by mechanisms other than simple competitive inhibition (Uchaipichat et al., 2004 and 2008) and the K_i value should be obtained using the relevant equation. Mechanism-based inactivation (MBI), which can be assessed using a pre- versus co-incubation strategy (Section 8.7.2) (Polasek and Miners, 2007), should be excluded since the determination of inhibition kinetic parameters requires a different experimental approach.

Knowledge of the kinetics of the victim drug also assumes importance for the design and interpretation of experiments. Although most studies of drug glucuronidation assume a hyperbolic model, there is increasing evidence for atypical (i.e., non-Michaelis–Menten) kinetics (Miners et al., 2004). Studies with recombinant UGTs have reported hyperbolic, positive- (autoactivation) and negative-homotropic cooperativity, and substrate inhibition (Stone et al., 2003; Uchaipichat et al., 2004). Autoactivation and substrate inhibition have also been observed with HLM as the enzyme source, for example (Kirkwood et al., 1998; Williams et al., 2002; Bauman et al., 2005). Atypical kinetics may be modeled empirically using the Hill or substrate inhibition equations or by mechanistic models based on the existence of multiple substrate and/or effector binding sites. Models that assume two equivalent substrate binding sites, that presumably arise from homodimerization, have been successfully applied to drug and xenobiotic glucuronidation (Stone et al., 2003; Uchaipichat et al., 2004, 2008).

8.5.2 Heterotropic Cooperativity

Heterotropic interactions, as might occur in screening studies to identify potential inhibitors of drug glucuronidation in vitro, add another level of complexity. Modifiers may act as an inhibitor or activator, or both. In the latter regard, a compound may activate enzyme activity at low concentrations but cause inhibition at higher concentrations depending on interactions occurring with the active site and the involvement of effector sites. For example, ethinylestradiol and anthraflavic acid were shown to activate and inhibit human liver microsomal UGT1A1-catalyzed estradiol 3-glucuronidation at low and high concentrations, respectively (Williams et al., 2002).

Effects can vary from substrate to substrate and modifier to modifier, which has been demonstrated with UGT2B7. 4-Methylumbelliferone (4MU) and 1-naphthol (1NP) are widely used as model substrates for inhibition screening studies. Whereas 1NP inhibited 4MU glucuronidation by UGT2B7, 1NP glucuronidation was activated by 4MU (Uchaipichat et al., 2008). Both 4MU and 1NP inhibited UGT2B7-catalyzed zidovudine glucuronidation, although in the presence of these modifiers zidovudine glucuronidation changed from hyperbolic to sigmoidal. In all cases, data were well described by multisite models that assume two equivalent substrate binding sites along with multiple effector sites. Differential effects of modifiers occur depending on the binding affinity of modifier in the active site, overlap between substrate/modifier "catalytic" binding sites, and the involvement of effector sites. The multiplicity of binding and effector sites results in complex kinetic interactions between the alternate UGT2B7 substrates, thereby complicating inhibition screening approaches.

8.5.3 Genotype-Dependent Effects

Treatment with indinavir is known to elevate total, conjugated, and unconjugated serum bilirubin in HIV patients. Indinavir was shown to inhibit a number of UGT enzymes in vitro including UGT1A1, the enzyme solely responsible for bilirubin glucuronidation (Boyd et al., 2006). However, the elevation in serum bilirubin was greatest in patients with mutant UGT1A1 genotypes due to already impaired bilirubin glucuronidation activity. Notably, the odds ratio for hyperbilirubinemia was 11-fold higher in patients with the combined *UGT1A1 *6* and **28* genotype compared to wild type. These data serve to demonstrate that an interaction may vary in severity and clinical significance between UGT genotypes, which is an important consideration given the highly polymorphic nature of *UGT1A1* and several other UGT enzymes.

8.6 Quantitative IV-IVE for DDI Involving Glucuronidated Drugs

8.6.1 Theoretical Considerations

Theoretical aspects of IV-IVE for the prediction of DDIs have been reviewed by Ito et al. (1998). For an orally administered drug at steady state that is metabolized by the liver along a single metabolic pathway (by a single enzyme), the ratio of the areas under the plasma concentration curve of the victim drug in the presence (AUC_i) and absence (AUC) of the inhibitor is given by

$$\frac{AUC_i}{AUC} = 1 + \frac{[I]}{K_i} \tag{8.1}$$

where [I] is the inhibitor concentration at the enzyme active site and assuming substrate concentration is much lower than K_m. Equation (8.1) further assumes that the inhibitor has no effect on the absorption and fraction unbound of the victim drug and the conditions of the well-stirred model of hepatic clearance apply. Although the use of Equation (8.1) is relatively widespread in the extrapolation of in vitro inhibition data, many drugs are metabolized along multiple pathways that involve multiple enzymes and may also be eliminated in part by renal or biliary excretion. In these circumstances, prediction of the AUC ratio in vivo is markedly improved by accounting for fraction metabolized by the inhibited pathway/enzyme (f_m) (Brown et al., 2005; Ito et al., 2005b). Here,

$$\frac{AUC_i}{AUC} = \frac{1}{\dfrac{f_m}{1+[I]/K_i} + (1-f_m)} \tag{8.2}$$

Equation (8.2) holds for both competitive and noncompetitive inhibition, but not uncompetitive inhibition and mechanism-based inactivation (Ito et al., 1998). As noted in Section 8.4, inhibitor probes are not yet available for all human UGT enzymes and accurate assessment of f_m may not be possible for all compounds. Nevertheless, it has been possible to link the metabolism of several glucuronidated drugs to a single UGT employing a combination of the techniques outlined in Section 8.4. For example, codeine, morphine, and zidovudine are glucuronidated predominantly, if not solely, by UGT2B7 although all are additionally cleared in part by other routes.

There is no consensus as to the value of [I] employed in Equation (8.2). However, prediction of in vivo DDI potential is generally improved by use of the maximum hepatic inlet concentration (Ito et al., 2005a), which takes into account input from the systemic circulation and portal vein (Ito et al., 1998):

$$[\mathrm{I}]_{\mathrm{inlet}} = [\mathrm{I}] + \frac{k_{\mathrm{a}} \times f_{\mathrm{a}} \times D}{Q_{\mathrm{H}}} \tag{8.3}$$

where [I] can be either the maximum or average concentration of inhibitor in the systemic circulation, k_{a} is the absorption rate constant, f_{a} is the fraction absorbed from the gastrointestinal tract, D is the inhibitor dose, and Q_{H} is hepatic blood flow. The unbound inlet concentration is calculated by multiplying the right-hand side of Equation (8.3) by fraction unbound. Studies of DDI involving inhibition of drug glucuronidation have generally obtained optimal prediction of the AUC ratio using inlet total drug concentration (Rowland et al., 2006).

8.6.2 Prediction of DDI Involving Glucuronidated Drugs

Numerous studies have investigated the inhibition of xenobiotic glucuronidation in vitro, primarily using HLM and recombinant UGTs as the enzyme source (Kiang et al., 2005). Many of these studies have employed UGT2B7-selective substrates as the victim drug, including codeine, morphine, and zidovudine. It should be noted in this regard that UGT2B7 appears to be the most important drug metabolizing UGT, glucuronidating epirubicin, nonsteroidal anti-inflammatory agents, opioids, valproic acid, zidovudine, and numerous other compounds. K_i values for inhibitors of UGT2B7 substrates in vitro typically exceed 100 μM and are not uncommonly >1 mM (Kiang et al., 2005). For example, reported K_{i}s for the inhibition of human liver microsomal zidovudine glucuronidation by fluconazole and valproic acid are 933 and 1600 μM, respectively (Ethell et al., 2003, Uchaipichat et al., 2006a). These values greatly exceed plasma concentrations observed with therapeutic dosing and DDIs would not be predicted on the basis of $[\mathrm{I}]/K_{\mathrm{i}}$ ratios (Equations 8.1 and 8.2) despite known interactions in vivo (Section 8.1). Relatively low K_i values (7.9–9.5 μM) have been reported for diclofenac inhibition of the microsomal glucuronidation of codeine and dimethylxanthenone acetic acid (Ammon et al., 2000; Miners et al., 1997), both UGT2B7 substrates, but again DDIs are not predicted given the low concentrations (total and unbound) of diclofenac observed in vivo. On the basis of the high K_{m} values generally observed for glucuronidated compounds in vitro and K_{i} values that generally exceed in vivo inhibitor concentrations, Williams et al. (2004) argued a low likelihood of DDIs between drugs cleared by glucuronidation. It is now known, however, that in vitro K_{m} and K_{i} values for substrates and inhibitors of the major drug metabolizing UGTs are overestimated due to an experimental artifact (Sections 8.6.3 and 8.6.4).

8.6.3 Endogenous Fatty Acids as Inhibitors of Drug Glucuronidation In Vitro

As noted in Section 8.3.1, there is a preference for the use of human hepatocytes over HLM for extrapolation of in vitro intrinsic clearance given the higher prediction bias observed with the latter enzyme source (Riley et al., 2005; Brown et al., 2007). Studies with the model glucuronidated drug zidovudine reported K_m values that were 6- to 9-fold higher for HLM and recombinant UGT2B7 compared to human hepatocytes (Engtrakul et al., 2005). As a result of the high K_m values observed with HLM, IV-IVE of human liver microsomal glucuronidation kinetic data typically underestimates in vivo intrinsic and hepatic clearances by an order of magnitude (Boase and Miners, 2002; Miners et al., 2004).

The mechanism responsible for the higher K_ms obtained with HLM and recombinant UGTs (and indeed many cytochrome P450 enzymes) has been elucidated only recently. It was observed that long-chain unsaturated fatty acids, including arachidonic, linoleic, and oleic acids, are potent inhibitors of UGT1A9, UGT2B7, and microsomal UGT activity (Tsoutsikos et al., 2004). For example, the K_i for arachidonic acid inhibition of microsomal 4-methylumbelliferone glucuronidation, a reaction primarily catalyzed by UGT1A9, was 0.15 μM. Subsequent work demonstrated that long-chain unsaturated fatty acids are released from membranes during the course of incubations of HLM and recombinant enzyme preparations. These fatty acids act as potent competitive inhibitors of UGT1A9 and UGT2B7 (and also P450s such as CYP2C9), resulting in overestimation of the "true" substrate K_m (Rowland et al., 2007, 2008a, 2008b). In contrast, fatty acids appear to have only a minor affect on the glucuronidation of UGT1A1, 1A4, and 1A6 substrates (Rowland et al., 2006, 2008a). Interestingly, reported K_m values for substrates of UGT1A1 (e.g., bilirubin) and UGT1A4 (e.g., trifluoperazine) are often lower than those of UGT1A9 and UGT2B7 substrates.

The inhibitory effects of long-chain unsaturated fatty acids may be eliminated by the addition of bovine serum albumin (BSA) or fatty acid free human serum albumin (HSAFAF) to incubation of HLM or recombinant enzymes. Albumin binds fatty acids with high affinity and thus sequesters these endogenous inhibitors as they are released from the membrane (Rowland et al., 2007, 2008a, 2008b). Consequently, K_m values of UGT1A9, UGT2B7, and CYP2C9 substrates determined from incubations conducted in the presence of albumin are approximately an order of magnitude lower than those generated from experiments performed in the absence of albumin. Importantly, the human liver microsomal K_ms reflect hepatocellular K_m values and extrapolated intrinsic and hepatic clearances either are in close agreement with those reported in vivo (UGT1A9 and CYP2C9) or give equivalent predictivity to data generated from experiments with hepatocytes (zidovudine) (Rowland et al., 2007, 2008a, b).

The "albumin effect" is predicted to apply to any UGT or CYP enzyme that is inhibited by long-chain unsaturated fatty acids and to date has been demonstrated for UGT1A9, UGT2B7, CYP2C9, and CYP1A2. As noted earlier, UGT1A1, 1A4, and 1A6 activities are minimally altered by albumin. Since many substrates of UGT (especially 1A9 and 2B7) and CYP enzymes are organic acids, they may bind extensively to albumin. It is therefore essential that albumin binding, along with nonspecific binding to HLM, is determined and taken into account in the calculation of in vitro K_m (and hence intrinsic clearance) and K_i values.

8.6.4 The Albumin Effect and DDI Prediction for Glucuronidated Drugs

Just as K_m values for substrates of UGT1A9 and UGT2B7 are overestimated due to the effects of long-chain unsaturated fatty acids, fatty acid inhibition leads to overestimation of K_is for several UGT inhibitors generated using HLM and recombinant enzymes. Thus, addition of BSA or HSAFAF to incubations provides "true" estimates of this parameter and potentially improves IV-IVE for DDIs. This has been demonstrated in two studies to date.

K_i values for fluconazole inhibition of zidovudine glucuronidation by both HLM and recombinant UGT2B7 were reduced by approximately an order of magnitude when BSA was included in incubation mixtures, from 1133 to 145 μM (HLM) and 527 to 73 μM (UGT2B7) (Uchaipichat et al., 2006a). The K_i values generated in the presence of BSA predicted the known mean increase in zidovudine AUC (1.92-fold) observed in vivo within $\pm$ 25% using Equations (8.2) and (8.3). Thus, where glucuronide formation is primarily due to a single UGT enzyme, it may be possible to predict DDI potential from studies with recombinant enzymes rather than HLM or hepatocytes. Similarly, the magnitude of the valproic acid–lamotrigine interaction (2.6-fold increase in lamotrigine AUC in patients receiving valproic acid 500 mg twice daily) was well predicted by the K_i value obtained for valproic acid from incubations of HLM supplemented with BSA (Rowland et al., 2006). Mean K_i values generated in the absence and presence of BSA were 2465 and 387 μM, respectively.

8.7 Inhibitory DDIs Due to Glucuronide Conjugates

8.7.1 Mechanism-Based Inactivation

Glucuronides may act as inhibitors of CYP-catalyzed drug biotransformation and hence perpetrators of pharmacokinetic DDIs. Although the extent to which this occurs is unknown, the serious interaction between gemfibrozil and cerivastatin is well documented and illustrates the potential clinical significance of this mechanism. Gemfibrozil is a more potent inhibitor of CYP2C9 than CYP2C8 in vitro, but the opposite is observed clinically. The explanation for this paradox is that glucuronidation of gemfibrozil shifts the selectivity of inhibition from CYP2C9 to CYP2C8. Under experimental conditions suitable for detecting reversible inhibition, gemfibrozil 1-*O*-β-glucuronide is a stronger inhibitor of CYP2C8-catalyzed cerivastatin metabolism than is gemfibrozil (IC_{50} = 4 and 28 μM, respectively) (Shitara et al., 2004). Furthermore, gemfibrozil 1-*O*-β-glucuronide, but not its parent, is a selective mechanism-based inactivator of CYP2C8. The kinetic constants of inactivation, K_I (the concentration required for half-maximal inactivation) and k_{inact} (the maximal inactivation rate), were estimated as 20–52 μM and 0.21 min^{-1}, respectively (Ogilvie et al., 2006). Although the exact molecular mechanism remains to be determined, hydroxylation of the dimethylphenoxy moiety at the opposite end to the glucuronide appears to be important in the formation of a reactive intermediate(s) that inactivates CYP2C8.

8.7.2 Assessment of Mechanism-Based Inactivation by Glucuronides

In vitro assessment of glucuronides as inhibitors of CYP should therefore consider reversible and irreversible inhibitory mechanisms. These are readily differentiated by time-dependent inhibition, a distinguishing feature of irreversible inhibition (sometimes called mechanism-based inactivation). Briefly, the degree of CYP inhibition following co-incubation of the glucuronide with a CYP-selective substrate is compared with the degree of inhibition following a pre-incubation period prior to the addition of selective substrate. Since catalytic turnover is required, glucuronides that exhibit greater inhibition of CYP with time (i.e., following pre-incubation) are likely to be mechanism-based inactivators. To characterize K_I and k_{inact}, an experimental protocol involving two steps is typically employed. The glucuronide is pre-incubated with a high concentration of CYP for various times prior to dilution into a second incubation containing a saturating concentration of selective substrate. A plot of the log of remaining CYP activity versus pre-incubation time allows the observed inactivation rate at each glucuronide concentration to be estimated, from which K_I and k_{inact} are solved by nonlinear least squares regression analysis (Polasek and Miners, 2007). Importantly, Ogilvie et al. (2006) demonstrated that the inactivation of CYP2C8 by gemfibrozil 1-*O*-β-glucuronide could also be observed using an experimental design starting with gemfibrozil. To form the glucuronide, gemfibrozil was initially incubated with alamethicin-activated HLM in the presence of UDP-glucuronic acid ("glucuronidation step"). Samples were then incubated with NADPH-regenerating system prior to dilution and measurement of residual CYP activity ("CYP inhibition step"). In this example, the ability of gemfibrozil to inhibit CYP2C8 was increased after both steps, firstly by formation of the glucuronide, and subsequently via inactivation of CYP2C8 by the glucuronide which was NADPH and time dependent. This in vitro approach therefore serves as a universal method for determining whether test compounds are converted to glucuronides that inhibit CYP via reversible or irreversible mechanisms.

8.7.3 Potential Competitive Inhibition of CYP2C8 by Glucuronides

CYP2C8 is also known to metabolize diclofenac acyl glucuronide (Kumar et al., 2002) and estradiol-17β-glucuronide (Delaforge et al., 2005). Likewise, naproxen acyl glucuronide and testosterone-17β-glucuronide are subject to oxidative metabolism by HLM (Delaforge M, personal communication). Although formal studies have not been published, these and other drug glucuronides may act as an important source of CYP inhibition, the clinical significance of which is currently unknown.

8.8 Prospects

Reaction phenotyping for glucuronidated drugs is advancing. UGT enzyme-selective substrates are now available for the major hepatic xenobiotic glucuronidating enzymes. Although the availability of selective inhibitors is limited, this is an area of ongoing investigation in several laboratories and it is anticipated that the reaction phenotyping of glucuronidated substrates will eventually be as feasible as it is currently for drugs cleared by cytochrome P450. Complex homo- and heterotropic effects and the recent identification of certain glucuronides as inhibitors of CYP2C8 create further levels of complexity in screening for DDIs, but these difficulties can be overcome by careful experimental design. Similarly, experimental paradigms have been developed in recent years to optimize the calculation of kinetic parameters for glucuronidated drugs. In particular, the addition of BSA or HSAFAF to incubations of HLM or recombinant UGTs may permit the accurate prediction of in vivo clearance parameters and DDI potential, although further work is required to define the scope of the albumin effect as it applies to glucuronidated drugs.

References

Ammon S, von Richter O, Hofmann U, Thon K-P, Eichelbaum M and Mikus G (2000) In vitro interaction of codeine and diclofenac. *Drug Metab Disp* **28**:1149–1152.

Austin RP, Barton P, Cockroft SL, Wenlock MC and Riley RJ (2002) The influence of non-specific microsomal binding on apparent intrinsic clearance, and its prediction from physicochemical properties. *Drug Metab Disp* **30**:1497–1503.

Bauman JN, Goosen TC, Tugnait M, Peterkin V, Hurst DI, Menning LC, Milad M, Court MH and Williams JA (2005) UDP-Glucuronosyltransferase 2B7 is the major enzyme responsible for gemcabene glucuronidation in human liver microsomes. *Drug Metab Disp* **33**:1349–1354.

Benoit-Biancamano M-O, Connelly J, Villeneuve L, Caron P and Guillemette C (2009) Deferiprone glucuronidation by human tissues and recombinant UDP-glucuronosyltransferase 1A6: an in vitro investigation of genetic and splice variants. *Drug Metab Disp* **37**: 322–329.

Boase S and Miners JO (2002) In vitro – in vivo correlations for drugs eliminated by glucuronidation: investigations with the model substrate zidovudine. *Br J Clin Pharmacol* **54**:493–503.

Boyd MA, Srasuebkul P, Ruxrunggtham K, Mackenzie PI, Uchaipichat V, Stek M, Lange JMA, Phanuphak P, Cooper DA, Udomuksorn W and Miners JO (2006) Relationship between hyperbilirubinemia and UDP-glucuronosyltransferase 1A1 polymorphism in HIV-infected Thai patients treated with indinavir. *Pharmacogenet Genomics* **16**:321–329.

Bowalgaha K, Elliot DJ, Mackenzie PI, Knights KM and Miners JO (2007) The glucuronidation of Δ^4-3-keto-hydroxysteroids by human liver microsomal UDP-glucuronosyltransferases: 6α- and 21- hydroxyprogesterone are metabolized selectively by UGT2B7. *Drug Metab Disp* **35**:363–370.

Brown HS, Griffin M and Houston JB (2007) Evaluation of cryopreserved human hepatocytes as an alternative in vitro system to microsomes for the prediction of metabolic clearance. *Drug Metab Disp* **35**:293–301.

Brown HS, Ito K, Galetin A and Houston JB (2005) Prediction of in vivo drug – drug interactions from in vitro data: impact of incorporating parallel pathways of drug elimination and inhibitor absorption rate constant. *Br J Clin Pharmacol* **60**:508–518.

Corona G, Vaccher E, Sandron S, Sartor I, Tirelli U, Innocenti F and Toffoli G (2008) Lopinavir-ritonavir dramatically affects the pharmacokinetics of irinotecan in HIV patients with Kaposi's sarcoma. *Clin Pharmacol Ther* **83**:601–606.

Court MH (2005) Isoform-selective probe substrates for in vitro studies of human UDP-glucuronosyltransferases. *Methods Enzymol* **400**:104–116.

Court MH, Duan SX, von Moltke LL, Greenblatt DJ, Patten CJ, Miners JO and Mackenzie PI (2001) Interindividual variability in acetaminophen glucuronidation by human liver microsomes: identification of relevant acetaminophen UDP-glucuronosyltransferase isoforms. *J Pharmacol Exp Ther* **299**:998–1006.

Delaforge M, Pruvost A, Perrin L and André F (2005) Cytochrome P450-mediated oxidation of glucuronide derivatives: example of estradiol-17β-glucuronide oxidation to 2-hydroxy-estradiol-17β-glucuronide by CYP 2C8. *Drug Metab Dispos* **33**:466–473.

Engtrakul JJ, Foti RS, Strelevitz TJ and Fisher MB (2005) Altered AZT glucuronidation kinetics in liver microsomes as an explanation for underprediction of in vivo clearance. *Drug Metab Disp* **33**:1621–1627.

Ethell BT, Anderson GD and Burchell B (2003) The effect of valproic acid on drug and steroid glucuronidation by expressed UDP-glucuronosyltransferases. *Biochem Pharmacol* **65**: 1441–1449.

Fisher MB, Campanale K, Ackermann BL, Vandenbranden M and Wrighton SA (2000) In vitro glucuronidation using human liver microsomes and the pore-forming peptide alamethicin. *Drug Metab Disp* **28**:560–566.

Gertz M, Kilford PJ, Houston JB and Galetin A (2008) Drug lipophilicity and microsomal protein concentration as determinants in the prediction of the fraction unbound in microsomal incubations. *Drug Metab Disp* **36**:535–542.

Guillemette C (2003) Pharmacogenetics of human UDP-glucuronosyltransferase enzymesM. *Pharmacogenet J* **3**:136–158.

Ito K, Brown HS and Houston JB (2005a) Database analyses for the prediction of in vivo drug–drug interactions from in vitro data. *Br J Clin Pharmacol* **57**:473–486.

Ito K, Hallifax D, Obach RS and Houston JB (2005b) Impact of parallel metabolic pathways of drug elimination and multiple cytochrome P450 involvement on drug – drug interactions: CYP2D6 paradigm. *Drug Metab Disp* **33**:837–844.

Ito K, Iwatsubo T, Kanamitsu S, Ueda K, Suzuki H and Sugiyama Y (1998) Prediction of pharmacokinetic alterations caused by drug – drug interactions: metabolic interaction in the liver. *Pharmacol Rev* **50**:387–411.

Kerdpin O, Elliot DJ, Mackenzie PI and Miners JO (2006) Sulfinpyrazone C-glucuronidation is catalyzed selectively by human UDP-glucuronosyltransferase 1A9 (UGT1A9). *Drug Metab Disp* **34**:1950–1953.

Kiang TKL, Ensom MHH and Chang TKH (2005) UDP-Glucuronosyltransferases and clinical drug–drug interactions. *Pharmacol Ther* **106**:97–132.

Kirkwood LC, Nation RL and Somogyi AA (1998) Glucuronidation of dihydrocodeine by human liver microsomes and the effect of inhibitors. *Clin Exp Pharmacol Physiol* **25**: 266–270.

Kumar S, Samuel K, Subramanian R, Braun MP, Stearns RA, Chiu SH, Evans DC and Baillie TA (2002) Extrapolation of diclofenac clearance from in vitro microsomal metabolism data: role of acyl glucuronidation and sequential oxidative metabolism of the acyl glucuronide. *J Pharmacol Exp Ther* **303**:969–978.

Lewis LD, Benin A, Szumlanski CL, Otterness DM, Lennard L and Weinshilboum RM (1997) Olsalazine and 6-mercaptopurine related bone marrow suppression: a possible drug–drug interaction. *Clin Pharmacol Ther* **62**:464–475.

Mackenzie PI, Bock KW, Burchell B, Guillemette C, Iyanagi T, Miners JO, Owens IS and Nebert DW (2005) Nomenclature update for the mammalian UDP glycosyltransferase gene superfamily. *Pharmacogenet Genomics* **15**:677–685.

McLure JA, Miners JO and Birkett DJ (2000) Nonspecific binding of drugs to human liver microsomes. *Br J Clin Pharmacol* **49**:453–461.

Miners JO, Knights KM, Houston JB and Mackenzie PI (2006) In vitro – in vivo correlation for drugs cleared by glucuronidation: pitfalls and promises. *Biochem Pharmacol* **71**:1531–1539.

Miners JO, Lillywhite KJ, Yoovathaworn K, Pongmarutai M and Birkett DJ (1990) Characterisation of paracetamol UDP-glucuronosyltransferase activity in human liver microsomes. *Biochem Pharmacol* **40**:595–600.

Miners JO and Mackenzie PI (1991) Drug glucuronidation in humans. *Pharmacol Ther* **51**: 347–369.

Miners JO, McKinnon RA and Mackenzie PI (2002) Genetic polymorphism of UDP-glucuronosyltransferases and their functional significance. *Toxicology* **181–182**:453–456.

Miners JO, Smith PA, Sorich MJ, McKinnon RA and Mackenzie PI (2004) Predicting human drug glucuronidation parameters: applications of in vitro and in silico modeling approaches. *Annu Rev Pharmacol Toxicol* **44**:1–25.

Miners JO, Valente L, Lillywhite KJ, Mackenzie PI, Burchell B, Baguley BC and Kestell P (1997) Preclinical prediction of factors influencing the elimination of dimethylxanthenone-4-acetic acid, a new anticancer drug. *Cancer Res* **57**:284–289.

Ogilvie BW, Zhang D, Li W, Rodrigues AD, Gipson AE, Holsapple J, Toren P and Parkinson A (2006) Glucuronidation converts gemfibrozil to a potent, metabolism-dependent inhibitor of CYP2C8: implications for drug–drug interactions. *Drug Metab Disp* **34**:191–197.

Oleson L and Court MH (2008) Effect of the ß-glucuronidase inhibitor saccharolactone on glucuronidation by human liver microsomes and recombinant UDP-glucuronosyltransferases. *J Pharm Pharmacol* **60**:1175–1182.

Polasek TM and Miners JO (2007) In vitro approaches to investigate mechanism-based inactivation of cytochromes P450. *Exp Opin Drug Metab Toxicol* **3**:321–329.

Riley RJ, McGinnity DF and Austin RP (2005) A unified model for predicting human hepatic, metabolic clearance from in vitro clearance data in hepatocytes and microsomes. *Drug Metab Disp* **33**:1304–1311.

Rowland A, Elliot DJ, Knights KM, Mackenzie PI, and Miners JO (2008b) The albumin effect and in vitro - in vivo extrapolation: sequestration of long chain unsaturated fatty acids enhances phenytoin hydroxylation by human liver microsomes and recombinant cytochrome P450 2C9. *Drug Metab Disp* **36**:870–877.

Rowland A, Elliot DJ, Williams JA, Mackenzie PI, Dickinson RG and Miners JO (2006) In vitro characterization of lamotrigine N2-glucuronidation and the lamotrigine – valproic acid interaction. *Drug Metab Disp* **34**:1055–1063.

Rowland A, Gaganis P, Elliot DJ, Mackenzie PI, Knights KM and Miners JO (2007) Binding of inhibitory fatty acids is responsible for enhancement of UDP-glucuronosyltransferase 2B7 (UGT2B7) activity by albumin. *J Pharmacol Exp Ther* **321**:137–147.

Rowland A, Knights KM, Mackenzie PI, and Miners JO (2008a) The albumin effect and drug glucuronidation: bovine serum albumin and fatty acid free human serum albumin enhance the glucuronidation of UGT1A9 substrates but not UGT1A1 and UGT1A6 activities. *Drug Metab Disp* **36**:1056–1062.

Sahai J, Gallicano K, Pakuts A and Cameron DW (1994) Effect of fluconazole on zidovudine pharmacokinetics in patients infected with human immunodeficiency virus. *J Infec Dis* **169**:1103–1107.

Shitara Y, Hirano M, Sato H and Sugiyama Y (2004) Gemfibrozil and its glucuronide inhibit the organic anion transporting peptide 2 (OATP2/OATP1B1:*SLC21A6*)-mediated hepatic uptake and CYP2C8-mediated metabolism of cerivastatin: analysis of the mechanism of the clinically relevant drug–drug interaction between cerivastatin and gemfibrozil. *J Pharmacol Expl Ther* **311**:228–236.

Soars MG, Burchell B and Riley RJ (2002) In vitro analysis of human drug glucuronidation and prediction of in vivo metabolic clearance. *J Pharmacol Exp Ther* **301**:382–390.

Soars MG, Ring BJ and Wrighton SA (2003) The effect of incubation conditions on the enzyme kinetics of UDP-glucuronosyltransferases. *Drug Metab Disp* **31**:762–767.

Stone AN, Mackenzie PI, Galetin A, Houston JB and Miners JO (2003) Isoform selectivity and kinetics of morphine 3- and 6- glucuronidation by human UDP-glucuronosyltransferases: evidence for atypical glucuronidation kinetics by UGT2B7. *Drug Metab Disp* **31**: 1086–1089.

Sykes MJ, Sorich MJ and Miners JO (2006) Molecular modelling approaches for the prediction of the non-specific binding of drugs to human liver microsomes. *J Chem Inf Model* **46**: 2661–2673.

Tsoutsikos P, Miners JO, Stapleton A, Thomas T, Sallustio BC and Knights KM (2004) Evidence that unsaturated fatty acids are potent inhibitors of renal UDP-glucuronosyltransferases: kinetic studies using human kidney cortical microsomes and recombinant UGT1A9 and UGT2B7. *Biochem Pharmacol* **67**:191–199.

Tukey RH and Strassburg CP (2000) Human UDP-glucuronosyltransferases: metabolism, expression and disease. *Ann Rev Pharmacol Toxicol* **40**:581–616.

Uchaipichat V, Elliot DJ, Mackenzie PI and Miners JO (2006b) Selectivity of substrate (trifluoperazine) and inhibitor (amitriptyline, androsterone, canrenoic acid, hecogenin, phenylbutazone, quinidine, quinine and sulfinpyrazone) probes for human UDP-glucuronosyltransferases. *Drug Metab Disp* **34**:449–456.

Uchaipichat V, Galetin A, Houston JB, Mackenzie PI, Williamd JA and Miners JO (2008) Kinetic modeling of the interactions between 4-methylumbelliferone, 1-naphthol and zidovudine glucuronidation by UDP-glucuronosyltransferase 2B7 (UGT2B7) provides evidence for multiple substrate and effector binding sites. *Molec Pharmacol* **74**:1152–1162.

Uchaipichat V, Mackenzie PI, Guo X-H, Gardner-Stephen D, Galetin A, Houston JB and Miners JO (2004) Human UDP-glucuronosyltransferases: isoform selectivity and kinetics of 4-methylumbelliferone and 1-naphthol glucuronidation, effects of organic solvents, and inhibition by diclofenac and probenecid. *Drug Metab Disp* **32**:413–423.

Uchaipichat V, Winner LK, Elliot DJ, Mackenzie PI, Williams JA and Miners JO (2006a) Quantitative prediction of inhibitory interactions involving glucuronidated drugs from in vitro data: the effect of fluconazole on zidovudine glucuronidation. *Br J Clin Pharmacol* **61**:427–439.

Udomuksorn W, Elliot DJ, Lewis BC, Mackenzie PI, Yoovathaworn K and Miners JO (2007) Influence of mutations associated with the Gilbert and Crigler-Najjar type II syndromes on the glucuronidation of bilirubin and other UDP-glucuronosyltransferase 1A substrates. *Pharmacogenet Genomics* **17**:1017–1029.

Wen Z, Tallman MN, Ali SY and Smith PC (2007) UDP-Glucuronosyltransferase 1A1 is the principal enzyme responsible for etoposide glucuronidation in human liver and intestinal microsomes: structural characterization of phenolic and alcoholic glucuronides of etoposide and estimation of enzyme kinetics. *Drug Metab Disp* **35**:371–380.

Williams JA, Hyland R, Jones BC, Smith DA, Hurst S, Goosen TC, Peterkin V, Koup JR and Ball SE (2004) Drug–drug interactions for UDP-glucuronosyltransferase substrates: a

pharmacokinetic explanation for typically observed low exposure (AUC_i/AUC) ratios. *Drug Metab Disp* **32**:1201–1208.
Williams JA, Ring BJ, Cantrell VE, Campanale K, Jones DR, Hall SD and Wrighton SA (2002) Differential modulation of UDP-glucuronosyltransferase 1A1 (UGT1A1) catalyzed estradiol 3-glucuronidation by the addition of UGT1A1 substrates and other compounds to human liver microsomes. *Drug Metab Disp* **30**:1266–1273.

Chapter 9
In Vitro Techniques to Study Transporter-Based DDI

Kelly Bleasby, Xiaoyan Chu, and Raymond Evers

Abstract In recent years numerous examples have been published where pharmacokinetic drug–drug interactions could be ascribed to inhibition of uptake or efflux drug transporters in, for instance, the liver and kidney. In drug discovery and development, it therefore has become increasingly important to identify the propensity of drug candidates to cause such interactions, either as a victim or perpetrator. In this chapter, we describe the status of in vitro methodologies currently applied to predict the propensity of drug candidates to be a victim or perpetrator in DDIs due to inhibition of transporter activity. Assay systems discussed are recombinant cell lines expressing transporters, primary cells such as hepatocytes, and membrane vesicles. Special focus is on transporters expressed in liver and kidney, with the exception of ABCB1 which is covered elsewhere in this book. In addition, we present the current understanding of how data generated in these systems can be used to predict DDI potential of drug candidates.

9.1 Introduction

Based on current knowledge, transporters most relevant to drug transport in liver and kidney are depicted in Fig. 9.1 Probe substrates and prototypical inhibitors are summarized in Table 9.1. The molecular identification of transporters involved in the disposition of drugs has resulted in the development of various in vitro assay systems which can be used to monitor drug transport. These systems are being used to elucidate pathways by which drugs are absorbed, distributed, and excreted by the body. Using this information, it is possible to rationalize pharmacokinetic drug–drug interactions (DDIs) which have been observed in the clinic and to predict

R. Evers (✉)
Department of Drug Metabolism and Pharmacokinetics, Merck & Co., In Vitro Technologies, Rahway, NJ, USA
e-mail: raymond_evers@merck.com

K.S. Pang et al. (eds.), *Enzyme- and Transporter-Based Drug–Drug Interactions*,
DOI 10.1007/978-1-4419-0840-7_9,

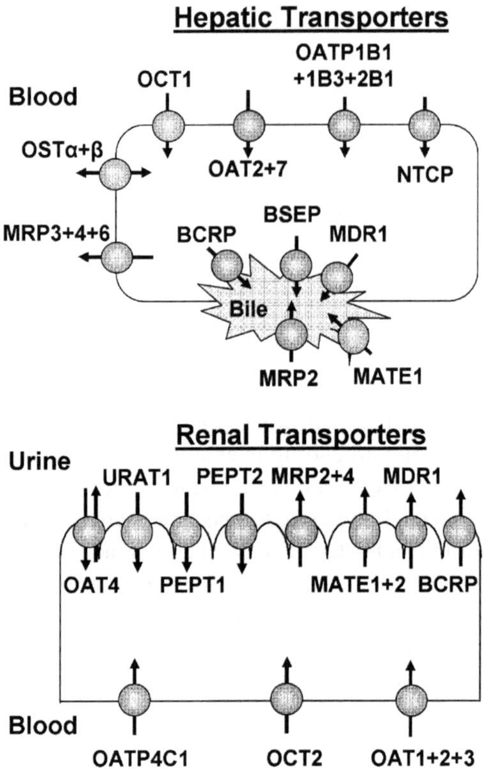

Fig. 9.1 Drug transporters expressed in human liver and kidney

semi-quantitatively whether drug candidates in clinical development have the potential to cause transporter-mediated DDIs, either as a victim or as a perpetrator.

Several examples have been found where DDIs and cases of toxicity could be ascribed to reduced transport, due to polymorphisms in transporter genes or inhibition of transporter activity by coadministration of two drugs that used a shared transporter mechanism for uptake or efflux. An elegant example of how polymorphisms can explain a toxicological response was obtained in a large correlation study which demonstrated that the incidence of a non-synonymous polymorphism (V174A) in the liver uptake organic anion transporting polypeptide 1B1 (OATP1B1) correlated with the incidence of myopathy in individuals taking a 40 or 80 mg dose of the cholesterol lowering drug simvastatin (Link et al., 2008). In another example, it was found that in individuals carrying a reduced function polymorphism of the organic cation transporter OCT1, the liver active antihyperglycemic agent metformin had a significantly reduced efficacy (Shu et al., 2007).

Examples of DDIs explained by liver transporter inhibition are, for instance, the coadministration of the nonspecific transporter inhibitor cyclosporine A (CsA) with statins. Coadministration of CsA with pravastatin resulted in a 5- to 10-fold

Table 9.1 Example in vitro probe substrates and inhibitors for major human drug transporters

Transporter	Probe substrates	Prototypic inhibitors	References
MDR1 (ABCB1)	Digoxin, quinidine, verapamil	Cyclosporine A, GF120918	Rautio et al. (2006)
BCRP (ABCG2)	Prazosin, cimetidine	Ko143	Allen et al. (2002)
MRP2 (ABCC2)	Estradiol 17β-D-glucuronide, ethacrynic acid glutathione	Probenecid, MK-571, benzbromarone, indomethacin, sulfinpyrazone	Zhou et al. (2008), Chu et al. (2004)
MRP3 (ABCC3)	Estradiol 17β-D-glucuronide, leukotrien C4	MK-571, probenecid, benzbromarone, indomethacin, sulfinpyrazone	Zeng et al. (2000), Zelcer et al. (2001)
MRP4 (ABCC4)	Folic acid	MK-571, Probenecid	Chen et al. (2002), Chu et al. (2009)
NTCP (SLC10A2)	Taurocholate	BSP	Hagenbuch and Meier (1994)
BSEP (ABCB11)	Taurocholate, glycocholate	Cyclosporine A, rifampicin	Byrne et al. (2002)
OCT1 (SLC22A1)	MPP	Decynium-22, quinidine	Zhang et al. (1997), Hayer-Zillgen et al. (2002)
OCT2 (SLC22A2)	TEA, MPP, cimetidine	Decynium-22, quinidine	Gorboulev et al. (1997), Hayer-Zillgen et al. (2002)
OAT1 (SLC22A6)	Cidofovir, PAH	Probenecid	Cihlar et al. (1999)
OAT3 (SLC22A8)	Cimetidine, estrone sulfate	Probenecid	Tahara et al. (2005)
MATE1 (SLC47A1)	TEA, cimetidine, metformin, fexofenadine	Cimetidine	Terada and Inui (2007), Matsushima et al. (2009)
MATE2 (SLC47A2)	TEA, cimetidine, metformin		Terada and Inui (2007)
PEPT1(SLC15A1)	Gly-Sar, cephalexin	β-lactam antibiotics	Russel et al. (2002), Lee and Kim (2004)
PEPT2 (SLC15A2)	Gly-Sar, cephalexin	β-lactam antibiotics	Russel et al. (2002), Lee and Kim (2004)
OATP1B1 (SLCO1B1)	Pitavastatin, estradiol 17β-D-glucuronide	Cyclosporine A, BSP, rifampicin	König et al. (2000), Hirano et al. (2004)
OATP1B3 (SLCO1B3)	Pitavastatin, estradiol 17β-D-glucuronide, CCK-8	BSP, rifampicin	König et al. (2000), Hirano et al. (2004)
OATP2B1 (SLCO1B1)	Estrone Sulfate, BSP	BSP, rifamycin SV	Tamai et al. (2001), Vavricka et al. (2002)
OATP4C1 (SLCO4C1)	Digoxin, ouabain, sitagliptin	Digitoxin, digoxigenin, thyroxine	Mikkaichi et al. (2004), Chu et al. (2007)

increased plasma exposure of pravastatin (Regazzi et al., 1993). Other interactions are explained by inhibition of organic anion and cation transporters in the kidney such as the cimetidine–metformin interaction. The dual OAT3, OCT2 substrate cimetidine increased the AUC of the OCT2 substrate metformin by ∼50% and reduced the CL_{renal} by 28%, whereas the pharmacokinetics for cimetidine were not altered significantly (Somogyi et al., 1987). Recently, a number of excellent reviews have been published on the topic of transporter-mediated DDIs, to which the reader is referred for a complete overview of this area (Li et al., 2006; El-Sheikh et al., 2008; Neuvonen et al., 2006).

The aim of this chapter is to provide a summary of the in vitro methodologies currently available to monitor drug transport by transporters localized in the liver and kidney and how data obtained in these models can be used to assess the potential for substrates to act as perpetrators or victims in transporter-mediated DDIs.

9.2 Hepatocytes

9.2.1 Hepatocytes in Suspension

Freshly isolated or cryopreserved hepatocyte suspensions are a valuable tool to study hepatic uptake transporters (Hewitt et al., 2007; Shitara et al., 2003b). However, internalization of canalicular efflux transporters during hepatocyte isolation limits the use of suspended hepatocytes to study the function of efflux transporters (Roelofsen et al., 1995; Ghibellini et al., 2006).

A commonly used method to determine the contribution of transporter-mediated uptake to total uptake using hepatocyte suspensions is to perform uptake studies at 37°C and 4°C, as transporter-mediated uptake is highly temperature dependent (Hewitt et al., 2007). The difference of rates of uptake at 37°C and 4°C can be used to distinguish active and passive transport with the assumption that the active transport is not functional at 4°C. However, membrane fluidity, and therefore passive diffusion, is also temperature dependent to some degree. Therefore, it may be difficult to distinguish the contribution of carrier-mediated transport and passive diffusion for compounds with a high passive permeability (Poirier et al., 2008). Preferably, temperature dependence studies should be combined with a concentration dependence study to assess whether transport is saturated at higher drug concentrations, or in the presence of transporter inhibitors. Concentration dependence studies will give a detailed analysis of the contribution of transporter-mediated vs. passive diffusion, and the K_m and V_{max} values obtained will aid in the understanding of the affinity and capacity of carrier-mediated uptake. CL_{int} calculated from K_m and V_{max} can potentially be used to predict hepatic clearance *in vivo* (Webborn et al., 2007).

The kinetic parameters for the uptake of a compound into hepatocytes can be calculated using Equation (9.1) (Hirano et al., 2004).

$$Vo = \frac{V_{\max} \times S}{K_m + S} + P_{dif} \times S \qquad (9.1)$$

where V_0 is the initial uptake rate (pmol/min/mg protein), S is the substrate concentration (μM), K_m is the Michaelis constant (μM), V_{max} is the maximum uptake rate (pmol/min/mg protein), and P_{dif} is the non-saturable uptake clearance (μl/min/mg protein).

Uptake of compounds into suspended hepatocytes can be measured by a rapid centrifugation/filtration method. To test uptake of radiolabeled compounds, after incubating hepatocytes with a test compound, uptake can be stopped by separating the cells from the buffer by rapid centrifugation of the cells through a silicon cushion to prevent the loss of the compound back into the medium (Fig. 9.2). The amount of compound retained in hepatocytes is measured by liquid scintillation counting (Hirano et al., 2006). Uptake studies can also be conducted in 96 deep well plates with either radiolabeled or non-labeled compounds. Uptake is stopped by the addition of ice-cold phosphate buffered saline (PBS), followed by immediate centrifugation at 4°C and washing of the cell pellet with PBS. Cell pellets are resuspended in 50% acetonitrile and the amount of compound associated with the cell pellets can be quantified by liquid scintillation counting or LC/MS/MS (Fig. 9.3).

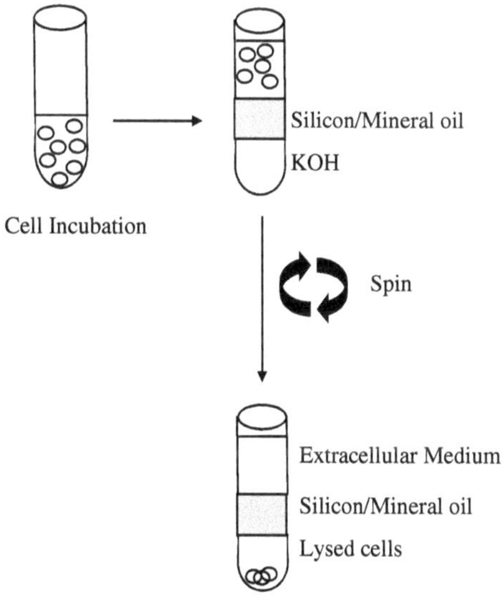

Fig. 9.2 Rapid centrifugation method for measuring uptake into suspended hepatocytes

9.2.2 Sandwich-Cultured Hepatocytes

Sandwich-cultured hepatocytes have become another useful tool to evaluate both uptake and efflux transporters (LeCluyse et al., 1994; Liu et al., 1999a,b). Sandwich-cultured hepatocytes maintain the hepatocyte architecture, including tight junctions, a canalicular biliary network, and functional transporters. Because depletion of Ca^{2+} will open tight junctions, accumulation of compounds in hepatocytes and

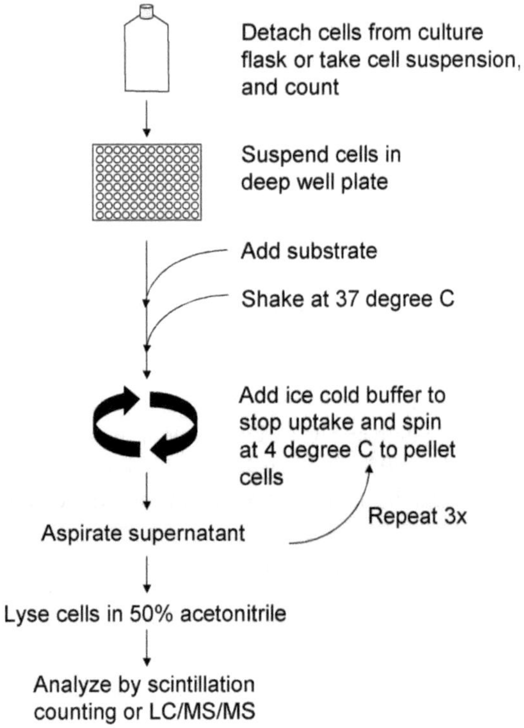

Fig. 9.3 Rapid centrifugation method for measuring uptake into cell suspensions

bile ducts, or in hepatocytes only can be determined in Ca^{2+} and Ca^{2+}-free buffer, respectively. Biliary elimination is then calculated as the difference of these two measurements. Utilizing sandwich-cultured rat hepatocytes, good correlations have been demonstrated between in vitro and in vivo biliary clearance for a number of compounds (Liu et al., 1999a,b). In addition, the potential for prediction of in vivo biliary clearance in humans using sandwich-cultured human hepatocytes has also been investigated with Tc-99m sestamibi, mebrofenin, and piperacillin as probe compounds (Ghibellini et al., 2007). Biliary excretion of these compounds was determined in seven healthy volunteers using an oroenteric catheter to aspirate duodenal secretions, and gamma scintigraphy to determine gallbladder contraction. The number of reports successfully using this model with human hepatocytes is limited, however. A limitation of the sandwich-cultured hepatocyte model is the absence of a constant bile flow, leading to characteristics of cholestasis like the upregulation of MRP3 and intrahepatic accumulation of bile salts.

Overall, hepatocytes are a valuable model to study hepatic transport and can be used as a tool to estimate the potential for drug–drug interactions. However, their utility is confined by the limited capability to study the relative contribution of drug uptake and efflux transporters. Additional disadvantages include the limited supply of fresh and cryopreserved human hepatocytes, relatively high cost, and inter-donor variation due to transporter polymorphisms and expression levels of transporters.

One way to address the latter points is the use of pooled hepatocytes from various donors.

9.3 Uptake into Recombinant Cell Lines

Recombinant cell lines expressing drug uptake transporters from the SLC22 (OAT, OCT), SLC10 (NTCP), and SLCO (OATP) families have been described in the literature (see Table 9.1 for references). To determine if a test compound is a substrate of an uptake transporter and thus has potential to be the victim of a transporter DDI, direct accumulation can be measured in cell lines transiently or stably expressing the transporter of interest. Uptake experiments can be performed with cells attached to permeable membrane supports (Fig. 9.4), wells of plastic plates, or with cells in suspension (in deep well plates, or individual tubes; Fig. 9.3).

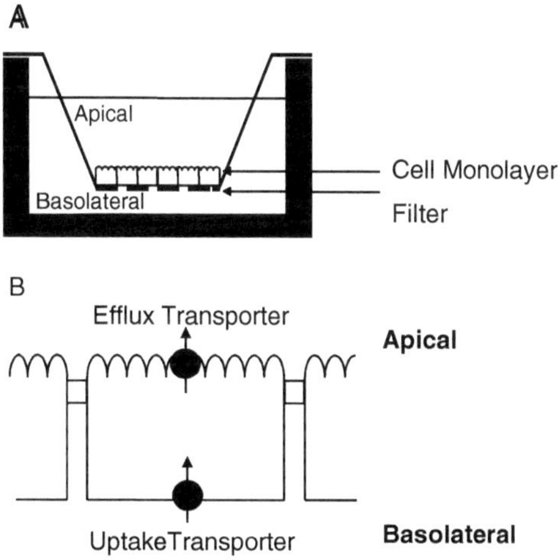

Fig. 9.4 Bidirectional transport assay. (**A**) Cell monolayers are grown on permeable filters, forming apical and basolateral compartments of the well. (**B**) Cell monolayers may express an efflux transporter or both uptake and efflux transporters

When measuring uptake in polarized cell monolayers, it is important to ensure that the membrane domain the transporter is expressed on is accessible to the substrate solution. For example, in MDCKII cells, OATP1B1 localizes to the basolateral membrane (König et al., 2000, Sasaki et al., 2002), and so when measuring compound uptake, the cells must either be in suspension, on a permeable support, or be sub-confluent on the plate, to ensure that the compound can access OATP1B1.

In a typical experiment, cells are rinsed in buffer and incubated with substrate for varying amounts of time, at 37°C. Compound uptake is stopped by centrifugation of the cell suspension, or aspiration of the substrate for the attached cells, and washing the cells several times in ice-cold buffer. The cells are then lysed and the amount of drug in the cells can be analyzed by liquid scintillation counting or by LC/MS/MS (Fig. 9.3). Uptake is usually calculated as amount of compound accumulated per amount of protein or number of cells, over time (e.g., pmole/mg protein or 10^6 cells/minute). Uptake in the transfected cell line is compared to the parental (untransfected) cell line to determine if the test compound is a substrate of the transporter of interest. Once this is established, experiments can be performed with inhibitors added to the substrate solution in order to assess the potential for DDIs. The transporter-mediated uptake is calculated (uptake into transporter expressing cells minus uptake into parental cells) at each concentration of inhibitor and, when expressed as percent of control (without inhibitor), an IC_{50} value can be calculated (Chu et al., 2007).

Although measurement of direct uptake into recombinant cell lines is often an efficient means to determine if a compound is a substrate for an individual transporter, limitations do exist. In particular, for compounds that bind to cell membranes, or are substrates for endogenous uptake transporters, it may be difficult to detect a clear difference between parental and recombinant cells. Furthermore, care should be taken when extrapolating recombinant cell data to the in vivo situation, as transporter expression levels can be very different from the native tissue, and the interplay between multiple transporters with overlapping substrate specificities is not accounted for.

9.4 Bidirectional Transport in Recombinant Cell Lines

Bidirectional transport assays are used to study efflux transporters and the interplay between uptake and efflux transporters. These assays are performed with polarized cultured cell monolayers (e.g., LLC-PK1 or MDCKII), grown on permeable membrane supports (Fig. 9.4). Cells are stably or transiently transfected with efflux and/or uptake transporters (Smit et al., 1998; Sasaki et al., 2002), which are localized to the apical and basolateral cell membranes, respectively.

In recent years, several cell lines expressing two or more transporters have been developed; for example, in order to model hepatic vectorial transport, MDCKII cells expressing OATP1B1 in the basolateral membrane and MRP2 in the apical membrane were established (Sasaki et al., 2002). Although this second generation of recombinant cell lines has greatly enhanced our understanding of the interplay between multiple transporters, quantitative extrapolation of data from these systems to the in vivo situation is complex, due to differences in transporter expression levels, compensatory transport mechanisms and, for the OATP/Oatp family in particular, a lack of clear orthologs between species for some transporters (Sasaki et al., 2004; Evers and Chu, 2008). Table 9.2 summarizes the recombinant cell lines expressing two or more transporters currently described in the literature.

In a typical bidirectional transport experiment to establish whether a compound is a substrate of the transporter and thus a potential DDI victim, the test compound is added to the apical (A) compartment, with buffer in the basolateral (B) compartment (A–B transport), and in separate wells, test compound is added to the basolateral compartment, with buffer in the apical compartment (B–A transport). The monolayers are incubated at 37°C, with samples removed for quantification from each

Table 9.2 Polarized cell monolayers expressing two or more drug transporters

Cell line	Basolateral transporter(s)	Apical transporter	References
MDCKII	OATP1B1	MRP2	Sasaki et al. (2002)
MDCKII	OATP1B1	BCRP	Matsushima et al. (2005)
MDCKII	OATP1B1	MDR1	Matsushima et al. (2005)
MDCKII	OATP1B3	MRP2	Cui et al. (2001)
MDCKII	OATP1B3	BCRP	Ishiguro et al. (2008)
MDCKII	OATP1B3	MDR1	Ishiguro et al. (2008)
MDCKII	OATP1B1, OATP1B3, OATP2B1	MRP2	Kopplow et al. (2005)
MDCKII	Rat Oatp1b2	Rat Mrp2	Sasaki et al. (2004)
MDCKII	OCT1	MDR1	Nies et al. (2008)
MDCKII	OCT1	MATE1	Sato et al. (2008)
MDCKII	OCT2	MATE2	Sato et al. (2008)
MDCKII	Rat Ntcp	Rat BSEP	Mita et al. (2005)

compartment at various time points. The apparent permeability (P_{app}) is calculated (Equation (9.2)) for each direction of transport, and data are reported as the P_{app} B–A/A–B ratio. Typically, the P_{app} B–A/A–B ratio is compared between control (untransfected) monolayers and those containing the transporter(s) of interest (Yamazaki et al., 2001; Rautio et al., 2006).

$$P_{app} = \frac{\text{Volume of receptor chamber (mL)}}{[\text{Area of membrane (cm}^2)][\text{initial concentration }(\mu\text{M})]} \times \frac{\Delta \text{ in concentration }(\mu\text{M})}{\Delta \text{ in time (s)}} \quad (9.2)$$

Once it is established that the test compound is the substrate of the transporter of interest, further experiments can be performed to assess the potential for transport to be inhibited by prototypic inhibitors that may be coadministered in the clinic. In these experiments, the inhibitor is added at various concentrations to both apical and basolateral compartments. Similar experiments can be performed to assess whether the test compound is an inhibitor of a known probe substrate and thus may be a perpetrator of a transporter-based DDI. An IC_{50} can be calculated in several ways and the method that is most predictive of in vivo DDI observations is the subject of ongoing discussion (Troutman and Thakker, 2003; Acharya et al., 2008; Balimane et al., 2008, Kalvass and Pollack, 2007). Figure 9.5 shows how IC_{50} values can differ within the same data set, when calculated from either changes in net transport (P_{app} B–A – P_{app} A–B) or changes in P_{app} B–A/A–B ratio.

Bidirectional transport assays are a sensitive method for determining transport of compounds that tend to bind to cell membranes as only compound which is fluxed through the cell monolayer, either *trans-* or *para*-cellularly, is measured, and therefore transport measurements are less confounded by compound binding to cell membranes (as in direct cell uptake studies). However, this method does have some limitations: For cells expressing only efflux transporters, transport will be limited

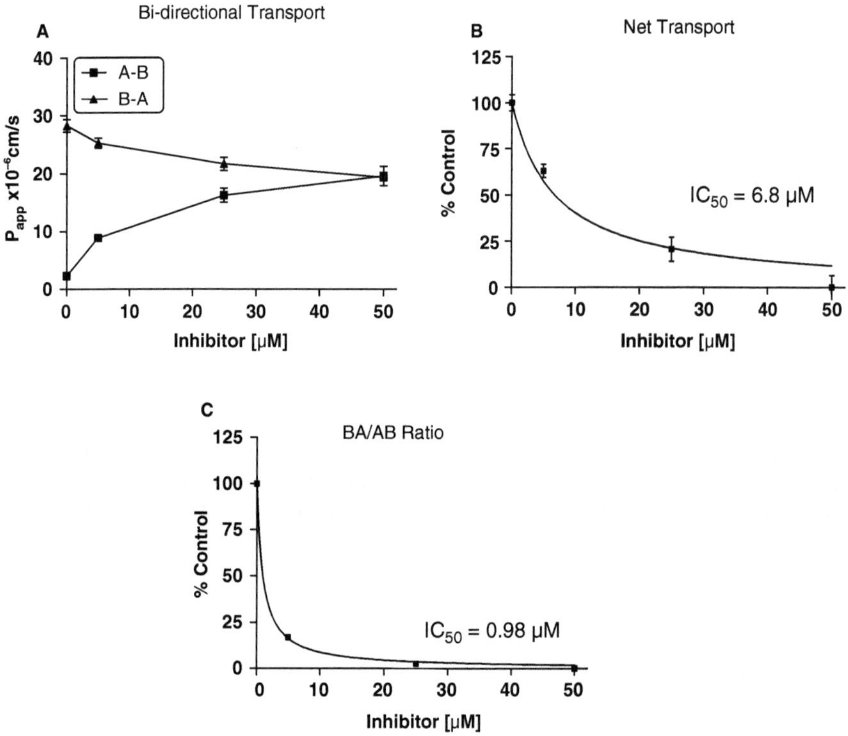

Fig. 9.5 Inhibition of bidirectional transport. (**A**) Inhibition of A–B and B–A transport of probe substrate. (**B**) Inhibition of net transport (P_{app} (B–A) – P_{app} (A–B)) of probe substrate. (**C**) Inhibition of P_{app} B–A/A–B ratio of probe substrate

by the diffusion rate of the compound across the basolateral membrane and so for low-permeability compounds, the P_{app} B–A/A–B ratio may be underestimated. In some cases, this can be overcome by using cell lines expressing both uptake and efflux transporters (Table 9.2); however, it must be noted that the commonly used cell lines MDCKII and LLC-PK1 express significant amounts of canine Pgp and porcine Bcrp, respectively, and thus careful controls such as single transfected cells must be included to account for any contribution they may make to overall transport. Furthermore, it is recommended to measure recovery of compound at the end of the experiment to account for adhesion to the lab wear and, although not commonly observed, the potential for compound metabolism.

9.5 The Vesicular Transport Assay

The vesicular transport assay is a valuable tool to study functional activity of ABC-transporters and some SLC transporters (Glavinas et al., 2008). Membranes

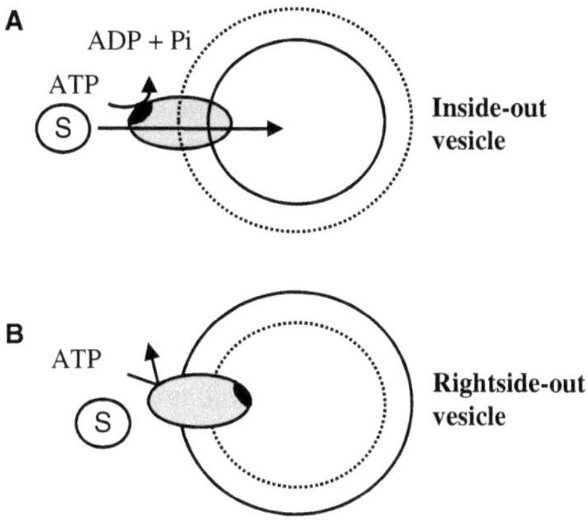

Fig. 9.6 Vesicular uptake assay and the orientation of membrane vesicles. (**A**) Inside-out membrane vesicles and ATP-dependent vesicular uptake. (**B**) Rightside-out membrane vesicles

prepared under suitable conditions contain inside-out-oriented vesicles, with the ATP-binding site and substrate-binding site of the transporter facing the buffer (Fig. 9.6). The difference of the uptake of a substrate in the presence or absence of ATP or other driving force such as Na^+ and H^+ is attributed to ABC-transporter or uptake transporter-mediated transport, respectively. The rapid filtration method can be used to separate the vesicles from the incubation solution (Ishikawa et al., 1990; Saito et al., 2006). After filtration, the membrane vesicles are retained on the filter and the amount of test compound trapped inside the vesicles can be measured by LC/MS/MS, fluorescence detection, or liquid scintillation counting.

The membranes used in vesicular transport assay are typically prepared from baculovirus-infected insect cells, transfected or selected mammalian cell lines, tissues expressing the transporter of interest, or transfected yeast cells (Glavinas et al., 2008; Cai and Gros, 2003), using a number of membrane preparation methods (Steck et al., 1970). Membranes are partially purified or crude, and typically contain a mixture of lamellae, inside-out and rightside-out vesicles. In vesicular uptake studies, only the activity of transporters present in inside-out vesicles is detected, as in these vesicles the ATP-binding domain is located extra-vesicularly (Figure 9.6). Although membrane preparations can be purified to enrich for inside-out vesicles, crude membranes have been used extensively to study ABC-transporters. Isolated membrane vesicles can be suspended in a hypotonic or isotonic buffer and stored at –80°C or in liquid nitrogen for many months. The commercial availability of transporter-expressing membrane vesicles makes this assay feasible and available for routine use, especially to monitor transport by ABC-transporters.

The vesicular transport assay is most suitable for studying transport of compounds with low membrane permeability and low nonspecific binding, such as glutathione and glucuronide conjugates (Jedlitschky et al., 2006). These low-permeable substrates are difficult to study in cell-based assays, as they do not penetrate the cell membrane without the presence of an uptake transporter. The availability of membrane vesicles isolated from cells over-expressing a single transporter makes it an efficient and useful tool, which can be applied to (i) directly measure transport of substrate compounds across the cell membrane, (ii) conduct detailed kinetic analysis of transported substrates, such as determination of K_m and V_{max} values, (iii) study the interaction of test compounds with a prototypical substrate of the transporter of interest to obtain K_i or IC_{50} values for inhibitors, and (iv) study the driving force of transport, or the requirement for the presence of cotransported molecules such as glutathione. For instance, some substrates of members of the multidrug resistance protein family are associated with the cotransport of reduced glutathione (Cole and Deeley, 2006; Borst et al., 2007). A limitation of vesicular uptake studies is that this assay may give false-negative results for highly lipophilic compounds due to high nonspecific binding to lipid membranes or high passive diffusion. Compounds with medium-to-high passive permeability are not retained inside the vesicles, which makes transport measurements for this class of compounds difficult to perform.

9.6 Utility of In Vitro Transporter Assays in Assessing Transporter-Based DDIs

Clinically relevant interactions mediated by transporters are one of the critical issues in drug development. The in vitro transporter models described above provide important tools to understand the contribution of transporters to drug absorption, disposition, and elimination. Together with information on therapeutic concentrations of the interacting drugs, the data obtained can be used to predict the propensity for transporter-mediated DDIs. Recently, several studies have illustrated how in vitro transporter assays can be applied to predict the potential for transporter based DDIs.

The HMG-CoA reductase inhibitor cerivastatin was withdrawn from the market in 2001 due to severe toxicity and DDIs. After coadministration with CsA, the *C*max and plasma AUC of cerivastatin in kidney transplant patients increased 5-fold and 3.8-fold, respectively (Mück et al., 1999). The mechanism of this drug–drug interaction was investigated in vitro (Shitara et al., 2003a). Uptake of cerivastatin by human hepatocytes and OATP1B1-transfected MDCKII cells was saturable, and inhibited by CsA with K_i values of 0.3–0.7 and 0.2 μM, respectively, which is lower than the estimated maximum unbound concentration of CsA at the inlet of the liver. This suggested that the drug–drug interaction between cerivastatin and CsA could be attributed, at least in part, to inhibition of OATP1B1-mediated cerivastatin transport by CsA.

Sitagliptin, a selective dipeptidyl peptidase 4 inhibitor for the treatment of type 2 diabetes, is primarily excreted into the urine via active tubular secretion in humans. In vitro transporter studies indicated that sitagliptin is a substrate for the human renal uptake transporter OAT3 and efflux transporter MDR1 Pgp (Chu et al., 2007). This suggested that OAT3 and MDR1 Pgp might play a role in transporting sitagliptin into and out of renal proximal tubule cells, respectively. Based on these findings,

Table 9.3 Effect of various drugs on OAT3-mediated uptake of sitagliptin (Adapted from Chu et al., 2007, and references therein)

Compound	IC_{50} (μM)	Plasma C_{max}	Dose (mg)
Cimetidine	79 ± 20	7.0[a]	400
Enalapril	~100	0.83[a]	10
Enalaprilat	>100	0.16[a]	10
Fenofibric acid	2.2 ± 0.1	0.26–0.81[b]	67 (Fenofibrate)
Furosemide	1.7 ± 0.4	5.1[a] 0.07[b]	40
Gabapentin	>100	28[a]	400
Ibuprofen	3.7 ± 0.3	135[a] 0.7[b]	400
Indapamide	11 ± 1.7	0.91[c]	5
Probenecid	5.6 ± 1.4	12–44[b]	500–2000
Quinapril	6.2 ± 1.7	1.31[d]	20

[a]Maximal total concentration in plasma.
[b]Maximal concentration of unbound drug.
[c]Maximal total concentration in blood.
[d]Maximal total concentration of active carboxylic acid.

the propensity of sitagliptin to cause DDIs at the level of OAT3 and Pgp was evaluated in vitro. By comparing IC_{50} values obtained from OAT3 transfected cells with unbound or total plasma concentration of clinically used drugs (Table 9.3), the potential of sitagliptin to act as a victim when coadministered with these drugs, including some OAT3 substrates and inhibitors, was examined. The results suggested that renal secretion of sitagliptin could be inhibited by probenecid, a nonspecific inhibitor for organic anion transporters. However, the magnitude of the interactions would be expected to be low, and the effects might not be clinically meaningful, due to the high safety margin of sitagliptin. In vitro studies also suggested that sitagliptin is unlikely to be a perpetrator of drug–drug interactions with Pgp, OAT1, or OAT3 substrates at clinically relevant concentrations.

Varenicline is a selective nicotinic acetylcholine receptor partial agonist aiding in smoking cessation (Feng et al., 2008). It is predominantly eliminated unchanged in urine, and active tubular secretion partially contributes to its renal elimination. In vitro studies using cell lines stably transfected with human renal transporters were performed to determine the mechanism underlying the renal elimination of varenicline and to evaluate its potential to be involved in DDIs with other renally excreted drugs, such as cimetidine (Feng et al., 2008). Varenicline was identified as a moderate-affinity substrate for OCT2, and its OCT2-mediated uptake was partially inhibited by cimetidine. Varenicline at high concentrations inhibited OCT2-mediated uptake, while its inhibitory effect on other transporters, including OAT1, OAT3, OCTN1, and OCTN2 was not observed in vitro. These in vitro data were further confirmed by a clinical DDI study showing that coadministration of cimetidine (1,200 mg/day) reduced the renal clearance of varenicline in 12 smokers, resulting in a 29.0% increase in systemic exposure. This was not considered clinically relevant, however, due to the safety margin of varenicline.

9.7 Conclusions and Perspective

The currently available in vitro test systems for transporters allow the investigation of the molecular mechanism by which drugs are eliminated via the kidney and liver. Although more transporters may be identified in the future, the currently identified ones provide a valuable framework to rationalize many of the transporter-mediated DDIs qualitatively and in some instances semi-quantitatively. In the pharmaceutical industry, a current challenge is to predict the susceptibility of development candidates to cause DDIs more quantitatively. Development of compounds with low liabilities in this regard is preferred from a safety, marketing, and regulatory perspective.

Despite the considerable progress made in the field of transporter research in the last decade, predicting transporter-mediated DDIs prospectively is still associated with a number of significant challenges, some of which can be summarized as follows: (i) Currently, there is a lack of and/or general access to specific, or at least selective, inhibitors for most transporters of interest. Availability of these would be valuable to dissect the relative contribution of transporters in drug uptake in for instance hepatocytes, which would allow more quantitative predictions of whether a drug could be a victim in DDIs. (ii) The lack of the availability of absolute protein concentration levels in recombinant systems, isolated primary cells, and human tissues. Attempts have been made to obtain this information by Western blotting but in most cases the information acquired was at best semi-quantitative and no absolute amounts could be determined due to a variety of technical limitations. Recently, several groups have been successful in measuring transporter levels using LC-MS/MS technologies. For instance, a quantitative body atlas was generated for a range of drug transporters in various mouse tissues (Kamiie et al., 2008), and using a similar approach, MRP2/Mrp2 protein levels were determined in liver from various species (Li et al., 2009). Correlating protein levels measured in tissues to those measured in in vitro systems can be used to calculate scaling factors which will be useful to more quantitatively assess the relative contribution of transporters to drug disposition both in vitro and in vivo, and therefore increase the capability to predict the likelihood that DDIs will be found. (iii) The examples provided in this chapter where DDI potential could be assessed were all for compounds excreted mainly as parent drug. The situation is far more complex for compounds undergoing both metabolism and transport such as, for example, atorvastatin (Lau et al., 2006). Availability of systems able to predict such complex interactions therefore is needed. (iv) In the expression systems available thus far, it is difficult to assess the relative contribution of uptake and efflux transporters in the *trans*-epithelial flux of drugs and therefore the rate-limiting step in excretion. Although progress has been made for human hepatocytes by the development of the sandwich culture system and double transfected cell lines, significant challenges still need to be overcome. A technology that holds promise is based on micropatterned cocultures of hepatocytes with 3T3 fibroblasts (Khetani and Bhatia, 2008). Hepatocytes cultured under these conditions are able to maintain phase I/II metabolism and canalicular transport over a period of several weeks. Potentially, therefore, microscale cultures may provide transporter data more representative for in vivo.

In conclusion, improvements are needed in our methodologies to predict transporter-mediated DDIs, but based on the examples provided above, it is likely that significant progress will be made in the coming years. Therefore, it is to be expected that exciting updates can be provided in future editions of this book.

References

Acharya P, O'Connor MP, Polli JW, Ayrton A, Ellens H and Bentz J (2008) Kinetic identification of membrane transporters that assist P-glycoprotein-mediated transport of digoxin and loperamide through a confluent monolayer of MDCKII-hMDR1 cells. *Drug Metab Dispos* **36**: 452–460

Allen JD, van Loevezijn A, Lakhai JM, van der Valk M, van Tellingen O, Reid G, Schellens JH, Koomen GJ and Schinkel AH (2002) Potent and specific inhibition of the breast cancer resistance protein multidrug transporter in vitro and in mouse intestine by a novel analogue of fumitremorgin C. *Mol Cancer Ther* **1**:417–425

Balimane PV, Marino A and Chong S (2008) P-gp inhibition potential in cell-based models: which "calculation" method is the most accurate?. *AAPS J* **10**:577–586

Borst P, Zelcer N, van de Wetering K and Poolman B (2007) On the putative cotransport of drugs by multidrug resistance proteins. *FEBS Lett* **580**:1085–1093

Byrne JA, Strautnieks SS, Mieli-Vergani G, Higgins CF, Linton KJ and Thompson RJ (2002) The human bile salt export pump: characterization of substrate specificity and identification of inhibitors. *Gastroenterology* **123**:1649–1658

Cai J and Gros P (2003) Overexpression, purification and functional characterization of ATP-binding cassette transporters in the yeast Pichia Pastoris. *Biochem Biophys Acta* **1610**:63–76

Chen ZS, Lee K, Walther S, Raftogianis RB, Kuwano M, Zeng H and Kruh GD (2002) Analysis of methotrexate and folate transport by multidrug resistance protein 4 (ABCC4): MRP4 is a component of the methotrexate efflux system. *Cancer Res* **62**:3144–3150

Chu XY, Bleasby K, Yabut J, Cai X, Chan GH, Hafey MJ, Xu S, Bergman AJ, Braun MP, Dean DC and Evers R (2007) Transport of the dipeptidyl peptidase-4 inhibitor sitagliptin by human organic anion transporter 3, organic anion transporting polypeptide 4C1, and multidrug resistance P-glycoprotein. *J Pharmacol Exp Ther* **321**:673–683

Chu XY, Huskey SE, Braun MP, Sarkadi B, Evans DC and Evers R (2004) Transport of ethinylestradiol glucuronide and ethinylestradiol sulfate by the multidrug resistance proteins MRP1, MRP2, and MRP3. *J Pharmacol Exp Ther* **309**:156–164

Chu XY, Liang Y, Cai X, Cuevas-Licea K, Rippley RK, Kassahun K, Shou M, Braun MP, Doss GA, Anari MR and Evers R (2009) Metabolism and renal elimination of gaboxadol in humans: role of UDP-glucuronosyltransferases and transporters. *Pharm Res* **26**:459–468

Cihlar T, Lin DC, Pritchard JB, Fuller MD, Mendel DB and Sweet DH (1999) The antiviral nucleotide analogs cidofovir and adefovir are novel substrates for human and rat renal organic anion transporter 1. *Mol Pharmacol* **56:**:570–580

Cole SP and Deeley RG (2006) Transport of glutathione and glutathione conjugates by MRP1. *Trends Pharmacol Sci* **27**:438–446

Cui Y, König J and Keppler D (2001) Vectorial transport by double-transfected cells expressing the human uptake transporter SLC21A8 and the apical export pump ABCC2. *Mol Pharmacol* **60**:934–943

El-Sheikh AA, Masereeuw R, and Russel FG. (2008) Mechanisms of renal anionic transport. *Eur J Pharamcol* **585**:245–255

Evers R and Chu XY (2008) Role of the murine organic anion-transporting polypeptide 1b2 (Oatp1b2) in drug disposition and hepatotoxicity. *Mol Pharmacol* **74**:309–311

Feng B, Obach RS, Burstein AH, Clark DJ, de Morais SM and Faessel HM (2008) Effect of human renal cationic transporter inhibition on the pharmacokinetics of varenicline, a new therapy for smoking cessation: an in vitro-in vivo study. *Clin Pharmacol Ther* **83**: 567–576

Ghibellini G, Leslie EM and Brouwer KL (2006) Methods to evaluate biliary excretion of drugs in humans: an updated review. *Mol Pharm* **3**:198–211

Ghibellini G, Vasist LS, Leslie EM, Heizer WD, Kowalsky RJ, Calvo BF and Brouwer KL (2007) In vitro-in vivo correlation of hepatobiliary drug clearance in humans. *Clin Pharmacol Ther* **81**:406–413

Glavinas H, Méhn D, Jani M, Oosterhuis B, Herédi-Szabó K and Krajcsi P (2008) Utilization of membrane vesicle preparations to study drug-ABC transporter interactions. *Expert Opin Drug Metab Toxicol* **4**:721–732

Gorboulev V, Ulzheimer JC, Akhoundova A, Ulzheimer-Teuber I, Karbach U, Quester S, Baumann C, Lang F, Busch AE and Koepsell H (1997) Cloning and characterization of two human polyspecific organic cation transporters. *DNA Cell Biol* **16**:871–881

Hagenbuch B and Meier PJ (1994) Molecular cloning, chromosomal localization, and functional characterization of a human liver Na+/bile acid cotransporter. *J Clin Invest* **93**:1326–1331

Hayer-Zillgen M, Brüss M and Bönisch H (2002) Expression and pharmacological profile of the human organic cation transporters hOCT1, hOCT2 and hOCT3. *Br J Pharmacol* **136**: 829–836

Hewitt NJ, Lechón MJ, Houston JB, Hallifax D, Brown HS, Maurel P, Kenna JG, Gustavsson L, Lohmann C, Skonberg C, Guillouzo A, Tuschl G, Li AP, LeCluyse E, Groothuis GM and Hengstler JG (2007) Primary hepatocytes: current understanding of the regulation of metabolic enzymes and transporter proteins, and pharmaceutical practice for the use of hepatocytes in metabolism, enzyme induction, transporter, clearance, and hepatotoxicity studies. *Drug Metab Rev* **39**:159–234

Hirano M, Maeda K, Shitara Y and Sugiyama Y (2004) Contribution of OATP2 (OATP1B1) and OATP8 (OATP1B3) to the hepatic uptake of pitavastatin in humans. *J Pharmacol Exp Ther* **311**:139–146

Hirano M, Maeda K, Shitara Y and Sugiyama Y (2006) Drug-drug interaction between pitavastatin and various drugs via OATP1B1. *Drug Metab Dispos* **34**:1229–1236

Ishiguro N, Maeda K, Saito A, Kishimoto W, Matsushima S, Ebner T, Roth W, Igarashi T and Sugiyama Y (2008) Establishment of a set of double transfectants coexpressing organic anion transporting polypeptide 1B3 and hepatic efflux transporters for the characterization of the hepatobiliary transport of telmisartan acylglucuronide. *Drug Metab Dispos* **36**:796–805

Ishikawa T, Müller M, Klünemann C, Schaub T and Keppler D (1990) ATP-dependent primary active transport of cysteinyl leukotrienes across liver canalicular membrane. Role of the ATP-dependent transport system for glutathione S-conjugates. *J Biol Chem* **265**:19279–19286

Jedlitschky G, Hoffmann U and Kroemer HK (2006) Structure and function of the MRP2 (ABCC2) protein and its role in drug disposition. *Expert Opin Drug Metab Toxicol* **2**:351–366

Kalvass JC and Pollack GM (2007) Kinetic considerations for the quantitative assessment of efflux activity and inhibition: implications for understanding and predicting the effects of efflux inhibition. *Pharm Res* **24**:265–276

Kamiie J, Ohtsuki S, Iwase R, Ohmine K, Katsukura Y, Yanai K, Sekine Y, Uchida Y, Ito S and Terasaki T (2008) Quantitative atlas of membrane transporter proteins: development and application of a highly sensitive simultaneous LC/MS/MS method combined with novel in-silico peptide selection criteria. *Pharm Res* **25**:1469–1483

Khetani SR and Bhatia SN (2008) Microscale culture of human liver cells for drug development. *Nat Biotechnol* **26**:120–126

Kopplow K, Letschert K, König J, Walter B and Keppler D (2005) Human hepatobiliary transport of organic anions analyzed by quadruple-transfected cells. *Mol Pharmacol* **68**: 1031–1038

König J, Cui Y, Nies AT and Keppler D (2000) A novel human organic anion transporting polypeptide localized to the basolateral hepatocyte membrane. *Am J Physiol Gastrointest Liver Physiol* **278**:G156–G164

Lau YY, Huang Y, Frasetto L and Benet LZ (2006) Effect of OATP1B1 transporter inhibition on the pharmacokinetics of atorvastatin in healthy volunteers. *Clin Pharmacol Ther* **81**:194–204

LeCluyse EL, Audus KL and Hochman JH (1994) Formation of extensive canalicular networks by rat hepatocytes cultured in collagen-sandwich configuration. *Am J Physiol* **266**:C1764–C1774

Lee W and Kim RB (2004) Transporters and renal drug elimination. *Ann Rev Pharmacol Toxicol* **44**:137–166

Li M, Anderson GD and Wang J (2006) Drug-drug interactions involving membrane transporters in the human kidney. *Expert Opin Drug Metab Toxicol* **2**:505–532

Li N, Zhang Y, Hua F and Lai Y (2009) Absolute difference of hepatobiliary transporter multidrug resistance-associated protein (MRP2/Mrp2) in liver tissues and isolated hepatocytes from rat, dog, monkey and human. *Drug Metab Dispos* **37**:66–73

Link E, Parish S, Armitage J, Bowman L, Heath S, Matsuda F, Gut I, Lathrop M and Collins R (2008) SLCO1B1 variants and statin-induced myopathy – a genome wide study. *N Engl J Med* **359**:789–799

Liu X, Chism JP, LeCluyse EL, Brouwer KR and Brouwer KL (1999a) Correlation of biliary excretion in sandwich-cultured rat hepatocytes and in vivo in rats. *Drug Metab Dispos* **27**: 637–644

Liu X, LeCluyse EL, Brouwer KR, Gan LS, Lemasters JJ, Stieger B, Meier PJ and Brouwer KL (1999b) Biliary excretion in primary rat hepatocytes cultured in a collagen-sandwich configuration. *Am J Physiol* **277**:G12–G21

Matsushima S, Maeda K, Inoue K, Ohta KY, Yuasa H, Kondo T, Nakayama H, Horita S, Kusuhara H and Sugiyama Y (2009) The inhibition of human multidrug and toxin extrusion 1 is involved in the drug-drug interaction caused by cimetidine. *Drug Metab Dispos* **37**:555–559

Matsushima S, Maeda K, Kondo C, Hirano M, Sasaki M, Suzuki H and Sugiyama Y (2005) Identification of the hepatic efflux transporters of organic anions using double-transfected Madin-Darby canine kidney II cells expressing human organic anion-transporting polypeptide 1B1 (OATP1B1)/multidrug resistance-associated protein 2, OATP1B1/multidrug resistance 1, and OATP1B1/breast cancer resistance protein. *J Pharmacol Exp Ther* **314**:1059–1067

Mikkaichi T, Suzuki T, Onogawa T, Tanemoto M, Mizutamari H, Okada M, Chaki T, Masuda S, Tokui T, Eto N, Abe M, Satoh F, Unno M, Hishinuma T, Inui K, Ito S, Goto J and Abe T (2004) Isolation and characterization of a digoxin transporter and its rat homologue expressed in the kidney. *Proc Natl Acad Sci USA* **101**:3569–3574

Mita S, Suzuki H, Akita H, Stieger B, Meier PJ, Hofmann AF and Sugiyama Y (2005) Vectorial transport of bile salts across MDCK cells expressing both rat Na+-taurocholate cotransporting polypeptide and rat bile salt export pump. *Am J Physiol Gastrointest Liver Physiol* **288**: G159–G167

Mück W, Mai I, Fritsche L, Ochmann K, Rohde G, Unger S, Johne A, Bauer S, Budde K, Roots I, Neumayer HH and Kuhlmann J (1999) Increase in cerivastatin systemic exposure after single and multiple dosing in cyclosporine-treated kidney transplant recipients. *Clin Pharmacol Ther* **65**:251–261

Neuvonen PJ, Niemi M and Backman J (2006) Drug interactions with lipid- lowering drugs: Mechanism and clinical relevance. *Clin Pharmacol Ther* **80**:565–581

Nies AT, Herrmann E, Brom M and Keppler D (2008) Vectorial transport of the plant alkaloid berberine by double-transfected cells expressing the human organic cation transporter 1 (OCT1, SLC22A1) and the efflux pump MDR1 P-glycoprotein (ABCB1).. *Naunyn Schmiedebergs Arch Pharmacol* **376**:449–461

Poirier A, Lave T, Portmann R, Brun ME, Senner F, Kansy M, Grimm HP and Funk C (2008) Design, data analysis, and simulation of in vitro drug transport kinetic experiments using a mechanistic in vitro model. *Drug Metab Dispos* **36**:2434–2444

Rautio J, Humphreys JE, Webster LO, Balakrishnan A, Keogh JP, Kunta JR, Serabjit-Singh CJ and Polli JW (2006) In vitro P-glycoprotein inhibition assays for assessment of clinical drug interaction potential of new drug candidates: a recommendation for probe substrates. *Drug Metab Dispos* **34**:786–792

Regazzi MB, Iacona I, Campana C, Raddato V, Lesi C, Perani G, Gavazzi A and Vigano M (1993) Altered disposition of pravastatin following concomitant drug therapy with cyclosporin A in transplant recipients. *Transplant Proc* **25**:2732–2734

Roelofsen H, Bakker CT, Schoemaker B, Heijn M, Jansen PL and Elferink RP (1995) Redistribution of canalicular organic anion transport activity in isolated and cultured rat hepatocytes. *Hepatology* **21**:1649–1657

Russel FG, Masereeuw R and van Aubel RA (2002) Molecular aspects of renal anionic drug transport. *Ann Rev Physiol* **64**:563–594

Saito H, Hirano H, Nakagawa H, Fukami T, Oosumi K, Murakami K, Kimura H, Kouchi T, Konomi M, Tao E, Tsujikawa N, Tarui S, Nagakura M, Osumi M and Ishikawa T (2006) A new strategy of high-speed screening and quantitative structure-activity relationship analysis to evaluate human ATP-binding cassette transporter ABCG2-drug interactions. *J Pharmacol Exp Ther* **317**:1114–1124

Sasaki M, Suzuki H, Aoki J, Ito K, Meier PJ and Sugiyama Y (2004) Prediction of in vivo biliary clearance from the in vitro transcellular transport of organic anions across a double-transfected Madin-Darby canine kidney II monolayer expressing both rat organic anion transporting polypeptide 4 and multidrug resistance associated protein 2. *Mol Pharmacol* **66:**:450–459

Sasaki M, Suzuki H, Ito K, Abe T and Sugiyama Y (2002) Transcellular transport of organic anions across a double-transfected Madin-Darby canine kidney II cell monolayer expressing both human organic anion-transporting polypeptide (OATP2/SLC21A6) and Multidrug resistance-associated protein 2 (MRP2/ABCC2. *J Biol Chem* **277**:6497–6503

Sato T, Masuda S, Yonezawa A, Tanihara Y, Katsura T and Inui K (2008) Transcellular transport of organic cations in double-transfected MDCK cells expressing human organic cation transporters hOCT1/hMATE1 and hOCT2/hMATE1. *Biochem Pharmacol* **76**:894–903

Shitara Y, Itoh T, Sato H, Li AP and Sugiyama Y (2003a) Inhibition of transporter-mediated hepatic uptake as a mechanism for drug-drug interaction between cerivastatin and cyclosporin A. *J Pharmacol Exp Ther* **304**:610–616

Shitara Y, Li AP, Kato Y, Lu C, Ito K, Itoh T and Sugiyama Y (2003b) Function of uptake transporters for taurocholate and estradiol 17beta-D-glucuronide in cryopreserved human hepatocytes. *Drug Metab Pharmacokinet* **18**:33–41

Shu Y, Sheardown SA, Brown C, Owen RP, Zhang S, Castro RA, Ianculescu AG, Yue L, Lo JC, Burchard EG, Brett CM, and Giacomini KM. (2007) Effect of genetic variation in the organic cation transporter 1(OCT1) in metformin action. *J Clin Invest* **117**:1422–1431

Smit JW, Weert B, Schinkel AH and Meijer DK (1998) Heterologous expression of various P-glycoproteins in polarized epithelial cells induces directional transport of small (type 1) and bulky (type 2) cationic drugs. *J Pharmacol Exp Ther* **286**:321–327

Somogyi A, Stockley C, Keal J, Rolan P and Bochner F (1987) Reduction of metformin renal tubular secretion by cimetidine in man. *Br J Clin Pharmacol* **23**:545–551

Steck TL, Weinstein RS, Straus JH and Wallach DF (1970) Inside-out red cell membrane vesicles: preparation and purification. *Science* **168**:255–257

Tahara H, Kusuhara H, Endou H, Koepsell H, Imaoka T, Fuse E and Sugiyama Y (2005) A species difference in the transport activities of H2 receptor antagonists by rat and human renal organic anion and cation transporters. *J Pharmacol Exp Ther* **315:**:337–345

Tamai I, Nozawa T, Koshida M, Nezu J, Sai Y and Tsuji A (2001) Functional characterization of human organic anion transporting polypeptide B (OATP-B) in comparison with liver-specific OATP-C. *Pharm Res* **18**:1262–1269

Terada T and Inui K (2007) Physiological and pharmacokinetic roles of H+/organic cation antiporters (MATE/SLC47A. *Biochem Pharmacol* **75**:1689–1696

Troutman MD and Thakker DR (2003) Novel Experimental Parameters to quantify the modulation of absorptive and secretory transport of compounds by P-glycoprotein in cell culture models of intestinal epithelium. *Pharm Res* **20**:1210–1224

Vavricka SR, Van Montfoort J, Ha HR, Meier PJ and Fattinger K (2002) Interactions of rifamycin SV and rifampicin with organic anion uptake systems of human liver. *Hepatology* **36**:164–172

Webborn PJ, Parker AJ, Denton RL and Riley RJ (2007) In vitro-in vivo extrapolation of hepatic clearance involving active uptake: theoretical and experimental aspects. *Xenobiotica* **37**: 1090–1109

Yamazaki M, Neway WE, Ohe T, Chen I, Rowe JF, Hochman JH, Chiba M and Lin JH (2001) In vitro substrate identification studies for p-glycoprotein-mediated transport: species difference and predictability of in vivo results. *J Pharmacol Exp Ther* **296**:723–735

Zelcer N, Saeki T, Reid G, Beijnen JH and Borst P (2001) Characterization of drug transport by the human multidrug resistance protein 3 (ABCC3. *J Biol Chem* **276**:46400–46407

Zeng H, Liu G, Rea PA and Kruh GD (2000) Transport of amphipathic anions by human multidrug resistance protein 3. *Cancer Res* **60**:4779–4784

Zhang L, Dresser MJ, Gray AT, Yost SC, Terashita S and Giacomini KM (1997) Cloning and functional expression of a human liver organic cation transporter. *Mol Pharmacol* **51**:913–921

Zhou SF, Wang LL, Di YM, Xue CC, Duan W, Li CG and Li Y (2008) Substrates and inhibitors of human multidrug resistance associated proteins and the implications in drug development. *Curr Med Chem* **15**:1981–2039

Chapter 10
In Vitro Techniques to Study Drug–Drug Interactions Involving Transport: Caco-2 Model for Study of P-Glycoprotein and Other Transporters

William R. Proctor, Xin Ming, and Dhiren R. Thakker

Abstract Transporters play a pivotal role in defining disposition, therapeutic outcome, and adverse reactions of medicines. P-glycoprotein (P-gp) is an efflux transporter that is constitutively expressed in barrier tissues and excretory organs, and has wide substrate specificity; consequently, it may be responsible for wide-ranging drug–drug interactions (DDIs). This chapter critically evaluates the utility and limitations of Caco-2 cells, a model for human intestinal epithelium that expresses most intestinal transporters, in assessing DDIs mediated by P-gp and other transporters.

10.1 Introduction

Efficacy of therapeutic agents depends not only on the potency and selectivity of their interactions with biological targets but also on their ability to attain certain threshold target concentrations for a desired period of time. The processes that define the ability of compounds to attain and sustain the desired systemic and target organ concentrations, i.e., absorption, distribution, metabolism, and excretion, play a pivotal role in therapeutic outcome of drugs. For orally administered drugs, which account for the vast majority of marketed drugs, absorption is the first critical step for therapeutic efficacy. Intestinal epithelium presents a formidable barrier to drug absorption (Madara and Trier, 1994).

Intestinal epithelium contains a wide variety of transporters that mediate absorption of nutrients (reviewed in (Oh et al., 1999). Drugs that can act as substrates for these transporters are absorbed across the intestinal epithelium even if they are not able to permeate the cell membrane. Unlike the absorptive transporters, the efflux transporters in the apical membrane of enterocytes, most notably P-glycoprotein

D.R. Thakker (✉)
Division of Molecular Pharmaceutics, Eshelman School of Pharmacy, The University of North Carolina at Chapel Hill, Chapel Hill, NC, USA
e-mail: dthakker@unc.edu; dhiren_thakker@unc.edu

K.S. Pang et al. (eds.), *Enzyme- and Transporter-Based Drug–Drug Interactions*, 257
DOI 10.1007/978-1-4419-0840-7_10,

[P-gp; multidrug resistance 1 (MDR1); ABCB1], attenuate absorption of their substrates. First discovered by Juliano and Ling (1976) as a trans-membrane protein that is overexpressed in cells that are treated with high doses of cytotoxic compounds, this efflux pump was found to confer multidrug resistance (MDR) in mammalian cell lines treated with high doses of chemotherapeutic agents (Kartner et al., 1983). It was discovered that P-gp is constitutively expressed in many barrier tissues including the blood–brain barrier and the intestinal epithelium (Thiebaut et al., 1987) as well as in excretory organelle such as bile canaliculus in hepatocytes (Chandra and Brouwer, 2004). Therefore, it is not surprising that P-gp and related transporters can play a significant role in the disposition of drugs in the body and can contribute to serious drug–drug interactions (DDIs) (Troutman et al., 2008).

Caco-2 cells, which originated from human colon adenocarcinoma cells, were developed as a cell-based model to study transport and absorption across human intestinal epithelium (Hidalgo et al., 1989; Artursson, 1990). These cells, like enterocytes in the intestinal epithelium, express P-gp on their apical membrane (Augustijns et al., 1993; Hunter et al., 1993). Because of the ease with which various transport characteristics of drugs can be studied with Caco-2 cell monolayers grown in a Transwell™ setup, these cells have been extensively used to study mechanisms of drug transport, including the role of P-gp and other efflux transporters in drug disposition (Gan and Thakker, 1997). In this chapter, the use of Caco-2 cell model to study the role of P-gp and other transporters in absorption-related and other DDIs is critically examined.

10.2 Role of P-gp in Drug Disposition

After the withdrawal of terfenadine from market in the late 1990s, which was attributed to CYP3A4-mediated DDI (Huang et al., 2004), there has been a widespread awareness of the role of cytochrome P450 in DDIs. The attention given to drug-metabolizing enzymes as mediators of DDIs is justified based on the available data on over 400 marketed drugs in the United States and Europe, which show that cytochrome P450-mediated metabolism accounts for approximately 55% of the all the clearance mechanisms of all marketed drugs (Clarke and Jones, 2008). However, there is increasing awareness about the role played by transporters in renal and biliary excretion, and in intestinal absorption of drugs (Chandra and Brouwer, 2004; Shitara et al., 2006). Hence, it is reasonable to expect that transporters would play an important role in causing DDIs. Among many transporters that play a role in the disposition of drugs, P-gp has been the most extensively investigated transporter (Troutman et al., 2008).

10.2.1 Tissue Distribution of P-gp

P-gp is constitutively expressed in barrier tissues such as intestinal epithelium, placenta, capillary endothelium in the blood–tissue barriers (e.g., brain, testis), and in

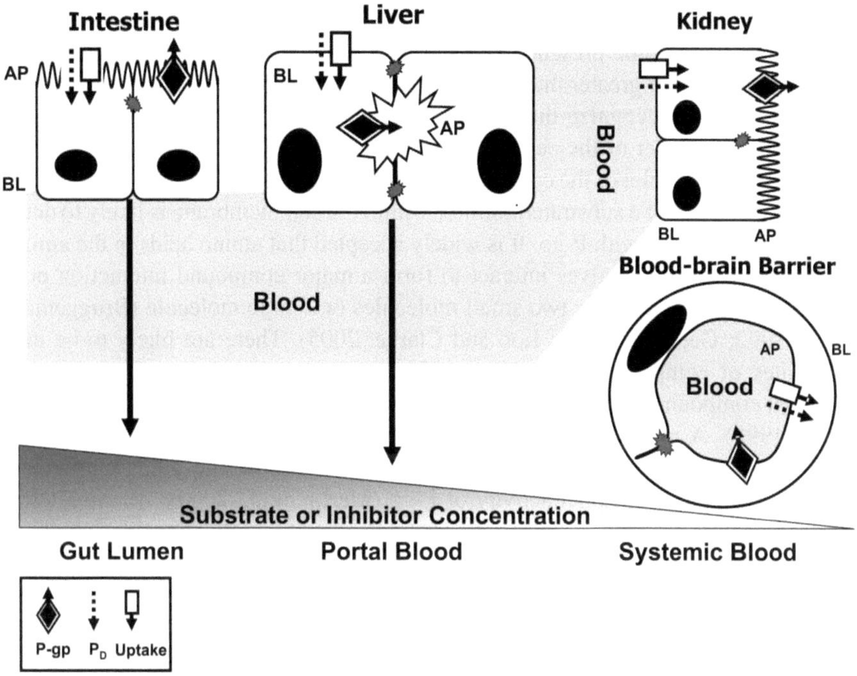

Fig. 10.1 Relative substrate/inhibitor concentrations experienced by P-gp in the intestine, the liver, the kidney, and the blood–brain barrier. A compound can enter the cell via an uptake transporter and/or through passive diffusion (P_D); P-gp attenuates its entry into barrier tissues (e.g., the intestine and the blood–brain barrier) or accelerates its excretion by the liver and the kidney

organs of excretion (canalicular membrane of hepatocytes and epithelium in renal proximal tubules) (Thiebaut et al., 1987; Cordon-Cardo et al., 1990). The pattern of localization at the tissue/organ level and at the cellular level reveals two major functions of P-gp, i.e., protection against entry of xenobiotics into critical tissues/organs and excretion of compounds via the biliary or the renal route (Fig. 10.1). Based on the observations regarding pharmacokinetic behavior of certain drugs (Gramatte et al., 1996; Mayer et al., 1996; Terao et al., 1996), it is now believed that its presence in the intestinal epithelium may serve both these roles of preventing absorption from intestinal lumen and secretion of compounds and metabolic products into the intestinal lumen.

10.2.2 Substrates and Inhibitors of P-gp

P-gp displays broad substrate specificity with respect to the chemical space and size of the substrates. In fact, the broad substrate specificity makes it difficult to define structural characteristics of good substrates and/or inhibitors of P-gp. Hence, only broad structural attributes can be used to describe P-gp's substrates/inhibitors, such as the presence of aromatic groups or other planar domains, amphiphilic or lipophilic behavior, the presence of a nitrogen group or a net positive charge, and molecular weight of greater than 300 Da (Litman et al., 1997; Etievant et al., 1998). It is important to recognize that substrates bind to P-gp while they are associated with the inner leaflet of the cell membrane (Homolya et al., 1993) or at the interface of the inner leaflet of the cell membrane and the cytosol (Altenberg et al., 1994). Thus the behavior of a substrate/inhibitor within the cell membrane is likely to determine its interactions with P-gp. It is widely accepted that amino acids in the amino- and carboxy-terminal halves interact to form a major compound interaction pore, capable of accommodating two small molecules or a large molecule (Bruggemann et al., 1992; Germann, 1996; Loo and Clarke, 2005). There are likely to be multiple sites of compound recognition within this compound interaction pore, and different compounds can bind to distinct, perhaps allosterically coupled sites (Ferry et al., 1995). A minimum of four distinct binding domains have been proposed, three of which are substrate-binding sites and the fourth one binds compounds that modulate the functional activity of P-gp (Martin et al., 2000a; b). An alternative explanation for multiple substrate-binding/substrate-modulating sites invokes a concept of substrate-induced binding, in which the substrates could affect conformational changes by interactions with amino acid residues within the binding pore to enhance the substrate–P-gp interactions (Loo and Clarke, 2000). Our understanding of how P-gp binds and affects their efflux continues to evolve as evidenced by a recent commentary (Sheps, 2009).

10.2.3 P-gp and Drug Disposition

With its wide substrate specificity and its widespread expression, P-gp can have a major influence on trafficking of drug molecules in the body. P-gp can affect absorption of compounds across the intestinal epithelium, distribution into organs, tissues, and cells, metabolism in the enterocytes and hepatocytes by influencing the availability of substrates to intracellular metabolic enzymes, and excretion (secretion) from the liver, the kidney, and the intestine.

Evidence has accumulated over the past several years that P-gp can efflux orally administered drugs back into the intestinal lumen during their absorption across the intestinal epithelium and therefore reduce their oral absorption (Relling, 1996; Sparreboom et al., 1997). For example, AUC values of paclitaxel obtained in *mdr1a*(–/–) animals after 10 mg/kg oral dose were several-fold greater than those obtained in the *mdr1a*(+/+) mice (Sparreboom et al., 1997). The oral bioavailability increased from 11% in the P-gp-competent mice to 35% in P-gp-deficient mice (Sparreboom et al., 1997).

The distribution of drugs into those tissues that are protected by a blood–tissue barrier (e.g., blood–brain barrier) is significantly influenced by P-gp. For example, vinblastine concentrations in the brains of *mdr1a*(–/–) mice after 1 mg/kg dose were ~20-fold higher than those in the *mdr1a*(+/+) mice (Schinkel et al., 1994). In the same study, the brain concentrations of ivermectin were found to be ~100-fold higher in *mdr1a(–/–)* mice compared to those in the wild-type mice. In a separate study, the brain tissue concentrations of 2,5-D-penicillamine enkephalin were found to be two- to fourfold higher in *mdr1a(–/–)* mice compared to those in the wild-type mice, and the dose required to elicit a comparable pharmacodynamic response was ~30-fold lower in the P-gp knockout mice (Chen and Pollack, 1998).

In the intestinal tissue, where P-gp can effectively influence the access of its substrates to intracellular drug-metabolizing enzymes, it can have a significant effect on the metabolism of dual P-gp/metabolic enzyme substrate. This was first demonstrated in Caco-2 cells when it was observed that oxidative metabolism of cyclosporin A was significantly higher during apical to basolateral transport than during basolateral to apical transport (Gan et al., 1996). The modulation of CYP3A-mediated metabolism of compounds by P-gp in intestinal tissue or CYP3A-competent Caco-2 cells for dual substrates has been demonstrated in several studies (Gan et al., 1996; Fisher et al., 1999; Cummins et al., 2002). However, whether P-gp contributes to increase or decrease in the CYP3A-mediated metabolism of compounds remains to be elucidated, as conflicting views have been expressed in the literature (Benet et al., 1999; Johnson et al., 2003; Tam et al., 2003; Knight et al., 2006).

While P-gp's role in hepatic and renal excretion of drugs and metabolites is well established (Chandra and Brouwer, 2004; Shitara et al., 2006), its importance in secretion/excretion of compounds into intestinal lumen (Arimori and Nakano, 1998) is not as well recognized. To the extent that intestine can act as an excretory organ, P-gp is likely to play a significant role in this function. For example, after intravenous administration of vinblastine (1 mg/kg) to P-gp-competent mice, 25% of the dose was eliminated as intact drug in the feces over a 24-hr period, whereas in the P-gp knockout mice only 9% of the dose was recovered in the feces over the same time period (van Asperen et al., 2000). Similar decrease in intestinal excretion occurred in *mdr1a(–/–)* compared to wild-type mice for paclitaxel, digoxin, and doxorubicin (Mayer et al., 1996; van Asperen et al., 1996; Sparreboom et al., 1997).

10.3 P-gp and DDI

P-gp-mediated efflux is a saturable process, which raises the potential for a DDI when two substrates or a substrate and an inhibitor are co-administered. However, unlike the DDIs involving drug-metabolizing enzymes, the DDIs with P-gp are likely to be much more complex and therefore more difficult to predict based on the information on the binding affinities, structure–transport relationships, or even in vitro experimental determinations of DDIs. First, P-gp exhibits broad substrate specificity due to either promiscuous binding site and/or multiple binding sites. Second, P-gp can be inhibited by interfering with substrate binding or by interfering with ATP binding and/or hydrolysis. Third, P-gp is widely distributed in the body and can affect absorption, distribution, metabolism, excretion, or any combination of these processes for an orally administered drug that is a P-gp substrate.

Of all the sites where P-gp operates, intestinal epithelium is the most likely site for DDIs because P-gp is likely to be exposed to relatively high concentrations of drugs achieved in the intestinal lumen after an oral dose. It is difficult to alter the disposition of a P-gp substrate by modulating P-gp activity at other sites with a co-administered substrate or inhibitor because it would be difficult to achieve high enough concentrations to saturate the efflux transporter (or inhibit majority of its activity) at these sites (Fig. 10.1). This is borne out by the fact that most of the attempts in the clinic to improve therapeutic efficacy of chemotherapeutic agents by inhibiting overexpressed P-gp in the tumor tissue with a co-administered P-gp inhibitor have not succeeded (see (Troutman et al., 2008)).

A few examples of clinically observed DDIs that can be attributed to P-gp do exist. For example, renal clearance of digoxin was decreased by co-administered clarithromycin. In vitro studies using kidney epithelial cell lines confirmed that the decrease in digoxin clearance by clarithromycin was due to inhibition of P-gp in the kidney (Wakasugi et al., 1998). Similarly, the P-gp modulator verapamil also decreased renal clearance of digoxin (Ito et al., 1993). Cyclosporin A, a P-gp modulator, caused significant increase in the systemic toxicity of the chemotherapeutic regimen involving three P-gp substrates: vincristine/dactinomycin/cyclophosphamide. The increased toxicity was attributed to decreased clearance and increased systemic exposure to the chemotherapeutic agents caused by P-gp inhibition due to co-administered cyclosporin A (Theis et al., 1998). Whether these DDIs could have been predicted using an in vitro assay for P-gp inhibition remains to be determined.

An additional factor that could modulate P-gp activity, and therefore alter drug disposition and resultant change in pharmacodynamic response or toxicity, involves genetic polymorphisms of the *MDR1* gene. As many as 28 single nucleotide polymorphisms have been identified on the *MDR1* gene, and a functional analysis of many of these polymorphism has been performed (Hoffmeyer et al., 2000; Sakaeda et al., 2003; Troutman et al., 2008). These studies have revealed a wide-ranging effect of these polymorphisms on the expression, localization, and function of P-gp (see Troutman et al., 2008). When a P-gp-related DDI is suspected for a drug, it is important to recognize that the unanticipated change in the disposition or pharmacodynamic response may be due to a polymorphism in the *MDR1* gene and not due to a DDI.

Clearly, with a high potential for P-gp to affect the disposition and therapeutic outcome of drug candidates, good in vitro tools need to be developed to assess the DDI potential involving P-gp during the lead identification and/or the lead optimization phase. As discussed in this chapter, it is extremely difficult to anticipate how modulation of P-gp by a co-administered substrate and/or inhibitor will influence the disposition of a P-gp substrate. However, well-designed in vitro studies can provide valuable information about the DDI potential of a drug candidate. The following section describes the Caco-2 cell-based model that can be used to assess the potential of drug candidates to encounter a P-gp-related DDI and provides a critical evaluation of the utility and pitfalls of this model.

10.4 In Vitro Assessment of Transporter-Related DDI

Cellular models have been extensively used to determine if a compound is a substrate or an inhibitor of P-gp. These models also determine if its interactions with P-gp will affect its absorption across intestinal epithelium, or excretion into the bile, or distribution into the brain. However, in this section, only Caco-2 cell model will be discussed.

10.4.1 Caco-2 Cell Model

Drugs can permeate across the intestinal epithelium via two pathways: paracellular and transcellular. The paracellular pathway is size limited due to the presence of the tight junction (TJ) and is typically favored by small hydrophilic drugs and ions (Knipp et al., 1997). When establishing a Caco-2 cell model that mimics intestinal epithelium, it is critical to characterize the "leakiness" of the monolayer in order to adequately account for this pathway in the overall transport. Electrophysiological measurements, such as transepithelial electrical resistance (TEER), are commonly used to assess monolayer integrity prior to and immediately following a transport experiment. TEER values can vary significantly across different laboratories (Walter and Kissel, 1995) and across different Caco-2 cell clones (Fleet et al., 2002). In addition, TEER values are dependent not only on monolayer integrity but also on the relative expression of a class of TJ proteins, claudins. Specific claudin isoforms can gate or facilitate sodium flux, in turn altering the resistance across the monolayer while not altering the paracellular transport of small molecules (Colegio et al., 2003; Van Itallie and Anderson, 2006). A more direct and reliable measurement of paracellular permeability is performed by measuring flux of a paracellular probe compound (e.g., mannitol, inulin, dextran, lucifer yellow, and PEG 4000).

Transcellular transport occurs by passive diffusion across apical and basolateral cell membranes and can be enhanced or attenuated by transporters at both membranes. In addition to P-gp, Caco-2 cells express many intestinal transporters that include both uptake and efflux processes. A brief description of the transporters and metabolic enzymes expressed in Caco-2 cell monolayers is provided below as this subject matter is covered in an excellent recent review by Sun et al. (2008). Although mRNA has been detected for many intestinal transporters in Caco-2 cells (Englund et al., 2006; Seithel et al., 2006; Hilgendorf et al., 2007; Maubon et al., 2007; Hayeshi et al., 2008), protein expression, localization, and function have yet to be determined for the majority of the transporters identified in these studies. The transporters that exhibit message (mRNA), protein expression, and/or functional activity in Caco-2 cells are depicted in Fig. 10.2. On the apical membrane, the organic anion-transporting polypeptide 2B1 (OATP2B1) (Sai et al., 2006), organic cation transporter 3 (OCT3) (Muller et al., 2005), organic cation/carnitine transporter (OCTN2) (Elimrani et al., 2003), H^+-di–tripeptide transporter 1 (PEPT1) (Saito et al., 1997), monocarboxylic acid transporter 1 (MCT1)

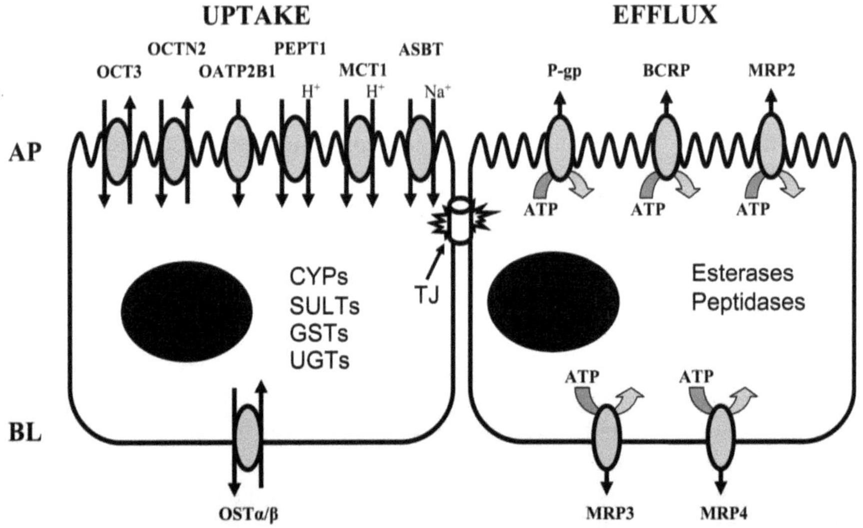

Fig. 10.2 Apical and basolateral membrane transporters and metabolizing enzymes present in Caco-2 cell monolayers

(Tamai et al., 1995), and the apical Na^+-dependent bile acid transporter (ASBT) (Hidalgo and Borchardt, 1990) facilitate uptake of drugs into the cells. The apical membrane also contains several ATP-binding cassette efflux transporters: P-gp (Anderle et al., 1998; Engman et al., 2001), breast cancer resistance protein (BCRP) (Xia et al., 2005), and multidrug resistance-associated protein 2 (MRP2) (Gutmann et al., 1999; Prime-Chapman et al., 2004). The relative expression of P-gp in Caco-2 cells is comparable (within twofold) to human intestine and other polarized cell models such as MDCK-MDR1 cell line (Troutman and Thakker, 2003b); therefore, this model can be used to predict potential intestinal P-gp interactions. The presence of BCRP in Caco-2 cells has been controversial. Xia et al. (2005) have provided evidence for functional activity of BCRP in Caco-2 cells, whereas several other studies have reported significantly lower expression (mRNA) in Caco-2 cells than in the human intestine (Hilgendorf et al., 2007; Maubon et al., 2007) with one study reporting almost 100-fold lower relative expression (Taipalensuu et al., 2001). A recent report confirmed the finding by Xia et al. (2005) of the functional activity of BCRP in Caco-2 cell monolayers by creating a stable knockdown BCRP Caco-2 cell line without altering the expression of P-gp or MRP2 (Zhang et al., 2009). It appears that certain Caco-2 cell clones, but not all clones, contain functionally active BCRP on the apical membrane. Hence, the role of this transporter in drug absorption or in DDIs may be underestimated by Caco-2 cell model unless a BCRP-expressing clone is selected for these studies.

The basolateral membrane contains MRP3 (Prime-Chapman et al., 2004), MRP4 (Ming and Thakker, 2009), and potentially MRP6 (mRNA only (Prime-Chapman et al., 2004)) that facilitate efflux of anionic and/or zwitterionic compounds from the cytosol to the basolateral compartment. In addition, the organic solute transporters (OSTα/β) are present in Caco-2 cells (Okuwaki et al., 2007) and function on the basolateral membrane. On the other hand, uptake transporters that may facilitate the secretion of cations from the serosal to the lumenal side have yet to be identified. Although there are some reports of the presence of OCT1 (mRNA) (Muller et al., 2005; Englund et al., 2006), direct localization and functional activity have yet to be determined. There does not appear to be any functionally active organic anion transporters (OATs) that facilitate uptake of anions across the basolateral membrane, although one cannot rule out the presence of anion secretory processes across Caco-2 cell monolayers.

In addition to the various transporters, several drug-metabolizing enzymes are also present in Caco-2 cell monolayers, including sulfotransferases (SULTs) (Meinl et al., 2008), cytochrome P450 1A1 (CYP1A1) (Schreiber et al., 2006), glutathione *S*-transferases (GSTs) (Peters and Roelofs, 1989), and UDP-glucuronosyltransferases (UGTs) (Munzel et al., 1999). Caco-2 cells also express several esterases, including carboxylesterase-1 and carboxylesterase-2, that have been implicated in prodrug conversion in vitro (Imai et al., 2005). Under normal culture conditions, Caco-2 cells do not express functionally active cytochrome P450 3A4 (CYP3A4), the predominant drug-metabolizing enzyme in the intestine (Paine et al., 2006). CYP3A4 can be induced by treatment with the active form of vitamin D_3 (1α,25-diydroxyviamin D_3) (Schmiedlin-Ren et al., 1997) or can be stably expressed through transfection of the *CYP3A4* gene (Brimer et al., 2000). These modified Caco-2 cell lines have been used to study first-pass intestinal metabolism (Fisher et al., 1999) or the interplay between transporters and intestinal metabolism (Cummins et al., 2002; Paine et al., 2002).

10.4.2 Cell Culture- and Age-Dependent Changes to Caco-2 Cells

Culture conditions for Caco-2 cells can influence the expression and functional activity of transporters, making it difficult to interpret and compare data obtained from different laboratories when culture conditions are significantly different. The differences observed across different Caco-2 cultures and laboratories are due, in part, to the heterogeneity of the parental cell line, where sub-populations of cells can be selectively passaged to become more prominent. Alternatively, this heterogeneity has been exploited to isolate more homogenous clones of the parental cell line that have specific functional traits or phenotypes. Understanding the specific Caco-2 clone and culture conditions used is paramount to making meaningful interpretations of the transport data obtained. The following section will focus on the morphology and protein expression of the parental clone of Caco-2 cells (ATCC HTB-37) and the effects culture conditions have on these properties.

Transporter expression, in particular on the apical membrane, increases as a function of culture time. For example, P-gp, MRP, BCRP, and PEPT1 expression increases during the culture time and plateaus at around 21 days (Behrens and Kissel, 2003; Xia et al., 2005; Englund et al., 2006). Consequently, transport experiments with Caco-2 cells are typically performed at 21 days post seeding. The age or the passage number of fully differentiated Caco-2 cells can also affect the barrier properties and protein expression. TEER increases with increasing passage number (Lu et al., 1996; Briske-Anderson et al., 1997; Yu et al., 1997). Higher monolayer resistance in late passage cells correlated with a reduction in the paracellular permeability of mannitol (Briske-Anderson et al., 1997; Yu et al., 1997), although in another study an increase in TEER associated with late passage cells was not accompanied by a decrease in mannitol permeability (Lu et al., 1996). This example clearly shows the importance of using a paracellular permeability probe (e.g., mannitol) to estimate paracellular transport, as relying solely on the TEER value of the cell monolayer may be misleading.

Increasing passage number appears to alter the expression of transport proteins. Although the majority of the work in this area has focused on nutrient transporters (e.g., glucose transporters), there has been some recent work looking at the changes in the expression of the xenobiotic transporters such as P-gp and BCRP. Although Caco-2 is generally known to express P-gp stably over a wide range in passage numbers, it has been shown that in both the wild-type and Caco-2 cells treated with the P-gp inducing ligand vinblastine, P-gp expression (mRNA) and function were significantly reduced between passages 29 and 49 (Siissalo et al., 2007). Similarly, the expression of BCRP was also affected by passage number, where the dimer protein expression decreased approximately tenfold between passage 36 and passage 56 of Caco-2 cells (Xia et al., 2005). In short, passage number can potentially alter both barrier function and transporter expression; therefore, functional activity of the transporter of interest in the passage range employed needs to be properly characterized.

As stated earlier, Caco-2 cells are made up of a heterogeneous population of epithelial cells that form a confluent monolayer. Many clones have been isolated to increase homogeneity or to enhance particular phenotypes. Different Caco-2 clones have distinct morphologic and phenotypic differences in relation to the parental Caco-2 cell clone (ATCC clone HTB-37) and need to be properly characterized. For example, the Caco-2 clone TC7 has high AP sucrase–isomaltase activity, higher relative TEER, and similar mannitol permeability to the parental line (Carriere et al., 1994; Ranaldi et al., 2003). Unlike the parental clone, which does not have basal functional CYP3A activity, Caco-2/TC7 cells express stable levels of CYP3A5 (Raeissi et al., 1997). For more detailed information regarding the different Caco-2 clones and the affects of culture conditions on monolayer functional and morphological characteristics, please refer to the review by Sambuy et al. (2005).

10.4.3 Experimental Approaches to Study Transporter-Mediated DDI in Caco-2 Cell Monolayers

Caco-2 cell monolayers are grown on permeable porous membrane supports that readily allow for solute and ion flux between the apical and the basolateral compartments through a monolayer of cells (Fig. 10.3). The appearance of drug into the

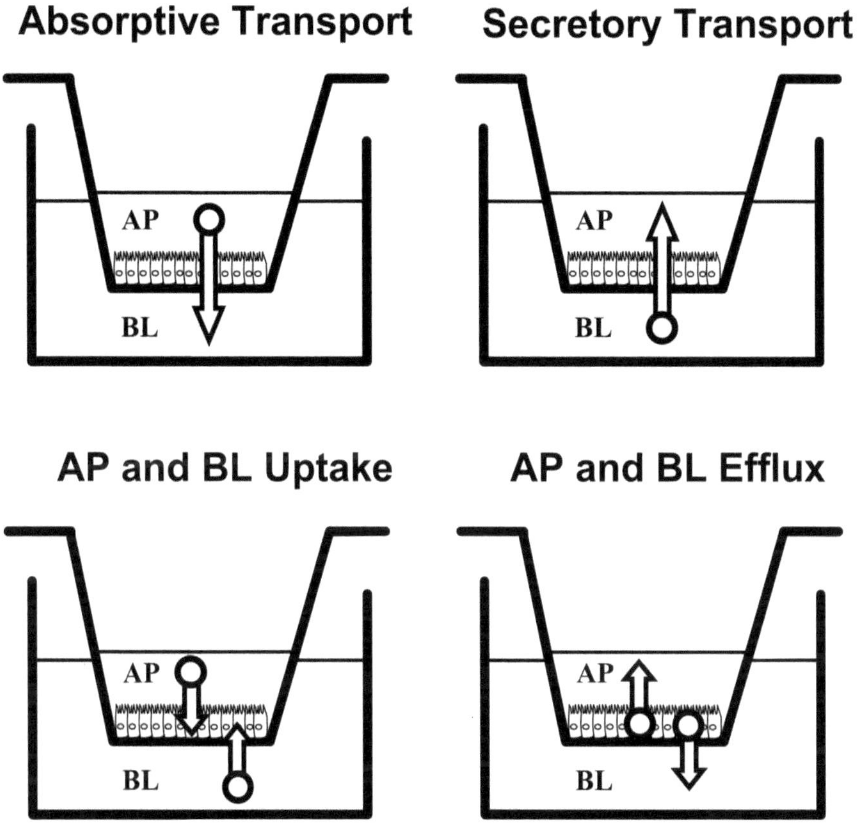

Fig. 10.3 Schematic of experimental approaches to study drug–transporter interactions in the Caco-2 Transwell^{TM} model

receiver compartment (the basolateral compartment for absorptive transport and the apical compartment for secretory transport) is monitored as a function of time. A transport clearance, corrected for the surface area of Transwell^{TM} insert, referred to as the apparent permeability (P_{app}), is calculated using the following equation:

$$P_{app} = \frac{J}{C_0} = \frac{dX/dt}{A^*C_o} \tag{10.1}$$

where J represents the flux of the mass of drug (X) transported over time (t) corrected for the porous surface area (A) and C_0 is the initial concentration in the donor compartment. P_{app} values are commonly measured in both absorptive (AP to BL) and secretory directions (BL to AP) and over a range of donor concentrations. Changes in P_{app} with donor concentrations indicate the presence of saturable processes and the involvement of metabolic enzymes and/or transporters. Kinetic constants (J_{max}, K_m, $K_{d,\,transport}$) can be obtained for transport data by fitting a model incorporating one saturable and one nonsaturable component to the overall transport using the following equation:

$$J = \frac{(J_{\text{max}}{}^{*}C)}{(K_{\text{m}} + C)} + K_{\text{d, transport}}{}^{*}C \qquad (10.2)$$

where C is the donor concentration of the drug, J_{max} is the maximal flux, K_{m} is the Michaelis–Menten constant, and $K_{\text{d, transport}}$ is the nonsaturable component of transport. The relative contribution of saturable process to the overall transport can be estimated by stripping out the nonsaturable component (e.g., $K_{\text{d, transport}}{}^{*}C$) from the overall transport data (see Bourdet and Thakker, 2006). Although this approach can indicate the presence of saturable processes, it does not allow for elucidation of the specific transporters responsible and is not amenable to high-throughput screening approaches due to the sample burden of assessing flux across a donor concentration range.

A common screening approach to determine if a transporter is involved is to measure absorptive transport (AP to BL or A–B) and secretory transport (BL to AP or B–A) at a fixed concentration. The efflux ratio (EfR) can then be calculated using the following equation:

$$\text{EfR} = \frac{P_{\text{app,B-A}}}{P_{\text{app,A-B}}} \qquad (10.3)$$

A compound whose efflux ratio deviates from unity implies that there is a transport mechanism acting on the drug. Typically, EfR values > 2 are deemed to be substrates for apical efflux transporters, and the magnitude of EfR is related to the efficacy of an efflux transporter in attenuating its absorptive transport. It is important to note that compounds that have high passive permeability may produce low EfR despite being good substrates for an apical efflux transporter (Troutman and Thakker, 2003a). EfR determination does not provide information about a particular transporter involved. Specific chemical inhibitors, knockout cell lines, or antibodies, all of which would move the EfR value toward unity by removing the effect of a specific transporter, can be used to tease out which transporter is responsible. A list of specific substrates and inhibitors for the major drug transporters is provided in Table 9.1 (Chapter 9). EfR can be misleading sometimes in that it does not inform on whether the efflux transporter is attenuating A–B transport or accelerating B–A transport; (Troutman and Thakker, 2003a). To assess the affect of P-gp in each direction, the absorptive quotient (AQ) parameter has been proposed, which is represented in the following equation:

$$\text{AQ} = \frac{(P_{\text{PD,A}\rightarrow\text{B}}) - (P_{\text{app}},\text{A} \rightarrow \text{B})}{(P_{\text{PD,A}\rightarrow\text{B}})} \qquad (10.4)$$

where P_{app} and P_{PD} represent the apparent permeability in the presence and absence of functional transport activity (e.g., in the absence and presence of the specific transporter inhibitor), respectively (Troutman and Thakker, 2003b). The AQ value ranges from 0 (no attenuation of absorptive transport by P-gp) to 1 (complete attenuation by P-gp); a drug is considered a reasonable substrate for the efflux transporter when the AQ value is greater than 0.3 (Thiel-Demby et al., 2004). This unidirectional approach provides a better assessment of specific efflux transporter involvement in absorptive transport rather than whether a drug is a substrate for any efflux transporter. Furthermore, it is more amenable to high-throughput screen due to limited number of samples generated. Although this method was developed to better characterize interactions of P-gp with its substrates, it can also be applied to other AP efflux transporters in Caco-2 such as BCRP and MRP2.

Overall transport across the cell monolayer accounts for multiple processes that are occurring at each membrane and through the cell. The polarized monolayer of the Caco-2 cell model allows for not only bidirectional transport studies but also for studying uptake and efflux transport processes at each membrane (Fig. 10.3). Uptake studies are performed by dosing either compartment and then monitoring the accumulation or uptake into the cells as a function of time and donor concentration. Akin to transport studies, saturable initial uptake (e.g., uptake occurring during the initial linear phase with respect to time) indicates the presence of carrier-mediated transport processes. Initial uptake data plotted as a function of donor concentration are commonly fitted to a model describing one saturable and one nonsaturable component (10.5)

$$\text{Uptake rate} = \frac{(V_{\text{max}}{}^{*}C)}{(K_{\text{m}} + C)} + K_{\text{d,uptake}}{}^{*}C \tag{10.5}$$

where C is the donor concentration, V_{max} is the maximal velocity, K_{m} is the Michaelis–Menten constant, and $K_{\text{d, uptake}}$ is the nonsaturable component of uptake (see Bourdet and Thakker, 2006). Specific chemical inhibitors (refer to Table 9.1) can be used to identify the particular uptake transporter class or specific isoform involved. Efflux experiments can also be performed in the Caco-2 cell model, in which drug is loaded into the cells to achieve pseudo-steady-state concentrations. The loaded monolayer is repeatedly but rapidly washed with ice-cold buffer, replaced with warm buffer, and appearance of drug into each compartment can be monitored as a function of time (Fig. 10.3). Specific inhibitors can be placed in the buffer following the washing to inhibit the efflux and tease out the role of a specific transporter in facilitating the efflux of the compound. Conversely, experiments have been conducted in Caco-2 cell monolayers, where labeled drug is loaded into the cells and unlabeled drug is present in the buffer, causing an increase in efflux. This classical pulse-chase experiment or trans-stimulation approach has been used to show functional evidence for bidirectional transport (Villalobos and Braun, 1998).

The methods outlined above demonstrate how to determine if a drug is a substrate for a transporter, the net effect that transporter has on the overall transport across the Caco-2 cell monolayer, and also information regarding what transporter(s) are involved at each membrane. A drug can be a transporter substrate and consequently an inhibitor; conversely, a drug may be a potent inhibitor and not a substrate for the transport. Selective probe substrates for many of the transporters are known, so that transport of these probe substrates can be evaluated in the presence and absence of test compounds to identify transporter inhibitors. In this chapter, a brief description of IC_{50} determination will be presented in which an inhibitory E_{max} model will be described, although thorough description of different methodologies to determine the extent of transporter inhibition in the Caco-2 cell model can be found in a recent review by Sun et al. (2008). Concentration-dependent inhibition of a probe substrate is used to determine the potency of a compound in inhibiting a specific transporter. Inhibitory potency, or IC_{50} value, can be determined by using nonlinear regression on the inhibitory E_{max} model using a modified Hill equation:

$$v = \frac{V_{max}{}^{*}[I]^n}{IC_{50}{}^n + [I]^n} \quad (10.6)$$

where v is the rate of transport in nmol/min/cm^2, V_{max} is the maximal rate of unidirectional transport of the probe substrate, I is the concentration of the inhibitor, n is the Hill coefficient, and IC_{50} is the inhibitor concentration that produces 50% reduction in the rate of transport of the probe substrate (Rautio et al., 2006). Equation (10.6) was adapted to fit bidirectional transport data (e.g., EfR), which the US Food and Drug Administration (FDA) proposed in its guidance for studying P-gp-mediated DDIs (Guidance for Industry, 2006), to yield the following expression:

$$\frac{EfR^{I}}{EfR^{W/O}} = 1 - \frac{I_{max}{}^{*}[I]^n}{[I]^n + IC_{50}^n} \quad (10.7)$$

where I_{max} is the maximal inhibitory effect and the efflux ratios, EfR^{I} and $EfR^{W/O}$, represent the inhibited and uninhibited efflux ratio, respectively. Efflux ratios are preferred measurements over flux or permeability values to determine IC_{50} values, for the latter two measurements can underestimate the extent of efflux inhibition and overestimate the IC_{50} value (Kalvass and Pollack, 2007). The inhibitory E_{max} model proposed by the FDA (Eq. (10.7)) not only can be used to identify transporter inhibition in the Caco-2 Transwell™ model but also can be applied to other polarized epithelial cell models, such as MDCK and LLC-PK_1 cells. For a description of these models and other in vitro assays to assay transporter DDIs, please refer to Chapter 9.

10.5 In Vitro/In Vivo Relationships

Because P-gp can potentially affect drug disposition of a large number of compounds, several in vitro assays have been developed to assess if compounds being considered as drug candidates are substrates or modulators of P-gp. The general approaches used in these assays involve measuring transport of test compounds in both directions across polarized cell monolayers, or activation by test compounds of ATPase activity associated with P-gp-mediated efflux, or inhibition of efflux of a pre-loaded fluorescent P-gp substrate (e.g., calcein-AM) by test compounds (Polli et al., 2001; Zhang et al., 2003; Guidance for Industry, 2006). Each of these assays has its benefits and pitfalls. As described previously in this chapter, the Caco-2 cell model, in which the polarized transport of test compounds is measured, can provide valuable information about a compound's interactions with P-gp. For example, these studies can (i) reveal if a compound is a substrate and/or an inhibitor of P-gp, (ii) provide relative potency of the test compound as a substrate and/or inhibitor of P-gp in relation to known P-gp substrates and inhibitors, (iii) inform if P-gp may attenuate absorptive transport of the test compound across an epithelial or an endothelial barrier, and (iv) assess if the test compound can increase absorptive transport of co-administered P-gp substrates and/or if other P-gp substrates/inhibitors can increase absorptive transport of the compound across an epithelial or an endothelial barrier. The goal of obtaining this information is to assess and predict if P-gp will influence absorption and other drug disposition parameters for the test compound and to predict if a co-administered P-gp substrate/inhibitor would increase exposure (AUC or C_{max}) to the test compound or vice versa.

The data generated with this model show that the efficacy of P-gp to attenuate absorptive transport of a compound depends not only on the kinetic parameters that define the interaction between the compound and the P-gp but also on the rate of passive diffusion across the cell membrane (Polli et al., 2001; Troutman and Thakker, 2003a). Further, it has been proposed that the behavior of compounds in the BCS system can provide insight on whether an efflux transporter like P-gp would affect disposition of these compounds (Wu and Benet, 2005). For example, it is proposed that for compounds in class II (high permeability, low solubility) P-gp-mediated efflux would play a significant role in intestinal absorption, whereas compounds in class I (high permeability, high solubility) would readily overwhelm P-gp in the cell membrane and therefore would not be affected by P-gp.

The emerging recognition of the importance of transporters in drug disposition and of the potential for a transporter-based DDI is evidenced by the inclusion of P-gp-related in vitro studies in the Guidance to Industry issued by the FDA on drug interaction studies. This guidance includes "criteria for determining whether a test drug is a substrate for P-gp, and whether an in vivo interaction study is needed." The draft guidance proposes that an in vivo drug interaction study may be warranted for a test compound if it meets the following criteria: (i) the test compound yields an efflux ratio of >2 in bidirectional transport studies using Caco-2 or MDCK-MDR1 cell monolayers and (ii) the efflux ratio can be decreased by a specific inhibitor for P-gp (list of inhibitors provided in the draft guidance). Similarly, the guidance proposes that an in vivo drug interaction study may be warranted for a test compound (i) if it decreases efflux ratio of probe substrates for P-gp in a bidirectional transport study as above, and (ii) if $[I]/IC_{50}$ (or K_i) > 0.1, where $[I]$ is the steady-state C_{max} value for total drug and IC_{50} (or K_i) represents the P gp inhibitory potency of the test compound.

It is understandable that the draft guidance may reflect the fact that our understanding of transporter-based DDIs is in its early phase of evolution. However, the draft guidance singles out just P-gp among many uptake and efflux transporters that are likely to play a significant role in DDIs (see Section 10.6). Furthermore, the draft guidance does not address a very real possibility that many test compounds are likely to be substrates for multiple transporters, present on the same or different cellular membrane. In such a case, the in vitro test sequence proposed will yield ambiguous results at best and will not provide guidance for the design of an in vivo DDI study. Also, the criteria proposed that would trigger a clinical DDI study do not appear to be grounded in reality. For example, an efflux ratio of 2 would suggest a fairly weak substrate activity for P-gp, not worthy of triggering a clinical DDI study. On the other hand, many of the specific P-gp inhibitors listed in the draft guidance interact with one or more other transporters and may not be appropriate as P-gp inhibitor probes. Finally, a recent study by Fenner et al. (2009) in which well over 100 clinical studies involving digoxin were analyzed, the authors make several observations that highlight the inadequacy of the current draft guidance. The authors observe that the pharmacokinetics of digoxin, a P-gp substrate that yields efflux ratios significantly greater than 2, did not change greatly when it was co-administered with widely varied drugs, many of which were potent P-gp inhibitors. On the other hand, in addressing the draft guidance that $[I]/IC_{50} > 0.1$ would trigger a clinical digoxin DDI study, the authors pointed out that 41% false-negative rate was revealed for compounds with $[I]/IC_{50} < 0.1$. These results indicated that in vitro data under-predicted the in vivo DDI involving digoxin and P-gp inhibitors. It appears that the draft guidance for in vitro studies to guide decisions regarding P-gp-mediated DDIs is quite premature at this point in time. More critical analysis like the one reported by Fenner et al. (2009) will be needed before a rational approach can be developed for in vitro studies that can guide decisions regarding clinical DDI studies.

10.6 Other Uptake and Efflux Transporters in Caco-2 Cells

Due to high expression and broad substrate specificity, P-gp can play an important role in mediating DDIs in the intestine. However, the knowledge of the role played by other transporters, including uptake and efflux transporters, on DDIs in the intestine is rapidly emerging. Many nutrients as well as hydrophilic drugs utilize vectoral transport systems and achieve higher bioavailability than that expected from their physicochemical properties. Vectoral transport systems for intestinal drug absorption involve uptake transporters in the apical membrane and efflux transporters in the basolateral membrane of the enterocytes (Fig. 10.2). Depending on their relative efficiency and capacity, apical uptake transporters or basolateral efflux transporters may control the transepithelial movement when the corresponding transport step becomes rate limiting (Cheeseman, 1992). Conceivably, intestinal basolateral transporters may also mediate efflux of hydrophilic compounds generated in the enterocytes, being either the active drugs generated from their prodrugs or the metabolites of the parent drugs.

In this section, current advances in our understanding of the intestinal transporters other than P-gp are summarized. The potential utility of Caco-2 cells, which express most of the transporters expressed by the intestinal epithelium, in studying the transporters other than P-gp is highlighted. Also discussed in the section is the complexity of interpreting transport data generated in the Caco-2 model when a compound is a substrate for P-gp as well as other uptake and/or efflux transporters present in Caco-2 cells.

10.6.1 Apical Uptake Transporters

Apical uptake transporters that may share substrate specificity with P-gp are OATP transporters, mediating bidirectional, Na^+-independent, pH-dependent transport of anionic and zwitterionic drugs (Hagenbuch and Meier, 2004). Recent studies using real-time PCR showed that the expression of OATP2B1 is much higher than other subtypes (Sai et al., 2006; Hilgendorf et al., 2007). OATP2B1 protein is localized at the apical surface of both human small intestine (Kobayashi et al., 2003) and Caco-2 cells (Sai et al., 2006). Functional studies demonstrated that OATP2B1 transports bile acids and endogenous sulfate conjugates such as estrone-3-sulfate (Nozawa et al., 2004), and the greater activity at lower pH broadens its specificity to include some clinically used drugs such as the HMG-CoA reductase inhibitor pravastatin, the antihistamine fexofenadine, and the antidiabetic glibenclamide (Kobayashi et al., 2003). Recently, OATP2B1 was shown to aid in the apical uptake of fexofenadine at pH 6.0 in Caco-2 cells (Ming and Thakker, unpublished). These studies clearly suggest a potential role for OATP2B1 in the intestinal absorption of anionic or zwitterionic drugs such as pravastatin and fexofenadine. Citrus juices and constituents are able to inhibit OATP2B1 activity, identifying OATP2B1 as a potential site for diet–drug interactions (Satoh et al., 2005).

10.6.2 Other Apical Efflux Transporters

BCRP is expressed in several tissues including the intestine, the liver, and the placenta and localized to the brush-border membrane in the intestine (Maliepaard et al., 2001). An increasing number of diverse drugs have been identified as substrates for BCRP including anticancer compounds (e.g., mitoxantrone, daunorubicin, doxorubicin) (Theis et al., 1998), sulfated conjugates of steroids and xenobiotics (Suzuki et al., 2003), and the H_2-receptor antagonist cimetidine (Pavek et al., 2005). Similar to P-gp, BCRP likely attenuates the oral absorption of drug substrates, thus limiting their overall oral bioavailability. BCRP knockout mice exhibited strong sensitivity to the dietary chlorophyll breakdown product (and BCRP substrate) pheophorbide a, suggesting that the intestine limits systemic exposure to this toxin (Wang et al., 2002). BCRP appears to reduce the oral bioavailability of the anticancer agent topotecan. Oral administration of topotecan in the presence of a BCRP/P-gp inhibitor (GF120918) in P-gp-deficient mice resulted in a sixfold increase in bioavailability compared to vehicle treated mice, clearly suggesting a role for BCRP in limiting topotecan absorption (Jonker et al., 2000). As stated in Section 10.4.1, the relative expression of BCRP in Caco-2 cells is low in comparison to human intestine and variable depending on the clone employed; therefore, the role of this transporter in drug absorption or in DDIs may be underestimated by the Caco-2 cell model.

MRP2 is localized exclusively to the apical membrane of polarized cells, such as hepatocytes, renal proximal tubule epithelia, and intestinal epithelia (Buchler et al., 1996). Compared to the extensively studied and established role of MRP2 in detoxification of drugs in the liver, its role in the intestine has been less well studied. MRP2 is highly expressed in human jejunum and Caco-2 cells (Hilgendorf et al., 2007). Studies utilizing wild-type and MRP2-deficient rats have clearly demonstrated a significantly reduced intestinal excretion of a glutathione conjugate DNP-SG, an MRP2 substrate, in the deficient rat after intravenous administration of the unconjugated parent (Gotoh et al., 2005). However, MRP2 may be less important than P-gp in limiting the absorption of drugs, because an 8.5-fold increase in oral bioavailability of paclitaxel in *mdr1a/b*(–/–) mice was observed compared to the wild-type mice, whereas change was not detected in *mrp2*(–/–) mice, despite the fact that Mrp2 plays an important role in biliary excretion of paclitaxel (Lagas et al., 2006).

10.6.3 Basolateral Efflux Transporters

MRP3 transports organic compounds conjugated to glutathione, sulfate, or glucuronate, as well as bile salts and unconjugated drugs such as methotrexate (Sparreboom et al., 1997). MRP3 mRNA is highly expressed in human jejunum and in Caco-2 cells (Taipalensuu et al., 2001). Rat Mrp3 protein and human MRP3 protein are localized in the basolateral membrane in rat intestine (Rost et al., 2002) and Caco-2 cells (Ming and Thakker, 2009), respectively. Mrp3 plays a minor role (<20%) in the basolateral efflux of bile acids (Zelcer et al., 2006; Suzuki et al., 2008). Recent study showed that MRP3 mediates the basolateral efflux of fexofenadine and that overall absorptive transport of this drug is dependent on relative activities of P-gp in the apical membrane and MRP3 in the basolateral membrane (Ming and Thakker, unpublished).

MRP4, the fourth member of the ABCC family, was initially identified as a homolog of MRP1 by screening databases of human expressed sequence tags (Kool et al., 1997). MRP4 is the only transporter in ABCC family that shows tissue-specific localization; it assumes basolateral localization in most organs such as liver (Rius et al., 2003), whereas in kidney proximal tubules it is located at the apical membranes (van Aubel et al., 2002). Recent results showed that MRP4 is localized in the basolateral membrane of Caco-2 cells and mediates the basolateral efflux of adefovir generated intracellularly from its prodrug, adefovir dipivoxil (Ming and Thakker, 2009). This study suggested that the oral bioavailability of adefovir can be significantly affected if an MRP4 inhibitor is co-administered with adefovir dipivoxil.

10.7 Conclusions

P-gp is an important efflux transporter that can influence disposition of drugs that are its substrates. Its wide substrate specificity and extensive tissue distribution would suggest that it affects disposition of a large number of drugs. It follows therefore that P-gp may contribute significantly to DDIs. While this chapter mainly deals with the use of Caco-2 and other related in vitro cell culture models in assessing DDI potential of P-gp substrates and inhibitors, an attempt has been made to convey an important message regarding P-gp and other transporters. That is, our understanding about transporters is in its infancy, and accordingly, the current in vitro and in vivo models to elucidate and predict roles of transporters in drug disposition and DDIs are quite inadequate. Therefore, the results derived from these models should be interpreted with much care and caution.

References

Altenberg GA, Vanoye CG, Horton JK and Reuss L (1994) Unidirectional fluxes of rhodamine 123 in multidrug-resistant cells: evidence against direct drug extrusion from the plasma membrane. *Proc Natl Acad Sci U S A* **91**:4654–4657.

Anderle P, Niederer E, Rubas W, Hilgendorf C, Spahn-Langguth H, Wunderli-Allenspach H, Merkle HP and Langguth P (1998) P-Glycoprotein (P-gp) mediated efflux in Caco-2 cell monolayers: the influence of culturing conditions and drug exposure on P-gp expression levels. *J Pharm Sci* **87**:757–762.

Arimori K and Nakano M (1998) Drug exsorption from blood into the gastrointestinal tract. *Pharm Res* **15**:371–376.

Artursson P (1990) Epithelial transport of drugs in cell culture. I: a model for studying the passive diffusion of drugs over intestinal absorptive (Caco-2) cells. *J Pharm Sci* **79**:476–482.

Augustijns PF, Bradshaw TP, Gan LS, Hendren RW and Thakker DR (1993) Evidence for a polarized efflux system in CACO-2 cells capable of modulating cyclosporin A transport. *Biochem Biophys Res Commun* **197**:360–365.

Behrens I and Kissel T (2003) Do cell culture conditions influence the carrier-mediated transport of peptides in Caco-2 cell monolayers? *Eur J Pharm Sci* **19**:433–442.

Benet LZ, Izumi T, Zhang Y, Silverman JA and Wacher VJ (1999) Intestinal MDR transport proteins and P-450 enzymes as barriers to oral drug delivery. *J Control Release* **62**:25–31.

Bourdet DL and Thakker DR (2006) Saturable absorptive transport of the hydrophilic organic cation ranitidine in Caco-2 cells: role of pH-dependent organic cation uptake system and P-glycoprotein. *Pharm Res* **23**:1165–1177.

Brimer C, Dalton JT, Zhu Z, Schuetz J, Yasuda K, Vanin E, Relling MV, Lu Y and Schuetz EG (2000) Creation of polarized cells coexpressing CYP3A4, NADPH cytochrome P450 reductase and MDR1/P-glycoprotein. *Pharm Res* **17**:803–810.

Briske-Anderson MJ, Finley JW and Newman SM (1997) The influence of culture time and passage number on the morphological and physiological development of Caco-2 cells. *Proc Soc Exp Biol Med* **214**:248–257.

Bruggemann EP, Currier SJ, Gottesman MM and Pastan I (1992) Characterization of the azidopine and vinblastine binding site of P-glycoprotein. *J Biol Chem* **267**:21020–21026.

Buchler M, Konig J, Brom M, Kartenbeck J, Spring H, Horie T and Keppler D (1996) cDNA cloning of the hepatocyte canalicular isoform of the multidrug resistance protein, cMrp, reveals a novel conjugate export pump deficient in hyperbilirubinemic mutant rats. *J Biol Chem* **271**:15091–15098.

Carriere V, Lesuffleur T, Barbat A, Rousset M, Dussaulx E, Costet P, de Waziers I, Beaune P and Zweibaum A (1994) Expression of cytochrome P-450 3A in HT29-MTX cells and Caco-2 clone TC7. *FEBS Lett* **355**:247–250.

Chandra P and Brouwer KL (2004) The complexities of hepatic drug transport: current knowledge and emerging concepts. *Pharm Res* **21**:719–735.

Cheeseman C (1992) Role of intestinal basolateral membrane in absorption of nutrients. *Am J Physiol* **263**:R482–R488.

Chen C and Pollack GM (1998) Altered disposition and antinociception of [D-penicillamine(2,5)] enkephalin in mdr1a-gene-deficient mice. *J Pharmacol Exp Ther* **287**:545–552.

Clarke SE and Jones BC (2008) Human cytochromes P450 and their role in metabolism-based drug–durg interactions, in *Durg to Durg Interactions* (Rodrigues DA ed) pp 53–87, Informa Healthcare, New York.

Colegio OR, Van Itallie C, Rahner C and Anderson JM (2003) Claudin extracellular domains determine paracellular charge selectivity and resistance but not tight junction fibril architecture. *Am J Physiol Cell Physiol* **284**:C1346–C1354.

Cordon-Cardo C, O'Brien JP, Boccia J, Casals D, Bertino JR and Melamed MR (1990) Expression of the multidrug resistance gene product (P-glycoprotein) in human normal and tumor tissues. *J Histochem Cytochem* **38**:1277–1287.

Cummins CL, Jacobsen W and Benet LZ (2002) Unmasking the dynamic interplay between intestinal P-glycoprotein and CYP3A4. *J Pharmacol Exp Ther* **300**:1036–1045.

Elimrani I, Lahjouji K, Seidman E, Roy MJ, Mitchell GA and Qureshi I (2003) Expression and localization of organic cation/carnitine transporter OCTN2 in Caco-2 cells. *Am J Physiol Gastrointest Liver Physiol* **284**:G863–G871.

Englund G, Rorsman F, Ronnblom A, Karlbom U, Lazorova L, Grasjo J, Kindmark A and Artursson P (2006) Regional levels of drug transporters along the human intestinal tract: co-expression of ABC and SLC transporters and comparison with Caco-2 cells. *Eur J Pharm Sci* **29**:269–277.

Engman HA, Lennernas H, Taipalensuu J, Otter C, Leidvik B and Artursson P (2001) CYP3A4, CYP3A5, and MDR1 in human small and large intestinal cell lines suitable for drug transport studies. *J Pharm Sci* **90**:1736–1751.

Etievant C, Schambel P, Guminski Y, Barret JM, Imbert T and Hill BT (1998) Requirements for P-glycoprotein recognition based on structure–activity relationships in the podophyllotoxin series. *Anticancer Drug Des* **13**:317–336.

Fenner KS, Troutman MD, Kempshall S, Cook JA, Ware JA, Smith DA and Lee CA (2009) Drug–drug interactions mediated through P-glycoprotein: clinical relevance and in vitro–in vivo correlation using digoxin as a probe drug. *Clin Pharmacol Ther* **85**:173–181.

Ferry DR, Malkhandi PJ, Russell MA and Kerr DJ (1995) Allosteric regulation of [3H]vinblastine binding to P-glycoprotein of MCF-7 ADR cells by dexniguldipine. *Biochem Pharmacol* **49**:1851–1861.

Fisher JM, Wrighton SA, Watkins PB, Schmiedlin-Ren P, Calamia JC, Shen DD, Kunze KL and Thummel KE (1999) First-pass midazolam metabolism catalyzed by 1alpha,25-dihydroxy vitamin D3-modified Caco-2 cell monolayers. *J Pharmacol Exp Ther* **289**:1134–1142.

Fleet JC, Eksir F, Hance KW and Wood RJ (2002) Vitamin D-inducible calcium transport and gene expression in three Caco-2 cell lines. *Am J Physiol Gastrointest Liver Physiol* **283**:G618–G625.

Gan LS, Moseley MA, Khosla B, Augustijns PF, Bradshaw TP, Hendren RW and Thakker DR (1996) CYP3A-like cytochrome P450-mediated metabolism and polarized efflux of cyclosporin A in Caco-2 cells. *Drug Metab Dispos* **24**:344–349.

Gan LS and Thakker D (1997) Applications of the Caco-2 model in the design and development of orally active drugs: elucidation of biochemical and physical barriers posed by the intestinal epithelium. *Adv Drug Deliv Rev* **23**:77–98.

Germann UA (1996) P-glycoprotein – a mediator of multidrug resistance in tumour cells. *Eur J Cancer* **32A**:927–944.

Gotoh Y, Kamada N and Momose D (2005) The advantages of the using chamber in drug absorption studies. *J Biomol Screen* **10**:517–523.

Gramatte T, Oertel R, Terhaag B and Kirch W (1996) Direct demonstration of small intestinal secretion and site-dependent absorption of the beta-blocker talinolol in humans. *Clin Pharmacol Ther* **59**:541–549.

Guidance for Industry (2006). Drug interaction studies–study design, data analysis and implications for dosing and labeling. FDA/CDER and CBER, 2006. Available from: http://www.fda.gov/cder/Guidance/6695dft.htm [Last accessed 4 May 2009]

Gutmann H, Fricker G, Torok M, Michael S, Beglinger C and Drewe J (1999) Evidence for different ABC-transporters in Caco-2 cells modulating drug uptake. *Pharm Res* **16**: 402–407.

Hagenbuch B and Meier PJ (2004) Organic anion transporting polypeptides of the OATP/ SLC21 family: phylogenetic classification as OATP/ SLCO superfamily, new nomenclature and molecular/functional properties. *Pflugers Arch* **447**:653–665.

Hayeshi R, Hilgendorf C, Artursson P, Augustijns P, Brodin B, Dehertogh P, Fisher K, Fossati L, Hovenkamp E, Korjamo T, Masungi C, Maubon N, Mols R, Mullertz A, Monkkonen J, O'Driscoll C, Oppers-Tiemissen HM, Ragnarsson EG, Rooseboom M and Ungell AL (2008) Comparison of drug transporter gene expression and functionality in Caco-2 cells from 10 different laboratories. *Eur J Pharm Sci* **35**:383–396.

Hidalgo IJ and Borchardt RT (1990) Transport of bile acids in a human intestinal epithelial cell line, Caco-2. *Biochim Biophys Acta* **1035**:97–103.

Hidalgo IJ, Raub TJ and Borchardt RT (1989) Characterization of the human colon carcinoma cell line (Caco-2) as a model system for intestinal epithelial permeability. *Gastroenterology* **96**:736–749.

Hilgendorf C, Ahlin G, Seithel A, Artursson P, Ungell AL and Karlsson J (2007) Expression of thirty-six drug transporter genes in human intestine, liver, kidney, and organotypic cell lines. *Drug Metab Dispos* **35**:1333–1340.

Hoffmeyer S, Burk O, von Richter O, Arnold HP, Brockmoller J, Johne A, Cascorbi I, Gerloff T, Roots I, Eichelbaum M and Brinkmann U (2000) Functional polymorphisms of the human multidrug-resistance gene: multiple sequence variations and correlation of one allele with P-glycoprotein expression and activity in vivo. *Proc Natl Acad Sci U S A* **97**:3473–3478.

Homolya L, Hollo Z, Germann UA, Pastan I, Gottesman MM and Sarkadi B (1993) Fluorescent cellular indicators are extruded by the multidrug resistance protein. *J Biol Chem* **268**: 21493–21496.

Huang W, Lin YS, McConn DJ, 2nd, Calamia JC, Totah RA, Isoherranen N, Glodowski M and Thummel KE (2004) Evidence of significant contribution from CYP3A5 to hepatic drug metabolism. *Drug Metab Dispos* **32**:1434–1445.

Hunter J, Hirst BH and Simmons NL (1993) Transepithelial secretion, cellular accumulation and cytotoxicity of vinblastine in defined MDCK cell strains. *Biochim Biophys Acta* **1179**: 1–10.

Imai T, Imoto M, Sakamoto H and Hashimoto M (2005) Identification of esterases expressed in Caco-2 cells and effects of their hydrolyzing activity in predicting human intestinal absorption. *Drug Metab Dispos* **33**:1185–1190.

Ito S, Woodland C, Harper PA and Koren G (1993) The mechanism of the verapamil–digoxin interaction in renal tubular cells (LLC-PK1. *Life Sci* **53**:PL399–PL403.

Johnson BM, Charman WN and Porter CJ (2003) Application of compartmental modeling to an examination of in vitro intestinal permeability data: assessing the impact of tissue uptake, P-glycoprotein, and CYP3A. *Drug Metab Dispos* **31**:1151–1160.

Jonker JW, Smit JW, Brinkhuis RF, Maliepaard M, Beijnen JH, Schellens JH and Schinkel AH (2000) Role of breast cancer resistance protein in the bioavailability and fetal penetration of topotecan. *J Natl Cancer Inst* **92**:1651–1656.

Juliano RL and Ling V (1976) A surface glycoprotein modulating drug permeability in Chinese hamster ovary cell mutants. *Biochim Biophys Acta* **455**:152–162.

Kalvass JC and Pollack GM (2007) Kinetic considerations for the quantitative assessment of efflux activity and inhibition: implications for understanding and predicting the effects of efflux inhibition. *Pharm Res* **24**:265–276.

Kartner N, Riordan JR and Ling V (1983) Cell surface P-glycoprotein associated with multidrug resistance in mammalian cell lines. *Science* **221**:1285–1288.

Knight B, Troutman M and Thakker DR (2006) Deconvoluting the effects of P-glycoprotein on intestinal CYP3A: a major challenge. *Curr Opin Pharmacol* **6**:528–532.

Knipp GT, Ho NF, Barsuhn CL and Borchardt RT (1997) Paracellular diffusion in Caco-2 cell monolayers: effect of perturbation on the transport of hydrophilic compounds that vary in charge and size. *J Pharm Sci* **86**:1105–1110.

Kobayashi D, Nozawa T, Imai K, Nezu J, Tsuji A and Tamai I (2003) Involvement of human organic anion transporting polypeptide OATP-B (SLC21A9) in pH-dependent transport across intestinal apical membrane. *J Pharmacol Exp Ther* **306**:703–708.

Kool M, de Haas M, Scheffer GL, Scheper RJ, van Eijk MJ, Juijn JA, Baas F and Borst P (1997) Analysis of expression of cMOAT (MRP2), MRP3, MRP4, and MRP5, homologues of the multidrug resistance-associated protein gene (MRP1), in human cancer cell lines. *Cancer Res* **57**:3537–3547.

Lagas JS, Vlaming ML, van Tellingen O, Wagenaar E, Jansen RS, Rosing H, Beijnen JH and Schinkel AH (2006) Multidrug resistance protein 2 is an important determinant of paclitaxel pharmacokinetics. *Clin Cancer Res* **12**:6125–6132.

Litman T, Zeuthen T, Skovsgaard T and Stein WD (1997) Structure-activity relationships of P-glycoprotein interacting drugs: kinetic characterization of their effects on ATPase activity. *Biochim Biophys Acta* **1361**:159–168.

Loo TW and Clarke DM (2000) Drug-stimulated ATPase activity of human P-glycoprotein is blocked by disulfide cross-linking between the nucleotide-binding sites. *J Biol Chem* **275**:19435–19438.

Loo TW and Clarke DM (2005) Recent progress in understanding the mechanism of P-glycoprotein-mediated drug efflux. *J Membr Biol* **206**:173–185.

Lu S, Gough AW, Bobrowski WF and Stewart BH (1996) Transport properties are not altered across Caco-2 cells with heightened TEER despite underlying physiological and ultrastructural changes. *J Pharm Sci* **85**:270–273.

Madara JL and Trier JS (1994) The functional morphology of the mucosa of the small intestine, in *Physiology of the Gastrointestinal Tract* (Johnson LR ed) pp 1577–1622, Raven Press, New York.

Maliepaard M, Scheffer GL, Faneyte IF, van Gastelen MA, Pijnenborg AC, Schinkel AH, van De Vijver MJ, Scheper RJ and Schellens JH (2001) Subcellular localization and distribution of the breast cancer resistance protein transporter in normal human tissues. *Cancer Res* **61**: 3458–3464.

Martin C, Berridge G, Higgins CF, Mistry P, Charlton P and Callaghan R (2000a) Communication between multiple drug binding sites on P-glycoprotein. *Mol Pharmacol* **58**: 624–632.

Martin C, Berridge G, Mistry P, Higgins C, Charlton P and Callaghan R (2000b) Drug binding sites on P-glycoprotein are altered by ATP binding prior to nucleotide hydrolysis. *Biochemistry* **39**:11901–11906.

Maubon N, Le Vee M, Fossati L, Audry M, Le Ferrec E, Bolze S and Fardel O (2007) Analysis of drug transporter expression in human intestinal Caco-2 cells by real-time PCR. *Fundam Clin Pharmacol* **21**:659–663.

Mayer U, Wagenaar E, Beijnen JH, Smit JW, Meijer DK, van Asperen J, Borst P and Schinkel AH (1996) Substantial excretion of digoxin via the intestinal mucosa and prevention of long-term digoxin accumulation in the brain by the mdr 1a P-glycoprotein. *Br J Pharmacol* **119**: 1038–1044.

Meinl W, Ebert B, Glatt H and Lampen A (2008) Sulfotransferase forms expressed in human intestinal Caco^{-2} and TC7 cells at varying stages of differentiation and role in benzo[a]pyrene metabolism. *Drug Metab Dispos* **36**:276–283.

Ming X and Thakker DR (2009) Role of Basolateral Efflux Transporter MRP4 in the Intestinal Absorption of the Antiviral Drug Adefovir Dipivoxil. *Biochem Pharmacol* (in press).

Muller J, Lips KS, Metzner L, Neubert RH, Koepsell H and Brandsch M (2005) Drug specificity and intestinal membrane localization of human organic cation transporters (OCT). *Biochem Pharmacol* **70**:1851–1860.

Munzel PA, Schmohl S, Heel H, Kalberer K, Bock-Hennig BS and Bock KW (1999) Induction of human UDP glucuronosyltransferases (UGT1A6, UGT1A9, and UGT2B7) by t-butylhydroquinone and 2,3,7,8-tetrachlorodibenzo-p-dioxin in Caco-2 cells. *Drug Metab Dispos* **27**:569–573.

Nozawa T, Imai K, Nezu J, Tsuji A and Tamai I (2004) Functional characterization of pH-sensitive organic anion transporting polypeptide OATP-B in human. *J Pharmacol Exp Ther* **308**: 438–445.

Oh DM, Han HK and Amidon GL (1999) Drug transport and targeting: intestinal transport, in *Membrane Transporters as Drug Targets* (Amidon GL and Sadee W eds) pp 59–88, Kluwer Academic and Plenum Publishers, New York.

Okuwaki M, Takada T, Iwayanagi Y, Koh S, Kariya Y, Fujii H and Suzuki H (2007) LXR alpha transactivates mouse organic solute transporter alpha and beta via IR-1 elements shared with FXR. *Pharm Res* **24**:390–398.

Paine MF, Hart HL, Ludington SS, Haining RL, Rettie AE and Zeldin DC (2006) The human intestinal cytochrome P450 "pie". *Drug Metab Dispos* **34**:880–886.

Paine MF, Leung LY, Lim HK, Liao K, Oganesian A, Zhang MY, Thummel KE and Watkins PB (2002) Identification of a novel route of extraction of sirolimus in human small intestine: roles of metabolism and secretion. *J Pharmacol Exp Ther* **301**:174–186.

Pavek P, Merino G, Wagenaar E, Bolscher E, Novotna M, Jonker JW and Schinkel AH (2005) Human breast cancer resistance protein: interactions with steroid drugs, hormones, the dietary carcinogen 2-amino-1-methyl-6-phenylimidazo(4,5-b)pyridine, and transport of cimetidine. *J Pharmacol Exp Ther* **312**:144–152.

Peters WH and Roelofs HM (1989) Time-dependent activity and expression of glutathione S-transferases in the human colon adenocarcinoma cell line Caco-2. *Biochem J* **264**:613–616.

Polli JW, Wring SA, Humphreys JE, Huang L, Morgan JB, Webster LO and Serabjit-Singh CS (2001) Rational use of in vitro P-glycoprotein assays in drug discovery. *J Pharmacol Exp Ther* **299**:620–628.

Prime-Chapman HM, Fearn RA, Cooper AE, Moore V and Hirst BH (2004) Differential multidrug resistance-associated protein 1 through 6 isoform expression and function in human intestinal epithelial Caco-2 cells. *J Pharmacol Exp Ther* **311**:476–484.

Raeissi SD, Guo Z, Dobson GL, Artursson P and Hidalgo IJ (1997) Comparison of CYP3A activities in a subclone of Caco-2 cells (TC7) and human intestine. *Pharm Res* **14**: 1019–1025.

Ranaldi G, Consalvo R, Sambuy Y and Scarino ML (2003) Permeability characteristics of parental and clonal human intestinal Caco-2 cell lines differentiated in serum-supplemented and serum-free media. *Toxicol In Vitro* **17**:761–767.

Rautio J, Humphreys JE, Webster LO, Balakrishnan A, Keogh JP, Kunta JR, Serabjit-Singh CJ and Polli JW (2006) In vitro p-glycoprotein inhibition assays for assessment of clinical drug interaction potential of new drug candidates: a recommendation for probe substrates. *Drug Metab Dispos* **34**:786–792.

Relling MV (1996) Are the major effects of P-glycoprotein modulators due to altered pharmacokinetics of anticancer drugs? *Ther Drug Monit* **18**:350–356.

Rius M, Nies AT, Hummel-Eisenbeiss J, Jedlitschky G and Keppler D (2003) Cotransport of reduced glutathione with bile salts by MRP4 (ABCC4) localized to the basolateral hepatocyte membrane. *Hepatology* **38**:374–384.

Rost D, Mahner S, Sugiyama Y and Stremmel W (2002) Expression and localization of the multidrug resistance-associated protein 3 in rat small and large intestine. *Am J Physiol Gastrointest Liver Physiol* **282**:G720–G726.

Sai Y, Kaneko Y, Ito S, Mitsuoka K, Kato Y, Tamai I, Artursson P and Tsuji A (2006) Predominant contribution of organic anion transporting polypeptide OATP-B (OATP2B1) to apical uptake of estrone-3-sulfate by human intestinal Caco-2 cells. *Drug Metab Dispos* **34**:1423–1431.

Saito H, Motohashi H, Mukai M and Inui K (1997) Cloning and characterization of a pH-sensing regulatory factor that modulates transport activity of the human H+/peptide cotransporter, PEPT1. *Biochem Biophys Res Commun* **237**:577–582.

Sakaeda T, Nakamura T and Okumura K (2003) Pharmacogenetics of MDR1 and its impact on the pharmacokinetics and pharmacodynamics of drugs. *Pharmacogenomics* **4**:397–410.

Sambuy Y, De Angelis I, Ranaldi G, Scarino ML, Stammati A and Zucco F (2005) The Caco-2 cell line as a model of the intestinal barrier: influence of cell and culture-related factors on Caco^{-2} cell functional characteristics. *Cell Biol Toxicol* **21**:1–26.

Satoh H, Yamashita F, Tsujimoto M, Murakami H, Koyabu N, Ohtani H and Sawada Y (2005) Citrus juices inhibit the function of human organic anion-transporting polypeptide OATP-B. *Drug Metab Dispos* **33**:518–523.

Schinkel AH, Smit JJ, van Tellingen O, Beijnen JH, Wagenaar E, van Deemter L, Mol CA, van der Valk MA, Robanus-Maandag EC, te Riele HP, et al. (1994) Disruption of the mouse mdr1a P-glycoprotein gene leads to a deficiency in the blood–brain barrier and to increased sensitivity to drugs. *Cell* **77**:491–502.

Schmiedlin-Ren P, Thummel KE, Fisher JM, Paine MF, Lown KS and Watkins PB (1997) Expression of enzymatically active CYP3A4 by Caco-2 cells grown on extracellular matrix-coated permeable supports in the presence of 1alpha,25-dihydroxyvitamin D3. *Mol Pharmacol* **51**:741–754.

Schreiber TD, Kohle C, Buckler F, Schmohl S, Braeuning A, Schmiechen A, Schwarz M and Munzel PA (2006) Regulation of CYP1A1 gene expression by the antioxidant *tert*-butylhydroquinone. *Drug Metab Dispos* **34**:1096–1101.

Seithel A, Karlsson J, Hilgendorf C, Bjorquist A and Ungell A (2006) Variability in mRNA expression of ABC- and SLC-transporters in human intestinal cells: comparison between human segments and Caco-2 cells. *Eur J Pharm Sci* **28**:291–299.

Sheps JA (2009) Biochemistry. Through a mirror, differently. *Science* **323**:1679–1680.

Shitara Y, Horie T and Sugiyama Y (2006) Transporters as a determinant of drug clearance and tissue distribution. *Eur J Pharm Sci* **27**:425–446.

Siissalo S, Laitinen L, Koljonen M, Vellonen KS, Kortejarvi H, Urtti A, Hirvonen J and Kaukonen AM (2007) Effect of cell differentiation and passage number on the expression of efflux proteins in wild type and vinblastine-induced Caco-2 cell lines. *Eur J Pharm Biopharm* **67**:548–554.

Sparreboom A, van Asperen J, Mayer U, Schinkel AH, Smit JW, Meijer DK, Borst P, Nooijen WJ, Beijnen JH and van Tellingen O (1997) Limited oral bioavailability and active epithelial excretion of paclitaxel (Taxol) caused by P-glycoprotein in the intestine. *Proc Natl Acad Sci U S A* **94**:2031–2035.

Sun H, Chow EC, Liu S, Du Y and Pang KS (2008) The Caco-2 cell monolayer: usefulness and limitations. *Expert Opin Drug Metab Toxicol* **4**:395–411.

Suzuki T, Seth A and Rao R (2008) Role of Phospholipase C{gamma}-induced Activation of Protein Kinase C{epsilon} (PKC{epsilon}) and PKC{beta}I in Epidermal Growth Factor-mediated Protection of Tight Junctions from Acetaldehyde in Caco-2 Cell Monolayers. *J Biol Chem* **283**:3574–3583.

Suzuki M, Suzuki H, Sugimoto Y and Sugiyama Y (2003) ABCG2 transports sulfated conjugates of steroids and xenobiotics. *J Biol Chem* **278**:22644–22649.

Taipalensuu J, Tornblom H, Lindberg G, Einarsson C, Sjoqvist F, Melhus H, Garberg P, Sjostrom B, Lundgren B and Artursson P (2001) Correlation of gene expression of ten drug efflux proteins of the ATP-binding cassette transporter family in normal human jejunum and in human intestinal epithelial Caco-2 cell monolayers. *J Pharmacol Exp Ther* **299**:164–170.

Tam D, Sun H and Pang KS (2003) Influence of P-glycoprotein, transfer clearances, and drug binding on intestinal metabolism in Caco-2 cell monolayers or membrane preparations: a theoretical analysis. *Drug Metab Dispos* **31**:1214–1226.

Tamai I, Takanaga H, Maeda H, Sai Y, Ogihara T, Higashida H and Tsuji A (1995) Participation of a proton-cotransporter, MCT1, in the intestinal transport of monocarboxylic acids. *Biochem Biophys Res Commun* **214**:482–489.

Terao T, Hisanaga E, Sai Y, Tamai I and Tsuji A (1996) Active secretion of drugs from the small intestinal epithelium in rats by P-glycoprotein functioning as an absorption barrier. *J Pharm Pharmacol* **48**:1083–1089.

Theis JG, Chan HS, Greenberg ML, Malkin D, Karaskov V, Moncica I, Koren G and Doyle J (1998) Increased systemic toxicity of sarcoma chemotherapy due to combination with the P-glycoprotein inhibitor cyclosporin. *Int J Clin Pharmacol Ther* **36**:61–64.

Thiebaut F, Tsuruo T, Hamada H, Gottesman MM, Pastan I and Willingham MC (1987) Cellular localization of the multidrug-resistance gene product P-glycoprotein in normal human tissues. *Proc Natl Acad Sci U S A* **84**:7735–7738.

Thiel-Demby VE, Tippin TK, Humphreys JE, Serabjit-Singh CJ and Polli JW (2004) In vitro absorption and secretory quotients: practical criteria derived from a study of 331 compounds to assess for the impact of P-glycoprotein-mediated efflux on drug candidates. *J Pharm Sci* **93**:2567–2572.

Troutman M, Luo G, Knight BM and Thakker DR (2008) The role of P-glycoprotein in drug disposition: signifcance to drug development, in *Durg to Durg Interactions* (Rodrigues DA ed) pp 359–434, Informa Healthcare, New York.

Troutman MD and Thakker DR (2003a) Efflux ratio cannot assess P-glycoprotein-mediated attenuation of absorptive transport: asymmetric effect of P-glycoprotein on absorptive and secretory transport across Caco-2 cell monolayers. *Pharm Res* **20**:1200–1209.

Troutman MD and Thakker DR (2003b) Novel experimental parameters to quantify the modulation of absorptive and secretory transport of compounds by P-glycoprotein in cell culture models of intestinal epithelium. *Pharm Res* **20**:1210–1224.

van Asperen J, Schinkel AH, Beijnen JH, Nooijen WJ, Borst P and van Tellingen O (1996) Altered pharmacokinetics of vinblastine in Mdr1a P-glycoprotein-deficient Mice. *J Natl Cancer Inst* **88**:994–999.

van Asperen J, van Tellingen O and Beijnen JH (2000) The role of mdr1a P-glycoprotein in the biliary and intestinal secretion of doxorubicin and vinblastine in mice. *Drug Metab Dispos* **28**:264–267.

van Aubel RA, Smeets PH, Peters JG, Bindels RJ and Russel FG (2002) The MRP4/ABCC4 gene encodes a novel apical organic anion transporter in human kidney proximal tubules: putative efflux pump for urinary cAMP and cGMP. *J Am Soc Nephrol* **13**:595–603.

Van Itallie CM and Anderson JM (2006) Claudins and epithelial paracellular transport. *Annu Rev Physiol* **68**:403–429.

Villalobos AR and Braun EJ (1998) Substrate specificity of organic cation/H+ exchange in avian renal brush-border membranes. *J Pharmacol Exp Ther* **287**:944–951.

Wakasugi H, Yano I, Ito T, Hashida T, Futami T, Nohara R, Sasayama S and Inui K (1998) Effect of clarithromycin on renal excretion of digoxin: interaction with P-glycoprotein. *Clin Pharmacol Ther* **64**:123–128.

Walter E and Kissel T (1995) Heterogeneity in the human intestinal cell line Caco-2 leads to differences in transepithelial transport. *Eur J Pharm Sci* **3**:215–230.

Wang D, Jonker JW, Kato Y, Kusuhara H, Schinkel A and Sugiyama Y (2002) Involvement of organic cation transporter 1 in hepatic and intestinal distribution of metformin. *J Pharmacol Exp Ther* **302**:510–515.

Wu CY and Benet LZ (2005) Predicting drug disposition via application of BCS: transport/absorption/ elimination interplay and development of a biopharmaceutics drug disposition classification system. *Pharm Res* **22**:11–23.

Xia CQ, Liu N, Yang D, Miwa G and Gan LS (2005) Expression, localization, and functional characteristics of breast cancer resistance protein in Caco-2 cells. *Drug Metab Dispos* **33**: 637–643.

Yu H, Cook TJ and Sinko PJ (1997) Evidence for diminished functional expression of intestinal transporters in Caco-2 cell monolayers at high passages. *Pharm Res* **14**:757–762.

Zelcer N, Wetering K, Waart R, Scheffer GL, Marschall HU, Wielinga PR, Kuil A, Kunne C, Smith A, Valk M, Wijnholds J, Elferink RO and Borst P (2006) Mice lacking Mrp3 (Abcc3) have normal bile salt transport, but altered hepatic transport of endogenous glucuronides. *J Hepatol* **44**:768–775.

Zhang W, Li J, Allen SM, Weiskircher EA, Huang Y, George RA, Fong RG, Owen A and Hidalgo IJ (2009) Silencing the breast cancer resistance protein expression and function in caco-2 cells using lentiviral vector-based short hairpin RNA. *Drug Metab Dispos* **37**:737–744.

Zhang Y, Bachmeier C and Miller DW (2003) In vitro and in vivo models for assessing drug efflux transporter activity. *Adv Drug Deliv Rev* **55**:31–51.

Chapter 11
Use of In Vivo Animal Models to Assess Drug–Drug Interactions

Thomayant Prueksaritanont

Abstract In this chapter, theoretical basis and specific examples are presented to illustrate the utility of the animal models in assessing and understanding the underlying mechanisms of DDIs. In vivo assessments in an appropriate animal model are considered key to help verify in vivo relevance of in vitro studies and substantiate a basis for extrapolating in vitro human data to clinical outcomes. From a pharmacokinetic standpoint, an important consideration for successful selection of the animal model is based on broad similarities to humans in key physiological and biochemical parameters governing drug absorption, distribution, metabolism, or excretion (ADME) process of interest for both the interacted and the interacting drugs. Also equally important are specific in vitro and/or in vivo experiments demonstrating animal–human similarities, usually both qualitative and quantitative, in the ADME property/process under investigation. Additional insights can also be gained with the use of knockout animals lacking specific drug transporters or drug-metabolizing enzymes and/or transgenic animal models with humanized mouse lines expressing specific drug transporters and/or metabolizing enzymes of interest.

11.1 Introduction

Pharmacokinetic drug interactions, typically characterized by alterations of plasma concentration–time curves, could be mediated via changes in processes of absorption, distribution, metabolism, and/or excretion (ADME) of a drug substance by another compound given concomitantly. In drug discovery and development processes, the assessment of drug–drug interaction (DDI) potential of a drug candidate usually encompasses two main objectives: (1) to help select/design a new chemical entity with least DDI liability potential in humans and (2) to help understand the

T. Prueksaritanont (✉)
Merck Research Laboratories, Department of Drug Metabolism and Pharmacokinetics, West Point, PA, USA
e-mail: thomayant_prueksaritanont@merck.com

K.S. Pang et al. (eds.), *Enzyme- and Transporter-Based Drug–Drug Interactions*,
DOI 10.1007/978-1-4419-0840-7_11,

underlying mechanism of DDI observed in preclinical and clinical studies. The former is the main focus during lead optimization in early discovery phase, while the latter is accomplished throughout from discovery to post-marketing. In both cases, in vitro studies using human tissue preparations are widely utilized and critical for DDI assessments (Weaver, 2001; Worboys and Carlile, 2001). However, uses of the in vitro models are to be successful only when their in vivo relevance can be established. One of the approaches to help bridge this gap involves using a preclinical model to demonstrate an in vitro–in vivo relationship, under conditions reflective of a likely therapeutic scenario. A proper in vitro–in vivo preclinical assessment will not only help form a basis for extrapolating in vitro human data to clinical outcomes but also provide a mechanistic insight for the interpretation of interactions observed clinically (Kanazu et al., 2004; Prueksaritanont et al., 2002).

Conceptually, an appropriate animal model for DDI studies should possess similarities to humans with respect to key physiological or biochemical factors which govern specific ADME characteristics of interest for *both* the drug candidate and an interacting drug under study. Examples of these important factors are organ blood flow, volume or pH, and tissue distribution/localization of drug transporters or metabolizing enzymes. In order to maximize the outcome of the in vivo animal DDI studies, there are three important considerations in choosing the animal model: (1) a thorough understanding of underlying mechanisms of a specific ADME process of both a drug candidate and the interacting drug in humans and the animal model of choice, (2) information regarding similarities between humans and the animal model in physiological and biochemical parameters relevant to the ADME process of interest, and (3) evidence or specific experiments, to show animal–human similarities, ideally both qualitative and quantitative, of a key factor governing the ADME property under investigation. The first and the last considerations are compound specific, necessitating conducting in vivo or in vitro experiments, while the second one is more general and may be accomplished based on literature data. It is important to emphasize that while qualitative similarities to humans with respect to each of the ADME properties under investigation are "must have" to qualify an animal model, quantitative differences are allowed and anticipated between the animal model and humans regarding a specific governing physiological factor, such as expression levels of drug transporters or drug-metabolizing enzymes. Assessment of the magnitude difference is needed for proper interpretations in gauging potential differences between the magnitude of DDI observed in the animal model and anticipated in humans.

In the subsequent part (Section 11.2) of this chapter, animal models known to possess AME-related physiological and biochemical properties common to humans are described. Animal models and DDI cases specific to distribution process have been scarce and thus will not be covered in this chapter. Detailed illustrations on the utility of these animal models in assessing DDI and understanding its associated underlying mechanisms, especially at the levels of drug-metabolizing enzymes and drug transporters, are presented in a following section (Section 11.3). Recent advancements in transgenic animal models with humanized mouse lines expressing and knockout animals lacking specific drug transporters and/or metabolizing enzymes also are expected to provide additional valuable mechanistic insights into the underlying mechanisms of DDIs. For the sake of completeness, general concepts of the utility of these transgenic and knockout mouse models are included in this chapter (Section 11.4); readers are referred to a recent excellent review article on this topic (Lin, 2008) for more details.

11.2 Native Animal Models

11.2.1 Absorption Model

Although extent of drug absorption in humans can generally be extrapolated reasonably well from animal data (Lin, 1995; Lin and Lu, 1998), there have been reports of marked species differences in absorption due to species differences in physiological and/or biochemical parameters. Among the important factors known to cause species differences in absorption include gastrointestinal pH and expression and localization of drug transporters. It is important to emphasize that the term "absorption" is different from "oral bioavailability." Oral bioavailability, which is often used as an indirect measure of drug absorption in vivo, is dependent on both the extent of absorption and the first-pass metabolism/elimination in the intestine and the liver. While high oral bioavailability would be indicative of good absorption, the opposite is not true for compounds subjected to extensive intestinal and/or hepatic metabolism. For such a compound, study designs and data interpretations of in vivo DDI studies are more complicated and are briefly discussed under Sections 11.2.2 and 11.3. Described below are generalities around species similarities and differences in physiological properties to aid in selecting an animal model for studying DDI mediated via absorption.

Dogs: the dog is generally considered as an absorption animal model for humans (Lin, 1995; Lin and Lu, 1998). This is attributable to the well-established similarities in major gastrointestinal physiological features. Exceptions include compounds with pH-sensitive solubility profiles, primarily due to differences in gastric pH (higher in dogs than in humans) between dogs and humans (Lin, 1995). In addition, small, hydrophilic, and passively transported drugs have also been shown to be better absorbed in dogs than in humans (Lenneras, 2007). To date, information has been sparse regarding similarities between dogs and humans in drug transporters expressed in the gastrointestinal tract. Among commonly utilized preclinical species, dogs have been most frequently used for studying absorption-related interactions, primarily between food and drugs (Lentz et al., 2007). However, results from these studies should be interpreted with cautions, considering that food may affect the gastric emptying time and intestinal transit time in dogs differently than in humans (Paulson et al., 2001).

Rats: unlike dogs, rodents are categorized as gastric acid secretor and therefore are used for studying pH-sensitive absorption. McConnell et al. (McConnell et al., 2008) have recently reported that a mean intestinal pH in mice and rats was lower than that in man (<pH 5.2 in the mouse; <pH 6.6 in the rat). Also, the water content in the gastrointestinal tract, when normalized for body weight, was higher in mice and rats than in man. In theory, these physiological differences may have an impact on the extent of drug absorption, but thus far experimental evidence supporting this is lacking. Lenneras (2007) has recently shown a high correlation for drug intestinal permeability between human and rat small intestine ($R^2 = 0.8$–0.95) with both carrier-mediated absorption and passive diffusion mechanisms. Moderate correlation between the two species ($R^2 > 0.56$) was also found for the expression levels of transporters in the duodenum, which provides evidence for a similarity in the molecular mechanism of drug absorption. Transport properties (permeability) for different compounds were also highly correlated between rats and humans when using rat intestinal specimens in the Ussing chamber model. Despite these similarities, reports are limited with respect to successful DDI studies in rats, at the absorption level, due presumably to differences in other key physiological parameters, including gastric emptying and intestinal transit time.

Monkeys: Cynomolgus monkeys (*Macaca fascicularis*) have been suggested to have both the gastric pH during fasted state and the gastric emptying rate similar to humans. However, they have also been shown to express higher levels of MDR1, MRP2, and BCRP in the intestine than those in human intestine (Takahashi et al., 2008). The latter has been substantiated based on apparently lower intestinal permeability (P_{app}) of several compounds known to be substrates of these effluxes in cynomolgus monkey intestine than in human intestine. To date, limited absorption-mediated DDI studies have been reported in cynomolgus monkeys, but based on these known differences, it is foreseeable that the cynomolgus monkeys could be used to provide some insights into transporter-mediated interactions at the intestinal absorption level.

11.2.2 Metabolism Model

Species differences/similarities in drug-metabolizing enzymes, particularly CYPs, have been extensively studied. In the early 1990s, most comparisons were made at enzyme functional activities (Sharer et al., 1995; Prueksaritanont et al., 1996). No single animal species is completely similar to man with regard to functional activities of all drug-metabolizing enzymes. In general, higher species exhibit higher degree of sequence identity to humans in the amino acid sequences. Recent advancements in molecular biology have provided additional insights into the observed differences/similarities at the molecular level. It is important to note that differences in enzyme catalytic specificity have been reported even when a high degree of amino acid sequence identity exists between isoforms; readers are referred to a more comprehensive review by Martignoni et al. on this topic (Martignoni et al., 2006). Additionally, species differences have been reported in enzyme susceptibility to inducers and inhibitors. Most notably, marked species differences between rodents and humans are well documented in response to CYP inducers, including rifampin and omeprazole (Gibson et al., 2002; Graham and Lake, 2008). Complicating this matter further is the finding that there are also species differences between animals and humans in the expression levels and tissue localization of drug-metabolizing enzymes. In this regard, Komura and Iwaki (Komura and Iwaki, 2008) showed that as is the case with hepatic CYP3A, species differences exist in intestinal CYP3A enzymes. Namely, identical CYP3A4 enzyme is expressed in human intestine and liver, but different CYP3A isoforms are expressed in intestines (CYP3A62) and liver (CYP3A1/2) of the rat. Additionally, CYP3A-mediated activities are also much higher in human and/or monkey than rat intestines (Komura and Iwaki, 2008; Roller et al., 2009). These differences could lead to differences in the pharmacokinetic profiles not only following systemic administration but also after oral administration of compounds known to undergo extensive intestinal first-pass metabolism.

In the pharmaceutical industry, both rhesus monkeys (*Macaca mulatta*) and cynomolgus monkeys have generally been considered to be more appropriate animal models than are rodents for drug metabolism studies, based on the sequence and functional activity similarities of drug-metabolizing enzymes (Kanazu et al., 2004; Sharer et al., 1995). Recently, rhesus monkey CYP3A64 has been cloned and in vitro characterization results showed that CYP3A64 was most similar, in protein sequences and functional activities, to human CYP3A4 (Carr et al., 2006). Several CYPs of cynomolgus monkeys have also been cloned and characterized; CYP2A23, CYP2A24, CYP2E1, CYP3A5, and CYP3A8 showed a high sequence identity (93–96%) to the homologous human CYP cDNAs and metabolized typical substrates for human CYPs in the corresponding subfamilies. In contrast, cynomolgus monkey CYP2C76 does not have a corresponding ortholog in humans – this has been shown to be at least partly responsible for differences between cynomolgus monkeys and humans in the metabolism of pivastatin (Uno et al., 2006). Interestingly, we have also recently observed species differences between rhesus monkeys and humans in CYP2C-mediated metabolism of one of our compounds – it is metabolized extensively by CYP2C75 in rhesus monkeys (Tang et al., 2008) but CYP3A4 in humans (unpublished information). Also, unlike human CYP2C9, rhesus CYP2C75 was not sensitive to the inhibitory effect of sulfaphenazole, a potent inhibitor of human CYP2C9-mediated diclofenac 4′-hydroxylation (Tang et al., 2007). These findings underscore the need to have a thorough understanding of the catalyzing enzyme of a drug candidate in the animal model vs. in humans before conducting preclinical in vivo metabolic DDI studies.

The chimpanzee (*Pan troglodytes*) has been characterized as a surrogate for drug oxidation and glucuronidation in humans and as a pharmacokinetic model for the selection of drug candidates (Wong et al., 2004; 2006). Chimpanzees have levels of CYP3A (testosterone 6β-hydroxylation)- and 2C9 (tolbutamide methylhydroxylation)-like enzyme activity similar to those of humans. However, levels of CYP2D (dextromethorphan *O*-demethylation)- and 1A (phenacetin *O*-deethylation)-like enzyme activity were shown to be higher (10-fold) in the chimpanzee than in humans. With respect to the glucuronidation pathway, similarities in the in vivo and/or in vitro metabolism of acetaminophen, estradiol, and morphine have been reported between chimpanzees and humans, also consistent with Western immunoblot analysis of chimpanzee liver microsomes, which revealed a single immunoreactive band when probed with anti-human UGT1A1, anti-human UGT1A6, and anti-human UGT2B7 (Wong et al., 2006). In a more recent study, chimpanzee CYP3A67 has been found to be closely related to human CYP3A7, with the mRNA expression of CYP3A67 comparable to the expression of CYP3A4 (Williams et al., 2007). Also, 99.7% nucleotide similarity of CYP3A5 clone has been reported between chimpanzee and humans (Williams et al., 2007). It is conceivable that the chimpanzee would serve as an animal model for assessment of DDI potential of a drug candidate undergoing CYP3A- and possibly UGT-mediated metabolism.

11.2.3 Excretion Model

Biliary excretion – It is well known that the amount of xenobiotics, especially those with the molecular size less than 700 Da, excreted in bile varies widely among species (Lin, 1995). In general, the underlying mechanism for the species differences is poorly understood, and consequently, no animal species is considered as a representative model for studying biliary excretion in humans. However, this may be changing in the next decade, considering rapidly growing knowledge and understanding of the role and identity of drug transporters. For example, recent data have revealed that rat liver contains much more (~10-fold) multidrug resistance-associated proteins 2 (Mrp2), consistent with a much higher capacity for the biliary excretion of some of organic anions in rats than in humans or other preclinical species (Zamek-Gliszczynski et al., 2006). More information is also becoming available on functional differences and similarities of drug transporters between animals and humans (Herédi-Szabó et al., 2009), which should help provide basis for choosing an animal model for biliary excretion studies.

Renal excretion – The rate of renal excretion (renal clearance) is dependent on renal blood flow, glomerular filtration rate (GFR), and tubular secretion and reabsorption. The GFR values vary considerably among species, depending on the number of nephrons. It is generally well accepted that a good allometric relationship holds for both the GFR and the number of nephrons and that for compounds with GFR and passive reabsorption as major mechanisms for renal excretion, any animal species could be considered as good animal models for humans (Lin, 1995; Lin and Lu, 1998). However, due to species differences in drug transporters, marked species differences in renal excretion may be observed for compounds subjected to significant tubular secretion. In this regard, species differences in the functionality of organic anion transporters (OATs) and tissue localization of organic cation transporters (OCTs) between rats and humans have been reported to be associated with species differences in the renal elimination of cimetidine, as well as a number of organic anions and cations (Tahara et al., 2005a, b). Recent studies showed that, unlike the findings in rats, species similarities have been demonstrated between cynomolgus monkeys and humans with respect to the function and tissue localization of these transporters (Tahara et al., 2006). The authors also reported a successful use of cynomolgus monkeys to study the renal DDI between famotidine/cimetidine and probenecid. In addition, the cynomolgus monkey may also be useful in studying DDI mediated via P-glycoprotein (P-gp), based on high homology in amino acid sequence (>96%) and functionality between cynomolgus P-gp and human P-gp (Lin, 2004; Xia et al., 2006; Tang et al., 2009).

11.3 Case Examples – Use of Native Animal Models for DDI Studies

In this section, we present a number of examples where an animal model has been used to either assess DDI potential during lead optimization or help understand the underlying mechanism of DDI observed for a drug candidate during efficacy/safety evaluations of a drug candidate. The emphasis is given to common DDI cases, namely metabolic and renal excretion interactions, mediated at the levels of drug-metabolizing enzymes and drug transporters, respectively. Although not covered in this chapter, it is important to note that DDI mediated through the interplay of uptake and efflux transporters with metabolic enzymes has also been frequently demonstrated (Custodio et al., 2008; Lam et al., 2007). In addition, our focus is on recently reported DDI examples; readers are referred to a review by Marathe and Rodrigues (2006) for more comprehensive and earlier preclinical case studies on CYP3A- and P-gp-mediated DDIs. Where applicable, we also highlighted potential limitations of the chosen animal model, as well as key experiments connecting findings between animals and humans for meaningful interpretations.

11.3.1 Metabolic Drug Interactions

11.3.1.1 Rhesus Model to Assess CYP3A-Mediated DDI

In our recent study (Prueksaritanont et al., 2006a), the rhesus monkey was chosen as an animal model for CYP3A-mediated DDI assessments based on our previous work (Carr et al., 2006), showing similarities between rhesus CYP3A64 and human CYP3A4 with respect to their ability to metabolize known human CYP3A substrates, including midazolam (MDZ). To further qualify the rhesus monkey for an enzyme induction and inhibition study, we compared between rhesus CYP3A64 and human CYP3A4, and between rhesus monkey and human hepatocytes, their susceptibility to induction and inhibition by a known CYP3A4 inducer, rifampin, and a known inhibitor, compound A. We also showed that in rhesus monkeys, MDZ was metabolized extensively in vivo, with 1′-hydroxy MDZ as the major metabolite, similar to humans. However, rhesus monkeys exhibited a relatively higher metabolic clearance and a lower hepatic availability (Fh − 16%), as compared to humans (Fh ∼40%). Consistent with the induction of hepatic metabolism of a high-clearance compound, pretreatment with rifampin (for 5 days to achieve plasma concentrations comparable to therapeutic concentrations in humans) did not significantly affect the intravenous (i.v.) kinetics of MDZ but caused a pronounced reduction (∼10-fold) in the systemic exposure to MDZ and consequently its Fh following intra-hepatic portal vein (i.pv.) administration of MDZ. The magnitude of DDI agreed well with the finding that MDZ is a higher clearance compound in rhesus than in humans. It is important to note that the i.pv. administration was used in this study to allow a direct comparison with the in vitro hepatocyte studies and avoid potential complications from incomplete absorption and intestinal enzyme induction or inhibition. The latter consideration deduced from the fact that there is little information on the exact identity of CYP3A enzymes in the monkey intestine, unlike the liver, except for comparative functional enzyme activities using CYP3A probes between human and monkey intestines (Komura and Iwaki, 2008). Overall, our results suggested that the rhesus could be used as an animal model to evaluate propensity of a compound to induce CYP3A substrate or be susceptible to induction by a potent CYP3A inducer, provided that appropriate key experiments connecting animal to human findings are done to aid in proper interpretations. Similar in vitro–in vivo findings were also observed with compound A, suggesting the potential applicability of the rhesus monkey model for studying CYP3A inhibition studies in humans. Table 11.1 summarizes the key considerations and information acquired in this study to qualify the rhesus monkey as the animal model for studying DDI between MDZ and rifampin. The same principle could be applied for studying other compound pairs of interest.

Tang et al. (2008) used the rhesus monkey as an animal model to help understand the underlying mechanism of autoinduction observed after oral treatment (once daily) for 4 weeks with a potent bradykinin B_1 receptor antagonist. This compound was primarily eliminated via biotransformation in rhesus monkeys, with oxidation of the chlorophenyl ring as one of the major metabolic pathways (M11 and M13). The formation rate of these two metabolites determined in liver microsomes from compound-treated groups was ≥twofold greater than the value of a control group. Studies with recombinant rhesus P450s and monoclonal antibodies against human P450 enzymes suggested that CYP2C75 played an important role in the formation of M11 and M13. The induction of this enzyme by the compound was further confirmed by a concentration-dependent increase of its mRNA in rhesus hepatocytes and the enhanced CYP2C proteins and catalytic activities toward CYP2C75 probe substrates, as well as M11 and M13 formation in liver microsomes from compound-treated animals. The authors concluded that the compound, both a substrate and an inducer for CYP2C75, caused autoinduction of its own metabolism in rhesus monkeys by increasing the expression of this enzyme. The results from this work, together with additional studies using human hepatocytes, helped to rationalize the autoinduction potential of this compound in clinic (unpublished data).

11.3.1.2 Rhesus Model for Evaluating Diclofenac (DF) as a Valid CYP2C9 Probe in Humans

Based on their in vitro metabolism studies, Kumar et al. (2002) have recently proposed that the direct glucuronidation of DF is a more important component to the in vivo clearance than the oxidation pathway in rats, dogs, and humans. If confirmed,

Table 11.1 Key consideration examples for choosing an animal model for studying metabolic DDI*

	MDZ – interacted drug		Rifampin – interacting drug	
	Human	Monkey	Human	Monkey
Relevant ADME properties	Oxidative metabolism – major clearance mechanism	Same	At therapeutic dose, C_{max} ~10 μM Plasma protein binding ~30%	At the dose used in the study, C_{max} ~15 μM Plasma protein binding ~20%
Key metabolizing enzyme	CYP3A4	CYP3A64 – previously shown to have similar amino acid sequence and catalytic activity to CYP3A4	Not critical	Not critical
Additional supporting evidence (qualitative and quantitative)	Moderate clearance (CL); hepatic extraction ratio (Eh) ~40%	High CL; Eh ~80%	In hepatocytes, induced CYP3A4 mRNA by ~6- to 14-fold and MDZ 1′-hydroxylation by ~2-fold	In hepatocytes, induced CYP3A64 mRNA by ~5- to 10-fold and MDZ 1′-hydroxylation by ~3- to 10-fold
Outcome of DDI studies	~2-fold change in i.v. MDZ CL, consistent with moderate CL characteristics ~20-fold change in oral AUC	No change in i.v. MDZ pharmacokinetics, consistent with high CL characteristic ~10-fold change in AUC after hepatic portal vein administration; expected much higher magnitude following oral administration	N/A	N/A

N/A = Not applicable.
*Extracted from Prueksaritanont et al., 2006a.

the utility of DF as an in vivo CYP2C9 probe in humans is no longer valid. We subsequently used the rhesus monkey as an animal model to show that the CYP2C-mediated oxidative metabolism of DF is not the major determinant for its in vivo clearance in monkeys, and likely in humans (Prueksaritanont et al., 2006b). This conclusion was based on a couple of in vitro and in vivo experiments. First, in both monkey and human liver microsomes (and later hepatocytes), DF underwent predominantly glucuronidation and modestly oxidation; the intrinsic clearance value for the glucuronidation pathway accounted for >90% (vs. about 75% in human liver microsomes) of the total (glucuronidation + hydroxylation) intrinsic clearance value. Second, effects of rifampin on in vitro oxidative metabolism and in vivo pharmacokinetics of DF were investigated in rhesus monkeys. Although rifampin markedly induced DF 4′-hydroxylase activity in monkey hepatocytes (as well as human hepatocytes), pretreatment with rifampin did not alter the pharmacokinetics of DF obtained following either i.v. or i.pv. administration of DF to monkeys. At the dose studied, plasma concentrations of rifampin reached 10 μM, far exceeding the in vitro EC_{50} values for the DF 4′-hydroxylase activity (0.2–0.4 μM). Finally, under similar treatment conditions, rifampin was previously shown to induce MDZ 1′-hydroxylation in rhesus monkey hepatocytes and markedly affected the in vivo pharmacokinetics of MDZ in this animal species (Prueksaritanont et al., 2006a). Based on these in vitro and in vivo studies in rhesus, together with the in vitro findings in humans, it is concluded that rifampin may also elicit modest effects on the DF pharmacokinetics, via induction of CYP2C9, in humans, as was the case in monkeys. Taken together, the results provided a convincing case that DF is not a suitable in vivo probe for CYP2C9-mediated DDI studies in humans.

11.3.1.3 Rat Model for Assessing CYP3A-Mediated DDI

Mandlekar et al. (2007) used the rat as an in vivo screening model to rank order compounds for their potential liability to interact with ketoconazole, a potent CYP3A inhibitor. Based on the relative magnitude of pharmacokinetic interaction observed with ketoconazole in the rat, the compounds were prioritized for further preclinical development. To qualify the rat as an appropriate animal model, they conducted in vitro reaction phenotyping using individual human and rat cDNA-expressed CYP enzymes and human or rat liver microsomes in the presence of ketoconazole to demonstrate similarities between rats and humans with respect to the main drug-metabolizing enzyme CYP3A. The authors acknowledged that the degree of pharmacokinetic interaction with ketoconazole would also be dependent on the fraction of compound metabolized (fm) in the rat relative to other disposition pathways and that fm may be different between rats and humans. It is also important to note that, as highlighted above, species differences exist in intestinal CYP3A proteins between rats and humans and that while oral administration was used in this study, comparative in vitro metabolism was not conducted with rat and human intestinal tissues. These factors may impact different compounds to varying extents, and therefore successful uses of the rat in vivo screening model may be limited to a selected group of compounds, and not applicable for an early broad screening.

11.3.2 Transporter-Mediated Drug Interaction

11.3.2.1 Cynomolgus Monkey Model to Assess Renal DDI

Tahara et al. (2006) proposed the use of cynomolgus monkeys to assess renal DDI mediated by OATs. The basis for choosing the cynomolgus as the animal model was based on (1) a good correlation of the renal transport activities with respect to that of reference compounds between monkey (mk) and human (h) OAT3, as opposed to poor correlation between rat and human OAT3 (Zamek-Gilszczynski et al., 2006), and (2) high expression levels of mkOCT1 and mkOCT2 in the liver and the kidney, respectively, a pattern similar to humans. They further showed that the transport activities of famotidine, a H_2 receptor antagonist by mkOAT3, were comparable to those by hOAT3. While famotidine was transported only by mkOAT3, cimetidine and ranitidine, other H_2 receptor antagonists, are substrates of mkOAT1 and mkOAT3. In monkeys, both famotidine and cimetidine are predominantly excreted into the urine, with almost identical relative contribution between tubular secretion and glomerular filtrate rates to their respective renal clearances. In monkeys, probenecid reduced the renal tubular secretion clearance of famotidine, resulting in a twofold increase in the AUC, consistent with the previous findings in humans. In contrast, the plasma concentration and renal clearance of cimetidine were not affected by probenecid in this species, also similar to the finding in humans. The authors concluded that monkeys, rather than rodents, can be used to predict drug–drug interactions involving tubular secretion, particularly when multiple transporters are involved.

11.3.2.2 Rat and Rhesus Models to Assess Potential Renal Transporter-Mediated DDI

In this example, the mechanism of renal excretion of compound A, a potent and selective $\alpha_v\beta_3$ integrin antagonist, and its renal transporter-mediated DDI potential were investigated (Prueksaritanont et al., 2004). In both rats and rhesus, renal excretion of compound A involved tubular secretion; ratios between renal clearance, corrected for unbound fraction in plasma (CLr,u), and GFR were greater than unity. In rats, the tubular secretion of compound A was inhibited significantly, although modestly (~twofold) by relatively high plasma concentrations of the organic anion PAH and the cation cimetidine but not by the P-gp inhibitor quinidine (~50 μM). In rhesus monkeys, the renal secretion of compound A was not affected by either cimetidine or PAH. In both species, compound A had a minimal effect on the renal tubular secretion of both cimetidine and PAH. In vitro, compound A was not a substrate for P-gp in the Caco-2, human *MDR1*, and mouse *mdr1a*-transfected LLC-PK1 cell lines but was shown using rOAT1- and rOAT3-transfected HEK cell lines to be a substrate for rat OAT3 (K_m = 15 μM), but not rat OAT1. These results suggest that the tubular secretion of compound A is not mediated by P-gp but rather is mediated, at least in part, via the organic anion transporter OAT3, the renal transporter shown to be capable of transporting both the organic anion PAH and the organic cation cimetidine. Unfortunately, no information is available in the literature regarding renal transporters in rhesus monkeys and species differences in the transporters between rats and monkeys. Nevertheless, given the relatively low magnitude of interaction observed in both species, we concluded that the magnitude of interaction between compound A and substrates or inhibitors of OAT3, at the renal excretion level, would likely be modest in humans at clinically relevant doses.

11.4 Transgenic and Knockout Animal Models

With the recent breakthroughs in molecular biology, a number of genetically modified animals have been established as models for evaluating the metabolism and transport of drugs, allowing better understanding of specific roles played by each drug transporter or drug-metabolizing enzyme. Currently, there are many humanized and knockout mouse models of CYPs and other drug-metabolizing enzymes, but mainly knockout mouse models are available for studying functional roles of efflux or influx transporters. Very recently, van de Steeg et al. (2009) have generated and characterized a transgenic mouse model with specific and functional expression of human OATP1B1 (SLCO1B1) in the liver. Although these humanized and/or knockout mouse models are powerful tools for scientific understanding of the function and regulation of particular genes in ADME processes, their uses for quantitative prediction of human pharmacokinetics and DDI assessment are not fully validated. Significant limitations of the genetically modified mouse models include the observations that modification of a given gene does not always result in the anticipated phenotype. Specific examples include compensatory mechanisms with increased sinusoidal membrane Mrp3 expression and activity observed in Mrp2 mutant rats (Konig et al., 1999; Johnson et al., 2006) and increased Mrp4 mRNA and protein in liver and kidney in $Mrp2^{-/-}$ mice (Chu et al., 2006). In addition, differences in liver cytochrome P450 and UGT1a levels between wild-type and mutant rats were detected (Newton et al., 2005; Chu et al., 2006). Thus, interpretations of the significance of the findings from studies using genetically modified mouse models should be done with caution. An in-depth review of this topic with specific examples is provided by Lin (2008).

11.5 Conclusions

Uses of preclinical models to assess DDI are complicated by species differences commonly encountered in the expression level, the functional activity, and the tissue distribution of drug-metabolizing enzymes and drug transporters, major determinants of ADME processes. Although this issue can theoretically be addressed by utilizing in vitro systems using human tissue preparations, the in vivo relevance of such in vitro systems is uncertain and needs to be validated. In this chapter, we describe an in vivo animal model approach to help bridge this gap. An appropriate animal model, when chosen and used properly, could be a valuable tool to provide a basis for extrapolating in vitro human data to clinical outcomes, as well as a mechanistic insight for the interpretation of interactions observed clinically. Other complementary tools for additional insights include knockout animals lacking specific drug transporters or drug-metabolizing enzymes and/or transgenic animal models with humanized mouse lines expressing specific drug transporters and/or metabolizing enzymes of interest. Although quantitative assessments using these animal models are currently limited, it is conceivable that in the next decade they could become more valuable in DDI assessments during drug discovery and early development processes.

References

Carr B, Norcross R, Fang YL, Lu P, Rodrigues AD, Shou MG, Rushmore T and Booth-Genthe C (2006) Characterization of the rhesus monkey CYP3A64 enzyme: Species comparisons of CYP3A substrate specificity and kinetics using baculovirus-expressed recombinant enzymes. *Drug Metab Dispos* **34**:1703–1712.

Chu XY, Strauss JR, Mariano MA, Li J, Newton DJ, Cai X, Wang RW, Yabut J, Hartley DP, Evans DC and Evers R (2006) Characterization of mice lacking the multidrug resistance protein Mrp2 (Abcc2). *J Pharmacol Exp Ther* **317**:579–589.

Custodio J, Wu C and Benet LZ (2008) Predicting drug disposition, absorption/elimination/transporter interplay and the role of food on drug absorption. *Adv Drug Delivery Rev* **60**:717–733.

Gibson GG, Plant NJ, Swales KE, Ayrton A and El-Sankary W (2002) Receptor-dependent transcriptional activation of cytochrome P450 3A genes: induction, mechanisms, species differences and interindividual variations in man. *Xenobiotica* **32**:165–206.

Graham MJ and Lake BG (2008) Induction of drug metabolism: species differences and toxicological relevance. *Toxicology* **254**:184–191.

Herédi-Szabó K, Glavinas H, Kis E, Méhn D, Báthori G, Veres Z, Kóbori L, von Richter O, Jemnitz K and Krajcsi P (2009) Multidrug Resistance Protein 2-mediated estradiol-17β-D-glucuronide transport potentiation: In vitro-in vivo correlation and species specificity. *Drug Metab Dispos* **37**:794–801.

Johnson BM, Zhang P, Schuetz JD and Brouwer KL (2006) Characterization of transport protein expression in multidrug resistance-associated protein (Mrp) 2-deficient rats. *Drug Metab Dispos* **34**:556–562.

Kanazu T, Yamaguchi Y, Okamura N, Baba T and Koike M (2004) Model for the drug–drug interaction responsible for CYP3A enzyme inhibition. I: evaluation of cynomolgus monkeys as surrogates for humans. *Xenobiotica* **34**:391–402.

Komura H and Iwaki M (2008) Species differences in in vitro and in vivo small intestinal metabolism of CYP3A substrates. *J Pharm Sci* **97**:1775–1800.

Konig J, Rost D, Cui Y and Keppler D (1999) Characterization of the human multidrug resistance protein isoform MRP3 localized in the basolateral hepatocyte membrane. *Hepatology* **29**:1156–1163.

Kumar S, Samuel K, Subramanian R, Braun MP, Stearns RA, Chiu SL, Evans DC and Baillie TA (2002) Extrapolation of diclofenac clearance from in vitro microsomal metabolism data: role of acyl glucuronidation and sequential oxidative metabolism of the acyl glucuronide. *J Pharmacol Exp Ther* **303**:969–978.

Lam J, Shugarts S, Okochi H and Benet LZ (2007) Elucidating the effect of final-day dosing of rifampin in induction studies on hepatic drug disposition and metabolism. *J Pharmacol Exp Ther* **319**:864–870.

Lenneräs H (2007) Animal data: the contributions of the Ussing chamber and perfusion systems to predicting human oral drug delivery in vivo. *Adv Drug Delivery Rev* **59**:1103–1120.

Lentz KA, Quitko M, Morgan DG, Grace JE, Gleason C and Marathe PH (2007) Development and validation of a preclinical food effect model. *J Pharm Sci* **96**:459–472.

Lin JH (1995) Species similarities and differences in pharmacokinetics. *Drug Metab Dispos* **23**:1008–1021.

Lin JH (2004) How significant is the role of P-glycoprotein in drug absorption and brain uptake? *Drugs Today* **40**:5–22.

Lin JH (2008) Applications and limitations of genetically modified mouse models in drug discovery and development. *Curr Drug Metab* **9**:419–438.

Lin JH and Lu AHY (1998) Role of pharmacokinetics and metabolism in drug discovery and development. *Pharmacol Rev* **49**:403–449.

Mandlekar SV, Rose AV, Cornelius G, Sleczka B, Caporuscio C, Wang J and Marathe PH (2007) Development of an in vivo rat screen model to predict pharmacokinetic interactions of CYP3A4 substrates. *Xenobiotica* **37**:923–942.

Marathe PH and Rodrigues AD (2006) In vivo animal models for investigating potential CYP3A- and Pgp-mediated drug–drug interactions. *Curr Drug Metab* **7**:687–704.

Martignoni M, Groothuis GMM and de Kanter R (2006) Species differences between mouse, rat, dog, monkey and human CYP-mediated drug metabolism, inhibition and induction. *Expert Opin Drug Metab Toxicol* **2**:875–894.

McConnell EL, Basit AW and Murdan S (2008) Measurements of rat and mouse gastrointestinal pH fluid and lymphoid tissue, and implications for in-vivo experiments. *J Pharm Pharmacol* **60**:63–70.

Newton DJ, Wang RW and Evans DC (2005) Determination of phase I metabolic enzyme activities in liver microsomes of Mrp2 deficient TR^- and EHBR rats. *Life Sci* **77**:1106–1115.

Paulson SK, Vaughn MB, Jessen SM, Lawal Y, Gresk CJ, Yan B, Maziasz TJ, Cook CS and Karim A (2001) Pharmacokinetics of celecoxib after oral administration in dogs and humans: effect of food and site of absorption. *J Pharmacol Exp Ther* **297**:638–645.

Prueksaritanont T, Gorham LM, Hochman JH, Tran L and Vyas KP (1996) Comparative studies of drug-metabolizing enzymes in dog, monkey, and human small intestines, and in Caco-2 cells. *Drug Metab Dispos* **24**:634–642.

Prueksaritanont T, Hochman JH, Meng Y, Pudvah NT, Barrish A, Ma B, Yamazaki M, Fernandez-Metzler C and Lin JH (2004) Renal elimination of a novel and potent $\alpha_v\beta_3$ antagonist in animals. *Xenobiotica* **34**:1059–1074.

Prueksaritanont T, Kuo Y, Tang C, Li C, Qiu Y, Lu B, Strong-Basalyga K, Richards K, Carr B and Lin JH (2006a) In vitro and in vivo CYP3A64 induction and inhibition studies in rhesus monkeys: a preclinical approach for CYP3A-mediated drug interaction studies. *Drug Metab Dispos* **34**:1546–1555.

Prueksaritanont T, Li C, Kuo Y, Tang C, Strong-Basalyga K and Carr B (2006b) Rifampin induces the in vitro oxidative metabolism, but not the in vivo clearance of Diclofenac in rhesus monkeys. *Drug Metab Dispos* **34**:1806–1810.

Prueksaritanont T, Subramanian R Fang X, Ma B, Qiu, Y, Lin JH, Pearson PG, and Baillie TA (2002) Glucuronidation of statins in animals and humans: a novel mechanism of statin lactonization. *Drug Metab Dispos* **30**:505–512.

Roller S, Cui D, Laspina C, Miller-Stein C, Rowe J, Wong B and Prueksaritanont T (2009) Preclinical pharmacokinetics of MK-0974, an orally active calcitonin-gene related peptide (CGRP)-receptor antagonist, mechanism of dose dependency and species differences. *Xenobiotica* **39**:33–45.

Sharer JE, Shipley LA, Vandenbranden MR, Binkley SN and Wrighton SA (1995) Comparisons of phase I and phase II in vitro hepatic enzyme activities of human, dog, rhesus monkey, and cynomolgus monkey. *Drug Metab Dispos* **23**:1231–1241.

Tahara H, Kusuhara H, Chida M, Fuse E and Sugiyama Y (2006) Is the monkey an appropriate animal model to examine drug–drug interactions involving renal clearance? Effect of probenecid on the renal elimination of H-2 receptor antagonists. *J Pharmacol Exp Ther* **316**: 1187–1194.

Tahara H, Kusuhara H, Endou H, Koepsell H, Imaoka T, Fuse E and Sugiyama Y (2005a) A species difference in the transport activities of H-2 receptor antagonists by rat and human renal organic anion and cation transporters. *J Pharmacol Exp Ther* **315**:337–345.

Tahara H, Shono M, Kusuhara H, Kinoshita H, Fuse E, Takadate A, Otagiri M and Sugiyama Y (2005b) Molecular cloning and functional analyses of OAT1 and OAT3 from cynomolgus monkey kidney. *Pharm Res* **22**:647–660.

Takahashi M, Washio T, Suzuki N, Igeta K, Fujii Y, Hayashi M, Shirasaka Y and Yamashita S (2008) Characterization of gastrointestinal drug absorption in cynomolgus monkeys. *Mol Pharmaceutics* **5**:340–348.

Tang C, Carr BA, Poignant F, Ma B, Polsky-Fisher S, Kuo Y, Strong-Basalyga K, Norcross A, Richards K, Eisenhandler R, Carlini E, Ng C, Kuduk S, Yu N, Raab C, Rushmore TH, Frederick CB, Bock MG and Prueksaritanont T (2008) CYP2C75-involved autoinduction of metabolism in rhesus monkeys of MK-0686, a Bradykinin B1 receptor antagonist. *J Pharmacol Exp Ther* **325**:935–946.

Tang C, Fang Y, Booth-Genthe C, Kuo Y, Kuduk S, Rushmore T and Carr BA (2007) Diclofenac hydroxylation in monkeys: efficiency, regioselectivity, and response to inhibitors. *Biochem Pharmacol* **73**:880–890.

Tang C, Kuo Y, Pudvah NT, Ellis JD, Michener MS, Egbertson M, Graham SL, Cook JJ, Hochman J and Prueksaritanont T (2009) Effect of P-Glycoprotein-mediated Efflux on Cerebrospinal Fluid Concentrations in Rhesus Monkeys. *Biochem Pharmacol* **78**:642–647.

Uno Y, Hiroko S, Shotaro U, Takayuki K, Kiyomi M, Chika N, Go K, Tetsuya K and Ryoichi N (2006) A null allele impairs function of CYP2C76 gene in cynomolgus monkeys: a possible genetic tool for generation of a better animal model in drug metabolism. *Mol Pharmacol* **70**:477–486.

van de Steeg E, van der Kruijssen CMM, Wagenaar E, Burggraaff JEC, Mesman E, Kenworthy KE and Schinkel AH (2009) Methotrexate pharmacokinetics in transgenic mice with liver-specific expression of human organic anion-transporting polypeptide 1B1 (*SLCO1B1*). *Drug Metab Dispos* **37**:277–281.

Weaver RJ (2001) Assessment of drug–drug interactions: concepts and approaches. *Xenobiotica* **31**:499–538.

Williams ET, Schouest KR, Leyk M and Strobel HW (2007) The chimpanzee cytochrome P450 3A subfamily: is our closest related species really that similar? *Comp Biochem Physiol D-Genom & Proteom* **2**:91–100.

Wong H, Grace JE, Wright MR, Browning MR, Grossman SJ, Bai SA and Christ DD (2006) Glucuronidation in the chimpanzee (*Pan troglodytes*): studies with acetaminophen, oestradiol and morphine. *Xenobiotica* **36**:1178–1190.

Wong H Grossman SJ, Bai SA, Diamond S, Wright MR, Grace JE, Qian MX, He K, Yeleswaram K and Christ DD (2004) The chimpanzee (*Pan troglodytes*) as a pharmacokinetic model for selection of drug candidates: model characterization and application. *Drug Metab Dispos* **32**:1359–1369.

Worboys PD and Carlile DJ (2001) Implications and consequences of enzyme induction on preclinical and clinical drug development. *Xenobiotica* **31**:539–556.

Xia CQ, Xiao GQ, Liu N, Pimprale S, Fox L, Patten CJ, Crespi CL, Miwa G and Gan LS (2006) Comparison of species differences of P-glycoproteins in beagle dog, rhesus monkey, and human using ATPase activity assays. *Mol Pharmaceutics* **3**:78–86.

Zamek-Gliszczynski MJ, Hoffmaster KA, Nezasa K, Tallman MN and Brouwer KLR (2006) Integration of hepatic drug transporters and phase II metabolizing enzymes: mechanisms of hepatic excretion of sulfate, glucuronide, and glutathione metabolites. *Eur J Pharm Sci* **27**:447–486.

Chapter 12
Extrapolation of In Vitro Metabolic and P-Glycoprotein-Mediated Transport Data to In Vivo by Modeling and Simulations

Motohiro Kato, Yoshihisa Shitara, Masato Kitajima, Tatsuhiko Tachibana, Masaki Ishigai, Toshiharu Horie, and Yuichi Sugiyama

Abstract Recently, a prediction method using in vivo K_i values for inhibitors of cytochrome P450 with a physiologically based pharmacokinetic modeling was proposed to improve the accuracy of the prediction. Also, a method to predict the alterations caused by drug–drug interactions mediated by intestinal cytochrome P450 3A4 or P-glycoprotein was introduced. In this chapter, these methods and computerized simulation method are shown.

12.1 Prediction of Drug–Drug Interactions in Hepatic Metabolism Using Physiologically Based Pharmacokinetic (PBPK) Modeling

12.1.1 Introduction

Most drugs are cleared by cytochrome P450 (CYP)-mediated metabolism. In clinical practice, combination therapies using multiple drugs are routinely applied. Thus, one drug can sometimes inhibit the metabolism of others, consequently increasing their plasma concentrations, which may cause severe adverse reactions. For example, a combination of cerivastatin and gemfibrozil caused severe adverse effects and has even resulted in some deaths (Shitara et al., 2004). Half of the withdrawals of drugs for safety reasons from the US market between 1999 and 2003 were associated with important drug–drug interactions (DDIs) (Huang et al., 2008). It should be

Y. Sugiyama (✉)
Department of Molecular Pharmacokinetics, Graduate School of Pharmaceutical Sciences, The University of Tokyo, Bunkyo-ku, Tokyo, Japan
e-mail: sugiyama@mol.f.u-tokyo.ac.jp

Motohiro Kato and Yoshihisa Shitara are equally contributed to this chapter.

K.S. Pang et al. (eds.), *Enzyme- and Transporter-Based Drug–Drug Interactions*,
DOI 10.1007/978-1-4419-0840-7_12,

noted that, among withdrawn drugs, only mibefradil is an inhibitor, whereas astemizole, cerivastatin, cisapride, and terfenadine are victim drugs (substrate drugs). Thus, to avoid clinically relevant DDIs, new drug candidates should be evaluated for their possible interactions with other drugs, not only as inhibitors but also as victim drugs. In particular, drug candidates with narrow therapeutic windows should be evaluated with precision.

The ratio of the inhibitor concentration for CYP to the inhibition constant (I/K_i) is generally used as an index of hepatic enzyme inhibition. When coadministered with inhibitor drugs, the metabolic rate of substrate drugs reduces to $1/(1 + I/K_i)$, under the condition that the substrate concentration is much lower than the K_m value. Therefore, when the inhibitor concentration is constant, one can predict the maximum degree of increase in the area under the concentration–time curve (AUC) by a factor of $(1 + I/K_i)$. Because human microsomes or human CYP expression systems are now commercially available, it is possible to evaluate the K_i values for human CYPs by in vitro studies. The I/K_i is useful as an index for predicting DDIs (Ito et al., 2002, 2004; Obach et al., 2006). However, false negative predictions, where the actual increased ratio of the AUC exceeds the predicted ratio, are inevitable and this may reduce the success rate of drug candidates. To avoid false negative predictions, the US Food and Drug Administration (FDA) draft guidance (2006) for drug interaction recommends the use of the steady-state total maximum plasma concentration ($I_{p,max}$) as the inhibitor concentration. In contrast, guidance from the Ministry of Health, Labour and Welfare of Japan (2001) recommends the use of the maximum unbound concentration at the inlet to the liver ($I_{u,max}$) as an inhibitor concentration (12.1)

$$I_{u,max} = f_u \cdot \left(I_{p,max} + \frac{k_a \cdot F_a \cdot \text{Dose}}{Q_H} \right) \tag{12.1}$$

where f_u, k_a, F_a, and Q_H represent the blood unbound fraction, the absorption rate constant, the fraction absorbed, and the hepatic blood flow rate, respectively. Both of these inhibitor concentrations should be sufficiently high to avoid false negative predictions. However, these predictions overestimate the extent of DDIs, possibly leading to false negative predictions. Thus, by these methods, a promising candidate drug may be excluded, although it would not actually interact with other drugs. Therefore, the development of a prediction method with a higher accuracy is desired. Predictions using physiologically based pharmacokinetic (PBPK) modeling may be more accurate because the time profile of inhibitor concentrations in the liver is also simulated. In addition, the plasma concentration–time profile of substrates coadministered with an inhibitor has been fitted to the PBPK model to estimate the in vivo K_i values to produce more accurate predictions (Kato et al., 2008). In this chapter, the prediction of DDIs with the PBPK model using in vivo K_i values will be introduced.

12.1.2 False Negative and False Positive Predictions Using the I/K_i Method

To clarify which concentration of inhibitor, i.e., $I_{p,max}$ or $I_{u,max}$, should be used, the prediction using $1 + I/K_i$ and the extent of DDIs were evaluated by Monte Carlo simulation using a PBPK model (Kato et al., 2003b). Prediction using $I_{p,max}$ sometimes provides false negative predictions, whereas that using $I_{u,max}$ provides no false negative predictions (Table 12.1). Overall, the frequencies of false positive, true negative, and true positive predictions using $I_{p,max}$ and $I_{u,max}$ were comparable, suggesting that the accuracy of both methods is similar. A report by Ito et al. (2004) also supports this result. The relationship between $I_{u,max}/K_i$ and $I_{p,max}/K_i$ is shown in Fig. 12.1 based on data by Ito et al. (2004). This relationship represents a 1:1 correspondence. When $I_{u,max}/K_i$ or $I_{p,max}/K_i$ was less than 0.25, some inhibitors increased the AUC of substrates by more than 1.25. False negative predictions were observed when using both concentrations, even though maximum inhibitor concentrations were assumed at any time. Some of the false negative predictions can be explained by mechanism-based inhibition (MBI), whereas others remain unexplained. In vitro K_i values might not reflect in vivo K_i values. Thus, the plasma concentration–time profile of substrates when coadministered with inhibitors was fitted to the PBPK model to estimate the in vivo K_i values.

Table 12.1 Comparison of the predictions using $I_{u,max}$, $I_{p,max,u}$, and $I_{p,max}$[a]

	$I_{u,max}/K_i$	$I_{p,max,u}/K_i$	$I_{p,max}/K_i$
True negative	556	732	534
True positive	32	87	45
False negative	0	14	6
False positive	412	167	415

[a]Data are reproduced from the report by Kato et al. (2003b) with permission from the Japanese Society for the Study of Xenobiotics. $I_{u,max}$, $I_{p,max,u}$, and $I_{p,max}$ are the maximum unbound concentration at the inlet to the liver, the maximum unbound, and total concentrations in the circulating blood, respectively. A change in AUC of the substrate following coadministration of the inhibitor was simulated 1000 times using the physiologically based pharmacokinetic model.

12.1.3 The PBPK-Based Method

The in vivo K_i values of the inhibitors were estimated using a PBPK model (Fig. 12.2) from the blood concentration–time profile of a substrate with coadministration of inhibitors. The following mass balance equations for the substrate and the inhibitor were used.

For substrate drugs

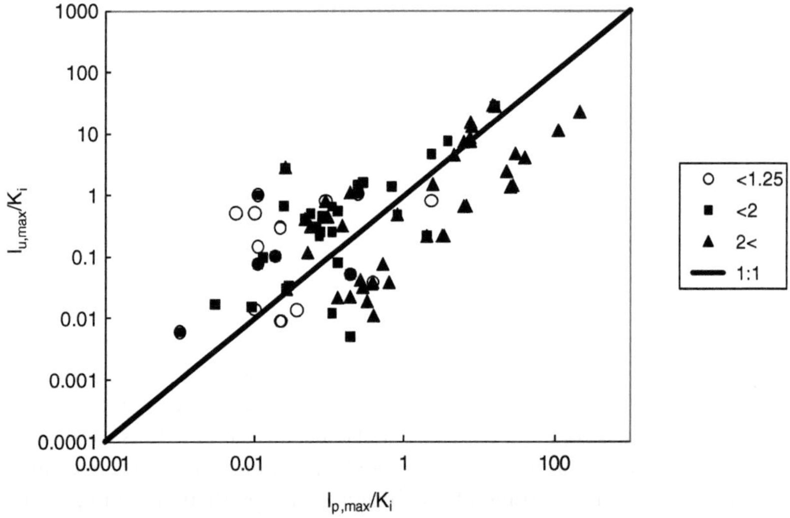

Fig. 12.1 Relationship between $I_{\mathrm{p,max}}/K_{\mathrm{i}}$ and $I_{\mathrm{u,max}}/K_{\mathrm{i}}$. $I_{\mathrm{p,max}}$ and $I_{\mathrm{u,max}}$ is the steady-state total maximum plasma concentration and the maximum unbound concentration at the inlet to the liver, respectively. *Open circle*, *closed square*, and *closed triangle* represent <125, 125–200, and > 200% increase of the control in the area under the plasma concentration–time curve (AUC), respectively. Data are modified from the report by Ito et al. (2004)

$$V_1\frac{\mathrm{d}C_\mathrm{b}}{\mathrm{d}t} = -Q_\mathrm{H}\cdot C_\mathrm{b} + Q_\mathrm{H}\cdot\frac{C_\mathrm{H}\cdot R_\mathrm{B}}{Kp_\mathrm{H}} - k_{12}\cdot V_1\cdot C_\mathrm{b} + k_{21}\cdot X_\mathrm{sub} - \mathrm{CL_{NH}}\cdot C_\mathrm{b} \quad (12.2)$$

$$V_\mathrm{H}\frac{\mathrm{d}C_\mathrm{H}}{\mathrm{d}t} = Q_\mathrm{H}\cdot C_\mathrm{b} - Q_\mathrm{H}\cdot\frac{C_\mathrm{H}\cdot R_\mathrm{B}}{Kp_\mathrm{H}} + k_\mathrm{a}\cdot\left(F_\mathrm{a}\cdot F_\mathrm{g}\right)\cdot\mathrm{Dose}\cdot\exp\left(-k_\mathrm{a}\cdot t\right)$$
$$-f_\mathrm{m}\cdot\frac{\mathrm{CL_{H,int}}}{1+\dfrac{f_{\mathrm{u,p,i}}\big/R_{\mathrm{B,i}}\cdot I_\mathrm{H}}{K_\mathrm{i}}}\cdot\frac{C_\mathrm{H}}{Kp_\mathrm{H}}\cdot f_{\mathrm{u,p}} - (1-f_\mathrm{m})\cdot\mathrm{CL_{H,int}}\cdot\frac{C_\mathrm{H}}{Kp_\mathrm{H}}\cdot f_{\mathrm{u,p}} \quad (12.3)$$

$$\frac{\mathrm{d}X_\mathrm{sub}}{\mathrm{d}t} = k_{12}\cdot C_b\cdot V_1 - k_{21}\cdot X_\mathrm{sub} \quad (12.4)$$

For the inhibitor drugs

$$V_{1,\mathrm{i}}\frac{\mathrm{d}I_\mathrm{b}}{\mathrm{d}t} = -Q_\mathrm{H}\cdot I_\mathrm{b} + Q_\mathrm{H}\cdot\frac{I_\mathrm{H}\cdot R_{\mathrm{B,i}}}{Kp_{\mathrm{H,i}}} - k_{12,\mathrm{i}}\cdot V_{1,\mathrm{i}}\cdot I_\mathrm{b} + k_{21,\mathrm{i}}\cdot X_{\mathrm{sub,i}} - \mathrm{CL_{NH,i}}\cdot I_\mathrm{b} \quad (12.5)$$

$$V_\mathrm{H}\frac{\mathrm{d}I_\mathrm{H}}{\mathrm{d}t} = Q_\mathrm{H}\cdot I_\mathrm{b} - Q_\mathrm{H}\cdot\frac{I_\mathrm{H}\cdot R_{\mathrm{B,i}}}{Kp_{\mathrm{H,i}}} + k_{\mathrm{a,i}}\cdot\left(F_{\mathrm{a,i}}\cdot F_{\mathrm{g,i}}\right)\cdot\mathrm{Dose_i}\cdot\exp\left(-k_{\mathrm{a,i}}\cdot(t+T)\right)$$
$$-\mathrm{CL_{H,int,i}}\cdot\frac{I_\mathrm{H}}{Kp_{\mathrm{H,i}}}\cdot f_{\mathrm{u,p,i}} \quad (12.6)$$

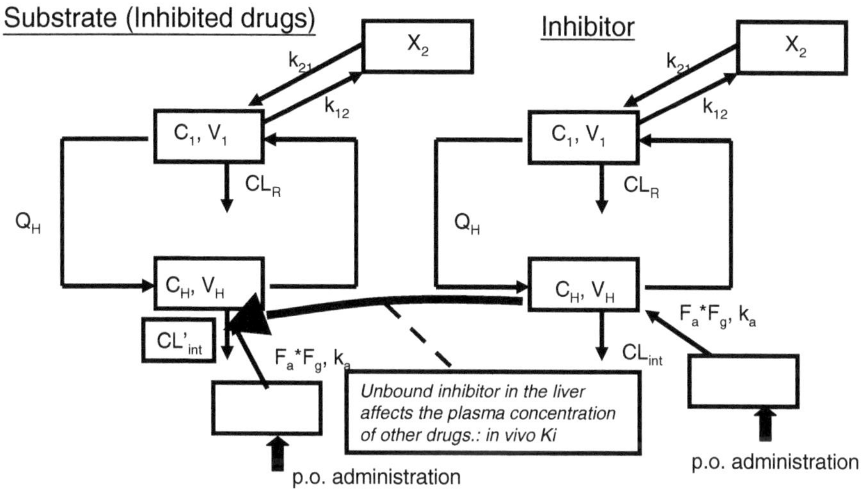

Fig. 12.2 Physiologically based pharmacokinetic model for the description of the time profiles of substrate and inhibitor concentrations. $CL_{H,int}$, V_1, k_a, k_{12}, and k_{21} represent the intrinsic hepatic clearance, the volume of systemic circulation, the absorption rate constant, the transfer rate constant from the systemic circulation to the tissue compartment, and the transfer rate constant from the tissue compartment to the systemic circulation, respectively. C_b, C_H, and X_{sub} are the blood and hepatic concentrations and the amount in the tissue compartment for a substrate, respectively. I_b, I_H, and $X_{sub,i}$ are the blood and hepatic concentrations and the amount in the tissue compartment for an inhibitor. The parameters with subscript "I" represent the parameters for an inhibitor. fm is the contribution of each cytochrome P450 (CYP) isoform to the total hepatic metabolism of a substrate. F_a, F_g, and Q_H are the fraction absorbed, the intestinal availability, and the hepatic blood flow rate, respectively, which was assumed to be 96.6 L/h. Q_H and the hepatic volume (V_H) were assumed to be 96.6 L/h and 1.4 L, respectively

$$\frac{dX_{sub,i}}{dt} = k_{12,i} \cdot I_b \cdot V_{1,i} - k_{21,i} \cdot X_{sub,i} \tag{12.7}$$

The fitting analyses of the itraconazole–midazolam interaction, which is a typical example, are shown in Fig. 12.3. Because itraconazole inhibits CYP3A4 in the liver and the intestine, this interaction was analyzed in two different ways: (i) estimation of K_i alone as a variable, assuming that enzymes only in the liver are inhibited and (ii) estimation of K_i and F_aF_g as variables, assuming that hepatic and intestinal enzymes are inhibited. When K_i was estimated alone, the fitted curve was not satisfactory; however, a good fit was obtained for the combined estimation of K_i and F_aF_g (Fig. 12.3). The values of F_aF_g for cyclosporine, midazolam, sildenafil, simvastatin, tacrolimus, and triazolam were estimated to have increased more than 1.3-fold by coadministration with CYP3A4 inhibitors. These drugs are metabolized by intestinal first-pass metabolism (Kato et al., 2003a; Kato, 2008). The geometric mean, maximum, and minimum values of the in vivo K_i values for 11 inhibitors are shown in Table 12.2. The estimated in vivo K_i values varied widely. For example, the in vivo K_i values of itraconazole were obtained from 12 independent studies

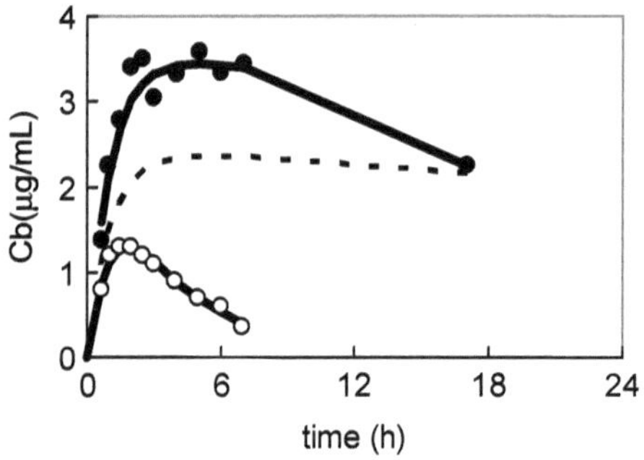

Fig. 12.3 Blood concentration (C_b)–time profiles of midazolam. *Open* and *closed circles* represent placebo and coadministration of itraconazole. *Solid lines* represent the fitted line. The *solid line* with coadministration represents the fitted line, assuming inhibition both in the liver and in the intestine. The *dotted line* represents the fitted line, assuming that there is inhibition only in the liver. Data are reproduced from the report by Kato et al. (2008b) with permission from Springer

and varied widely depending on the studies. The maximum K_i value was more than 100-fold greater than the minimum K_i value (Table 12.2).

12.1.4 Discrepancy Between In Vitro and In Vivo K_i

Figure 12.4 shows the relationship between the in vivo K_i and mean values of the in vitro K_i. There exists a tendency that in vitro K_i values are relatively higher than the in vivo K_i values when they are relatively low. The K_i value for paroxetine was about 4000-fold greater than its in vivo K_i value (Fig. 12.4) because it is a mechanism-based inhibitor (Bertelsen et al., 2003). The in vitro K_i value for itraconazole was about 600-fold greater than the in vivo K_i value (Fig. 12.4). However, the in vitro K_i values for itraconazole corrected by the unbound fraction in microsomes were reported to be 1.3–4.7 nM (Ishigam et al., 2001; Isoherranen et al., 2004), which is close to its in vivo K_i value (0.4 nM). Thus, the difference between the in vitro and the in vivo K_i values might be at least partly because of the binding of itraconazole to microsomes. A good correlation between the $c \log P$ and the ratio of the in vivo and the in vitro K_i values was observed, except for paroxetine and fluvoxamine (Fig. 12.5). This result suggests that, for lipophilic drugs, in vitro K_i values should not be used for the prediction of DDI unless these values are corrected. The correction of in vitro K_i values using the regression curve obtained from the $c \log P$ values and the ratio of the in vivo K_i to the in vitro K_i might be useful. Reported in vitro K_i values vary. The K_i values in the FDA draft guidance for drug interactions also have ranges. For example, the K_i value of ketoconazole is 0.0037–0.18 μM and the ratio of the maximum to the minimum value is about 50-fold. Some drugs are concentrated in the liver by hepatic transporters (Shitara et al., 2003, 2005; Shitara et al., 2006). The active hepatic uptake by transporters might also partly contribute to this discrepancy between in vivo and in vitro K_i values.

Table 12.2 In vitro and in vivo K_i values for several inhibitor drugs for cytochrome P450*

Inhibitor	Enzyme	AUC ratio (%)	n	Substrate	K_i (mean) (μg/L)	K_i (mean) (μM)	K_i (max) (μg/L)	K_i (min) (μg/L)	In vitro K_i (μM)
Azithromycin	3A	127	1	Midazolam	4641	(5.91)			29.5
Cimetidine	3A	132–177	3	Nifedipine, sildenafil, triazolam	13452	(53.3)	39386	6206	71
Fluconazole	2C9	184–217	2	Fluvastatin, glimepiride	12930	(42.2)	33372	5010	8
Fluconazole	3A	182–350	4	Cyclosporine, triazolam, tacrolimus	5270	(12.9)	32762	2570	9.5
Fluoxetine	2D6	225	1	Desipramine	1.76	(0.005)			0.71
Fluvoxamine	2C9	150–164	2	Chloroguanide, tolbutamide	8.74	(0.020)	79.75	0.957	13.3
Fluvoxamine	3A	141	1	Quinidine	28.3	(0.065)			14
Indinavir	3A	438	1	Sildenafil	0.264	(0.056)			0.67
Itraconazole	3A	132–2711	12	Alprazolam, atorvastatin, diazepam, dexamethasone, midazolam, simvastatin, triazolam, zolpidem	0.282	(0.0004)	1.9	0.014	0.187
Ketoconazole	3A	133–2200	5	Midazolam, tacrolimus, triazolam, tirilazad	0.821	(0.0015)	5.03	0.134	0.02
Paroxetine	2D6	175–453	2	Desipramine, imipramine	0.113	(0.0003)	0.412	0.0307	1.22
Propafenone	2D6	171	1	Metoprolol	5.01	(0.013)			0.58
Quinidine	2D6	133–266	4	Desimipramine, haloperidol, imipramine	5.47	(0.017)	32.5	0.600	0.064

*Data are reproduced from the report by Kato et al. (2008). In vivo K_i values were obtained from the PBPK modeling analyses based on the reported clinical drug-drug interaction data. The data of clinically reported drug-drug interactions are also shown in this table. Inhibition of intestinal metabolism was also considered for CYP3A4 substrates. In vitro K_i represents the reported values.

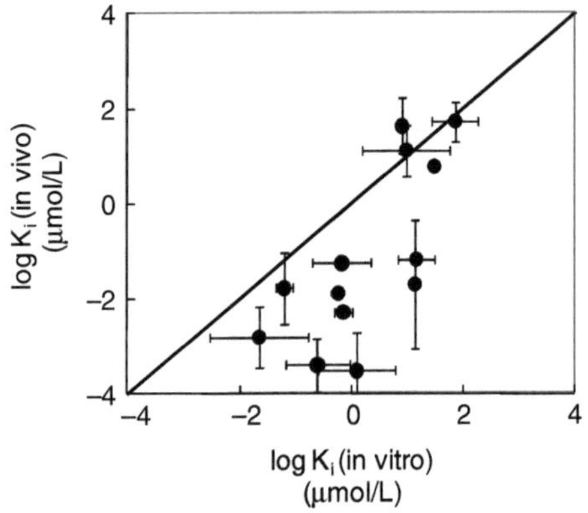

Fig. 12.4 Relationship between the in vitro and the in vivo K_i values. The in vitro K_i values were obtained by the PBPK modeling analyses and in vivo K_i values were obtained from literatures. The K_i values shown here are geometric mean values. Data are reproduced from the report by Kato et al. (2008b) with permission from Springer

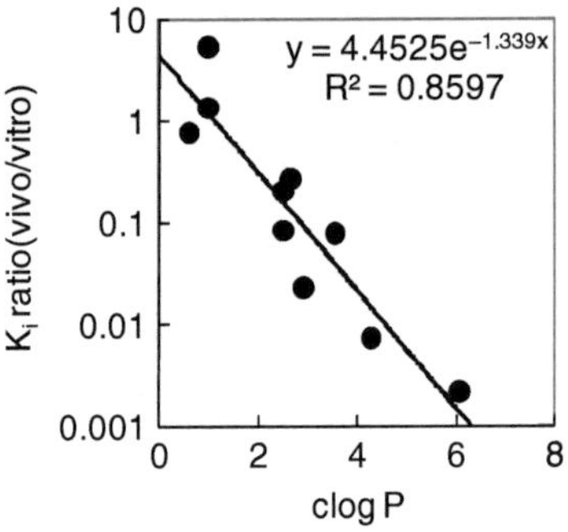

Fig. 12.5 Relationship between clog*P* and the ratio of in vitro K_i to in vivo K_i. The ratio was calculated from the geometric mean values of in vivo K_i values, which were obtained by the PBPK modeling analyses, and in vitro K_i values, which were obtained from literatures (Table 12.2). *c* log *P* is computer-calculated logarithm value of octanol–water partition as neutral. Data are reproduced from the report by Kato et al. (2008b) with permission from Springer

12.1.5 Prediction Result

To verify the accuracy of the predictions by the PBPK model, the results of drug interaction studies for drugs that had been approved in Japan between 1999 and 2004 were analyzed by this method. Among them, azithromycin is an inhibitor, whereas others are substrates. To evaluate predictability, the AUC ratios predicted by the PBPK model were compared with those obtained from $1 + I/K_i$ using the in vivo K_i values. Most of the predictions by $1 + I/K_i$ using the constant $I_{u,max}$ produced false positives, with some of them giving more than 100-fold overestimations (Fig. 12.6). This remarkable false positive prediction is possibly because of an assumption that there is one clearance pathway, although most drugs are cleared via multiple pathways. When substrate drugs are cleared via multiple pathways, even if an inhibitor acts on one of them, other pathways help their elimination and the degree of DDI should be low. It has been reported that predictions using I/K_i

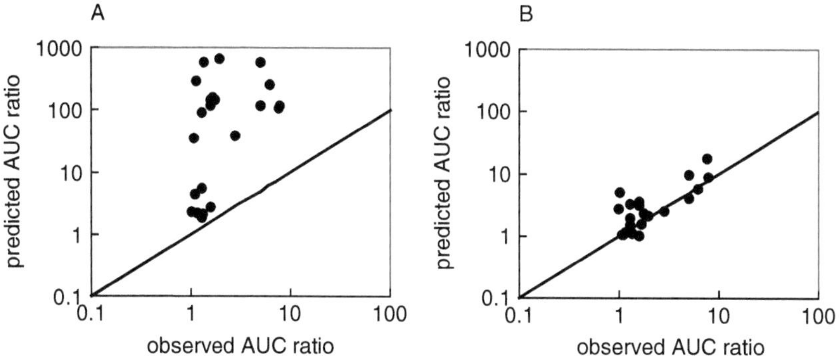

Fig. 12.6 Relationship between the observed and the predicted values by $1+I/K_i$ (A) and the physiologically based pharmacokinetic (PBPK) model using the in vivo K_i for the fold increases in the AUC caused by drug–drug interactions. Data are reproduced from the report by Kato et al. (2008b) with permission from Springer. For predictions with the PBPK model, the inhibition of intestinal enzymes was assumed to be maximal (i.e., $F_aF_g = 1$)

method that consider f_m, which represents the contribution of a single CYP enzyme to the total metabolism, are more accurate (Obach et al., 2006). Predictions using the PBPK model for each combination of inhibitor and substrate drugs, and simulations of hepatic and intestinal inhibition, were performed (Fig. 12.6B). The predictions using the PBPK model taking the time-dependent change of inhibitor concentrations into account were more accurate than predictions using the constant $I_{u,max}$ (Fig. 12.6). As mentioned above, drug candidates should be evaluated as "inhibitors" and "substrates." Although the I/K_i method is widely used as a prediction method, this method is unfavorable, especially when the drug of interest is a victim drug. The method overestimates the degree of DDIs. In contrast, predictions using the PBPK model are suitable for substrate and inhibitor drugs.

12.2 Prediction of DDIs in Intestinal Metabolism and Transport

12.2.1 Introduction

CYP3A4, the most abundant CYP isoform in the human liver, is also expressed in human intestines and mediates the intestinal first-pass metabolism of drugs (Thummel et al., 1996). CYP3A4 substrates and inhibitors interact in the intestine (Kato et al., 2003a). The analysis, using the PBPK model mentioned above, indicates that DDIs mediated by intestinal CYP3A4 affect intestinal metabolism. In addition, P-glycoprotein (P-gp) is also expressed in the intestine and causes DDIs. Oral administration of CYP3A4 or P-gp inhibitors causes a direct and high level of exposure in the intestine. Therefore, even at a low dose, CYP3A4 or P-gp inhibitors may cause DDIs in the intestine rather than in the liver when they are orally administered. To predict intestinal DDIs, a method that considers the liver or the systemic concentration of an inhibitor results in false negatives. Rostami-Hodjegan and Tucker (2004) proposed a prediction method for intestinal enzyme-mediated DDIs, although this has not been validated. To date, there is no standard method for the prediction of DDIs caused by CYP3A4 or P-gp in the intestine. One of the problems in validating a method to predict intestinal DDIs is that it is not possible to estimate the alterations in intestinal availability (F_g). Another problem is the inability to measure the inhibitor concentration in the intestine (I) (Lin et al., 1999). Thus, in this chapter, I/K_i is substituted with the dose of inhibitors divided by K_i (Dose, i/K_i), termed the drug interaction number (DIN), which indexes intestinal interactions because Dose, i divided by intestinal fluid volume (V_g) is independent of inhibitor type (Fig. 12.7) (Tachibana et al., 2009).

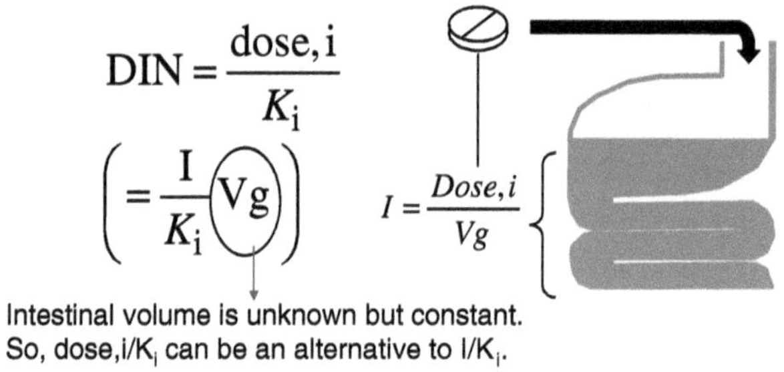

Fig. 12.7 The concept of drug interaction number (DIN). For the prediction of the extent of drug–drug interactions caused by inhibition of intestinal cytochrome P450, the dose of inhibitors divided by K_i (Dose, i/K_i) should be used instead of I/K_i. This value is termed as drug interaction number (DIN), which indexes the extent of drug–drug interactions caused by inhibition of intestinal cytochrome P450 because intestinal fluid volume (V_g) is independent of inhibitor type. This figure illustrates the concept of "DIN."

12.2.2 The Dose/K_i Method

The DIN is defined as an index for DDIs and is determined by the following equation:

$$\text{DIN} = \frac{\text{Dose},i}{K_i}\left(= \frac{I}{K_i}V_g\right) \tag{12.8}$$

Dose,i, the oral dose of inhibitors, equals the product of I and V_g in cases where the inhibitor dissolves completely in the intestine. Because V_g is independent of inhibitor type, the DIN is an appropriate index for intestinal DDIs. The relationship between the fold increase in the AUC of substrate drugs and the DIN was examined from DDI data for CYP3A4-specific substrates, which are not transported by

P-gp, including felodipine, midazolam, and triazolam, and for P-gp-specific substrates, which are not metabolized by CYP3A4, including digoxin, fexofenadine, and talinolol, to elicit a method for predicting the risk of intestinal DDIs. The prediction method was also applied to DDI data for dual CYP3A4/P-gp substrates including atorvastatin, celiprolol, cyclosporine, docetaxel, paclitaxel, saquinavir, and tacrolimus.

12.2.3 Results

The relationship between the DIN of inhibitor drugs and the AUC ratio for CYP3A4-specific substrates (felodipine, midazolam, and triazolam) is shown in Fig. 12.8. A greater DIN tends to correlate with a higher AUC ratio. The inhibitor with the smallest DIN, which increased the AUC of the substrate by more than 1.25-fold, was ranitidine (DIN = 2.8 L). Among the inhibitors that did not increase the AUC of the substrate by 1.25-fold but dissolved in the intestine (Do < 15.6), aprepitant had the highest DIN (9.4 L). Thus, the boundaries for risk of CYP3A4-mediated interactions in the intestine were defined – inhibitors with a DIN below 2.8 L have a low risk, those with a DIN between 2.8 and 9.4 L have a medium risk, and those with a DIN above 9.4 L have a high risk. The relationship between the DIN and the AUC ratio for each P-gp-specific substrate (digoxin, fexofenadine, and talinolol) is shown also in Fig. 12.8. The highest AUC ratio among the P-gp-specific substrates was 2.73-fold and was observed for fexofenadine coadministered with

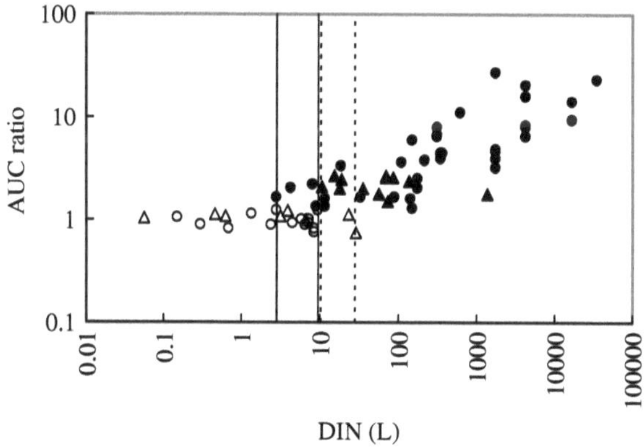

Fig. 12.8 The relationship between the DIN and the increase in the AUC ratio for CYP3A4 and P-glycoprotein (P-gp) substrates coadministered with inhibitors. *Circles* and *triangles* represent CYP3A4 substrates and P-gp substrates, respectively. *Open* and *closed* symbols represent less than and more than 1.25-fold increase in AUC, respectively. *Solid* and *dotted* lines represent DIN boundaries for CYP3A and P-gp inhibitors, respectively

itraconazole, which was lower than that of the CYP3A4-specific substrates. The inhibitor with the lowest DIN that increased the AUC of the substrate more than 1.25-fold was erythromycin (DIN = 10.8 L). The inhibitor with the largest DIN that did not increase the AUC of substrates by 1.25-fold was *R*-verapamil (DIN = 27.9 L). Thus, the boundaries for risk of P-gp interaction are as follows: inhibitors with a DIN below 10.8 L have a low risk, those with a DIN between 10.8 and 27.9 L have a medium risk, and those with a DIN above 27.9 L have a high risk. These DIN boundaries were also applied to the dual CYP3A4/P-gp substrates and were found to be applicable.

Here, a novel index, DIN, for the prediction of intestinal interaction has been introduced and applied to data from DDI studies of CYP3A4 and P-gp substrates and inhibitors. From the results, a set of empirical rules that make possible the prediction of the DDI risk regarding intestinal inhibition of CYP3A4 and/or P-gp will be introduced. These empirical rules are comparable with the index for intestinal DDI presented by the FDA (Zhang et al., 2008) as described below. For a new drug candidate that has an inhibitory effect on CYP3A4 and/or P-gp, this prediction method enables a safe clinical dose of the candidate drug to be determined by comparing K_i with the dose. A high DIN indicates a high risk; thus, researchers will be able to address the problem by selecting other candidates at an earlier stage. Because this method makes it possible to assess risk without defining inhibitor concentration in the liver and the intestine, it should be useful during the new drug development process.

12.2.4 Comparison with FDA Draft Guidance

In 2006, the FDA released draft guidance on transporter-mediated DDIs. In this draft guidance, I/IC_{50} (or K_i), which determines that the mean value of total maximum inhibitor concentration (protein-bound plus unbound) in the intestine following administration of the highest proposed clinical dose of drugs at a steady state divided by IC_{50} or K_i is higher than 0.1, was presented as a way of determining whether an investigational drug is an inhibitor for P-gp and whether an in vivo drug interaction study with a P-gp substrate such as digoxin is needed. Recently, this criterion was modified in an article reported by Zhang et al. (2008). In the report, they proposed that if I_1/IC_{50} (or K_i) > 0.1 or I_2/IC_{50} (or K_i) > 10, then an investigational drug is an inhibitor for P-gp and an in vivo drug interaction study with a P-gp substrate is needed, where I_2 is the mean maximum concentration of inhibitor drugs in the intestine at a steady state and I_2 is obtained by dividing the inhibitor dose by a volume of 250 mL. The volume estimate of 250 mL is derived from typical clinical study protocols that prescribe administration of a drug to subjects with a glass of water. The former condition, I_1/IC_{50} (or K_i) > 0.1, may be appropriate for DDIs in the liver, the kidney, or the blood–brain barrier; however, it is not appropriate for the DDIs in the intestine because it considers only the inhibitor concentration in circulating blood. The latter condition, I_2/IC_{50} (or K_i) > 10, elicited from clinical DDI data together with the I_2 value and the in vitro IC_{50} value is equivalent to the condition DIN > 2.5 L (= 10 × 250 mL). The DIN value, 2.5 L, is somewhat smaller than our DIN value of 10.8 L, which separates low- and medium-risk P-gp-mediated DDIs. The FDA may have adopted this smaller DIN to avoid false negative predictions. Thus, the method shown here and that by Zhang et al. have independently reached a comparable conclusion, indicating the appropriateness of these methods for predicting the risk of intestinal DDI (Zhang et al., 2008).

12.3 Development of a Computer Program for Predicting DDIs in the Liver and the Intestine

12.3.1 Introduction

It was mentioned above that the prediction of DDIs using the PBPK model is more accurate than conventional methods. In addition, we have proposed a method to predict the pharmacokinetic alterations caused by inhibition of intestinal CYPs, which suggests that a more accurate prediction is possible by combining PBPK-based prediction that includes intestinal CYP inhibition. However, to predict clinically relevant DDIs, the competitive inhibition of drug-metabolizing enzymes in the liver or the intestine, as mentioned above, is not sufficient and other mechanisms such as the MBI of CYP enzymes and/or transporter inhibition should also be considered. It should be noted that a remarkable increase in the AUC has been reported, especially in the case of MBI. The PBPK model-based analyses are also useful for the prediction of MBI. Our PBPK model-based prediction can be applied to MBI and/or transporter inhibition.

Although the prediction of DDI using the PBPK model is useful, its use is restricted to only some specialists because of its complexity, which requires some special skills. In addition, without the help of computer programs, such analyses can be complicated and time consuming. Model-based DDI-predicting software should make these analyses easier. Thus, we are developing a computer program that will help researchers to predict DDI using the PBPK model. To date, the program being developed has the following features (Fig. 12.9):

- Database function, which can be customized by users The database contains in vivo K_i and PK parameters of 11 CYP inhibitor drugs and the PK parameters of 39 substrate drugs. Users can easily add parameters of other drugs or new drug candidates.
- Simulation function, which enables PBPK model-based simulation of plasma concentration–time profile when coadministered with CYP inhibitors. The effect of competitive and mechanism-based inhibitors can be simulated. For CYP3A4 inhibitors, a model-based simulation assuming a complete inhibition of intestinal CYP3A4 (i.e., $F_aF_g = 1$) can be performed.

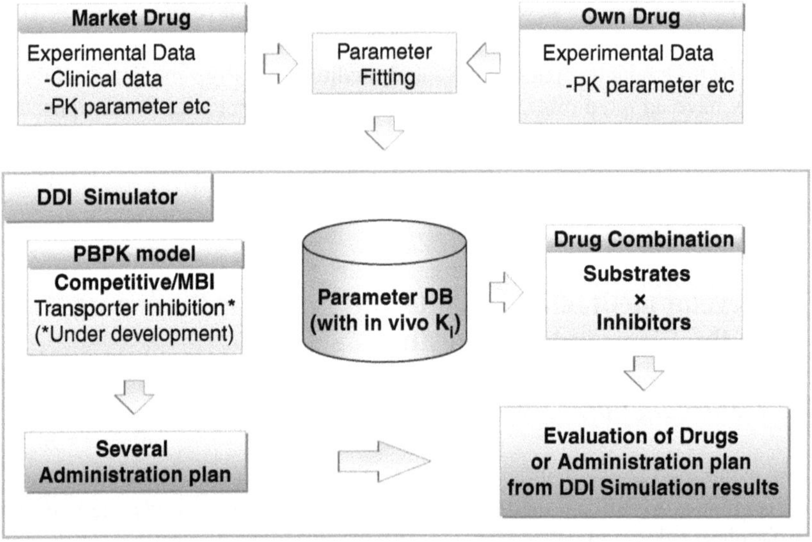

Fig. 12.9 Features of the computer program (the DDI Simulator). The DDI simulator has database and simulation functions. Users can easily add parameters of other drugs or new drug candidates obtained from experimental data or PBPK modeling analyses. This simulator enables the prediction of the extent of drug–drug interactions caused by competitive and mechanism-based inhibition of cytochrome P450. It also allows the simulation of different dosing regimens and helps in dose optimization

12.3.2 Importance of PBPK Model-Based Prediction

In this chapter, we have shown several advantages of PBPK model-based prediction over the conventional method using $1+I/K_i$. Although the conventional method predicts only the maximum increase in AUC, the PBPK method can quantitatively predict the plasma concentration–time profile. It enables the prediction of alterations in other parameters than AUC, including maximum concentration (C_{max}) and half-life ($t_{1/2}$) of substrate drugs. Moreover, it is possible to predict simultaneous inhibitions for several kinds of CYPs and transporters. There have been several reports regarding the prediction of MBI using the PBPK model although they have not been sufficiently evaluated. As mentioned above, substituting the in vivo K_i values calculated using the PBPK model with the in vitro K_i values can predict DDI for competitive inhibition more successfully. In the case of MBI, the turnover rate of enzymes as well as the $K_{i,app}$ value of the inhibitor can be optimized for accurate prediction using the PBPK model. Using these values, more accurate prediction should be obtained. In this analysis, an inhibitor may influence its own clearance, possibly leading to its delayed elimination. It should be noted that this effect will influence the accuracy of prediction. Because PBPK model-based simulations allow easy modification of several parameters and simulation conditions, it is possible to evaluate different parameters or conditions, including the self-inhibition of inhibitors' clearance, for more accurate prediction.

12.3.3 Importance of the Prediction of DDIs in Drug Development

Although a substantial number of pharmacotherapeutics are combined in clinical practices, only a limited number of drug combinations, including typical substrates or inhibitors for CYP enzymes, are examined in clinical studies. However, it is impossible to examine the risks associated with cyclopedic combinations of drugs although it is many times necessary. Ohno et al. (2007) proposed a method of cyclopedic prediction using the ratio of the contribution of CYP3A4 to the total oral clearance of substrates and the time-averaged apparent inhibition ratio of CYP3A4 calculated from the results of clinical trials. Although this method is based only on clinical results and does not require K_i values, it enables easy prediction of DDIs. However, it cannot be applied to new compounds which function as inhibitors because of the lack of in vivo information on inhibitory potencies. On the other hand, our program enables the prediction of DDIs caused by newly developed compounds with their in vitro K_i values by registering new drug information in the database. It also enables cyclopedic and accurate predictions, even for inhibitions involving several metabolic enzymes, with each of the in vitro K_i values for multiple enzymes. By using this program, plasma concentrations of multiple substrate drugs coadministered with inhibitors can be simulated at once (Fig. 12.10). If precise simulation by this program is validated, it may help make plans of clinical DDI studies, possibly resulting in waiver from clinically DDI studies, which are predicted to make low risks. Our program also allows the simulation of different dosing regimens and helps in dose optimization. For example, there are some cases where DDIs occur for drugs given in a single dose a day; however, this can be avoided by changing the administration plan to three times a day or changing the dosing schedule. Our program determines better dosing regimens by using comparative simulations. It can be used for a wide range of risk evaluation, from non-clinical to clinical phases.

PBPK model-based analyses show that the simulation using in vitro K_i values does not necessarily reflect the result observed in vivo. It would be desirable to calculate the in vivo K_i value from the results of DDIs in clinical studies; however, this calculation method is not easy. The program currently uses the method of predicting the in vivo K_i value based on correlations of $c \log P$ with the ratio of the in vivo to the in vitro K_i values. More accurate and easier calculation of in vivo K_i values will be implemented in the program in the future.

In conclusion, the computerized DDI simulator can help in the screening of a large number of potentially valuable drug candidates, with a low risk of DDIs, at an earlier stage in the drug development process. The DDI simulator would reduce the time and cost of the new drug development. In addition, the program will help determine a dosing regimen that reduces the risk of pharmacokinetic alterations caused by DDIs in clinical stages.

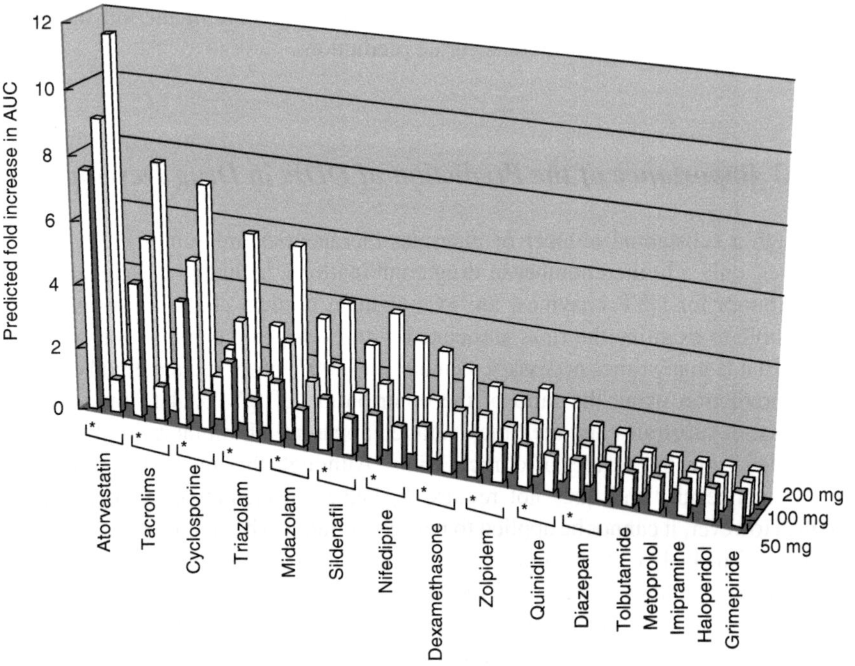

Fig. 12.10 Prediction of pharmacokinetic alterations of multiple drugs caused by coadministration of itraconazole (10, 50, or 200 mg). The DDI simulator can simultaneously predict the pharmacokinetic alterations of multiple drugs caused by coadministration of single or multiple inhibitors. Here, the effect of coadministration of 10, 50, or 200 mg itraconazole is shown. $*$ means pharmacokinetic alterations, assuming that intestinal metabolism is completely inhibited (i.e., $F_a F_g = 1$)

References

Bertelsen KM, Venkatakrishnan K, Von Moltke LL, Obach RS and Greenblatt DJ (2003) Apparent mechanism-based inhibition of human CYP2D6 in vitro by paroxetine: comparison with fluoxetine and quinidine. *Drug Metab Dispos* **31**:289–293.

Food and Drug Administration (2006) Guidance for industry: drug interaction studies – study design, data analysis, and implications for dosing and labeling (draft guidance).

Huang SM, Strong JM, Zhang L, Reynolds KS, Nallani S, Temple R, Abraham S, Habet SA, Baweja RK, Burckart GJ, Chung S, Colangelo P, Frucht D, Green MD, Hepp P, Karnaukhova E, Ko HS, Lee JI, Marroum PJ, Norden JM, Qiu W, Rahman A, Sobel S, Stifano T, Thummel K, Wei XX, Yasuda S, Zheng JH, Zhao H and Lesko LJ (2008) New era in drug interaction evaluation: US food and drug administration update on CYP enzymes, transporters, and the guidance process. *J Clin Pharmacol* **48**:662–670.

Ishigam M, Uchiyama M, Kondo T, Iwabuchi H, Inoue S, Takasaki W, Ikeda T, Komai T, Ito K and Sugiyama Y (2001) Inhibition of in vitro metabolism of simvastatin by itraconazole in humans and prediction of in vivo drug–drug interactions. *Pharm Res* **18**: 622–631.

Isoherranen N, Kunze KL, Allen KE, Nelson WL and Thummel KE (2004) Role of itraconazole metabolites in CYP3A4 inhibition. *Drug Metab Dispos* **32**:1121–1131.

Ito K, Brown HS and Houston JB (2004) Database analyses for the prediction of in vivo drug–drug interactions from in vitro data. *Br J Clin Pharmacol* **57**:473–486.

Ito K, Chiba K, Horikawa M, Ishigami M, Mizuno N, Aoki J, Gotoh Y, Iwatsubo T, Kanamitsu S, Kato M, Kawahara I, Niinuma K, Nishino A, Sato N, Tsukamoto Y, Ueda K, Itoh T and Sugiyama Y (2002) Which concentration of the inhibitor should be used to predict in vivo drug interactions from in vitro data?. *AAPS Pharm Sci* **4**:E25.

Kato M (2008) Intestinal first-pass metabolism of CYP3A4 substrates. *Drug Metab Pharmacokinet* **23**:87–94.

Kato M, Chiba K, Hisaka A, Ishigami M, Kayama M, Mizuno N, Nagata Y, Takakuwa S, Tsukamoto Y, Ueda K, Kusuhara H, Ito K and Sugiyama Y (2003a) The intestinal first-pass metabolism of substrates of CYP3A4 and P-glycoprotein-quantitative analysis based on information from the literature. *Drug Metab Pharmacokinet* **18**:365–372.

Kato M, Shitara Y, Sato H, Yoshisue K, Hirano M, Ikeda T and Sugiyama Y (2008) The quantitative prediction of CYP-mediated drug interaction by physiologically based pharmacokinetic modeling. *Pharm Res* **25**:1891–1901.

Kato M, Tachibana T, Ito K and Sugiyama Y (2003b) Evaluation of methods for predicting drug–drug interactions by Monte Carlo simulation. *Drug Metab Pharmacokinet* **18**:121–127.

Lin JH, Chiba M and Baillie TA (1999) Is the role of the small intestine in first-pass metabolism overemphasized?. *Pharmacol Rev* **51**:135–158.

Ministry of Health, Labour and Welfare, Japan (2001) Methods of drug interaction studies: notification No. 813 of the pharmaceutical affair bureau 2001 Japan.

Obach RS, Walsky RL, Venkatakrishnan K, Gaman EA, Houston JB and Tremaine LM (2006) The utility of in vitro cytochrome P450 inhibition data in the prediction of drug–drug interactions. *J Pharmacol Exp Ther* **316**:336–348.

Ohno Y, Hisaka A and Suzuki H (2007) General framework for the quantitative prediction of CYP3A4-mediated oral drug interactions based on the AUC increase by coadministration of standard drugs. *Clin Pharmacokinet* **46**:681–696.

Rostami-Hodjegan A, Tucker GT. (2004) "In silico" simulations to assess the "in vivo" consequences of "in vitro" metabolic drug–drug interactions. *Drug Discov Today: Technol* **1**:441–448.

Shitara Y, Hirano M, Sato H and Sugiyama Y (2004) Gemfibrozil and its glucuronide inhibit the organic anion transporting polypeptide 2 (OATP2/OATP1B1:SLC21A6)-mediated hepatic uptake and CYP2C8-mediated metabolism of cerivastatin: analysis of the mechanism of the clinically relevant drug–drug interaction between cerivastatin and gemfibrozil. *J Pharmacol Exp Ther* **311**:228–236.

Shitara Y, Horie T and Sugiyama Y (2006) Transporters as a determinant of drug clearance and tissue distribution. *Eur J Pharm Sci* **27**:425–446.

Shitara Y, Li AP, Kato Y, Lu C, Ito K, Itoh T and Sugiyama Y (2003) Function of uptake transporters for taurocholate and estradiol 17beta-D-glucuronide in cryopreserved human hepatocytes. *Drug Metab Pharmacokinet* **18**:33–41.

Shitara Y, Sato H and Sugiyama Y (2005) Evaluation of drug–drug interaction in the hepatobiliary and renal transport of drugs. *Annu Rev Pharmacol Toxicol* **45**:689–723.

Tachibana T, Kato M, Watanabe T, Mitsui T and Sugiyama Y (2009) A method for predicting the risk of drug–drug interactions involving inhibition of intestinal CYP3A4 and P-glycoprotein. *Xenobiotica* in press.

Thummel KE, O'Shea D, Paine MF, Shen DD, Kunze KL, Perkins JD and Wilkinson GR (1996) Oral first-pass elimination of midazolam involves both gastrointestinal and hepatic CYP3A-mediated metabolism. *Clin Pharmacol Ther* **59**:491–502.

Zhang L, Zhang YD, Strong JM, Reynolds KS and Huang SM (2008) A regulatory viewpoint on transporter-based drug interactions. *Xenobiotica* **38**:709–724.

Chapter 13
Translation of In Vitro Metabolic Data to Predict In Vivo Drug–Drug Interactions: IVIVE and Modeling and Simulations

Amin Rostami-Hodjegan

Abstract For many years, researchers have been able to study xenobiotic metabolism using several in vitro systems. However, quantitative translation of the information, which requires appropriate understanding of the link between these systems and the human body, is still in its infancy. Recent progress in systems biology, the advent of powerful computers capable of handling numerous complicated non-linear models and the realization that there is virtually no end to the number of various studies which would be required to cover all the possible permutations of clinical scenarios that can occur in real life have all contributed to the increased prominence of in vitro–in vivo extrapolation (IVIVE) and virtual clinical studies using modeling and simulation in drug discovery and development. This chapter provides an overview of these translational aspects and describes the prerequisite data, models and tools which can be applied for the prediction of metabolic drug–drug interactions (M-DDI).

13.1 Introduction

In previous chapters, the enzymatic basis of drug metabolism and commonly used in vitro techniques for studying metabolism were discussed in detail. The use of these techniques may help with anticipating and minimizing potential metabolic drug–drug interactions during discovery. Regulatory perspectives on the value of in vitro data and the extrapolation of such data, as a guide to determine the need for specific clinical M-DDI studies and their design, will be the subject of later chapters. It will also become evident from the discussions in Chapters 26 and 27 that M-DDI would have financially relevant consequences with respect to labelling, regulatory

A. Rostami-Hodjegan (✉)
Department of Human Metabolism (M129), The Medical School, University of Sheffield, Sheffield, South Yorkshire, UK
e-mail: a.rostami@sheffield.ac.uk

K.S. Pang et al. (eds.), *Enzyme- and Transporter-Based Drug–Drug Interactions*, 317
DOI 10.1007/978-1-4419-0840-7_13,

approval and competitive marketing. Thus, identifying, quantifying and managing M-DDI are not concerns solely for the prescribers engaged in the clinical and toxicological aspects of drug treatment. There are databases which collect reported M-DDI and provide records showing specific information on the "victim" or "perpetrator" drugs (Carlson et al., 2002); these provide evidence for the proportional role that certain pathways may play in the elimination of drugs and could satisfy the need to identify clinically important metabolic interactions for drugs on the market (see also Chapter 15). However, these databases cannot replace the in vitro–in vivo extrapolation (IVIVE) which is necessary for the prediction of M-DDI associated with the new drug candidates. Attempts have been made to use existing data from clinical observations and combine them with predictive models of M-DDI (Ohno et al., 2007, 2008; Kozawa et al., 2009); these are not the focus of this chapter as we concentrate on predictive methods based on in vitro data for drug candidates at early stages of development (i.e. IVIVE) when human data are sparse.

Although in vitro screening tools for assessing metabolic stability of xenobiotics have been used for a long time, the quantitative extrapolation of in vitro observations to clinical consequences is a relatively modern technique. Moreover, it is now well recognized that such efforts should not be limited to likely events in an "average" person but they should also explore certain sub-groups of patients who might be at higher risk (Krayenbuhl et al., 1999). The latter requires a good understanding of the biological elements involved in each scaling step of the IVIVE exercise and the inter-individual variability which stems from genetic as well as epigenetic factors.

Einolf, in 2007, proposed a classification to distinguish between various approaches used to predict M-DDIs due to CYP inhibition (Einolf, 2007). Predictions were categorized according to whether they were "pragmatic" and based purely on a static inhibitor concentration and the inhibition constant (e.g. "$[I]/K_i$" approach); based on static-mechanistic models if they utilized additional information on the characteristic of the "victim" drug such as the fraction of clearance via the inhibited pathway (fm) and availability across the gut wall (F_G); or based on a dynamic mechanistic model if they also considered time-variant inhibitor concentration. This has been extended recently by Almond et al. (2009) to describe the current status of investigations used to predict M-DDIs resulting from CYP induction. Since the assessment of changes in full kinetics is the ultimate goal and will arguably be the most reliable method to quantitatively predict in vivo M-DDI, this chapter summarizes the required data, models and infrastructure which are needed to translate in vitro metabolic information to in vivo expectations.

The information that concerns IVIVE and modeling and simulation efforts to assess M-DDI are provided in three sections in this chapter encompassing the considerations for characterization of concentration–time profiles of xenobiotics (as "perpetrator" or "victim" drugs), the general structure of an ideal "modeling and simulation" platform for IVIVE and prediction of ADME, and finally, the specific requirements for prediction of M-DDI.

13.2 Translation of In Vitro ADME Data to Characterize In Vivo Pharmacokinetics

Defining the plasma concentration–time course of a drug involves prediction of absorption (unless administered via parenteral routes), distribution, metabolism and excretion (ADME). Many "pragmatic" approaches may not appear to require the knowledge of full kinetics as they rely on a "single static" value of the average plasma concentration of the "perpetrator". However, in reality, assigning a single average steady-state value requires some knowledge of the dose as well as bioavailability and clearance. Such information is not available for new candidate drugs under development until late clinical phases. Thus, the value of so-called pragmatic approaches in assessing the M-DDI potential of a new drug candidate as a "perpetrator" is limited. Moreover, recent simulation studies have indicated that the timing of administration of the "perpetrator" and "victim" drugs can be a critical determinant of the level of inhibition (Yang et al., 2003; Zhao et al., 2009). These effects cannot be captured using "pragmatic" approaches which ignore time-varying concentrations of the interacting drugs. So, it is unsurprising that there is increasing recognition of the fact that there is not a single optimal design for drug interaction evaluation (Huang and Lesko, 2009). Thus, it is recommended that pharmacokinetic characteristics of both substrates and inhibitors should be considered when designing in vivo drug interaction studies (Huang and Lesko, 2009) despite the fact that regulatory guidelines may not have indicated this explicitly [Food and Drug Administration (FDA), 2006]. Modeling and simulation, combined with IVIVE, can be used to characterize concentration–time profiles and guide the best dosing strategy during the conduct of clinical M-DDI studies (Huang and Lesko, 2009).

Accurate IVIVE of ADME is possible when the in vitro data are reliable and all necessary extrapolation factors (physical chemistry, biology, physiology and genetics) are available for each in vitro system. Table 13.1 summarizes some of the factors which influence the various components of ADME. More popular use of IVIVE may be attributed to the ability to "integrate" various aspects related to the human body and the characteristics of drugs shown in Table 13.1 (Rostami-Hodjegan and Tucker, 2007), which has been in parallel to an increased perception of "model-based drug development" initiatives in the pharmaceutical industry (e.g. see Lalonde et al. 2007) and the regulatory agencies (e.g. see Buckman et al. 2007).

13.2.1 Prediction of Exposure

The area under the concentration–time curve (AUC) defines the overall (systemic) exposure, whereas other features of concentration–time curves are used to define early exposure (e.g. truncated AUC) and maximal exposure ($C_{\max}$) (Rostami-Hodjegan et al., 1994). AUC after any non-parenteral administration is dependent on the proportion of the dose that is absorbed and subsequently available in the systemic circulation. The most common route for drug intake is oral drug administration. Bioavailability (F), in the case of oral administration of solid dosage forms,

Table 13.1 Separation of drug-related factors determining ADME from those concerning the system (human body) and study design when considering translation of in vitro data to in vivo observations

	Influential factors		
ADME element	Inherent drug characteristics	Inherent physiology and biology	Study condition/design
Absorption	– Release from formulation – Solubility, dissolution, precipitation rate – Permeability (and lipophilicity) – Affinity to gut wall enzymes – Chemical stability – Molecular weight and pK_a	– Baseline pH and GI tract motility (may differ by age, sex, ethnicity, etc.) – Stomach and intestinal residence time (may differ by age, sex, etc.) – Enzyme levels and activity in gut wall (may differ by genetics, environment, age, sex, etc.) – Enzyme levels and activity in liver (may differ by genetics, environment, age, sex, etc.) – Liver (/gut) size and its blood flow (may differ by age, sex, etc.) – Plasma proteins (may differ by age, sex, etc.) – Transporter levels and activity in gut and liver (may differ by genetics, age, sex, environment, etc.)	– Size of the dose taken – Volume of fluid taken with the dose – Type and volume of food taken with the dose and its staggering from the dose – Posture and other conditions of study affecting stomach and intestinal residence time and blood flow – Composition of study population (sex, age, dietray habits) which affects distribution of physiological and biological values
Distribution	– Permeability (and lipophilicity) – Molecular weight and pK_a – Protein binding and affinity to transporters – Tissue and red blood cell partitioning	– Organ size and perfusion (may differ by age, sex, environment, etc.) – Protein levels (may differ by age, sex, etc.) – Hematocrit (may differ by age, sex, environment, etc.) – Transporters (may differ by genetics, age, sex, environment, etc.)	– Composition of study population (sex, age, race, ect) which affects distribution of physiological and biological values

Table 13.1 (continued)

	Influential factors		
ADME element	Inherent drug characteristics	Inherent physiology and biology	Study condition/design
Metabolism	– Molecular size, permeability (/polarity) and pK_a – Affinity to kidney transporters – Protein binding and – Red blood cell partitioning	– Transporter levels and activity in kidney (may differ by genetics, age, sex, environment, etc.) – Plasma proteins (may differ by age, sex, environment, etc.) – Kidney function (may differ by age, sex, environment, etc.) – Urine flow and its pH – Hematocrit (may differ by age, sex, environment, etc).	– Posture and other conditions of study affecting liver blood flow – Composition of study population (sex, age, dietray habits) which affects distribution of physiological and biological values
Excretion	– Affinity to liver enzymes – Affinity to liver transporters – Protein binding and – Red blood cell partitioning	– Enzyme levels and activity in liver (may differ by genetics, age, sex, environment, etc.) – Transporter levels and activity in liver (may differ by genetics, age, sex, environment, etc.) – Plasma proteins (may differ by age, sex, environment, etc.) – Liver size and perfusion (may differ by age, sex, environment, etc.) – Protein levels (may differ by age, sex, etc.) – Hematocrit (may differ by age, sex, environment, etc.) – Transporter (may differ by genetics, age, sex, environment, etc.)	– Fluid intake affecting urine flow – Diet affecting urine pH – Composition of study population (sex, age, dietray habits) which affects distribution of physiological and biological values

involves release of the drug from the formulation, dissolution, passage through the gut wall and subsequently through the liver. The bioavailability together with the clearance (CL) and the dose of the drug (*D*) determines the overall systemic exposure (AUC = *FD*/CL).

13.2.1.1 IVIVE to Determine Oral Bioavailability

Bioavailability can be split into three elements involving the fraction of the dose which enters the gut wall (fa), the fraction of the drug which escapes metabolism in the gut wall and enters the portal vein (F_G) and the fraction of the drug that enters the liver and escapes metabolism (F_H) (i.e. $F = \text{fa} \times F_G \times F_H$). Low "fa" could be associated with one, or several, of the following problems:

(a) Decomposition in the gut lumen
(b) Problems with the release of the active moiety from the formulation
(c) Failure of the solid form to dissolve within the window of absorption time
(d) Failure of dissolved molecules to permeate passively through the gut wall
(e) Efflux of drug in the enterocytes by gut wall transporters (acting as a barrier to absorption in the absence of adequate passive permeability)

Absolute or relative values of "fa" and each of its elements can be assessed by in vitro systems and subsequent IVIVE (Jamei et al., 2009b). However, validating IVIVE attempts for estimated "fa" is fraught with problems since it is not possible to obtain a separate estimate of "fa" and "F_G" from ordinary clinical data. Many reports in the literature erroneously refer to "fa" when the composite function of "F_G × fa" has been used in their assessments. Evidently, these values can be assumed to be the same only if there is no gut wall metabolism at all (i.e. $F_G = 1$).

The importance of gut wall metabolism (and consequently F_G) has been highlighted in Chapter 17. Although there are simple models to describe F_G based on some in vitro data (Yang et al., 2007), it is well understood that these are not based on fully mechanistic approaches (Yang et al., 2007). Nonetheless, they accommodate the two main drug-related parameters defining F_G, namely permeability and metabolism by gut wall enzymes, as well as some physiological parameters such as blood flow to the villi (Yang et al., 2007). Other more mechanistic models occasionally contain elements which cannot be measured in vitro (Pang, 2003; Tam et al., 2003). However, the use of such models is increasing as investigators use them for scientific exploration beyond IVIVE (Jamei et al., 2009b).

Assessment of F_H is an integral part of estimating liver clearance, which is discussed in the next section.

13.2.1.2 IVIVE to Determine Drug Clearance

Metabolism is a major route of elimination for the majority of drugs currently on the market (Wienkers and Heath, 2005). Full understanding of the contribution that each metabolic route makes to the overall elimination of a given drug helps with the assessment of M-DDI by assigning an important parameter to a "victim" drug – fractional elimination (fm) via inhibited pathway. In addition, the relative importance of genetic polymorphisms and the effects of environmental factors could be evaluated once the contributions of various routes are known. There are different in vitro systems to facilitate estimation of liver clearance including human liver microsomes, human hepatocytes and recombinantly expressed human enzymes. Each of these requires their own scaling factor (Barter et al., 2007) and has their own advantages and disadvantages. Many early screening methodologies for assessing metabolism are designed to evaluate the overall stability rather than identify proportional elimination by each route. In vivo hepatic intrinsic clearance and associated variability in human populations can be extrapolated from the in vitro systems as previously described (Rostami-Hodjegan and Tucker, 2007).

Intrinsic clearance can be converted to whole organ clearance using various models (Pang and Chiba, 1994) (e.g. "well-stirred", "parallel tube", "dispersion"). These models, in their original form, assume that the passage of the drug from the blood into the liver is dependent on liver blood flow (Q_H) and that only unbound drug in blood (f_{uB}) crosses the cell membrane and is available to be metabolized. Variations of these models are now used to accommodate the effects of transporters in hepatic uptake and efflux (see Chapter 16).

Although many drugs are mainly eliminated via metabolism, renal excretion plays a major role in elimination of a considerable number of drugs. Moreover, even when renal excretion is not a major elimination route, it plays a significant role in determining the maximum level of M-DDI that can be achieved by full inhibition of metabolic pathway (i.e. 1/[1–fm]). It is also important to note that most metabolites formed from parent drugs are removed by renal excretion. With increasing knowledge on the role of metabolites of "perpetrator" drugs on M-DDI (Isoherranen et al., 2004; Ogilvie et al., 2006), it is becoming important to consider the concentration–time profiles of the metabolites in addition to the parent compound when there is in vitro information on the inhibitory potency of such metabolites (Jamei et al., 2009a).

Many drug characteristics which determine the extent of renal elimination have been known for decades [including physical chemistry, lipophilicity and ionization (Tucker, 1981), plasma protein and erythrocyte binding (Levy, 1980)]. Nonetheless, there are neither in vitro systems nor any in silico quantitative-structure-kinetics models which can be used to estimate the potential renal clearance of xenobiotics. This could, in part, be related to specific affinity of various drugs to certain transporter proteins in the kidney (Song et al., 2008; Urban et al., 2008; Wang et al., 2008). Combining this information with the knowledge of physico-chemical determinants of renal clearance may present new opportunities to build predictive models for drug excretion by the kidneys (i.e. fractional tubular re-absorption, glomerular filtration and active secretion) (Levy, 1980; Janku, 1993; Rostami-Hodjegan, 2009). In the meantime, extrapolation from animal data, using allometry and after correction for differences in protein binding, might be the most reliable approach to estimate renal clearance in humans.

Recent IVIVE attempts to assess biliary excretion using sandwich-cultured hepatocytes have been successful (Ghibellini et al., 2006, 2007). However, experience in this area is limited and in vivo studies in humans which involve collecting bile and estimating its precise secretion rate are costly and may hamper validation studies in larger sets of compounds.

13.2.2 Prediction of Concentration–Time Profile

Similar AUC values are not necessarily indicative of similar concentration–time profiles. For single-dose studies, the rates of drug absorption, distribution and elimination define the shape of the concentration–time profile. Moreover, in the case of multiple dose studies where the overall daily dose is the same, dosing frequency also determines the level of fluctuations (i.e. swing). If the non-linearity in inhibitory effects of "perpetrator" drugs and their concentrations is considered, it becomes evident that evaluating the impact of various concentration–time profiles of a "perpetrator" is essential in simulating M-DDI studies. A prime example of such an assessment has recently been provided by Zhao et al. (2009).

Absorption rate is also influenced by the majority of parameters which determine bioavailability. Hence, various in vitro systems used for bioavailability assessment can also be used to evaluate the rate of absorption (Jamei et al., 2009b).

The rate of elimination is a hybrid function of the clearance and volume of distribution. Thus, to characterize concentration–time curves for any given dose, estimation of drug distribution is essential.

13.2.2.1 IVIVE to Determine Drug Distribution

Some factors which determine the distribution of drugs relate to characteristics of the individual (e.g. the perfusion rate of different tissues by blood, the concentration of plasma proteins, hematocrit, body composition, tissue density and genetic variants of transporter proteins). However, there are drug-related characteristics (such as ability of a drug to cross membranes, bind to plasma proteins, partition into red blood cells, tissues or fat, and become a substrate of specific influx or efflux transporter proteins) which can be measured in vitro and used for IVIVE purposes. As described above, drug distribution within the body influences the elimination rate and maximum exposure ($C_{\max}$). It is important to note that the proportion of the drug in different tissues changes with time (i.e. the tissue drug concentrations–time profiles are not necessarily changing in parallel to those in plasma), and hence volume of distribution is not a fixed term. The V_{ss} (volume of distribution at steady state) is considered a "purer distributional term" (Rodgers and Rowland, 2007) since other volume terms (such as central, V_c, and terminal, V_z, distribution volume) can be affected significantly by the relative speed of drug elimination and distribution. Current IVIVE approaches for estimating volume of distribution involve estimation of tissue partitions based on physicochemical characteristics combined with in vitro measurements of blood and plasma protein binding (Poulin and Theil, 2002; Berezhkovskiy, 2004; Rodgers and Rowland, 2007).

13.3 Requirements for an M-DDI "Modeling and Simulation" Platform

Modeling and simulation within the pharmaceutical industry has traditionally been a limited exercise carried out by only a few scientists working in selected departments. However, this is rapidly changing to a more comprehensive practice in all divisions from discovery to post-marketing (Lalonde et al., 2007). In parallel, the regulatory agencies (Buckman et al., 2007) are embracing the applications of modeling and simulation more than ever. IVIVE and prediction of M-DDI using modeling techniques has not been an exception. However, the more widespread use of simulations within organizations necessitates implementation of strategies that broaden access to the tools and models, ensure consistency of approaches, provide appropriate training, offer adequate flexibility for updating models and databases as scientific knowledge improves, allow compatibility between simulation systems used in other sections of the organization and make the models and assumptions transparent at a level required for regulatory decisions. A number of recent reports have commented on the advantages and disadvantages of in-house vs commercially available tools in the area of predictive ADME and M-DDI (Riley et al., 2007; Peters, 2008; Grime et al., 2009); however, an objective and comprehensive analysis is still lacking. Sugano (2009) has recently provided some general guidance, albeit in the specific area of predictive absorption models, which could be used as a fair outline of the advantages and disadvantages of each strategy. The author concluded that both in-house and commercial systems might be necessary due to complementary nature of their features which are listed in Table 13.2.

Table 13.2 Analysis of attributes for in-house and externally provided M&S tools

Commercial programs	An in-house template/program/platform
(1) A larger human physiology database (including ethnic, age, and disease differences) which enables virtual trials to be conducted in clinically relevant populations rather than a non-representative average subject (2) The opportunity to gather pre-competitive information from multiple (pharmaceutical) organizations which facilitates knowledge exchange (3) User-friendly graphical interfaces	(1) Highest flexibility for real-time changes in models (2) Tailor-made models for specific in-house experimental tools which are not available elsewhere (3) More insight within the organization as the people responsible for the coding learn the subject by developing the models!

Building on this general guidance of Sugano, an ideal system specifically designed for simulating M-DDI (whether external or internal) should satisfy several criteria. It should use the most mechanistic models, which are validated for routine use by non-modellers based on routine data generated in vitro or are designed for scientific explorations (i.e. asking "what if?" questions). It should contain data not just for a population of healthy volunteers but for a variety of disease populations. It should provide the flexibility to input data from various in vitro enzyme and cellular systems. The relevant attributes of compounds (i.e. affinity to various enzymes, non-enzymatic routes, protein binding, gut wall permeability) should be separated from the information on the system (i.e. population demography, physiological and genetic information on different patients, enzyme abundances and their turnover rates, level of plasma proteins, kidney function). User-friendly interfaces are generally preferable; however, there are specific requirements related to M-DDI including a clear distinction between the parameter entries with apparent similarity (e.g. inhibitory constants for competitive inhibition (K_i) and mechanism-based inhibition (K_I); see later sections for further details). Considering the absence of in vitro data on many old ("perpetrator") drugs, an ideal platform should also allow adequate flexibility to combine some in vivo values with in vitro observations if necessary. The outputs (reports) should be easy to interpret and flexible to manipulate. Conducting simulations in populations rather than single individuals is important to ensure more realistic virtual clinical trials; however, virtual trials of large populations without any ability to identify the outlier individuals may limit the power to identify covariates of M-DDI (Johnson et al., 2009), which is an essential feature of the ideal platform. The ability to process multiple cases (using parallel computing) is an advantage for drug discovery groups dealing with large numbers of compounds. Parallel computing can also become useful when the simulator is capable of optimizing values by fitting the models to observed data.

13.4 In Silico Simulations to Assess In Vivo M-DDIs Using In Vitro Data

This section summarizes the general pharmacokinetic basis of M-DDIs. Aspects of some specific M-DDI issues, such as "mechanism-based inactivation of enzymes", "interplay of transporters and metabolism", and "concomitant enzyme inhibition and induction", are covered in later chapters.

13.4.1 Static Models for Estimating M-DDI, Ignoring Gut Metabolism

The most commonly used mathematical models for predicting M-DDI due to "inhibition" of hepatic metabolism, in the absence of any complications by the perturbed gut wall metabolism, are derived under so-called static (non-time-variant) concentrations for the "perpetrator". These equations capture the two main elements defining the level of any M-DDI: (a) fractional metabolism ("fm") and (b) potency of interacting drug ("fold reduction or induction in $\mathrm{CL}u_{\mathrm{int}}$"). Equations (13.1) and (13.2) [after Rowland and Matin (1973), with expansions introduced by Rostami-Hodjegan and Tucker (2004) for multiple M-DDI] are the general forms of such static calculations:

$$\frac{\mathrm{AUC(inhibited)}}{\mathrm{AUC}} = \frac{1}{\sum\limits_{j=1}^{n} \frac{\mathrm{fm}_j}{\text{Fold reduction in } \mathrm{CL}u_{\mathrm{int},j}} + (1 - \sum\limits_{j=1}^{n} \mathrm{fm}_j)} \tag{13.1}$$

where fm_j is the fraction of substrate clearance mediated by the inhibited metabolic pathway "j" and $\mathrm{CL}u_{\mathrm{int\,j}}$ is the intrinsic metabolic clearance of substrate down pathway j. Similar analogous equations can be considered in the case of M-DDI involving enzyme induction:

$$\frac{\mathrm{AUC(induced)}}{\mathrm{AUC}} = \frac{1}{\sum\limits_{j=1}^{n} \left(\mathrm{fm}_j \times \text{Fold induction in } \mathrm{CL}u_{\mathrm{int},j}\right) + (1 - \sum\limits_{j=1}^{n} \mathrm{fm}_j)} \tag{13.2}$$

It should be noted that both the maximum increase in exposure to the "victim" drug following inhibition and the degree of reduction in exposure to the "victim" drug after induction of its metabolism are dependent on the fractional metabolism by the affected pathway(s) ("fm"). As shown in Equation (13.3), "1–fm" would be the ceiling of any inhibitory effect in the presence of the "strongest inhibitor" according to static models:

$$\text{Fold reduction} \rightarrow \infty \Rightarrow \frac{\mathrm{AUC(inhibited)}}{\mathrm{AUC(uninhibited)}} = \frac{1}{1 - \mathrm{fm}} \tag{13.3}$$

Thus, any observed inhibition that goes beyond this value (1/[1–fm]) strongly suggests the involvement of an additional mechanism(s) (e.g. inhibition of non-metabolic routes). However, it should be noted that "fm" varies within populations and using a typical average value to draw conclusions on the role of other mechanisms is not prudent. The sources of variable "fm" include, but are not restricted to, the following (Rostami-Hodjegan and Tucker, 2004):

- Genetic control of non-inhibited metabolic pathways
- Involvement of genetically regulated transporters in clearance
- Renal impairment influencing "1–fm"
- Variable rates of maturation of enzymatic and non-enzymatic routes

Although "fm" is an important determinant of the level of M-DDI involving enzyme induction (Equation (13.2)), maximum achievable interaction (and relative change in exposure to a "victim" drug) is not dependent on fm. Thus, enzyme activity can, theoretically, increase substantially if the "fold induction in CLu_{int}" is adequately high. This is in contrast to the impact of very large "fold reduction in CLu_{int}" which cannot modify exposure beyond 1/[1–fm].

Fold inhibition can be defined according to the type of inhibition that occurs (Rostami-Hodjegan and Tucker, 2004). Thus, for multiple ("*p*") competitive inhibitors acting *via* same mechanism to inhibit enzyme "j":

$$\text{Fold reduction in } \mathrm{CL}u_{\mathrm{int},j} = 1 + \sum_{k=1}^{p} \frac{[I_k]}{Ki_k} m \tag{13.4}$$

where $[I_k]$ is the concentration of inhibitor "*k*" at the enzyme site and Ki_k is the inhibition constant for inhibitor "*k*" obtained from in vitro studies after accounting for non-specific binding. The same equation applies for multiple non-competitive inhibitors acting at the same enzyme site. However, if the mechanisms of inhibition by multiple inhibitors are different (independent), the fold reduction in clearance would be greater:

$$\text{Fold reduction in} CLu_{\mathrm{int},j} = \prod_{k=1}^{p} \left(1 + \frac{[I_k]}{Ki_k}\right) \tag{13.5}$$

In the case of mechanism-based (suicidal) inactivation of enzymes, the ratio of basal degradation to the new accelerated rate of degradation determines the fold reduction in clearance:

$$\text{Fold reduction in } \mathrm{CL}u_{\mathrm{int},j} = \frac{k_{\mathrm{deg}}}{k_{\mathrm{deg}} + \frac{k_{\mathrm{inact}} \times [I]}{K_I + [I]}} \tag{13.6}$$

where k_{deg} is the natural degradation rate constant for the enzyme, k_{inact} is the maximum degradation rate constant in the presence of a very high concentration of inhibitor and K_{I} is the concentration of inhibitor associated with half maximal inactivation.

Assuming that the inhibited metabolic pathway is the only elimination pathway (i.e. "fm" = 1) for the "victim" drug, simplified versions of Equation (13.1) have been used to calculate M-DDI (Equation (13.7)); however, these should be avoided as they may lead to substantial over-prediction when assessing strong inhibitors:

$$[1-\mathrm{fm}] \to 0 \Rightarrow \frac{\mathrm{AUC(inhibited)}}{\mathrm{AUC(uninhibited)}} = \begin{cases} \text{Competitive:} & 1 + \frac{[I]}{K_i} \\ \text{Mechanism-based:} & \frac{k_{\mathrm{deg}}}{k_{\mathrm{deg}} + \frac{k_{\mathrm{inact}} \times [I]}{K_{\mathrm{I}} + [I]}} \end{cases} \tag{13.7}$$

Estimation of "fold induction in CLu_{int}" involves knowledge of the efficacy and potency of an inducing agent and the concentration of the inducer at the site of effect ($[I]$), according to Equation (13.8). Potency is related to ECu_{50}, the unbound concentration of the inducer that elicits 50% of the maximum response ($E_{\max}$), after correcting for any non-specific binding; the lower the ECu_{50} value, the higher the potency. E_{max} value is calculated using Equation (13.9):

$$\text{Fold induction } CLu_{\text{int},j} = 1 + \frac{E_{max} \times [I]}{ECu_{50} + [I]} \tag{13.8}$$

$$E_{max} = \frac{E_{ind} - E_0}{E_0} \tag{13.9}$$

where E_{ind} is the amount of enzyme at maximal induction and E_0 is its basal level. Hypothetically, in vitro endpoints to assess induction potential for various xenobiotics (nuclear factor activation, mRNA expression, protein expression and protein activity) are all correlated proportionally. Hence, the ECu_{50} value should be independent of the method used to assess induction (Almond et al., 2009). The latter is consistent with the recent report by McGinnity et al., who observed similar EC50 values using various systems and endpoints to measure induction. Different in vitro methods may not provide similar E_{max}, hence the fold increase in mRNA and nuclear receptor activation in response to an inducer might be much greater than the increase in protein level or activity (Luo et al., 2002; Martin et al., 2008). Therefore, any IVIVE may (Shou et al., 2008) or may not (Fahmi et al., 2008) assume equivalent potency and efficacy for in vitro and in vivo data. Deriving a scaling factor to calibrate the various in vitro systems requires an internal standard (e.g. rifampicin). Similarly, scaling factors to link in vitro data to in vivo manifestations involves calibrators. However, defining in vivo ECu_{50} and E_{max} values requires intensive clinical studies in which the time courses of both "perpetrator" and "victim" drugs are followed during sustained administration of a range of doses of the "perpetrator". Such studies are non-existent, resulting in a paucity of information on in vivo values defining the induction of metabolism. The limited information that is available can be obtained through various literature reports (e.g. see Kozawa et al., 2009 Ohno et al., 2008).

Various technical limitations (e.g. low compound solubility, cell toxicity) may hamper full characterization of the concentration–induction profile and obtaining E_{max} and ECu_{50} values. Analogous to the definition of intrinsic metabolic clearance at low substrate concentration [much lower than the Michaelis–Menten constant (K_m)], it is possible to define a linear relationship between concentration and induction based on the initial slope (Magnusson, 2007; Shou et al., 2008).

13.4.2 Static Model to Estimate M-DDI in Gut Wall

Gut wall metabolism may play a major role in reducing bioavailability as described in Chapter 4. Drugs with an appreciable gut wall extraction ratio (E_G), which include substrates of CYP3A and UGT, might be susceptible to an increased systemic exposure under inhibition of gut wall metabolism. The calculation of perturbed gut wall metabolism, and its impact on AUC values, should be considered by multiplying the right-hand sides of Equations (13.1) and (13.2) by the ratio F_G'/F_G, where F_G and F_G' describe the fraction of the drug that escapes first-pass gut metabolism in the absence and presence of any interacting drug, respectively. This ratio has been considered in the following form by a number of groups (Wang et al., 2004; Ernest et al., 2005; Obach et al., 2006):

$$\frac{F_G'}{F_G} = \frac{1}{F_G + \left[(1 - F_G) \times \text{Fold change in } \mathrm{CL}u_{\text{int,Gut}}\right]} \tag{13.10}$$

Although there are mechanistic (and dynamic) models for estimating F_G, such as "segmental and segregated intestinal transport and metabolism" described by Pang and colleagues (Pang, 2003; Tam et al., 2003) or ADAM model described by Jamei et al. (2009b), derivation of the above equation relies on more simplistic (and static) models such as those described by Yang et al. (2007). In the latter case, F_G is described by an operational "Q_{Gut}" model which retains the form of the "well-stirred" model with reference gut intrinsic clearance values ($\mathrm{CL}u_{\text{int, G}}$), but the flow term ($Q_{\text{Gut}}$) is a hybrid of both permeability through the enterocyte membrane and villous blood flow:

$$F_G = \frac{Q_{\text{Gut}}}{Q_{\text{Gut}} + \mathrm{fu}_G \cdot \sum_{j=1}^{n} \mathrm{CL}u_{\text{int,G}_j}} \tag{13.11}$$

Q_{Gut} can be described in terms of $\mathrm{CL}_{\text{perm}}$, a clearance term defining permeability through the enterocytes, and Q_{villi}, actual villous blood flow:

$$Q_{\text{Gut}} = \frac{Q_{\text{villi}} \cdot \mathrm{CL}_{\text{perm}}}{Q_{\text{villi}} + \mathrm{CL}_{\text{perm}}} \tag{13.12}$$

In the presence of an inhibitor that alters only intrinsic gut metabolic clearance, Equation (13.11) can be rewritten to include an estimate of the concentration of inhibitor in the enterocyte ($[I]_{\text{gut}}$) and its inhibitory constant for the affected pathway "j" (Rostami-Hodjegan and Tucker, 2004):

$$F_G' = \frac{'Q_{\text{Gut}}'}{'Q_{\text{Gut}}' + \mathrm{fu}_{\text{Gut}} \cdot \sum_{j=1}^{n} \frac{\mathrm{CL}u_{\text{int,G}_j}}{\text{Fold inhibition of pathway } j}} \tag{13.13}$$

where fold inhibition can be defined depending on the type of inhibition:

$$\text{Fold inhibition of } \mathrm{CL}u_{\text{int,Gut}} = \begin{cases} \text{Competitive:} & 1 + \frac{[I]_{\text{Gut}}}{K_i} \\ \text{Mechanism-based:} & \frac{k_{\text{deg,Gut}}}{k_{\text{deg,Gut}} + \frac{k_{\text{inact}} \times [I]_{\text{Gut}}}{K_I + [I]_{\text{Gut}}}} \end{cases}$$

Various assumptions can be made regarding an estimate of $[I]_{\text{gut}}$ including the unbound concentration of the inhibitor in the systemic circulation (when the inhibitor is not co-administered with the substrate), enterocytic concentrations based on blood flow to the tips of the villi (Q_{ent}) and the pre-hepatic absorption rate of the inhibitor according to Equation (13.15) (Rostami-Hodjegan and Tucker, 2004):

$$[I]_{\mathrm{gut}} = \frac{f_{\mathrm{a}} \times k_{\mathrm{a}}(I) \times \mathrm{Dose}(I)}{Q_{\mathrm{ent}}} \tag{13.15}$$

where f_{a} is the fraction of the inhibitor dose that is absorbed into the gut wall and $k_{\mathrm{a}}(I)$ and Dose(I) are the absorption rate constant and dose of inhibitor, respectively. The above equation assumes that the inhibitor itself is not subject to major first-pass gut metabolism.

Operating concentrations for induction of gut metabolism would be similar to those described above; however, Equation (13.16) is used instead of Equation (13.13) to account for the increase (rather than decrease) in activity:

$$F_{\mathrm{G}}' = \frac{'Q_{\mathrm{Gut}}'}{'Q_{\mathrm{Gut}}' + \mathrm{fu}_{\mathrm{Gut}} \cdot \sum_{j=1}^{n} \left(\mathrm{CL}u_{\mathrm{int,G}_j} \times \left[1 + \left(\frac{\mathrm{Ind}_{\max} \times [I]_{\mathrm{Gut}}}{\mathrm{Ind}C_{50} + [I]_{\mathrm{Gut}}}\right)\right]\right)} \tag{13.16}$$

Equations (13.11) together with Equations (13.13) and (13.16) can be used to derive Equation (13.10). All these equations highlight the non-linear relationship between the inhibitor concentrations (/dose) and the extent of the effect on F_{G}. For drugs with low basal F_{G} values, variations in $[I]_{\mathrm{Gut}}$ (e.g. changes in dose, absorption rate, variable blood flow) lead to variable inhibition of intestinal first-pass metabolism; this is not the case for drugs with high F_{G}.

13.4.3 Considerations for Operational Concentration of the "Perpetrator"

The "free drug hypothesis" assumes that unbound intracellular concentration of compounds should be used as the operational concentration to calculate any drug interactions. However, since "static" models ignore many of the mechanistic elements of predicting M-DDI, the outcome of predictions may not reflect the observed clinical data. Hence, many investigators have used various drug concentrations within the "static" framework to overcome shortcomings of these non-mechanistic approximations and assign an arbitrary concentration which is "most predictive" based on fitting to sets of clinically observed data. It should be noted that many of the conclusions from such exercises are model dependent and attempts to identify the best surrogate concentrations, whilst using minimalistic (abbreviated) models to define the M-DDI, could be viewed only as a compensatory mechanism related to the non-mechanistic model assumptions. Although it has been reported by some investigators that predicted M-DDI could be relatively insensitive to the choice of inhibitor concentration (Venkatakrishnan and Obach, 2005), others (Ito et al.,

2002, 2004) have observed variations in the outcome. It has been recommended that multiple options be examined and the risk of false negative M-DDIs be assessed in any analysis (Venkatakrishnan and Obach, 2007). Importantly, the findings (i.e. the best practice for various [*I*] values) shift as the models change. Although most reports which describe variation in assumption on inhibitor concentrations use static models, Kanamitsu et al. (2000) compared the use of a static maximum unbound inlet concentration with three different dynamic unbound concentrations (systemic blood, inlet and liver). The former tended to result in significant overprediction, particularly when intrinsic metabolic clearance was high. The accurate prediction of M-DDI between sulfaphenazole (500 mg) on the CYP2C9-mediated metabolism of tolbutamide using the static model was attributed to the fact that the concentration exceeded the threshold (I_u:K_i ratio) for complete inhibition, minimizing expected differences between the use of the static single concentration and full concentration–time profile.

In general, the investigations with complex and more comprehensive time-varying models for M-DDI often use the unbound blood concentrations as a more biological surrogate for intracellular unbound concentrations. However, this assumes rapid equilibration across cellular barriers and no active influx or efflux, which may not be the case for many compounds. Indeed, the influence of transporters on intracellular concentration in hepatocytes is described in Chapter 16.

Finally, it is important to distinguish between the drug concentration in the inlet blood supply to the liver and that in the hepatic vein leaving the liver (liver outlet concentration). According to the "well-stirred model" of hepatic clearance, unbound concentration in the liver outlet equates to the unbound concentration of the drug within the liver. The difference between the drug concentration values in the liver inlet and outlet is determined by the hepatic extraction ratio of the "perpetrator" drug. Such a difference has recently been considered when using a full concentration–time profile of the "perpetrator" drug and Fig. 13.1 shows an example of the potential impact of the assumption on inhibitory concentrations for a series of M-DDI studies simulated for five mechanism-based inhibitors: diltiazem, verapamil, clarithromycin, and azithromycin (Rowland-Yeo, in preparation).

13.4.4 Using Full Time Course of "Perpetrator" to Assess M-DDI

It is intuitive that more physiologically based models accounting for the time-varying concentration of "perpetrators" might offer advantages over "static" models and provide better insight into M-DDI observations. A clear example of the applications of these models from a regulatory point of view was recently demonstrated by an investigation into the appropriate study design for ketoconazole and CYP3A substrates (Zhao et al., 2009). Some of these will be discussed in later chapters by Huang and colleagues. The disadvantages of such models include their complexity, which may hamper common use unless user-friendly interfaces are provided and adequate training is given to users. As discussed in earlier sections,

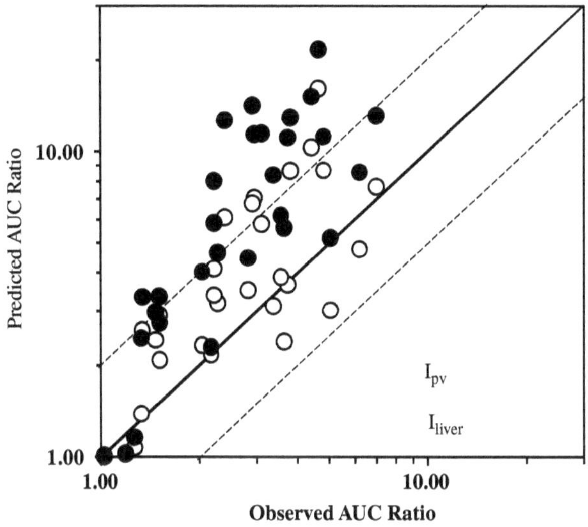

Fig. 13.1 The impact of assumptions on inhibitory concentrations operating on an enzyme could be significant when predicting M-DDI. A series of M-DDI ($n = 30$) involving mechanism-based inhibition of CYP3A4 were simulated using full concentration–time profiles of the "perpetrator" drugs in the liver or the portal vein within the Simcyp® Population-based ADME Simulator (V8). The inhibitory effects predicted from liver concentrations (*open circles*) were lower than those predicted based on portal concentrations (*solid circles*) and they were more consistent with the observed level of inhibition (turnover half-life of CYP3A4 in the gut wall and the liver were 24 and 70 h, respectively, courtesy of Dr Karen Rowland Yeo, Simcyp Limited 2008)

such complex models can be implemented within internal general purpose software or they can be obtained as specialized platforms from commercial vendors (e.g. the Simcyp® Population-based ADME Simulator; http://www.simcyp.com; Simcyp Ltd, Sheffield, UK). These models are more informative for drug interaction assessments particularly in relation to the impact of population variability (Rostami-Hodjegan and Tucker, 2004; Rostami-Hodjegan, 2009) and study design (Yang et al., 2003; Zhao et al., 2009). This follows since they can easily incorporate variables such as CYP expression level and genetic polymorphisms in hepatic as well as gut wall metabolism using a Monte Carlo approach. Moreover, they can define the drug concentration–time profiles of the "victim" and multiple "perpetrators", the combined effects of parent drug and its metabolite and time- and dose-dependent phenomena such as auto-induction and auto-inhibition, using differential equations. An account of various strategies for using qualitative, semi-quantitative and quantitative predictions of M-DDI has been provided recently by Einolf (2007), who indicated an improvement in the prediction of M-DDI magnitude using full concentration–time profiles of inhibitors compared to "static" methods. Einolf (2007) also commented on the ability to predict variability in magnitude of M-DDI, the automated prediction of the inhibition of parallel metabolic pathways, and simulations of different dosing regimens.

Another advantage of using time-course analysis rather than "static" calculations relates to multiple pharmacokinetic changes. Indeed, Orlando et al. recently emphasized the need to consider simultaneous changes in plasma binding and enzyme activity when predicting complex DDIs (Orlando et al., 2009). A further complication is that some compounds, such as ritonavir (Hsu et al., 1998; Foisy et al., 2008; Kenny et al., 2008), troleandomycin (Luo et al., 2002) and mifepristone (Kenny et al., 2008), cause simultaneous CYP inhibition and induction and the balance of these phenomena may be dose dependent. Clearly, these often opposing effects on drug exposure require appropriate time-based models to predict the net outcome of a DDI (Rostami-Hodjegan and Tucker, 2004; Fahmi et al., 2008).

The essential element of models involving the time course of drugs, and their ability to accommodate simultaneous events (e.g. competitive inhibition, mechanism-based inactivation, combined effects of parent compound and metabolite, induction and displacement from plasma proteins), is the reliance on differential equations. For instance, whilst the abundance of a particular enzyme in an individual can be defined as a single static value, differential equations can describe the balance between the rate of synthesis and degradation. Thus, the enzyme level at time t, E_t, can be described by the following equation:

$$\frac{\mathrm{d}E_t}{\mathrm{d}t} = R_{\mathrm{syn},t} - k_{\mathrm{deg},t} \times E_t \tag{13.17}$$

where $R_{\mathrm{syn},t}$ and $k_{\mathrm{deg},t}$ are the rate of enzyme synthesis and the degradation rate constant at time t, respectively. At the basal level and when the balance between degradation and synthesis is not disturbed (i.e. in the absence of any "perpetrator" that influences either synthesis via induction or degradation via mechanism-based inactivation), $R_{\mathrm{syn},t}$ can be described by the following equation:

$$R_{\mathrm{syn,base}} = E_{\mathrm{base}} \times k_{\mathrm{deg,base}} \tag{13.18}$$

where $R_{\mathrm{syn,base}}$, E_{base}, $k_{\mathrm{deg,base}}$ are the basal rate of enzyme synthesis, the basal level of enzyme and the basal degradation rate constant, respectively. Increased degradation or synthesis can be accommodated within Equation (13.17) based on quantitative relationships between the concentration of the "perpetrator" at the enzyme site ($[I]_t$) and its inactivating or inductive abilities at any given concentration as previously described by Equations (13.6) and (13.8). Hence,

$$\frac{\mathrm{d}E_t}{\mathrm{d}t} = \left[E_{\mathrm{base}} \times k_{\mathrm{deg,base}} \times \left(1 + \frac{E_{\mathrm{max}} \times [I]_t}{\mathrm{EC}_{50} + [I]_t}\right)\right] - \left[\left(k_{\mathrm{deg,base}} + \frac{k_{\mathrm{inact}} \times [I]_t}{K_{\mathrm{I}} + [I]_t}\right) \times E_t\right] \tag{13.19}$$

It is important to realize that the value of $[I]$ at the enzyme site is not a fixed ("static") value and it changes with time (using sets of differential equations). Moreover, as discussed in the previous sections, the assumptions regarding the impact of blood and protein binding (free drug hypothesis) and the affinity of the "perpetrator" to any efflux and influx proteins will determine the relationship between $[I]_t$ at the enzyme site and $[I]_t$ in the plasma. Combining the differential equation (13.19), which describes the changes in active enzyme level, with those defining the events in the liver (or any other eliminatory organ) extends the capability of the models to include the effects of competitive inhibition as well:

$$\frac{\mathrm{d}C_{\mathrm{Liv}}}{\mathrm{d}t} = \frac{1}{V_{\mathrm{Liv}}}\left[Q_{\mathrm{AR}}C_{\mathrm{AR}} + Q_{\mathrm{PV}}C_{\mathrm{PV}} - Q_{\mathrm{Liv}}C_{\mathrm{Liv}}\left(\frac{\mathrm{B{:}P}}{\mathrm{P}_{\mathrm{Liv{:}p}}}\right) - \mathrm{CL}u_{\mathrm{int},t} \times Cu_{\mathrm{Liv}}\right] \quad (13.20)$$

where V_{Liv}, Q_{AR}, Q_{PV}, Q_{Liv}, C_{AR}, C_{PV}, C_{Liv} are volume of the liver compartment, the arterial and portal blood flow to the liver, the blood flow out of the liver, the arterial and portal concentrations of the "victim" substrate and the concentration of the substrate within the liver compartment, respectively. The terms B:P and $\mathrm{P}_{\mathrm{Liv{:}p}}$ describe the blood to plasma and liver to plasma concentration ratios. $\mathrm{CL}u_{\mathrm{int},t}$ is the time-varying intrinsic clearance of the substrate by combination of multiple clearance values from each enzymatic elimination route described as

$$\mathrm{CL}u_{\mathrm{int},t} = \frac{V_{\mathrm{max}} \times E_t}{K_{\mathrm{m}} \times \left(1 + \frac{[I]_t}{K_{\mathrm{i}}}\right) + Cu_{\mathrm{Liv}}} \quad (13.21)$$

where V_{max} is the maximum rate of metabolism per unit of enzyme (so-called k_{cat}) and K_{m} in the Michaelis constant (concentration of substrate associated with half the maximal metabolic rate). Similar equations can be derived for other organs (e.g. intestine, kidney) and other chemical moieties (e.g. additional inhibitors, metabolites) in the circulation. A currently available commercial platform, the Simcyp® Population-based ADME Simulator (Jamei et al., 2009b), can accommodate up to three inhibitors and two metabolic profiles (one for the substrate and one for an inhibitor) creating a multiplex of self-induction, self-inactivation and mutual interactions (as shown in Fig. 13.2).

13.4.5 Assessing Variability in M-DDI

The final part of this chapter focuses on the issue of variability in M-DDI. A full account of this issue is given elsewhere (Rostami-Hodjegan and Tucker, 2004; Rostami-Hodjegan, 2009) and only a brief overview is provided here.

Variation in "fm", which depends on relative variability in metabolic routes, is a common cause of inter-individual differences in M-DDI even at a fixed level of a given inhibitor (i.e. fixed $[I]$ and $[I]_{\mathrm{Gut}}$). The pie charts in Fig. 13.3 show the proportional abundance of CYP enzymes as an average value of the population together with selected individuals within the population. This highlights problems with using values from an "average" individual. The role of "fm" in determining the variability in M-DDI is particularly dominant when strong inhibitors are used. Hypothetical examples have been considered by Rostami-Hodjegan and Tucker (2004) to show the inter-individual variability of M-DDI by a strong inhibitor for

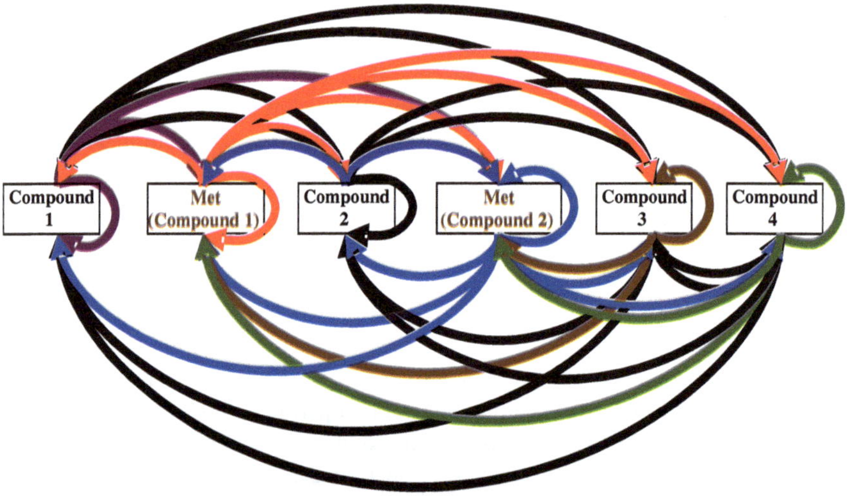

Fig. 13.2 A realistic assessment of M-DDI may require consideration of a multiplex of effects between various compounds in the human body as shown in this figure. The mutual inhibition of various compounds, the inhibitory or inducer effects of metabolites, self-induction and self-inhibition via mechanism-based inactivation of enzymes and simultaneous effects via competitive inhibition, mechanism-based inactivation and induction of enzymes are often ignored when using simple models. New modeling and simulations tools are available to incorporate all the complexities shown in the figure (Jamei et al., 2009a); however, the rate-limiting step in using these platforms is often availability of data on in vitro values for all compounds

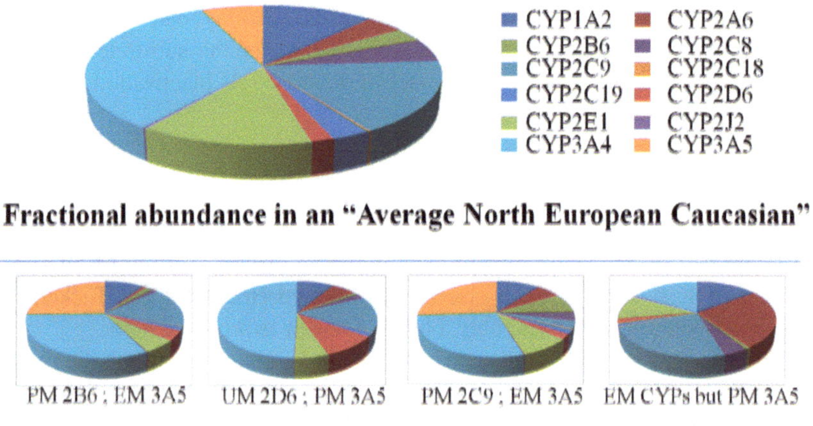

Fig. 13.3 The pie chart at the *top* shows the average fractional abundance of various CYPs (this may not represent any person as some genotypes lead to a lack of any expression). Thus an average value from the population corresponds neither to poor (PM) nor to extensive (EM) or ultra-rapid (UM) phenotype. The *bottom panel* of pie charts show the proportional abundance of CYPs in selected individuals. The charts highlight that the fractional metabolism of "victim" drugs may vary in different individuals. Most mechanistic prediction models for M-DDI use fm (see text); however, they ignore the variations that may exist in "fm". The variation in abundance (and hence fm) can have significant effect on the variability in M-DDI

a "victim" compound in sub-groups such as patients with renal failure or individuals with genetically impaired capacity of metabolism down a non-inhibited pathway (i.e. reduced capacity of alternative elimination routes). However, a recent report by Orlando et al. (2009) in cirrhotic patients clearly indicates variation in M-DDI when the relative "fm" changes in different populations (Fig. 13.3). Interestingly, the studies to investigate the covariates associated with increased risk of MDDI are rare and typical population data analyses to assess M-DDI are rarely powered to identify these covariates (Johnson et al., 2009).

Mechanism-based (suicidal) inactivation (Ghanbari et al., 2006) and induction (Shou et al., 2008) are time-dependent processes, where the basal degradation rate (k_{deg}) determines the time course of interaction. Inter-individual variability in degradation could be substantial (Yang et al., 2008). Moreover, k_{deg} determines the level of increased exposure to the victim drug as shown in Equation (13.6), making its inter-individual differences even more important. Further details on requirement for sensitivity analysis are given in chapter 19 by Obach et al.

The circulating concentration of the inhibitor (and hence concentration at the active site) is also subject to variability as described earlier in this chapter. Variations in the administered dose of the "perpetrator" or issues related to variable compliance (or absorption rates) may also affect the outcome of M-DDI as shown in Fig. 13.4 for inhibition of midazolam and ketoconazole.

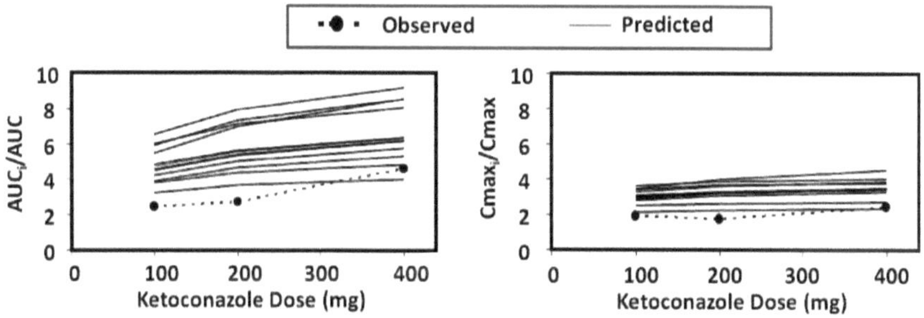

Fig. 13.4 The effect of dose escalation of ketoconazole on the midazolam interaction. Data points are the mean values for AUC*i*/AUC (*left-hand panel*) and $C_{max}i/C_{max}$ (*right-hand panel*) observed in vivo and each of the *solid lines* are the mean of the values obtained for the 10 trials of 8 subjects simulated with Simcyp Population-Based ADME Simulator (V8) (Ozdemir et al., in preparation)

Emerging issues with genetic differences in transporters responsible for efflux and active uptake into hepatocytes (Watanabe et al., 2008) are covered in chapters 12 and 21 and readers are referred to these chapters for appreciation of their impact. Many of these effects can be easily accommodated within the models. However, mechanistic scaling of transporter-related values from in vitro data, without some type of fitting procedure, is not possible yet.

13.5 Conclusion

The mechanistic prediction of M-DDI, and associated variability, is possible only by taking into account all physiological and drug-related knowledge (both "perpetrator" and "victim" drugs) together with information on study design. M-DDI remains the major source of drug–drug interactions; however, the role of transporters and their interplay with the enzymes, when relevant, can be easily incorporated into simulations if the appropriate data are gathered from in vitro studies. Whilst these considerations increase the complexity of modeling and simulation tasks, availability of common platforms has reduced the reliance on specialized modeling experts to carry out such efforts.

Acknowledgments The author wishes to thank Dr Lizzie Kingsley for useful comments and acknowledge helpful discussion with scientists within Simcyp Limited and the member companies of the Simcyp Consortium (www.simcyp.com).

References

Almond LM, Yang J, Jamei M, Tucker GT and Rostami-Hodjegan A (2009) Towards a quantitative framework for the prediction of DDIs arising from cytochrome P450 induction. *Curr Drug Metab* **10:** 420–432

Barter ZE, Bayliss MK, Beaune PH, Boobis AR, Carlile DJ, Edwards RJ, Houston JB, Lake BG, Lipscomb JC, Pelkonen OR, Tucker GT and Rostami-Hodjegan A (2007) Scaling factors for the extrapolation of in vivo metabolic drug clearance from in vitro data: reaching a consensus on values of human microsomal protein and hepatocellularity per gram of liver. *Curr Drug Metab* **8**:33–45.

Berezhkovskiy LM (2004) Volume of distribution at steady state for a linear pharmacokinetic system with peripheral elimination. *J Pharm Sci* **93**:1628–1640.

Buckman S, Huang SM and Murphy S (2007) Medical product development and regulatory science for the 21st century: the critical path vision and its impact on health care. *Clin Pharmacol Ther* **81**:141–144.

Carlson SP, Ragueneau-Majlessi I, Bougan TE and Levy RH (2002) Development of a metabolic drug interaction database at the University of Washington, in: *Drug–Drug Interactions* (Rodrigues AD ed), pp 650, Marcel Dekker, New York & Basel.

Einolf HJ (2007) Comparison of different approaches to predict metabolic drug–drug interactions. *Xenobiotica* **37**:1257–1294.

Ernest CS, 2nd, Hall SD and Jones DR (2005) Mechanism-based inactivation of CYP3A by HIV protease inhibitors. *J Pharmacol Exp Ther* **312**:583–591.

Fahmi OA, Maurer TS, Kish M, Cardenas E, Boldt S and Nettleton D (2008) A combined model for predicting CYP3A4 clinical net drug–drug interaction based on CYP3A4 inhibition, inactivation, and induction determined in vitro. *Drug Metab Dispos* **36**:1698–1708.

Foisy MM, Yakiwchuk EM and Hughes CA (2008) Induction effects of ritonavir: implications for drug interactions. *Ann Pharmacother* **42**:1048–1059.

Food and Drug Administration (FDA) (2006) Food and Drug Administration (FDA). Guidance for industry: drug interaction studies – study design, data analysis, and implications for dosing and labeling, in http://www.fda.gov/cber/gdlns/interactstud.pdf

Ghanbari F, Rowland-Yeo K, Bloomer JC, Clarke SE, Lennard MS, Tucker GT and Rostami-Hodjegan A (2006) A critical evaluation of the experimental design of studies of mechanism based enzyme inhibition, with implications for in vitro-in vivo extrapolation. *Curr Drug Metab* **7**:315–334.

Ghibellini G, Vasist LS, Hill TE, Heizer WD, Kowalsky RJ and Brouwer KL (2006) Determination of the biliary excretion of piperacillin in humans using a novel method. *Br J Clin Pharmacol* **62**:304–308.

Ghibellini G, Vasist LS, Leslie EM, Heizer WD, Kowalsky RJ, Calvo BF and Brouwer KL (2007) In vitro-in vivo correlation of hepatobiliary drug clearance in humans. *Clin Pharmacol Ther* **81**:406–413.

Grime KH, Bird J, Ferguson D and Riley RJ (2009) Mechanism-based inhibition of cytochrome P450 enzymes: an evaluation of early decision making in vitro approaches and drug–drug interaction prediction methods. *Eur J Pharm Sci* **36**:175–191.

Hsu A, Granneman GR and Bertz RJ (1998) Ritonavir. Clinical pharmacokinetics and interactions with other anti-HIV agents. *Clin Pharmacokinet* **35**:275–291.

Huang SM and Lesko LJ (2009) The need for multiple doses of 400 mg ketoconazole as a precipitant inhibitor of a CYP3A substrate in an in vivo drug–drug interaction study. *J Clin Pharmacol* **49**:370.

Isoherranen N, Kunze KL, Allen KE, Nelson WL and Thummel KE (2004) Role of itraconazole metabolites in CYP3A4 inhibition. *Drug Metab Dispos* **32**:1121–1131.

Ito K, Brown HS and Houston JB (2004) Database analyses for the prediction of in vivo drug–drug interactions from in vitro data. *Br J Clin Pharmacol* **57**:473–486.

Ito K, Chiba K, Horikawa M, Ishigami M, Mizuno N, Aoki J, Gotoh Y, Iwatsubo T, Kanamitsu S, Kato M, Kawahara I, Niinuma K, Nishino A, Sato N, Tsukamoto Y, Ueda K, Itoh T and Sugiyama Y (2002) Which concentration of the inhibitor should be used to predict in vivo drug interactions from in vitro data?. *AAPS Pharm Sci* **4**:E25.

Jamei M, Marciniak S, Feng K, Barnett A, Tucker GT and Rostami-Hodjegan A (2009a) The Simcyp® population-based ADME simulator. *Expert Opin Drug Metab Toxicol*:in press.

Jamei M, Turner D, Yang J, Neuhoff S, Polak S, Rostami-Hodjegan A and Tucker GT (2009b) Population-based mechanistic prediction of oral drug absorption. *AAPS J* 11:225–237.

Janku I (1993) Physiological modelling of renal drug clearance. *Eur J Clin Pharmacol* **44**:513–519.

Johnson TN, Kerbusch T, Jones B, Tucker GT, Rostami-Hodjegan A and Milligan PA (2009) Assessing the efficiency of mixed effects modelling in quantifying metabolism based drug–drug interactions: using in vitro data as an aid to assess study power. *Pharm Stat*:Mar 16. [Epub ahead of print].

Kanamitsu S, Ito K.,Suganiyama Y (2000) Quantitative prediction of in vivo drug-drug interactions from in vito data based on physiological pharmacokinetics: use of maximum unbound concentration of inhibitor at the inlet to the liver. *Pham Res* **17:** 336–343.

Kenny JR, Chen L, McGinnity DF, Grime K, Shakesheff KM, Thomson B and Riley R (2008) Efficient assessment of the utility of immortalized Fa2N-4 cells for cytochrome P450 (CYP) induction studies using multiplex quantitative reverse transcriptase-polymerase chain reaction (qRT-PCR) and substrate cassette methodologies. *Xenobiotica* **38**:1500–1517.

Kozawa M, Honma M and Suzuki H (2009) Quantitative prediction of in vivo profiles of CYP3A4 induction in humans from in vitro results with reporter gene assay. *Drug Metab Dispos* **37**: 1234–1241.

Krayenbuhl JC, Vozeh S, Kondo-Oestreicher M and Dayer P (1999) Drug–drug interactions of new active substances: mibefradil example. *Eur J Clin Pharmacol* **55**:559–565.

Lalonde RL, Kowalski KG, Hutmacher MM, Ewy W, Nichols DJ, Milligan PA, Corrigan BW, Lockwood PA, Marshall SA, Benincosa LJ, Tensfeldt TG, Parivar K, Amantea M, Glue P, Koide H and Miller R (2007) Model-based drug development. *Clin Pharmacol Ther* **82**:21–32.

Levy G (1980) Effect of plasma protein binding on renal clearance of drugs. *J Pharm Sci* **69**: 482–483.

Luo G, Cunningham M, Kim S, Burn T, Lin J, Sinz M, Hamilton G, Rizzo C, Jolley S, Gilbert D, Downey A, Mudra D, Graham R, Carroll K, Xie J, Madan A, Parkinson A, Christ D, Selling B, LeCluyse E and Gan LS (2002) CYP3A4 induction by drugs: correlation between a pregnane X receptor reporter gene assay and CYP3A4 expression in human hepatocytes. *Drug Metab Dispos* **30**:795–804.

Magnusson MO (2007) Pharmacodynamics of enzyme induction and its consequence for substrate elimination, in *Faculty of Pharmacy*, p 64, Uppsala University, Uppsala.

Martin P, Riley R, Back DJ and Owen A (2008) Comparison of the induction profile for drug disposition proteins by typical nuclear receptor activators in human hepatic and intestinal cells. *Br J Pharmacol* **153**:805–819.

Obach RS, Walsky RL, Venkatakrishnan K, Gaman EA, Houston JB and Tremaine LM (2006) The utility of in vitro cytochrome P450 inhibition data in the prediction of drug–drug interactions. *J Pharmacol Exp Ther* **316**:336–348.

Ogilvie BW, Zhang D, Li W, Rodrigues AD, Gipson AE, Holsapple J, Toren P and Parkinson A (2006) Glucuronidation converts gemfibrozil to a potent, metabolism-dependent inhibitor of CYP2C8: implications for drug–drug interactions. *Drug Metab Dispos* **34**: 191–197.

Ohno Y, Hisaka A and Suzuki H (2007) General framework for the quantitative prediction of CYP3A4-mediated oral drug interactions based on the AUC increase by coadministration of standard drugs. *Clin Pharmacokinet* **46**:681–696.

Ohno Y, Hisaka A, Ueno M and Suzuki H (2008) General framework for the prediction of oral drug interactions caused by CYP3A4 induction from in vivo information. *Clin Pharmacokinet* **47**:669–680.

Orlando R, De Martin S, Pegoraro P, Quintieri L and Palatini P (2009) Irreversible CYP3A inhibition accompanied by plasma protein-binding displacement: a comparative analysis in subjects with normal and impaired liver function. *Clin Pharmacol Ther* **85**:319–326.

Pang KS (2003) Modeling of intestinal drug absorption: roles of transporters and metabolic enzymes (for the Gillette Review Series). *Drug Metab Dispos* **31**:1507–1519.

Pang KS and Chiba M (1994) Metabolism: scaling up from in vitro to organ and whole body, in *Pharmacokinetics of Drugs* (Welling PG and Balant LP eds), pp 101–187, Springer-Verlag, New York.

Peters SA (2008) Evaluation of a generic physiologically based pharmacokinetic model for lineshape analysis. *Clin Pharmacokinet* **47**:261–275.

Poulin P and Theil FP (2002) Prediction of pharmacokinetics prior to in vivo studies. 1. Mechanism-based prediction of volume of distribution. *J Pharm Sci* **91**:129–156.

Riley RJ, Grime K and Weaver R (2007) Time-dependent CYP inhibition. *Expert Opin Drug Metab Toxicol* **3**:51–66.

Rodgers T and Rowland M (2007) Mechanistic approaches to volume of distribution predictions: understanding the processes. *Pharm Res* **24**:918–933.

Rostami-Hodjegan A (2009) Predicting inter-individual variability of metabolic drug–drug interactions: identifying the causes and accounting for them using systems approach, in *Enzyme Inhibition in Drug Discovery and Development: The Good and the Bad* (Lu C and Li AP eds), pp (in press), John WHey & Sons, Inc., New Jersey.

Rostami-Hodjegan A, Jackson PR and Tucker GT (1994) Sensitivity of indirect metrics for assessing "rate" in bioequivalence studies—moving the "goalposts" or changing the "game". *J Pharm Sci* **83**:1554–1557.

Rostami-Hodjegan A and Tucker GT (2004) 'In silico' simulations to assess the 'in vivo' consequences of 'in vitro' metabolic drug–drug interactions. *Drug Discov Today: Technol* **1**:441–448.

Rostami-Hodjegan A and Tucker GT (2007) Simulation and prediction of in vivo drug metabolism in human populations from in vitro data. *Nat Rev Drug Discov* **6**:140–148.

Rowland M and Matin SB (1973) Kinetics of drug–drug interactions. *J Pharmacokinet Biopharm* **1**:553–567.

Shou M, Hayashi M, Pan Y, Xu Y, Morrissey K, Xu L and Skiles GL (2008) Modeling, prediction, and in vitro in vivo correlation of CYP3A4 induction. *Drug Metab Dispos* **36**:2355–2370.

Song IS, Shin HJ, Shim EJ, Jung IS, Kim WY, Shon JH and Shin JG (2008) Genetic variants of the organic cation transporter 2 influence the disposition of metformin. *Clin Pharmacol Ther* **84**:559–562.

Sugano K (2009) Introduction to computational oral absorption simulation. *Expert Opin Drug Metab Toxicol* **5**:259–293.

Tam D, Tirona RG and Pang KS (2003) Segmental intestinal transporters and metabolic enzymes on intestinal drug absorption. *Drug Metab Dispos* **31**:373–383.

Tucker GT (1981) Measurement of the renal clearance of drugs. *Br J Clin Pharmacol* **12**: 761–770.

Urban TJ, Brown C, Castro RA, Shah N, Mercer R, Huang Y, Brett CM, Burchard EG and Giacomini KM (2008) Effects of genetic variation in the novel organic cation transporter, OCTN1, on the renal clearance of gabapentin. *Clin Pharmacol Ther* **83**:416–421.

Venkatakrishnan K and Obach RS (2005) In vitro-in vivo extrapolation of CYP2D6 inactivation by paroxetine: prediction of nonstationary pharmacokinetics and drug interaction magnitude. *Drug Metab Dispos* **33**:845–852.

Venkatakrishnan K and Obach RS (2007) Drug–drug interactions via mechanism-based cytochrome P450 inactivation: points to consider for risk assessment from in vitro data and clinical pharmacologic evaluation. *Curr Drug Metab* **8**:449–462.

Wang YH, Jones DR and Hall SD (2004) Prediction of cytochrome P450 3A inhibition by verapamil enantiomers and their metabolites. *Drug Metab Dispos* **32**:259–266.

Wang ZJ, Yin OQ, Tomlinson B and Chow MS (2008) OCT2 polymorphisms and in-vivo renal functional consequence: studies with metformin and cimetidine. *Pharmacogenet Genom* **18**:637–645.

Watanabe T, Kusuhara H, Maeda K, Shitara Y and Sugiyama Y (2009) Physiologically based pharmacokinetic modeling to predict transporter-mediated clearance and distribution of pravastatin in humans. *J Pharmacol Exp Ther* **328**:652–662.

Wienkers LC and Heath TG (2005) Predicting in vivo drug interactions from in vitro drug discovery data. *Nat Rev Drug Discov* **4**:825–833.

Yang J, Jamei M, Yeo KR, Tucker GT and Rostami-Hodjegan A (2007) Prediction of intestinal first-pass drug metabolism. *Curr Drug Metab* **8**:676–684.

Yang J, Liao M, Shou M, Jamei M, Yeo KR, Tucker GT and Rostami-Hodjegan A (2008) Cytochrome p450 turnover: regulation of synthesis and degradation, methods for determining rates, and implications for the prediction of drug interactions. *Curr Drug Metab* **9**:384–394.

Yang JS, Kjellsson M, Rostami-Hodjegan A and Tucker GT (2003) The effects of dose staggering on metabolic drug–drug interactions. *Eur J Pharm Sci* **20**:223–232.

Zhao P, Ragueneau-Majlessi I, Zhang L, Strong JM, Reynolds KS, Levy RH, Thummel KE and Huang SM (2009) Quantitative evaluation of pharmacokinetic inhibition of CYP3A substrates by ketoconazole: a simulation study. *J Clin Pharmacol* **49**:351–359.

Chapter 14
Absorption Models to Examine Bioavailability and Drug–Drug Interactions in Humans

Ahsan Naqi Rizwan and Kim L.R. Brouwer

Abstract Human drug absorption studies are warranted in many situations. The development of modified release formulations, the evaluation of different formulations designed to improve bioavailability by enhancing solubility or permeability, the investigation of mechanisms of drug-drug/food interactions in the gastrointestinal tract, and the determination of intestinal and biliary secretion in humans all require assessment of drug absorption/excretion in the human gastrointestinal tract. This chapter discusses in vivo approaches that can be used to assess absorption/bioavailability issues in the human gastrointestinal tract. The design and the application of remotely activated capsules and aspiration/perfusion catheters are reviewed, and the utility of human microdose studies is discussed. With the use of appropriate probes, these approaches also may be used to assess drug–drug or drug–food interactions, as well as interactions involving transporters and enzymes present in the gastrointestinal tract and/or the liver.

14.1 Introduction

Oral administration of drugs in the form of tablets or capsules is the most common route of drug administration because it is the safest and easiest, however, compared to other routes of administration (e.g., intravenous, topical, or inhalational), it is the route that takes the drug molecule through the most barriers prior to reaching the systemic circulation. The fraction of the administered dose of unchanged drug that reaches the systemic circulation, defined as the bioavailability, is one of the principal pharmacokinetic parameters of a drug product. Adequate bioavailability of a therapeutic dose ensures that appropriate concentrations of the drug reach the site of action to produce a therapeutic effect.

K.L.R. Brouwer (✉)
Division of Pharmacotherapy and Experimental Therapeutics, Eshelman School of Pharmacy, The University of North Carolina at Chapel Hill, Chapel Hill, NC, USA
e-mail: kbrouwer@unc.edu

K.S. Pang et al. (eds.), *Enzyme- and Transporter-Based Drug–Drug Interactions*, 343
DOI 10.1007/978-1-4419-0840-7_14,

To reach the systemic circulation, drug molecules must first pass from the gastrointestinal lumen through the epithelial barrier, and then through the liver in series before reaching the general circulation. Both the liver and the gastrointestinal tract contain numerous metabolizing enzymes that are capable of altering the drug molecule such that only a fraction of the dose administered actually reaches the systemic circulation. This is termed the first-pass effect. Bioavailability also can be modulated by factors such as solubility, other physicochemical properties of the drug, formulation effects, local pH in the gastrointestinal tract, the presence of food, intestinal microflora, and intestinal and hepatic influx/efflux transporters (Martinez and Amidon, 2002).

The reasons for performing drug absorption and disposition studies in humans are numerous, including the evaluation of lead compounds for IND submissions (Davis and Wilding, 2001), satisfying phase 4 commitments (FDA, Data standards manual phase 4 commitment categories, 1996), and in the life-cycle management of older drugs going off patent (Davis and Wilding, 2001; Fleming and Ma, 2002). Situations where human drug absorption studies may be warranted include the following:

1. Development of modified release formulations (e.g., *developing sustained release formulations for once-a-day therapy*)
2. Improvement of bioavailability by enhancing solubility or permeability (e.g., *exploiting transporters to promote active uptake of drugs*)
3. Generation of human data supporting an IND application (e.g., *microdosing pharmacokinetic/imaging studies to increase attrition rate in early development*)
4. Evaluation of mechanisms of drug–drug/food interactions
5. Evaluation of specific routes of elimination in humans (e.g., *measuring biliary excretion patterns in humans*)

Most drugs in development now require a once- or twice-daily dosing regimen because it makes them more marketable due to improved patient compliance resulting from reduced dosing frequency (Fleming and Ma, 2002). Therefore, understanding how a drug performs during transit through the human gastrointestinal tract provides the key to developing effective oral formulations. For older compounds that are already on the market, or compounds in late stages of clinical development, human drug absorption studies can provide useful and relevant information to address the situations mentioned above. For example, information about regional differences in drug absorption in the gastrointestinal tract can aid in the development of modified release formulations. The use of intelligent, remotely activated capsules capable of delivering drugs to specific regions of the gastrointestinal tract has made it possible to determine regional differences in drug absorption. Given the substantial species differences in metabolic enzymes and transport proteins, drugs whose bioavailability is thought to be affected by these systems can be evaluated in humans using intestinal aspiration/perfusion catheters and suitable probes. It is notable that drugs excreted in bile may undergo enterohepatic recirculation and exhibit multiple peaks in their drug absorption profile. Biliary excretion of drugs/metabolites can be measured in humans by specialized oro-enteric catheters.

The Biopharmaceutics Classification System (BCS) is currently the accepted method of classifying drugs in terms of their predicted intestinal permeability and solubility in humans. The FDA advocates the use of this system to gain bioequivalence waivers, thereby avoiding the need for expensive clinical bioequivalence studies. Poorly soluble and highly impermeable molecules are prone to reduced bioavailability. Molecules that undergo high first-pass metabolism or transporter-mediated absorption fall into the category of drugs whose bioavailability may be limited due to interactions with other drugs or food. Orally administered, immediate release (IR) drug products in the top 200 drug-product lists from the United States, Great Britain, Spain, and Japan have been provisionally classified based on the BCS. Oral IR drug products constituted more than 50% of the top 200 drug products on all four lists and ranged from 102 to 113 in number. In essence, more than 55% of the drug products are classified as high-solubility (class 1 and class 3) drugs in the four lists, suggesting that in vivo bioequivalence (BE) may be assured with a less expensive and more easily implemented in vitro dissolution test (Takagi et al., 2006). For molecules that do not fall in this category, in vivo drug absorption studies are necessary to understand intestinal permeability.

Microdosing is another strategy that has been employed to gain an understanding of the pharmacokinetics of a lead molecule before it is advanced to phase I. Human microdosing studies are designed to elucidate the pharmacokinetic profile following administration of a sub-pharmacological, sub-therapeutic dose of a novel drug candidate that has demonstrated efficacy in non-clinical models. Regulatory bodies have determined that because these exploratory IND studies present fewer potential risks than traditional phase I studies that evaluate dose-limiting toxicities, such limited exploratory investigations in humans can be initiated with less, or different, preclinical support than that required for traditional IND studies (Marchetti and Schellens, 2007).

Carefully conducted in vivo human experiments utilizing innovative absorption models can ensure a more informed approach for drugs that may be categorized as follows:

- Poorly soluble and poorly permeable compounds
- Compounds affected by metabolizing enzymes and carrier-mediated transport
- New chemical entities for which human absorption, distribution, metabolism, and elimination (ADME) data are needed
- Compounds exhibiting drug–drug/food interactions
- Drugs whose intestinal bioavailability needs to be demonstrated (e.g., pancreatic enzyme supplements)
- Drugs/formulations where regional absorption in the gastrointestinal tract needs to be defined
- Lead compounds where preclinical assessment is mandated to gain knowledge regarding disposition

This chapter discusses in vivo approaches that can be used to assess absorption/bioavailability issues in the human gastrointestinal tract. In principle, these approaches also may be used to assess drug–drug or drug–food interactions, and interactions involving transporters and enzymes present in the gastrointestinal tract and/or in the liver.

14.2 Remotely Activated Capsules

Remotely activated drug delivery systems provide a rapid and cost-effective method to assess the oral absorption/bioavailability of drugs in specific regions of the gastrointestinal tract. These delivery systems provide a realistic assessment of the drug's absorption/bioavailability profile and help to identify the site(s) of absorption so that controlled release dosage forms can be developed to achieve maximum bioavailability. Remotely activated capsules have been used to facilitate evaluation of performance modifiers, such as penetration enhancers or inhibitor agents that can be co-administered with the drug, thus defining their role in the overall drug absorption or degradation process. These devices avoid the need to subject volunteers to invasive procedures like oral intubation. An important advantage of these devices is that they can deliver many forms of the drug (dry powder, liquids or suspensions, and whole formulations) to key regions of the human gastrointestinal tract precisely where they are required. They can be activated without the need for invasive measures, which can disrupt local physiology and affect the accuracy of measurements.

14.2.1 HF-Capsule

14.2.1.1 Design

The high-frequency (HF)-capsule (Fig.14.1) was developed by Shuster and Hugemann in the early 1980s (Wouters, 1998). The HF-capsule is a remote controlled device for bolus delivery of a drug in different regions of the human gastrointestinal tract. It consists of a polyurethane hull (length 2.5 cm, diameter 0.7 cm) containing a small latex balloon as a drug reservoir and a simple mechanism for destruction of the balloon by an external electromagnetic high-frequency signal. The drug in solution penetrates the hull through small gaps after triggering the release, and the hull itself is excreted in the feces. Part of the hull is radio-opaque to enable localization in the gastrointestinal tract by radioscopy.

14.2.1.2 Application

Since its development, the HF-capsule has been employed in a number of studies to evaluate the absorption of drugs including theophylline, nitrendipine, nisoldipine, and ipsapirone (Bode et al., 1996). The HF-capsule was used to investigate the absorption profile of ciprofloxacin solution from different regions of the gastrointestinal tract. Pharmacokinetic data obtained from 24 h venous blood samples and urine collections were compared after ciprofloxacin release in the stomach, jejunum, ileum, ascending colon, and descending colon. These values were compared with data generated after the same dose of ciprofloxacin was administered as an oral solution, and after the administration of a commercially available tablet formulation. The capsule was activated successfully in most instances, and results demonstrated that the primary absorption site of ciprofloxacin was the upper gastrointestinal tract up

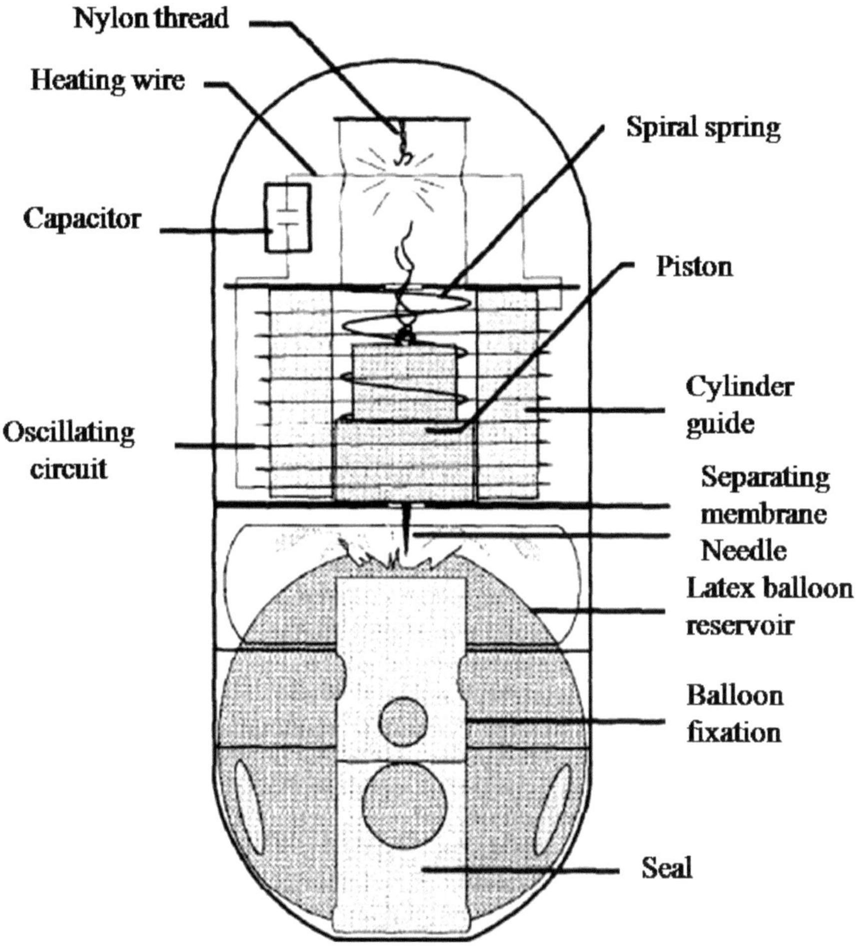

Fig. 14.1 Schematic diagram of the HF-capsule. With kind permission from Springer Science+ Business. (Staib et al. *Am J Med* **87**, 1989 Suppl. 5A, p. 67S)

to the jejunum. In addition, urinary excretion of the parent drug and metabolites was monitored in this study; although urinary recovery of ciprofloxacin and metabolites decreased due to regional differences in absorption along the GI tract, the metabolite pattern was unchanged, thereby ruling out differences in presystemic metabolism as a cause of the observed differences in absorption.

In a separate study, ipsapirone in solution was administered orally, rectally, and at specific gastrointestinal sites using the HF-capsule. The relative systemic bioavailability of the drug from the colon and the rectum was two- to three-fold greater than that from the upper regions of the gastrointestinal tract. Hence, these data provided a rationale for developing a prolonged release formulation of the drug (Fuhr et al., 1994).

Bode et al. (1996) used the HF-capsule to determine whether absorption windows for nifedipine existed along the gastrointestinal tract. The absence of absorption windows, coupled with the finding that no particular regions of the gastrointestinal tract exhibited significant impairment in the extent of nifedipine absorption, provided the rationale for developing nifedipine as a sustained release formulation.

14.2.2 InteliSite®

14.2.2.1 Design

The InteliSite® capsule (Fig. 14.2), an externally activated drug delivery system, is approximately the size of a "000" capsule (10 mm wide and 35 mm long) and can carry ~0.8 mL of solution in a storage chamber. Following oral administration, the InteliSite® capsule can be tracked through the gastrointestinal tract using gamma scintigraphy. Either indium (^{111}In) or technetium (^{99m}Tc) gamma isotopes may be incorporated into the capsule. Therefore, the capsule and the release of drug can be tracked simultaneously using gamma cameras with dual isotope detection capabilities. Two versions of the InteliSite® capsule are available. One version

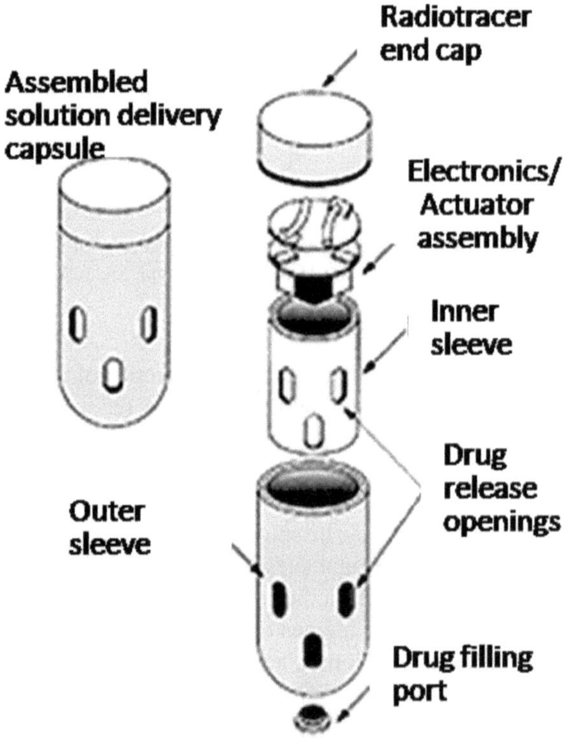

Fig. 14.2 Schematic diagram of the InteliSite® capsule. With kind permission from Springer Science+Business Media. (Parr et al., *Pharm Res*, 1999, **16**: p. 267, Fig. 4.1)

delivers solutions or suspensions; another version delivers powder formulations (Parr et al., 1999).

When the InteliSite® capsule reaches the desired location in the gastrointestinal tract, it is externally activated by a remote control device. Activation is accomplished by exposing the capsule to a radio frequency magnetic field that induces a small amount of heat in the capsule's activation assembly. This causes two shape-memory alloy wires to straighten, and this action rotates an inner sleeve of the capsule in relation to an outer sleeve. The rotation process aligns a series of slots in the sleeve surface, which permits the contents of the capsule to be released. Serial blood samples are taken to characterize drug absorption from the specific gastrointestinal region. Post-activation, the InteliSite® capsule passes intact through the gastrointestinal tract and is excreted in the feces.

14.2.2.2 Application

The safety and utility of the InteliSite® capsule was assessed by Pithavala et al. (1998) and Parr et al. (1999) in healthy human volunteers; the test substance, a radiotracer compound or ranitidine, respectively, was delivered to different regions of the human intestine. The purpose of these studies was to establish that this device could be used to administer drugs to specific sites in the human intestine, and to correlate release of the drug to measured plasma concentrations. The study by Parr et al. (1999) evaluated the safety and tolerance of the device; no active drug was included in the capsule. Instead, radiolabeled ^{99m}Tc-diethylenetriaminepentaacetic acid (DTPA) solution was added to the "drug" reservoir, and release of DTPA from the capsule was visualized by gamma scintigraphy. The position of the capsule in the correct anatomical region, before and after release, was verified by the second radioisotope, ^{111}In-DTPA, which was incorporated in a compartment in the cap of the capsule. The aim of this study was to selectively release the contents of the capsule from the "drug" reservoir into the stomach, jejunum, ileum, or colon, during separate activation phases of the study. Results demonstrated that subjects experienced no adverse events associated with the administration of the InteliSite® capsule, and that the device successfully released the contents in most instances at the desired location (Parr et al., 1999).

The study by Pithavala et al. (1998) evaluated the extent of ranitidine absorption from different regions of the human intestine including the jejunum, ileum, and colon. Ranitidine solution was administered to healthy adult volunteers via the InteliSite® capsule. Following capsule activation and release of ranitidine, blood samples were collected to measure the absorption and elimination profile. The data obtained correlated well with previous intubation studies showing that remote activation of the InteliSite® capsule was a non-invasive way to release ranitidine into the human intestine and measure site-specific absorption of drugs (Pithavala et al., 1998).

Having established the safety and utility of the device in delivering a drug to the desired location in the intestine, the application of the InteliSite® capsule was validated further by co-administering two model drugs, theophylline and furosemide, with known differences in absorption patterns along the intestine. Theophylline is well absorbed along the full length of the intestinal tract, whereas furosemide exhibits regional differences in absorption. Theophylline and furosemide were incorporated inside the InteliSite® capsule in the form of split immediate release tablets. In this three-way crossover study, eight subjects were administered either tablets orally, or the split tablets were inserted inside the capsule and the capsule was activated either in the small intestine or the colon after scintigraphic verification of the capsule location. Successful capsule activation was confirmed by including a soluble internal marker along with the split tablets. The capsule released the entire radioactive marker in the small intestine in all but one subject. Both theophylline and furosemide were well absorbed following InteliSite® activation in the small intestine, but near-complete release was observed for only two of the eight subjects in the colonic region, possibly due to low water content in the colon, which resulted in poor mixing and solubilization (Clear et al., 2001).

14.2.3 *Enterion™*

14.2.3.1 Design

The Enterion™ capsule (Fig. 14.3) is a remotely activated capsule that can be used to obtain detailed information on the human regional gut absorption window for all types of oral formulations, provided they fit in the drug reservoir. The round-ended capsule is the size of a "000" capsule, and its design incorporates a spring which actively expels its contents (up to 1 mL of formulation) at the target site. The Enterion™ capsule can be used to evaluate solutions, suspensions, particulates, pellets, or mini-tablets. The primary application of this device is to identify factors responsible for poor bioavailability in order to optimize oral drug delivery. Data from the Enterion™ capsule have been used in the development of extended-release versions of over 30 drugs currently on the market, including carvedilol,

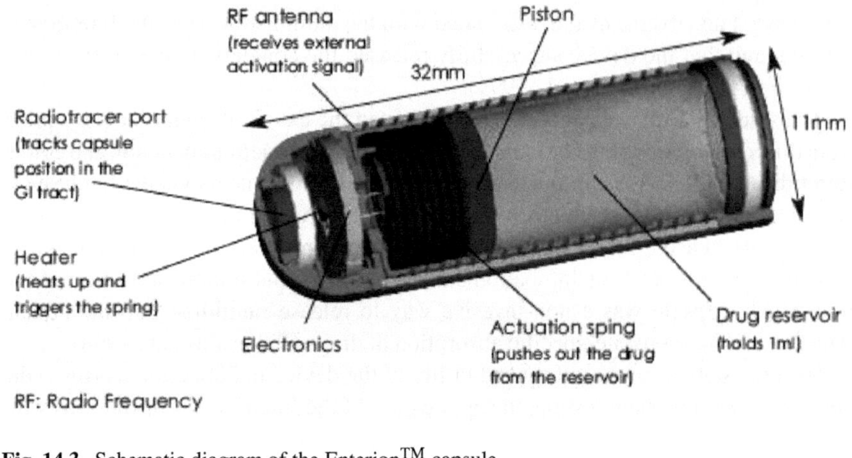

Fig. 14.3 Schematic diagram of the Enterion™ capsule

faropenem, oseltamivir, and acipimox (Enterion^{TM}, 2009; Martin et al., 2003). A notable advantage of the Enterion^{TM} capsule is that whole formulations can be tested, thus avoiding, at an early stage, the development of formulations that may not prove viable as drug products.

After the subject has swallowed the Enterion^{TM} capsule, it is tracked in real time as it passes through the gastrointestinal tract using a gamma camera which detects the gamma-emitting radionuclide sealed inside the tracer port of the capsule (Fig. 14.3). The subject also ingests a drink containing a second gamma-emitting radionuclide, which provides an outline of his/her stomach and colon. When the capsule has reached its target destination, the volunteer stands inside an activation unit, which sends an electromagnetic signal to the capsule and results in the instantaneous release of the actuation spring, expelling the drug formulation within milliseconds (Enterion^{TM}, 2009; Martin et al., 2003).

14.2.3.2 Application

Enterion^{TM} clinical studies are designed in a crossover format with regular blood sampling to assess the impact of the type of oral formulation, including the immediate release formulation, on bioavailability. These studies may also answer questions regarding the cause of poor absorption, which can be used to assess the advantages of developing a modified release formulation. The different treatment arms may include intravenous administration and an immediate release formulation to compare with Enterion^{TM} capsule activation and release of test drug/formulation in select regions of the gastrointestinal tract (e.g., stomach, proximal small bowel, distal small bowel, or ascending colon).

The anti-influenza prodrug oseltamivir phosphate (Tamiflu) exhibits high aqueous solubility and moderately low permeability. In order to evaluate the feasibility of developing oseltamivir as a once-daily modified release formulation, the Enterion^{TM} capsule was employed to administer oseltamivir phosphate into the stomach, the proximal small bowel, the distal small bowel, and the ascending colon. Both the parent prodrug and the active metabolite exhibited a good absorption profile from all regions of the gastrointestinal tract except for the ascending colon; although the absorption rate from the colon was slower, substantial C_{max} concentrations still were achieved. The results, therefore, supported the feasibility of developing a modified release formulation of oseltamivir phosphate (Oo et al., 2003).

Similarly, the regional absorption of the dual angiotensin-converting enzyme/neutral endopeptidase (ACE/NEP) inhibitor, M100240, was evaluated in three regions of the gastrointestinal tract using the Enterion^{TM} capsule and compared to intravenous and oral administration of the standard immediate release tablet formulation to determine the absolute and relative bioavailability of this compound. The estimates of relative bioavailability in the proximal small bowel, the distal small bowel, and the ascending colon, relative to the oral immediate release tablet, supported the development of a modified release formulation of the compound. These studies also provided a more accurate determination of absolute and relative bioavailability from different regions along the gastrointestinal tract (Martin et al., 2003).

Bevirimat, a first in class human immunodeficiency virus-1 (HIV-1) maturation inhibitor, is in clinical development for combination therapy for HIV. However, the immediate release formulation exhibits a relative bioavailability of less than 50% and higher inter-subject variability compared with an oral solution. Bevirimat has little or no interaction with CYP enzymes or P-glycoprotein. The Enterion^{TM} capsule was used to investigate whether a narrow permeability-dependent absorption window existed within the human gastrointestinal tract that could explain the observed inter-subject variability in bioavailability of bevirimat. Results demonstrated that bioavailability of the immediate release formulation relative to the oral solution was similar along different regions of the gastrointestinal tract, suggesting that the observed interindividual variability in bioavailability was probably due to other factors (Connor et al., 2009).

14.3 Intestinal Aspiration/Perfusion Catheters

The intestinal absorption of a drug is governed by its dissolution characteristics, solubility and permeability, and the stability of the drug during exposure to the gut wall and its contents following oral administration. The pharmacokinetic events following oral drug administration are often poorly predicted by current models of first-pass effects mediated by the gastrointestinal tract and liver, especially in first time in human studies. The use of multilumen aspiration/perfusion catheters offers one approach to study these processes in a very direct fashion in the most relevant species. Many versions of these catheters have been used over the past several decades (Sladen, 1968; Sladen and Dawson, 1968, 1970; von Richter et al., 2001; Glaeser et al., 2002; Lee et al., 2004) to address many questions: intestinal solubilization of drugs; in vivo dissolution and bioavailability; permeability of drugs in various regions of the intestine; absorption/metabolism of drugs in the intestine; differentiating between paracellular/transcellular and carrier-mediated intestinal transport; extent of intestinal secretion of drugs; regional single-pass jejunal perfusion; biliary clearance/excretion; induction of intestinal drug metabolism; absorption after oral, jejunal, cecal, or colonic administration; the effect of pancreatico-biliary secretions on drug absorption; and intestinal drug–drug interactions. These are summarized in Tables 14.1 and 14.2.

Although there has been considerable interest in studying intestinal permeability in humans, the feasibility of conducting such studies is limited due to the complexity and invasiveness of these procedures. Relatively few labs have developed expertise in the use of aspiration/perfusion catheters to obtain samples directly from the intestine or the stomach in humans. The advantages of using these techniques are many, and depending upon the design of the catheter, by adjusting the location of the occluding balloons and/or the function of aspiration/perfusion ports, many paradigms can be evaluated. The applications of two specific catheters the Loc-I-Gut® and CHOL-ect catheters are discussed below.

Table 14.1 Human intestinal aspiration/perfusion studies using the Loc-I-Gut® catheter

Application	Victim/drug	Perpetrator	References
Drug–drug interaction and jejunal permeability	Fexofenadine	Ketoconazole	Tannergren et al. (2003)
	Amoxicillin	Amiloride	Lennernäs et al. (2002)
	(*R*/*S*)-Verapamil	Ketoconazole	Sandstrom et al. (1999)
	Finasteride	St. John's Wort	Lundahl et al. (2009)
Jejunal permeability	Cimetidine, ranitidine, propranolol	–	Takamatsu et al. (2001)
	Terbutaline, antipyrine	–	Fagerholm et al. (1995)
	Levodopa, antipyrine	–	Nilsson et al. (1994)
	(*R*/*S*)-Verapamil	–	Hedeland et al. (2004), Sandstrom et al. (1998)
Influence of interindividual variability in gastric emptying/precipitation/dissolution of a drug on its pharmacokinetic profile	Gefitinib	–	Bergman et al. (2007)
Biliary secretion	Rosuvastatin	–	Bergman et al. (2006)

Table 14.2 Human intestinal aspiration studies using the CHOL-ect multilumen catheter

Application	Drug/probe	References
In vitro–in vivo correlation of biliary clearance; estimation of gall bladder ejection fraction	^{99m}Tc-mebrofenin, ^{99m}Tc-sestamibi, piperacillin	Ghibellini et al. (2007a, b)
Determination of in vitro and in vivo metabolism	Piperacillin	Ghibellini et al. (2007a)
Biliary clearance/excretion	^{99m}Tc-mebrofenin, piperacillin	Ghibellini et al. (2006b)
In vivo dissolution and bioavailability of a pancreatic enzyme formulation	Pancrelipase	ClinicalTrials.gov (NCT00744250)
Absorption after oral or colonic administration	Zafirlukast	Fischer et al. (2000)
Absorption after oral, jejunal or cecal administration	Sumatriptan	Warner et al. (1995)
Effect of pancreatico-biliary secretions on drug absorption	Ranitidine	Reynolds et al. (1998)

14.3.1 Loc-I-Gut®

The primary application of the Loc-I-Gut® is in measuring the effective permeability of a molecule across the wall of the small bowel (Table 14.1). Effective permeability P_{eff} is expressed in units of 10^4 cm/s and is a quantitative estimation of passive and carrier-mediated flux across membranes (Lennernäs, 2007). The effective permeability of a compound depends on its physicochemical characteristics, effects on membrane fluidity, and carrier-mediated uptake or efflux. Depending upon the compound, there may be substantial differences in P_{eff} between species due to differences in gastric volume and pH, gastrointestinal transit time, differences in drug-metabolizing enzymes, and/or influx/efflux transporters. Furthermore, permeability of a compound is not necessarily constant throughout the gastrointestinal tract; for some compounds, permeability may be greater in the jejunum than the colon, while for other compounds, permeability may be constant throughout the gastrointestinal tract.

14.3.1.1 Design

The Loc-I-Gut® catheter (Fig. 14.4; Synectics Medical, Sweden) is a multichannel polyvinyl tube with an external diameter of 5.3 mm. It is 175 cm long with two inflatable balloons attached 10 cm apart. Inflation or deflation of the two balloons allows this catheter to be used in three configurations: (1) open; (2) semi-open, with occlusion by inflating the proximal balloon; and (3) closed, with two inflated balloons occluding a 10-cm long intestinal segment. The tube often is used in conjunction with a naso-gastric tube to aid in gastric sampling and drug administration. The non-absorbable marker ^{14}C-polyethylene glycol (PEG) 4000 is used for normalization of the data. The tube can be positioned using a Teflon-coated guidewire and placement can be confirmed by fluoroscopy. Details are discussed in a comprehensive review (Lennernäs, 1998).

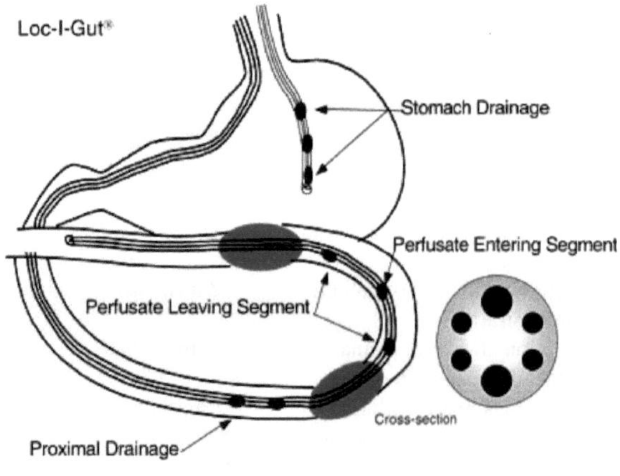

Fig. 14.4 The Loc-I-Gut® tube. A schematic view of the Loc-I-Gut® catheter positioned in the proximal human jejunum with an isolated segment created by inflation of the two balloons. With kind permission from Springer Science+Business. (Petri et al. *Drug Metab Dispos*. 2003, **31**:805–813)

When using the Loc-I-Gut®, a 10-cm segment can be created between two inflated balloons enabling single-pass perfusion of a well-defined region of the small intestine. One of the primary advantages of this technique is that occlusion of the segment between the two intra-luminal balloons minimizes contamination with luminal fluids both proximally and distally into the perfused segment. In addition, leakage from the segment over the balloons is small, so that the recovery of the non-absorbable marker is almost complete. These qualities enable control of the absorption conditions in the intestinal segment, and thus facilitate the study of mechanisms of transport and metabolism of xenobiotics and nutrients in the human intestine (Lennernäs, 1998).

14.3.1.2 Applications

Two important parameters that can be measured by using the Loc-I-Gut® catheter include the intestinal absorption rate (P_{eff}) and gut wall extraction (E_{G}). Bioavailability following oral administration can be described using the following equation:

$$F = f_{\mathrm{a}}(1 - E_{\mathrm{G}})(1 - E_{\mathrm{H}}) \tag{14.1}$$

where f_{a} is the fraction of the dose available for absorption; E_{G} is the first-pass extraction of the drug in the gut wall; and E_{H} is the first-pass extraction of the drug in the liver.

The amount of drug that disappears during a single passage through the intestinal segment is assumed to be absorbed and can be calculated as follows:

$$f_{\mathrm{a}} = 1 - \frac{C_{\mathrm{out}}}{C_{\mathrm{in}}} \times \frac{\mathrm{PEG_{in}}}{\mathrm{PEG_{out}}} \tag{14.2}$$

where C_{in} and C_{out} are the inlet and outlet concentrations of the drug, respectively; and $\mathrm{PEG_{in}}$ and $\mathrm{PEG_{out}}$ are the concentrations of ^{14}C-polyethylene glycol 4000 entering and leaving the segment, respectively.

The effective intestinal permeability can be calculated based on the following equation:

$$P_{\mathrm{eff}} = \frac{(C_{\mathrm{in}} - C_{\mathrm{out}})Q_{\mathrm{in}}}{C_{\mathrm{out}} \times 2\Pi rL} \tag{14.3}$$

where Q_{in} is the flow rate of the perfusion solution and L is the length of the perfused intestinal segment (usually 10 cm). The human intestinal radius (r) is assumed to be 1.7 cm.

The two parameters f_{a} and P_{eff} can be quantified by sampling through the tube. P_{eff} is a descriptor of the transport process across the intestinal barrier and is dependent upon the perfusion technique, the perfusion rate, and the intestinal motility. The fraction of the dose available for absorption (f_{a}) is an estimate of the fraction of the dose administered via the tube and transported across the apical cell membrane into the cellular space of the enterocytes. In Equation (14.1), the bioavailability (*F*) can be estimated from the ratio of dose normalized blood/plasma area under the curve (AUC) values determined following intestinal perfusion and after intravenous administration of the drug, and f_a can be determined by measuring the rate of disappearance of the drug from the perfusate using Equation (14.2). Hepatic extraction following intravenous administration can be calculated by dividing total blood clearance by hepatic blood flow. Thus, incorporating the value of f_a in Equation (14.1) and assuming equal hepatic extraction following intravenous administration and intestinal perfusion, the first-pass extraction from the gut can be estimated using Equation (14.4):

$$E_G = 1 - \frac{F}{f_a(1 - E_H)} \tag{14.4}$$

A major advantage of this approach is that extraction from the gut wall can be differentiated from hepatic extraction assuming that the liver is the primary organ for elimination of the drug after intravenous administration. The following sections describe how Loc-I-Gut® can be used and what questions can be answered.

14.3.1.3 Determination of Intestinal Presystemic Metabolism

Metabolic drug–drug interactions involving the CYP P450 enzyme systems can affect the amount of drug entering into the systemic circulation and in some cases have been demonstrated to play a more dominant role than hepatic enzymes (Thelen and Dressman, 2009). The Loc-I-Gut® was used in the double balloon configuration to determine the effective jejunal permeability of (*R*/*S*)-enantiomers of verapamil (a CYP3A4 substrate) along a 10-cm segment of the jejunum. The aim of this study was to investigate the effect of ketoconazole, a modulator of P-glycoprotein and a strong inhibitor of CYP3A enzymes, on the jejunal permeability and first-pass metabolism of verapamil. Results showed that ketoconazole did not affect the permeability of verapamil, suggesting that ketoconazole did not affect P-glycoprotein-mediated efflux of verapamil. However, inhibition of CYP3A4 led to a decreased appearance of both the (*R*) and (*S*) metabolites of verapamil, norverapamil, in the perfusate leaving the human jejunal segment. Inhibition of the gut wall metabolism of verapamil also may have been responsible for the observed increase in the overall flux into the systemic circulation (Sandstrom et al., 1999). This provided a specific mechanism for the drug–drug interaction between ketoconazole and verapamil, and established a method to measure the presystemic metabolism of a compound mediated by the gastrointestinal tract in humans.

14.3.1.4 Evaluation of Biliary Excretion

Biliary excretion of drugs often is overlooked or not taken into account because no convenient method exists to accurately detect this process in humans. Biliary excretion is, however, a significant confounder, if not taken into account, when secretion into bile is a major route of drug excretion. The Loc-I-Gut® technique permits simultaneous dosing and bile collection. It allows for bile samples to be collected from the distal duodenum/proximal jejunum during the absorption phase, i.e., during the first couple of hours of drug absorption and disposition. The Loc-I-Gut® was used to generate data for rosuvastatin, a statin that acts in the liver and exhibits toxicity after systemic exposure. The appearance of rosuvastatin in human bile was very rapid; peak biliary concentrations were reached within 42 min after rosuvastatin administration via a naso-gastric tube (Bergman et al., 2006).

St. John's Wort (SJW) is a commonly used over-the-counter herbal remedy for mild depression. SJW is a known inducer of CYP3A4. The effect of SJW on the plasma, biliary, and urinary pharmacokinetics of finasteride was investigated using a Loc-I-Gut® catheter in a semi-open configuration. The objective was to study the consequences of induced drug metabolism caused by SJW treatment on the plasma, urinary and biliary pharmacokinetics of finasteride and its two phase I metabolites (hydroxyl- and carboxy-finesteride). Finasteride was administered as a solution directly into the intestine via a gastric port distal to the inflated balloon. The tube was positioned such that the inflated balloon blocked bile passage down the intestinal lumen, and bile samples were collected from aspiration ports proximal to the balloon. Finasteride is known to undergo complete metabolism via CYP3A4 with 60% of the dose recovered in feces as metabolites. Thus, biliary excretion may be an important route of elimination for finasteride metabolites. This was the first study to investigate the excretion of finasteride in bile. The investigators quantified the biliary excretion of the metabolite carboxy-finesteride and demonstrated that biliary excretion increased following 2 weeks of induction with SJW. Furthermore, plasma finasteride AUC and C_{max} values were reduced by 50%. The interaction between SJW and finasteride was deemed to be clinically relevant, and dose adjustments were recommended in order to maintain efficacy in the treatment of benign prostatic hyperplasia, especially since a significant proportion of the male population uses SJW as an herbal supplement (Lundahl et al., 2009).

14.3.1.5 Impact of Modulators of Hepatic Transport

The Loc-I-Gut® was used to study the importance of membrane transport proteins in the pharmacokinetics of fexofenadine absorption. Fexofenadine is not metabolized extensively and exhibits low passive membrane permeability because of low P_{eff} and f_a values, both measured and predicted based on physicochemical properties. Therefore, fexofenadine probably relies on active transport to gain systemic exposure. The objective of this study was to investigate the in vivo transport mechanisms involved in the intestinal absorption and bioavailability of fexofenadine in humans. Comparison with intravenous pharmacokinetic data was not possible because an injectable form of fexofenadine was not available. Verapamil was used as a membrane transport inhibitor. A jejunal single-pass perfusion study was performed in healthy human volunteers and the drugs were administered by perfusing a 10-cm-long jejunal segment. Since the P_{eff} of fexofenadine was not affected by verapamil, it was hypothesized that simultaneous inhibition of the efflux transporter P-glycoprotein and influx transporter(s), possibly organic anion-transporting polypeptides (OATPs), by verapamil resulted in no net change in fexofenadine permeability. However, the plasma AUC of fexofenadine was increased by four-fold in the presence of verapamil. The intestinal permeability of verapamil was high, ensuring adequate liver exposure. The reason for the higher bioavailability of fexofenadine in the face of low intestinal absorption was attributed to either inhibition of OATP-mediated uptake of fexofenadine across the sinusoidal membrane and/or P-glycoprotein-mediated secretion across the canalicular membrane during first pass through the liver. This study showed that although OATPs and P-glycoprotein work in opposite directions in the intestine, their cooperative function in the liver to facilitate hepatic uptake and biliary secretion, respectively, would certainly affect first-pass hepatic extraction in the presence of a dual modulator (in this case verapamil) (Sandstrom et al., 1999).

14.3.1.6 Directly Detecting Drug–Food Interactions

Dietary constituents may interact with drugs and lead to interindividual variability in drug disposition. While it is challenging to delineate the mechanism(s) of drug–drug interactions, it is even more complex to establish mechanisms for interactions between drugs and dietary constituents. Petri et al. (2003) showed that components present in foods may rapidly induce phase II enzymes after entering enterocytes. Using the Loc-I-Gut®, the permeability and the metabolism of the phytochemicals sulforaphane and quercetin (from onion and broccoli, respectively) were shown for the first time. Additionally, for the first time, short-term changes in the gene expression levels of metabolic enzymes were demonstrated in viable shed enterocytes collected via this catheter. This novel work established a potential method for detecting direct drug–food interactions in the gastrointestinal tract (Petri et al., 2003).

14.3.2 CHOL-ect

14.3.2.1 Design

The CHOL-ect multilumen catheter (Fig. 14.5) is a versatile device that has multiple applications. It has been used to determine drug absorption following administration in different regions along the gastrointestinal tract, measure the effect of pancreatico-biliary secretions and gastrointestinal transit time on drug absorption and pharmacokinetics, determine biliary clearance of drugs in humans for assessment of in vitro and in vivo correlations, determine in vivo intestinal bioavailability of pancreatic enzyme supplements, and assess the influence of hepatic transporter modulation on hepatic exposure of drugs in humans (Ghibellini et al., 2004, 2006a, b, 2007a, 2008) (summarized in Table 14.2).

This customized PVC catheter was designed by Ghibellini, Johnson, Heizer, Brouwer, and co-workers in collaboration with Dentsleeve International Ltd., and currently is manufactured by Mui Scientific, Inc. The CHOL-ect catheter is geared toward establishing methods to examine biliary excretion of xenobiotics. It has an outer diameter of 7.1 mm for the first 100 cm, and 4.6 mm for the remaining 28 cm.

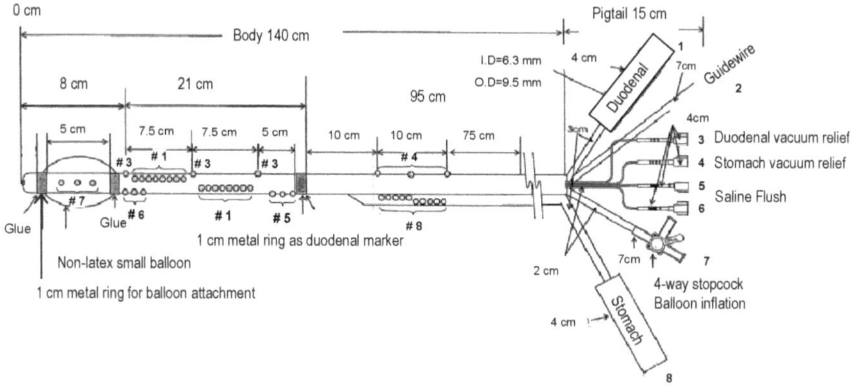

Fig. 14.5 The CHOL-ect tube. Cross-sectional diagram of the CHOL-ect multilumen catheter. Lumens are numbered according to function. Design copyrighted and manufactured by Mui Scientific, Mississauga, Ontario, Canada

An inflatable non-latex balloon is located at the tip with radio-opaque marker rings at three positions along the length of the tube. The lumen of the tube consists of eight channels for duodenal aspiration, Teflon-coated guidewire insertion, duodenal vacuum relief, stomach vacuum relief, proximal and distal duodenal saline flush, balloon inflation, and stomach aspiration. The large number and the distribution of aspiration/perfusion ports along the tube add to the versatility of the device for different applications (Fig. 14.5).

14.3.2.2 Applications

Although the CHOL-ect tube is amenable to studies evaluating the effect of site-specific absorption of drugs along the gastrointestinal tract (Warner et al., 1995; Fischer et al., 2000) or to the influence of pancreatico-biliary secretions on the absorption of a drug (Reynolds et al., 1998), the need to gain information on hepatobiliary drug disposition has led to more specialized applications of this technology.

The prominent role of the liver in the biotransformation and elimination of xenobiotics is well recognized. The liver is anatomically unique in that it receives blood from two sources: from the spleen, gastrointestinal tract and its associated organs, (portal vein), and from the general circulation (hepatic artery). The liver tissue is not vascularized with a capillary network as with most other organs but consists of blood-filled sinusoids surrounding the hepatic cells. Also surrounding the hepatocytes are the bile canaliculi, which drain into bile ducts. Substances entering the hepatocytes across the basolateral membrane can undergo biotransformation and either be excreted back into sinusoidal blood, or can be excreted apically into bile and end up in the duodenum following storage and subsequent expulsion from the gall bladder.

A major obstacle in accurately quantifying biliary excretion has been the interindividual variability in biliary secretion as a result of incomplete gall bladder emptying during the course of bile collection. One way to circumvent this is by pharmacologically inducing gall bladder contraction by infusing cholecystokinin-8 (CCK) or by stimulating physiological CCK secretion through olfactory and/or gastric stimuli. Ghibellini et al. (2004) circumvented these problems by developing a novel method to quantify biliary excretion of drugs in healthy human volunteers. Using ^{99m}Tc-mebrofenin (MEB), an imaging agent administered intravenously that is excreted primarily unchanged in bile, they were able to visualize gall bladder contractions and ejection of contents that were aspirated via the CHOL-ect tube. This allowed for accurate calculation of in vivo biliary clearance of ^{99m}Tc-MEB corrected for gall bladder ejection fraction (EF), which accounted for the amount of ^{99m}Tc-MEB remaining in the gall bladder due to incomplete gall bladder contraction (X_{GB}). The gall bladder ejection fraction was defined as follows:

$$E_{\mathrm{G}} = \frac{\mathrm{GB}^{t^{\mathrm{CCK}}} - \mathrm{GB}^{t^{\mathrm{last}}}}{\mathrm{GB}^{t^{\mathrm{CCK}}}} \tag{14.5}$$

where GB represents the counts over the gall bladder region, t^{CCK} is the time of CCK administration, and t^{last} is the last imaging time point.

In vivo biliary clearance, corrected for ejection fraction, was estimated using the following equation:

$$\text{in vivo}\,\mathrm{CL}_{\mathrm{biliary}} = \frac{X_{\mathrm{GB}_{0-t\,\mathrm{last}}}}{\mathrm{AUC}_{0-t\,\mathrm{last}}} \tag{14.6}$$

where X_{GB} is the amount excreted into the gall bladder, corrected for gall bladder ejection fraction, from time 0 to the last time point (Ghibellini et al. 2006b).

The gamma-emitting probe substrate ^{99m}Tc-MEB was employed initially to establish this method and quantify gall bladder ejection fraction. Collection of bile in the bile ducts and gall bladder, followed by expulsion of bile from the gall bladder, could be visualized by the gamma camera and quantified by measuring the ^{99m}Tc-MEB dose recovered in bile aspirated from the intestine. Bile collection was facilitated by using the CHOL-ect tube with a distally occluding balloon. Following proper placement of the catheter in the duodenum below the ampulla of Vater, baseline blood, urine, and bile samples were collected, and the occlusive balloon was inflated. Subjects were administered an intravenous bolus dose of ^{99m}Tc-MEB and blood samples were collected from the contralateral arm over 180 min. Bile samples also were obtained over 180 min and urine was collected at the end of 180 min. Anterior gamma scintigraphic images were acquired at 1-min intervals throughout the study. Imaging results clearly showed that bile was ejected from the gall bladder in pulses rather than in a sustained secretary response. Each of the four subjects studied showed a different pattern of bile secretion into the intestine, probably due to differences in spontaneous contraction patterns of the gall bladder. However, the gamma camera images proved conclusively that the catheter collected biliary and duodenal secretions and that the secretions were aspirated from the duodenal region via the CHOL-ect tube. Data clearly demonstrated that correction for gall bladder ejection fraction contributed substantially to the accuracy of the method to quantify biliary clearance of molecules (Ghibellini et al., 2004).

This technique was applied to evaluate the biliary excretion of piperacillin in healthy human volunteers. Piperacillin is an intravenously administered antibiotic used to treat biliary tract infections. Piperacillin concentrates in bile and achieves minimum inhibitory concentrations for target pathogens in tissues within 2–3 h following a 5 g intravenous bolus injection (Ghibellini et al., 2006b). The CHOL-ect tube was positioned fluoroscopically prior to starting the piperacillin infusion. About 120 min after piperacillin administration, ^{99m}Tc-MEB was administered as an intravenous bolus, and subjects were positioned under the gamma camera for gamma scintigraphic imaging of the gall bladder over 240 min. Gall bladder contraction was induced at 240 min by CCK administration. When CCK failed to induce gall bladder contraction, other gall bladder contraction stimulants were tried. Blood samples were collected for 600 min. The ejection fraction was estimated using Equation (14.5), and the total amount of piperacillin excreted in bile before and after inducing gall bladder emptying at 240 min, corrected for the EF, was obtained using the following equation:

$$\text{Piperacillin}\, X_{\text{GB}} = X_{\text{bile}}^{0-240\ \text{min}} + (X_{\text{bile}}^{240\ \text{min}-360\ \text{min}}/\text{EF}) \tag{14.7}$$

where X_{bile} is the amount collected from the CHOL-ect tube and X_{GB} is the estimated amount excreted into the gall bladder. Piperacillin is not a drug that is extensively secreted into bile, yet a sensitive estimate of in vivo biliary clearance was determined by factoring in interindividual variability due to differences in gall bladder emptying.

The CHOL-ect tube has been employed to validate in vitro–in vivo correlations in hepatobiliary drug clearance in humans. In preclinical species, piperacillin metabolism and biliary excretion varied considerably. This is often the case with many drugs due to species differences in drug transport and metabolizing enzymes. A better prediction of biliary clearance can be made by extrapolating data generated in human hepatocytes. However, for estimating biliary clearance, bile canaliculi must be present in the cultured hepatocytes. Hence, suspended human hepatocytes will not give an accurate estimate of biliary clearance. Hepatocytes cultured in a sandwich configuration develop extensive bile network formation, and the in vivo biliary clearance can be estimated using B-CLEAR® technology and compared with actual in vivo biliary clearance values measured using the CHOL-ect catheter. This approach published by Ghibellini et al. (2007b) was the first report of the comparison of predicted values for biliary clearance obtained in sandwich-cultured human hepatocytes to in vivo determinations in humans. When biliary clearance predictions were made by scaling in vitro data to kilogram of body weight, a good correlation was achieved.

As with the Loc-I-Gut®, the CHOL-ect catheter has been used to characterize the absorption profile of drugs following administration in various anatomical regions of the human gastrointestinal tract. For drugs with low bioavailability administered orally, understanding where in the gastrointestinal tract absorption/metabolism occurs can aid in the development of approaches to improve bioavailability. In the case of sumatriptan, a drug that undergoes rapid absorption after oral administration but has a low bioavailability (14%), oral versus colonic administration using the CHOL-ect tube established that sumatriptan was absorbed to a greater extent in the upper regions of the gastrointestinal tract rather than in the colon and that most of the metabolism occurred in this region (Warner et al., 1995).

The CHOL-ect multilumen catheter has been used to assess in vivo intestinal dissolution and bioavailability of a drug formulation that acts within the duodenum of humans. Patients with pancreatic insufficiency take exogenous porcine pancreatic enzymes which contain lipase, pancrease, and amylase necessary for digestion. These enzymes are prone to inactivation in the acidic pH of the stomach. Thus, pancreatic enzymes are formulated as enteric-coated and buffered microspheres because deficient bicarbonate secretion in patients with a compromised pancreas (such as patients with cystic fibrosis and chronic pancreatitis) results in insufficient neutralization of bile acids which can delay dissolution and increase the risk of enzyme inactivation; both situations lead to decreased intestinal bioavailability of pancreatic enzymes. In recent guidelines to the industry, the FDA has mandated that manufacturers demonstrate in vivo intestinal bioavailability of these products using intestinal aspiration techniques in humans. The CHOL-ect catheter has been used to evaluate the in vivo dissolution and release of these enzymes from a commercially available pancreatic enzyme supplement in subjects with pancreatic insufficiency. Samples from the duodenum were collected 45 min after administering the pancreatic enzyme supplement with a high-fat liquid meal to simulate a fed state. Differences in enzyme activity compared to placebo, measured in samples collected over 3 h, were used to assess intestinal bioavailability of these products (ClinicalTrials.gov., NCT00744250).

Another novel application of the CHOL-ect multilumen catheter may be to address the dilemma the pharmaceutical industry faces in explaining mechanisms of drug-induced liver injury and hepatic transporter-mediated drug–drug interactions. Multiple and poly-specific efflux and influx transporters in the liver, the absence of specific inhibitors/substrates for individual transporters, and species differences in hepatic transport processes have limited our ability to accurately predict interactions and altered hepatic disposition of drugs in humans. To increase our understanding of these events, specific probes need to be developed and validated in vivo so that a better extrapolation can be made from in vitro studies. For example, imaging agents may be used to assess hepatic exposure and the CHOL-ect tube can be used to quantify biliary excretion in humans. The influence of hepatic transporter modulation by drug–drug/food interactions, disease states or genetic variations in hepatic drug exposure, and systemic disposition and biliary excretion of drugs in humans could be assessed using the CHOL-ect tube and suitable probes.

14.4 Microdosing

Recently, the FDA and the European Medicines Agency (EMEA) have developed guidances for phase 0 microdosing or imaging studies designed to evaluate the pharmacokinetics, metabolism, or distribution of test compounds in humans. The impact of these guidelines has been reviewed recently by Marchetti and Schellens (Marchetti and Schellens, 2007). The aim of human microdosing studies is not to gain information about safety or efficacy, but rather to determine key pharmacokinetic parameters. This information can then be incorporated into in silico pharmacokinetic models to aid first-in-human studies, if possible, or used to make an early decision to stop development of the drug candidate in order to save resources. Thus, human microdosing primarily is directed toward reducing the cost of failure, once a new drug candidate is selected, in a highly competitive industry. Undesirable pharmacokinetic properties account for almost 40% of the drugs that are withdrawn after phase I. Existing guidelines, therefore, allow for significant flexibility in terms of the amount of data that need to be submitted with an IND application for microdosing studies.

Candidates for human microdosing studies are selected on the basis of demonstrated pharmacological activity in animal models and after some limited animal safety testing. This is followed by oral and/or intravenous administration of a microdose to human volunteers and analysis of relevant body fluids using sensitive bioanalytical methods such as accelerator mass spectrometry, ultrasensitive LC-MS/MS, tandem liquid chromatography/mass spectroscopy, or positron-emission tomography. A microdose is defined as less than 1/100th of the dose of a test substance calculated (based on animal data) to yield a pharmacologic effect of the test substance with a maximum dose of $\leq$100 μg (for imaging agents, the latter criterion applies). Due to differences in molecular weights as compared to synthetic drugs, the maximum dose for protein products is $\leq$30 nmol.

The scope of these exploratory studies involving very limited human exposure can extend to assessing whether a particular mechanism of action defined during drug development also can be observed in humans, to gaining pharmacokinetic information, and to selecting the most promising lead compound(s) from a group of candidates based on pharmacokinetic or pharmacodynamic properties. An obvious limitation with microdosing is ascertaining whether pharmacokinetic data obtained at these low doses are predictive of data that would be obtained at the clinically relevant dose.

Data demonstrating the utility of human microdosing studies are just beginning to emerge. Microdosing studies (excluding those studies aimed at imaging) published in the literature were reviewed in 2008 to assess how well microdose data have predicted the pharmacokinetics of drugs compared to administration of a therapeutic dose. Of the 18 drugs evaluated, 15 demonstrated linear pharmacokinetics within a factor of 2 between the microdose and the therapeutic dose (Lappin and Garner, 2008).

A trial was performed in healthy volunteers to compare the pharmacokinetics of five drugs [warfarin, ZK253 (Schering), diazepam, midazolam, and erythromycin] in order to determine the feasibility of using microdosing to predict the pharmacokinetics of these drugs at a therapeutic dose. The compounds were chosen to represent a situation in which prediction of pharmacokinetics from either animal or in vitro studies (or both) was likely to be complicated. In a crossover design, volunteers received: (1) one of the five compounds as a microdose labeled with ^{14}C (100 μg), (2) the corresponding ^{14}C-labeled therapeutic dose on a separate occasion, and (3) simultaneous administration of an intravenous ^{14}C-labeled microdose and an oral non-labeled therapeutic dose for ZK253, midazolam, and erythromycin. Analysis of ^{14}C-labeled drugs in plasma was performed by HPLC followed by accelerator mass spectrometry. Liquid chromatography–tandem mass spectrometry was used to measure plasma concentrations of ZK253, midazolam, and erythromycin at therapeutic concentrations, whereas HPLC–accelerator mass spectrometry was used to measure warfarin and diazepam concentrations. Results showed concordance between the pharmacokinetics of diazepam, midazolam, and ZK253 following the microdose and the therapeutic dose. However, a discrepancy was observed in warfarin's distribution that may have been attributed to high-affinity, low-capacity tissue binding, although the clearance was reasonably well predicted. The oral microdose of erythromycin failed to provide detectable plasma concentrations as a result of possible acid lability in the stomach. The absolute bioavailability of the three compounds examined yielded excellent concordance with data from the literature or generated data. Overall, it appears that microdosing, when used appropriately, offers the potential to aid in early drug candidate selection (Lappin et al., 2006).

14.5 Future Challenges

14.5.1 Need for Innovation

There is no doubt that considerable innovation and research is needed to gain a better understanding of drug absorption processes in humans. It has been estimated that currently a new molecular entity (NME) entering clinical phase I testing has only an 8% chance of reaching the market, and the probability is even lower for an anticancer drug (FDA, Critical Path Opportunities Report, 2006). The current cost of bringing a new medicine to market, estimated to be as high as US $0.8–1.7 billion, is a major barrier to investment in innovative, higher risk drugs or in therapies for uncommon diseases or diseases that predominantly afflict the poor (FDA, Innovation or Stagnation? Challenge and Opportunity on the Critical Path to New Medical Products, 2004). Identifying at an early stage lead molecules that are likely to fail at later stages of clinical development will enhance the efficiency of drug development. Currently, the ability of in silico models to predict the biopharmaceutical properties of a compound, and of animal models to predict ADME characteristics is limited. Due to the complexities of human drug absorption, current in vitro and preclinical in vivo models may not accurately predict bioavailability or drug–drug/food interactions in drug absorption. It is therefore essential that human absorption data are obtained using innovative technologies in order to enhance the attrition rate in early clinical testing (Wilding et al., 2000).

14.5.2 Identification of Probe Substrates

Clarifying the role of transport proteins and metabolic enzymes during the drug absorption process, determining mechanisms of modulation of these proteins, and understanding the therapeutic or toxicological implications of such alterations are key in drug development. Figure 14.6 shows an example of a decision tree for drug–drug interaction studies to evaluate the interaction profile of a candidate molecule with respect to in vitro metabolism to determine when clinical in vivo data are needed (FDA, MAPP: Clinical Pharmacology and Biopharmaceutics NDA Review

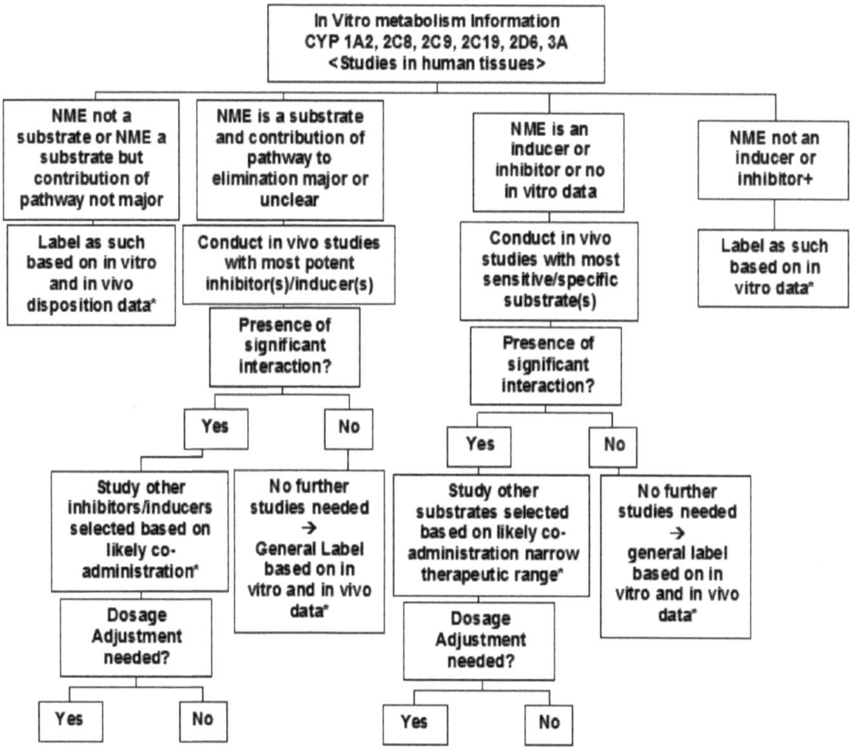

Fig. 14.6 A possible model for decision making: CYP-based drug–drug interaction studies (from FDA's CDER homepage). NME: New molecular entity. *Additional population pharmacokinetic analysis may assist the overall evaluation, + negative results from an in vivo cocktail study would preclude further evaluation to determine whether an NME is an inhibitor or an inducer of a particular CYP enzyme

Template, 2004). When a compound is determined to be a substrate for a particular enzyme or a transporter, and the contribution of that pathway to drug elimination is significant or unclear, in vivo studies in humans need to be carried out using the most potent inducer(s) and/or inhibitor(s), depending on the potential clinical interaction. With seemingly endless possibilities for clinically significant transporter- and enzyme-mediated drug interactions, the need to identify specific probe substrates and inhibitors is imperative prior to undertaking these expensive studies. The techniques described in this chapter may aid in delineating mechanisms of interactions in drug absorption and addressing bioavailability problems.

14.5.3 Better Detection and Quantification Methods

Non-invasive means of monitoring drug concentrations, for example, using molecular tags that can be located through imaging techniques, could markedly improve product development by enabling sponsors to correlate response with drug availability at the target site. Furthermore, the potential to assess the role of metabolizing enzymes, transporters, and their interplay in the clearance of a drug by a particular organ would be increased using these methods.

14.5.4 Understanding the Physiology of Drug Disposition Within Organs

Considering recent guidelines stressing the need to define mechanistically enzyme- and transporter-based drug interactions, now more than ever there is a need to accurately characterize a drug molecule in terms of its intestinal permeability and first-pass metabolism; to delineate its interaction with metabolizing enzymes; to understand whether its disposition is transporter dependent; and to predict the potential drug interaction profile. In vitro, in silico and animal in vivo and ex vivo models provide a significant amount of information during development of a lead molecule. Nevertheless, there are numerous instances when such molecules fail because of poor bioavailability, toxicity, drug interactions, or other factors when administered to humans. To bridge the gap between in vitro data and in vivo absorption and intestinal first-pass metabolism, accurate prediction of the intestinal availability will allow for a more precise assessment of the clearance of a molecule and its drug–drug interaction potential.

14.6 Conclusions

In conclusion, as our understanding of drug absorption, metabolism, and disposition processes increases, innovative new technologies are needed to evaluate potential new drugs, to optimize existing development strategies, and to ensure the safety and efficacy of medicines.

Acknowledgments Dr. Brouwer's research program is supported by grant R01 GM41935 from the National Institute of General Medical Sciences, National Institutes of Health. Ahsan N. Rizwan, M. Pharm., Ph.D. is a GlaxoSmithKline sponsored Clinical Pharmacokinetics and Pharmacodynamics Fellow at the UNC Eshelman School of Pharmacy, Division of Pharmacotherapy and Experimental Therapeutics.

References

Bergman E, Forsell P, Persson EM, Knutson L, Dickinson P, Smith R, Swaisland H, Farmer MR, Cantarini MV and Lennernäs H (2007) Pharmacokinetics of gefitinib in humans: the influence of gastrointestinal factors. *Int J Pharm* **341**:134–142.

Bergman E, Forsell P, Tevell A, Persson EM, Hedeland M, Bondesson U, Knutson L and Lennernäs H (2006) Biliary secretion of rosuvastatin and bile acids in humans during the absorption phase. *Eur J Pharm Sci* **29**:205–214.

Bode H, Brendel E, Ahr G, Fuhr U, Harder S and Staib AH (1996) Investigation of nifedipine absorption in different regions of the human gastrointestinal (GI) tract after simultaneous administration of ^{13}C- and ^{12}C-nifedipine. *Eur J Clin Pharmacol* **50**:195–201.

Clear NJ, Milton A, Humphrey M, Henry BT, Wulff M, Nichols DJ, Anziano RJ and Wilding I (2001) Evaluation of the Intelisite capsule to deliver theophylline and frusemide tablets to the small intestine and colon. *Eur J Pharm Sci* **13**:375–384.

ClinicalTrials.gov. NCT00744250-Intraduodenal Aspiration Study to Assess the Bioavailability of Oral Pancrecarb® Compared to Placebo Control (http://clinicaltrials.gov/ct2/show/NCT00744250?term=pancrecarb&rank=3). Accessed on April 27, 2009.

Connor A, Evans P, Doto J, Ellis C and Martin DE (2009) An oral human drug absorption study to assess the impact of site of delivery on the bioavailability of bevirimat. *J Clin Pharmacol* **49**:606–612.

Davis SS and Wilding IR (2001) Oral drug absorption studies: the best model for man is man!. *Drug Discov Today* **6**:127–130.

Enterion™. Fast, efficient development of oral drugs requires knowledge of human gastrointestinal absorption. http://www.enterion.co.uk/ (Accessed on April 27, 2009).

Fagerholm U, Borgstrom L, Ahrenstedt O and Lennernäs H (1995) The lack of effect of induced net fluid absorption on the in vivo permeability of terbutaline in the human jejunum. *J Drug Target* **3**:191–200.

FDA CDER Data standards manual phase 4 commitment categories manual of policies and procedures (1996). MAPP 6010.2 Attachment B, p.16. (http://www.fda.gov/Cder/mapp/6010-2.pdf) Accessed on April 27, 2009.

FDA Critical Path Opportunities Report (2006) (http://www.fda.gov/oc/initiatives/criticalpath/reports/opp_report.pdf) Accessed on April 27, 2009.

FDA Guidance to Industry (2000) Center for Drug Evaluation and Research (CDER). Waiver of in vivo bioavailability and bioequivalence studies for immediate-release solid oral dosage forms based on a biopharmaceutics classification system. (http://www.fda.gov/downloads/Drugs/GuidanceComplianceRegulatoryInformation/Guidances/UCM070246.pdf). Accessed on 31 August 2009.

FDA Innovation or Stagnation? Challenge and Opportunity on the Critical Path to New Medical Products (2004). (http://www.fda.gov/oc/initiatives/criticalpath/whitepaper.html) Accessed on April 27, 2009.

FDA MAPP: Clinical Pharmacology and Biopharmaceutics NDA Review Template (Issued 4/27/2004, Posted 6/24/2004): http://www.fda.gov/cder/ mapp/4000.4.pdf (Accessed on April 27, 2009).

Fischer JD, Song MH, Suttle AB, Heizer WD, Burns CB, Vargo DL and Brouwer KLR (2000) Comparison of zafirlukast (Accolate) absorption after oral and colonic administration in humans. *Pharm Res* **17**:154–159.

Fleming E and Ma P (2002) Drug life-cycle technologies. *Nat Rev Drug Discov* **1**:751–752.

Fuhr U, Staib AH, Harder S, Becker K, Liermann D, Schollnhammer G and Roed IS (1994) Absorption of ipsapirone along the human gastrointestinal tract. *Br J Clin Pharmacol* **38**:83–86.

Ghibellini G, Bridges AS, Generaux CN and Brouwer KLR (2007a) In vitro and in vivo determination of piperacillin metabolism in humans. *Drug Metab Dispos* **35**:345–349.

Ghibellini G, Johnson BM, Kowalsky RJ, Heizer WD and Brouwer KLR (2004) A novel method for the determination of biliary clearance in humans. *AAPS J* **6**(4):Article 33.

Ghibellini G, Leslie EM and Brouwer KLR (2006a) Methods to evaluate biliary excretion of drugs in humans: an updated review. *Mol Pharm* **3**:198–211.

Ghibellini G, Leslie EM, Pollack GM and Brouwer KLR (2008) Use of tc-99m mebrofenin as a clinical probe to assess altered hepatobiliary transport: integration of in vitro, pharmacokinetic modeling, and simulation studies. *Pharm Res* **25**:1851–1860.

Ghibellini G, Vasist LS, Hill TE, Heizer WD, Kowalsky RJ and Brouwer KLR (2006b) Determination of the biliary excretion of piperacillin in humans using a novel method. *Br J Clin Pharmacol* **62**:304–308.

Ghibellini G, Vasist LS, Leslie EM, Heizer WD, Kowalsky RJ, Calvo BF and Brouwer KLR (2007b) In vitro–in vivo correlation of hepatobiliary drug clearance in humans. *Clin Pharmacol Ther* **81**:406–413.

Glaeser H, Drescher S, van der Kuip H, Behrens C, Geick A, Burk O, Dent J, Somogyi A, Von Richter O, Griese EU, Eichelbaum M and Fromm MF (2002) Shed human enterocytes as a tool for the study of expression and function of intestinal drug-metabolizing enzymes and transporters. *Clin Pharmacol Ther* **71**:131–140.

Hedeland M, Fredriksson E, Lennernäs H and Bondesson U (2004) Simultaneous quantification of the enantiomers of verapamil and its N-demethylated metabolite in human plasma using liquid chromatography–tandem mass spectrometry. *J Chromatogr B Analyt Technol Biomed Life Sci* **804**:303–311.

InteliSite® A Proven Non-Invasive Method for Assessing the Oral Absorption of Drugs Within a Specific Region of the Gastrointestinal Tract (http://www./innovativedevices.com/intelli.html) Accessed on 31 August 2009.

Lappin G and Garner RC (2008) The utility of microdosing over the past 5 years. *Expert Opin Drug Metab Toxicol* **4**:1499–1506.

Lappin G, Kuhnz W, Jochemsen R, Kneer J, Chaudhary A, Oosterhuis B, Drijfhout WJ, Rowland M and Garner RC (2006) Use of microdosing to predict pharmacokinetics at the therapeutic dose: experience with 5 drugs. *Clin Pharmacol Ther* **80**:203–215.

Lee L, Hossain M, Wang Y and Sedek G (2004) Absorption of rivastigmine from different regions of the gastrointestinal tract in humans. *J Clin Pharmacol* **44**:599–604.

Lennernäs H (1998) Human intestinal permeability. *J Pharm Sci* **87**:403–410.

Lennernäs H (2007) Modeling gastrointestinal drug absorption requires more in vivo biopharmaceutical data: experience from in vivo dissolution and permeability studies in humans. *Curr Drug Metab* **8**:645–657.

Lennernäs H, Knutson L, Knutson T, Hussain A, Lesko L, Salmonson T and Amidon GL (2002) The effect of amiloride on the in vivo effective permeability of amoxicillin in human jejunum: experience from a regional perfusion technique. *Eur J Pharm Sci* **15**:271–277.

Lundahl A, Hedeland M, Bondesson U, Knutson L and Lennernäs H (2009) The effect of St. John's wort on the pharmacokinetics, metabolism and biliary excretion of finasteride and its metabolites in healthy men. *Eur J Pharm Sci* **36**:433–443.

Marchetti S and Schellens JH (2007) The impact of FDA and EMEA guidelines on drug development in relation to Phase 0 trials. *Br J Cancer* **97**:577–581.

Martin NE, Collison KR, Martin LL, Tardif S, Wilding I, Wray H and Barrett JS (2003) Pharmacoscintigraphic assessment of the regional drug absorption of the dual angiotensin-converting enzyme/neutral endopeptidase inhibitor, M100240, in healthy volunteers. *J Clin Pharmacol* **43**:529–538.

Martinez MN and Amidon GL (2002) A mechanistic approach to understanding the factors affecting drug absorption: a review of fundamentals. *J Clin Pharmacol* **42**: 620–643.

Nilsson D, Fagerholm U and Lennernas H (1994) The influence of net water absorption on the permeability of antipyrine and levodopa in the human jejunum. *Pharm Res* **11**: 1540–1547.

Oo C, Snell P, Barrett J, Dorr A, Liu B and Wilding I (2003) Pharmacokinetics and delivery of the anti-influenza prodrug oseltamivir to the small intestine and colon using site-specific delivery capsules. *Int J Pharm* **257**:297–299.

Parr AF, Sandefer EP, Wissel P, McCartney M, McClain C, Ryo UY and Digenis GA (1999) Evaluation of the feasibility and use of a prototype remote drug delivery capsule (RDDC) for non-invasive regional drug absorption studies in the GI tract of man and beagle dog. *Pharm Res* **16**:266–271.

Petri N, Tannergren C, Holst B, Mellon FA, Bao Y, Plumb GW, Bacon J, O'Leary KA, Kroon PA, Knutson L, Forsell P, Eriksson T, Lennernäs H and Williamson G (2003) Absorption/metabolism of sulforaphane and quercetin, and regulation of phase II enzymes, in human jejunum in vivo. *Drug Metab Dispos* **31**:805–813.

Pithavala YK, Heizer WD, Parr AF, O'Connor-Semmes RL and Brouwer KLR (1998) Use of the InteliSite capsule to study ranitidine absorption from various sites within the human intestinal tract. *Pharm Res* **15**:1869–1875.

Reynolds KS, Song MH, Heizer WD, Burns CB, Sica DA and Brouwer KLR (1998) Effect of pancreatico-biliary secretions and GI transit time on the absorption and pharmacokinetic profile of ranitidine in humans. *Pharm Res* **15**:1281–1285.

Sandstrom R, Karlsson A, Knutson L and Lennernäs H (1998) Jejunal absorption and metabolism of R/S-verapamil in humans. *Pharm Res* **15**:856–862.

Sandstrom R, Knutson TW, Knutson L, Jansson B and Lennernäs H (1999) The effect of ketoconazole on the jejunal permeability and CYP3A metabolism of (R/S)-verapamil in humans. *Br J Clin Pharmacol* **48**:180–189.

Sladen GE (1968) Perfusion studies in relation to intestinal absorption. *Gut* **9**:624–628.

Sladen GE and Dawson AM (1968) An evaluation of perfusion techniques in the study of water and electrolyte absorption in man: the problem of endogenous secretions. *Gut* **9**: 530–535.

Sladen GE and Dawson AM (1970) Further studies on the perfusion method for measuring intestinal absorption in man: the effects of a proximal occlusive balloon and a mixing segment. *Gut* **11**:947–954.

Staib AH, Beermann D, Harder S, Fuhr U and Liermann D (1989) Absorption differences of ciprofloxacin along the human gastrointestinal tract determined using a remote-control drug delivery device (HF-capsule). *Am J Med* **87**:66S–69S.

Takagi T, Ramachandran C, Bermejo M, Yamashita S, Yu LX and Amidon GL (2006) A provisional biopharmaceutical classification of the top 200 oral drug products in the United States, Great Britain, Spain, and Japan. *Mol Pharm* **3**:631–643.

Takamatsu N, Kim ON, Welage LS, Idkaidek NM, Hayashi Y, Barnett J, Yamamoto R, Lipka E, Lennernas H, Hussain A, Lesko L and Amidon GL (2001) Human jejunal permeability of two polar drugs: cimetidine and ranitidine. *Pharm Res* **18**:742–744.

Tannergren C, Knutson T, Knutson L and Lennernäs H (2003) The effect of ketoconazole on the in vivo intestinal permeability of fexofenadine using a regional perfusion technique. *Br J Clin Pharmacol* **55**:182–190.

Thelen K and Dressman JB (2009) Cytochrome P450-mediated metabolism in the human gut wall. *J Pharm Pharmacol* **61**:541–558.

von Richter O, Greiner B, Fromm MF, Fraser R, Omari T, Barclay ML, Dent J, Somogyi AA and Eichelbaum M (2001) Determination of in vivo absorption, metabolism, and transport of drugs by the human intestinal wall and liver with a novel perfusion technique. *Clin Pharmacol Ther* **70**:217–227.

Warner PE, Brouwer KLR, Hussey EK, Dukes GE, Heizer WD, Donn KH, Davis IM and Powell JR (1995) Sumatriptan absorption from different regions of the human gastrointestinal tract. *Pharm Res* **12**:138–143.

Wilding II, Hirst P and Connor A (2000) Development of a new engineering-based capsule for human drug absorption studies. *Pharm Sci Technol Today* **3**:385–392.

Wouters S (1998) *Electronically Controlled Drug Delivery*. Published by Informa Health Care. Editors Bret Berner, Steven M. Dinh. p 119.

Chapter 15
Management of Drug Interactions of New Drugs in Multicenter Trials Using the Metabolism and Transport Drug Interaction Database©

Houda Hachad, Isabelle Ragueneau-Majlessi, and René H. Levy

Abstract The Metabolism and Transport Drug Interaction Database© (DIDB) is a web-based research tool (www.druginteractioninfo.org) for scientists and clinicians working in the field of drug interactions (DIs). The DIDB enables users to access, manage, and analyze the scientific basis of clinical drug interactions for drugs, biologics, food, and herbal derivatives. Users can search the database using a number of queries that allow them to approach the data from different perspectives. This chapter illustrates how the DIDB can be used by clinical investigators to manage drug interactions of new molecular entities in the context of a disease and its comorbidities or during multicenter trials.

15.1 Introduction

Once the pharmacokinetic and metabolic characteristics of a new molecular entity (NME) have been established, scientists and clinicians encounter the need to assess its drug interaction (DI) potential. That potential is assessed using in vitro and in vivo studies. When the NME advances into clinical trials, the drug interaction "company" scientists must provide clinical investigators with recommendations regarding coadministration of the NME with a large array of marketed drugs, based on the therapeutic regimens of patients enrolled in the clinical trials. There is then a need to translate the drug interaction potential of the NME into patient-specific guidance. This requires rapid access to vast literature sets on metabolic isozymes, transporters, substrates, inducers, and inhibitors that allow judgements on dosage adjustments. In the first part of this chapter, it is shown how the Metabolism and Transport Drug Interaction Database (www.druginteractioninfo.org) (DIDB) can be used to provide

R.H. Levy (✉)
Department of Pharmaceutics, University of Washington, Seattle, WA, USA
e-mail: rhlevy@u.washington.edu

K.S. Pang et al. (eds.), *Enzyme- and Transporter-Based Drug–Drug Interactions*, DOI 10.1007/978-1-4419-0840-7_15,

scientifically based recommendations to clinicians in real time, during clinical trials. The DIDB, developed by the Drug Interaction Prediction Group at the University of Washington, includes one of the largest sets of comprehensive data pertaining to drug–drug, drug–food, and drug–herb interactions in humans. The DIDB was initially designed to serve as a tool for scientists in academia, regulatory agencies, and the pharmaceutical industry who need to build or evaluate a drug interaction program. As shown previously (Hachad et al., 2008), the DIDB has been used extensively by researchers and clinicians interested in correlating in vitro and in vivo findings on interactions associated with metabolic enzymes (phases I and II) and transporters.

The second part of this chapter addresses a new approach to DI evaluation based on a disease orientation. It introduces the use of a new section of the DIDB (*Disease-Oriented Database*) that allows novel queries regarding the DI potential of an NME in the context of a disease and its comorbidities.

15.2 Database Design and Content

Structure: The DIDB application has a typical multitier architecture in a Microsoft® .NET environment (the web part of the database, which is accessed by the user over the internet, is hosted on a Microsoft Windows 2003 server running IIS and version 2.0 of the ASP.NET framework. All data are stored on a Microsoft SQL Server 2005 database). The use of the Web facilitates worldwide access as well as upgrades and updates; the DIDB is updated daily.

Content: Currently the DIDB has data extracted from more than 7400 published articles referenced in *PubMed* (Pub Med) (1966–present), 60 New Drug Applications (NDA), and 360 product labels (1998–present). The information is being devoted to drug metabolism, transport and interactions.

The unit of information (*citation*) is either a published research article or the "NDA Clinical Reviews" section available from drugs@FDA website (drugs@FDA).

Detailed records are generated from each research article or NDA, highlighting study results as well as experimental conditions; the records are structured in the database according to a defined hierarchy. For example, relevant information collected from in vitro studies pertains to the role of particular metabolic enzymes in the various metabolic pathways of substrates and the inhibition and induction spectra of drugs toward metabolic enzymes. Particular attention is paid to experimental conditions used in the determination of enzyme kinetic parameters, including K_m, K_i, IC_{50}, $K_I - K_{inact}$, EC_{50}. In vivo studies include pharmacokinetic studies with blood level measurements, pharmacokinetic–pharmacodynamic studies, as well as case reports. In addition to research articles, the DIDB team has built original excerpts from Product Label of recently approved drugs (1998–2009) in the United States, available from the drugs@FDA website (drugs@FDA).

Recently, a new section analyzing DIs in the context of specific diseases and their comorbidities (*Disease-Oriented Database*) has been added to the DIDB. This section allows users to retrieve overall summaries on DIs related not only to drugs used to treat the disease but to drugs used to treat the main comorbidities of that disease.

Queries: The DIDB search interface utilizes a list of prestructured searches called by "*Queries*." These are set along intuitive themes such as drug, enzyme, therapeutic class, transporter and thus allow the user to quickly select the appropriate search without the need for extensive training (see *appendix* for more details).

The following section describes two examples of use of the DIDB highlighting the three-step logic used to perform a search:

- Defining the issue (background and question)
- Selecting the search (queries)
- Analyzing and interpreting the result output

15.3 Examples of Use

15.3.1 Example: Finding CYP3A Inhibitors in the Context of Clinical Trials

15.3.1.1 Background

In its last guidance document (FDA Guidance for Industry), the FDA proposed that CYP3A inhibitors be classified based on the magnitude of changes in plasma area under the curve (AUC) of oral midazolam or other sensitive CYP3A substrate. For instance, if the ratio $AUC_{inhibited}/AUC_{control}$ (AUC_R) of oral midazolam (or other CYP3A sensitive substrate) is 5 or higher, the inhibitor is considered a strong CYP3A inhibitor. If the ratio is between 2 and 5, the inhibitor is classified as moderate, and finally, if the ratio is between 1.25 and 2, it will be considered a weak inhibitor. A similar classification has been proposed for the other CYP enzymes. By using a clear and consistent categorization of drugs as substrates and inhibitors, the FDA hopes to facilitate analyses across DDI studies and to help health-care providers to safely administer these drugs through a consistent labeling language.

In the following example, an NME has been shown in early development to be mainly metabolized by CYP3A (fm_{CYP3A} of 0.94), and a drug interaction study performed with the strong CYP3A inhibitor ketoconazole yielded an AUC_R of this NME of 3.6. Considering that the therapeutic range of the NME is not wide, clinicians in charge of clinical trials need to identify all known strong inhibitors of CYP3A because they might be contraindicated. Similarly, these clinicians need to identify all moderate CYP3A inhibitors to recommend clinical monitoring for potential toxicity when they are coadministered with the NME.

15.3.1.2 Question

Before starting the program of phase II trials, the sponsor wishes to identify all drugs that behave as strong or moderate inhibitors of CYP3A, in order to provide clinical investigators with recommendations.

15.3.1.3 Search Strategy

It is possible to obtain all strong inhibitors from a single query if one includes all sensitive substrates at once: in the DIDB section "*AUC and CL Changes Queries*," the query "*Percent AUC or CL with Multiple Objects*" retrieves all changes in AUC that were measured in DI studies with multiple substrates. In the situation described above, the CYP3A probe substrate midazolam as well as all drugs listed by the FDA as sensitive CYP3A substrates (i.e., which exhibit an AUC_R of 5 or more when given concomitantly with a CYP3A inhibitor) will be used. These sensitive substrates are the following: budesonide, buspirone, eplerenone, eletriptan, felodipine, fluticasone, lovastatin, midazolam, saquinavir, sildenafil, simvastatin, triazolam, vardenafil (FDA Guidance for Industry). To find strong CYP3A inhibitors, a percent change in substrate's AUC of 400% or more (i.e., $AUC_R \geq 5$) will be selected (Fig. 15.1).

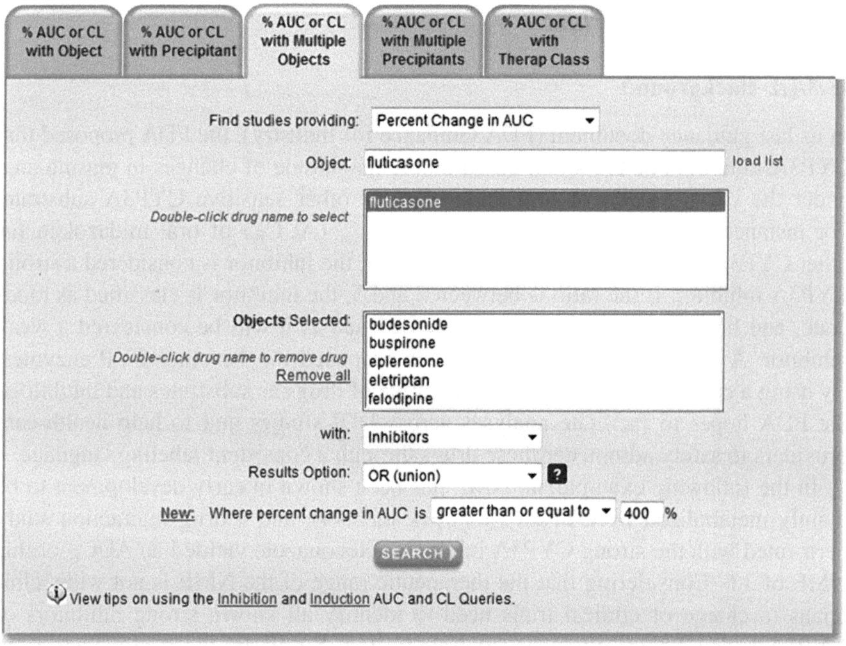

Fig. 15.1 Query labeled "Percent *AUC or CL with* Multiple *Objects"* used to find in vivo inhibitors of CYP3A using sensitive substrates. Display from the *Metabolism and Transport Drug Interaction Database*© (http://www.druginteractioninfo.org, accessed Apr 2009)

15.3.1.4 Result Outputs

The display shown in Fig. 15.2 has, for each sensitive substrate, an alphabetical list of inhibitors that have been shown to increase the AUC of the substrate by at least 400%.

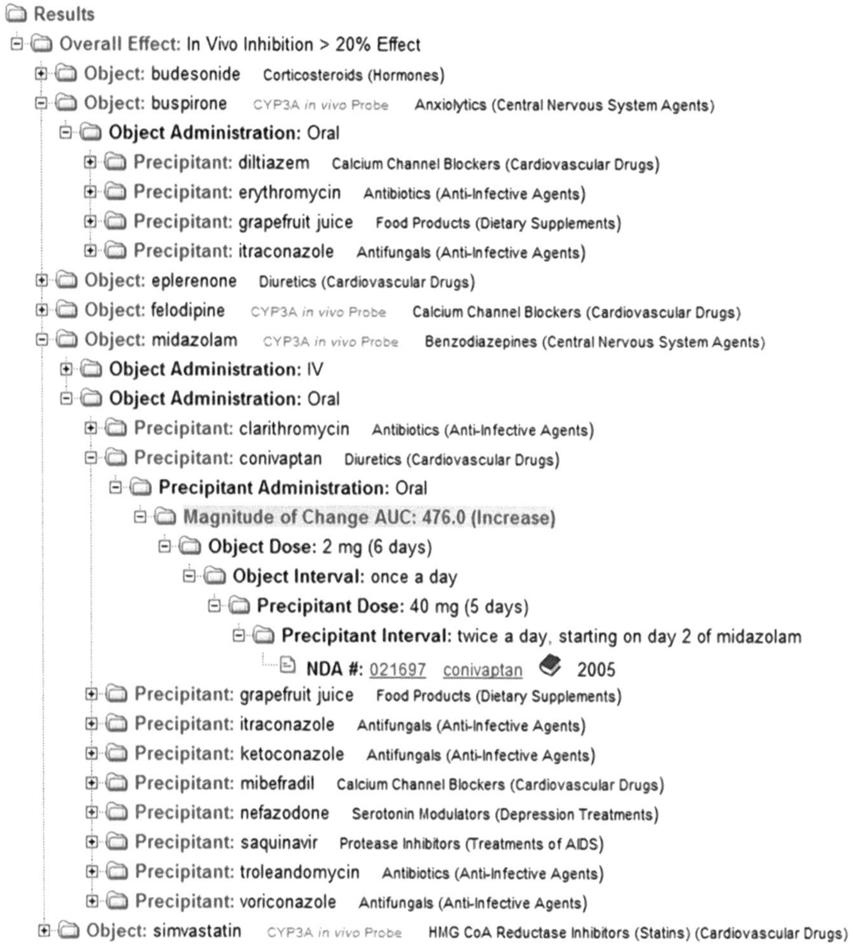

Fig. 15.2 List of *precipitants* evaluated with CYP3A sensitive substrates. These *precipitants* have increased the AUC of substrate(s) by 400% or more. Display from the *Metabolism and Transport Drug Interaction Database*© (http://www.druginteractioninfo.org, accessed Apr 2009)

Each *precipitant* (inhibitor) in the list has its own folder containing more detailed information: exact value of the AUC change observed in the study, dosing regimen of the *object* (substrate) and the *precipitant* (inhibitor), and a link to the source article identified by either an *Accession Number* (PMID number) or an *NDA Number*. By clicking directly on this number, the full description of the article can be retrieved (study design, population, drug dosing regimen, results of pharmacokinetic measurements, side effects. . .). Two additional features are available next to the Accession/NDA Number: *Abstract* of the article (visualized with the icon) and *Reference PK* parameters for drugs (retrieved by clicking on the icon).

There are several options for displaying the results in a table and performing filter operations as well as exporting capabilities into Microsoft Excel® (Fig. 15.3) or Microsoft Word®.

Object	Precipitant	% ↑ AUC	Precipitant Dosing regimen		Accession # or NDA #	Publication Date
budesonide	ketoconazole	580.9	200 mg (4 days)	once daily (8 am)	Accession #: 10945311	2000 Jul
buspirone	diltiazem	433.3	60 mg (2 days)	tid (total of 5 doses) (1 pm)	Accession #: 9663178	1998 Jun
buspirone	erythromycin	490.9	500 mg (4 days)	tid (7 am)	Accession #: 9333111	1997 Sep
buspirone	grapefruit juice	821.3	200 mL double-strength (2.5 days)	tid for 2 days, and at the same time, 0.5	Accession #: 9871430	1998 Dec
buspirone	itraconazole	1347.9	200 mg (4 days)	once daily	Accession #: 10068153	1999 Feb
buspirone	itraconazole	1815.2	100 mg (4 days)	bid (7 am)	Accession #: 9333111	1997 Sep
eplerenone	ketoconazole	438.7	200 mg (6 days)	BID	Accession #: 15204695	2004 Mar
felodipine	itraconazole	533.6	200 mg (4 days)	once daily (7 am)	Accession #: 9129558	1997 Apr
midazolam	clarithromycin	405.5	500 mg (7 days)	bid	Accession #: 9728893	1998 Aug
midazolam	clarithromycin	448.3	500 mg (7 days)	BID	Accession #: 17495878	2008 Jan
midazolam	clarithromycin	531.9	500 mg (7 days)	BID	Accession #: 17635500	2008 Jan
midazolam	clarithromycin	600	500 mg (7 days)	bid	Accession #: 9728893	1998 Aug
midazolam	clarithromycin	739.3	500 mg (7 days)	bid	Accession #: 16432272	2006 Feb
midazolam	clarithromycin	861	500 mg (7 days)	bid	Accession #: 9728893	1998 Aug
midazolam	conivaptan	476	40 mg (5 days)	twice a day, starting on day 2 of	NDA # 021697	2005
midazolam	grapefruit juice	495	240 mL (3 days)	double-strength given tid on Days 1-2	Accession #: 12953340	2003 Aug
midazolam	itraconazole	474.5	100 mg (4 days)	once daily (1 pm)	Accession #: 8527290	1995 Sep
midazolam	itraconazole	516.2	200 mg (4 days)	once daily (11 am)	Accession #: 9591931	1998 Mar
midazolam	itraconazole	564.4	200 mg (6 days)	once daily (1 pm)	Accession #: 8623953	1996 Mar
midazolam	itraconazole	976.9	200 mg (4 days)	once daily (2 pm)	Accession #: 8181191	1994 May
midazolam	ketoconazole	419.5	200 mg	single dose, 2 hours before midazolam	Accession #: 10586386	1999 Dec
midazolam	ketoconazole	545	200 mg	single dose (9 am)	Accession #: 10586386	1999 Dec
midazolam	ketoconazole	547.1	200 mg (4 days)	bid	Accession #: 15114429	2004 Jun
midazolam	ketoconazole	772	200 mg (12 days)	once daily	Accession #: 14551182	2003 Nov
midazolam	ketoconazole	851	400 mg (10 days)	once daily	Accession #: 16580903	2006 Apr
midazolam	ketoconazole	1261.6	200 mg (1.5 days)	bid (first dose 12 h before midazolam	Accession #: 10579473	1999 Nov
midazolam	ketoconazole	1489.7	400 mg (4 days)	once daily (2 pm)	Accession #: 8181191	1994 May
midazolam	mibefradil	786.3	100 mg	single dose (8 am)	Accession #: 14517191	2003 Oct
midazolam	nefazodone	444	100 to 200 mg (12 days)	bid (100 mg bid for 5 days and 200 mg	Accession #: 14551182	2003 Nov
midazolam	saquinavir	417.6	1200 mg (soft-gelatin capsule) (5	tid (8 am)	Accession #: 10430107	1999 Jul
midazolam	troleandomycin	1382.5	500 mg	single doses: 2 hours before midazolam	Accession #: 15536460	2004 Dec
midazolam	voriconazole	839.6	200-400 mg (2 days)	400 mg bid on the first day and 200 mg	Accession #: 16580904	2006 Apr
simvastatin	clarithromycin	895.5	500 mg (9 days)	BID on days 10-18	Accession #: 15518608	2004 Nov
simvastatin	conivaptan	489.5	20 mg (5 days)	twice a day	NDA # 021697	2005
simvastatin	conivaptan	1193.7	40 mg (5 days)	twice a day	NDA # 021697	2005
simvastatin	cyclosporine	696.9	236 ± 71 mg/day (7 months)	Not Provided	Accession #: 11583721	2001 Oct
simvastatin	erythromycin	521.5	500 mg (2 days)	tid (7 am)	Accession #: 9728898	1998 Aug
simvastatin	grapefruit juice	1249.5	200 mL double-strength (3 days)	tid for 2 days, and on day 3: with	Accession #: 11061578	2000 Oct
simvastatin	grapefruit juice	1513.7	200 mL (double-strength) (2.5	tid for 2 days, and 0.5 and 1.5 h after	Accession #: 9834039	1998 Nov
simvastatin	itraconazole	900	200 mg (4 days)	once daily (7 am)	Accession #: 9542477	1998 Mar
simvastatin	ketoconazole	1155.2	400 mg (10 days)	once daily	Accession #: 16580903	2006 Apr
simvastatin	nelfinavir	507.1	1250 mg (14 days)	bid	Accession #: 11709322	2001 Dec

Fig. 15.3 Excel download of the results of the query that retrieved strong CYP3A inhibitors (same results than Fig. 15.2). Display from the *Metabolism and Transport Drug Interaction Database*© (http://www.druginteractioninfo.org, accessed Apr 2009)

When the lists of inhibitors obtained for each substrate are compared and duplicates are removed, the following strong inhibitors remain (Table 15.1).

Using the AUC_R of the probe substrate midazolam when available, users can compare potencies of various inhibitors and extrapolate what would be observed if these inhibitors are coadministered with the NME of interest.

Repeating the same query and changing the AUC range to 100–400% (AUC_R of 2–5) retrieves the moderate inhibitors of CYP3A. These are presented in Table 15.2.

15.3.1.5 Interpretation

Considering the results of Table 15.1, it appears that most AUC_R were obtained with midazolam making it easier to compare the effects of ketoconazole to those of

Table 15.1 Strong CYP3A inhibitors obtained using sensitive substrates

Precipitant	Therapeutic class	*Object* (orally)	AUC_{ratio}	PMID or NDA #	Publication Year
Strong CYP3A inhibitors (yielding substrate AUC_r> 5)					
Ritonavir	Protease inhibitors	Triazolam	**40.7**	16513448	2006
Indinavir	Protease inhibitors	Vardenafil	**16.2**	NDA # 021400	2003
Ketoconazole	Antifungals	Midazolam	**15.9**	8181191	1994
Troleandomycin	Antibiotics	Midazolam	**14.8**	15536460	2004
Itraconazole	Antifungals	Midazolam	**10.8**	8181191	1994
Voriconazole	Antifungals	Midazolam	**9.4**	16580904	2006
Saquinavir/RIT[a]	Protease inhibitors	Maraviroc	**9.2**	18333863	2008
Mibefradil	Calcium channel blockers	Midazolam	**8.8**	14517191	2003
Clarithromycin	Antibiotics	Midazolam	**8.4**	16432272	2006
Lopinavir / RIT	Protease inhibitors	Aplaviroc	**7.7**	16934050	2006
Nelfinavir	Protease inhibitors	Simvastatin	**6.1**	11709322	2001
Telithromycin	Antibiotics	Midazolam	**6.0**	NDA # 021144	2004
Grapefruit juice DS[b]	Food products	Midazolam	**5.6**	12953340	2003
Conivaptan	Diuretics	Midazolam	**5.7**	NDA # 021697	2005
Nefazodone	Antidepressants	Midazolam	**5.4**	14551182	2003
Saquinavir	Protease inhibitors	Midazolam	**5.2**	10430107	1999

[a]RIT: ritonavir
[b]DS: double strength

other strong inhibitors. For example, since the NME under consideration exhibited an AUC_R of 3.6 with ketoconazole, other strong inhibitors such as telithromycin, conivaptan, nefazodone, and saquinavir are expected to produce smaller increases in AUC when they are coadministered with the NME. Additional insights can be obtained by searching for CYP3A substrates that exhibit AUC_R values close to 3.6 when coadministered with ketoconazole. In the present example, the lists of inhibitors generated become useful tools for investigators involved in early clinical studies since they provide a rational basis for dosing adjustments. Also, in the case of multicenter trials, clinicians forward queries about existing comedications in individual patients' regimens. While the DIDB would allow a rapid assessment of the in vivo inhibition potential of these comedications, clinicians are told that if these additional comedications are not found in the lists of CYP3A inhibitors provided, they do not need to be concerned about potential interactions since the NME is metabolized only by CYP3A.

While this question focused on an NME that is CYP3A substrate, the same approach can be easily reproduced with NMEs that are substrates of any other CYP enzyme and allows the identification of drugs that need to be considered carefully when coadministered with the NME.

Table 15.2 Moderate CYP3A inhibitors obtained using sensitive substrates

Precipitant	Therapeutic class	*Object* (orally)	AUC_{ratio}	PMID or NDA #	Publication year
Moderate CYP3A inhibitors ($AUC_r \geq 2$ and < 5)					
Fluconazole	Antifungals	Midazolam	**4.9**	16172184	2005
Atazanavir/RIT[a]	Protease inhibitors	Maraviroc	**4.9**	18333863	2008
Erythromycin	Antibiotics	Midazolam	**4.4**	8453848	1993
Darunavir/RIT	Protease inhibitors	Sildenafil	**4.0**	NDA # 021976	2006
Diltiazem	Calcium channel blockers	Midazolam	**3.7**	8198928	1994
Atazanavir	Protease inhibitors	Maraviroc	**3.6**	18333863	2008
Aprepitant	Antiemetics	Midazolam	**3.3**	12891225	2003
Imatinib	Antineoplastic agents	Simvastatin	**2.9**	14612892	2003
Verapamil	Calcium channel blockers	Midazolam	**2.9**	8198928	1994
Grapefruit juice	Food products	Midazolam	**2.4**	10546916	1999
Tofisopam	Benzodiazepines	Midazolam	**2.4**	17989974	2008
Ciprofloxacin	Antibiotics	Sildenafil	**2.1**	16372380	2005
Cimetidine	H-2 receptor antagonists	Midazolam	**2.0**	6152615	1984

[a]RIT: ritonavir

15.3.2 Example: Analysis of Drug Interactions in the Context of a Disease and Its Comorbidities

15.3.2.1 Background

Assessment of the DDI risk potential of an NME during drug development takes into consideration the clinical outcome of administration of the NME and focuses not only on drugs used to treat the primary disease but also on those used to treat comorbidities. Moreover, questions arise regarding the roles of environmental factors (food, herbal medications) and patient characteristics (phenotype, age, etc.) that may also alter drug disposition.

In the problem at hand, an NME is being developed for the treatment of anxiety. This NME is mainly metabolized by CYP1A2 with some contribution by CYP3A4. It was also found that this NME is a weak CYP3A4 inhibitor yielding an AUC ratio of midazolam of 2.3.

15.3.2.2 Question

Because anxiety symptoms are common in 60% of patients with major depression and sleep problems such as insomnia, the user wanted to evaluate whether antidepressants would have any clinically relevant impact on the disposition of this new anxiolytic; in addition, given the inhibitory profile of this NME, the user wanted to determine whether any drugs used in depression are likely to be affected by this new anxiolytic.

15.3.2.3 Search Strategy

First Query

Within the DIDB website, the disease monograph for depression will be used (Fig. 15.4). This depression monograph has been compiled using in-depth analyses of the metabolic profiles and effects of all major antidepressant classes (tricyclics, selective serotonin reuptake inhibitors (SSRIs), monoamine oxidase inhibitors (MAOIs), serotonin–norepinephrine reuptake inhibitors (SNRIs), and dopamine–norepinephrine reuptake inhibitors). The depression monograph is organized into summaries, 22 individual drug monographs, and queries.

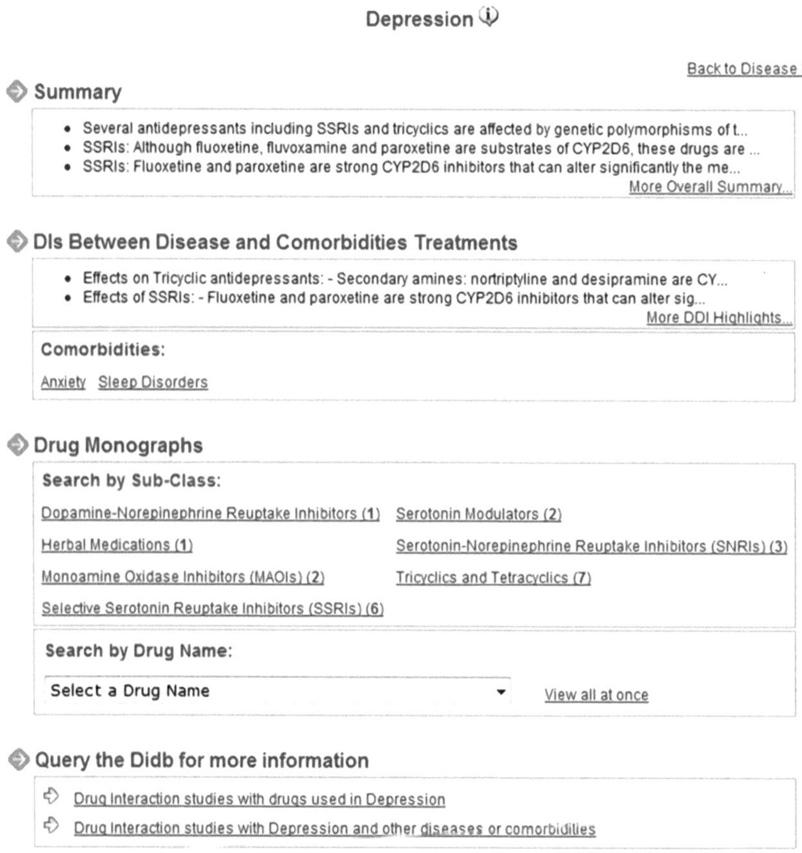

Fig. 15.4 One-page presentation of the different sections (summaries, individual drug monographs, and queries) within the monograph for depression. Display from the *Metabolism and Transport Drug Interaction Database©* (http://www.druginteractioninfo.org, accessed Apr 2009)

Result Output

For all the drugs considered within the five antidepressant classes cited above, complete profiles are presented within a table that highlights the main characteristics of each drug considered as an inhibitor/inducer (Fig. 15.5).

For each antidepressant shown in Fig. 15.5, that table provides the enzymes and /or transporters affected and a corresponding DDI risk level. Four risk levels have been created based on a combination of the following characteristics: (i) sensitivity to inhibition and induction of the involved enzymes and/or transporters; (ii) therapeutic range; (iii) documented clinical interactions.

Analyzing Results of the First Query

The inhibitory profiles of the 22 antidepressants listed in Fig. 15.5 show that these drugs inhibit mostly CYP2D6 and only one of them is a CYP1A2 inhibitor, namely

As Inhibitors / Inducers Export to: Excel Word

Drug Name	PK	Inhibition		Induction		DDI Risk Level
		Enzymes	Transporters	Enzymes	Transporters	
(S)-citalopram (escitalopram)		CYP2D6				III: (No or Low)
amitriptyline						III: (No or Low)
bupropion		CYP2D6				I (High)
citalopram		CYP2D6				III: (No or Low)
clomipramine						II (Intermediate)
desipramine						III: (No or Low)
desvenlafaxine (O-desmethylvenlafaxine)						III: (No or Low)
doxepin		CYP2D6				II (Intermediate)
duloxetine		CYP2D6				II (Intermediate)
fluoxetine		CYP2D6				I (High)
fluvoxamine		CYP1A2 CYP2C19 CYP2C9 CYP3A				I (High)
imipramine						III: (No or Low)
mirtazapine						III: (No or Low)
nefazodone		CYP3A				I (High)
nortriptyline						III: (No or Low)
paroxetine		CYP2D6				I (High)
phenelzine						Risk Unknown
sertraline		CYP2D6				III: (No or Low)
st Johns Wort extract (Hypericum perforatum)				CYP3A CYP2C9 CYP2C19 UGT1A1	P-gp (ABCB1)	I (High)
tranylcypromine						Risk Unknown
trazodone						Risk Unknown
venlafaxine						III: (No or Low)

Fig. 15.5 Complete DI profile of 22 antidepressants presented within a table that highlights the main characteristics of each drug considered as an inhibitor or an inducer. Display from the *Metabolism and Transport Drug Interaction Database*© (http://www.druginteractioninfo.org, accessed Apr 2009)

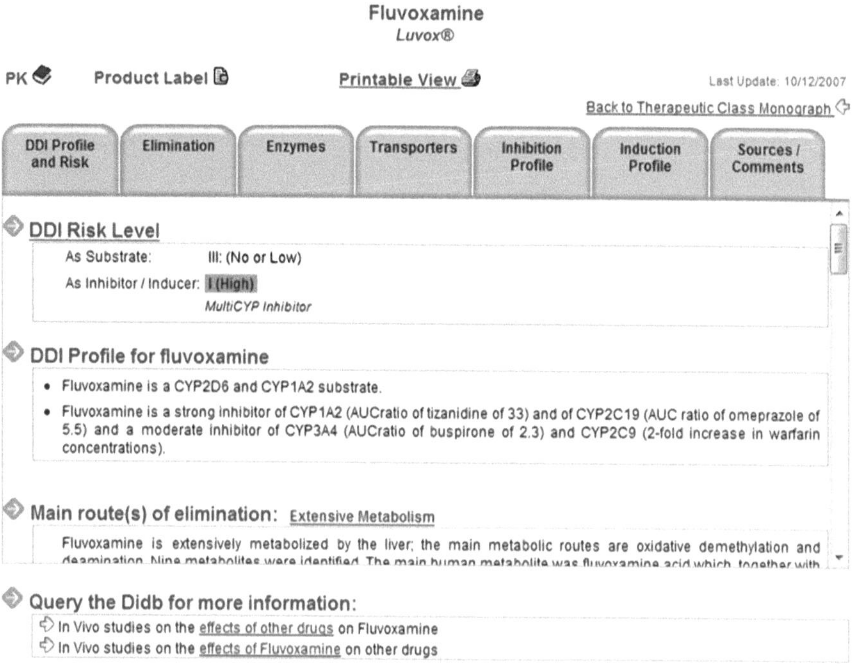

Fig. 15.6 Fluvoxamine monograph: complete DI profile for this drug which is organized into various sections: metabolism/elimination, enzymes/transporters and main associated interactions, inhibition/induction profiles. Display from the *Metabolism and Transport Drug Interaction Database*© (http://www.druginteractioninfo.org, accessed Apr 2009)

fluvoxamine. The fluvoxamine drug monograph (Fig. 15.6) shows the extent of CYP1A2 inhibition by this drug in a study with duloxetine (CYP1A2 sensitive substrate) yielding an AUC ratio of 6. This study can then be used to predict the outcome of a DI between the new NME (CYP1A2 substrate) and fluvoxamine.

Second Query

To address the CYP3A inhibitory potential of the NME and assess whether its weak inhibitory potency might affect certain antidepressants, the overall view that highlights the main characteristics of all antidepressant as substrates is used (Fig. 15.7).

Result Output

This table shows that nefazodone and trazodone are the only antidepressants metabolized principally by CYP3A4. Additionally, CYP3A is partially involved in mirtazapine disposition, along with CYP2A6 and CYP1A2.

Drugs used in Depression

Back to Disease Monograph

As Substrates Export to: Excel Word

Drug Name	PK	Routes of Elimination	Main Enzymes	Main Transporters	DDI Risk Level
(S)-citalopram (escitalopram)		Extensive Metabolism	CYP2C19		III: (No or Low)
amitriptyline		Extensive Metabolism	CYP2D6 CYP2C19		III: (No or Low)
bupropion		Extensive Metabolism	CYP2B6		III: (No or Low)
citalopram		Extensive Metabolism	CYP2C19 CYP2D6		II (Intermediate)
clomipramine		Extensive Metabolism	CYP2D6 CYP2C19		II (Intermediate)
desipramine CYP2D6 in vivo Probe		Extensive Metabolism	CYP2D6		I (High)
desvenlafaxine (O-desmethylvenlafaxine)		Renal Excretion (unchanged)			III: (No or Low)
doxepin		Extensive Metabolism	CYP2D6 CYP2C19		III: (No or Low)
duloxetine		Extensive Metabolism	CYP1A2		I (High)
fluoxetine		Extensive Metabolism	CYP2D6		III: (No or Low)
fluvoxamine		Extensive Metabolism	CYP2D6		III: (No or Low)
imipramine		Extensive Metabolism	CYP2D6 CYP2C19		II (Intermediate)
mirtazapine		Extensive Metabolism	CYP2D6 CYP1A2 CYP3A		II (Intermediate)
nefazodone		Extensive Metabolism	CYP3A		II (Intermediate)
nortriptyline		Extensive Metabolism	CYP2D6		I (High)
paroxetine		Extensive Metabolism	CYP2D6	P-gp (ABCB1)	II (Intermediate)
phenelzine		Extensive Metabolism			Risk Unknown
sertraline		Extensive Metabolism	CYP2C19		III: (No or Low)
st Johns Wort extract (Hypericum perforatum)					Risk Unknown
tranylcypromine		Extensive Metabolism			Risk Unknown
trazodone		Extensive Metabolism	CYP3A		II (Intermediate)
venlafaxine		Extensive Metabolism	CYP2D6		I (High)

Fig. 15.7 Complete DI profile of 22 antidepressants presented within a table that highlights the main characteristics of each drug considered as a substrate. Display from the *Metabolism and Transport Drug Interaction Database©* (http://www.druginteractioninfo.org, accessed Apr 2009)

Analyzing the Results of the Second Query

An analysis of the DDI profiles of these three antidepressants (Fig. 15.8) shows that only trazodone is susceptible to CYP3A inhibition as indicated by a 2.4 AUC ratio in the presence of the potent CYP3A inhibitor ritonavir (the three antidepressants are also sensitive to potent inducers such as carbamazepine). The ritonavir effect toward trazodone indicates that a weak CYP3A inhibitor such as the NME under consideration should have a limited effect on trazodone exposure.

Nefazodone
Serzone®

Main Enzyme(s) and Associated Interactions: CYP3A

In vitro studies have shown that nefazodone undergoes metabolism mediated by CYP3A.

Coadministration of cimetidine (a multiCYP inhibitor) did not have any significant effects on the pharmacokinetics of nefazodone.

The coadministration of carbamazepine with nefazodone resulted in 90% reduction in AUCs for nefazodone and hydroxynefazodone.

Trazodone
Desyrel®

Main Enzyme(s) and Associated Interactions: CYP3A

Studies in human liver microsomes show that trazodone is N-dealkylated to form m-chlorophenylpiperazine by CYP3A. Metabolic processes other than dealkylation might be mediated by CYP2D6 or CYP1A2.

Ritonavir, a strong CYP3A inhibitor, yields an AUCratio of 2.4 with trazodone. It is likely that ketoconazole, indinavir, and other CYP3A inhibitors such as itraconazole or nefazodone may lead to substantial increases in trazodone plasma concentrations.

Carbamazepine reduces plasma concentrations of trazodone by 76%.

Fluoxetine and thioridazine (CYP2D6 inhibitors) increase the plasma concentrations of trazodone by 60% and 40%, respectively.

Smokers have 20% lower plasma concentrations than nonsmokers, suggesting a role of CYP1A2 in trazodone metabolism.

Mirtazapine
Remeron®

Main Enzyme(s) and Associated Interactions: CYP2D6 CYP1A2 CYP3A

Cytochrome 2D6 and 1A2 are involved in the formation of the 8-hydroxy metabolite of mirtazapine, whereas CYP3A is responsible for the formation of the N-desmethyl and N-oxide metabolite.

There was a 50% difference between mirtazapine plasma levels in CYP2D6 poor metabolizers and CYP2D6 extensive metabolizers.
Sertraline, a weak CYP2D6 inhibitor, increases the plasma levels of mirtazapine by 33%. Other strong (paroxetine, fluoxetine, quinidine) and moderate (duloxetine, terbinafine) CYP2D6 inhibitors are likely to increase mirtazapine levels as well.

Carbamazepine decreases the AUC of mirtazapine by 60%.

Fig. 15.8 Example of the drug monograph section called: *main enzymes and associated interactions* for three antidepressants that are CYP3A substrates: nefazodone, trazodone, and mirtazapine. Display from the *Metabolism and Transport Drug Interaction Database©* (http://www.druginteractioninfo.org, accessed Apr 2009)

15.3.2.4 Interpretation

The new disease section has multiple uses and it allows a rapid assessment of the following:

- the DI potential of an NME in comparison with other marketed drugs used to treat the same disease,
- the DI potential of this NME with drugs used to treat comorbidities of that disease,

– the complete DI profile of a disease in a summarized tabulated view as shown in Figs. 15.5 and 15.7.

Clinical investigators may also be interested in using two novel queries that yield the DI profile resulting from the coexistence of any two diseases of interest (e.g., depression and diabetes) (Fig. 15.9).

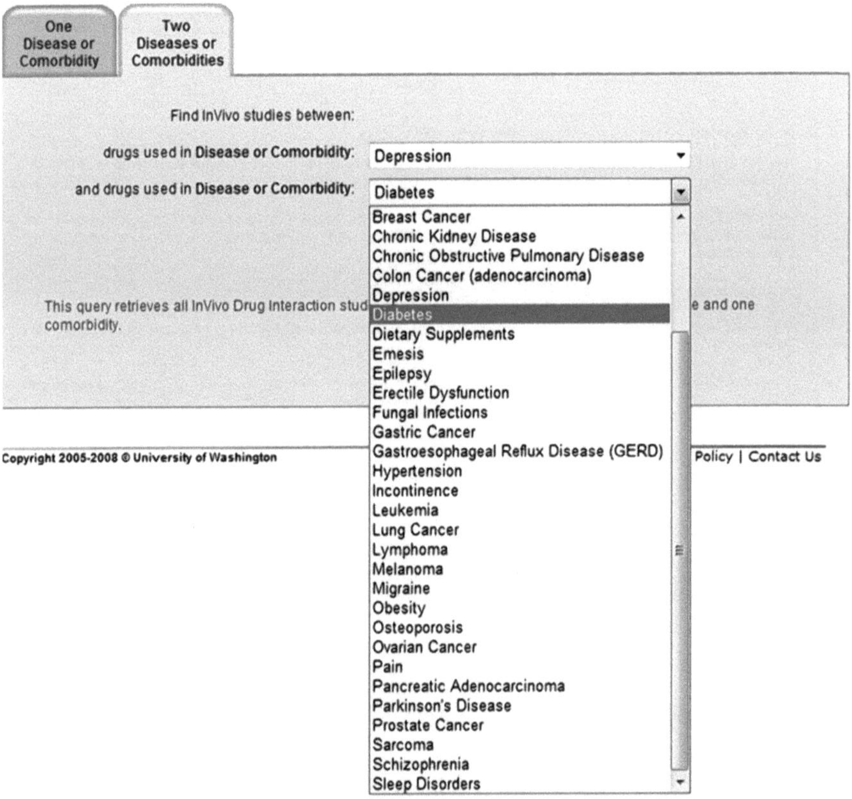

Fig. 15.9 Query labeled "*Two diseases or comorbidities*" used to retrieve all DI studies between antidepressants and antidiabetics. Display from the *Metabolism and Transport Drug Interaction Database*© (http://www.druginteractioninfo.org, *accessed Apr 2009*)

15.4 Ongoing Developments

The DIDB is a tool that is constantly being optimized as a result of feedback from a large base of users including requests for specific searches. These can be in the format of new queries or special reports tailored by the DIDB team. New features currently being developed include the addition of data sets pertaining to emerging areas (transporters, nuclear receptors). The DIDB is also enhanced with tools that allow the users to rapidly focus on important DI reports and sort through the large body of literature. Examples of such tools include: graphical displays of extent of DI (AUC_R, changes in clearance of substrates) and "flagging" important drug characteristics (narrow therapeutic range drugs, probe substrates, potent inhibitors, or inducers).

Appendix

The only terminology specific to the DIDB pertains to drugs/compounds which appear as object or precipitant depending on their role in specific drug interaction. *Object* refers to a compound that acts as the modified agent (i.e., substrate) and *precipitant* refers to a compound that acts as the causative agent. A *precipitant* can be an inhibitor, inducer, or activator but it may have no effect at all.

The DIDB search interface utilizes a list of prestructured searches called by "*Queries*." These are set along intuitive themes such as drug, enzyme, therapeutic class, and transporter and thus allow the user to quickly select the appropriate search without the need for extensive training. The ten sets of queries can be categorized into qualitative or quantitative as shown below:

Qualitative

Drug	Search by *Drug* (using generic names)
Enzyme	Search by *Enzyme*
Nuclear receptor and induction	Search by *Nuclear Receptor*
Therapeutic class	Search by *Therapeutic Class*
Transporters	Search by *Transporter*
QT interval	Search for data summarizing *QTc effects* of drugs
Other	Search for articles using journal, author name, etc.

Quantitative

AUC/CL changes	Search for AUC and CL changes (%) observed in drug interaction studies
In vitro parameters	Search for *in vitro* parameters: K_m, K_i, IC_{50}, K_I , K_{inact}, etc.
Pharmacokinetics	Search for pharmacokinetic parameters of selected drugs

References

Drugs@FDA: http://www.accessdata.fda.gov/scripts/cder/drugsatfda/ (accessed Apr 2009)

FDA Guidance for Industry: Drug Interaction Studies – Study Design, Data Analysis, and Implications for Dosing and Labeling. Draft guidance (Released September 2006). Available at: http://www.fda.gov/cder/guidance/6695dft.htm (accessed Apr 2009)

Hachad H, Ragueneau-Majlessi I and Levy RH. Metabolism and transport drug interaction database: a web-based research and analysis tool, in *Drug–Drug Interactions* Second Edition (David Rodrigues A ed), pp 567–579, Informa Healthcare, New York (2008).

PubMed: http://www.ncbi.nlm.nih.gov/sites/entrez (accessed Apr 2009)

Chapter 16
Web-Based Database as a Tool to Examine Drug–Drug Interactions Involving Transporters

Kazuya Maeda, Yoshihisa Shitara, Toshiharu Horie, and Yuichi Sugiyama

Abstract The clinical importance of drug–drug interaction mediated by drug transporters has been gradually recognized; its quantitative prediction and an in silico database of drug–drug interaction have been much wanted. In this chapter, examples of, and prediction methods for, transporter-mediated drug–drug interactions are shown, and a Web-based transporter-mediated drug–drug interaction database in TP-search (http://www.TP-Search.jp/) is also described.

16.1 Drug Transporters as a Target of Drug–Drug Interaction

16.1.1 General Features of Transporter-Mediated Drug–Drug Interaction

Currently, various kinds of drug transporters in several organs have been cloned and characterized, and the importance of drug transporters in the in vivo pharmacokinetics of their substrate drugs has also been clarified by the use of gene-knockout or gene-deficient animals and in human clinical studies demonstrating the effects of genetic polymorphisms or interacting drugs on the pharmacokinetics of substrates. As shown in Fig. 16.1, many kinds of transporters are located on both sides of polarized epithelial or endothelial cells in several organs. Among these drug transporters, SLC (solute carrier)-type transporters are mainly involved in the uptake of drugs, while ABC (ATP-binding cassette)-type transporters are important for the cellular efflux of drugs. Coordination of uptake and efflux transporters enables compounds to penetrate these cells efficiently in so-called vectorial transport (Giacomini and Sugiyama, 2005). Vectorial transport in several organs can be a determinant of

Y. Sugiyama (✉)
Department of Molecular Pharmacokinetics, Graduate School of Pharmaceutical Sciences, The University of Tokyo, Bunkyo-ku, Tokyo, Japan
e-mail: sugiyama@mol.f.u-tokyo.ac.jp

K.S. Pang et al. (eds.), *Enzyme- and Transporter-Based Drug–Drug Interactions*, DOI 10.1007/978-1-4419-0840-7_16,

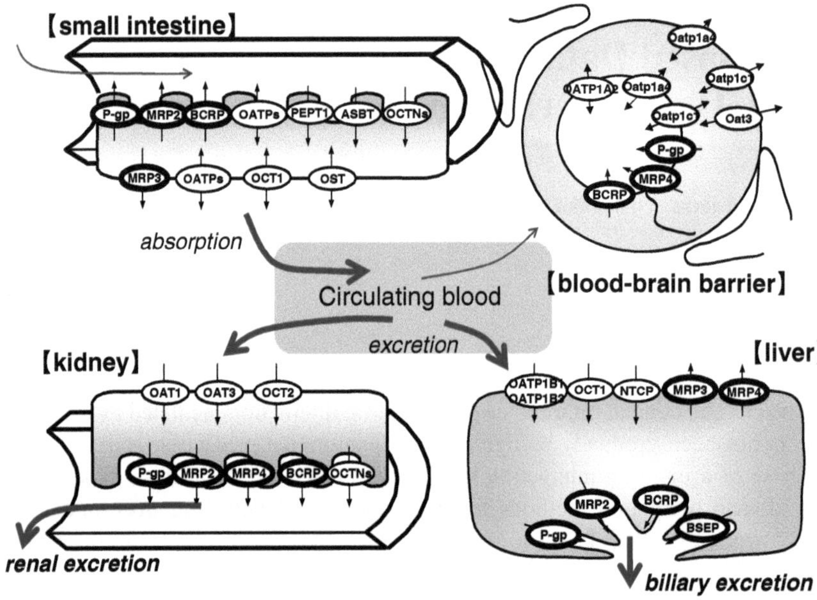

Fig. 16.1 Drug transporters expressed in various tissues. *Bold circles* represent the ATP-binding cassette (ABC) family of transporters and others are the solute carrier (SLC) family of transporters

the itinerary of substrate drugs in the body by effecting systemic clearance in the liver and kidney, intestinal absorption, and brain distribution across the blood–brain barrier (BBB). For example, in human liver, SLC-type transporters, organic anion transporting polypeptide (OATP) 1B1 and OATP1B3, are thought to be responsible for the uptake of anionic compounds, while ABC-type transporters, multidrug resistance-associated protein (MRP) 2 and breast cancer resistance protein (BCRP), play important roles in their biliary excretion. Though the protein sequences and structures of SLC and ABC transporters are very different, their substrate specificities often overlap, which results in the efficient biliary excretion of a wide variety of organic anions. Because drug transporters can generally accept several structurally unrelated compounds, including endogenous compounds such as bilirubin and bile acids and clinically important drugs, we must notice that inhibitors or inducers for a specific transporter can affect the pharmacokinetics of multiple compounds, which cannot presently be easily determined by their chemical structure alone. The impact of functional change in the uptake and efflux transporters on the pharmacokinetics of drugs and their subsequent pharmacological or toxicological effects depends on the circumstances. For transporters expressed in the liver, kidney, and intestine, functional alterations directly affect the bioavailability and systemic clearance of drugs. For example, a decrease in the transport function of an uptake or efflux transporter in a clearance organ reduces the organ clearance, which results in an increase of the drug concentration in the systemic circulation and its exposure to all the organs in the body. In this case, a routine survey of drug concentration in blood such as therapeutic drug monitoring (TDM) greatly helps us to understand easily any functional change in the transport function by coadministered drugs. On the other hand, for transporters expressed at pharmacological or toxicological targets or at tightly connected endothelial cell monolayer barriers between the blood and important tissues (including brain, fetus, and testis), if the distribution volume of the target tissue is much smaller than that of whole body, their functional change does not affect the plasma concentration but dramatically alters the drug's local tissue concentration. In this case, we cannot easily detect the drug–drug interaction, because in humans, we can only measure the plasma concentration of drugs. Moreover, the inhibition of transporters for endogenous compounds by drugs causes an abnormal increase in their plasma and tissue concentration. For example, hepatobiliary transport of bile acids is achieved by their hepatic uptake by Na^+-taurocholate cotransporting polypeptide (NTCP) and biliary excretion by the bile salt export pump (BSEP). Some drugs, such as troglitazone sulfate, glyburide, bosentan, rifampicin, and cyclosporin A, potently inhibit the function of NTCP and BSEP, which sometimes causes drug-induced cholestasis in clinical situations (Fattinger et al., 2001; Funk et al., 2001; Kostrubsky et al., 2003). Bilirubin is taken up into hepatocytes mainly via OATP1B1, conjugated with glucuronic acid, and then excreted into the bile via MRP2 as bilirubin glucuronide. A previous report has indicated that indinavir, cyclosporin A, and rifamycin SV, which have been clinically reported to induce hyperbilirubinemia, inhibited the function of human OATP1B1 at therapeutically relevant concentrations (Campbell et al., 2004).

16.1.2 Examples of Clinically Relevant Drug–Drug Interaction Mediated by Transporters

Recently the number of clinical studies demonstrating the putative transporter-mediated drug–drug interaction has gradually increased, and even for drugs that are thought to be eliminated from the body by extensive metabolism, transporters sometimes play important roles in their drug–drug interactions. In this section, some clinically relevant examples of the transporter-meditated drug interactions are briefly introduced.

16.1.2.1 Multidrug Resistance (MDR) 1 (P-glycoprotein)

MDR1 is expressed in various kinds of organs, such as intestine, liver, kidney, and the BBB, and it works to effect the cellular efflux of various kinds of cationic and neutral compounds with relatively high hydrophobicity. MDR1 restricts intestinal absorption of, and facilitates the excretion of, these sorts of compounds from liver and kidney; thus the inhibition of MDR1 function leads to an increase in the concentration of these substrates in the circulating blood with consequent potentiation of their pharmacological or toxicological effects. Digoxin is a typical MDR1 substrate with minimal metabolism and is generally recognized as a drug probe for MDR1. Because the therapeutic window of digoxin is very narrow and even a small fluctuation of its blood concentration can cause severe side effects, many clinical reports regarding its drug–drug interaction have been published. For example, many reports have previously indicated that coadministration of quinidine, verapamil (antiarrhythmic drugs), or ritonavir (an anti-HIV drug) with digoxin results in an increase in the blood concentration of digoxin and the risk of side effects (Fromm et al., 1999; Verschraagen et al., 1999; Kusuhara and Sugiyama, 2001). This interaction is thought to be caused by a combination of increased intestinal absorption and decreased hepatic and renal clearance of digoxin by the inhibition of MDR1 in the small intestine, liver, and kidney. Because the substrate specificities of MDR1 often overlap those of cytochrome P450 (CYP) 3A4, the relative contribution of MDR1 and CYP3A4 in the drug–drug interaction of MDR1 substrates at the small intestine cannot be easily separated in most cases. MDR1 is also localized on the blood side of the BBB and plays an important role in limiting the brain distribution of substrate drugs. Using MDR1 knockout mice, the brain concentration of some MDR1 substrates was drastically increased, whereas their blood concentration was not significantly changed (Schinkel et al., 1994). Thus, it is very difficult to determine the types of drug–drug interaction as mentioned above. A previous clinical study has indicated that coadministration of loperamide (an antidiarrheal) and quinidine causes respiratory suppression in humans (Sadeque et al., 2000). Normally, MDR1 actively pumps loperamide out from the brain and acts peripherally as an agonist of opioid receptors in the small intestine. However, coadministered quinidine can increase the brain distribution of loperamide by its potent inhibition of MDR1-mediated efflux, and it exerts its opioid receptor agonistic effects in the central nervous system (CNS). The expression of side effects occurs earlier than any increase in the blood concentration of loperamide, suggesting that its CNS side effect is not determined by its blood concentration.

Some drugs upregulate mRNA expression of MDR1 mainly via pregnane X receptor (PXR). A previous report has indicated that repeated administration of rifampicin, a typical potent PXR agonist, increased the protein expression of MDR1 in human small intestine biopsy samples and significantly decreased the blood concentration of orally administered digoxin (Greiner et al., 1999).

16.1.2.2 OATP Family Transporters

OATP1B1 and OATP1B3 are almost exclusively expressed in the basolateral membrane of hepatocytes and are responsible for the hepatic uptake of a wide variety of anionic compounds. The inhibition of OATP1B1 and OATP1B3 decreases the hepatic clearance of substrate drugs and subsequently increases their blood concentration. For example, a clinical report indicated that coadministration of cyclosporin A drastically increased the plasma concentration of cerivastatin (Muck et al., 1999). Because cerivastatin is metabolized mainly by CYP2C8 and partly by CYP3A4, many investigators have suspected the inhibition of CYP-mediated metabolism of cerivastatin by cyclosporin A. However, Shitara et al. (2003) have demonstrated that cerivastatin was taken up into hepatocytes by OATP1B1 and that cyclosporin

A potently inhibited OATP1B1-mediated uptake of cerivastatin in human hepatocytes, even at therapeutic doses, while its inhibitory effect on the metabolism of cerivastatin was less potent (Shitara et al., 2003). After this report, several drug interaction studies with cyclosporin A have been performed and indicated that the plasma concentration of many OATP drug substrates that are eliminated from the liver by extensive metabolism, such as repaglinide (2C8), nateglinide (2C9), glibenclamide (2C9 and 3A4), bosentan (2C9 and 3A4), and atorvastatin (3A4), is increased by the coadministration of cyclosporin A, while OATP substrates with minimal metabolism such as rosuvastatin, pitavastatin, and pravastatin also interact with cyclosporin A to the same extent (Shitara et al., 2005; Shitara et al., 2006). Thus, this evidence indicated that the hepatic clearance of these drugs is mainly determined by an OATP-mediated hepatic uptake process. (The theoretical consideration from the viewpoint of pharmacokinetics will be discussed below.) On the other hand, repeated doses of rifampicin and efavirenz decreased the plasma concentration of pravastatin, which is thought to be cleared from the body by OATP1B1 without any metabolism, suggesting that the expression of OATP1B1 may be induced by these drugs, probably via PXR (Kyrklund et al., 2004; Gerber et al., 2005). This hypothesis is supported by in vitro experiments demonstrating that exposure of rifampicin increased the mRNA level of OATP1B1 in cultured human hepatocytes (Sahi et al., 2006), though the elements responsible for PXR binding have not yet been clarified.

In the small intestine, OATP1A2 and OATP2B1 are reported to be expressed, though expression of OATP1A2 remains to be discussed. Recently, the plasma concentration of fexofenadine and several β-blockers (including celiprolol, atenolol, and talinolol) was significantly reduced by the coadministration of fruit juice made from grapefruits, oranges, or apples (Dresser et al., 2002; Lilja et al., 2003, 2005; Schwarz et al., 2005). In vitro experiments indicated that fruit juice, especially a major constituent of grapefruit juice, naringin, can inhibit the uptake mediated by OATP1A2 and OATP2B1 as well as Oatp1a5, which is expressed in the small intestine in rodents, and several animal experiments apparently reproduced the inhibitory effect of fruit juice on the intestinal absorption of several drugs, which includes drugs interacting with fruit juice in humans (Dresser et al., 2002; Lilja et al., 2003, 2005; Schwarz et al., 2005). Currently we do not know which transporters are involved in this type of drug–drug interaction in humans, and further studies will be needed to demonstrate directly the role of OATP transporters in the intestinal absorption of organic anions by the use of the genetic polymorphisms or drugs interacting with these transporters.

16.1.2.3 Organic Anion Transporter (OAT) 1 and OAT3

OAT1 and OAT3 are expressed on the blood side of kidney epithelial cells, and several reports have suggested that these transporters are important for the renal secretion of organic anions. OAT3 accepts various kinds of bulky hydrophobic anions that are very similar to substrates of OATP transporters in the liver, while OAT1 can transport relatively hydrophilic small molecules, such as nucleoside analogues. Previous reports have indicated that the coadministration of probenecid and some nonsteroidal anti-inflammatory drugs (NSAIDs) decreases renal clearance of cephalosporin antibiotics and methotrexate by the inhibition of OAT family transporters (Maiche, 1986; McLeod, 1998). In in vitro experiments, many drugs can inhibit OAT-mediated uptake, but most of them should not interact with OAT-mediated drug uptake in clinical situations because their clinical plasma concentrations are far below the K_i values for OATs in most cases, except for *p*-aminohippurate (PAH), probenecid, indomethacin, and salicylic acid. Interestingly, coadministration of probenecid with famotidine results in a reduction of the renal clearance of famotidine in humans but not in rats. To clarify the interspecies difference, Tahara et al. have demonstrated the effect of probenecid on the uptake of famotidine mediated by renal transporters (Tahara et al., 2005). In rats, the relative contribution of organic cation transporters (Octs) to the overall renal uptake of famotidine is thought to be high, and inhibition of Oat3-mediated uptake by probenecid may not have a significant effect on the renal clearance of famotidine, whereas in the case of humans, the relative contribution of OAT3 is high compared with that in rats, and the inhibition of OAT3-mediated uptake of famotidine by probenecid causes a drug–drug interaction.

16.1.2.4 Multidrug and Toxin Extrusion (MATE)

MATE family transporters are mainly expressed on the apical membrane of hepatocytes and kidney epithelial cells. MATEs are thought to be involved in the cellular efflux of compounds to bile or urine. Cimetidine was reported to decrease the renal clearance of fexofenadine and metformin in humans, and organic cation transporter (OCT) 2 in the kidney is thought to be one of the target transporters for these interactions (Yasui-Furukori et al., 2005). However, the plasma protein-unbound concentration of cimetidine is lower than the K_i value for OCT2. On the other hand, cimetidine can inhibit MATE function more potently than OCT2; thus, this interaction can be explained by the inhibition of MATE-mediated efflux by cimetidine (Matsushima et al., 2009; Tsuda et al., 2009). The number of substrates for MATEs has been increasing, and the inhibition of MATEs is expected to decrease the biliary excretion and increase the hepatic distribution of certain substrates. Thus, the importance of drug–drug interaction for MATE family transporters should be further investigated.

16.2 What Should Be Considered When Predicting a Transporter-Mediated Drug–Drug Interaction?

As indicated above, the number of examples of drug–drug interaction at transporters has been increasing, and its prediction is important for the development of new drugs as well as for the appropriate clinical use of currently existing drugs. Therefore, in this section, we will discuss what kinds of factors influence the extent of drug–drug interaction based on pharmacokinetic theory.

16.2.1 The Effect of Inhibitors on the Decrease in the Intrinsic Clearance of Substrate Drugs

If inhibitors inhibit the transport of substrates in a competitive and noncompetitive manner, the transport clearance can be described as follows:

$$(Competitive)\ CL_{\mathrm{int}}(+I) = \frac{V_{\mathrm{max}}}{K_{\mathrm{m}}\left(1 + \frac{I}{K_{\mathrm{i}}}\right) + S} \tag{16.1}$$

and

$$(Noncompetitive)\ CL_{\mathrm{int}}(+I) = \frac{V_{\mathrm{max}}/\left(1 + \frac{I}{K_{\mathrm{i}}}\right)}{K_{\mathrm{m}} + S} \tag{16.2}$$

where $CL_{\mathrm{int}}(+I)$, K_{m}, V_{max}, S, I, and K_{i} represent intrinsic clearance in the presence of inhibitors, the Michaelis constant of substrates, maximum transport velocity of substrates, substrate concentration, inhibitor concentration, and inhibition constant of inhibitors, respectively. Thus, the degree of inhibition (*R* value) can be determined by inhibitor concentration and K_{i} values for inhibitors:

$$R = \frac{CL_{\mathrm{int}}(+I)}{CL_{\mathrm{int}}(-I)} = \frac{1}{1 + \frac{I}{K_{\mathrm{i}}}} \tag{16.3}$$

where $CL_{\mathrm{int}}(-I)$ represents the intrinsic clearance in the absence of inhibitors. Thus, the accurate estimation of the inhibitor concentration and K_{i} values is one of the key points for the quantitative prediction of the drug–drug interactions. K_{i} values can be obtained from various kinds of in vitro experimental systems, such as transporter-expressing cell lines, *Xenopus* oocytes, and membrane vesicles, simply by plotting the data for transport clearance of substrates in the presence of several concentrations of inhibitors. *I* should be the inhibitor concentration in the vicinity of transporters, but especially for humans, with the exception of plasma concentrations, it is very difficult to measure their true concentrations. Moreover, the inhibitor concentration changes continuously with time. For the early phase of drug development, researchers want to know whether new drug candidates have the potential to interact with other drugs via transporters as well as metabolic enzymes. In that case, to avoid false-negative prediction, the maximum concentration of inhibitors can be used for the sensitive detection of the putative drug–drug interactions. Following the classical assumption in which only the protein-unbound form of drugs can inhibit the transporters, to consider the effect of inhibitors on the uptake transporters, the initial protein-unbound concentrations of inhibitors that are intravenously administered should be the maximum values. On the other hand, if inhibitors are orally administered, maximum concentrations of inhibitors (C_{max}) can be regarded as maximum values in most cases, whereas in the liver, because orally administered drugs enter the liver not only from blood circulation but also from the small intestine, the maximum input of inhibitors from both pathways should be considered. Ito et al. proposed the following equation for the estimation of maximum protein-unbound concentration at the inlet to the liver ($I_{in,max,u}$) after oral administration of drugs (Ito et al., 1998):

$$I_{in,max,u} = \left(I_{max} + \frac{k_a \cdot D \cdot F_a}{Q_H} \right) \times f_B \tag{16.4}$$

where I_{max}, k_a, D, F_a, Q_H, and f_B represent the maximum concentration of inhibitors in blood, absorption rate constant, dose, intestinal availability, hepatic blood flow (= 1.4 L/min in humans) and protein-unbound fraction in blood. If we do not know the exact k_a and F_a values for the calculation of maximum concentration, maximum k_a (= 0.1 min^{-1}; gastric emptying rate) and F_a (= 1) can also be assigned to Equation (16.4). In the interpretation of the results from this approach, we should keep in mind that even if the R value for the combination of drugs is calculated to be less than 1, drug–drug interaction does not always occur. On the other hand, if the R value is nearly equal to 1, we are not concerned that there is a drug–drug interaction for that drug combination.

For the efflux transporters, if the protein-unbound form of inhibitors can only passively pass into the intracellular compartment, unbound blood concentration of inhibitors is the same as that near transporters inside the cells. However, if the inhibitors are actively taken up by transporters, intracellular unbound concentration ($I_{u,tissue}$) must be higher than the blood unbound concentration ($I_{u,blood}$) and we may underestimate the inhibition potency by the prediction using $I_{u,blood}$ values. Thus

$$I_{u,tissue} = q \times I_{u,blood} \tag{16.5}$$

where q represents the ratio of the unbound concentration in cells to that in blood. The concentrative uptake of inhibitors can be predicted with in vitro experiments using isolated organ cells. The steady-state cell-to-plasma concentration ratio of inhibitors is measured in the presence and absence of metabolic inhibitors such as *p*-trifluoromethoxyphenylhydrazone (FCCP), rotenone, and sodium azide, and the ratio of these values minus 1 corresponds to the q value. If these data are not available, for the safety margin, intracellular unbound concentration is estimated by using 5–10 as a q value. For the prediction of drug–drug interactions at transporters in the small intestine, I should be the concentration of drugs in the intestinal lumen; however, it is very difficult to estimate. Recently, Tachibana et al. have proposed a new approach to estimating the apparent intestinal volume to predict drug–drug interaction in the small intestine (Tachibana et al., 2009). The details are described in Chapter 12. Efflux transporters such as MDR1 and BCRP are often suspected to be candidate targets for drug–drug interaction in the small intestine. In this case, the inhibition of efflux transporters results in an increase of intestinal availability ($F_a F_g$). However, we must keep in mind that if $F_a F_g$ is close to one under normal conditions, the inhibition of efflux transporters no longer causes the enhancement of intestinal absorption.

16.2.2 The Influence of the Pharmacokinetic Properties of Substrate Drugs on the Extent of Drug–Drug Interaction

The severity of drug–drug interaction largely depends on not only the concentration of inhibitors but also the pharmacokinetic properties of substrate drugs. In this section, the influence of the pharmacokinetic properties of substrate drugs on the extent of drug–drug interaction will be discussed.

16.2.2.1 Contribution of Each Transporter to the Overall Membrane Transport of Substrates

Several transporters are usually expressed on the same membrane of the same tissue. Because of the broad substrate specificities of transporters, one compound is often recognized by multiple transporters. The substrate specificities of transporters often overlap, and so one compound can frequently be recognized by multiple transporters. For example, OATP1B1 expressed on the basal side of hepatocytes can transport similar types of compounds such as OATP1B3 and OATP2B1. In that case, information regarding the contribution of specific transporters to the overall hepatic uptake is required to evaluate quantitative changes in the pharmacokinetics of drugs when the function of specific transporters is changed. Moreover, especially for lipophilic compounds that can easily penetrate the plasma membrane, passive diffusion is not negligible. Therefore, the intrinsic membrane transport clearance ($CL_{\mathrm{int,membrane}}$) is described as follows:

$$CL_{\mathrm{int,membrane}} = \sum_i CL_{\mathrm{transport},i} + CL_{\mathrm{passive}} \qquad (16.6)$$

where $CL_{\mathrm{transport},i}$ and CL_{passive} represent the active transport clearance mediated by "transporter i" and passive transport clearance, respectively. We can roughly discriminate the active transport clearance from passive clearance by measuring the decrease in the transport activity in the presence of metabolic inhibitors or at low temperatures.

When multiple transporters move substrate drugs in the same direction in parallel, if an inhibitor can inhibit multiple transporters with different inhibition potencies, the net degree of inhibition (R) is given by the following equation:

$$R = \frac{\sum_n PS_{\text{int},n}(+I)}{\sum_n PS_{\text{int},n}(-I)} = \sum_n R_n \cdot \frac{PS_{\text{int},n}(-I)}{\sum_n PS_{\text{int},n}(-I)} = \sum_n R_n \cdot f_n = \sum_n \frac{f_n}{1 + \frac{I_u}{K_{i,n}}} \tag{16.7}$$

where $PS_{\text{int},n}(+I)$, $PS_{\text{int},n}(-I)$, R_n, and f_n represent the intrinsic membrane transport clearance mediated by "transporter n" in the presence and absence of an inhibitor, R value for "transporter n," and the contribution of "transporter n" to the overall transport clearance, respectively. Therefore, considering this equation, when we predict the clinical drug–drug interaction, we must obtain information regarding the contribution of each transporter to the overall membrane transport (f_n), protein-unbound concentration of an inhibitor (I_u), and its inhibition constant for each transporter ($K_{i,n}$). Various compounds are generally accepted by several multispecific drug transporters when considering the f_n value. For example, OATP1B1 and OATP1B3 are responsible for the hepatic uptake of drugs, and OAT1 and OAT3 play an important role in their renal uptake, but the importance of each transporter in the uptake process depends on the substrates. Methods for estimating the contribution of each uptake transporter to the overall hepatic and renal uptake in humans have been established. Hirano et al. have proposed three different methods for estimating the relative contribution of OATP1B1 and OATP1B3 to the hepatic uptake of various drugs (Hirano et al., 2004, 2006). The first approach is to use reference compounds that are recognized by a specific transporter, and this is called the "RAF (relative activity factor) method." E-sul (estrone-3-sulfate) and CCK-8 (cholecystokinin octapeptide) can be used as reference compounds for OATP1B1 and OATP1B3, respectively. Thus, the ratio of the uptake clearance of reference compounds in human hepatocytes to that in expression systems is calculated and defined as "R_{act}" for OATP1B1 and 1B3, and by multiplying the R_{act} value by the uptake clearance of test compounds in expression systems (CL_{test}), we could estimate the uptake clearance of test compounds mediated by specific transporter in human liver. Assuming that the hepatic uptake clearance (CL_{hep}) can be explained by OATP1B1 and OATP1B3-mediated transport, the following equation should be correct:

$$CL_{\text{hep}} = R_{\text{act,OATP1B1}} \times CL_{\text{test,OATP1B1}} + R_{\text{act,OATP1B3}} \times CL_{\text{test,OATP1B3}} \tag{16.8}$$

Using the same methodology, the contribution of OAT1 and OAT3 to the renal uptake of compounds in rats has been estimated by using PAH and pravastatin as reference compounds for OAT1 and OAT3, respectively (Hasegawa et al., 2003). This type of approach has also been applied to human kidney slices (Nozaki et al., 2004).

The second approach is to estimate directly the ratio of the expression level of OATP1B1, 1B3, and 2B1 in human hepatocytes to that in expression systems by comparing their band densities in Western blot analyses and by estimating their relative contributions using that ratio instead of using R_{act} values as shown above. The third approach is to estimate the inhibitable portion of the uptake of test compounds in human hepatocytes in the presence of a specific inhibitor for each transporter. E-sul can be used as a specific inhibitor for OATP1B1. For example, the uptake of pitavastatin was completely inhibited by 30 μM E-sul, indicating a major role for OATP1B1 in its hepatic uptake, whereas that of telmisartan (OATP1B3-specific substrate) was not inhibited by E-sul. We have attempted to estimate the contribution of OATP1B1 and OATP1B3 to the hepatic uptake of multiple anionic drugs, and though some anionic drugs shared the same pharmacokinetic properties by which they are efficiently accumulated in the liver, the relative contribution of each transporter depends on individual substrates. From our estimation, pitavastatin and rosuvastatin are taken up mainly via OATP1B1 (Hirano et al., 2004; Kitamura et al., 2008), and fexofenadine and telmisartan are transported predominantly by OATP1B3 (Shimizu et al., 2005; Ishiguro et al., 2006).

16.2.2.2 Importance of the Transport Process in Overall Intrinsic Clearance

The detoxification process in each organ consists of some intrinsic processes such as uptake of compounds from blood to tissue cells, metabolism at intracellular compartment, backflux from cells to blood, and excretion from cells into another compartment such as bile (liver) and urine (kidney) (Fig. 16.2). Then, the overall intrinsic organ clearance ($CL_{\text{int},all}$) can be described by the following equation (Shitara et al., 2006):

$$CL_{\text{int},all} = CL_{\text{inf}} \times \frac{CL_{\text{metab}} + CL_{\text{bile}}}{(CL_{\text{metab}} + CL_{\text{bile}}) + CL_{\text{back}}} \tag{16.9}$$

where CL_{inf}, CL_{metab}, CL_{bile}, and CL_{back} represent the intrinsic clearance for influx from blood, intracellular metabolism, efflux from cells to bile or urine, and backflux to blood, respectively. If the sum of the intrinsic metabolic clearance and efflux clearance (CL_{metab} + CL_{bile}) is much higher than intrinsic backflux clearance (CL_{back}), overall intrinsic organ clearance is very close to CL_{inf}, and influx clearance is solely determined by the overall intrinsic clearance:

$$CL_{\text{int},all} \sim CL_{\text{inf}} \tag{16.10}$$

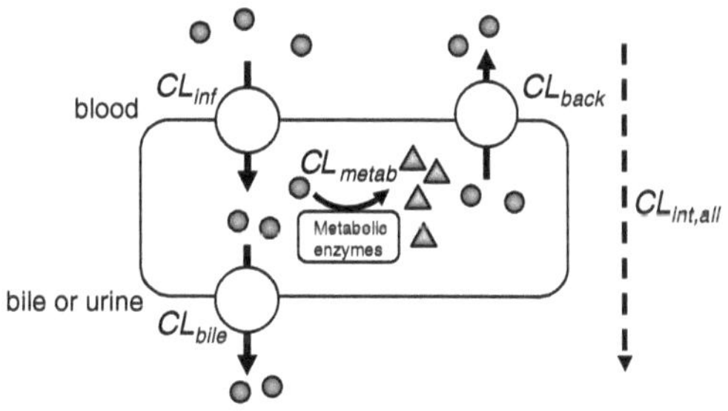

Fig. 16.2 The elementary steps comprising overall intrinsic clearance

On the other hand, if ($CL_{\text{metab}} + CL_{\text{bile}}$) is much lower than CL_{back}, overall intrinsic clearance is determined by the activity of all processes as shown below:

$$CL_{\text{int},all} \sim CL_{\text{inf}} \times \frac{CL_{\text{metab}} + CL_{bile}}{CL_{\text{back}}} \tag{16.11}$$

When the membrane transport clearance is very rapid and the membrane transport on the blood side is symmetric ($CL_{\text{inf}} = CL_{\text{back}}$), Equation (16.9) can be approximated by the following equation:

$$CL_{\text{int},all} \sim CL_{\text{metab}} + CL_{\text{bile}} \tag{16.12}$$

This equation is applicable to lipophilic drugs with rapid membrane permeability. Considering these equations, when a drug is recognized by influx transporters, it is important to recognize that even if a compound is finally eliminated by extensive metabolism, overall intrinsic organ clearance is determined not by metabolic intrinsic clearance but by influx clearance under certain conditions as discussed above. Figure 16.3 shows the simulation curves showing the impact of the intrinsic clearance for each process on a decrease in the overall hepatic intrinsic clearance when metabolism or an uptake process is inhibited. Assuming that this drug is taken up into hepatocytes by an uptake transporter and then extensively metabolized, if the sum of intrinsic metabolic clearance and efflux clearance ($CL_{\text{metab}} + CL_{\text{bile}}$) is higher than intrinsic backflux clearance (CL_{back}), the inhibition of uptake clearance (CL_{inf}) directly affects the decrease in the overall intrinsic clearance ($CL_{\text{int},\,all}$), but the inhibition of metabolism does not greatly change the $CL_{\text{int},\,all}$ value. On the other hand, if ($CL_{\text{metab}} + CL_{\text{bile}}$) is much lower than CL_{back}, the inhibition of uptake clearance and metabolism equally affects the decrease in the $CL_{\text{int},\,all}$ value. Assuming that an orally administered drug is completely absorbed from the small intestine and predominantly excreted by the liver, its plasma area under the curve (AUC) is described as follows:

$$\text{AUC}_{\text{plasma}} = \frac{\text{Dose}}{f_{\text{B}} \cdot CL_{\text{int},all}} = \frac{\text{Dose}}{f_{\text{B}} \cdot CL_{\text{inf}} \cdot \frac{CL_{\text{metab}}+CL_{\text{bile}}}{CL_{\text{back}}+CL_{\text{metab}}+CL_{\text{bile}}}} \tag{16.13}$$

where f_{B} represents protein-unbound fraction in blood.

AUC in the liver is estimated by the following equation:

$$\begin{aligned} \text{AUC}_{\text{liver}} &= \frac{CL_{\text{inf}}}{CL_{\text{back}} + CL_{\text{metab}} + CL_{\text{bile}}} \cdot \text{AUC}_{\text{plasma}} \\ &= \frac{CL_{\text{inf}}}{CL_{\text{back}} + CL_{\text{metab}} + CL_{\text{bile}}} \cdot \frac{\text{Dose}}{f_{\text{B}} \cdot CL_{\text{inf}} \cdot \frac{CL_{\text{metab}}+CL_{\text{bile}}}{CL_{\text{back}}+CL_{\text{metab}}+CL_{\text{bile}}}} = \frac{\text{Dose}}{f_{\text{B}} \cdot (CL_{\text{metab}} + CL_{\text{bile}})} \end{aligned} \tag{16.14}$$

Therefore, if the function of uptake intrinsic clearance (CL_{inf}) is decreased, plasma AUC is inversely proportional to CL_{inf}, while liver AUC is not affected.

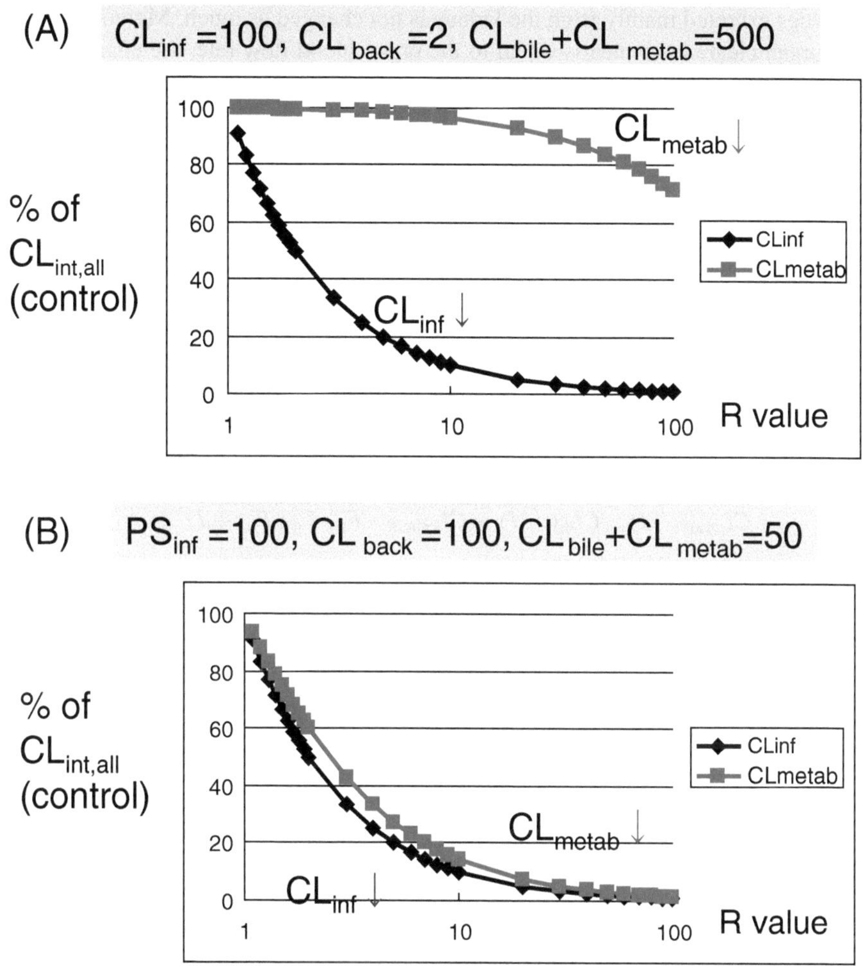

Fig. 16.3 The impact of a change in the intrinsic clearance for uptake and metabolism on the decrease in the overall hepatic intrinsic clearance. *X*- and *Y*-axes represent the *R* value for uptake (CL_{inf}) or metabolism (CL_{metab}) and the percentage of $CL_{\mathrm{int},all}$ value in the absence of inhibitors, respectively

On the other hand, a decrease in the function of metabolism or biliary sequestration directly causes an increase in liver AUC, but if drug uptake is the rate-determining process of the hepatic intrinsic clearance, plasma AUC is not changed.

16.2.2.3 Effect of the Absolute Value of Organ Clearance and the Fraction of Substrates Excreted from the Liver and Kidney on the Extent of Drug–Drug Interaction

Drugs are mainly eliminated through the liver and kidney, and thus, even if hepatic clearance is greatly decreased, it is intuitively accepted that the pharmacokinetics of drugs excreted mainly from the kidney is not changed as much. Moreover, when the organ clearance is nearly equal to the organ blood flow rate, the change in the intrinsic organ clearance does not affect the total organ clearance. The contribution of hepatic clearance to the total clearance is defined as f_h:

$$f_h = \frac{CL_h}{CL_h + CL_{others}} \tag{16.15}$$

where CL_h and CL_{others} represent hepatic clearance and extrahepatic (e.g., renal) clearance, respectively.

When a drug is administered orally, the first-pass effect in the intestine and liver must be considered. Assuming that only hepatic clearance is inhibited by inhibitors, the ratio of oral clearance in the presence of inhibitors ($CL_{oral}(+I)$) to that in their absence ($CL_{oral}(-I)$) can be calculated using Equation (16.16).

$$\begin{aligned}\frac{CL_{oral}(+I)}{CL_{oral}(-I)} &= \frac{CL_h(+I) + CL_{others}}{CL_h(-I) + CL_{others}} \cdot \frac{F_a \cdot F_g \cdot F_h(-I)}{F_a \cdot F_g \cdot F_h(+I)} \\ &= \left(f_h \cdot \frac{CL_h(+I)}{CL_h(-I)} + (1 - f_h)\right) \cdot \frac{F_h(-I)}{F_h(+I)}\end{aligned} \tag{16.16}$$

Here, F_a, F_g, and F_h are the fractions absorbed through the gastrointestinal tract (F_a), the intestinal availability (F_g), and the hepatic availability in the presence of inhibitors ($F_h(+I)$) and in their absence ($F_h(-I)$).

If a substrate drug exhibits high intrinsic hepatic clearance ($Q_h << f_{uB} \cdot CL_{h,int}$, Q_h: hepatic blood flow, f_{uB}: fraction unbound in the blood) even in the presence of inhibitors, Equation (16.16) would be approximated to Equation (16.17), because $CL_h \approx Q_h$ and $F_h \approx Q_h/(f_{uB} \cdot CL_{h,int})$:

$$\frac{CL_{oral}(+I)}{CL_{oral}(-I)} = \frac{CL_{h,int}(+I)}{CL_{h,int}(-I)} \tag{16.17}$$

On the other hand, if a substrate exhibits low intrinsic hepatic clearance ($Q_h >> f_{uB} \cdot CL_{h,int}$), Equation (16.16) would be approximated to Equation (16.18), because $CL_h \approx f_{uB} \cdot CL_{h,int}$, and $F_h \approx 1$:

$$\begin{aligned}\frac{CL_{oral}(+I)}{CL_{oral}(-I)} &= \left(f_h \cdot \frac{CL_h(+I)}{CL_h(-I)} + (1 - f_h)\right) \cdot \frac{F_h(-I)}{F_h(+I)} \\ &= f_h \cdot \frac{CL_{h,int}(+I)}{CL_{h,int}(-I)} + (1 - f_h)\end{aligned} \tag{16.18}$$

Figure 16.4 shows a curve to simulate the influence of f_h on the change in the plasma AUC of substrate drugs whose hepatic clearance is low compared with hepatic blood flow in the presence of inhibitors of hepatic clearance. As shown in this graph, it is noted that a small difference in the f_h value sometimes produces

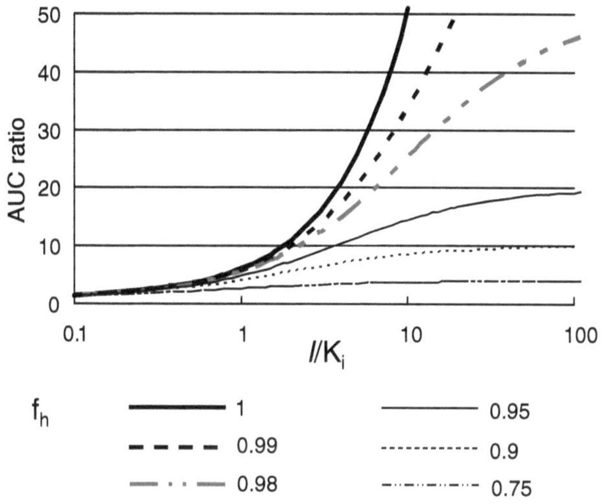

Fig. 16.4 The influence of f_h on the change in the plasma AUC of substrate drugs whose hepatic clearance is low compared with hepatic blood flow in the presence of inhibitors for hepatic clearance. *X*- and *Y*-axes represent the I/K_i value and the ratio of plasma AUC in the presence of inhibitors to that in their absence when the f_h value is changed from 0.75 to 1, respectively

a large variation in the change in the AUC; thus, the estimation of the f_h value is also important for accurate prediction of the extent of drug–drug interaction.

As mentioned previously, if the overall intrinsic organ clearance of transporter substrates approximates their uptake intrinsic clearance, their hepatic and renal clearance can be predicted from in vitro uptake experiments using isolated hepatocytes and kidney slices. Watanabe et al. have recently attempted such a type of prediction for several transporter substrates in rats and have demonstrated that hepatic clearance, renal clearance, and the fraction excreted into urine are well predicted from in vitro uptake assays for most test compounds except those with low renal clearance, possibly because of the involvement of reabsorption (Watanabe et al., 2009b). This prediction method can also be applied to humans by using human cryopreserved hepatocytes and human kidney slices (Watanabe et al., unpublished observation).

16.2.3 The Impact of the Inhibition of Uptake and Efflux Processes on Change in the Pharmacokinetics of Substrate Drugs

Some drugs are excreted into bile as an unchanged form without any metabolism such as some kinds of HMG-CoA reductase inhibitors [so-called statins (pitavastatin, rosuvastatin, pravastatin)] and angiotensin II receptor antagonists [so-called sartans (olmesartan, valsartan)]. In that case, both uptake and efflux transporters are cooperatively involved in their transcellular biliary transport in hepatocytes and so it may be possible that the function of both uptake and efflux transporters is inhibited by drugs. For avoiding the false-negative prediction, the maximum effect of drug–drug interaction mediated by uptake and efflux transporters should be considered (Maeda and Sugiyama, 2007). If the hepatic clearance is very close to the hepatic blood flow ($f_B CL_{int,h} >> Q_h$), the inhibitory effect of hepatic intrinsic clearance does not affect the hepatic clearance so much, but in the opposite case ($f_B CL_{int,h} << Q_h$), the hepatic clearance approximates $f_B CL_{int,h}$ and the inhibition of hepatic intrinsic clearance directly affects the decrease in the hepatic clearance. Thus, for estimating the maximum inhibitory effect, decrease in the hepatic intrinsic clearance is considered to be equal to that in the hepatic clearance. Then, as mentioned above, the hepatic intrinsic clearance can be described as in Equation (16.9). If the hepatic uptake is the rate-determining process for overall intrinsic clearance, the intrinsic clearance approximates hepatic influx clearance (Equation 16.10) and the inhibition of efflux transport does not affect the overall intrinsic clearance. But if not the case (Equation 16.11), the inhibition of both uptake and efflux processes directly influence the decrease in the overall intrinsic clearance. When the inhibition of uptake and efflux transport simultaneously occurs and R values ($=1+I/K_i$) for the inhibition of uptake and efflux transport are R_{uptake} and $R_{excretion}$, respectively, the degree of inhibition of overall intrinsic clearance (R_{net}) is described as follows;

$$R_{uptake} \leq R_{net} \leq R_{uptake} \times R_{excretion} \tag{16.19}$$

Ueda et al. have predicted the extent of drug–drug interaction between methotrexate and probenecid in rats by using the method mentioned above (Ueda et al., 2001). They estimated the R values for uptake and efflux processes by using the results of inhibition study using rat isolated hepatocytes and bile canalicular membrane vesicles, respectively. Then, the maximum R_{net} value was estimated as $R_{uptake} \times R_{excretion}$. As a result, the degree of reduction in the hepatic clearance was overestimated by a simple calculation of the maximum R_{net} value.

16.3 *Examples of the Prediction of Drug–Drug Interaction from* In Vitro *Data*

Hirano et al. have investigated the inhibition potencies of several drugs for OATP1B1 by using a gene expression system and have calculated R values for each drug using clinical concentrations of drugs in humans, assuming that substrates are cleared only via OATP1B1 and that uptake is a rate-limiting step for overall hepatic intrinsic clearance (Hirano et al., 2006). This type of analysis was also performed for OATP1B3 (Matsushima et al., 2008). They found candidate inhibitors for OATP1B1 with an R value > 2.0, which are likely to inhibit OATP1B1 in clinical situations, include rifamycin SV, rifampicin, cyclosporin A, clarithromycin, indinavir, and ritonavir, while those for OATP1B3 are rifampicin and cyclosporin A.

Cyclosporin A inhibits both multiple metabolic enzymes and transporters. Thus, the main target of drug–drug interaction with cyclosporin A and several transporter substrates is discussed by using calculated R values for each molecule. Table 16.1 indicates reported drug–drug interactions with cyclosporin A in humans

Table 16.1 The reported drug–drug interactions with cyclosporin A in humans and predicted *R* value for OATP1B1 and major metabolic enzymes

Substrate drugs	Dose of cyclosporin A (mg)	Observed AUC ratios (fold increase)	Predicted *R* values for OATP1B1	Predicted *R* values for major CYPs
Pitavastatin	150	4.6	3.46	–
Rosuvastatin	200	7.1	4.28	–
Cerivastatin	200	3.8	4.28	~1 (2C8)[a] 1.6–6.6 (3A4)
Fluvastatin	344	1.9–3.5	6.64	1.05–1.48 (2C9)[a] 2.0–11 (3A4)
Pravastatin	300	5–7.9	5.92	–
Atorvastatin	100	7.4	2.64	1.3–3.8 (3A4)[a]
Repaglinide	100	2.44	2.64	~1 (2C8)[a] 1.3–3.8 (3A4)
Bosentan	300	1.97	5.92	1.04–1.42 (2C9)[a] 1.8–9.4 (3A4)

Predicted *R* values are calculated by using the maximum unbound concentration of cyclosporin A at each dose based on Equation (16.4) and the in vitro K_i value for each molecule.
[a]The most dominant metabolic pathway for each substrate drug.

and predicted *R* values for OATP1B1 and major metabolic enzymes. In the calculation of *R* values, maximum protein-unbound concentration of cyclosporin A calculated by Equation (16.4) and reported in vitro K_i values for OATP1B1 (0.24 μM), CYP3A4 (1.4 μM), CYP2C8 (no inhibition), and CYP2C9 (28 μM) were used. When calculating the *R* values for metabolic enzymes, to provide a safety margin, the ratio of the unbound concentration in cells to that in blood (*q* value) is assumed to be 1 to 10. In the case of statins without any metabolism (pitavastatin, rosuvastatin, and pravastatin), predicted *R* values are almost comparable or even lower than the observed increase in the plasma AUC. Cerivastatin and repaglinide are thought to be mainly metabolized by CYP2C8, but cyclosporin A cannot inhibit CYP2C8-mediated metabolism, and the predicted *R* value for OATP1B1 can explain the increase in their plasma AUC; thus, OATP1B1 can be a major target for the drug–drug interaction of cerivastatin and repaglinide. Fluvastatin and bosentan are mainly metabolized by CYP2C9. In each case, the predicted *R* value for CYP2C9 does not exceed 1.5 even if the *q* value is set to 10, suggesting that CYP2C9 is not a main target for these interactions. The *R* values for OATP1B1 are larger than the observed increase in plasma AUC. This is maybe caused by the contribution of passive diffusion and transporters other than OATP1B1 to their hepatic uptake. On the other hand, atorvastatin is mainly metabolized by CYP3A4, and the predicted *R* value for CYP3A4 is very similar to that for OATP1B1; thus, we cannot conclude which is a possible target for this interaction from this data alone. However, previous reports have indicated that itraconazole, a potent CYP3A4 inhibitor, largely increased the plasma AUC of simvastatin and lovastatin, but it slightly increased that of atorvastatin, suggesting that CYP3A4 is not a major target. The observed increase in the AUC is larger than that for the predicted *R* value for OATP1B1, probably because the

F_aF_g value for atorvastatin is relatively low among these drugs, and the inhibition of intestinal CYP3A4 and MDR1 results in an increase in F_aF_g.

16.4 Web-Based Transporter-Mediated DDI Database

Now that many human transporters, as well as those in rodents, have been cloned and characterized, and the clinical importance of these transporters has been well indicated by several reports presenting the effect of genetic polymorphisms and interacting drugs on the pharmacokinetics of substrate drugs, researchers in drug discovery and development and clinical pharmacists wish for updated information regarding drug transporters. However, because many papers regarding transporters are continuously being published and many experimental methods for functional characterization of the same transporters (including cellular uptake, uptake into membrane vesicles, transcellular transport, ATPase assay) and methods of data presentation (including percentage of inhibition, percentage of uptake in control cells, K_i, IC_{50}) are reported, this can cause confusion for researchers without extensive knowledge of multiple transporters, which restricts the use of variable information for the efficient development and appropriate clinical use of drugs. Therefore, we have constructed a Web-based comprehensive database for drug transporters, named "TP-search" (http://www.TP-Search.jp/) (Ozawa et al., 2004). In this database, the following information for more than 90 kinds of transporters in humans and rodents can be searched. Currently, the data collected from various international journals published until 2007 are available in this database:

1. Substrates/inhibitors/inducers
2. Tissue distribution of drug transporters
3. Genetic polymorphisms of drug transporters
4. Genetic disease caused by drug transporters
5. Clinical drug–drug interactions mediated by transporters
6. Knockout mice and gene-deficient mice for transporters
7. Expression and function of transporters under pathophysiological condition
8. Gender difference in the expression of transporters

Substrates, inhibitors, and inducers for each transporter can be searched for by transporter name or compound name. When information regarding compounds for specific transporters is required, transporter name, species, and category of compound are selected. After sending a query to this database, one can obtain a list of compounds with chemical structure, K_m, K_i or IC_{50} values (if available), and experimental methods and materials (Fig. 16.5). A link to the PubMed database is also attached to each record so that detailed information may be readily obtained. This database includes information regarding nonsubstrates or noninhibitors for each transporter which have been experimentally confirmed. To determine whether a specific compound is a substrate, inhibitor, or inducer of multiple drug transporters, the compound name is input, and the species and category of compound are selected. Thus, the same information described above can be obtained. Currently, more than 6700 records are included in this database.

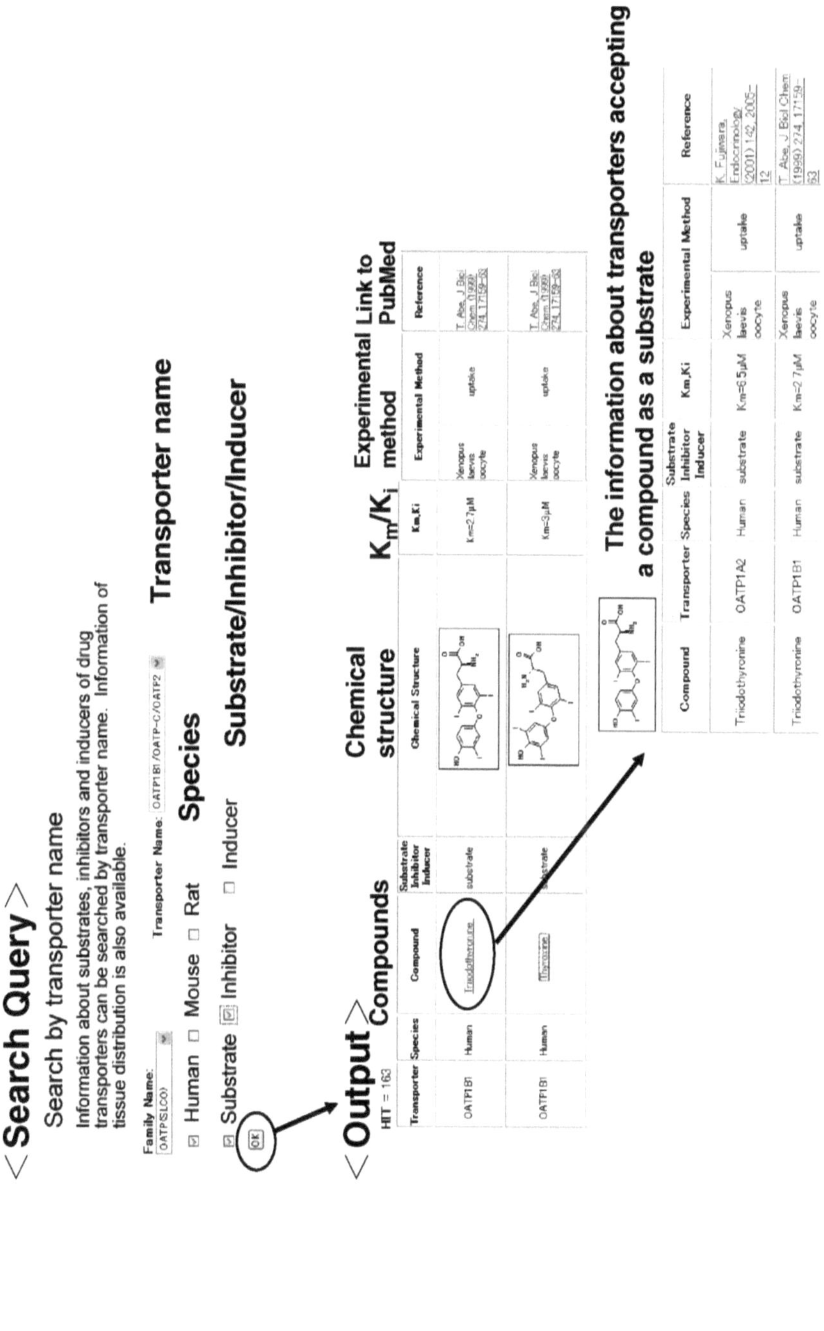

Fig. 16.5 The display image of a search query and data output for substrates, inhibitors, and inducers of a transporter in TP-search

Information regarding in vivo drug–drug interactions that are thought to be mediated by transporters can also be searched in TP-search. The compound name is input, and the species and category of compound are selected. After sending the query to this database, one can obtain a list for a pair of compounds used in each drug–drug interaction study, the change in the pharmacokinetic parameters, a possible interaction site as suggested by the authors, and a link to the PubMed database (Fig. 16.6). If both metabolic enzymes and transporters are involved in the drug–drug interaction, the names of metabolic enzymes and transporters are shown together. Currently, information regarding detailed dosing conditions such as dose and interval is lacking in this database. In the near future, we hope to include detailed dosing conditions and the predicted results for the potency of transporter-mediated drug–drug interaction from in vitro data (e.g., *R* value) in this database to facilitate the prediction of drug–drug interactions mediated by transporters.

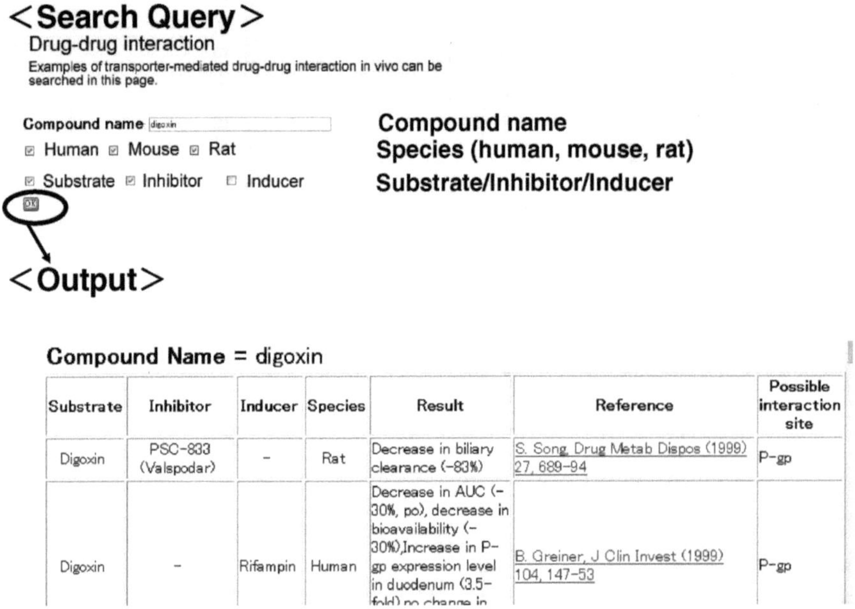

Fig. 16.6 The display image of a search query and data output for in vivo transporter-mediated drug–drug interactions in TP-search

16.5 In Silico Prediction of the Time Profiles of Substrate Drugs in the Presence of Inhibitors Using Physiologically Based Pharmacokinetic Modeling

Recently, a method for quantitatively predicting a pharmacokinetic alteration caused by inhibition of hepatic CYP enzymes based on a physiologically based pharmacokinetic (PBPK) model was introduced [see also Chapter 12]. This method can

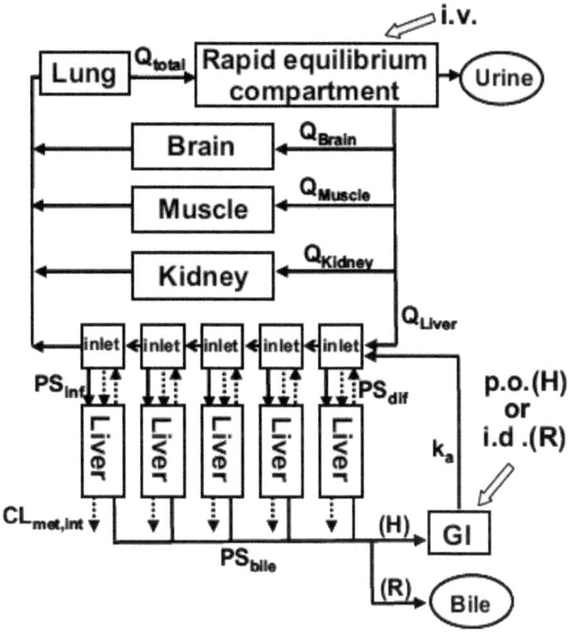

Fig. 16.7 A physiologically based pharmacokinetic model to predict the concentration–time profiles of pravastatin. The liver compartment was divided into five compartments to mimic the dispersion model. In this model, Q represents the blood flow rate, PS_{inf}, PS_{dif}, $CL_{met,int}$, and PS_{inf} represent the active hepatic uptake clearance, the passive diffusion clearance, the metabolic clearance, and the biliary clearance, respectively

also be applied to transporter-mediated DDI. Here a method to predict transporter-mediated DDI by the same method will be introduced.

Watanabe et al. (2009a) reported a simulation analysis of the transporter-mediated clearance of pravastatin in rats and humans using a PBPK model and attempted to simulate a pharmacokinetic alteration caused by inhibitions of hepatic uptake and biliary excretion via transporters. Figure 16.7 shows the PBPK model used in this analysis. The compartments for the liver and the inlet to the liver are divided into five, which produces the closest hepatic availability to that obtained in the dispersion model. Among parameters shown in Fig. 16.7, PS_{bile} and $CL_{met,int}$ were calculated using in vivo data, while PS_{dif} was assumed to be the same as the in vitro-estimated PS_{dif}. PS_{inf} was a hybrid parameter of in vivo-estimated parameters and in vitro-estimated PS_{dif}. In addition, the initial distribution volume was estimated from the in vivo study, and the observed tissue-to-plasma concentration ratios were used. Subsequently, the observed plasma concentration of pravastatin was fitted to the PBPK model to obtain its absorption rate constant. By using these parameters, the plasma concentration and biliary excretion of pravastatin after intravenous or oral administration were simulated, and simulated curves were well fitted to the observed data, although deviation was observed between the simulated and observed values for its plasma concentration at the terminal phase after oral administration. The simulated tissue concentration was also close to the in vivo-observed concentration. By the comparison of in vivo-estimated parameters and in vitro-obtained parameters, scaling factors, which are used for the in vitro–in vivo extrapolation as the ratios of in vivo intrinsic clearances to in vitro-estimated clearances, were obtained. By using these scaling factors and in vitro-estimated parameters using human cryopreserved hepatocytes and bile canalicular membrane vesicles, the plasma and tissue concentrations of pravastatin in humans can be simulated. Thus, it was shown that model-based analysis can help to simulate plasma concentrations of drugs accurately with in vitro-obtained parameters using human tissue samples and the scaling factors obtained from animal studies.

These findings also suggest that it is possible to simulate pharmacokinetic alterations caused by transporter inhibition using PBPK model-based analyses. We also showed the simulation of the plasma concentration of pravastatin when its hepatic uptake or biliary excretion is affected (Fig. 16.8). This simulation shows that the alteration in PS_{inf} markedly changes the plasma concentration with a minimal effect on the liver concentration. On the other hand, the altered PS_{bile} results in a marked change in the liver concentration with a minor effect on the plasma concentration. Thus, using the PBPK model, a sensitivity analysis can be performed. This analysis suggests that it is possible to predict the pharmacokinetic alterations caused by transporter inhibition.

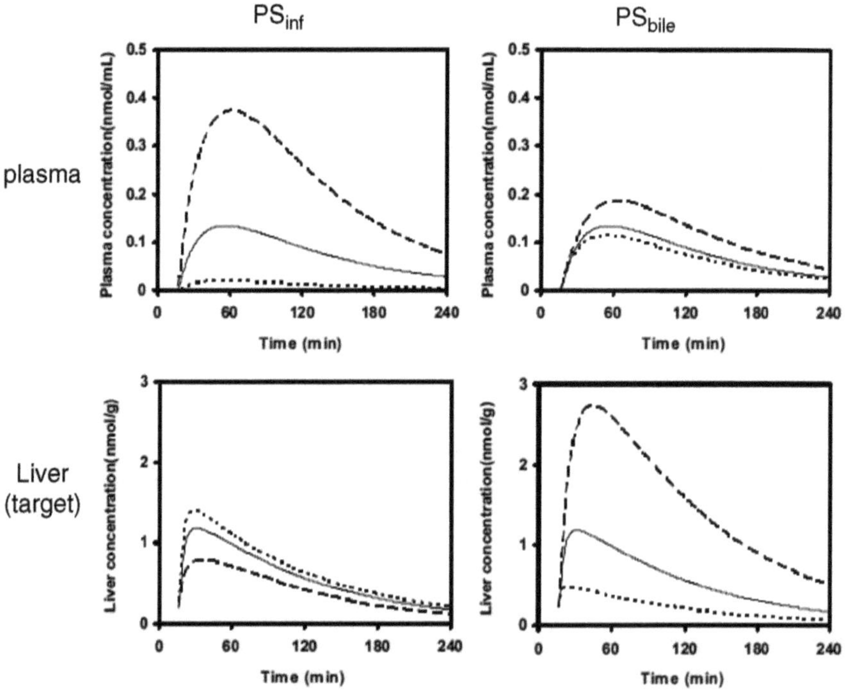

Fig. 16.8 Effects of changes in transporter activity on the time profiles of plasma and liver concentrations of pravastatin in humans. Plasma and liver concentrations after oral administration (40 mg) of pravastatin were simulated using the PBPK model with varying hepatic transport activities over a 1/3- to 3-fold range of the initial values. *Solid*, *dotted*, and *dashed lines* represent ×1, ×1/3, and ×3 the initial transporter activity for uptake and biliary excretion by the liver

Here, an attempt to predict quantitatively the extent of transporter-mediated DDI based on a PBPK model will be introduced. For this prediction, the PBPK model shown in Fig. 16.9 can be used. As mentioned above, a model-based simulation of the transporter-mediated clearance is possible. Thus, a database of transporter substrates and inhibitors, which contains their pharmacokinetic parameters and transporter inhibition constants, will help the quantitative prediction of transporter-mediated DDIs. At the present time, this method has not been validated because there is scant information regarding transporter-mediated DDIs. However, recently, increasing numbers of transporter-mediated DDIs have been reported. Thus, this method may be validated using multiple combinations of substrates and inhibitors of transporters and utilized for drug development. In future, this type of prediction using PBPK modeling may be performed on the Web site using the database for pharmacokinetic parameters and inhibition constants of substrates and inhibitors to generate the parameters for PBPK modeling.

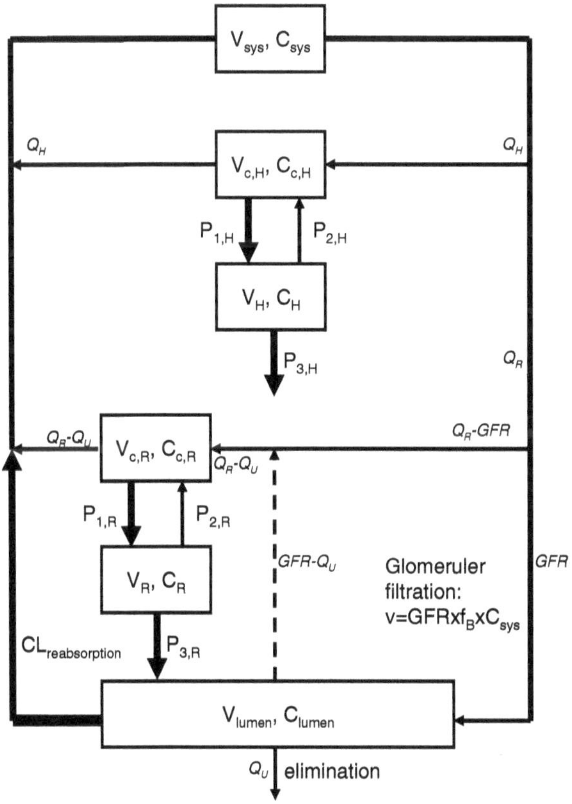

Fig. 16.9 A proposed physiologically based pharmacokinetic model for the simulation of time profile of plasma concentration of drugs considering the involvement of transporters. A physiologically based pharmacokinetic model to simulate a plasma concentration–time profile of drugs whose clearance is affected by hepatic and renal transporters. Transporter-mediated processes are shown by *bold arrows*

References

Campbell SD, de Morais SM and Xu JJ (2004) Inhibition of human organic anion transporting polypeptide OATP 1B1 as a mechanism of drug-induced hyperbilirubinemia. *Chem Biol Interact* **150**:179–187.

Dresser GK, Bailey DG, Leake BF, Schwarz UI, Dawson PA, Freeman DJ and Kim RB (2002) Fruit juices inhibit organic anion transporting polypeptide-mediated drug uptake to decrease the oral availability of fexofenadine. *Clin Pharmacol Ther* **71**:11–20.

Fattinger K, Funk C, Pantze M, Weber C, Reichen J, Stieger B and Meier PJ (2001) The endothelin antagonist bosentan inhibits the canalicular bile salt export pump: a potential mechanism for hepatic adverse reactions. *Clin Pharmacol Ther* **69**:223–231.

Fromm MF, Kim RB, Stein CM, Wilkinson GR and Roden DM (1999) Inhibition of P-glycoprotein-mediated drug transport: A unifying mechanism to explain the interaction between digoxin and quinidine. *Circulation* **99**:552–557.

Funk C, Ponelle C, Scheuermann G and Pantze M (2001) Cholestatic potential of troglitazone as a possible factor contributing to troglitazone-induced hepatotoxicity: in vivo and in vitro interaction at the canalicular bile salt export pump (Bsep) in the rat. *Mol Pharmacol* **59**: 627–635.

Gerber JG, Rosenkranz SL, Fichtenbaum CJ, Vega JM, Yang A, Alston BL, Brobst SW, Segal Y and Aberg JA (2005) Effect of efavirenz on the pharmacokinetics of simvastatin, atorvastatin, and pravastatin: results of AIDS Clinical Trials Group 5108 Study. *J Acquir Immune Defic Syndr* **39**:307–312.

Giacomini KM and Sugiyama Y (2005) Membrane transporters and drug response, in: *Goodman & Gilman's The Pharmacological Basis of Therapeutics* 11th edition (Brunton LL, Lazo JS and Parker KL eds), pp 41–70, McGraw-Hill, New York.

Greiner B, Eichelbaum M, Fritz P, Kreichgauer HP, von Richter O, Zundler J and Kroemer HK (1999) The role of intestinal P-glycoprotein in the interaction of digoxin and rifampin. *J Clin Invest* **104**:147–153.

Hasegawa M, Kusuhara H, Endou H and Sugiyama Y (2003) Contribution of organic anion transporters to the renal uptake of anionic compounds and nucleoside derivatives in rat. *J Pharmacol Exp Ther* **305**:1087–1097.

Hirano M, Maeda K, Shitara Y and Sugiyama Y (2004) Contribution of OATP2 (OATP1B1) and OATP8 (OATP1B3) to the hepatic uptake of pitavastatin in humans. *J Pharmacol Exp Ther* **311**:139–146.

Hirano M, Maeda K, Shitara Y and Sugiyama Y (2006) Drug–drug interaction between pitavastatin and various drugs via OATP1B1. *Drug Metab Dispos* **34**:1229–1236.

Ishiguro N, Maeda K, Kishimoto W, Saito A, Harada A, Ebner T, Roth W, Igarashi T and Sugiyama Y (2006) Predominant contribution of OATP1B3 to the hepatic uptake of telmisartan, an angiotensin II receptor antagonist, in humans. *Drug Metab Dispos* **34**:1109–1115.

Ito K, Iwatsubo T, Kanamitsu S, Ueda K, Suzuki H and Sugiyama Y (1998) Prediction of pharmacokinetic alterations caused by drug–drug interactions: metabolic interaction in the liver. *Pharmacol Rev* **50**:387–412.

Kitamura S, Maeda K, Wang Y and Sugiyama Y (2008) Involvement of multiple transporters in the hepatobiliary transport of rosuvastatin. *Drug Metab Dispos* **36**:2014–2023.

Kostrubsky VE, Strom SC, Hanson J, Urda E, Rose K, Burliegh J, Zocharski P, Cai H, Sinclair JF and Sahi J (2003) Evaluation of hepatotoxic potential of drugs by inhibition of bile-acid transport in cultured primary human hepatocytes and intact rats. *Toxicol Sci* **76**: 220–228.

Kusuhara H and Sugiyama Y (2001) Drug–drug interactions involving the membrane transport process, in: *Drug–drug interactions* (Rodrigues AD ed), pp 123–188, Marcel Dekker, Basel.

Kyrklund C, Backman JT, Neuvonen M and Neuvonen PJ (2004) Effect of rifampicin on pravastatin pharmacokinetics in healthy subjects. *Br J Clin Pharmacol* **57**:181–187.

Lilja JJ, Backman JT, Laitila J, Luurila H and Neuvonen PJ (2003) Itraconazole increases but grapefruit juice greatly decreases plasma concentrations of celiprolol. *Clin Pharmacol Ther* **73**:192–198.

Lilja JJ, Raaska K and Neuvonen PJ (2005) Effects of orange juice on the pharmacokinetics of atenolol. *Eur J Clin Pharmacol* **61**:337–340.

Maeda K and Sugiyama Y (2007) In vitro-in vivo scale-up of drug transport activities, in: *Drug transporters* (You G and Morris ME eds), pp 557–588, John Wiley & Sons, Inc., New Jersey.

Maiche AG (1986) Acute renal failure due to concomitant action of methotrexate and indomethacin. *Lancet* **1**:1390.

Matsushima S, Maeda K, Inoue K, Ohta KY, Yuasa H, Kondo T, Nakayama H, Horita S, Kusuhara H and Sugiyama Y (2009) The inhibition of human multidrug and toxin extrusion 1 is involved in the drug–drug interaction caused by cimetidine. *Drug Metab Dispos* **37**: 555–559.

Matsushima S, Maeda K, Ishiguro N, Igarashi T and Sugiyama Y (2008) Investigation of the inhibitory effects of various drugs on the hepatic uptake of fexofenadine in humans. *Drug Metab Dispos* **36**:663–669.

McLeod HL (1998) Clinically relevant drug–drug interactions in oncology. *Br J Clin Pharmacol* **45**:539–544.

Muck W, Mai I, Fritsche L, Ochmann K, Rohde G, Unger S, Johne A, Bauer S, Budde K, Roots I, Neumayer HH and Kuhlmann J (1999) Increase in cerivastatin systemic exposure after single and multiple dosing in cyclosporine-treated kidney transplant recipients. *Clin Pharmacol Ther* **65**:251–261.

Nozaki Y, Kusuhara H, Endou H and Sugiyama Y (2004) Quantitative evaluation of the drug–drug interactions between methotrexate and nonsteroidal anti-inflammatory drugs in the renal uptake process based on the contribution of organic anion transporters and reduced folate carrier. *J Pharmacol Exp Ther* **309**:226–234.

Ozawa N, Shimizu T, Morita R, Yokono Y, Ochiai T, Munesada K, Ohashi A, Aida Y, Hama Y, Taki K, Maeda K, Kusuhara H and Sugiyama Y (2004) Transporter database, TP-Search: a web-accessible comprehensive database for research in pharmacokinetics of drugs. *Pharm Res* **21**:2133–2134.

Sadeque AJ, Wandel C, He H, Shah S and Wood AJ (2000) Increased drug delivery to the brain by P-glycoprotein inhibition. *Clin Pharmacol Ther* **68**:231–237.

Sahi J, Sinz MW, Campbell S, Mireles R, Zheng X, Rose KA, Raeissi S, Hashim MF, Ye Y, de Morais SM, Black C, Tugnait M and Keller LH (2006) Metabolism and transporter-mediated drug–drug interactions of the endothelin-A receptor antagonist CI-1034. *Chem Biol Interact* **159**:156–168.

Schinkel AH, Smit JJ, van Tellingen O, Beijnen JH, Wagenaar E, van Deemter L, Mol CA, van der Valk MA, Robanus-Maandag EC, te Riele HP, et al. (1994) Disruption of the mouse mdr1a P-glycoprotein gene leads to a deficiency in the blood-brain barrier and to increased sensitivity to drugs. *Cell* **77**:491–502.

Schwarz UI, Seemann D, Oertel R, Miehlke S, Kuhlisch E, Fromm MF, Kim RB, Bailey DG and Kirch W (2005) Grapefruit juice ingestion significantly reduces talinolol bioavailability. *Clin Pharmacol Ther* **77**:291–301.

Shimizu M, Fuse K, Okudaira K, Nishigaki R, Maeda K, Kusuhara H and Sugiyama Y (2005) Contribution of OATP (organic anion-transporting polypeptide) family transporters to the hepatic uptake of fexofenadine in humans. *Drug Metab Dispos* **33**:1477–1481.

Shitara Y, Horie T and Sugiyama Y (2006) Transporters as a determinant of drug clearance and tissue distribution. *Eur J Pharm Sci* **27**:425–446.

Shitara Y, Itoh T, Sato H, Li AP and Sugiyama Y (2003) Inhibition of transporter-mediated hepatic uptake as a mechanism for drug–drug interaction between cerivastatin and cyclosporin A. *J Pharmacol Exp Ther* **304**:610–616.

Shitara Y, Sato H and Sugiyama Y (2005) Evaluation of drug–drug interaction in the hepatobiliary and renal transport of drugs. *Annu Rev Pharmacol Toxicol* **45**:689–723.

Tachibana T, Kato M, Watanabe T, Mitsui T and Sugiyama Y (2009) A method for predicting the risk of drug–drug interactions involving inhibition of intestinal CYP3A4 and P-glycoprotein. *Xenobiotica* **39**:430–443.

Tahara H, Kusuhara H, Endou H, Koepsell H, Imaoka T, Fuse E and Sugiyama Y (2005) A species difference in the transport activities of H2 receptor antagonists by rat and human renal organic anion and cation transporters. *J Pharmacol Exp Ther* **315**:337–345.

Tsuda M, Terada T, Ueba M, Sato T, Masuda S, Katsura T and Inui K (2009) Involvement of human multidrug and toxin extrusion 1 in the drug interaction between cimetidine and metformin in renal epithelial cells. *J Pharmacol Exp Ther* **329**:185–191.

Ueda K, Kato Y, Komatsu K and Sugiyama Y (2001) Inhibition of biliary excretion of methotrexate by probenecid in rats: quantitative prediction of interaction from in vitro data. *J Pharmacol Exp Ther* **297**:1036–1043.

Verschraagen M, Koks CH, Schellens JH and Beijnen JH (1999) P-glycoprotein system as a determinant of drug interactions: the case of digoxin-verapamil. *Pharmacol Res* **40**:301–306.

Watanabe T, Kusuhara H, Maeda K, Shitara Y and Sugiyama Y (2009a) Physiologically based pharmacokinetic modeling to predict transporter-mediated clearance and distribution of pravastatin in humans. *J Pharmacol Exp Ther* **328**:652–662.

Watanabe T, Maeda K, Kondo T, Nakayama H, Horita S, Kusuhara H and Sugiyama Y (2009b) Prediction of the hepatic and renal clearance of transporter substrates in rats using in vitro uptake experiments. *Drug Metab Dispos* **37**:1471–1479 (2009).

Yasui-Furukori N, Uno T, Sugawara K and Tateishi T (2005) Different effects of three transporting inhibitors, verapamil, cimetidine, and probenecid, on fexofenadine pharmacokinetics. *Clin Pharmacol Ther* **77**:17–23.

Part III
Impact of Drug–Drug Interactions

Chapter 17
Drug Disposition and Drug–Drug Interactions: Importance of First-Pass Metabolism in Gut and Liver

Catherine K. Yeung, Ping Zhao, Danny D. Shen, and Kenneth E. Thummel

Abstract Drugs with high intestinal and hepatic extraction that undergo biotransformation are often involved in significant metabolic drug–drug interactions when co-administered with enzyme inhibitors or inducers. These interactions can result in sub- or supra-therapeutic drug levels which can lead to major adverse events. We present an overview of clinically relevant examples of metabolically based, first-pass drug–drug interactions, as well as an introduction to the pharmacokinetic principles of sequential first-pass intestinal and hepatic metabolism. We conclude with pharmacokinetic simulations that illustrate the complex relationship between inhibitor-induced changes in organ intrinsic clearances and systemic drug exposure as a first step in the development of pharmacokinetic modeling techniques that will facilitate the prediction and avoidance of metabolically based drug–drug interactions.

17.1 Clinical Aspects of First-Pass Drug Interactions

17.1.1 Background

In 1990, reports of sudden deaths or cardiovascular collapse due to *torsade de pointes* were associated with the combined use of the nonsedating antihistamine terfenadine (trade name Seldane) and either macrolide antibiotics or azole antifungals (Paris et al., 1994). This resulted in warnings from the manufacturer to doctors and other healthcare providers of the potential for fatal cardiac arrhythmia caused by a metabolically based drug–drug interaction between terfenadine, a CYP3A4 substrate, and CYP3A4 inhibitors. Subsequent investigation revealed that terfenadine metabolism by CYP3A4 was inhibited by erythromycin and ketoconazole,

K.E. Thummel (✉)
Departments of Pharmaceutics, University of Washington, Seattle, WA, USA
e-mail: thummel@u.washington.edu

K.S. Pang et al. (eds.), *Enzyme- and Transporter-Based Drug–Drug Interactions*,
DOI 10.1007/978-1-4419-0840-7_17,

which resulted in an estimated 30- to 100-fold elevation in terfenadine plasma levels (Shen et al., 2000; Petty and Vega, 2008). In vitro studies further showed that high levels of terfenadine and not its active metabolite, fexofenadine, block cardiac potassium channels, prolong cardiac repolarization, and could in turn precipitate a *torsade de pointes*-type polymorphic ventricular arrhythmia (Woosley et al., 1993). The impressive magnitude of metabolic interactions with terfenadine is attributed to its very extensive first-pass metabolism (>99%), after a typical antihistaminic dose, and the dramatic increase in systemic bioavailability when a potent inhibitor of CYP3A is co-administered (Shen et al., 2000). As a result of this potentially fatal adverse drug interaction, terfenadine, once among the top 10 best selling prescription drugs in the United States, was pulled from the market in 1997 (Petty and Vega, 2008). This was a wake-up call to the drug industry that metabolically based drug–drug interactions could cause life-threatening adverse drug responses and sparked new research focused on drug interactions involving first-pass drug metabolism processes in humans.

Since that time, other noteworthy examples of drug–drug interactions involving modulation of first-pass metabolism have surfaced. Several of the clinically relevant and prototypic interactions are presented below, namely, the interactions between grapefruit juice and simvastatin (CYP3A4 inhibition), rifampin and cyclosporine (3A4 induction), ketoconazole and cyclosporine along with ritonavir and saquinavir (beneficial interaction), and mycophenolate mofetil and tacrolimus (UGT inhibition). These examples are meant to provide illustration of the physiological mechanisms and pharmacokinetic principles governing metabolic drug interactions that occur during oral absorption, that is, lessons that we have learned in the nearly one and a half decades since the advent of the terfenadine experience.

17.1.2 CYP3A-Based Interactions

17.1.2.1 Enzymology

A full complement of drug-metabolizing enzymes is expressed in the human intestinal epithelium (Hickman et al., 1998; Tukey and Strassburg, 2001; Coles et al., 2002; Tam et al., 2003; Paine et al., 2006a). The most important Phase 1 enzymes in the context of drug–drug interactions are the cytochrome P450 enzymes. The specific P450 content of microsomes isolated from mucosal epithelium of the human proximal small intestine is roughly 1/6–1/8 of that found in liver microsomes (Paine et al., 1997; Zhang et al., 1999). The most important component of the intestinal P450 pool is the human CYP3A subfamily: CYP3A4, CYP3A5, CYP3A7, and CYP3A43. The first three of these enzymes share at least 85% amino acid sequence homology (Kolars et al., 1994), although they are not identical regarding substrate selectivity or tissue localization. CYP3A43 expression in the liver and small intestine is relatively low and, hence, its contribution to drug clearance is considered minimal. CYP3A7 is found mainly in fetal liver (Dresser et al., 2000). As a result, CYP3A4 and CYP3A5 are the dominant hepatic and intestinal CYP3A isoforms.

The specific CYP3A4 content of the liver and small intestine generally exceeds that of CYP3A5 (Lown et al., 1994), although CYP3A5 expression can be comparable to or greater than CYP3A4 in individuals carrying the wild-type *CYP3A5*1* allele (Lin et al., 2001). In terms of cellular CYP3A content, the level found within enterocytes of the upper (duodenal) region of the small intestine can be equivalent to or exceed that found in hepatocytes (von Richter et al., 2004). Due to their localization, CYP3A4/5 play a prominent role in first-pass drug metabolism and contribute to low oral drug bioavailability (Thummel et al., 2008; Thummel and Wilkinson, 1998).

The following sections present adverse drug–drug interactions that occur as a result of concurrent treatment with CYP3A4/5 inhibitors and inducers. Inhibition of intestinal CYP3A4/5 has also been exploited to increase oral bioavailability of first-pass drug substrates to enhance potency, achieve dose sparing, and effect cost savings.

17.1.2.2 Examples of Intestinal CYP3A Inhibition

Grapefruit constituents (dihydroxybergamottin and bergamottin) have been shown to cause irreversible, mechanism-based inactivation of intestinal CYP3A4 and CYP3A5 (Edwards et al., 1996; Schmiedlin-Ren et al., 1997). These furanocoumarins undergo metabolism by CYP3A to form reactive metabolites that covalently bind the enzyme, rendering it irreversibly inactivated (Schmiedlin-Ren et al., 1997; Paine et al., 2006b). Restoration of enzymatic activity does not occur until biosynthesis of new CYP3A protein (Lown et al., 1997; Greenblatt et al., 2003). Existing data indicate that ingestion of one glass (200–300 mL) of regular strength grapefruit juice results in localized inactivation of intestinal CYP3A enzymes, whereas the same quantity of double-strength grapefruit juice can lead to expanded inhibition of both intestinal and liver enzymes (Thummel et al., 2008). In fact, a decrease in the first-pass metabolism of a CYP3A substrate following consumption of a usual quantity of grapefruit juice may provide an estimate of the contribution of intestinal CYP3A to its first-pass effect.

HMG-CoA reductase inhibitors, colloquially referred to as "statins," are a commonly used group of cholesterol-lowering agents. While generally safe and effective, the use of these drugs has been associated with complications including myalgia and creatine kinase elevations. Severe adverse reactions are rare; however, renal failure secondary to statin-associated rhabdomyolysis has been reported (Ayanian et al., 1988; Corpier et al., 1988; Biesenbach et al., 1996; Hino et al., 1996). CYP3A4 dominates the metabolism of lovastatin, simvastatin, atorvastatin, and cerevistatin (Dresser et al., 2000); in contrast, it plays a minor role in the metabolism of pravastatin (Ayanian et al., 1988). A randomized crossover study investigated the effect of regular consumption of grapefruit juice on the pharmacokinetics of simvastatin and demonstrated that co-administration of 200 mL of grapefruit juice with simvastatin increased the AUC_{0-24} of simvastatin 3.6-fold over control (Lilja et al., 2004). These results suggest that moderate consumption of grapefruit juice can elevate the plasma concentrations of simvastatin and increase the risk of myotoxicity. Moreover, it suggests that simvastatin undergoes significant first-pass metabolism in the intestinal mucosa.

As a general rule, low oral bioavailability drugs like simvastatin, lovastatin, saquinavir, as well as others for which intestinal CYP3A-dependent first-pass metabolism is significant are all subject to profound changes in systemic AUC when co-administered with potent CYP3A inhibitors. For example, midazolam (Paine et al., 1996), triazolam (Masica et al., 2004), buspirone (Mahmood and Sahajwalla, 1999), felodipine (Lundahl et al., 1997), verapamil (Fromm et al., 1996; von Richter et al., 2001), nifedipine (Holtbecker et al., 1996), and tirilazad (Fleishaker et al., 1996) all appear to undergo significant intestinal CYP3A-dependent first-pass metabolism and they are all markedly affected by co-administration with potent inhibitors such as ketoconazole and itraconazole (Strolin Benedetti and Bani, 1999; Dresser et al., 2000; Pelkonen et al., 2008).

17.1.2.3 Examples of Intestinal CYP3A Induction

Induction of CYP3A4 expression in the liver and intestine can result in subtherapeutic drug concentrations, loss of drug efficacy, and disease exacerbation. Multiple case reports (Allen et al., 1985; Cassidy et al., 1985; Modry et al., 1985; Offermann et al., 1985; Vandevelde et al., 1991) have confirmed a serious adverse interaction between rifampin, a known CYP3A inducer, and cyclosporine A. In order to prevent rejection following solid organ transplantation, patients are necessarily immunosuppressed with a combination of drugs, frequently including cyclosporine A, a CYP3A4 substrate. Immunosuppression can lead to reactivation of latent tuberculosis or fungal aspergillosis, both of which are commonly treated with complex regimens that include rifampin. In one case report, addition of rifampin caused a decrease in the serum trough cyclosporine level from a therapeutic 478 to 31 ng/mL (Modry et al., 1985). As a result, the patient developed severe acute graft rejection of the transplanted heart and required several rounds of treatment with methylpredisolone and antithymocyte globulin to resolve the rejection episode. The dose of cyclosporine was gradually increased, and this patient eventually required 30 mg/kg/day to reach target serum levels, a >20-fold increase from pre-rifampin levels (1.2–1.4 mg/kg/day) (Modry et al., 1985). Other case reports have described cyclosporine dosage adjustments of $\sim$4-fold with the addition of rifampin. These case reports are supported by a clinical pharmacokinetics study in which subjects on stable cyclosporine regimens following kidney transplantation were administered rifampin for treatment of active *Mycobacterium tuberculosis* (Kim et al., 1998). In this group, mean $C_{\max}$ decreased from 630 to 446 ng/mL, and a similar decrease was seen in AUC (4582.95–2790.48 ng/mL/h). To achieve optimal levels in these patients, cyclosporine doses had to be increased from 2.5- to 3-fold (Kim et al., 1998). Increases in oral cyclosporine clearance were greater than that expected if only hepatic CYP3A4 was involved, leading the authors to conclude that rifampin also induced intestinal CYP3A4 and decreased the bioavailability of cyclosporine, a claim which is indirectly supported by a significant increase in Vd/F (1.76–4.28 L/kg) (Kim et al., 1998). Clearly, induction of intestinal CYP3A4 by a precipitant drug (such as rifampin) can lead to a substantial decrease in oral bioavailability of an object drug (cyclosporine), resulting in subtherapeutic plasma concentrations and the subsequent clinical consequences.

Other important inducers of intestinal and hepatic CYP3A-dependent first-pass metabolism include phenytoin, dexamethasone, and phenobarbital (Pelkonen et al., 2008). Interestingly, some CYP3A inducers such as efavirenz appear to be selective for the liver and do not affect the intestine, possibly as a consequence of its selective activation of CAR and not PXR (Faucette et al., 2007); the former nuclear receptor appears active only in the liver.

17.1.2.4 Clinical Utility of CYP3A Inhibition

The ability to manipulate CYP3A4 activity has been cleverly exploited to decrease cost and enhance efficacy of poorly bioavailable drugs. In 1995, the average cost of immunosuppressive therapy in a cardiac transplantation patient was around $6,640 per year (Keogh et al., 1995). Oral ketoconazole, an inexpensive imidazole antifungal drug (200 mg daily) was added to the regimen to inhibit intestinal and hepatic CYP3A4, thereby increasing the bioavailability and slowing the metabolic clearance of cyclosporine. As a result, cyclosporine doses were decreased as much as 80% after 12 months of treatment, realizing a cost savings of $5,200 per patient per year (Keogh et al., 1995), with no increase in rates of organ rejection or adverse effects.

A similar strategy is currently being used to enhance the efficacy of saquinavir, a poorly bioavailable antiretroviral drug used in the treatment of HIV-1 infection (Dresser et al., 2000). Saquinavir, along with other protease inhibitors, is metabolized by and inhibits CYP3A4 (Eagling et al., 1997). Additionally, the protease inhibitors are also substrates and inhibitors of P-glycoprotein (Gutmann et al., 1999). Addition of ritonavir to saquinavir therapy causes a 33-fold increase in $C_{\max}$ and a 58-fold increase in AUC of saquinavir (Merry et al., 1997), which results in enhanced antiviral activity. The magnitude of this interaction suggests that both CYP3A4 and P-glycoprotein are inhibited and that both hepatic and intestinal proteins may be involved; this supposition is supported by additional research showing that the oral bioavailability of saquinavir is doubled when co-administered with grapefruit juice (Kupferschmidt et al., 1998). These findings have been translated to clinical practice, as saquinavir is almost always prescribed in combination with ritonavir. Ritonavir is now also used to boost the AUC and to prolong the elimination half-life of other protease inhibitors to achieve dose sparing and to improve compliance (Chandwani and Shuter, 2008).

17.1.3 UGT-Based Interactions

A number of Phase 2 enzymes are also present in human intestinal mucosa (Her et al., 1996; Radominska-Pandya et al., 1998; Strassburg et al., 1999), of which UDP-glucuronosyltransferases (UGTs) are the conjugating enzymes most often involved in clinically significant drug–drug interactions.

Both families of UGTs, UGT1 and UGT2, are expressed in the human gastrointestinal tract. In the small intestine, the most prominent members of the 1A family (in the order of mRNA expression) are UGT1A10, 1A1, 1A6, 1A5, and 1A8 (Strassburg et al., 2000; Nakamura et al., 2008; Ohno and Nakajin, 2009). In fact, UGT1A8 and UGT1A9 are exclusively expressed in the gut mucosa. For the 2B family, UGT2B17, 2B7, and 2B15 are the predominant members present in human small intestine. Almost all published UGT expression data are based on mRNA quantification. An inability to generate UGT isoform-specific antibodies has been a significant handicap in characterizing the protein expression and catalytic function of the individual UGTs expressed in various human tissues including the gastrointestinal tract. The considerable overlap in substrate specificity of UGTs also means the lack of selective chemical inhibitors that can serve as functional probes for assessing the catalytic contribution of multiple UGTs to the conjugation of a drug substrate. As a result, our understanding of the biochemistry and in vivo functioning of UGTs, particularly the role of intestinal UGTs, lags well behind current knowledge of the drug-metabolizing CYP enzymes. This situation extends to our understanding of the pharmacokinetics of drug interactions involving UGTs. When drug interactions involving first-pass metabolism mediated by CYPs and UGTs are compared, it becomes fairly obvious that inhibitory interactions involving UGTs rarely show an elevation in AUC exceeding 2- to 3-fold. The relatively modest in vivo inhibition with UGTs is probably attributed to the fact that multiple UGTs are involved in most cases and the in vivo potency of UGT inhibitors is generally modest, with K_i rarely below 10 μM and mostly above 100 μM (Kiang et al., 2005). Nonetheless, clinically significant inhibitory interactions involving UGTs have been noted for drugs with a narrow therapeutic index. Of course, induction of UGTs as with CYPs can still be quite remarkable and lead to significant loss in drug efficacy.

Clinically relevant drug interactions involving glucuronidation are well recognized with the immunosuppressant mycophenolate mofetil (MMF), which has become a key element of post-transplantation regimens. UGT metabolizes MMF to its inactive 4-hydroxyphenyl-β-glucuronide (MPAG) metabolite and active, but minor, acyl-glucuronide (AcMPAG) metabolite (Bullingham et al., 1998; Shipkova et al., 1999). MMF plasma levels must be tightly controlled to avoid underexposure and loss of immunosuppressant effect or overexposure and toxic side effects (anemia, leucopenia, and diarrhea) (Kuypers et al., 2005). A recent report concluded that concurrent administration of rifampin, a well-known UGT inducer, results in an increase in daily dose requirements from 2 to 6 g in order to achieve a desired trough concentration of 2.5 μg/mL (Kuypers et al., 2005). Due to an increased risk for the development of adverse effects from this high dose of MMF, transplant physicians have discontinued rifampin therapy, relying on dual antimycobacterial therapy instead of the preferred triple drug protocol. As expected, withdrawal of rifampin caused an increase in AUC_{0-12}, C_0, and C_{max} and necessitated a 50% decrease in dose (Kuypers et al., 2005). The authors speculate that UGT1A9, present in intestines, kidney, and liver, (Basu et al., 2004; Bernard and Guillemette, 2004; Picard et al., 2005) bears major responsibility for the metabolism of MMF to MPAG. Under noninduced conditions, UGT1A9 is responsible for ~40% of intestinal MPAG formation (Picard et al., 2005). UGT1A7 and UGT1A8, predominantly located in the gastrointestinal tract, are also involved in the first-pass metabolism of MMF (Bernard and Guillemette, 2004). While major adverse effects as a result of this drug–drug interaction have not yet been reported, a subtherapeutic plasma level of MMF, as part of an immunosuppression regimen, clearly has the potential to precipitate acute allograph rejection.

17.2 Prediction of First-Pass Drug Interactions

17.2.1 Overview

The potential for adverse drug reactions caused by alterations in first-pass metabolism is a challenge to predict quantitatively. In general, drugs that are subject to extensive metabolism and have high intestinal and hepatic extraction are most susceptible to metabolic drug–drug interactions when co-administered with enzyme inhibitors or inducers (Thummel et al., 2008). A major change in bioavailability can result in significant clinical consequences, particularly when drugs with narrow therapeutic indices (including, but not limited to, protease inhibitors, immunosuppressants, anticoagulants, antihypertensives, sedative/hypnotics, and cardiac glycosides) are involved. Fortunately, screening for metabolic drug–drug interactions is becoming an integral part of the drug development process, and interactions caused by older drugs are being minimized with therapeutic drug monitoring (TDM) programs. Nevertheless, predictive tools such as mathematical pharmacokinetic models that adequately describe both intestinal and hepatic first-pass metabolism would provide a cost-effective approach for predicting the in vivo impact of metabolic drug interactions and developing ways to circumvent or minimize adverse drug–drug interactions.

17.2.2 Pharmacokinetic Principles of First-Pass Metabolism

In order to predict the dynamics and extent of interaction between an enzyme modulator and an orally administered drug, it is necessary to consider the integration of sequential processes of intestinal and hepatic first-pass metabolism and systemic clearance. The theory and pharmacokinetic models necessary for predicting changes in hepatic first-pass metabolism and hepatic clearance are well established. However, at the level of intestinal metabolism, one needs to understand the dynamic relationship between biotransformation and drug absorption, along with consideration of the physiological characteristics unique to the small intestine. The pharmacokinetic theory behind intestinal drug–drug interactions is evolving and current approaches for evaluating such interactions will be presented. In order to remove the added complexity of drug release kinetics associated with solid formulations, we have assumed that drugs are administered orally via solutions.

17.2.3 Organ Intrinsic Clearance and Bioavailability

According to the physiological model presented in Fig. 17.1, the fraction of an oral drug dose that successfully traverses through the gut wall and liver and then into the systemic circulation can be expressed as the product of three bioavailability (F) terms:

$$F_{\text{po}} = F_{\text{a}} \cdot F_{\text{gm}} \cdot F_{\text{h}} \tag{17.1}$$

where F_{po} is the absolute oral bioavailability, F_{a} is the fraction absorbed from the gut lumen into the enterocyte, and F_{gm} and F_{h} are the fractions that escape intestinal and hepatic elimination before entering the systemic circulation for the first time (Rowland and Tozer, 1995). Assuming that a drug is metabolized in the liver and intestine, has high cell barrier permeability, and is not subject to intestinal or hepatic efflux transport, one can derive a blood flow-limited, "well-stirred" model for organ

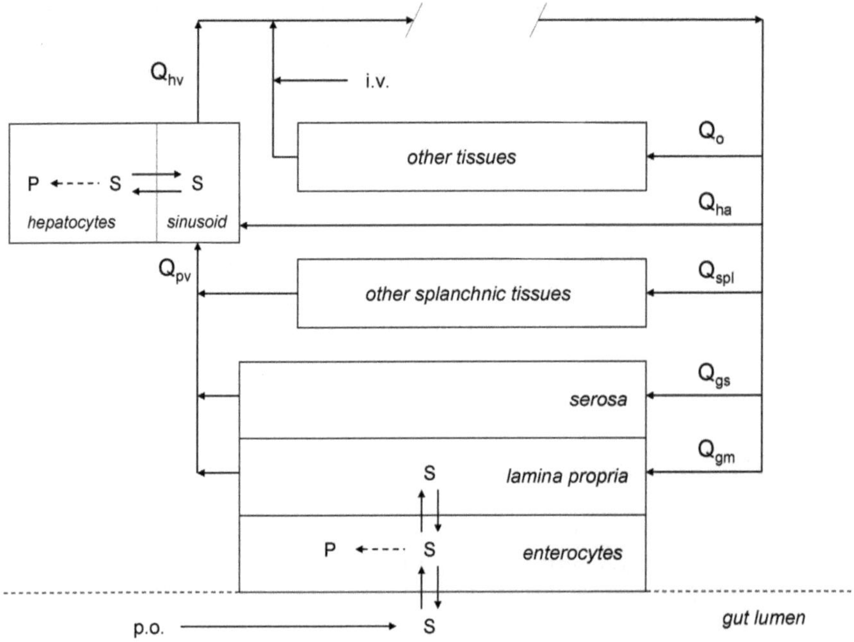

Fig. 17.1 Physiological model for sequential intestinal and hepatic first-pass metabolism. Blood flow to the small intestine is divided functionally into mucosal (Q_{gm}) and serosal (Q_{gs}) components. Mucosal blood flow in the lamina propria that perfuses the enterocyte epithelium. Portal blood flow (Q_{pv}) which perfuses the liver is comprised of blood leaving the small intestine (serosal and mucosal) and other splanchnic organs such as the stomach and spleen. Blood flow leaving the liver (Q_{hv}) represents the sum of hepatic arterial flow (Q_{ha}) and Q_{pv}. First-pass metabolism of an orally administered substrate (S) to product (P) may occur in the enterocyte or hepatocyte compartment. Reproduced with permission from (Thummel et al., 2008)

extraction, with first-order liver and intestinal metabolic elimination. This yields nonlinear relationships between organ bioavailability (F_{gm} or F_{h}) and the intrinsic metabolic capacity of the organ (Paine et al., 1997; Thummel et al., 1997):

$$F_{\mathrm{gm}} = 1 - ER_{\mathrm{gm}} = \frac{Q_{\mathrm{gm}}}{f_{\mathrm{ub}} \cdot CL_{\mathrm{gm}}^{\mathrm{int}} + Q_{\mathrm{gm}}} \tag{17.2}$$

and

$$F_{\mathrm{h}} = 1 - ER_{\mathrm{h}} = \frac{Q_{\mathrm{h}}}{f_{\mathrm{ub}} \cdot CL_{\mathrm{h}}^{\mathrm{int}} + Q_{\mathrm{h}}} \tag{17.3}$$

where ER, Q, f_{ub}, and CL^{int} represent the organ extraction ratio, blood flow, free fraction in blood, and intrinsic clearance, respectively. Organ blood flow is denoted by Q_{h} for the liver and Q_{gm} for the gut mucosa (Fig. 17.1). The organ intrinsic clearances (i.e., $V_{\mathrm{max}}/K_{\mathrm{m}}$ in biochemical terms) are values that can be modified by an interacting drug (inhibitor or inducer). Note that a term for blood-free fraction can be found in both hepatic and intestinal relationships. In the case of a well-stirred model for the liver, binding to blood constituents acts to "restrict" drug uptake into the hepatocytes, although this can be overcome by a very high unbound intrinsic clearance that can essentially shift the equilibrium at the cellular and plasma-binding sites to release free drug. In the case of the intestine, drug binding to blood components is assumed to act as a sink that facilitates diffusion of drug molecules out of the enterocyte compartment, i.e., removal from the metabolic site. Again, this process is in competition with intestinal intrinsic clearance; when the latter is sufficiently high, first-pass loss would be high and systemic bioavailability would necessarily be low.

17.2.4 Organ Bioavailability and Drug Exposure

According to basic pharmacokinetic principles, drug exposure measured by the area under the concentration–time curve (AUC) after oral administration is a function of the systemic clearance (CL) and absolute bioavailability (F_{po}):

$$\mathrm{AUC}_{\mathrm{po}} = \frac{\left(F_{\mathrm{a}} \cdot F_{\mathrm{gm}} \cdot F_{\mathrm{h}}\right) \cdot \mathrm{Dose}_{\mathrm{po}}}{CL} \tag{17.4}$$

It is generally accepted that intestinal metabolism plays a minor role in systemic clearance, given the relatively small fraction of blood flow through the mucosa (Q_{gm} or gut mucosal blood flow is approximately 10–15% of Q_{spl} or splanchnic blood flow) (Thummel et al., 1997; Pang, 2003). This factor limits the capacity of enterocytes to clear drug molecules from the systemic circulation. Conversely, since drug passage (Thummel et al., 1997; Pang, 2003) through the enterocytes is obligatory (for transcellularly absorbed drugs), mucosal flow simply competes with the intrinsic metabolic clearance and helps define the first-pass extraction efficiency.

Assuming that (i) organ extractions are blood flow limited and (ii) the drug exhibits complete absorption (F_a=1), Equation (17.5) can be derived from Equations (17.2), (17.3), and (17.4):

$$\frac{\text{Dose}_{\text{po}}}{\text{AUC}_{\text{po}}} = \frac{f_{\text{ub}} \cdot CL_{\text{h}}^{\text{int}}}{\dfrac{Q_{\text{gm}}}{f_{\text{ub}} \cdot CL_{\text{gm}}^{\text{int}} + Q_{\text{gm}}}} \tag{17.5}$$

When Equation (17.5) is applied to drug–drug interactions, the AUC ratio in the presence (*) and absence of an interacting drug can be expressed as

$$\frac{\text{AUC}_{\text{po}}^{*}}{\text{AUC}_{\text{po}}} = \left(\frac{Q_{\text{gm}} + f_{\text{ub}} \cdot CL_{\text{gm}}^{\text{int}}}{Q_{\text{gm}} + f_{\text{ub}} \cdot CL_{\text{gm}}^{\text{int}*}}\right) \cdot \left(\frac{f_{\text{ub}} \cdot CL_{\text{h}}^{\text{int}}}{f_{\text{ub}} \cdot CL_{\text{h}}^{\text{int}*}}\right) \tag{17.6}$$

The above equation illustrates the multiplicative effect that a simultaneous change in both the hepatic and intestinal intrinsic clearance can have on the systemic AUC of an orally administered drug. However, it is not useful by itself for quantitative resolution of the factors that govern the AUC change, as it requires prior knowledge of the altered intrinsic clearances. Fortunately that information can be generated from in vitro experimentation.

In the case of rapidly reversible enzyme inhibitors, the change in the intrinsic clearance for the liver and intestine in the presence of the inhibitor can be expressed as a function of the in vivo K_i and the respective unbound inhibitor concentration in the liver and intestine (I_h and I_{gm}):

$$\frac{f_{\text{ub}} \cdot Cl_{\text{h}}^{\text{int}}}{f_{\text{ub}} \cdot Cl_{\text{h}}^{\text{int}*}} = 1 + \frac{I_{\text{h}}}{K_{\text{i}}} \tag{17.7}$$

$$\frac{f_{\text{ub}} \cdot CL_{\text{gm}}^{\text{int}}}{f_{\text{ub}} \cdot CL_{\text{gm}}^{\text{int}*}} = 1 + \frac{I_{\text{gm}}}{K_{\text{i}}} \tag{17.8}$$

The K_i in vivo can be estimated from incubations of substrate and inhibitor with hepatic and intestinal subcellular fractions or intact cells, although several assumptions about access of the ligands to the enzyme and its function must be made. At both sites of metabolism, as the inhibitor concentration exceeds the K_i, a reduction in the intrinsic clearance of 50% or more is expected.

For prediction of enzyme inhibition, we generally assume that I_h and I_{gm} are equal to the unbound inhibitor concentration in plasma. However, the unbound inhibitor concentration in the intestinal epithelia may or may not be equivalent to that in plasma and the liver. For example, I_{gm} may well exceed the unbound portal plasma concentration during the inhibitor absorption phase, and it may be less than the unbound portal concentration in post-absorption, if there is not a rapid equilibrium between the intracellular and portal plasma compartments (i.e., a basolateral membrane diffusional barrier exists). Even under the most optimal diffusion conditions, a gradient between the intracellular and vascular compartments may exist. When there is active uptake or efflux of drug from the enterocyte, the discordance between the two compartments may be even greater. A similar possibility of active inhibitor uptake and efflux can occur in the liver. Thus, it is this uncertainty about the inhibitor concentration at the intestinal and hepatic sites of metabolism that makes quantitative predictions of a drug–drug interaction challenging. In general, we can only obtain measurements from the vascular compartment and must estimate the parameters in the metabolic compartments, based on the best guess estimate of inter-compartmental concentration ratio.

For reversible competitive inhibitors, Equations (17.7) and (17.8) are quantitatively accurate only when the substrate concentration at the enzyme active site is below its respective K_m. Saturation of metabolic enzymes by the substrate is generally not an issue for hepatic elimination of most drugs, but it is more likely to occur in the enterocyte while the substrate is being absorbed. If there is saturation of the intestinal or hepatic enzyme, the metabolic efficiency will be reduced and the organ bioavailability will increase. More to the point, competition between the substrate and inhibitor for the metabolic enzyme will decrease the extent of inhibition that is observed with the concomitantly administered drug.

Substitution of Equations (17.7) and (17.8) into Equation (17.6) yields the expression below (Equation 17.9), which clearly illustrates the multiplicative effect that an enzyme inhibitor can have on systemic exposure to an orally administered drug:

$$\frac{\mathrm{AUC}^*_{\mathrm{po}}}{\mathrm{AUC}_{\mathrm{po}}} = \left[\frac{Q_{\mathrm{gm}} + f_{\mathrm{ub}} \cdot CL^{\mathrm{int}}_{\mathrm{gm}}}{Q_{\mathrm{gm}} + \left(f_{\mathrm{ub}} \cdot CL^{\mathrm{int}}_{\mathrm{gm}} \Big/ 1 + \frac{I_{\mathrm{gm}}}{K_{\mathrm{i}}}\right)}\right] \cdot \left(1 + \frac{I_{\mathrm{h}}}{K_{\mathrm{i}}}\right) \tag{17.9}$$

The relationship is complex and while the effect of the inhibitor in the liver is independent of blood flow, this is not the case for intestinal metabolism. In this situation, the relative magnitude of mucosal blood flow compared to the baseline mucosal intrinsic clearance and the apparent intrinsic clearance in the presence of inhibitor must be considered. At one extreme, when the baseline mucosal intestinal intrinsic clearance is negligible compared to mucosal blood flow (i.e., negligible mucosal extraction), Equation (17.9) will collapse into the much simpler and well-recognized equation for a hepatic inhibitory interaction (Rowland and Matin, 1973).

$$\frac{\mathrm{AUC}^*_{\mathrm{po}}}{\mathrm{AUC}_{\mathrm{po}}} = 1 + \frac{I_{\mathrm{h}}}{K_{\mathrm{i}}} \tag{17.10}$$

This is illustrated in Fig. 17.2, when the basal condition for $CL_{\mathrm{int,gm}}/Q_{\mathrm{gm}}$ is near 0. Conversely, when the basal intestinal intrinsic clearance is very high relative to

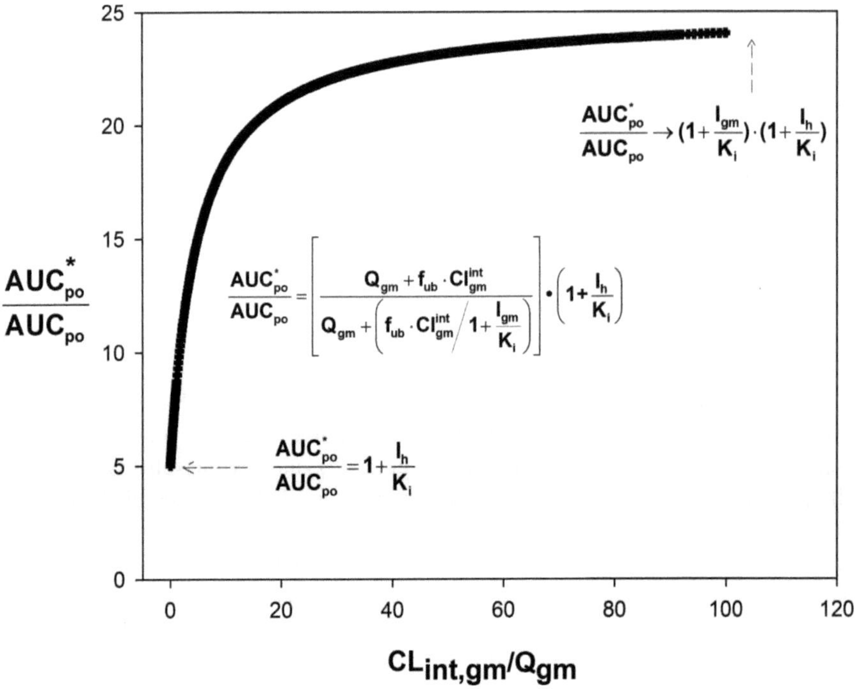

Fig. 17.2 Relationship between the degree of drug inhibition (AUC^*_{po}/AUC_{po}) and the intestinal metabolic extraction according to Equation (17.9). Assumptions are $Q_{gm} = 18$ L/h (Yang et al., 2007); $I_h/K_i = 4$; $I_{gm}/K_i = 4$

mucosal blood flow, the AUC ratio may be approximated as the product of the ($1+ I/K_i$) terms (i.e., the asymptote for the hyperbolic curve).

For an orally administered drug that is completely absorbed from the gastrointestinal tract and subject to first-pass intestinal extraction, one can assign a lower limit to the impact of change in mucosal extraction ratio on systemic AUC. For example, complete inhibition of an intestinal extraction that is 50% in the control state would result in a 2-fold increase in AUC, independent of any changes in the hepatic intrinsic clearance. If the mucosal extraction ratio is lower, the impact of intestinal inhibition will be reduced and eventually become negligible as the intestinal availability approaches unity. As the basal intestinal metabolic extraction ratio approaches 100%, the potential impact of inhibition of intestinal enzyme activity can be profound. If only 5% escapes first-pass intestinal metabolism in the basal state, this can be increased by as much as 20-fold (i.e., 100/5) with administration of a sufficiently high dose of a potent inhibitor. Considering that the gut enzyme may be more susceptible to first-pass inhibition than the liver because of the high inhibitor concentrations surrounding the enterocytes that exist during absorption of the inhibitor, one can envision a drug–drug interaction that is dominated by action at the intestinal mucosa. Of course potent inhibition of both intestinal and hepatic first-pass metabolism of a low bioavailability drug can lead to even more remarkable changes in systemic AUC, such as that observed for the ketoconazole–terfenadine interaction as mentioned earlier (Hönig et al., 1993).

Finally, it should be noted that one can still achieve very profound increases in AUC of a drug that does not undergo intestinal first-pass metabolism (Equation 17.10); i.e., when the substrate is metabolized by enzymes expressed specifically in the liver, the inhibitor concentration is high enough relative to the K_i, and there are no alternative pathways for systemic drug elimination.

Potent inducers (e.g., rifampin) of intestinal and hepatic drug metabolism can have an equally profound effect on systemic exposure to active compound as a potent inhibitor, but obviously in the opposite direction. When the basal intestinal extraction ratio is in the medium range, as in the case of midazolam and nifedipine, addition of an inducer can reduce systemic exposure by 90% or more (Niemi et al., 2003). Clearly an inducer will exert its effect on enzyme expression, independent of the basal extraction ratio, but as the basal intrinsic clearance decreases the net systemic effect of the inducers will be greatly minimized. For example, if the basal metabolic intrinsic clearance is low, such that the intestinal extraction ratio is 5%, the basal organ availability will be 95%. In this case, a 6-fold increase in the intestinal intrinsic clearance will reduce the organ bioavailability to 76%. Systemic exposure to the drug will be reduced but to a relatively modest degree.

17.2.5 Added Complexity of Intestinal Drug Interactions

17.2.5.1 Diffusion Barrier Effects

The relationship between an altered organ intrinsic clearance by an interacting drug and the degree of in vivo drug interaction becomes less direct when concurring events significantly influence the degree of metabolism as the drug molecule travels through the enterocytes (Pang, 2003). The pharmacokinetic relationship depicted in Equation (17.2) (for F_{gm}) assumes that permeability at either the apical or basolateral membrane of the epithelium (or both) is not rate limiting (i.e., permeability being much higher than mucosal flow). A more complex model that features the joint actions of permeability and metabolism has been proposed by Mizuma (2002), Mizuma et al. (2004) and Yang et al. (2007). As permeability becomes rate limiting (CL_{perm}/Q_{gm} in Equation 17.11), the relationship between drug–drug interactions at the enzyme level and the in vivo outcome becomes more complex as exemplified by the changing ratio of F_{gm} in Fig. 17.3:

$$F_{gm} = 1 - ER_{gm} = \frac{Q_{gm}}{\left[\left(f_{ub} \cdot CL_{gm}^{int}\right) \cdot \left(1 + Q_{gm}/CL_{perm}\right)\right] + Q_{gm}} \tag{17.11}$$

In this relationship, CL_{perm} represents the permeability clearance across both the apical and basolateral membranes once drug has gained access (i.e., been absorbed)

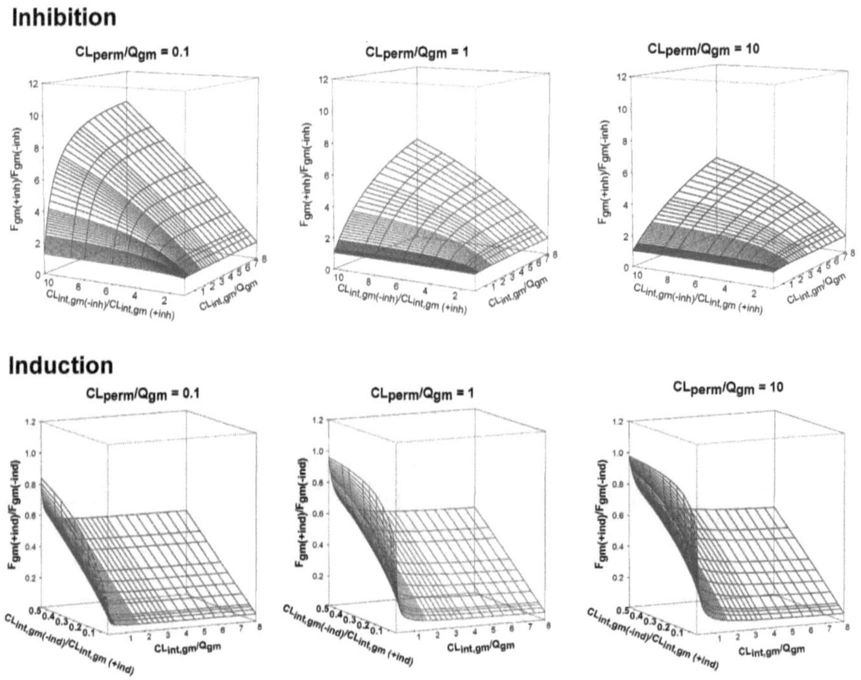

Fig. 17.3 Effect of drug permeability on the changes of intestinal bioavailability (F_{gm}) by drug interactions. *Upper panel*: inhibition; *Lower panel*: induction. Equation (17.11), from Yang et al. (2007), was used to calculate F_{gm}. For each panel, $CL_{\mathrm{perm}}/Q_{\mathrm{gm}}$ values are 0.1, 1, and 10 (from *left to right*) representing low-permeability (permeability limited), intermediate permeability, and high-permeability (blood flow limited) scenarios, respectively. One horizontal axis is the ratio of intrinsic clearances without and with interacting drugs, indicating the degree of drug interaction at the enzyme level; the other horizontal axis is the ratio of basal level intrinsic clearance to Q_{gm}, a measurement of the extent of metabolic extraction

into the intracellular compartment for the enterocyte. Again, for our purposes we have assumed that there is no active efflux of drug back into the intestinal lumen via P-glycoprotein or another membrane transporter. We acknowledge, this is an oversimplification for some drugs and more complex models that incorporate this phenomenon can be considered (Tam et al., 2003).

As seen from the 3D surface plots in Fig. 17.3, the most significant change in the ratio of intestinal bioavailability (F_{gm}) between the presence and absence of an enzyme inhibitor (upper panels) will occur when the intrinsic clearance is high relative to mucosal blood flow. With greater reductions in the intrinsic clearance in the presence of the inhibitor, the drug bioavailability across the organ increases markedly. Of course when the basal intrinsic clearance is low, further inhibition has little effect on the already high organ bioavailability; that is, the 3D surface approaches the 2D plane of $CL_{\mathrm{int,gm(-Inh)}}/CL_{\mathrm{int,gm(+Inh)}}$ versus the ratio of F_{gm}. What is equally interesting is the effect that a permeability barrier has on the inhibitory interaction. All things being equal, a reduction in the intrinsic clearance with an inhibitor has a greater effect on the organ bioavailability when cell permeability is limiting, than when blood flow is limiting. This is because a basolateral membrane permeability barrier tends to enhance the extraction efficiency at any given level of intrinsic clearance (de Lannoy and Pang, 1987; Gwilt et al., 1988). As seen in Fig. 17.3, for drugs with $CL_{\text{perm}}/ Q_{\text{gm}} = 0.1$ (low permeability) and $CL_{\text{int,gm}}/ Q_{\text{gm}} = 10$ (high intestinal extraction), a 10-fold decrease in $CL_{\text{int,gm}}$ by the inhibitor translates into an approximately 10-fold increase in F_{gm}; whereas for drugs with $CL_{\text{perm}}/ Q_{\text{gm}} = 10$ (high permeability) and high intestinal extraction ($CL_{\text{int,gm}}/ Q_{\text{gm}}$ approaching 10), a 10-fold decrease in $CL_{\text{int,gm}}$ by the inhibitor translates into an approximately 6-fold increase in F_{gm}.

With regard to enzyme induction (Fig. 17.3, lower panels), the impact of the permeability barrier is most noticeable when the basal intrinsic clearance is low. In this case, the organ bioavailability declines more rapidly with increasing enzyme induction for the low-permeability state compared to the high-permeability state. That is, for substrates with $CL_{\text{int,gm}}/Q_{\text{gm}} = 0.1$ (low extraction), a 10-fold increase in $CL_{\text{int,gm}}$ by the inducer translates into an F_{gm} ratio of approximately 0.3, 0.7, and 0.8 for $CL_{\text{perm}}/\text{Q}_{\text{gm}}$ of 0.1, 1.0, and 10, respectively. Again, the permeability barrier enhances the efficiency of drug extraction by the enterocyte at a given level of intrinsic clearance.

Besides permeability, factors such as gastric empting time, gastrointestinal transit time, luminal pH, and concomitant active processes by drug transporters may theoretically affect the extent of intestinal first-pass metabolism. Also, in the context of first-pass metabolism after oral administration, it is important to define the region of the gastrointestinal tract in which the majority of the drug dose will be absorbed when assigning values to the mucosal blood flow and unbound intrinsic clearance. For example, many CYP3A substrates are absorbed predominantly in the small intestine and, thus, one can define the total unbound intrinsic clearance and blood flow for the mucosa of a segment of the small intestine (Paine et al., 1997; Tam et al., 2003).

17.2.5.2 Inhibitor Dose and Temporal Factors

It is possible to have a drug–drug interaction that is confined to the intestine during first pass. As was discussed earlier, components of regular strength grapefruit juice appears to selectively inhibit CYP3A and drug transporters in the intestinal mucosa, but not those of the liver (Veronese et al., 2003). The level of exposure of the liver and intestinal mucosa to an inhibitor or inducer need not be identical, particularly during the peri-absorptive phase, when the modulator concentration at the intestinal mucosa may be much greater than that in the portal blood. It is also important to recognize that the intracellular mechanism underlying an interaction (e.g., *trans*-acting elements for induction or time-dependent inhibition) may not occur simultaneously in the intestine and the liver, or if it were to take place in both organs, they do not follow the same temporal course or occur to the same degree because of differential expression of accessory proteins, such as coactivators or corepressors of CYP3A (Mouly et al., 2002; Zhou et al., 2004) between enterocytes and hepatocytes. Consequently, the extent of induction or inhibition at each site of metabolism/transport following acute or chronic administration of an interacting drug could be quite different.

As discussed below, the magnitude of an AUC change observed in vivo will be dependent on the basal intestinal extraction ratio for any given individual. The change in oral AUC following administration of an inhibitor can be quite appreciable when the intestinal first-pass extraction is extensive. Conversely, if E_{gm} is already low, an inhibitor such as grapefruit juice will cause little change to the intestinal availability and to the oral AUC. Also, it has also been suggested that there is a ceiling effect for enzyme induction, whereby an individual with a pre-existing high level of enzyme expression will undergo a more modest degree of induction than will someone with low basal expression (Gorski et al., 2003).

A thorough understanding of the pharmacokinetic characteristics of both precipitating drug and substrate is critical for the design of drug interaction studies to maximize the interaction potential. By definition, the intestinal and hepatic concentration of an inhibitor or inducer will change over time following pulsed administration and, thus, the acute effect of the modulator on enzyme function should be time-dependent. Accordingly, administration of an oral ketoconazole dose approximately 2 h before the oral dose of a CYP3A substrate appears to be important to ensure optimal enzyme inhibition in both the liver and the small intestine (Yang et al., 2003). The relationships depicted in Equations (17.8) and (17.9) for enzyme inhibition take a static approach by allowing for the introduction of a stable inhibitor concentration. For prediction purposes, often an expected time average or maximum inhibitor concentration can be employed. However, a rigorous dynamic approach that considers the fluctuation in inhibitor concentration over time is technically more appropriate; there have been recent attempts to feature dynamic changes in the inhibitor concentration in the prediction of AUC changes (Chien et al., 2006; Zhang et al., 2009).

17.3 Conclusions

In conclusion, to fully assess the contribution of metabolic interaction at the intestinal mucosa during first pass to the overall drug interaction potential (e.g., AUC^*_{po}/AUC_{po}), one needs to account for multiple physiological factors that may affect the degree of first-pass metabolism. Physiologically based pharmacokinetic modeling approaches outlined in Chapters 5 and 26 and illustrated in Fig. 17.1 represent one way to evaluate such complexity. This could be one future direction for further refinement in our quantitative understanding of clinically relevant drug–drug interaction involving sequential first pass in the intestine and liver.

References

Allen RD, Hunnisett AG and Morris PJ (1985) Cyclosporin and rifampicin in renal transplantation. *Lancet* **1**:980.

Ayanian JZ, Fuchs CS and Stone RM (1988) Lovastatin and rhabdomyolysis. *Ann Intern Med* **109**:682–683.

Basu NK, Kole L, Kubota S and Owens IS (2004) Human UDP-glucuronosyltransferases show atypical metabolism of mycophenolic acid and inhibition by curcumin. *Drug Metab Dispos* **32**:768–773.

Bernard O and Guillemette C (2004) The main role of UGT1A9 in the hepatic metabolism of mycophenolic acid and the effects of naturally occurring variants. *Drug Metab Dispos* **32**: 775–778.

Biesenbach G, Janko O, Stuby U and Zazgornik J (1996) Myoglobinuric renal failure due to long-standing lovastatin therapy in a patient with pre-existing chronic renal insufficiency. *Nephrol Dial Transplant* **11**:2059–2060.

Bullingham RE, Nicholls AJ and Kamm BR (1998) Clinical pharmacokinetics of mycophenolate mofetil. *Clin Pharmacokinet* **34**:429–455.

Cassidy MJ, Van Zyl-Smit R, Pascoe MD, Swanepoel CR and Jacobson JE (1985) Effect of rifampicin on cyclosporin A blood levels in a renal transplant recipient. *Nephron* **41**:207–208.

Chandwani A and Shuter J (2008) Lopinavir/ritonavir in the treatment of HIV-1 infection: a review. *Ther Clin Risk Manag* **4**:1023–1033.

Chien JY, Lucksiri A, Ernest CS, 2nd, Gorski JC, Wrighton SA and Hall SD (2006) Stochastic prediction of CYP3A-mediated inhibition of midazolam clearance by ketoconazole. *Drug Metab Dispos* **34**:1208–1219.

Coles BF, Chen G, Kadlubar FF and Radominska-Pandya A (2002) Interindividual variation and organ-specific patterns of glutathione S-transferase alpha, mu, and pi expression in gastrointestinal tract mucosa of normal individuals. *Arch Biochem Biophys* **403**:270–276.

Corpier CL, Jones PH, Suki WN, Lederer ED, Quinones MA, Schmidt SW and Young JB (1988) Rhabdomyolysis and renal injury with lovastatin use. Report of two cases in cardiac transplant recipients. *JAMA* **260**:239–241.

de Lannoy IA and Pang KS (1987) Effect of diffusional barriers on drug and metabolite kinetics. *Drug Metab Dispos* **15**:51–58.

Dresser GK, Spence JD and Bailey DG (2000) Pharmacokinetic-pharmacodynamic consequences and clinical relevance of cytochrome P450 3A4 inhibition. *Clin Pharmacokinet* **38**:41–57.

Eagling VA, Back DJ and Barry MG (1997) Differential inhibition of cytochrome P450 isoforms by the protease inhibitors, ritonavir, saquinavir and indinavir. *Br J Clin Pharmacol* **44**:190–194.

Edwards DJ, Bellevue FH, 3rd and Woster PM (1996) Identification of 6′,7′-dihydroxybergamottin, a cytochrome P450 inhibitor, in grapefruit juice. *Drug Metab Dispos* **24**:1287–1290.

Faucette SR, Zhang TC, Moore R, Sueyoshi T, Omiecinski CJ, LeCluyse EL, Negishi M and Wang H (2007) Relative activation of human pregnane X receptor versus constitutive androstane receptor defines distinct classes of CYP2B6 and CYP3A4 inducers. *J Pharmacol Exp Ther* **320**:72–80.

Fleishaker JC, Pearson PG, Wienkers LC, Pearson LK and Peters GR (1996) Biotransformation of tirilazad in human: 2. Effect of ketoconazole on tirilazad clearance and oral bioavailability. *J Pharmacol Exp Ther* **277**:991–998.

Fromm MF, Busse D, Kroemer HK and Eichelbaum M (1996) Differential induction of prehepatic and hepatic metabolism of verapamil by rifampin. *Hepatology* **24**:796–801.

Gorski JC, Vannaprasaht S, Hamman MA, Ambrosius WT, Bruce MA, Haehner-Daniels B and Hall SD (2003) The effect of age, sex, and rifampin administration on intestinal and hepatic cytochrome P450 3A activity. *Clin Pharmacol Ther* **74**:275–287.

Greenblatt DJ, von Moltke LL, Harmatz JS, Chen G, Weemhoff JL, Jen C, Kelley CJ, LeDuc BW and Zinny MA (2003) Time course of recovery of cytochrome p450 3A function after single doses of grapefruit juice. *Clin Pharmacol Ther* **74**:121–129.

Gutmann H, Fricker G, Drewe J, Toeroek M and Miller DS (1999) Interactions of HIV protease inhibitors with ATP-dependent drug export proteins. *Mol Pharmacol* **56**:383–389.

Gwilt PR, Comer S, Chaturvedi PR and Waters DH (1988) The influence of diffusional barriers on presystemic gut elimination. *Drug Metab Dispos* **16**:521–526.

Her C, Szumlanski C, Aksoy IA and Weinshilboum RM (1996) Human jejunal estrogen sulfotransferase and dehydroepiandrosterone sulfotransferase: immunochemical characterization of individual variation. *Drug Metab Dispos* **24**:1328–1335.

Hickman D, Pope J, Patil SD, Fakis G, Smelt V, Stanley LA, Payton M, Unadkat JD and Sim E (1998) Expression of arylamine N-acetyltransferase in human intestine. *Gut* **42**:402–409.

Hino I, Akama H, Furuya T, Ueda H, Taniguchi A, Hara M and Kashiwazaki S (1996) Pravastatin-induced rhabdomyolysis in a patient with mixed connective tissue disease. *Arthritis Rheum* **39**:1259–1260.

Holtbecker N, Fromm MF, Kroemer HK, Ohnhaus EE and Heidemann H (1996) The nifedipine-rifampin interaction. Evidence for induction of gut wall metabolism. *Drug Metab Dispos* **24**:1121–1123.

Hönig PK, Wortham DC, Zamani K, Conner DP, Mullin JC and Cantilena LR (1993) Terfenadine-ketoconazole interaction. Pharmacokinetic and electrocardiographic consequences. *JAMA* **269**:1513–1518.

Keogh A, Spratt P, McCosker C, Macdonald P, Mundy J and Kaan A (1995) Ketoconazole to reduce the need for cyclosporine after cardiac transplantation. *N Engl J Med* **333**:628–633.

Kiang TK, Ensom MH and Chang TK (2005) UDP-glucuronosyltransferases and clinical drug–drug interactions. *Pharmacol Ther* **106**:97–132.

Kim YH, Yoon YR, Kim YW, Shin JG and Cha IJ (1998) Effects of rifampin on cyclosporine disposition in kidney recipients with tuberculosis. *Transplant Proc* **30**:3570–3572.

Kolars JC, Lown KS, Schmiedlin-Ren P, Ghosh M, Fang C, Wrighton SA, Merion RM and Watkins PB (1994) CYP3A gene expression in human gut epithelium. *Pharmacogenetics* **4**: 247–259.

Kupferschmidt HH, Fattinger KE, Ha HR, Follath F and Krahenbuhl S (1998) Grapefruit juice enhances the bioavailability of the HIV protease inhibitor saquinavir in man. *Br J Clin Pharmacol* **45**:355–359.

Kuypers DR, Verleden G, Naesens M and Vanrenterghem Y (2005) Drug interaction between mycophenolate mofetil and rifampin: possible induction of uridine diphosphate-glucuronosyltransferase. *Clin Pharmacol Ther* **78**:81–88.

Lilja JJ, Neuvonen M and Neuvonen PJ (2004) Effects of regular consumption of grapefruit juice on the pharmacokinetics of simvastatin. *Br J Clin Pharmacol* **58**:56–60.

Lin YS, Lockwood GF, Graham MA, Brian WR, Loi CM, Dobrinska MR, Shen DD, Watkins PB, Wilkinson GR, Kharasch ED and Thummel KE (2001) In-vivo phenotyping for CYP3A by a single-point determination of midazolam plasma concentration. *Pharmacogenetics* **11**: 781–791.

Lown KS, Bailey DG, Fontana RJ, Janardan SK, Adair CH, Fortlage LA, Brown MB, Guo W and Watkins PB (1997) Grapefruit juice increases felodipine oral availability in humans by decreasing intestinal CYP3A protein expression. *J Clin Invest* **99**:2545–2553.

Lown KS, Kolars JC, Thummel KE, Barnett JL, Kunze KL, Wrighton SA and Watkins PB (1994) Interpatient heterogeneity in expression of CYP3A4 and CYP3A5 in small bowel. Lack of prediction by the erythromycin breath test. *Drug Metab Dispos* **22**:947–955.

Lundahl J, Regardh CG, Edgar B and Johnsson G (1997) Effects of grapefruit juice ingestion—pharmacokinetics and haemodynamics of intravenously and orally administered felodipine in healthy men. *Eur J Clin Pharmacol* **52**:139–145.

Mahmood I and Sahajwalla C (1999) Clinical pharmacokinetics and pharmacodynamics of buspirone, an anxiolytic drug. *Clin Pharmacokinet* **36**:277–287.

Masica AL, Mayo G and Wilkinson GR (2004) In vivo comparisons of constitutive cytochrome P450 3A activity assessed by alprazolam, triazolam, and midazolam. *Clin Pharmacol Ther* **76**:341–349.

Merry C, Barry MG, Mulcahy F, Ryan M, Heavey J, Tjia JF, Gibbons SE, Breckenridge AM and Back DJ (1997) Saquinavir pharmacokinetics alone and in combination with ritonavir in HIV-infected patients. *Aids* **11**:F29–F33.

Mizuma T (2002) Kinetic impact of presystemic intestinal metabolism on drug absorption: experiment and data analysis for the prediction of in vivo absorption from in vitro data. *Drug Metab Pharmacokinet* **17**:496–506.

Mizuma T, Tsuji A and Hayashi M (2004) Does the well-stirred model assess the intestinal first-pass effect well?. *J Pharm Pharmacol* **56**:1597–1599.

Modry DL, Stinson EB, Oyer PE, Jamieson SW, Baldwin JC and Shumway NE (1985) Acute rejection and massive cyclosporine requirements in heart transplant recipients treated with rifampin. *Transplantation* **39**:313–314.

Mouly S, Lown KS, Kornhauser D, Joseph JL, Fiske WD, Benedek IH and Watkins PB (2002) Hepatic but not intestinal CYP3A4 displays dose-dependent induction by efavirenz in humans. *Clin Pharmacol Ther* **72**:1–9.

Nakamura A, Nakajima M, Yamanaka H, Fujiwara R and Yokoi T (2008) Expression of UGT1A and UGT2B mRNA in human normal tissues and various cell lines. *Drug Metab Dispos* **36**:1461–1464.

Niemi M, Backman JT, Fromm MF, Neuvonen PJ and Kivisto KT (2003) Pharmacokinetic interactions with rifampicin: clinical relevance. *Clin Pharmacokinet* **42**:819–850.

Offermann G, Keller F and Molzahn M (1985) Low cyclosporin A blood levels and acute graft rejection in a renal transplant recipient during rifampin treatment. *Am J Nephrol* **5**:385–387.

Ohno S and Nakajin S (2009) Determination of mRNA expression of human UDP-glucuronosyltransferases and application for localization in various human tissues by real-time reverse transcriptase-polymerase chain reaction. *Drug Metab Dispos* **37**:32–40.

Paine MF, Hart HL, Ludington SS, Haining RL, Rettie AE and Zeldin DC (2006a) The human intestinal cytochrome P450 "pie". *Drug Metab Dispos* **34**:880–886.

Paine MF, Khalighi M, Fisher JM, Shen DD, Kunze KL, Marsh CL, Perkins JD and Thummel KE (1997) Characterization of interintestinal and intraintestinal variations in human CYP3A-dependent metabolism. *J Pharmacol Exp Ther* **283**:1552–1562.

Paine MF, Shen DD, Kunze KL, Perkins JD, Marsh CL, McVicar JP, Barr DM, Gillies BS and Thummel KE (1996) First-pass metabolism of midazolam by the human intestine. *Clin Pharmacol Ther* **60**:14–24.

Paine MF, Widmer WW, Hart HL, Pusek SN, Beavers KL, Criss AB, Brown SS, Thomas BF and Watkins PB (2006b) A furanocoumarin-free grapefruit juice establishes furanocoumarins as the mediators of the grapefruit juice-felodipine interaction. *Am J Clin Nutr* **83**:1097–1105.

Pang KS (2003) Modeling of intestinal drug absorption: roles of transporters and metabolic enzymes (for the Gillette Review Series). *Drug Metab Dispos* **31**:1507–1519.

Paris DG, Parente TF, Bruschetta HR, Guzman E and Niarchos AP (1994) Torsades de pointes induced by erythromycin and terfenadine. *Am J Emerg Med* **12**:636–638.

Pelkonen O, Turpeinen M, Hakkola J, Honkakoski P, Hukkanen J and Raunio H (2008) Inhibition and induction of human cytochrome P450 enzymes: current status. *Arch Toxicol* **82**:667–715.

Petty KJ and Vega JM (2008) Drug–drug interactions: marketing perspectives, in *Drug–drug Interactions* (Rodrigues AD ed) pp 711–713, Informa Healthcare, New York.

Picard N, Ratanasavanh D, Premaud A, Le Meur Y and Marquet P (2005) Identification of the UDP-glucuronosyltransferase isoforms involved in mycophenolic acid phase II metabolism. *Drug Metab Dispos* **33**:139–146.

Radominska-Pandya A, Little JM, Pandya JT, Tephly TR, King CD, Barone GW and Raufman JP (1998) UDP-glucuronosyltransferases in human intestinal mucosa. *Biochim Biophys Acta* **1394**:199–208.

Rowland M and Matin SB (1973) Kinetics of drug–drug interactions. *J Pharmacokinet Biopharm* **1**:553–567.

Rowland M and Tozer TN (1995) *Clinical Pharmacokinetics: Concepts and Applications*. Lippincott, Philadelphia.

Schmiedlin-Ren P, Edwards DJ, Fitzsimmons ME, He K, Lown KS, Woster PM, Rahman A, Thummel KE, Fisher JM, Hollenberg PF and Watkins PB (1997) Mechanisms of enhanced oral availability of CYP3A4 substrates by grapefruit constituents. Decreased enterocyte CYP3A4 concentration and mechanism-based inactivation by furanocoumarins. *Drug Metab Dispos* **25**:1228–1233.

Shen DD, Madani S, Banfield C and Clement RP (2000) H1-receptor antagonists, in *Metabolic Drug Interactions* (Levy RH, Thummel KE, Trager WF, Hansten PD and Eichelbaum M eds) pp 435–455, Lippincott Williams & Wilkins, Philadelphia.

Shipkova M, Armstrong VW, Wieland E, Niedmann PD, Schutz E, Brenner-Weiss G, Voihsel M, Braun F and Oellerich M (1999) Identification of glucoside and carboxyl-linked glucuronide conjugates of mycophenolic acid in plasma of transplant recipients treated with mycophenolate mofetil. *Br J Pharmacol* **126**:1075–1082.

Strassburg CP, Kneip S, Topp J, Obermayer-Straub P, Barut A, Tukey RH and Manns MP (2000) Polymorphic gene regulation and interindividual variation of UDP-glucuronosyltransferase activity in human small intestine. *J Biol Chem* **275**:36164–36171.

Strassburg CP, Nguyen N, Manns MP and Tukey RH (1999) UDP-glucuronosyltransferase activity in human liver and colon. *Gastroenterology* **116**:149–160.

Strolin Benedetti M and Bani M (1999) Metabolism-based drug interactions involving oral azole antifungals in humans. *Drug Metab Rev* **31**:665–717.

Tam D, Sun H and Pang KS (2003) Influence of P-glycoprotein, transfer clearances, and drug binding on intestinal metabolism in Caco-2 cell monolayers or membrane preparations: a theoretical analysis. *Drug Metab Dispos* **31**:1214–1226.

Thummel KE, Kunze KL and Shen DD (1997) Enzyme-catalyzed processes of first-pass hepatic and intestinal drug extraction. *Adv Drug Deliv Rev* **27**:99–127.

Thummel KE, Shen DD and Isoherranen N (2008) Role of the gut mucosa in metabolically based drug–drug interactions, in *Drug–drug Interactions* (Rodrigues AD ed) pp 472–473, Informa Healthcare, New York.

Thummel KE and Wilkinson GR (1998) In vitro and in vivo drug interactions involving human CYP3A. *Annu Rev Pharmacol Toxicol* **38**:389–430.

Tukey RH and Strassburg CP (2001) Genetic multiplicity of the human UDP-glucuronosyltransferases and regulation in the gastrointestinal tract. *Mol Pharmacol* **59**:405–414.

Vandevelde C, Chang A, Andrews D, Riggs W and Jewesson P (1991) Rifampin and ansamycin interactions with cyclosporine after renal transplantation. *Pharmacotherapy* **11**:88–89.

Veronese ML, Gillen LP, Burke JP, Dorval EP, Hauck WW, Pequignot E, Waldman SA and Greenberg HE (2003) Exposure-dependent inhibition of intestinal and hepatic CYP3A4 in vivo by grapefruit juice. *J Clin Pharmacol* **43**:831–839.

von Richter O, Burk O, Fromm MF, Thon KP, Eichelbaum M and Kivisto KT (2004) Cytochrome P450 3A4 and P-glycoprotein expression in human small intestinal enterocytes and hepatocytes: a comparative analysis in paired tissue specimens. *Clin Pharmacol Ther* **75**: 172–183.

von Richter O, Greiner B, Fromm MF, Fraser R, Omari T, Barclay ML, Dent J, Somogyi AA and Eichelbaum M (2001) Determination of in vivo absorption, metabolism, and transport of drugs by the human intestinal wall and liver with a novel perfusion technique. *Clin Pharmacol Ther* **70**:217–227.

Woosley RL, Chen Y, Freiman JP and Gillis RA (1993) Mechanism of the cardiotoxic actions of terfenadine. *JAMA* **269**:1532–1536.

Yang J, Jamei M, Yeo KR, Tucker GT and Rostami-Hodjegan A (2007) Prediction of intestinal first-pass drug metabolism. *Curr Drug Metab* **8**:676–684.

Yang J, Kjellsson M, Rostami-Hodjegan A and Tucker GT (2003) The effects of dose staggering on metabolic drug–drug interactions. *Eur J Pharm Sci* **20**:223–232.

Zhang QY, Dunbar D, Ostrowska A, Zeisloft S, Yang J and Kaminsky LS (1999) Characterization of human small intestinal cytochromes P-450. *Drug Metab Dispos* **27**:804–809.

Zhang X, Jones DR and Hall SD (2009) Prediction of the effect of erythromycin, diltiazem, and their metabolites, alone and in combination, on CYP3A4 inhibition. *Drug Metab Dispos* **37**:150–160.

Zhou C, Tabb MM, Sadatrafiei A, Grun F and Blumberg B (2004) Tocotrienols activate the steroid and xenobiotic receptor, SXR, and selectively regulate expression of its target genes. *Drug Metab Dispos* **32**:1075–1082.

Chapter 18
Transporter-Based Drug–Drug Interactions and Their Effect on Distribution Volumes

Anita Grover and Leslie Z. Benet

Abstract Recently, drug transporters have emerged as significant modifiers of a patient's pharmacokinetics. In cases where the functioning of drug transporters is altered, such as by drug–drug interactions, the resulting change in volume of distribution can lead to a significant change in drug effect or likelihood of toxicity, as well as a change in half-life independent of a change in clearance. Here, we discuss pharmacokinetic interactions at the transporter level that have been investigated in animals and humans and reported in literature, with a focus on the changes in distribution volume. Trends are discussed as they may be used to predict volume changes given the function of a transporter and the primary location of the interaction.

18.1 Introduction

Significant attention has been paid to pharmacokinetic drug–drug interactions (DDIs) resulting from inhibition or induction of metabolizing enzymes in the gut and the liver. Such interactions are accepted to cause changes in bioavailability, clearance, and the associated half-life, but generally considered not to have an effect on volume of distribution. Less consideration has been given, therefore, to changes in distribution volume following potential DDIs. However, with the recent advances in the understanding of the role of drug transporters in pharmacokinetics, it has become critical to also understand drug–drug interactions that are rooted

---

L.Z. Benet (✉)
Department of Biopharmaceutical Sciences, University of California, San Francisco, CA, USA
e-mail: leslie.benet@ucsf.edu

K.S. Pang et al. (eds.), *Enzyme- and Transporter-Based Drug–Drug Interactions*, DOI 10.1007/978-1-4419-0840-7_18,

in transporters. Unlike enzyme-mediated interactions, these may markedly change distribution volume.

18.1.1 Volume of Distribution: V_1, V_{area}, and V_{ss}

The pharmacokinetic parameter volume of distribution describes the relationship between the measured systemic concentrations and the amount of drug in the body. It is a measure of the extent of tissue distribution, and it usually does not represent any physiological volume. Instead, it is considered a theoretical parameter that is dependent on a variety of drug properties: traditionally, lipophilicity (a measure of tissue affinity) and plasma protein binding (Øie, 1986) and, more recently, a drug's status as a substrate for transporters (Grover and Benet, 2009).

The relationship between systemic concentrations and amount of drug in the body can differ depending on the dosing regimen and time at which the parameters are measured. In particular, V_1, the initial dilution volume, defined as the dose divided by the initial plasma concentration following an intravenous bolus dose, is likely to be relatively small because equilibration to the other tissue spaces has not yet occurred. The volume at the terminal phase of elimination, V_{area} (also known as V_z), will be greater. Most commonly, this represents the phase when distribution is complete and elimination from the plasma is predominant, so drug is re-entering the circulation from the tissue spaces. V_{area} is defined as the clearance divided by the terminal rate of elimination and is therefore heavily dependent on the terminal rate of elimination. This rate is often a more difficult parameter to estimate experimentally because it requires concentration data for a long time period following the dose, when concentrations may begin to fall below the limits of sensitivity for some analytical methodology. Finally, the volume at steady state, V_{ss}, is the sum of the distribution volumes of all the compartments in a pharmacokinetic model. It can also be calculated from a single dose as the product of clearance and the mean residence time in the body following non-compartmental analysis. Its value will be between V_1 and V_{area}, and it is considered to be a more "accurate" measure of whole body distribution volume, as it is less directly dependent on changes in the elimination processes, a characteristic of V_{area}.

It is also important to note that calculations of volume are highly model dependent. The multicompartment volume parameters as defined above assume elimination from the central compartment of a pharmacokinetic model (Nakashima and Benet, 1988). The typical compartmental model assumes the liver and kidneys are part of the central compartment, as they are highly perfused organs and assumed to be in rapid equilibrium with the plasma, thus providing for central compartment elimination. However, when elimination does not occur from the central compartment, V_{ss} and V_{area} will significantly under-predict distribution volume (Yates and Arundel, 2008a, b). Therefore, it may be important to consider a possible case where transporter dysfunction means the liver and kidneys are not in rapid equilibrium with the plasma, such that elimination occurs from a peripheral compartment. We will return to this topic.

18.1.2 Drug Transporters and Drug–Drug Interactions

Drug transporters are found at numerous tissues in the body, implicating them as players in drug distribution. While a variety of transporters, including P-glycoprotein (P-gp), BCRP, and some members of the OATP family, are heavily expressed at the intestinal epithelium, they should not affect volume of distribution, as volume terms are related to the behavior of the drug once it has entered the systemic circulation. Within the body, the liver and kidneys express the greatest variety and level of drug transporters. At these two organs, transporters modulate access to metabolizing enzymes and excretion processes, both biliary and renal. Consequently, they are likely to also have an effect on other pharmacokinetic parameters, particularly clearance and half-life. The majority of published reports, therefore, focus on primary transporter interactions at either the liver or the kidneys.

Drug transporters can be loosely characterized as either uptake or efflux, denoting whether they facilitate drug entry into a cell or efflux out of a cell. Thus, an uptake transporter with reduced function prevents drug accumulation in the tissue expressing the transporter, while an efflux transporter with decreased function increases accumulation in the tissue expressing the transporter. The effect on total distribution volume depends on the tissue expressing the transporter, whether it is an uptake or efflux transporter and where the transporter is expressed in this tissue.

In those cases where the functioning of drug transporters is altered by a drug–drug interaction, the resulting change in volume of distribution can lead to a significant change in drug effect or likelihood of toxicity, as well as a change in half-life independent of a change in clearance. Generally, because many transporters have a wide range of substrates, DDIs in this consideration are rooted in the inhibition of a transporter, leading to decreased functionality. Genetic polymorphisms in the transporter gene or relevant genetic control elements may also affect the functioning of a transporter and, similarly, most polymorphisms will result in a reduced function transporter. Thus, here we primarily discuss decreased transporter function and its effects on the distribution volume. It is possible, however, for an increase in transporter function to occur either by up-regulation of the transporter gene under multiple dosing conditions or by a theoretical polymorphism that creates a transport protein with increased effectiveness or an increased amount of protein. Such transporter induction interactions have been reported to affect bioavailability in the gut; however, these should not affect distribution volume. While in vitro induction interactions have been reported at other tissues, including the liver and kidney (Jette et al., 1996; Brady et al., 2002; Cherrington et al., 2002), clinical, pharmacokinetic interactions with increased functioning transporters have not yet been reported outside the gut.

18.2 Trends in Transporter Effects on Distribution Volume

Grover and Benet (2009) reported a number of interactions at the transporter level that have been clinically investigated and reported in the literature. A summary of 37 such interactions, including the effects on the primary pharmacokinetic parameters, is included in Table 18.1. Further analysis of these interactions revealed a few interesting trends:

1. The magnitude of transporter-mediated change in volume of distribution may differ depending on which measure of volume is used.
2. A transporter-mediated change in volume of distribution may be independent of or correlated to a change in the drug's clearance and the associated half-life.
3. In general, interactions at uptake transporters at the liver lead to a significant decrease in volume of distribution, while those at the renal tubules do not lead to a change in volume of distribution, although there are exceptions.
4. Interactions with efflux transporters at the liver generally lead to a decrease in volume of distribution, while those at the renal tubules lead to an increase in volume of distribution.
5. The primary location of the interaction (liver or kidneys) is a more important determinant of the change in distribution volume than the secondary change in tissue distribution, as evidenced by interactions that affect the integrity of the blood–brain barrier.

Each of these trends will be discussed in relation to the interactions presented in Table 18.1.

18.2.1 Measures of Volume

The magnitude of transporter-mediated change in volume of distribution may differ depending on which measure of volume is used.

As discussed above, there are three measures of distribution volume. The relative contribution of changes in transporter function to these three measures of volume may differ. Grover and Benet (2009) extracted the plasma concentration–time data from four studies (Yagi et al., 2003; Breedveld et al., 2004; Lam et al., 2006; Yamashiro et al., 2006) and reanalyzed the pharmacokinetics. For each study, the three measures of volume were calculated under both control and decreased transporter functionality conditions, as shown in Table 18.2. V_1 changes less markedly than V_{ss} and V_{area}. This is expected since full tissue distribution is not likely to have occurred at the initial time points, and transporters not directly associated with very rapidly equilibrating organs may not have had the chance to exert their effects. In contrast, after all organs are in distribution equilibrium, V_{ss} and V_{area} show the full effect of transporter inhibition, exhibiting bigger changes than seen for V_1. With respect to transporter effects on equilibrium volume measures, little difference is seen between the effects on V_{ss} and V_{area}.

Table 18.1 Summary of transporter-based interactions and their effect on volume, clearance, and half-life after Grover and Benet (2009)

Drug	Interaction	*V*	*CL*	$t_{1/2}$	Mechanism of clearance in reference organism	Related transporter comments	References
Adriamycin	Verapamil	↑ 31.2% (V_{ss})	↓ 32.6%	↑ 37.7%	Metabolism > Biliary > Renal	Inhibition of P-gp efflux in humans after iv dosing; in rats, renal transport is saturable but biliary is not, indicating a transporter is involved in renal elimination	Kerr et al. (1986) and Tavoloni and Guraino (1980)
Atorvastatin	Rifampin	↓ 94.3% (V_{ss}/F)	↓ 87.0% (CL/F)	↓ 63.1%	Metabolism	Inhibition of OATP1B1 uptake into the liver in humans after oral dosing	Lau et al. (2007)
Atorvastatin	OATP1B1 reduced function allele	↓ 57.6% (V_{area}/F)	↓ 59.2% (CL/F)	↔	Metabolism	Reduced OATP1B1 uptake into the liver in humans after oral dosing	Pasanen et al. (2007)
Cefazolin	Probenecid	↔ (V_1)	↓ 32.1%	↑ 68.8%	Renal	Inhibition of OAT-mediated uptake into the renal tubules in humans after intramuscular or iv dosing, likely OAT2, OAT3	Brown et al. (1993), Shitara et al. (2005), Sakurai et al. (2005), and Khamdang et al. (2003)

Table 18.1 (continued)

Drug	Interaction	*V*	*CL*	$t_{1/2}$	Mechanism of clearance in reference organism	Related transporter comments	References
Cerivastatin	Cyclosporine	↓ 66.7% (V_1/F)	↓ 73.3% (CL/F)	↔	Metabolism	Inhibition of OATP1B1 uptake into the liver in humans after oral dosing	Muck et al. (1999) and Shitara et al. (2006)
Cetirizine	Pilsicainide	↑ 102% (V_{ss}/F)	↓ 28.9% (CL/F); ↓ 37.6% (CL_r)	↑ 188%	Renal >> Metabolism	Inhibition of P-gp efflux in humans after oral dosing, possible inhibition of OCT2 as well, but P-gp effect is more significant	Tsuruoka et al. (2006)
Ciprofloxacin	Probenecid	↔ (V_{ss})	↓ 41.2% (CL); ↓ 64.0% (CL_r)	↑ 51.5%	Renal > Metabolism	Inhibition of OAT-mediated uptake into the renal tubules in humans after iv dosing	Jaehde et al. (1995)
Colchicine	SDZ PSC 833	↔ (V_{area})	↓ 47.0%	↔	Biliary > Renal > Metabolism	Inhibition of P-gp efflux in rats after iv dosing	Desrayaud et al. (1997) and Speeg et al. (1992)
Daunomycin	Verapamil	↓ 63.7% (V_{area})	↓ 89.1%	↑ 232%	Metabolism > Renal ~ Biliary	Inhibition of P-gp efflux in rats after iv dosing	Nooter et al. (1987) and Yesair et al. (1972)
Digoxin	Rifampin	↓ 70.8% (V_{ss})	↓ 54.2%	↔	Metabolism	Inhibition of Oatp1a4 uptake into the liver under Cyp3a-induced conditions in rats after iv dosing	Lam et al. (2006)

Table 18.1 (continued)

Drug	Interaction	V	CL	$t_{1/2}$	Mechanism of clearance in reference organism	Related transporter comments	References
Digoxin	Ritonavir	↑ 76.7% (V_{ss})	↓ 41.8%	↑ 156%	Renal	Inhibition of P-gp efflux in humans after iv dosing	Ding et al. (2004)
Famotidine	Probenecid	↓ 39.5% (V_1/F)	↓ 64.0% (CL_r)	↔	Renal	Inhibition of uptake into the renal tubules, likely OAT3, in humans after oral dosing	Inotsume et al. (1990) and Tahara et al. (2006)
Glyburide	Rifampin	↓ 67.4% (V_{ss}/F)	↓ 54.6% (CL/F)	↔	Metabolism	Reduced OATP2B1 uptake into the liver in humans after oral dosing	Zheng et al. (2009)
Metformin	OCT1 reduced function allele	↓ 53.9% (V_{area}/F)	↓ 37.5% (CL/F)	↔	Renal	Reduced OCT1 uptake into the liver in humans after oral dosing; OCT1 does not mediate elimination at the renal tubules	Shu et al. (2008)
Methotrexate	Pantoprazole	↓ 21.6% (V_{area})	↓ 45.7%	↑ 44.4%	Biliary > Renal	Inhibition of Bcrp-mediated efflux in mice after iv dosing	Breedveld et al. (2004)
Penicillin G	oat3–/– mice	↓ 33.5% (V_{ss})	↓ 44.8%	↑ 72.9%	Renal > Metabolism	Decreased renal clearance in mice after iv dosing	VanWert et al. (2007)

Table 18.1 (continued)

Drug	Interaction	V	CL	$t_{1/2}$	Mechanism of clearance in reference organism	Related transporter comments	References
Pilsicainide	Cetirizine	$\uparrow$ 106% (V_{ss}/F)	$\downarrow$ 26.6% (CL/F); $\downarrow$ 41.3% (CL_r)	$\uparrow$ 206%	Renal >> Metabolism	Inhibition of P-gp efflux after oral dosing in humans, possible inhibition of OCT2 as well, but P-gp effect is more significant	Tsuruoka et al. (2006)
Pilsicainide	Cimetidine	$\leftrightarrow$ (V_1/F)	$\downarrow$ 26.4% (CL/F); $\downarrow$ 28.0% (CL_r)	$\uparrow$ 24.5%	Renal >> Metabolism	Inhibition of uptake into the renal tubules, likely OCT2, in humans after oral dosing	Tsuruoka et al. (2006) and Shiga et al. (2000)
Pitavastatin	OATP1B1 reduced function allele	$\downarrow$ 27.6% (V_{area}/F)	$\downarrow$ 44.3% (CL/F)	$\uparrow$ 29.9%	Biliary	Reduced OATP1B1 uptake into the liver in humans after oral dosing	Chung et al. (2005) and Hirano et al. (2005)
Procainamide	Cimetidine	$\downarrow$ 12.4% (V_{area}/F)	$\downarrow$ 30.5% (CL/F); $\downarrow$ 43.4% (CL_r)	$\uparrow$ 26.0%	Renal > Metabolism	Inhibition of OCT-mediated uptake in humans after oral dosing; not likely a gut interaction	Bauer et al. (2005) and Somogyi et al. (1983)
Procainamide	Ciprofloxacin	$\leftrightarrow$ (V_{area} and V_{ss})	$\leftrightarrow$ (CL); $\downarrow$ 14.8% (CL_r)	$\leftrightarrow$	Renal > Metabolism	Inhibition of OCT-mediated uptake in humans after iv dosing	Bauer et al. (2005) and Somogyi et al. (1983)

Table 18.1 (continued)

Drug	Interaction	V	CL	$t_{1/2}$	Mechanism of clearance in reference organism	Related transporter comments	References
Procainamide	Levofloxacin	↔ (V_{area} and V_{ss})	↓ 17.2% (CL); ↓ 25.9% (CL_r)	↑ 18.5%	Renal > Metabolism	Inhibition of OCT-mediated uptake in humans after iv dosing	Bauer et al. (2005) and Somogyi et al. (1983)
Repaglinide	Cyclosporine	↓ 59.0% (V_{area}/F)	↓ 59.0% (CL/F)	↔	Metabolism	Inhibition of OATP1B1 uptake into the liver in humans after oral dosing	Kajosaari et al. (2005)
Rosuvastatin	Cyclosporine	↓ 90.6% (V_{area}/F)	↓ 80.1% (CL/F)	↓ 52.7%	Biliary >> Metabolism	Inhibition of OATP1B1 uptake into the liver in humans after oral dosing; $t_{1/2}$ is normally reported as lower than it is in this study's baseline; decrease in $t_{1/2}$ might be artifactual	Simonson et al. (2004)
Rosuvastatin	Gemfibrozil	↓ 27.5% (V_{area}/F)	↓ 46.8% (CL/F)	↑ 36.3%	Biliary >> Metabolism	Inhibition of OATP1B1 uptake into the liver in humans after oral dosing	Schneck et al. (2004)
Rosuvastatin	OATP1B1 reduced function allele	↓ 50.9% (V_{area}/F)	↓ 38.3% (CL/F)	↔	Biliary >> Metabolism	Reduced OATP1B1 uptake into the liver in humans after oral dosing	Pasanen et al. (2007)

Table 18.1 (continued)

Drug	Interaction	V	CL	$t_{1/2}$	Mechanism of clearance in reference organism	Related transporter comments	References
Sotalol	Cimetidine	↔ (V_{ss})	↓ 26.7%	↔	Renal	Inhibition of OCT-mediated uptake in rats after iv dosing, likely OCT1, OCT2	Carr et al. (1996) and Ullrich (1999)
Tacrolimus	mdr1a–/– mice	↔ (V_{ss})	↓ 65.4%	↑ 99.4%	Metabolism >> Biliary > Renal	Lack of P-gp-mediated biliary clearance in mice after iv dosing	Yokogawa et al. (1999)
Telmisartan	Nisoldipine	↓ 62.2% (V_{area}/F)	↓ 51.2% (CL/F)	↔	Metabolism	Inhibition of P-gp efflux of metabolites from the liver in humans after oral dosing; the majority of the effect is likely a bioavailability consideration	Bajcetic et al. (2007)
Tetracycline	Diclofenac	↔ (V_{ss})	↓ 60.7% (CL); ↓ 61.0% (CL_r)	↑ 88.2%	Renal	Inhibition of OAT-mediated uptake in rats after iv dosing, likely OAT1, OAT2, OAT4	Khamdang et al. (2003), Oh and Han (2006), and Dresser et al. (2001)
Tetracycline	Naproxen	↔ (V_{ss})	↓ 75.0% (CL); ↓ 72.0% (CL_r)	↑ 400%	Renal	Inhibition of OAT-mediated uptake in rats after iv dosing, likely OAT1, OAT2, OAT4	Khamdang et al. (2003), Oh and Han (2006), and Dresser et al. (2001)

Table 18.1 (continued)

Drug	Interaction	V	CL	$t_{1/2}$	Mechanism of clearance in reference organism	Related transporter comments	References
Tezosentan	Cyclosporine	↓ 65.2% (V_{ss})	↓ 74.8%	↔	Biliary	Likely inhibition of P-gp efflux in humans after iv dosing	van Giersbergen et al. (2002)
Topotecan	Probenecid	↔ (V_{ss})	↔ (CL); ↓ 29.6% (CL_r)	↔	Renal > Biliary >> Metabolism	Inhibition of OAT-mediated uptake in mice after iv dosing	Zamboni et al. (1998)
Topotecan	Novobiocin	↑ 254% (V_{ss})	↓ 33.7%	↑ 341%	Renal > Biliary >> Metabolism	Inhibition of Bcrp efflux in rats after iv dosing	Su et al. (2007)
Ulifloxacin	Cyclosporine	↓ 31.4% (V_{ss})	↓ 32.1% (CL); ↓ 66.5% (CL_b)	↔	Biliary ~ Renal	Inhibition of Oat/Oatp-mediated uptake into the liver in rats after iv dosing	Yagi et al. (2003, 2007) and Nakashima et al. (1994)
Ulifloxacin	EHBR rats	↓ 34.1% (V_{ss})	↔	↓ 35.0%	Biliary ~ Renal	Lack of Mrp2 decreases biliary excretion of glucuronide; no change in clearance or percent of dose eliminated in the bile or urine after iv dosing	Yagi et al. (2003, 2007) and Nakashima et al. (1994)
Valsartan	EHBR rats	↓ 35.8% (V_1)	↓ 94.3% (CL); ↓ 97.7% (CL_b)	↑ 1030%	Biliary >> Metabolism > Renal	Lack of Mrp2 decreases biliary clearance, leading to increased plasma and liver concentrations after iv dosing	Yamashiro et al. (2006)

CL_r: renal clearance, CL_b: biliary clearance, EHBR: Eisai hyperbilirubinemic rats

Table 18.2 Effect of change in transporter function on the distribution volume parameters

	(mL/kg)	**Control**	**+Rifampin**	**% Decrease**	**References**
Digoxin (Oatp1a4)	V_1	933	454	51.2	Lam et al. (2006)
	V_{ss}	7140	1640	77.1	
	V_{area}	7360	1790	75.7	
	(L/kg)	**Control**	**+Pantoprazole**	**% Decrease**	
Methotrexate (Bcrp)	V_1	417	315	24.4	Breedveld et al. (2004)
	V_{ss}	933	651	30.3	
	V_{area}	1010	689	31.9	
	(mL/kg)	**Control**	**+Cyclosporine**	**% Decrease**	
Ulifloxacin (Oat/Oatp)	V_1	831	702	15.6	Yagi et al. (2003)
	V_{ss}	4450	3070	31.0	
	V_{area}	4880	3320	31.9	
	(mL/kg)	**Control**	**In EHBR rats**	**% Decrease**	
Valsartan (Mrp2)	V_1	51.0	44.7	12.4	Yamshiro et al. (2006)
	V_{ss}	258	111	57.2	
	V_{area}	420	118	71.9	

As presented by Grover and Benet (2009) the plasma concentration–time data from four studies were extracted, and the three measures of volume were calculated under both control and decreased transporter functionality conditions. EHBR: Eisai hyperbilirubinemic rats

18.2.2 Inter-relationship Between Volume, Clearance, and Half-Life

A transporter-mediated change in volume of distribution can be independent of or correlated to a change in the drug's clearance and the associated half-life.

Half-life is considered the most important parameter to the clinician for determining dosing changes due to drug–drug interactions or pharmacogenomic variability, as it is considered the parameter most closely associated with dosing interval and duration of drug effect. In the simplest relation, half-life, clearance, and volume of distribution are related by Equation (18.1):

$$t_{1/2} \approx \frac{\ln(2) \cdot V}{CL} \tag{18.1}$$

Therefore, the change in half-life is proportional to the change in distribution volume and inversely related to the change in clearance. In this simple single-phase approximation, there will only be one volume term ($V = V_1 = V_{ss} = V_{area}$). In reality, most drugs exhibit multiple phases of distribution and/or elimination and may have many half-lives (Sahin and Benet, 2008). However, changes in this single-phase approximation are still indicative of a general pharmacokinetic trend. Further, as noted by Sahin and Benet (2008), many different, single value half-lives can be reported for a drug that almost assuredly exhibits multicompartment kinetics, including the single-phase approximation or the half-life for the terminal phase. Therefore, for a number of drugs in Table 18.1, the relationship between clearance, volume, and half-life will not follow Equation (18.1).

For example, in rats, digoxin is primarily metabolized in the liver by Cyp3a. It is a substrate for Oatp1a4 uptake and P-glycoprotein efflux in hepatocytes. When rats were dosed with dexamethasone, Cyp3a, Oatp1a4, and P-gp were induced. Following administration of a single dose of the Oatp-inhibitor rifampin to these dexamethasone-induced rats, a decrease in steady-state volume of distribution of 70.8% was observed together with a decrease in clearance of 54.2%, while no change in half-life was evident in comparison to the induced, but not rifampin inhibited, controls (Lam et al., 2006). From previous studies, it is known that the concentration of inhibitor achieved after the single dose had minimal effects on Cyp3a and P-gp. Therefore, inhibition of the uptake transporter led to the pharmacokinetic changes observed. Inhibition of Oatp1a4 prevents liver accumulation, decreasing distribution volume. Preventing liver entry also prevents metabolism, leading to the decrease in clearance. Possibly because Oatp1a4 is also expressed at the blood–brain barrier, choroid plexus, ciliary bodies, and retina, in addition to the liver (Hagenbuch and Meier, 2004), the decrease in volume is greater than the decrease in clearance. In this case, the changes in volume and clearance appear to be correlated.

In humans, however, digoxin is predominantly excreted unchanged in the urine. This process is mediated by P-gp. In patients concomitantly dosed with ritonavir, a P-gp inhibitor, steady-state volume increased 76.7%, clearance decreased 41.8%, and half-life increased 156% as compared to controls (Ding et al., 2004). Here, inhibition of P-gp prevents efflux of drug from the renal epithelial cells into the urine, decreasing clearance. Inhibition of P-gp also prevents efflux of drug from other tissues protected by P-gp, such as the brain and heart. Therefore, the drug is more widely distributed in the body, and there is less drug in the systemic circulation, making less available to be cleared by the kidneys. Because the clearance rate is also decreased, both factors work toward increasing half-life. In this case, the changes in volume and clearance are not correlated.

These examples elucidate mechanisms by which transporter inhibition can lead to significantly different pharmacokinetic patterns for the clinician to consider.

18.2.3 Interactions Involving Hepatic and Renal Uptake

In general, interactions at uptake transporters at the liver lead to a significant decrease in volume of distribution, while those at the renal tubules do not lead to a change in volume of distribution, although there are exceptions.

Of 24 interactions that involved uptake transporters with decreased function, 9 did not cause a significant change in distribution volume. Each of these involved interactions documented at uptake transporters at the renal tubules for drugs that are primarily excreted unchanged in the urine. Three of twelve renal interactions did exhibit decreased volume. Conversely, the 12 interactions attributed to the liver all led to a decreased volume of distribution. A decrease in volume of distribution would be expected in these interactions, as inhibiting an uptake transporter prevents tissue accumulation. These uptake interactions are retabulated as either hepatic or renal in Table 18.3.

From a physiological perspective, the liver is significantly more massive than the kidneys: In the average man, the kidneys weigh about 150 g each, and the liver weighs about 1.5 kg (Guyton and Hall, 2006). The liver also contains more cellular space available for transporter expression, while considerable kidney mass is interstitial fluid and tubule volume. Similarly, hepatocytes are more available to drug sequestration and storage than kidney epithelial cells. Therefore, preventing drug from entering the hepatocytes will have a greater relative effect on the entire body volume of distribution than will preventing drug from entering the epithelial cells at the renal tubules. Despite substantial decreases in renal clearance and associated increases in systemic concentrations, it appears that volume change due to interactions at the kidney level is not observable.

A few further explanations for this disparity are possible. For one, if the kidney transporters were unique to the renal epithelium, while liver transporters were also expressed at other tissues in the body, inhibition of liver transporters would cause a more significant pharmacokinetic change. However, it seems like the opposite may be true: Table 18.4 shows the tissue distribution of the transporters highlighted in this chapter. A second possibility is that the transporters that are relatively uniquely expressed at the liver are also more specific for their substrates, while the renal transporters act on a wider range of substrates. In that case, inhibition of a renal transporter would not have much of an effect because another transporter could

Table 18.3 Summary of uptake transporter-based interactions at the liver and renal tubules

Drug	Interaction	V	CL	$t_{1/2}$	Mechanism of clearance in reference organism	Transporter-related comments	References
Hepatic interactions							
Atorvastatin	Rifampin	↓ 94.3% (V_{ss}/F)	↓ 87.0% (CL/F)	↓ 63.1%	Metabolism	Inhibition of OATP1B1 uptake into the liver in humans after oral dosing	Lau et al. (2007)
Atorvastatin	OATP1B1 reduced function allele	↓ 57.6% (V_{area}/F)	↓ 59.2% (CL/F)	↔	Metabolism	Reduced OATP1B1 uptake into the liver in humans after oral dosing	Pasanen et al. (2007)
Cerivastatin	Cyclosporine	↓ 66.7% (V_1/F)	↓ 73.3% (CL/F)	↔	Metabolism	Inhibition of OATP1B1 uptake into the liver in humans after oral dosing	Muck et al. (1999) and Shitara et al. (2006)
Digoxin	Rifampin	↓ 70.8% (V_{ss})	↓ 54.2%	↔	Metabolism	Inhibition of Oatp1a4 uptake into the liver under Cyp3a-induced conditions in rats after iv dosing	Lam et al. (2006)
Glyburide	Rifampin	↓ 67.4% (V_{ss}/F)	↓ 54.6% (CL/F)	↔	Metabolism	Inhibition of OATP2B1 uptake into the liver in humans after oral dosing	Zheng et al. (2009)

Table 18.3 (continued)

Drug	Interaction	V	CL	$t_{1/2}$	Mechanism of clearance in reference organism	Transporter-related comments	References
Metformin	OCT1 reduced function allele	$\downarrow$ 53.9% (V_{area}/F)	$\downarrow$ 37.5% (CL/F)	$\leftrightarrow$	Renal	Reduced OCT1 uptake into the liver in humans after oral dosing; OCT1 does not mediate elimination at the renal tubules	Shu et al. (2008)
Pitavastatin	OATP1B1 reduced function allele	$\downarrow$ 27.6% (V_{area}/F)	$\downarrow$ 44.2% (CL/F)	$\uparrow$ 29.9%	Biliary	Reduced OATP1B1 uptake into the liver in humans after oral dosing	Chung et al. (2005) and Hirano et al. (2005)
Repaglinide	Cyclosporine	$\downarrow$ 59.0% (V_{area}/F)	$\downarrow$ 59.0% (CL/F)	$\leftrightarrow$	Metabolism	Inhibition of OATP1B1 uptake into the liver in humans after oral dosing	Kajosaari et al. (2005)
Rosuvastatin	Cyclosporine	$\downarrow$ 90.6% (V_{area}/F)	$\downarrow$ 80.1% (CL/F)	$\downarrow$ 52.7%	Biliary >> metabolism	Inhibition of OATP1B1 uptake into the liver in humans after oral dosing; $t_{1/2}$ is normally reported as lower than it is in this study's baseline; decrease in $t_{1/2}$ might be artifactual	Simonson et al. (2004)

Table 18.3 (continued)

Drug	Interaction	V	CL	$t_{1/2}$	Mechanism of clearance in reference organism	Transporter-related comments	References
Rosuvastatin	Gemfibrozil	↓ 27.5% (V_{area}/F)	↓ 46.8% (CL/F)	↑ 36.3%	Biliary >> metabolism	Inhibition of OATP1B1 uptake into the liver in humans after oral dosing	Schneck et al. (2004)
Rosuvastatin	OATP1B1 reduced function allele	↓ 50.9% (V_{area}/F)	↓ 38.3% (CL/F)	↔	Biliary >> metabolism	Reduced OATP1B1 uptake into the liver in humans after oral dosing	Pasanen et al. (2007)
Ulifloxacin	Cyclosporine	↓ 31.4% (V_{ss})	↓ 32.1% (CL); ↓ 66.5% (CL_b)	↔	Renal ~ biliary	Inhibition of Oat/Oatp-mediated uptake into the liver in rats after iv dosing	Yagi et al. (2003, 2007) and Nakashima et al. (1994)
Renal interactions							
Cefazolin	Probenecid	↔ (V_1)	↓ 32.1%	↑ 68.8%	Renal	Inhibition of OAT-mediated uptake into the renal tubules in humans after intramuscular or iv dosing, likely OAT2, OAT3	Brown et al. (1993), Shitara et al. (2005), Sakurai et al. (2005) and Kamdang et al. (2003)
Ciprofloxacin	Probenecid	↔ (V_{ss})	↓ 41.2% (CL); ↓ 64.1% (CL_r)	↑ 51.5%	Renal > metabolism	Inhibition of OAT-mediated uptake into the renal tubules in humans after iv dosing	Jaehde et al. (1995)

Table 18.3 (continued)

Drug	Interaction	V	CL	$t_{1/2}$	Mechanism of clearance in reference organism	Transporter-related comments	References
Famotidine	Probenecid	$\downarrow$ 39.5% (V_1/F)	$\downarrow$ 52.4% (CL); $\downarrow$ 64.0% (CL_r)	$\leftrightarrow$	Renal	Inhibition of uptake into the renal tubules, likely OAT3, in humans after oral dosing	Inotsume et al. (1990) and Tahara et al. (2006)
Penicillin G	oat3-/- mice	$\downarrow$ 33.5% (V_{ss})	$\downarrow$ 44.8%	$\uparrow$ 72.9%	Renal > metabolism	Decreased renal clearance in mice after iv dosing	VanWert et al. (2007)
Pilsicainide	Cimetidine	$\leftrightarrow$ (V_1/F)	$\downarrow$ 26.4% (CL/F); $\downarrow$ 28.0% (CL_r)	$\uparrow$ 24.5%	Renal >> metabolism	Inhibition of uptake into the renal tubules, likely OCT2, in humans after oral dosing	Tsuruoka et al. (2006) and Speeg et al. (1992)
Procainamide	Cimetidine	$\downarrow$ 12.4% (V_{area}/F)	$\downarrow$ 30.5% (CL/F); $\downarrow$ 43.4% (CL_r)	$\uparrow$ 26.0%	Renal > metabolism	Inhibition of OCT-mediated uptake in humans after oral dosing; not likely a gut interaction	Bauer et al. (2005) and Somogyi et al. (1983)
Procainamide	Ciprofloxacin	$\leftrightarrow$ (V_{area} and V_{ss})	$\leftrightarrow$ (CL); $\downarrow$ 14.8% (CL_r)	$\leftrightarrow$	Renal > metabolism	Inhibition of OCT-mediated uptake in humans after iv dosing	Bauer et al. (2005) and Somogyi et al. (1983)
Procainamide	Levofloxacin	$\leftrightarrow$ (V_{area} and V_{ss})	$\downarrow$ 17.2% (CL); $\downarrow$ 25.9% (CL_r)	$\uparrow$ 18.5%	Renal > metabolism	Inhibition of OCT-mediated uptake in humans after iv dosing	Bauer et al. (2005) and Somogyi et al. (1983)

Table 18.3 (continued)

Drug	Interaction	V	CL	$t_{1/2}$	Mechanism of clearance in reference organism	Transporter-related comments	References
Sotalol	Cimetidine	$\leftrightarrow$ (V_{ss})	$\downarrow$ 26.7%	$\leftrightarrow$	Renal	Inhibition of OCT-mediated uptake in rats after iv dosing, likely OCT1, OCT2	Carr et al. (1996) and Ullrich (1999)
Tetracycline	Diclofenac	$\leftrightarrow$ (V_{ss})	$\downarrow$ 60.7% (CL); $\downarrow$ 61.0% (CL_r)	$\uparrow$ 88.2%	Renal	Inhibition of OAT-mediated uptake in rats after iv dosing, likely OAT1, OAT2, OAT4	Kamdang et al. (2003), Oh and Han (2006) and Dresser et al. (2001)
Tetracycline	Naproxen	$\leftrightarrow$ (V_{ss})	$\downarrow$ 75.0% (CL); $\downarrow$ 72.0% (CL_r)	$\uparrow$ 400%	Renal	Inhibition of OAT-mediated uptake in rats after iv dosing, likely OAT1, OAT2, OAT4	Kamdang et al. (2003), Oh and Han (2006) and Dresser et al. (2001)
Topotecan	Probenecid	$\leftrightarrow$ (V_{ss})	$\leftrightarrow$ (CL); $\downarrow$ 29.6% (CL_r)	$\leftrightarrow$	Renal > biliary >> metabolism	Inhibition of OAT-mediated uptake in mice after iv dosing	Su et al. (2007)

Table 18.4 Tissue distribution of transporters for which volume modulation has been investigated after Grover and Benet (2009)

Transporter	Species	Tissue expression	References
A. Efflux			
P-gp	Human	Adrenal, sweat glands, blood vessels, liver, kidney, lung > muscle, mammary glands, spleen, gall bladder, heart	van der Valk et al. (1990)
P-gp (mdr1a/b)	Rat	Brain > kidney > lung > liver	Brady et al. (2002)
P-gp (mdr1a)	Mouse	Adrenal > placenta> kidney, heart > liver, uterus, muscle, spleen, brain, lung	Croop et al. (1989)
Mrp2	Rat	Liver > kidney, brain	Cherrington et al. (2002)
Bcrp	Rat	Kidney > liver > gonads > brain > thymus, spleen	Tanaka et al. (2005)
Bcrp	Mouse	Kidney > liver > gonads > brain > spleen, muscle, lung	Tanaka et al. (2005)
B. OATPs			
OATP1B1	Human	Liver	Hagenbuch and Meier (2004)
OATP2B1	Human	Liver, placenta, ciliary body	Hagenbuch and Meier (2004)
Oatp1a4	Rat	Liver, brain; eye	Dresser et al. (2001) and Hagenbuch and Meier (2004)
C. OATs			
OAT1	Human	Kidney > brain	Dresser et al. (2001)
OAT2	Human	Liver > kidney	Pavlova et al. (2000)
OAT3	Human	Kidney > brain	Dresser et al. (2001)
OAT4	Human	Kidney; placenta	Dresser et al. (2001) and Sekine et al. (2000)
Oat1	Rat	Kidney > brain	Dresser et al. (2001)
Oat2	Rat	Liver > kidney	Dresser et al. (2001)
Oat3	Rat	Liver ~ kidney, brain; eye	Dresser et al. (2001), Sekine et al. (2000), and Sweet et al. (2002)
Oat4	Rat	Kidney, likely placenta	Cha et al. (2000)
Oat1	Mouse	Kidney > brain	Dresser et al. (2001)
Oat2	Mouse	Kidney > liver (female > male)	Kobayashi et al. (2002)
Oat3	Mouse	Kidney > brain	Sweet et al. (2002)
Oat4	Mouse	Kidney, placenta	Cha et al. (2000)
D. OCTs			
OCT1	Human	Liver > kidney	Dresser et al. (2001)
OCT2	Human	Kidney > brain	Dresser et al. (2001)
OCT3	Human	Liver > kidney > brain	Dresser et al. (2001)
Oct1	Rat	Kidney, liver > brain	Dresser et al. (2001)
Oct2	Rat	Kidney	Dresser et al. (2001)
Oct3	Rat	Kidney, brain	Dresser et al. (2001)

restore the activity of the dysfunctional transporter. In support of this is the fact that many of the studies focused on the kidney report only that the inhibited transporter is a member of the OAT or OCT family. Further, Sweet et al. (2002) report that the OAT transporters, which predominantly mediate clearance at the renal tubules, have significant overlap in substrate specificities. An inhibitor would then also inhibit multiple transporters. On the other hand, liver studies often report a specific transporter that is affected. Alternatively, the renal epithelium may be a "looser" barrier than the hepatocyte membranes, implying the transporters may simply have less importance at the renal epithelium. In support of this is the fact that many drugs have different permeability characteristics at the enterocytes of the intestine and at the hepatocytes; certain drugs, such as atorvastatin, may diffuse passively into the intestine, but require an uptake transporter at the liver. Thus, some drugs may also have different permeability characteristics at the renal epithelium and hepatocyte membranes. However, the measured changes in clearance contradict these possibilities: If a single transporter were less important in the kidneys, clearance, in addition to volume, would be unaffected.

Three interactions at the renal tubules presented in Table 18.3 do lead to a decrease in volume. For the interaction between procainamide and cimetidine (Somogyi et al., 1983), V_{area} was calculated by Grover and Benet (2009) from the reported data assuming the reported half-life was the terminal half-life. As discussed, the change in V_{area} is often more extensive than the change in the other volume parameters. However, the decrease in penicillin G V_{ss} in oat3–/– knockout mice (VanWert et al., 2007) and in famotidine V_1 in healthy human volunteers upon co-administration with probenecid (Inotsume et al., 1990) seem to be exceptions to the trend.

18.2.4 Interactions Involving Hepatic and Renal Efflux

Interactions with efflux transporters at the liver generally lead to a decrease in volume of distribution, while those at the renal tubules lead to an increase in volume of distribution.

Efflux transporters serve a protective purpose preventing drug distribution at some of the most sensitive tissue sites, such as the brain, lungs, and heart. They are also expressed at the liver canalicular membrane and renal epithelia to facilitate clearance. An increase in distribution volume would be expected after inhibiting an efflux transporter, by increasing penetration to tissues protected by the transporters. Table 18.5 highlights the interactions attributed to efflux transporters. Of these 13 interactions, 5 lead to an increase in volume, and they are all interactions at the renal tubules. The remaining eight that do not cause a change or lead to a decrease in volume are interactions at the liver.

For example, methotrexate and topotecan are both substrates of the efflux transporter Bcrp, distributed through the blood–brain barrier, liver, and kidneys, among other tissues. When methotrexate was dosed with the Bcrp inhibitor pantoprazole in mice, V_{area} decreased by 21.6%, clearance decreased by 45.7%, and the half-life

Table 18.5 Summary of efflux transporter-based interactions at the liver and renal tubules

Drug	Interaction	V	CL	$t_{1/2}$	Mechanism of clearance in reference organism	Transporter-related comments	References
Hepatic interactions							
Colchicine	SDZ PSC 833	$\leftrightarrow$ (V_{area})	$\downarrow$ 47.0%	$\leftrightarrow$	Biliary > renal > metabolism	Inhibition of P-gp efflux in rats after iv dosing	Desrayaud et al. (1997) and Speeg et al. (1992)
Daunomycin	Verapamil	$\downarrow$ 63.7% (V_{area})	$\downarrow$ 89.1%	$\uparrow$ 232%	Metabolism > renal ~ biliary	Inhibition of P-gp efflux in rats after iv dosing	Nooter et al. (1987) and Yesair et al. (1972)
Methotrexate	Pantoprazole	$\downarrow$ 21.6% (V_{area})	$\downarrow$ 45.7%	$\uparrow$ 44.4%	Biliary > renal	Inhibition of Bcrp efflux into the feces in mice after iv dosing	Breedveld et al. (2004)
Tacrolimus	mdr1a–/– mice	$\leftrightarrow$ (V_{ss})	$\downarrow$ 65.4%	$\uparrow$ 99.4%	Biliary >> renal > metabolism	Lack of P-gp-mediated biliary clearance in mice after iv dosing	Yokogawa et al. (1999)
Telmisartan	Nisoldipine	$\downarrow$ 62.2% (V_{area}/F)	$\downarrow$ 51.2% (CL/F)	$\leftrightarrow$	Metabolism	Inhibition of P-gp efflux of metabolites from the liver in humans after oral dosing; the majority of the effect is likely a bioavailability consideration	Bajcetic et al. (2007)
Tezosentan	Cyclosporine	$\downarrow$ 65.2% (V_{ss})	$\downarrow$ 74.8%	$\leftrightarrow$	Biliary	Likely inhibition of P-gp efflux in humans after iv dosing	van Giersbergen et al. (2002)

Table 18.5 (continued)

Drug	Interaction	V	CL	$t_{1/2}$	Mechanism of clearance in reference organism	Transporter-related comments	References
Ulifloxacin	EHBR rats (Mrp2–/–)	↓ 34.1% (V_{ss})	↔	↓ 34.9%	Biliary ~ renal	Lack of Mrp2 decreases biliary excretion of glucuronide; no change in clearance or percent of dose eliminated in the bile or urine after iv dosing	Yagi et al. (2003, 2007) and Nakashima et al. (1994)
Valsartan	EHBR rats (Mrp2–/–)	↓ 35.8% (V_1)	↓ 94.3% (CL); ↓ 97.7% (CL_b)	↑ 1030%	Biliary >> metabolism > renal	Lack of Mrp2 decreases biliary clearance, leading to increased plasma and liver concentrations after iv dosing	Yamashiro et al. (2006)
Renal interactions							
Adriamycin	Verapamil	↑ 31.2% (V_{ss})	↓ 32.6%	↑ 37.7%	Metabolism > Biliary > Renal	Inhibition of P-gp efflux in humans after iv dosing; in rats, renal clearance is saturable but hepatic is not, indicating a transporter is involved in renal elimination	Kerr et al. (1986) and Tavoloni and Guarino (1980)

Table 18.5 (continued)

Drug	Interaction	V	CL	$t_{1/2}$	Mechanism of clearance in reference organism	Transporter-related comments	References
Cetirizine	Pilsicainide	↑ 102% (V_{ss}/F)	↓ 28.91% (CL/F); ↓ 37.57% (CL_r)	↑ 188.4%	Renal >> Metabolism	Inhibition of P-gp efflux in humans after oral dosing, possible inhibition of OCT2 as well, but P-gp effect is more significant	Tsuruoka et al. (2006)
Digoxin	Ritonavir	↑ 76.7% (V_{ss})	↓ 41.8%	↑ 156%	Renal	Inhibition of P-gp efflux in humans after iv dosing	Ding et al. (2004)
Pilsicainide	Cetirizine	↑ 106% (V_{ss}/F)	↓ 26.6% (CL/F); ↓ 41.3% (CL_r)	↑ 206%	Renal >> Metabolism	Inhibition of P-gp efflux in humans after oral dosing, possible inhibition of OCT2 as well, but P-gp effect is more significant	Tsuruoka et al. (2006)
Topotecan	Novobiocin	↑ 254% (V_{ss})	↓ 33.7%	↑ 341%	Renal > Biliary >> Metabolism	Inhibition of Bcrp efflux in rats after iv dosing	Su et al. (2007)

CL_r: renal clearance, CL_b: biliary clearance, EHBR: Eisai hyperbilirubinemic rats increased by 44.4%. Methotrexate is primarily cleared via the bile, where Bcrp has a modulating role (Breedveld et al., 2004). On the other hand, when topotecan was dosed with the Bcrp inhibitor novobiocin in rats, V_{ss} increased by 254%, clearance decreased by 33.7%, and half-life increased by 341%. Topotecan is primarily eliminated unchanged in the urine, again mediated by Bcrp. In this case, the authors note that increased brain concentrations of topotecan could lead to the increased volume of distribution (Su et al., 2007). While increased peripheral tissue distribution is likely in both cases, the effect is not apparent in the liver interaction.

Similarly, the anti-cancer agents daunomycin and adriamycin are both substrates of P-glycoprotein. As daunomycin is eliminated predominantly in the liver by metabolism (Yesair et al., 1972), when dosed with the P-gp inhibitor verapamil in rats, V_{area} decreased by 63.7%, clearance decreased by 89.1%, and half-life increased by 232% (Nooter et al., 1987). On the other hand, adriamycin is eliminated through both the liver and the urine. In rats, Tavoloni and Guarino (1980) found that urinary elimination of adriamycin is saturable, while biliary excretion is not. This indicates that P-gp may play a more important role in the kidney than the liver. Upon co-administration with verapamil in humans, V_{ss} increased by 31.2%, clearance decreased by 32.6%, and half-life increased by 37.7%. Further, while the values were not reported, the authors do note that the volume of the central compartment decreased, and the volume of the peripheral compartments increased after P-gp inhibition (Kerr et al., 1986). Because V_{ss} is the sum of the volumes of all the compartments, the change in central compartment volume, which most likely includes the kidneys, must be minor compared to the increase in volume of the peripheral compartments.

Therefore, it seems that efflux transporter inhibition leads to a decrease in distribution volume for the central compartment and an increase in distribution volume for the peripheral tissue compartments. The magnitude of the increase in peripheral distribution is greater than the magnitude of the decrease in central compartment volume for a renal interaction, but it is less than the magnitude of the decrease in central compartment volume for a hepatic interaction. So, an increase in total distribution volume is evident for a kidney interaction, but a decrease in total distribution volume is evident for a liver interaction. This conclusion follows the above analysis on the difference in volume changes following uptake transporter interactions in the kidney and liver: The liver is again a greater contributor to distribution volume than the kidneys. Because peripheral distribution does not seem to be a factor in uptake interactions, it is also clear that efflux transporters play a larger role than uptake transporters outside the liver and kidney, despite the fact that uptake transporters are expressed at these other tissues.

A mechanism for a decrease in central compartment volume, however, is not immediately clear. As mentioned above, it is possible that when the efflux transporter is inhibited, the eliminating organ is no longer quickly equilibrating with the plasma. Effectually, the plasma concentrations do not reflect the amounts of drug at the elimination site because drug is now so highly sequestered in the hepatocytes. As Yates and Arundel (2008b) derived for a two-compartment model, the steady-state volume is under-predicted by the value of $V_1 \cdot (k_{10}/k_{21})$, when elimination is actually from the peripheral compartment, where k_{10} and k_{21} are defined for the central compartment elimination model shown in Fig. 18.1. Thus, the decrease in steady-state volume might be a consequence of the pharmacokinetic calculations and may not reflect a "real" volume change. It remains to be elucidated in which cases the central compartment elimination model does not hold, as it is foreseeable that uptake transporter dysfunction will also change the equilibration properties of the eliminating organs.

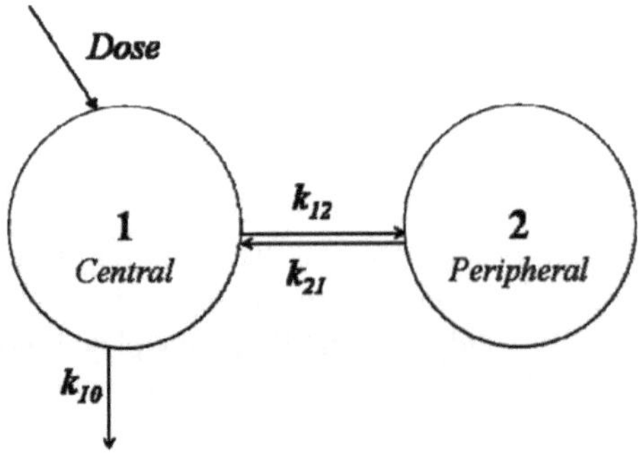

Fig. 18.1 Two-compartment pharmacokinetic model with elimination from the central compartment. Plasma concentrations reflect the central compartment, compartment 1

18.2.5 Secondary Interactions at the Blood–Brain Barrier

The primary location of the interaction (liver or kidneys) is a more important determinant of the change in distribution volume than the secondary change in tissue distribution is, as evidenced by interactions that affect the integrity of the blood–brain barrier.

While transporters are operative at almost all the major tissues in the body, including the heart, lungs, and muscle, they have been most studied, beside the liver and kidney, at the blood–brain barrier. Here, efflux transporters dominate, where they serve to protect the brain from xenobiotic penetration. Table 18.6 highlights the interactions that are associated with an increased distribution of drug to the brain. These five interactions involve P-gp and BCRP, the two transporters most highly implicated in maintaining the integrity of the blood–brain barrier. Despite the increase in brain concentrations in these five studies, there is no common increase in volume of distribution. Instead, it appears the trends discussed above for efflux transporters generally hold true. That is, interactions attributed to the renal transporters lead to an increased volume, while interactions at hepatic transporters lead to either a decrease or no change in volume of distribution.

At the kidney tubules, as discussed, when the Bcrp substrate topotecan was dosed with novobiocin in rats, volume of distribution increased 254%, with increased distribution to the brain (Su et al., 2007). At the liver, tacrolimus, a P-gp substrate, was dosed to wild-type and mdr1–/– (P-gp knockout) mice. In these mice, there was no significant change in volume, clearance decreased 65.4%, and half-life increased

Table 18.6 Summary of transporter-based interactions at the blood–brain barrier

Drug	Interaction	V	CL	$t_{1/2}$	Mechanism of clearance in reference organism	Transporter-related comments	References
Colchicine	SDZ PSC 833	$\leftrightarrow$ (V_{area})	$\downarrow$ 47.0%	$\leftrightarrow$	Biliary > Renal > Metabolism	Increased brain distribution with P-gp inhibition without significant effect on volume in rats after iv dosing	Desrayaud et al. (1997) and Speeg et al. (1992)
Rhodamine-123	Cyclosporine	$\leftrightarrow$ (V_1)	$\leftrightarrow$	$\leftrightarrow$	Renal > Biliary	Increased distribution to the brain with P-gp inhibition in rats, without significant effect on volume or clearance after iv dosing	Wang et al. (1995) and Kunihara et al. (1998)
Tacrolimus	mdr1a–/– mice	$\leftrightarrow$ (V_{ss})	$\downarrow$ 65.4%	$\uparrow$ 99.4%	Biliary >> Renal > Metabolism	Increased brain distribution in mice lacking P-gp without significant effect on volume after iv dosing	Yokogawa et al. (1999)
Tezosentan	Cyclosporine	$\downarrow$ 65.2% (V_{ss})	$\downarrow$ 74.8%	$\leftrightarrow$	Biliary	Likely inhibition of P-gp efflux into the bile in humans after iv dosing; side effects possibly caused by increased brain distribution	van Giersbergen et al. (2002)
Topotecan	Novobiocin	$\uparrow$ 254% (V_{ss})	$\downarrow$ 33.7%	$\uparrow$ 341%	Renal > Biliary >> Metabolism	Dramatically increased distribution to brain with Bcrp inhibition after iv dosing in rats	Zamboni et al. (1998)

99.4% as compared to wild-type mice. Knockout mice also exhibited a 33-fold increase in brain concentrations of tacrolimus. Minor increases in liver concentrations were also evident. In mice, tacrolimus is predominantly excreted in the bile (Yokogawa et al., 1999). Finally, tezosentan, also eliminated into the bile, was also dosed with cyclosporine for inhibition of P-gp in humans. In this study, volume of distribution decreased 65.2%, clearance decreased 74.8%, and half-life did not change. The authors note that an increased incidence of adverse events, including headache, hot flushes, and nausea, may have been caused by increased brain distribution of the drug (van Giersbergen et al., 2002).

Thus, while brain distribution may change, even dramatically as in the case of tacrolimus in P-gp knockout mice, these changes do not necessarily manifest in a total body volume of distribution change. It is possible, however, that changes at the other tissues expressing transporters might offer a different conclusion.

18.3 Pharmacodynamic Considerations

It is possible to predict the direction of the change in pharmacological effect given the mechanisms of action of the drug and the location of the interaction. For example, glyburide, metformin, and atorvastatin are substrates for uptake transporters in the liver. Following uptake inhibition either via polymorphism or concomitant medication, subjects in the three studies exhibit significantly reduced distribution volumes. However, the direction of the resulting change in pharmacological effect is different.

Glyburide is a hypoglycemic agent indicated for patients with type 2 diabetes. Its main effect is at the pancreatic beta cells, where it stimulates insulin secretion. It is primarily eliminated via metabolism by CYP2C9 and, to a lesser degree, by CYP3A4 in the liver. It is a substrate for uptake mediated by OATP2B1 at the hepatocytes, and subjects show a decrease in steady-state volume of 67.4%, a decrease in clearance of 54.6%, without a change in half-life, following uptake inhibition via concomitant dosing with rifampin. As would be predicted, inhibition of liver uptake decreases elimination, increasing plasma concentrations. This increases pancreatic beta cells' access to the drug, increasing the pharmacologic effect. Following a single dose of glyburide and a single dose of rifampin, subjects exhibited significantly decreased blood glucose AUCs over a 12-h period (Zheng et al., 2009).

Similarly, metformin is the first-line therapy for patients with type 2 diabetes. Its pharmacological effect is in the liver hepatocytes, where it prevents gluconeogenesis, effectively decreasing blood glucose levels. It is primarily eliminated via excretion at the renal tubules, a process mediated by OCT2. However, at the liver, it is a substrate for uptake by OCT1. In this unique case, the transporter interaction is not at the primary site of elimination, but because the drug is a substrate for hepatic uptake and it is highly distributed to the liver, the interaction still causes marked pharmacokinetic changes. This further attests to the importance of the liver in the determination of distribution volume in consideration of transporter dysfunction. In patients with polymorphisms in one of their OCT1 alleles, volume of distribution is decreased by 53.9% and clearance is reduced by 37.5%, without a change in half-life (Shu et al., 2008). In this case, a reduced function OCT1 allele decreases hepatocyte access to the drug, decreasing the pharmacologic effect.

Along the same lines, Pasanen et al. (2007) and Tachibana-Iimori et al. (2004) both studied a single nucleotide polymorphism (SNP) at position 521 in the SLCO1B1 (OATP1B1) gene. Both groups looked at atorvastatin, among other statins, to measure the effect of this polymorphism. Atorvastatin is primarily metabolized by CYP3A4 and by CYP2C9 to a lesser extent, and it is a substrate for OATP1B1 uptake. The drug and its active metabolites are less than 1% excreted in the urine (Lau et al., 2007). The statins, HMG-CoA reductase inhibitors, decrease cholesterol levels by preventing cholesterol synthesis and increasing clearance of LDL, or "bad," cholesterol at the hepatocytes. From the pharmacokinetic perspective, Pasanen et al. (2007) showed a decrease in V_{area} of 57.6% and a decrease in clearance of 59.2%, without a change in half-life, between patients homozygous for either the wild-type or mutant alleles. From a pharmacodynamic perspective, Tachibana-Iimori et al. (2004) showed the same patterns held for patients beginning atorvastatin, pravastatin, or simvastatin therapy, all OATP1B1 substrates. Analysis of patients on any of these three drugs found that patients homozygous for the wild-type alleles showed a decrease in total cholesterol of 22.3%, while patients heterozygous for the wild-type and polymorphic allele showed a decrease in total cholesterol of 16.5%, indicating a decreased pharmacological benefit. The difference between wild-type and polymorphic patients is likely to be greater for patients homozygous for the mutant alleles.

In this regard, while the pharmacokinetic consequences of an interaction are important for the clinician to understand, the pharmacodynamic change is also critical to consider before changes to the dosing regimen are made. Within these three examples, although the direction of pharmacokinetic change is the same, a glyburide–rifampin interaction would require a decreased dosing rate to maintain the same pharmacological effect, while patients with polymorphisms in OCT1 or OATP1B1 would require an increased dosing rate of metformin or atorvastatin, respectively, to maintain effect. The potential for toxicity when higher dosing rates are required complicates this issue and may lead to alternative therapies for patients with such pharmacogenetic variation.

18.4 Experimental Considerations

As with any pharmacokinetic study, it is important to understand the experimental conditions and variability that complicate the conclusions that are drawn from transporter interaction studies.

First, there are wide interspecies differences in drugs' elimination pathways, the expression of transporters, and transporter substrate profiles. For instance, as noted, digoxin is almost completely metabolized in rats, where it is a substrate for uptake mediated by Oatp1a4 at the liver (Lam et al., 2006). In humans, however, digoxin is predominantly eliminated in the urine. Similarly, as shown in Table 18.4, in humans OAT3 is expressed primarily at the kidney, and, to a lesser degree, in the brain, while in rats Oat3 is highly expressed in the liver in addition to the kidneys and brain. For substrates common to both rat and human Oat3/OAT3, this will most likely lead to different tissue distributions and volume calculations between the two species and indicates that results from an Oat3-focused pharmacokinetic study conducted in rats may not scale to humans. Finally, the interaction between famotidine and probenecid (Inotsume et al., 1990) resulting from the inhibition of OAT3 transport at the renal tubules in humans is not reproducible in rats (Lin et al., 1988). This is likely due to the increased expression of Oct1 in the rat kidney: Because famotidine is also a substrate of Oct1 and probenecid does not inhibit Oct1, the transporter serves as an alternate, compensatory route for renal clearance in rats concomitantly dosed with famotidine and probenecid (Shitara et al., 2005). Briefly, these few examples attest to the importance of considering interspecies differences before clinical extrapolations are made from animal data.

Further, while clearance is relatively easily extrapolated from in vitro data to in vivo relevance, the same is not true of volume of distribution. Because volume is focused on the entire body, even ex situ techniques, such as the isolated perfused rat liver (IPRL) or isolated perfused rat kidney, can lead to incorrect approximations of the direction of volume changes. Table 18.7 highlights these discrepancies. Although the published data are sparse, it appears the ex situ results for inhibited uptake transporters in the liver and kidney follow the analysis above (Lau, 2006; Kugler et al., 1996). However, the IPRL data for inhibited efflux transporters, in particular P-glycoprotein, show an increase in steady-state volume of distribution, while the in vivo trend predicts either a decrease or no change in this parameter (Wu, 2003; Booth et al., 1998). A mechanism for this discrepancy remains to be elucidated.

Table 18.7 V_{ss} values from ex situ studies in rat

Interaction	Transporter	Organ	V_{ss} Change (%)	Prediction from trends	References
Digoxin + Rifampin	Oatp1a4 uptake	Liver	↓ 17.9	↓	Lau (2006)
Quinapril + PAH	Oat uptake	Kidney	↓ 13.7	↔, (↓)	Kugler et al. (1996)
Digoxin + Quinidine	*P-gp efflux*	*Liver*	↑ 95.9	↓	Wu (2003)
Tacrolimus + GG918	*P-gp efflux*	*Liver*	↑ 30.1	↓	Wu (2003)
Talinolol + GG918	*P-gp efflux*	*Liver*	↑ 74.2	↓	Wu (2003)
Doxorubicin + GG918	*P-gp efflux*	*Liver*	↑ 70.2	↓	Booth et al. (1998)

Italicized interactions are those where the data do not match the in vivo predictions of Grover and Benet (2009).
PAH: p-aminohippurate, GG918: GF120918 (Elacridar)

18.5 Conclusions

Through the above discussion, it is apparent that active drug transporters that modulate tissue distribution act as modifiers of distribution volume. Because transporters can be significantly affected by drug–drug interactions or genetic polymorphisms, changes in drug transporter activity as they affect distribution volume require attention. The above analysis indicates that the primary location of the interaction, at the kidneys or the liver, serves as the major predictor of change in distribution volume. Figure 18.2 summarizes the trends in effects of transporter dysfunction on distribution volume as discussed above. As knowledge pertaining to the location and function of drug transporters and the substrate status of drugs for these transporters becomes more available, the present analysis provides a framework for understanding future pharmacokinetic interactions rooted in active drug transporters.

	Kidney	**Liver**
Uptake	↔, (↓)	↓
Efflux	↑	↓, ↔

Fig. 18.2 Summary of trends for predicting effects of transporter dysfunction on volume of distribution, depending on whether it is an uptake or efflux transporter and the primary location of the interaction (Reprinted with permission from Grover A and Benet LZ (2009). Effects of drug transporters on volume of distribution. *AAPS J* **11**:250–261. Copyright 2009, American Association of Pharmaceutical Scientists.)

References

Bajcetic M, Benndorf RA, Appel D, Schwedhelm E, Schulze F, Riekhof D, Maas R and Boger R (2007) Pharmacokinetics of oral doses of telmisartan and nisoldipine, given alone and in combination, in patients with essential hypertension. *J Clin Pharmacol* **47**:295–304.

Bauer LA, Black DJ, Lill JS, Garrison J, Raisys VA and Hooton TM (2005) Levofloxacin and ciprofloxacin decrease procainamide and N-acetylprocainamide renal clearances. *Antimicrob Agents Chemother* **49**:1649–1651.

Booth CL, Brouwer KR and Brouwer KL (1998) Effect of multidrug resistance modulators on the hepatobiliary disposition of doxorubicin in the isolated perfused rat liver. *Cancer Res* **58**: 3641–3648.

Brady JM, Cherrington NJ, Hartley DP, Buist SC, Li N and Klaassen CD (2002) Tissue distribution and chemical induction of multiple drug resistance genes in rats. *Drug Metab Dispos* **30**: 838–844.

Breedveld P, Zelcer N, Pluim D, Sonmezer O, Tibben MM, Beijnen JH, Schinkel AH, van Tellingen O, Borst P and Schellens JH (2004) Mechanism of the pharmacokinetic interaction between methotrexate and benzimidazoles: potential role for breast cancer resistance protein in clinical drug–drug interactions. *Cancer Res* **64**:5804–5811.

Brown G, Zemcov SJ and Clarke AM (1993) Effect of probenecid on cefazolin serum concentrations. *J Antimicrob Chemother* **31**:1009–1011.

Carr RA, Pasutto FM and Foster RT (1996) Influence of cimetidine coadministration on the pharmacokinetics of sotalol enantiomers in an anaesthetized rat model: evidence supporting active renal excretion of sotalol. *Biopharm Drug Dispos* **17**:55–69.

Cha SH, Sekine T, Kusuhara H, Yu E, Kim JY, Kim DK, Sugiyama Y, Kanai Y and Endou H (2000) Molecular cloning and characterization of multispecific organic anion transporter 4 expressed in the placenta. *J Biol Chem* **275**:4507–4512.

Cherrington NJ, Hartley DP, Li N, Johnson DR and Klaassen CD (2002) Organ distribution of multidrug resistance proteins 1, 2, and 3 (Mrp1, 2, and 3) mRNA and hepatic induction of Mrp3 by constitutive androstane receptor activators in rats. *J Pharmacol Exp Ther* **300**: 97–104.

Chung JY, Cho JY, Yu KS, Kim JR, Oh DS, Jung HR, Lim KS, Moon KH, Shin SG and Jang IJ (2005) Effect of OATP1B1 (SLCO1B1) variant alleles on the pharmacokinetics of pitavastatin in healthy volunteers. *Clin Pharmacol Ther* **78**:342–350.

Croop JM, Raymond M, Haber D, Devault A, Arceci RJ, Gros P and Housman DE (1989) The three mouse multidrug resistance (mdr) genes are expressed in a tissue-specific manner in normal mouse tissues. *Mol Cell Biol* **9**:1346–1350.

Desrayaud S, Guntz P, Scherrmann JM and Lemaire M (1997) Effect of the P-glycoprotein inhibitor, SDZ PSC 833, on the blood and brain pharmacokinetics of colchicine. *Life Sci* **61**:153–163.

Ding R, Tayrouz Y, Riedel KD, Burhenne J, Weiss J, Mikus G and Haefeli WE (2004) Substantial pharmacokinetic interaction between digoxin and ritonavir in healthy volunteers. *Clin Pharmacol Ther* **76**:73–84.

Dresser MJ, Leabman MK and Giacomini KM (2001) Transporters involved in the elimination of drugs in the kidney: organic anion transporters and organic cation transporters. *J Pharm Sci* **90**:397–421.

Grover A and Benet LZ (2009) Effects of drug transporters on volume of distribution. *AAPS J* **11**:250–261.

Guyton AC and Hall JE (2006) *Textbook of Medical Physiology*, Elsevier Saunders, Philadelphia.

Hagenbuch B and Meier PJ (2004) Organic anion transporting polypeptides of the OATP/ SLC21 family: phylogenetic classification as OATP/ SLCO superfamily, new nomenclature and molecular/functional properties. *Pflugers Arch* **447**:653–665.

Hirano M, Maeda K, Matsushima S, Nozaki Y, Kusuhara H and Sugiyama Y (2005) Involvement of BCRP (ABCG2) in the biliary excretion of pitavastatin. *Mol Pharmacol* **68**:800–807.

Inotsume N, Nishimura M, Nakano M, Fujiyama S and Sato T (1990) The inhibitory effect of probenecid on renal excretion of famotidine in young, healthy volunteers. *J Clin Pharmacol* **30**:50–56.

Jaehde U, Sorgel F, Reiter A, Sigl G, Naber KG and Schunack W (1995) Effect of probenecid on the distribution and elimination of ciprofloxacin in humans. *Clin Pharmacol Ther* **58**: 532–541.

Jette L, Beaulieu E, Leclerc JM and Beliveau R (1996) Cyclosporin A treatment induces overexpression of P-glycoprotein in the kidney and other tissues. *Am J Physiol* **270**: F756–765.

Kajosaari LI, Niemi M, Neuvonen M, Laitila J, Neuvonen PJ and Backman JT (2005) Cyclosporine markedly raises the plasma concentrations of repaglinide. *Clin Pharmacol Ther* **78**:388–399.

Kerr DJ, Graham J, Cummings J, Morrison JG, Thompson GG, Brodie MJ and Kaye SB (1986) The effect of verapamil on the pharmacokinetics of adriamycin. *Cancer Chemother Pharmacol* **18**:239–242.

Khamdang S, Takeda M, Babu E, Noshiro R, Onozato ML, Tojo A, Enomoto A, Huang XL, Narikawa S, Anzai N, Piyachaturawat P and Endou H (2003) Interaction of human and rat organic anion transporter 2 with various cephalosporin antibiotics. *Eur J Pharmacol* **465**:1–7.

Kobayashi Y, Ohshiro N, Shibusawa A, Sasaki T, Tokuyama S, Sekine T, Endou H and Yamamoto T (2002) Isolation, characterization and differential gene expression of multispecific organic anion transporter 2 in mice. *Mol Pharmacol* **62**:7–14.

Kugler AR, Olson SC and Smith DE (1996) Tubular transport mechanisms of quinapril and quinaprilat in the isolated perfused rat kidney: effect of organic anions and cations. *J Pharmacokinet Biopharm* **24**:349–368.

Kunihara M, Nagai J, Murakami T and Takano M (1998) Renal excretion of rhodamine 123, a P-glycoprotein substrate, in rats with glycerol-induced acute renal failure. *J Pharm Pharmacol* **50**:1161–1165.

Lam JL, Shugarts SB, Okochi H and Benet LZ (2006) Elucidating the effect of final-day dosing of rifampin in induction studies on hepatic drug disposition and metabolism. *J Pharmacol Exp Ther* **319**:864–870.

Lau YY (2006) *Examining the Regulation of Hepatic Drug Disposition and Metabolism by Organic Anion Transporting Peptide, P-Glycoprotein, and Multidrug Resistance-Associated Protein 2 [dissertation]*, University of California, San Francisco.

Lau YY, Huang Y, Frassetto L and Benet LZ (2007) Effect of OATP1B transporter inhibition on the pharmacokinetics of atorvastatin in healthy volunteers. *Clin Pharmacol Ther* **81**: 194–204.

Lin JH, Los LE, Ulm EH and Duggan DE (1988) Kinetic studies on the competition between famotidine and cimetidine in rats. evidence of multiple renal secretory systems for organic cations. *Drug Metab Dispos* **16**:52–56.

Muck W, Mai I, Fritsche L, Ochmann K, Rohde G, Unger S, Johne A, Bauer S, Budde K, Roots I, Neumayer HH and Kuhlmann J (1999) Increase in cerivastatin systemic exposure after single and multiple dosing in cyclosporine-treated kidney transplant recipients. *Clin Pharmacol Ther* **65**:251–261.

Nakashima E and Benet LZ (1988) General treatment of mean residence time, clearance, and volume parameters in linear mammillary models with elimination from any compartment. *J Pharmacokinet Biopharm* **16**:475–492.

Nakashima M, Uematsu T, Kosuge K, Okuyama Y, Morino A, Ozaki M and Takebe Y (1994) Pharmacokinetics and safety of NM441, a new quinolone, in healthy male volunteers. *J Clin Pharmacol* **34**:930–937.

Nooter K, Oostrum R and Deurloo J (1987) Effects of verapamil on the pharmacokinetics of daunomycin in the rat. *Cancer Chemother Pharmacol* **20**:176–178.

Oh YH and Han H (2006) Pharmacokinetic interaction of tetracycline with non-steroidal anti-inflammatory drugs via organic anion transporters in rats. *Pharmacol Res* **53**:75–79.

Øie S (1986) Drug distribution and binding. *J Clin Pharmacol* **26**:583–586.

Pasanen MK, Fredrikson H, Neuvonen PJ and Niemi M (2007) Different effects of SLCO1B1 polymorphism on the pharmacokinetics of atorvastatin and rosuvastatin. *Clin Pharmacol Ther* **82**:726–733.

Pavlova A, Sakurai H, Leclercq B, Beier DR, Yu AS and Nigam SK (2000) Developmentally regulated expression of organic ion transporters NKT (OAT1), OCT1, NLT (OAT2), and roct. *Am J Physiol Renal Physiol* **278**:F635–643.

Sahin S and Benet LZ (2008) The operational multiple dosing half-life: a key to defining drug accumulation in patients and to designing extended release dosage forms. *Pharm Res* **25**: 2869–2877.

Sakurai Y, Motohashi H, Ogasawara K, Terada T, Masuda S, Katsura T, Mori N, Matsuura M, Doi T, Fukatsu A and Inui K (2005) Pharmacokinetic significance of renal OAT3 (SLC22A8) for anionic drug elimination in patients with mesangial proliferative glomerulonephritis. *Pharm Res* **22**:2016–2022.

Schneck DW, Birmingham BK, Zalikowski JA, Mitchell PD, Wang Y, Martin PD, Lasseter KC, Brown CD, Windass AS and Raza A (2004) The effect of gemfibrozil on the pharmacokinetics of rosuvastatin. *Clin Pharmacol Ther* **75**:455–463.

Sekine T, Cha SH and Endou H (2000) The multispecific organic anion transporter (OAT) family. *Pflugers Arch* **440**:337–350.

Shiga T, Hashiguchi M, Urae A, Kasanuki H and Rikihisa T (2000) Effect of cimetidine and probenecid on pilsicainide renal clearance in humans. *Clin Pharmacol Ther* **67**:222–228.

Shitara Y, Horie T and Sugiyama Y (2006) Transporters as a determinant of drug clearance and tissue distribution. *Eur J Pharm Sci* **27**:425–446.

Shitara Y, Sato H and Sugiyama Y (2005) Evaluation of drug–drug interaction in the hepatobiliary and renal transport of drugs. *Annu Rev Pharmacol Toxicol* **45**:689–723.

Shu Y, Brown C, Castro RA, Shi RJ, Lin ET, Owen RP, Sheardown SA, Yue L, Burchard EG, Brett CM and Giacomini KM (2008) Effect of genetic variation in the organic cation transporter 1, OCT1, on metformin pharmacokinetics. *Clin Pharmacol Ther* **83**:273–280.

Simonson SG, Raza A, Martin PD, Mitchell PD, Jarcho JA, Brown CD, Windass AS and Schneck DW (2004) Rosuvastatin pharmacokinetics in heart transplant recipients administered an antirejection regimen including cyclosporine. *Clin Pharmacol Ther* **76**:167–177.

Somogyi A, McLean A and Heinzow B (1983) Cimetidine-procainamide pharmacokinetic interaction in man: evidence of competition for tubular secretion of basic drugs. *Eur J Clin Pharmacol* **25**:339–345.

Speeg KV, Maldonado AL, Liaci J and Muirhead D (1992) Effect of cyclosporine on colchicine secretion by a liver canalicular transporter studied in vivo. *Hepatology* **15**:899–903.

Su Y, Hu P, Lee SH and Sinko PJ (2007) Using novobiocin as a specific inhibitor of breast cancer resistant protein to assess the role of transporter in the absorption and disposition of topotecan. *J Pharm Pharmaceut Sci* **10**:519–536.

Sweet DH, Miller DS, Pritchard JB, Fujiwara Y, Beier DR and Nigam SK (2002) Impaired organic anion transport in kidney and choroid plexus of organic anion transporter 3 (Oat3 (Slc22a8)) knockout mice. *J Biol Chem* **277**:26934–26943.

Tachibana-Iimori R, Tabara Y, Kusuhara H, Kohara K, Kawamoto R, Nakura J, Tokunaga K, Kondo I, Sugiyama Y and Miki T (2004) Effect of genetic polymorphism of OATP-C (SLCO1B1) on lipid-lowering response to HMG-CoA reductase inhibitors. *Drug Metab Pharmacokinet* **19**:375–380.

Tahara H, Kusuhara H, Chida M, Fuse E and Sugiyama Y (2006) Is the monkey an appropriate animal model to examine drug–drug interactions involving renal clearance? Effect of probenecid on the renal elimination of H_2 receptor antagonists. *J Pharmacol Exp Ther* **316**:1187–1194.

Tanaka Y, Slitt AL, Leazer TM, Maher JM and Klaassen CD (2005) Tissue distribution and hormonal regulation of the breast cancer resistance protein (Bcrp/Abcg2) in rats and mice. *Biochem Biophys Res Commun* **326**:181–187.

Tavoloni N and Guarino AM (1980) Biliary and urinary excretion of adriamycin in anesthetized rats. *Pharmacology* **20**:256–267.

Tsuruoka S, Ioka T, Wakaumi M, Sakamoto K, Ookami H and Fujimura A (2006) Severe arrhythmia as a result of the interaction of cetirizine and pilsicainide in a patient with renal insufficiency: first case presentation showing competition for excretion via renal multidrug resistance protein 1 and organic cation transporter 2. *Clin Pharmacol Ther* **79**:389–396.

Ullrich K (1999) Affinity of drugs to the different renal transporters for organic anions and organic cations, in *Membrane Transporters as Drug Targets* (Amidon GL and Sadee W eds) pp 159–179, Kluwer Academic/Plenum Publishers, New York.

van der Valk P, van Kalken CK, Ketelaars H, Broxterman HJ, Scheffer G, Kuiper CM, Tsuruo T, Lankelma J, Meijer CJ and Pinedo HM (1990) Distribution of multi-drug resistance-associated P-glycoprotein in normal and neoplastic human tissues. Analysis with 3 monoclonal antibodies recognizing different epitopes of the P-glycoprotein molecule. *Ann Oncol* **1**:56–64.

van Giersbergen PL, Bodin F and Dingemanse J (2002) Cyclosporin increases the exposure to tezosentan, an intravenous dual endothelin receptor antagonist. *Eur J Clin Pharmacol* **58**: 243–245.

VanWert AL, Bailey RM and Sweet DH (2007) Organic anion transporter 3 (Oat3/Slc22a8) knockout mice exhibit altered clearance and distribution of penicillin G. *Am J Physiol* **293**:F1332–F1341.

Wang Q, Yang H, Miller DW and Elmquist WF (1995) Effect of the P-glycoprotein inhibitor, cyclosporin A, on the distribution of rhodamine-123 to the brain: an in vivo microdialysis study in freely moving rats. *Biochem Biophys Res Commun* **211**:719–726.

Wu C-Y (2003) *The Interactive Roles of P-glycoprotein and Cytochrome P-450 3A in Intestinal and Hepatic Drug Disposition [dissertation]*, University of California, San Francisco.

Yagi Y, Aoki M, Iguchi M, Shibasaki S, Kurosawa T, Kato Y and Tsuji A (2007) Transporter-mediated hepatic uptake of ulifloxacin, an active metabolite of a prodrug-type new quinolone antibiotic prulifloxacin, in rats. *Drug Metab Pharmacokinet* **22**:350–357.

Yagi Y, Shibutani S, Hodoshima N, Ishiwata K, Okudaira N, Li Q, Sai Y, Kato Y and Tsuji A (2003) Involvement of multiple transport systems in the disposition of an active metabolite of a prodrug-type new quinolone antibiotic, prulifloxacin. *Drug Metab Pharmacokinet* **18**:381–389.

Yamashiro W, Maeda K, Hirouchi M, Adachi Y, Hu Z and Sugiyama Y (2006) Involvement of transporters in the hepatic uptake and biliary excretion of valsartan, a selective antagonist of the angiotensin II AT1-receptor, in humans. *Drug Metab Dispos* **34**:1247–1254.

Yates JW and Arundel PA (2008a) Oral and IV dosing: a method to determine the compartment of drug elimination for two-compartment models. *J Pharm Sci* **97**:2036–2040.

Yates JW and Arundel PA (2008b) On the volume of distribution at steady state and its relationship with two-compartmental models. *J Pharm Sci* **97**:111–122.

Yesair DW, Schwartzbach E, Shuck D, Denine EP and Asbell MA (1972) Comparative pharmacokinetics of daunomycin and adriamycin in several animal species. *Cancer Res* **32**:1177–1183.

Yokogawa K, Takahashi M, Tamai I, Konishi H, Nomura M, Moritani S, Miyamoto K and Tsuji A (1999) P-glycoprotein-dependent disposition kinetics of tacrolimus: studies in mdr1a knockout mice. *Pharm Res* **16**:1213–1218.

Zamboni WC, Houghton PJ, Johnson RK, Hulstein JL, Crom WR, Cheshire PJ, Hanna SK, Richmond LB, Luo X and Stewart CF (1998) Probenecid alters topotecan systemic and renal disposition by inhibiting renal tubular secretion. *J Pharmacol Exp Ther* **284**:89–94.

Zheng HX, Huang Y, Frassetto LA and Benet LZ (2009) Elucidating rifampin's inducing and inhibiting effects on glyburide pharmacokinetics and blood glucose in healthy volunteers: unmasking the differential effects of enzyme induction and transporter inhibition for a drug and its primary metabolite. *Clin Pharmacol Ther* **85**:78–85.

Chapter 19
Inactivation of Human Cytochrome P450 Enzymes and Drug–Drug Interactions

R. Scott Obach, Odette A. Fahmi, and Robert L. Walsky

Abstract Inactivation of human P450 enzymes represents an important mechanism of drug–drug interactions (DDIs). Inactivators are distinct from other inhibitors in that the affected enzyme is responsible for bioactivating an otherwise inert drug into an intermediate that can irreversibly damage the enzyme, and recovery of activity in vivo requires the biosynthesis of new enzyme. Whether a new drug will be a mechanism-based inactivator depends on the identity of chemical substituents present in the substrate and their metabolism by the P450 enzyme. Experimental approaches used to define new drugs as possible time-dependent inhibitors and mechanism-based inactivators are described. Finally, it has been demonstrated that inactivation kinetic parameters generated in vitro can be used to predict DDI. The methods used to do this are described along with existing uncertainties in the input parameters needed for accurate predictions.

19.1 Introduction

Drug–drug interactions (DDIs) represent a continuing problem in clinical practice, particularly as the use of multiple drugs simultaneously to treat either one or concurrent conditions occurs with an aging world population. While it is sometimes the case that DDIs arise due to one drug affecting the pharmacological effect of a second drug, in most cases DDIs arise because one drug affects the activity of a drug-metabolizing enzyme responsible for the clearance of a second drug. If the pharmacokinetics of the affected drug do not elicit a pharmacologically meaningful result (i.e., increase in side effects or toxicity or loss of beneficial effect), the DDI will not be important and it may go unnoticed.

R.S. Obach (✉)
Pharmacokinetics, Pharmacodynamics, and Drug Metabolism, Pfizer Inc., Global Research & Development, Groton, CT, USA
e-mail: r.scott.obach@pfizer.com

K.S. Pang et al. (eds.), *Enzyme- and Transporter-Based Drug–Drug Interactions*,
DOI 10.1007/978-1-4419-0840-7_19,

The cytochrome P450 family is most often the target among drug-metabolizing enzymes involved in DDI. This is due to the fact that majority of drugs are cleared by P450 catalyzed metabolism (Williams et al. 2003). Furthermore, there are several drugs known to inhibit, inactivate, or induce the expression of P450 enzymes in vitro, some of which are dosed at high enough levels to attain in vivo concentrations that can cause these effects. Our understanding of the human P450 enzymes, their involvement in drug clearance, and an evolving knowledge of how to relate in vitro data to in vivo DDI has resulted in the regular application of in vitro approaches in drug research with the goal of designing and developing new drugs that lack DDI issues.

Other chapters in this book focus on inhibition and induction; the focus of this chapter is on inactivation of human P450 enzymes. Inactivators (also referred to as mechanism-based inactivators (MBI) or suicide substrates) are distinct from other inhibitors in that the affected enzyme is responsible for bioactivating an otherwise inert substrate into an intermediate that can irreversibly damage the enzyme. Recovery of activity in vivo requires the biosynthesis of new enzyme. The determination of whether a compound is an inactivator usually involves an initial demonstration of time-dependent inhibition (TDI), that is, a kinetic experiment that shows that when the compound is incubated with the enzyme, its inhibitory activity increases with time. However, in addition to data that show time-dependent inhibition, other biochemical evidence is required in order to demonstrate that a compound is an inactivator. Examples of drugs that have been demonstrated to be TDI and also have been shown to elicit DDI are in Table 19.1. In this chapter, inactivation of P450 will be discussed with regard to biochemical mechanisms, experimental approaches to measure inactivation, and how to relate these in vitro data to in vivo DDI. Several recent reviews of P450 inactivation have been published and the interested reader can refer to these for even greater detail (Zhou et al., 2005; Riley et al., 2007; Kalgutkar et al., 2007; Venkatakrishnan et al., 2007; Venkatakrishnan and Obach, 2007; Hollenberg et al., 2008).

19.2 Biochemical Aspects of Inactivation of Cytochrome P450

Inactivation of P450 enzymes can occur via one of three main mechanisms: (a) formation of a quasi-irreversible heme–ligand complex, (b) adduct formation to the heme, and (c) adduct formation to the protein. Despite mechanistic distinction, all three can ultimately yield the same kinetic outcome, both in vitro and in vivo (i.e., observation of TDI and in vivo DDI). Whether a new drug will be a mechanism-based inactivator depends on the identity of chemical substituents present in the substrate, whether the substituent can be metabolized by the P450 enzyme, and whether the activated entity is properly oriented in the enzyme active site to react with the enzyme or form a complex with the heme. For example, many drugs contain primary and secondary aliphatic amines, a substituent that can be activated to a nitroso metabolite capable of forming a complex to the heme iron. However, despite many of these drugs being metabolized at the amine, only a subset will actually

Table 19.1 Drugs that have been shown to be both time-dependent inhibitors in vitro and cause DDI in vivo

Enzyme/drug	Chemical substituent involved in inactivation	Example of DDI
CYP1A2		
Furafylline	Alkylimidazole	Caffeine
Enoxacin	Unknown	Theophylline
Rofecoxib	Unknown	Tizanidine
Zileuton	Thiophene?	Theophylline
CYP2A6		
Methoxsalen	Furan	Nicotine
CYP2B6		
Clopidogrel	Thiophene	Bupropion
thioTEPA	Unknown	Cyclophosphamide
Ticlopidine	Thiophene	Bupropion
CYP2C8		
Gemfibrozil[a]	Benzyl group	Repaglinide
CYP2C9		
Tienilic acid	Thiophene	Warfarin
CYP2D6		
Cimetidine	Unknown	Desipramine
MDMA	Methylenedioxyphenyl	Dextromethorphan
Paroxetine	Methylenedioxyphenyl	Desipramine
CYP2E1		
Disulfiram[a]	Diethyldithiocarbamate metabolite?	Chlorzoxazone
CYP3A		
Clarithromycin	Alkylamine	Midazolam
Delavirdine	Unknown	Buprenorphine
Diltiazem[a]	Alkylamine	Midazolam
Erythromycin	Alkylamine	Midazolam
Fluoxetine[a]	Alkylamine	Alprazolam
Irinotecan	Unknown	Sorafenib
Isoniazid	Hydrazine?	Triazolam
Mibefradil	Unknown	Midazolam
Nefazodone	Quinoneimine	Midazolam
Nelfinavir	Unknown	Simvastatin
Ritonavir	Unknown	Midazolam
Saquinavir	Unknown	Midazolam
Troleandomycin	Alkylamine	Midazolam
Verapamil[a]	Alkylamine	Buspirone

[a]Evidence suggests that a metabolite of the parent drug is the substrate that is the TDI.

cause inactivation. Thus, the presence of a structure alert for MBI of P450 enzymes does not guarantee that a compound possessing such an alert will be an inactivator, and at present there is no established approach whereby MBI could be predicted from structure alone – in vitro testing is necessary.

19.2.1 Quasi-irreversible Complex Formation

Several substituents can be metabolized by P450 enzymes to generate intermediates that can form tight non-covalent interactions with the heme iron. The term "quasi-irreversible" stems from the fact that many of these heme–ligand complexes can be reversed under specific in vitro conditions to restore active enzyme. For example, treatment of a quasi-irreversible complex with ferricyanide can restore the activity of the enzyme. However, despite being non-covalent and "reversible" under specific in vitro conditions, drugs that inactivate P450 by this mechanism are essentially irreversible in vivo. The formation of the complex (sometimes referred to as a metabolite–intermediate complex, abbreviated as MIC) occurs at the last step in the P450 catalytic cycle (Correia and Ortiz de Montellano, 2005). The initially formed product can ligand to the ferrous heme iron to form a complex Fe(II)-M that will not dissociate under ordinary conditions. Depending on the specific type of substituent the complex will demonstrate an absorbance maximum in the 448–458 nm region, and measurement of the increase in absorption at this region during incubation of P450, inactivator, and NADPH is evidence for the formation of MIC.

19.2.2 Covalent Modification of Heme and Protein

While it is commonly held that cytochrome P450 enzymes metabolize xenobiotics in order to detoxicate them, to provide functional groups able to be conjugated, and to make them more polar so as to be more readily excreted, in some cases the metabolism can result in the generation of a reactive electrophile or radical species. These latter metabolites are frequently associated with toxicity (e.g., drug allergy, carcinogenesis). They are also frequently the underlying mechanism for inactivation of P450 enzymes. The reactivity of the electrophile or radical will dictate whether it is long-lived enough to dissociate from the P450 active site and react with other nucleophiles (water, glutathione, or tissue macromolecule nucleophiles) or whether it will be so reactive that it will rapidly react with nucleophiles present on the P450 enzyme before it has an opportunity to dissociate. The latter can result in enzyme inactivation. The reaction can occur with groups on the apoprotein or with the porphyrin. Experimentally, this can be distinguished by determining whether the inactivated enzyme can still bind CO (which indicates that the heme is unmodified). In either case, the reaction results in an irreversible covalent bond to the enzyme which renders it inactive. If reaction occurs to the porphyrin, a change in the heme absorbance spectrum may be observed, depending on the structure of the adduct.

19.2.3 Examples of Functional Groups That Inactivate Cytochrome P450

There are numerous examples of drugs and other xenobiotics that have been demonstrated to inactivate P450 enzymes. In many cases, the mechanisms are unique to the specific structure of the drug and a description of every case is beyond the scope of this chapter. In this section, some of the most common inactivating substituents are described. A greater level of detail can be found in articles by Kalgutkar et al. (2007), Correia and Ortiz de Montellano (2005).

19.2.3.1 Alkylamines

Aliphatic amines are an extremely common substituent present in drugs and unavoidable in general drug design since a positively charged nitrogen center is required for binding to many drug receptor sites (e.g., neurotransmitter receptors). There are far more drugs that possess this substituent that are not inactivators than those that are inactivators; however, an understanding of which aliphatic amines will be inactivators and which will not is not known a priori. Depending on the substitution pattern on the nitrogen, there can be several metabolic steps needed to generate the intermediate that can form a quasi-irreversible complex with the heme iron in P450. A primary amine is necessary, and drugs that are secondary or tertiary amines must first undergo N-dealkylation to be able to be an inactivator. A nitroso intermediate is the one hypothesized to form the complex with the heme iron:

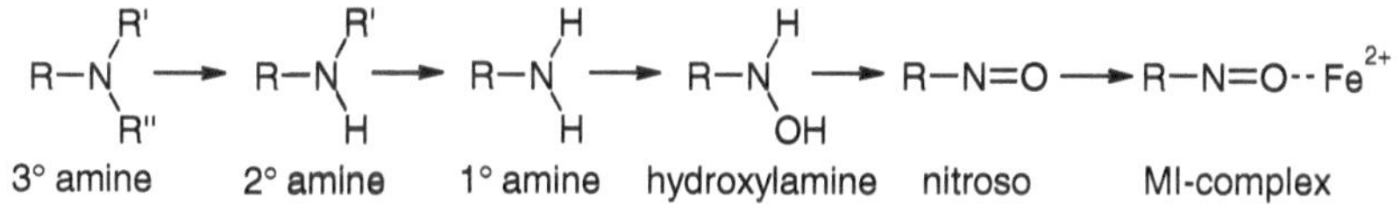

Examples of drugs that undergo this mechanism include diltiazem (Jones et al., 1999), verapamil (Wang et al., 2004), and several macrolide antibiotic drugs (e.g., troleandomycin, clarithromycin) (McConn et al., 2004).

19.2.3.2 Methylenedioxyphenyl

The methylenedioxyphenyl group is common in natural products and in some cases has been included in drugs. Hydroxylation at the bridging methylene group results in the formation of an unstable intermediate that either loses the methylene group resulting in a catechol or loses water to yield a carbene. While the former product can ultimately generate a reactive quinone metabolite with further oxidation, it is the carbene product that is important for the mechanism of P450 inactivation. The carbene will form an MI complex with the heme iron and is marked by formation of a Soret absorption peak around 456 nm.

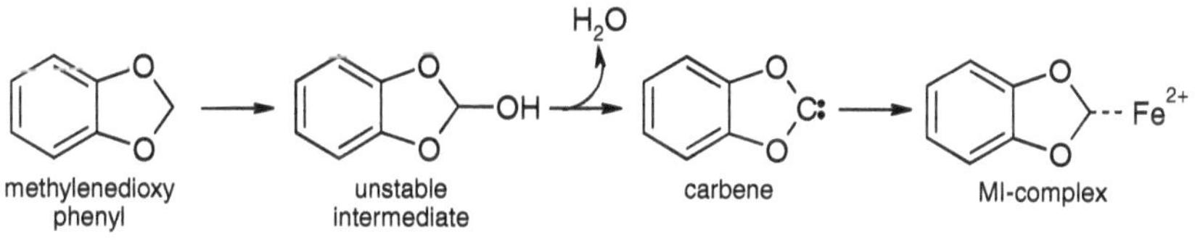

Examples of drugs that inactivate P450 enzymes by this mechanism include paroxetine (Bertelsen et al., 2003) and methylenedioxymethamphetamine (Van et al., 2006).

19.2.3.3 Alkynes

Alkynes have been shown to be inactivators of P450 enzymes, with different mechanisms possible. Addition of the activated iron-bound oxygen across the triple bond can result in a radical intermediate that can form an adduct with the porphyrin. Alternately, completion of the oxidation reaction of an alkyne will yield an electrophilic ketene. Reaction of the ketene with water yields an acetic acid metabolite (commonly observed for alkynes), or the ketene can react with nucleophiles present in the active site of the P450 enzyme, resulting in activation.

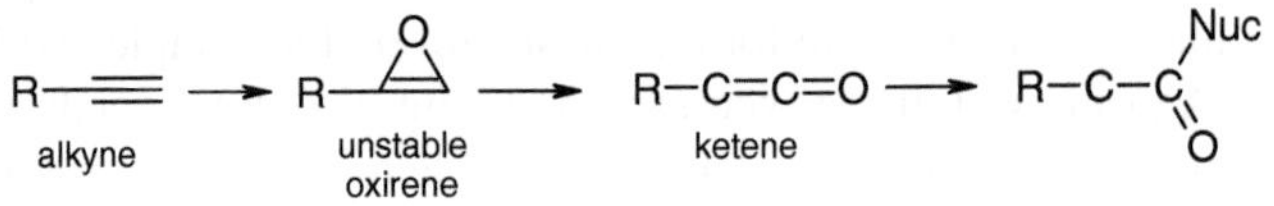

Ethinyl estradiol serves as an example of a drug containing an alkyne that is metabolized in this manner and which inactivates CYP3A4 (Lin et al., 2002).

19.2.3.4 Progenitors of Michael Acceptors

Michael acceptors are electrophiles containing conjugated double bond systems that can readily react with soft nucleophiles. Examples include α,β-unsaturated aldehydes, ketones, and amides, quinones, quinoneimines, and quinonemethides. Phenols can be readily oxidized to hydroquinones and quinones. A compelling example of this is raloxifene, which contains an extended hydroquinone system and has been shown to inactivate CYP3A4, although there are no reported in vivo DDI caused by this inactivation (Pearson et al., 2007; Baer et al., 2007). Nefazodone is the best example of an MBI that likely occurs via generation of a Michael acceptor that also causes in vivo DDI. The 3-chlorophenylpiperazine portion can undergo hydroxylation of the 4-position, and a second oxidation yields a quinoneimine intermediate(Kalgutkar et al., 2005).

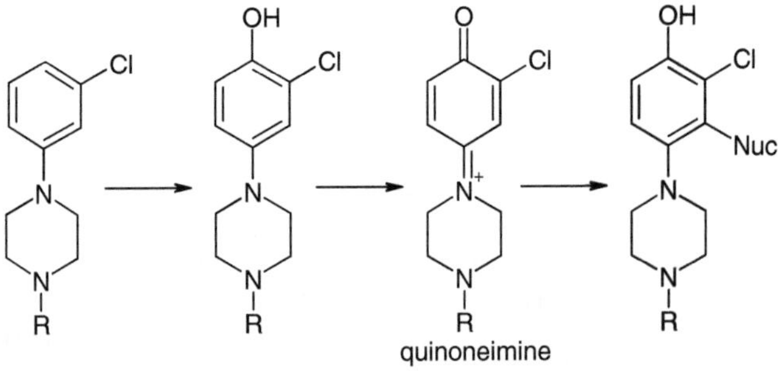

Nefazodone is hepatotoxic, and the same reaction, but occurring with other tissue nucleophiles, may be an underlying mechanism behind toxicity. Thus, bioactivation of drugs to reactive intermediates that can inactivate P450 enzymes can sometimes serve as a harbinger of other types of deleterious phenomena. However, there is not a 1:1 correlation since many drugs can inactivate P450 enzymes without causing such types of toxicity.

19.2.3.5 Five-Membered Aromatic Heterocyclic Compounds

There are numerous examples of furan- and thiophene-containing drugs, some of which are MBI for human P450 enzymes and cause DDI. (Furans and thiophenes generally inactivate P450 by formation of Michael acceptors, which were described in the preceding sub section. However, their generation from 5-membered aromatic heterocycles merits a separate discussion.) Menthofuran, a metabolite of a natural product, is a furan shown to be an MBI for CYP2A6 and adduction of menthofuran-related material to both CYP2A6 and P450 reductase was shown by electrophoresis and immunoblotting supporting a mechanism whereby the reactive intermediate covalently binds to the enzyme (Khojasteh-Bakht et al., 1998). Reduction of binding by added nucleophiles (e.g., glutathione and methoxylamine) showed a reduction of binding to reductase but not CYP2A6, suggesting the reactive intermediate reacts with CYP2A6 prior to release from the enzyme. Such a reactive metabolite is proposed to arise by P450 catalyzed epoxidation of the furan ring. Studies on L-745394, which contains a furanopyridine ring, also support a role for furans as MBI for P450 enzymes (Sahali-Sahly et al., 1996; Lightning et al., 2000). Findings support a mechanism of

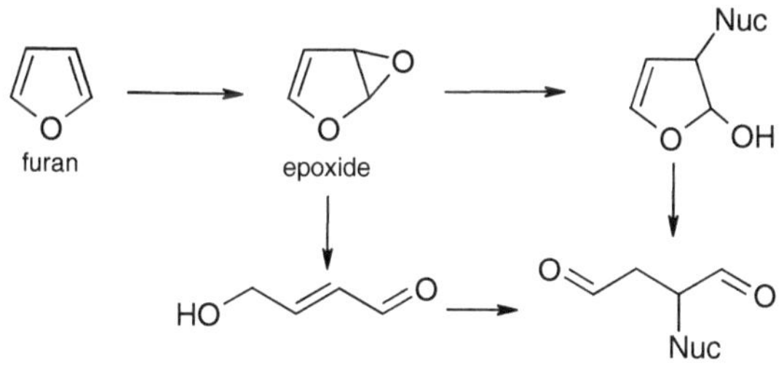

The antipsoriasis agent methoxsalen is a drug that inactivates CYP2A6 and CYP1A2 and causes DDI in vivo (Koenigs et al., 1997).

Drugs that contain thiophene rings and have been shown to be MBI for human P450 enzymes include ticlopidine, clopidogrel, suprofen, and tienilic acid (Lopez-Garcia et al., 1994; O'Donnell et al., 2003; Richter et al., 2004; Turpeinen et al., 2005; Walsky and Obach, 2007). The mechanism of bioactivation that leads to MBI of P450 could occur in a manner similar to furans (i.e., epoxidation) or via oxygenation of the thiophene sulfur. Adduction occurs to nucleophiles on the protein, and tienilic acid adducts to CYP2C9 have been shown (Lopez-Garcia et al., 1994). Possible mechanisms for thiophene ring MBI are

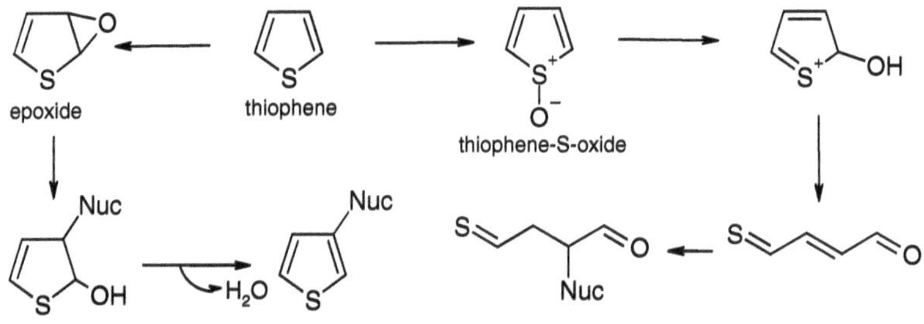

Both ticlopidine and clopidogrel have been shown to cause a reduction of CYP2B6 activity in vivo, and tienilic acid has been shown to inactivate CYP2C9 in vivo through a marked reduction in warfarin clearance (O′Reilly, 1982; Turpeinen et al., 2005). The mechanism of inactivation of CYP1A2 by zileuton has not been determined, but could be due to the benzothiophene substituent (Lu et al., 2003).

19.2.3.6 Alkyl-Substituted 5-Member Aromatic Azaheterocyclics

Noteworthy among structures of this type is furafylline, a frequently employed reagent to inactivate CYP1A2 in vitro and which had been shown to cause a marked increase in exposure to caffeine in humans (Kunze and Trager, 1993). Interestingly, despite possessing a freely accessible furan ring in its structure, this substituent is not responsible for the MBI demonstrated by furafylline. Rather, metabolism of the 8-methyl substituent is proposed to yield a reactive iminomethide intermediate which can either leave the active site and react with water (to yield the hydroxylated metabolite) or with a nucleophilic group on the P450 protein(Racha et al., 1998).

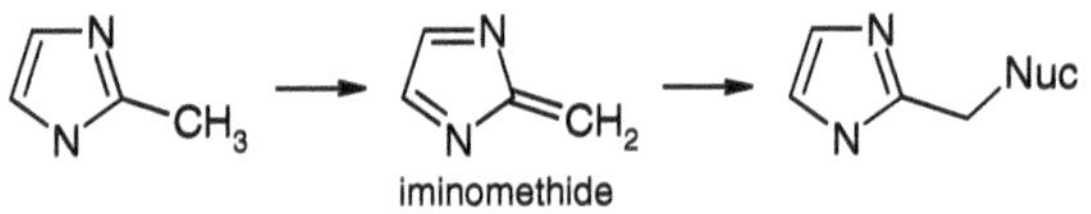

It is possible that midazolam, which despite being a very commonly used probe substrate activity of CYP3A4 that shows autoinactivation kinetics, could cause inactivation by this same mechanism during generation of the 1′-hydroxy metabolite (Khan et al., 2002)

The generation of reactive intermediates from 3-alkylindoles may be related to the bioactivation of 2-alkylimidazoles. The bioactivation of 3-methylindole has been well characterized in studies directed toward understanding its mechanism of pneumotoxicity in grazing animals, but it has also been shown to inactivate human CYP2F enzymes (Kartha and Yost, 2008). The mechanism involves oxidation of the 3-methyl group and dehydration to yield a reactive iminomethide intermediate:

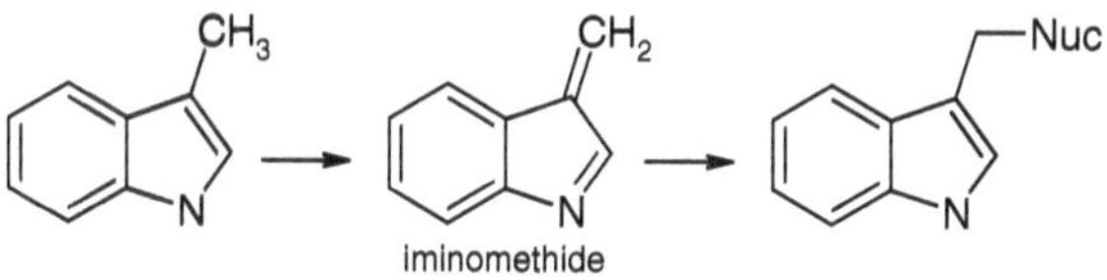

The iminomethide forms an adduct with the apoprotein, since the CO binding properties of the heme are unobstructed. The MBI exhibited by zafirlukast on CYP3A4 has been proposed to occur via this mechanism as well (Kassahun et al., 2005).

There are several other chemical substituents that have been shown to cause mechanism-based inactivation of P450 enzymes, but have not been associated with inactivation of human enzymes to cause in vivo DDI. These include alkenes (e.g., secobarbital) which can form adducts to porphyrin; haloalkanes (e.g., chloramphenicol) which form acyl halides that acylate the protein; cyclopropylamines (e.g., tranylcypromine) and hydrazines (e.g., phenelzine, dihydralazines) which generate radicals to form adducts to P450 protein, among others.

19.3 Experimental Approaches to Determination of Time-Dependent Inhibition

Time-dependent inhibition is the in vitro enzyme kinetic observation that is made that differentiates between reversible inhibition and irreversible/quasi-irreversible inhibition. The assessment of potential drugs for time-dependent inhibition of P450 enzymes using in vitro techniques has become common in the drug discovery and development processes and remains an area of active research (Venkatakrishnan et al., 2007; Ghanbari et al., 2006; Obach et al., 2007). A determination that a candidate is a time-dependent inhibitor in a kinetic experiment does not in itself confirm that it is also a mechanism-based inactivator. Proof that a compound is a mechanism-based inactivator requires gathering additional biochemical evidence. Depending on the mechanism of inactivation, this can include spectral demonstration of a metabolite–intermediate complex (MIC), demonstration of irreversible covalent binding to the P450 protein, a lack of recovery of activity after dialysis of inactivated enzyme, among other approaches (listed in Table 19.2).

Pooled human liver microsomes are the most often used in vitro test system for time-dependent inhibition study of P450 enzymes. The use of this human-derived system has been shown to be reasonably successful in the prediction of clinical DDI due to inactivation of CYP1A2 by furafylline, CYP2B6 by ticlopidine, CYP2D6 by paroxetine, and CYP3A by diltiazem, erythromycin, fluoxetine, ritonavir, verapamil (Obach et al., 2007; Venkatakrishnan and Obach, 2005; Wang et al., 2004; Mayhew et al., 2000). The use of recombinant P450 heterologous expression systems, while able to readily detect TDI, can be problematic when using the data to predict the magnitude of DDI. It has been suggested that inactivation data generated

Table 19.2 Biochemical experiments that can ascertain mechanism-based inactivation of P450 enzymes

Experiment	Use(s)	Approach
Spectral measurements of MIC	Detect MIC complex formation	Incubate TDI with P450 and NADPH; monitor formation of MIC by spectral scanning in the range of 445–460 nm (Franklin, 1974, 1991)
Ferricyanide reactivation	Detect MIC complex formed from alkylamine activation to nitroso	After treatment with TDI, add 10–50 mM potassium ferricyanide and measure reemergence of enzyme activity (Franklin, 1991; Delaforge et al., 1984)
P450 CO-difference spectral measurement	Detects adduction to porphyrin or apoprotein (such that CO-binding is disrupted)	After treatment with TDI, bubble CO in the enzyme solution and add a few grains of sodium dithionite; measure difference spectrum at 450 nm compared to enzyme solution without dithionite (Schenkman and Jansson, 1998)
Dialysis	Demonstrates irreversible inactivation	After treatment with TDI, subject enzyme to dialysis to remove the TDI; test for activity to see if it reemerges, as compared to non-inactivated control(Kunze and Trager, 1993)
Washing microsomes	Demonstrates irreversible inactivation	After treatment with TDI, spin microsomes in an ultracentrifuge at 100 k×*g* to remove the TDI; test for activity to see if it reemerges, as compared to non-inactivated control (Sinal and Bend, 1996)
Characterization of adducted porphyrin	Demonstrates adduct to porphyrin	After treatment with TDI, denature enzyme and analyze the porphyrin by HPLC, as compared to untreated control (Kunze et al., 1983)
Covalent binding	Demonstrates adduct to apoprotein	Treat enzyme with radiolabeled TDI, denature protein and exhaustively wash away excess TDI, determine radioactivity in protein pellet compared to a control lacking NADPH (De Matteis, 1974; Halpert, 1981)
Substrate protection	Demonstrates irreversible inactivation	During preincubation of enzyme and TDI, include a low K_s substrate to determine if inactivation can be competed away with increasing substrate concentration (Polasek et al., 2006)

in such systems may suffer from their unnatural molar ratio of NADPH/P450 oxidoreductase (OR) to P450 being 10- to 20-fold greater than that found in human liver microsomes and therefore are more likely to form reactive intermediates which may bias determination of k_{inact} values higher than normal (Polasek and Miners, 2007).

19.3.1 Abbreviated Approaches to Identify Time-Dependent Inhibition

Conduct of experiments to measure K_I and k_{inact} can be laborious. Thus, abbreviated approaches to identifying compounds that are TDI before investing the effort to do a complete inactivation enzyme kinetic experiments are worthwhile, provided that such approaches do not yield false-negative results. Abbreviated approaches to identifying TDI involve the comparison of samples preincubated in the presence and absence of NADPH followed by a conventional incubation with substrate to determine if a relative change in enzyme activity has occurred. Among the simplest of TDI experiments is the single concentration TDI assay (Venkatakrishnan et al., 2007; Obach et al., 2007; Walsky and Boldt, 2008). In this assay a single test compound concentration (IC_{25}) is used to maximize to possibility of observing a significant change in activity over the control (Fig. 19.1). The IC25 concentration can be determined from the data generated during an IC_{50} experiment. (Compounds which are not reversible inhibitors could still be TDI and may be evaluated at the highest concentration tested during the reversible inhibition assessment.) The evaluation of TDI following a reversible inhibition experiment represents a highly effective iterative approach. Test compound is assessed at 10× its IC_{25} concentration in microsomes at 10× their normal incubation concentration in the presence and absence of NADPH. After a 30 min preincubation, aliquots are diluted 10× into substrate at K_M-containing NADPH and a normal incubation conducted. At this point the test compound, if unconsumed during the preincubation, would be at its theoretical IC_{25} concentration (75% control activity), substrate concentration at K_M, and microsomes at their normal reversible inhibition experiment concentration. When results are compared to vehicle controls and percent control activities calculated, a direct comparison of percent change between them is made to determine if TDI is of concern or not. The IC_{25} concentration is used because it represents

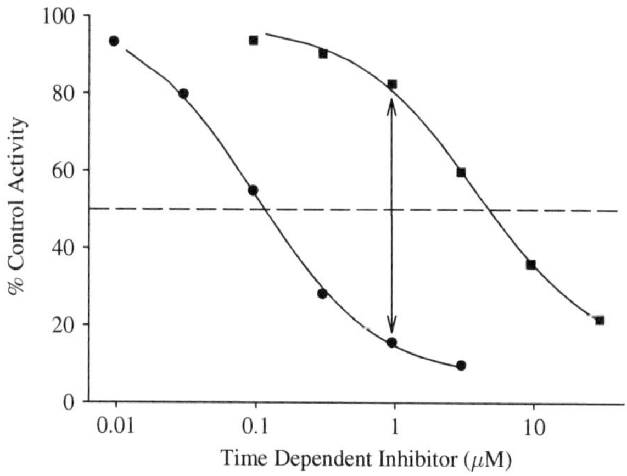

Fig. 19.1 IC_{50} plots in the presence (■) and absence (•) of 30 min NADPH preincubation. *Arrow* indicates the change in activity at the IC_{25}

a point on an inhibition curve near the top for which a small change in concentration will yield the greatest change in activity, hence the most sensitive point. (If concentrations are used that are too high or too low, changes with preincubation may not be observed and yield false-negative results.) Typically a loss of more than ~20% activity warrants further investigation. This experiment only indicates that the test compound has shown measurable time-dependent inhibition that requires further evaluation to determine K_I and k_{inact}. A second strategy is to perform the same type of experiment in the presence and absence of NADPH following the same dilution steps indicated above, but with multiple test concentrations. This provides IC_{50} values after preincubation which have been empirically shown to correlate to k_{inact}/K_I and these can be used in the prediction of DDI (Obach et al., 2007; Grime et al., 2009). In the earlier phases of drug research, greater emphasis is usually given toward the potential of a new compound to cause TDI for CYP3A due to the large number of drugs this enzyme metabolizes. However, mechanism-based inactivators of CYP1A2, CYP2B6, CYP2C8, CYP2C19, CYP2D6, and CYP2E1 have been identified and a thorough TDI assessment of all relevant enzymes should be conducted during the drug development process (Venkatakrishnan et al., 2007; Grimm, et al.,2009).

19.3.2 Determination of K_I *and* k_{inact}

Inactivation kinetics experiments are conducted similar to the TDI experiments described above (Obach et al., 2007) to assess the maximal rate of inactivation (k_{inact}) and the concentration of test compound that causes inactivation at half of the maximal rate of inactivation (K_I). The ratio of k_{inact}/K_I is sometimes used to rank order inactivation potential. A good correlation between this k_{inact}/K_I ratio and the degree of the shifted IC_{50} values obtained in the above experiment has been reported (Obach et al., 2007; Grime et al., 2009). In k_{inact}/K_I experiments, preincubation of the test compound at several concentrations with NADPH is performed at several preincubation time points, ranging from 0 to 10 min for strong, or 0 to 30 min for moderate or weak TDIs, then aliquots are diluted 10- to 20-fold into incubations containing high substrate concentrations such that velocity in the control incubations (i.e., no inactivator) is near V_{max}. The combination of high substrate concentration and the large dilution employed reduces the interference of reversible inhibition when conducting this assay. At each compound concentration (including a vehicle control), a k_{obs} value is determined by plotting the decrease in natural logarithm of activity over time and then determining the negative slopes of each line (or using nonlinear regression for first-order decay curves). Typically, the rate observed is greatest at the first non-zero time point taken, especially at the highest test compound concentration, and is attenuated at the longer preincubation time point(s). If the first non-zero time point is ignored, often a biphasic relationship is observed and the time points from zero to the point of that inflection are used for determining k_{obs}. Figure 19.2 illustrates a plot of natural log of percent control activity vs. preincubation time which illustrates this phenomena. The initial high slope described above

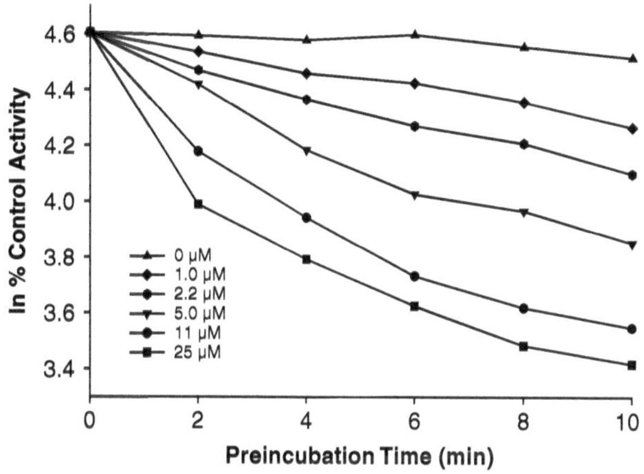

Fig. 19.2 Plot of ln % control activity vs. preincubation time (min)

Fig. 19.3 Plot of $k_{\text{obs,app}}$ vs. inactivator concentration ($K_{\text{I}}/k_{\text{inact}}$)

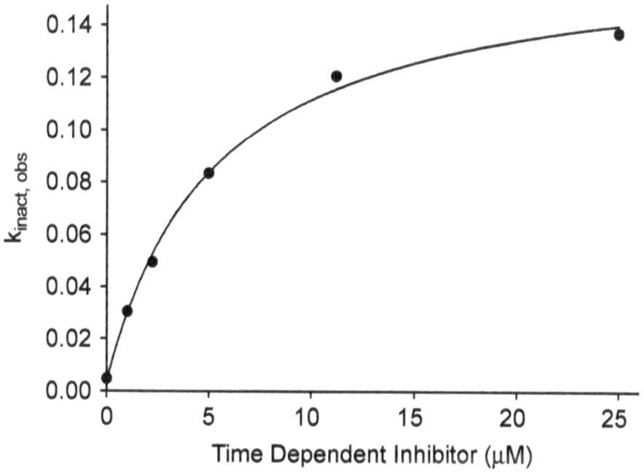

can be observed at the highest concentration at the 2 min preincubation time point. In this example, time points from 0 to 8 or 10 min could be used to determine the $K_{\text{I}}/k_{\text{inact}}$ values shown in Fig. 19.3 using the following equation:

$$k_{obs} = k_{obs[I=0]} + \frac{k_{inact} \cdot [I]}{K_I + [I]}$$

The $k_{\text{obs}[I=0]}$ value represents the apparent inactivation observed in samples containing only vehicle which can be due to self-inactivation of P450 enzymes in the presence of NADPH and absence of inactivator. It is used so that the curve is not erroneously forced through zero, which would yield inaccurate K_{I} and k_{inact} values. Other investigators have corrected the k_{obs} values by subtracting $k_{\text{obs}[I=0]}$ first from each inactivation curve and then forcing the relationship through the origin.

19.4 Correlation of In Vitro P450 Inactivation Data to In Vivo Drug–Drug Interactions

Understanding the risk for DDI associated with a new chemical entity is a key component of both the drug discovery and development processes. Identifying that risk early for new chemical entities under consideration as potential drugs gives a greater probability of removing that risk via drug design efforts. But it is important to distinguish those TDIs that have a high likelihood of causing DDI vs. those that show TDI but will not cause DDI. Several models have been proposed to predict DDI caused by inactivation. While these models can not yet replace clinical studies in all instances, they are extremely useful in identifying the level of DDI risk in early drug development and can be used for decision making.

19.4.1 Prediction of Drug Interactions for Time-Dependent Inhibitors

An equation for the prediction of drug–drug interactions due to time-dependent inhibition was originally developed to predict the magnitude of in vivo metabolic drug–drug interactions caused by CYP3A4 inactivators using in vitro data (Mayhew et al., 2000). The majority of known drugs that show time-dependent inactivation are targets for CYP3A4 interaction and there are only limited examples of drugs that are inactivators of other P450 enzymes (e.g., zileuton for CYP1A2, tienilic acid for CYP2C9, ticlopidine for CYP2C19, and paroxetine for CYP2D6). Nevertheless, the concepts used to develop predictions for DDI caused by TDI of CYP3A4 were extended to include the other P450 enzyme targets (Obach et al., 2007).

The equation that can be used to predict the magnitude of a DDI, i.e., the ratio of AUC values with the inhibitor relative to control (i.e., AUC′/AUC) of the probe substrate, is described below:

$$\frac{\mathrm{AUC}'}{\mathrm{AUC}} = \frac{1}{\left(\dfrac{f_m}{1+\left(\dfrac{k_{inact}\cdot[I]}{k_{deg}\cdot([I]+K_I)}\right)}\right)+(1-f_m)}$$

The terms k_{inact} and K_{I} are as described in the previous section; k_{deg} is the first-order degradation rate constant for P450 enzymes in vivo, f_{m} is the fraction of the affected drug that is cleared by the inactivated enzyme, and [*I*] is the in vivo free plasma concentration of the inactivator. For CYP3A4 inactivators, the impact of inactivation of intestinal enzyme is also necessary (Wang et al., 2004):

$$\frac{\text{AUC}'}{\text{AUC}} = \frac{1}{\left(\frac{f_m}{1+\left(\frac{k_{inact}\cdot[I]}{k_{deg}\cdot([I]+K_I)}\right)}\right)+(1-f_m)}\cdot$$

$$\frac{1}{F_G+(1-F_G)\cdot\left(\frac{1}{1+\left(\frac{k_{inact}\cdot[I]_G}{k_{deg}\cdot(K_I+[I]_G)}\right)}\right)}$$

wherein F_G is the fraction of the probe substrate that evades intestinal metabolism under the control condition, $[I]_G$ is the concentration of the inactivator in the enterocytes during absorption, and all other terms are as described above. To estimate $[I]_G$, the equation

$$[I]_G = \frac{k_a \cdot F_a \cdot Dose}{Q_g}$$

can be leveraged in which k_a, F_a, and Q_g are absorption rate constant, fraction absorbed, and intestinal blood flow, respectively (Rostami-Hodjegan and Tucker, 2007).

19.4.2 Combined Mathematical Equation

Although the magnitude of DDI predicted via the above approach is often close to that observed in the clinic for drugs known to inhibit CYP3A via time-dependent inactivation, it is important to acknowledge discrepancies that have been observed with some of the drugs tested, e.g., erythromycin, ethinyl estradiol, indinavir, mibefradil, oleandomycin, rifampicin, and verapamil (Lamberg et al., 1998). This is due mainly to the assumption that the amount of active CYP3A is decreased by the time-dependent inactivation and that the inhibitor has no effect on the rate of CYP3A synthesis. However, some drugs are also inducers, causing increased enzyme concentration that might counteract the inhibitory effect, and the predicted AUC increase will be less than that which is observed. While the above model is effective when only one mechanism (inactivation) is present, it cannot predict well when multiple effects are present (induction or reversible inhibition).

More recently, a combined mathematical model approach that accounts for the simultaneous influences of competitive inhibition, time-dependent inactivation, and induction of CYP3A in both the liver and the intestine has been proposed to provide a net drug–drug interaction prediction in terms of AUC ratio (Fahmi et al., 2008). This model provides a framework by which readily obtained in vitro values for competitive inhibition, time-dependent inactivation, and induction for the precipitant compound as well as literature values for f_m and F_G for the object drug can be used to provide quantitative predictions of DDIs. Using this model, DDIs arising via inactivation (e.g., erythromycin) continue to be well predicted, while those arising via mixed mechanisms (e.g., ritonavir) are also predicted within the ranges reported in the clinic (Fahmi et al., 2008). This comprehensive model quantitatively predicts clinical observations with reasonable accuracy and proven as a valuable tool to evaluate candidate drugs and rationalize clinical DDIs.

The equation used to predict the magnitude of DDI was previously reported and is shown below, expressed as the ratio of area under the exposure–time curve in the presence (AUC'_{po}) and absence (AUC_{po}) of a pharmacokinetic drug–drug interaction (Fahmi et al., 2008). This combined mathematical model is based on calculating the net effect of competitive inhibition, inactivation, and induction in both the intestine and the liver, as presented below:

$$\frac{AUC'_{po}}{AUC_{po}} = \left(\frac{1}{[A \times B \times C] \times f_m + (1 - f_m)}\right) \times \left(\frac{1}{[X \times Y \times Z] \times (1 - F_G) + F_G}\right),$$

in which

"A" is the term for TDI for the liver: $A = \dfrac{k_{deg,H}}{k_{deg,H} + \dfrac{[I]_H \times k_{inact}}{[I]_H + K_I}}$

"B" is the term for induction for the liver: $B = 1 + \dfrac{d \cdot E_{max} \cdot [I]_H}{[I]_H + EC_{50,I}}$

"C" is the term for reversible inhibition in the liver: $C = \dfrac{1}{1 + \dfrac{[I]_H}{K_i}}$

"X" is the term for TDI for the intestine: $X = \dfrac{k_{deg,G}}{k_{deg,G} + \dfrac{[I]_G \times k_{inact}}{[I]_G + K_I}}$

"Y" is the term for induction for the intestine: $Y = 1 + \dfrac{d \cdot E_{max} \cdot [I]_G}{[I]_G + EC_{50,I}}$

"Z" is the term for reversible inhibition for the intestine: $Z = \dfrac{1}{1 + \dfrac{[I]_G}{K_i}}$

$[I]_G$ and $[I]_H$ represent concentrations of inhibitor relevant for the intestine and liver, respectively (Obach et al., 2007). For the intestine, an estimate for $[I]_G$ was made as described above. For liver, free systemic C_{max} was used for the inactivation (term A) and induction (term B) portions of the expression, while free portal C_{max} was used for the reversible inhibition portion of the expression (term C). k_{deg} and f_m are as described above. E_{max}, EC_{50}, and "d" represent the maximum fold induction observed in cultured human hepatocytes, the concentration of inducer associated with half-maximum induction, and a calibration factor "d" (0.3), respectively, as described previously (Fahmi et al., 2008)

19.4.3 Computer Simulations

It is important to note that the simple and combined models described above only address the extent of DDI under the equilibrium, steady-state condition, using a representative number of the human population by applying an average of the systemic blood concentrations at steady state. In this regard, in silico approaches which link the time course of predicted precipitant and object liver and intestine exposure to the known CYP3A4 interaction mechanism is likely to address the aforementioned limitation.

With the use of the Simcyp®, population-based ADME application, simulating clinical study where the risk to individuals and the ability of simulating a clinical study with its unique identity are important features within this computer-simulated program. This population pharmacokinetic modeling provides not only point estimates of average DDI magnitudes but also simulation of DDI across different individuals and groups (Einolf, 2007; Rostami-Hodjegan and Tucker, 2007). The ability to use multiple subjects in the simulation representing the variable content of their drug-metabolizing enzymes in the human population is an important feature in Simcyp. This population model also allows one to consider subgroups of the population that are at greater risk for clinically relevant interactions to assess if they have a level of risk different from the rest of the population (e.g., poor metabolizers, impaired renal function). Also, the ability to input key details of clinical study design and simulation dynamic concentrations are other advantages of Simcyp. It should be noted that predictions with Simcyp require more comprehensive input data such as physiochemical properties, absorption, distribution, and elimination before accurate predictions are made.

19.4.4 Factors That Impact on Prediction Accuracy

The use of k_{inact} and K_{I} in predicting DDIs for inactivators has been the most frequently reported approach. However, careful attention to the design of the in vitro experiments to obtain accurate kinetic parameters is necessary for a reliable prediction, as previously described (Ghanbari et al., 2006; Yang et al., 2007). There are several parameters that are most important to making accurate predictions of DDIs caused by inactivators (Obach et al., 2006). Some are unique to predictions for inactivators while others are relevant for DDI prediction for any type of mechanism.

19.4.4.1 Factors Specifically Required for Predicting DDI for Inactivators

Enzyme Degradation Rate (k_{deg}). The true numerical value of k_{deg} can make a huge impact on the prediction and has not been adequately estimated in humans in vivo for any CYP, leaving this parameter as one important source of potential error in DDI predictions for inactivators. It has been known that the k_{deg} values could be very different for different individuals, disease states, and different isoforms.

In previous reports, the value used for k_{deg} for P450 enzymes frequently was a value derived from rat ($k_{deg} = 0.00083$ min^{-1} based on $t_{1/2} = 14$ h) (Mayhew et al., 2000). However, more recently, values for k_{deg} were derived from modeling the time course of reversal of DDIs caused by induction or inactivation of P450 enzymes in human study subjects. The current most used values for the degradation rates for CYP3A4 (k_{deg}) are 0.00032 min^{-1} based on $t_{1/2} = 36$ h and 0.00048 min^{-1} based on $t_{1/2} = 24$ h for the liver and intestine, respectively (Fahmi et al., 2008; Obach et al., 2007). For CYP1A2, as previously reported (Faber and Fuhr 2004), the k_{deg} used is 0.0003 min^{-1} based on $t_{1/2} = 38$ h. For CYP2D6, the k_{deg} used is 0.0002 min^{-1} based on $t_{1/2} = 51$ h (Obach et al., 2007; Venkatakrishnan and Obach, 2005). There has not been much investigation in the area of determining k_{deg} values for CYP2C9, CYP2C19, and CYP2B6. It is worth mentioning that there is considerable scatter among k_{deg} values that have been used for DDI predictions (Yang et al., 2007).

Estimation of In Vitro Inactivation Parameters k_{inact} *and* K_I. The Kitz–Wilson plot has been frequently applied for reporting the kinetics of time-dependent inactivators, where both K_I and k_{inact} parameters can be estimated. However, this method requires data linearization and parameter estimation by a two-step linear regression, which may adversely influence the accuracy and precision of estimation, particularly when the data contain substantial variabilities. Also, the inactivation rate (k_{inact}) and incubation time significantly influence the resultant K_I estimate. In addition, the observed form of the plot resulting from enzyme degradation may erroneously be interpreted as resulting from a noncompetitive inhibitory mechanism.

On the other hand, a nonlinear method can provide a relatively more robust method (Maurer and Fung, 2000) for the estimation of the K_I parameter than the classical Kitz–Wilson method in the presence of mechanism-based enzyme inactivation. The Kitz–Wilson method was developed before the widespread availability of computers and software designed for nonlinear parameter estimation. Now that these resources are readily available, nonlinear methods should be used to obtain more accurate and precise estimates of kinetic parameters.

19.4.4.2 Factors Which Are Important for Predicting DDI for Inactivators but Also Important for Other Mechanisms of DDI

In Vivo Inactivator Concentration $[I]_{in\ vivo}$. One area of DDI risk assessment that has not reached consensus is the determination of the in vivo precipitant concentration driving the DDI. Projected clinically relevant concentrations at the estimated therapeutic dose have included single-dose vs. steady-state values, total plasma vs. free plasma values, and systemic vs. estimated portal vein values. Different types of interactions have had success using different input values for $[I]$ with successful predictions of DDI caused by reversible inhibition utilizing unbound hepatic portal concentrations while predictions of induction DDI utilize free systemic concentrations. For predicting DDI for inactivators, estimated free systemic C_{max} generally provides a more accurate parameter to use for $[I]$ in vivo than portal vein C_{max} and was utilized by several authors (Kanamitsu et al., 2000; Obach et al., 2006). The use of estimated unbound hepatic inlet concentrations of inactivators occurring during the absorption phase, as surrogates for free concentrations in the liver, generally led to overpredictions of the magnitude of DDIs.

Fraction of Clearance Contributed by the Affected Enzyme $f_{CL(enz)}$. The fraction contribution of the specific P450 enzyme to the clearance of substrate drug in vivo is defined as f_m. Several approaches have been utilized to estimate the f_m value, including reaction phenotyping using human liver microsomes or the recombinant system with radiolabeled compound and selective inhibitors. It has been also estimated based on the estimates of urinary recovery of unchanged drugs, but they are not regarded as definitive and represent an upper estimate of the P450 contribution (Galetin et al., 2006; Gorski et al., 1998). Also, (Shou et al., 2008) utilized clinical DDIs associated with CYP3A4 inhibition data to estimate f_m values. Therefore, a slightly different range of values had been reported by several publications for midazolam (Shou et al., 2008; Gorski et al., 1998; Hallifax et al., 2008). Probably the best values for f_m are those estimated from pharmacokinetic studies conducted in subjects with and without expression of functional enzyme, i.e., poor metabolizers. In these instances, comparison of the intrinsic clearance of a probe substrate in subjects that are homozygous for a null allele for a P450 enzyme relative to those possessing the enzyme will show an overall mean contribution of the enzyme to clearance. This has been done to estimate f_m values for desipramine, metoprolol, omeprazole, tolbutamide, and (S)-warfarin. However, the limitation of this approach is that there has to be a genetic polymorphism such that some subjects lack the enzyme of interest. Clearly, prediction success can be highly impacted on the potential error in estimating f_m value and a challenge is gaining reliable values for this parameter.

Microsomal Binding. Nonspecific binding in enzyme preparations can have a significant impact on the inhibition potency, as reported previously (Margolis and Obach, 2003). Since inactivation experimental design often uses high concentration of microsomal protein in the preincubation portion of the experiment to enable an adequate dilution into the activity assay (thereby reducing reversible inhibition), estimation of K_I of highly bound compounds can be impacted. Therefore, the extent of nonspecific binding to microsomes is an important parameter to measure when attempting to relate K_I values measured in vitro to observations of drug–drug interactions in vivo.

19.5 Conclusions

Throughout the past decade, our ability to generate in vitro cytochrome P450 inactivation data and utilize it for predictions of DDI in the clinic has greatly advanced. Such data are routinely generated in the search for new drugs that will lack the liability of causing pharmacokinetic-based DDI. The sophistication with which in vitro TDI data are used has advanced to the point where not only predictions of simple point estimates of clinical DDI can be reliably made but computer simulation of population variability in DDI can also be conducted. This can aid in the clinical development of new drugs and appropriately direct effort to the conduct of only those clinical studies for which a DDI is anticipated.

Our increasing knowledge of the types of chemical substituents that can be bioactivated by P450 enzymes and cause enzyme inactivation can be used in drug design (Kalgutkar et al., 2007). While not all substituents that cause inactivation can be expunged from the repertoire of chemical substituents needed by the medicinal chemist to design potent drugs (e.g., alkylamines) some can be avoided (e.g., thiophene rings). And in those events where a potential inactivating substituent is incorporated into a molecule of interest, in vitro approaches can be readily applied to experimentally determine whether the compound will cause inactivation of P450 enzymes. Nevertheless, there will probably be instances in the future in which a new compound that contains no presently known P450-inactivating substituents will be shown to be an inactivator; this will lead us into new knowledge of the bioorganic mechanisms of P450-mediated bioactivation.

References

Baer BR, Wienkers LC and Rock DA (2007) Time-dependent inactivation of P450 3A4 by raloxifene: identification of Cys239 as the site of apoprotein alkylation. *Chem Res Toxicol* **20**:954–964.

Bertelsen KM, Venkatakrishnan K, Von Moltke LL, Obach RS and Greenblatt DJ (2003) Apparent mechanism-based inhibition of human CYP2D6 in vitro by paroxetine: comparison with fluoxetine and quinidine. *Drug Metab Dispos* **31**:289–293.

Correia M and Ortiz de Montellano PR (2005) Inhibition of cytochrome P450 enzymes, in *Cytochrome P450 Structure, Mechanism, and Biochemistry* (Ortiz de Montellano PR ed) pp 247–322, Kluwer Academic/Plenum Publishers, New York.

Delaforge M, Jaouen M, and Mansuy D (1984) The cytochrome P-450 metabolite complex derived from troleandomycin: properties in vitro and stability in vivo. *Chem-Biol Int.* **51**:371–376.

De Matteis F (1974) Covalent binding of sulfur to microsomes and loss of cytochrome P-450 during the oxidative desulfuration of several chemicals. *Mol. Pharmacol.* **10**:849–854.

Einolf HJ (2007) Comparison of different approaches to predict metabolic drug–drug interactions. *Xenobiotica* **37**:1257–1294.

Faber MS and Fuhr U (2004) Time response of cytochrome P450 1A2 activity on cessation of heavy smoking. *Clin. Pharmacol. Ther.* **76**:178–184.

Fahmi OA, Maurer TS, Kish M, Cardenas E, Boldt S and Nettleton D (2008) A combined model for predicting CYP3A4 clinical net drug–drug interaction based on CYP3A4 inhibition, inactivation, and induction determined in vitro. *Drug Metab Dispos* **36**: 1698–1708.

Franklin MR (1974) Complexes of metabolites of amphetamines with hepatic cytochrome P-450. *Xenobiotica* **4**:133–142.

Franklin MR (1991) Cytochrome P450 metabolic intermediate complexes from macrolide antibiotics and related compounds. *Meth. Enzymol.* **206**:559–573.

Galetin A, Burt H, Gibbons L and Houston JB (2006) Prediction of time-dependent CYP3A4 drug–drug interactions: impact of enzyme degradation, parallel elimination pathways, and intestinal inhibition. *Drug Metab Dispos* **34**:166–175.

Ghanbari F, Rowland-Yeo K, Bloomer JC, Clarke SE, Lennard MS, Tucker GT and Rostami-Hodjegan A (2006) A critical evaluation of the experimental design of studies of mechanism based enzyme inhibition, with implications for in vitro-in vivo extrapolation. *Curr Drug Metab* **7**:315–334.

Gorski JC, Jones DR, Haehner-Daniels BD, Hamman MA, O'Mara EM, Jr. and Hall SD (1998) The contribution of intestinal and hepatic CYP3A to the interaction between midazolam and clarithromycin. *Clin Pharmacol Ther* **64**:133–143.

Grime KH, Bird J, Ferguson D and Riley RJ (2009) Mechanism-based inhibition of cytochrome P450 enzymes: an evaluation of early decision making in vitro approaches and drug–drug interaction prediction methods. *Eur J Pharm Sci* **36**:175–191.

Grimm SW, Einolf HJ, Hall SD, et al., (2009). The conduct of in vitro studies to address time-dependent inhibition of drug-metabolizing enzymes: a perspective of the Pharmaceutical Research and Manufacturers of America. *Drug Metab. Dispos.* **37**:1355–1370.

Hallifax D, Galetin A and Houston JB (2008) Prediction of metabolic clearance using fresh human hepatocytes: comparison with cryopreserved hepatocytes and hepatic microsomes for five benzodiazepines. *Xenobiotica* **38**:353–367.

Halpert J. (1981) Covalent modification of lysine during the suicide inactivation of rat liver cytochrome P-450 by chloramphenicol. *Biochem. Pharmacol.* **30**:875–881.

Hollenberg PF, Kent UM and Bumpus NN (2008) Mechanism-based inactivation of human cytochromes p450s: experimental characterization, reactive intermediates, and clinical implications. *Chem Res Toxicol* **21**:189–205.

Jones DR, Gorski JC, Hamman MA, Mayhew BS, Rider S and Hall SD (1999) Diltiazem inhibition of cytochrome P-450 3A activity is due to metabolite intermediate complex formation. *J Pharmacol Exp Ther* **290**:1116–1125.

Kalgutkar AS, Obach RS and Maurer TS (2007) Mechanism-based inactivation of cytochrome P450 enzymes: chemical mechanisms, structure-activity relationships and relationship to clinical drug–drug interactions and idiosyncratic adverse drug reactions. *Curr Drug Metab* **8**:407–447.

Kalgutkar AS, Vaz AD, Lame ME, Henne KR, Soglia J, Zhao SX, Abramov YA, Lombardo F, Collin C, Hendsch ZS and Hop CE (2005) Bioactivation of the nontricyclic antidepressant nefazodone to a reactive quinone-imine species in human liver microsomes and recombinant cytochrome P450 3A4. *Drug Metab Dispos* **33**:243–253.

Kanamitsu SI, Ito K, Okuda H, Ogura K, Watabe T, Muro K and Sugiyama Y (2000) Prediction of in vivo drug–drug interactions based on mechanism-based inhibition from in vitro data: inhibition of 5-fluorouracil metabolism by (E)-5-(2-Bromovinyl)uracil. *Drug Metab Dispos* **28**:467–474.

Kartha JS and Yost GS (2008) Mechanism-based inactivation of lung-selective cytochrome P450 CYP2F enzymes. *Drug Metab Dispos* **36**:155–162.

Kassahun K, Skordos K, McIntosh I, Slaughter D, Doss GA, Baillie TA and Yost GS (2005) Zafirlukast metabolism by cytochrome P450 3A4 produces an electrophilic alpha,beta-unsaturated iminium species that results in the selective mechanism-based inactivation of the enzyme. *Chem Res Toxicol* **18**:1427–1437.

Khan KK, He YQ, Domanski TL and Halpert JR (2002) Midazolam oxidation by cytochrome P450 3A4 and active-site mutants: an evaluation of multiple binding sites and of the metabolic pathway that leads to enzyme inactivation. *Mol Pharmacol* **61**:495–506.

Khojasteh-Bakht SC, Koenigs LL, Peter RM, Trager WF and Nelson SD (1998) (R)-(+)-Menthofuran is a potent, mechanism-based inactivator of human liver cytochrome P450 2A6. *Drug Metab Dispos* **26**:701–704.

Koenigs LL, Peter RM, Thompson SJ, Rettie AE and Trager WF (1997) Mechanism-based inactivation of human liver cytochrome P450 2A6 by 8-methoxypsoralen. *Drug Metab Dispos* **25**:1407–1415.

Kunze KL, Mangold BLK, Wheeler C, Beilan HS, and Ortiz de Montellano PR (1983) The cytochrome P-450 active site. Regiospecificity of prosthetic heme alkylation by olefins and acetylenes. *J Biol. Chem.* **258**:4202–4207.

Kunze KL and Trager WF (1993) Isoform-selective mechanism-based inhibition of human cytochrome P450 1A2 by furafylline. *Chem Res Toxicol* **6**:649–656.

Lamberg TS, Kivisto KT and Neuvonen PJ (1998) Effects of verapamil and diltiazem on the pharmacokinetics and pharmacodynamics of buspirone. *Clin Pharmacol Ther* **63**: 640–645.

Lightning LK, Jones JP, Friedberg T, Pritchard MP, Shou M, Rushmore TH and Trager WF (2000) Mechanism-based inactivation of cytochrome P450 3A4 by L-754,394. *Biochemistry* **39**: 4276–4287.

Lin HL, Kent UM and Hollenberg PF (2002) Mechanism-based inactivation of cytochrome P450 3A4 by 17 alpha-ethynylestradiol: evidence for heme destruction and covalent binding to protein. *J Pharmacol Exp Ther* **301**:160–167.

Lopez-Garcia MP, Dansette PM and Mansuy D (1994) Thiophene derivatives as new mechanism-based inhibitors of cytochromes P-450: inactivation of yeast-expressed human liver cytochrome P-450 2C9 by tienilic acid. *Biochemistry* **33**:166–175.

Lu P, Schrag ML, Slaughter DE, Raab CE, Shou M and Rodrigues AD (2003) Mechanism-based inhibition of human liver microsomal cytochrome P450 1A2 by zileuton, a 5-lipoxygenase inhibitor. *Drug Metab Dispos* **31**:1352–1360.

Margolis JM and Obach RS (2003) Impact of nonspecific binding to microsomes and phospholipid on the inhibition of cytochrome P4502D6: implications for relating in vitro inhibition data to in vivo drug interactions. *Drug Metab Dispos* **31**:606–611.

Maurer T and Fung HL (2000) Comparison of methods for analyzing kinetic data from mechanism-based enzyme inactivation: application to nitric oxide synthase. *AAPS Pharm Sci* **2**:E8.

Mayhew BS, Jones DR and Hall SD (2000) An in vitro model for predicting in vivo inhibition of cytochrome P450 3A4 by metabolic intermediate complex formation. *Drug Metab Dispos* **28**:1031–1037.

McConn DJ, Lin YS, Allen K, Kunze KL and Thummel KE (2004) Differences in the inhibition of cytochromes P450 3A4 and 3A5 by metabolite-inhibitor complex-forming drugs. *Drug Metab Dispos* **32**:1083–1091.

Obach RS, Walsky RL and Venkatakrishnan K (2007) Mechanism-based inactivation of human cytochrome p450 enzymes and the prediction of drug–drug interactions. *Drug Metab Dispos* **35**:246–255.

Obach RS, Walsky RL, Venkatakrishnan K, Gaman EA, Houston JB and Tremaine LM (2006) The utility of in vitro cytochrome P450 inhibition data in the prediction of drug–drug interactions. *J Pharmacol Exp Ther* **316**:336–348.

O'Donnell JP, Dalvie DK, Kalgutkar AS and Obach RS (2003) Mechanism-based inactivation of human recombinant P450 2C9 by the nonsteroidal anti-inflammatory drug suprofen. *Drug Metab Dispos* **31**:1369–1377.

O'Reilly RA (1982) Ticrynafen-racemic warfarin interaction: hepatotoxic or stereoselective?. *Clin Pharmacol Ther* **32**:356–361.

Pearson JT, Wahlstrom JL, Dickmann LJ, Kumar S, Halpert JR, Wienkers LC, Foti RS and Rock DA (2007) Differential time-dependent inactivation of P450 3A4 and P450 3A5 by raloxifene: a key role for C239 in quenching reactive intermediates. *Chem Res Toxicol* **20**: 1778–1786.

Polasek TM, Elliot DJ, Somogyi AA, Gillam EMJ, Lewis BC, and Miners JO (2006) An evaluation of potential mechanism-based inactivation of human drug metabolizing cytochromes P450 by monoamine oxidase inhibitors, including isoniazid. *Br. J. Clin. Pharmacol.* **61**:570–584.

Polasek TM and Miners JO (2007) In vitro approaches to investigate mechanism-based inactivation of CYP enzymes. *Expert Opin Drug Metab Toxicol* **3**:321–329.

Racha JK, Rettie AE and Kunze KL (1998) Mechanism-based inactivation of human cytochrome P450 1A2 by furafylline: detection of a 1:1 adduct to protein and evidence for the formation of a novel imidazomethide intermediate. *Biochemistry* **37**:7407–7419.

Richter T, Murdter TE, Heinkele G, Pleiss J, Tatzel S, Schwab M, Eichelbaum M and Zanger UM (2004) Potent mechanism-based inhibition of human CYP2B6 by clopidogrel and ticlopidine. *J Pharmacol Exp Ther* **308**:189–197.

Riley RJ, Grime K and Weaver R (2007) Time-dependent CYP inhibition. *Expert Opin Drug Metab Toxicol* **3**:51–66.

Rostami-Hodjegan A and Tucker GT (2007) Simulation and prediction of in vivo drug metabolism in human populations from in vitro data. *Nat Rev Drug Discov* **6**:140–148.

Sahali-Sahly Y, Balani SK, Lin JH and Baillie TA (1996) In vitro studies on the metabolic activation of the furanopyridine L-754,394, a highly potent and selective mechanism-based inhibitor of cytochrome P450 3A4. *Chem Res Toxicol* **9**:1007–1012.

Schenkman JB and Jansson I (1998) Spectral analyses of cytochromes P450. *Meth. Molec. Biol.* **107**:25–33.

Sinal C and Bend JR (1996) Kinetics and selectivity of mechanism-based inhibition of guinea pig hepatic and pulmonary cytochrome P450 by N-benzyl-1-aminobenzotriazole and N-.alpha.-methylbenzyl-1-aminobenzotriazole. *Drug Metab. Dispos.* **24**:996–1001.

Shou M, Hayashi M, Pan Y, Xu Y, Morrissey K, Xu L and Skiles GL (2008) Modeling, prediction, and in vitro in vivo correlation of CYP3A4 induction. *Drug Metab Dispos* **36**: 2355–2370.

Turpeinen M, Tolonen A, Uusitalo J, Jalonen J, Pelkonen O and Laine K (2005) Effect of clopidogrel and ticlopidine on cytochrome P450 2B6 activity as measured by bupropion hydroxylation. *Clin Pharmacol Ther* **77**:553–559.

Van LM, Heydari A, Yang J, Hargreaves J, Rowland-Yeo K, Lennard MS, Tucker GT and Rostami-Hodjegan A (2006) The impact of experimental design on assessing mechanism-based inactivation of CYP2D6 by MDMA (Ecstasy). *J Psychopharmacol* **20**:834–841.

Venkatakrishnan K and Obach RS (2005) In vitro-in vivo extrapolation of CYP2D6 inactivation by paroxetine: prediction of nonstationary pharmacokinetics and drug interaction magnitude. *Drug Metab Dispos* **33**:845–852.

Venkatakrishnan K and Obach RS (2007) Drug–drug interactions via mechanism-based cytochrome P450 inactivation: points to consider for risk assessment from in vitro data and clinical pharmacologic evaluation. *Curr Drug Metab* **8**:449–462.

Venkatakrishnan K, Obach RS and Rostami-Hodjegan A (2007) Mechanism-based inactivation of human cytochrome P450 enzymes: strategies for diagnosis and drug–drug interaction risk assessment. *Xenobiotica* **37**:1225–1256.

Walsky RL and Boldt SE (2008) In vitro cytochrome P450 inhibition and induction. *Curr Drug Metab* **9**:928–939.

Walsky RL and Obach RS (2007) A comparison of 2-phenyl-2-(1-piperidinyl)propane (ppp), 1,1′,1″-phosphinothioylidynetrisaziridine (thioTEPA), clopidogrel, and ticlopidine as selective inactivators of human cytochrome P450 2B6. *Drug Metab Dispos* **35**:2053–2059.

Wang YH, Jones DR and Hall SD (2004) Prediction of cytochrome P450 3A inhibition by verapamil enantiomers and their metabolites. *Drug Metab Dispos* **32**:259–266.

Williams JA, Hurst SI, Bauman J, Jones BC, Hyland R, Gibbs JP, Obach RS and Ball SE (2003) Reaction phenotyping in drug discovery: moving forward with confidence?. *Curr Drug Metab* **4**:527–534.

Yang J, Jamei M, Yeo KR, Tucker GT and Rostami-Hodjegan A (2007) Theoretical assessment of a new experimental protocol for determining kinetic values describing mechanism (time)-based enzyme inhibition. *Eur J Pharm Sci* **31**:232–241.

Zhou S, Yung CS, Cher GB, Chan E, Duan W, Huang M and McLeod HL (2005) Mechanism-based inhibition of cytochrome P450 3A4 by therapeutic drugs. *Clin Pharmacokinet* **44**:279–304.

Chapter 20
Allosteric Enzyme- and Transporter-Based Interactions

Murali Subramanian and Timothy S. Tracy

Abstract Allosterism in enzymes and transporters can result in alterations in kinetic profiles and in rates of metabolism and transport. Cooperativity due to allosteric interactions has been observed with drug-metabolizing enzymes such as the cytochromes P450 and the uridine glucuronosyltransferases. In addition, transporters such as P-glycoprotein and breast cancer resistance protein have been demonstrated to also exhibit allosteric interactions resulting in cooperativity. Kinetic profiles such as autoactivation (sigmoidal profile), biphasic and substrate inhibition have been observed to occur due to homotropic cooperativity, and heteroactivation has been observed to occur due to heterocooperativity. Numerous examples of all types of allosteric interactions have been observed in vitro but in vivo examples are limited. The correct kinetic equation should be applied to kinetic profiles to properly estimate kinetic parameters for use in in vitro–in vivo correlations.

Abbreviations

P450s	Cytochrome P450s
UGTs	UDP-glucuronosyltransferases
SULTs	sulfotransferases
P-gp	P-glycoprotein
CBZ	carbamazepine
CBZ-E	carbamazepine–epoxide
DHEA	dehydroepiandrosterone
UDPGA	uridine diphosphate glucuronic acid
PhIP	2-amino-1-methyl-6-phenylimidazo[4,5-*b*]pyridine
MDCK-MDR	Madin Darby canine kidney-multiple drug resistance
BCRP	breast cancer resistance protein
HLMs	human liver microsomes

T.S. Tracy (✉)
Department of Experimental and Clinical Pharmacology, College of Pharmacy, University of Minnesota, Minneapolis, MN, USA
e-mail: tracy017@umn.edu

K.S. Pang et al. (eds.), *Enzyme- and Transporter-Based Drug–Drug Interactions*,
DOI 10.1007/978-1-4419-0840-7_20,

20.1 Atypical Kinetic Profiles Resulting from Allosterism

Allosterism can be defined as a change in the activity and conformation of an enzyme resulting from the binding of a compound at a site on the enzyme other than the active binding site. For the purposes of this chapter, we will assume that this allosteric binding site could be a second site within the enzyme active site but not the one from which substrate is undergoing metabolism at that instant. Allosteric interactions of enzyme proteins can result in atypical kinetic profiles that differ from the usual hyperbolic profile that is characterized by Michaelis–Menten kinetics. Detailed analyses of atypical kinetic data have been conducted and equations have been derived to model the four types of atypical kinetics most often observed – biphasic kinetics, autoactivation, heteroactivation, and substrate inhibition (Houston and Galetin, 2005; Korzekwa et al., 1998; Shou et al., 2001). In addition, the phenomena of partial inhibition and substrate-dependent inhibition are also included in this chapter since these likely also result from the biochemical/biophysical occurrences described above. The occurrence of atypical enzyme kinetics has been observed in numerous enzymes, including the cytochrome P450s (P450s), the UDP-glucuronosyltransferases (UGTs), the sulfotransferases (SULTs), and transporter proteins.

It is generally believed that these atypical profiles occur either due to the presence of multiple binding regions with the enzyme active site allowing a second or even a third ligand to bind (Ekroos and Sjogren, 2006), an allosteric site on the enzyme external to the active site (Schoch et al., 2004; Williams et al., 2004), or finally due to the existence of multiple conformers of the enzyme (Atkins et al., 2001) with the possibility that these conformers can be allosterically modulated by sites external to the active site and altering binding affinity within the catalytic site (Davydov and Halpert, 2008).

20.2 Kinetic Models to Describe Atypical Kinetics

Korzekwa et al. provided some of the first mathematical models for describing the various types of atypical kinetics (Korzekwa et al., 1998). Shou et al. extended this work by deriving equations that could describe differential kinetic rates for the formation of an SE complex versus an ES complex (where S is the substrate and E is the enzyme) (Shou et al., 2001). Either set of models appears to adequately describe atypical kinetic data. Further analyses have extended these concepts to provide both more generic and specialized models (Houston and Galetin, 2005). Several reviews have been published summarizing the various models (e.g., Hutzler and Tracy, 2002; Tracy, 2003; Tracy and Hummel, 2004; Tracy, 2006) and the reader is directed to these manuscripts for a more detailed description of these various kinetic profiles. For this chapter, the types of profiles and the appropriate equations have been summarized in Table 20.1 and partial inhibition and substrate-dependent inhibition are discussed below.

Table 20.1 Types of atypical kinetics and mathematical equations to describe the data

Type of atypical kinetics	Mathematical equation	Notes
Biphasic kinetics	$v = \frac{(V_{\max 1} \cdot [S]) + (\mathrm{CL_{int}}) \cdot [S]^2}{(K_{m1} + [S])}$	The $\mathrm{CL_{int}}$ term in the equation describes the linear portion of the kinetic profile and is separate from any other intrinsic clearance that might be calculated from the estimated K_m and $V_{\max}$
Autoactivation	$v = \frac{\frac{V_{\max 1} \cdot [S]}{K_{m1}} + \frac{V_{\max 2} \cdot [S]^2}{K_{m1} \cdot K_{m2}}}{1 + \frac{[S]}{K_{m1}} + \frac{[S]^2}{K_{m1} \cdot K_{m2}}}$	Because this equation requires a large number of data points to adequately estimate the four kinetic parameters, it may be easier to use the Hill equation in cases where insufficient data points are derived
Substrate inhibition	$v = \frac{V_{\max}}{1 + \frac{K_m}{[S]} + \frac{[S]}{K_i}}$	This is the most simple equation form that can be used to describe substrate inhibition. A more extensive form with a number of parameters can be found in Shou et al. (1999)
Heteroactivation	$v = \frac{V_{\max} \cdot [S]}{K_m \frac{\left(1 + \frac{[B]}{K_s}\right)}{\left(1 + \frac{\beta[B]}{\alpha K_s}\right)} + [S] \frac{\left(1 + \frac{[B]}{\alpha K_s}\right)}{\left(1 + \frac{\beta[B]}{\alpha K_s}\right)}}$	This equation describes a two binding site model. More complex models can be used to describe three binding sites and various types of cooperativity (see Galetin et al. (2002), Kenworthy et al. (2001))

Partial inhibition occurs when saturating concentrations of an inhibitor are unable to completely inhibit metabolism of a substrate (Tracy, 2003). This is usually indicative of mixed inhibition wherein a productive enzyme–substrate–inhibitor (ESI) complex is formed. This complex can continue to form products, hence resulting in partial inhibition. In this case, the inhibitor can either bind to the protein in an allosteric site or in the catalytic pocket of the enzyme, causing conformational changes or steric hindrance, thus affecting the binding and metabolism of the substrate.

Substrate-dependent inhibition is observed when an inhibitor exhibits different K_i values depending on the substrate. This is counter to what is normally expected in the case of traditional competitive inhibition wherein the K_i value for an inhibitor is constant regardless of the substrate. Since probe substrates and chemical inhibitors are used to assess drug–drug interactions in vivo, substrate-dependent inhibition can cofound interpretation of results. For example, substrate-dependent inhibition has been observed for CYP2C9, CYP3A4, and CYP2D6 where K_i values have been found to differ by orders of magnitude depending on the substrate (Kumar et al., 2006; Ramamoorthy et al., 2001; Stresser et al., 2000; Wang et al., 2000c).

20.3 Cytochrome P450-Based Allosteric Interactions

20.3.1 CYP3A4

CYP3A4 has been reported to make up 30% of the hepatic P450 content (Shimada et al., 1994), and estimates suggest it may responsible for the turnover of more than 50% of marketed drugs. Interestingly, with its broad substrate specificity comes an array of atypical kinetics phenomena with this P450 isoform displaying the greatest number of allosteric interaction examples including homo and heterocooperativity involving a number of substrates and effectors (Houston and Galetin, 2005; Shou et al., 2001; Niwa et al., 2008a).

In a recent review, Niwa et al. provided an updated table listing the heterotropic interactions involving CYP3A4 (Niwa et al., 2008a). Examples of heteroactivation were provided for 41 combinations of substrate, effector, and enzyme source; in some instances the same effector–substrate pair showed activation in one enzyme source and inhibition in a different enzyme source. 7,8-Benzoflavone has been reported to result in the heteroactivation of multiple CYP3A4 and CYP3A5 substrates including aflatoxin, carbamazepine (CBZ), phenanthrene, benzo(a)pyrene, progesterone, 17β-estradiol, diazepam, phenacetin, nifedipine, and testosterone (Ueng et al., 1997; Domanski et al., 2001; Niwa et al., 2003; Andersson et al., 1994; Emoto et al., 2001; Shou et al., 1994; Domanski et al., 2000). Quinidine has been implicated in the activation of meloxicam, diclofenac, and warfarin (Ngui et al., 2000; Ludwig et al., 1999; Ngui et al., 2001). In addition, many other effector–substrate pairs such as androstenedione–CBZ, estradiol–CBZ, testosterone–midazolam, and testosterone–diazepam have been identified (Nakamura et al., 2003; Wang et al., 2000c; Kenworthy et al., 2001). Heteroactivation of triazolam, alprazolam, midazolam, flunitrazepam, and diazepam by testosterone has been demonstrated with fresh human hepatocytes as well (Hallifax et al., 2008). Aldosterone, testosterone, progesterone, androstenedione, dehydroepiandrosterone (and its sulfate), cortisol, flavone, and α-naphthoflavone have been reported to exhibit heteroactivation of the metabolism of CBZ to carbamazepine–epoxide (CBZ-E) in human liver microsomes, hepatocytes, and expressed and purified CYP3A4 and CYP3A5 (Henshall et al., 2008).

In heterotropic activation of substrates exhibiting sigmoidal kinetics, at high enough concentrations of the activator, formation of predominantly [EA] (where E is the enzyme and A the activator) complex can occur, and hence the kinetics of substrate metabolism change from sigmoidal to hyperbolic ([ESS] to [EAS]). However, this is not always the case as seen for the heteroactivation of testosterone (a sigmoidal kinetics substrates), where high concentrations of the activators diazepam, quinidine, haloperidol, and azoles did not result in normalization of kinetics (Kenworthy et al., 2001; Galetin et al., 2002). This led Houston et al. to propose a three-site model for CYP3A4, wherein two testosterone molecules could occupy two sites in the active region, while the activator occupies a separate site (Houston and Galetin, 2005). The possibility of three binding regions within the active site of CYP3A4 has further been suggested in experiments using Nanodisc™ technology (Baas et al., 2004). Heteroactivation of CYP3A has not been limited to CYP3A4, as CYP3A7 heteroactivation has been observed with the effector–substrate pairs androstenedione–CBZ, CBZ–dehydroepiandrosterone (DHEA), and estradiol–CBZ (Nakamura et al., 2003).

Autoactivation is also commonly observed with CYP3A4. Diazepam, testosterone, aflatoxin, amitriptyline, CBZ, desmethyladinazolam, estradiol, nifedipine, nordiazepam, and progesterone have been demonstrated to exhibit sigmoidal kinetics (Ueng et al., 1997; Schmider et al., 1995; Shaw et al., 1997; Korzekwa et al., 1998; Kerr et al., 1994; Venkatakrishnan et al., 1998; Andersson et al., 1994; Kenworthy et al., 1999; Shou et al., 1999; Shou et al., 1999; Kerlan et al., 1992; Lee et al., 1995; Domanski et al., 1998; Kronbach et al., 1989; Ghosal et al., 1996). Autoactivation (exhibited by sigmoidal kinetics) has also been observed in human hepatocyte preparations with several benzodiazepines, including flunitrazepam, alprazolam, and triazolam (Hallifax et al., 2008).

Lastly, substrate inhibition and biphasic kinetics have also been observed in CYP3A4 enzyme systems, albeit less frequently than either heteroactivation or autoactivation. Benzyloxyresorufin, progesterone, testosterone, and triazolam show substrate inhibition, while levo-α-acetylmethadol, naphthalene, and nor-levo-α-acetylmethadol have been observed to exhibit biphasic kinetics (Lin et al., 2001; Schrag and Wienkers, 2001). Interestingly, pyrene also displayed biphasic kinetics with CYP3A4, although the addition of cytochrome b5 changed the profile to sigmoidal (autoactivation) (Jushchyshyn et al., 2005), while the addition of testosterone resulted in a hyperbolic kinetic profile. Regardless of the causative factor, the large active site of CYP3A4, which is postulated to accommodate up to three molecules simultaneously, may undergo significant conformational changes upon the binding of a ligand or redox transfer protein resulting in observation of allosteric interactions.

20.3.2 CYP2C9

The CYP2C9 enzyme has also been demonstrated to exhibit allosteric interactions, including heteroactivation, autoactivation, substrate inhibition, and biphasic kinetics. Dapsone is a classic effector of CYP2C9, resulting in the heteroactivation of flurbiprofen, naproxen, and piroxicam metabolism, while dapsone itself exhibits an autoactivation (sigmoidal) kinetic profile with CYP2C9 (Hutzler et al., 2001b; Korzekwa et al., 1998). The mechanisms by which dapsone heteroactivates CYP2C9 have been extensively studied. Co-incubation with dapsone decreases uncoupling of the P450 catalytic cycle by reducing the generation of peroxide and excess water and causing the more efficient use of NADPH reducing equivalents in formation of product. Interestingly, the presence of dapsone also physically causes the substrate (e.g., flurbiprofen) to be located closer to the reactive heme (Hummel et al., 2004; Hutzler et al., 2003, 2001a). Hence, multiple mechanisms appear to govern the action of dapsone heteroactivation. Substrate inhibition and biphasic kinetics have also been observed in CYP2C9 with celecoxib and naproxen, respectively (Korzekwa et al., 1998; Lin et al., 2001). A recent paper testing a library of molecules against CYP2C9 reported that 18% of the tested molecules exhibited atypical kinetics (McMasters et al., 2007).

20.3.3 CYP1A2

Though less extensively studied for atypical kinetics, CYP1A2 also has been demonstrated to exhibit this phenomenon. Rabbit CYP1A2 exhibits heteroactivation with 1-methoxy-4-nitrobenzene as the effector and 1,4-phenyldiisocyanide as the substrate (Miller and Guengerich, 2001). Interestingly, equilibrium-binding studies demonstrated that relatively minor changes in the oxygen-containing substituent of 4-nitrobenzene can dramatically alter the effect. For example, the 1-ethoxy-4-nitrobenzene compound is not a heteroactivator, while the 1-isopropoxy-4-nitrobenzene compound is an inhibitor of 1,4-phenyldiisocyanide. A substrate inhibition profile was observed when ethoxyresorufin was incubated with baculovirus-expressed CYP1A2, while biphasic kinetics were observed with 1-methoxy-4-nitrobenzene incubated with expressed CYP1A2 (Lin et al., 2001; Miller and Guengerich, 2001).

20.3.4 Other CYPs

Other cytochromes P450s have been demonstrated to exhibit atypical kinetics of substrate metabolism suggesting that they also may accommodate two substrate molecules (either of the same ligand or different ligands) in the active site at the same time. For example, substrate inhibition has been observed with both dextromethorphan and fluoxetine metabolism by CYP2D6 and for halothane metabolism by CYP2E1. Biphasic kinetics have been observed with aminopyrine and 7-ethoxycoumarin metabolism by CYP1A1 (Stevens and Wrighton, 1993; Ring et al., 2001; Lin et al., 2001; Inouye et al., 2000). In the only report to date of heteroactivation of CYP2D6, fluoxetine was observed to activate the metabolism of allopregnanolone and progesterone (Niwa et al., 2008b).

20.4 Conjugating Enzyme-Based Allosteric Interactions

20.4.1 Glucuronosyltransferases (UGTs)

Atypical kinetics with xenobiotics metabolized by the UGTs were first reported by Miners and colleagues in 1998 (Miners et al., 1988). While biphasic, substrate inhibition, and autoactivation kinetics are the most common atypical kinetic profiles observed, a few cases of heteroactivation have been reported. Atypical kinetic profiles of substrate metabolism have been observed with UGT1A1, 2B7, 1A9, 1A10, and 1A4. Though not examined to the same degree as the P450s, it is likely that large, flexible binding sites can accommodate the presence of multiple substrate molecules simultaneously. This may not be so surprising given that the UGTs are naturally designed to accommodate at least two co-substrates, the aglycone and the uridine diphosphate glucuronic acid (UDPGA) sugar. Table 20.2 lists a summary of the particular UGT isoform, the substrate, whether an effector molecule is involved, and the type of kinetic profile observed.

One of the classic compounds exhibiting atypical kinetics with UGTs is estradiol. Estradiol-17-glucuronidation exhibits autoactivation in human liver microsomes, but in the presence of 50 μM of the effector 17-α-ethinyl estradiol, normal hyperbolic kinetics are observed. This suggests two distinct substrate-binding regions within the active site, each independently occupied by estradiol and 17-α-ethinyl estradiol (Williams et al., 2002). Based on this finding, a number of potential effectors have been tested against this enzyme–substrate pair. Low concentrations of 17-α-ethinyl estradiol exerted a stimulatory effect, while high concentrations caused inhibition of estradiol 17-glucuronidation. In contrast 2-amino-1-methyl-6-phenylimidazo[4,5-*b*]pyridine (PhIP) exhibited a stimulatory effect (heteroactivation) even at its highest concentration of 100 μM and was inhibitory only at the highest estradiol concentration tested (100 μM).

Changes in regioselectivity of the site of glucuronidation have also been observed with the UGTs when a substrate and effector have been co-incubated. For example, the 3- and 17-glucuronidation of estradiol has been studied with a variety of flavone effectors. At low concentrations of the substrate (estradiol), daidzein activated 3-glucuronidation whereas genistein inhibited this metabolic pathway (Pfeiffer et al., 2005). However, either compound did not affect the 17-glucuronidation of estradiol.

Recently, both two-binding site and three-binding site models have been proposed to explain the interactions between zidovudine, 4-methylumbelliferone, and 1-naphthol glucuronidation by UGT2B7 (Uchaipichat et al., 2008). 1-Naphthol and 4-methylumbelliferone exhibit autoactivation (sigmoidal) kinetics, while zidovudine exhibits hyperbolic kinetics. However, autoactivation was observed with zidovudine when co-incubated with either 1-naphthol or 4-methylumbelliferone. A two-site binding model adequately described this change in zidovudine kinetics from hyperbolic to autoactivation. However, inhibition of the metabolism of both 1-naphthol and 4-methylumbelliferone was observed, as well. In this case, a three-binding site model was required to explain these interactions. This three-site model is similar to the one proposed by Houston et al. to explain the heteroactivation of

Table 20.2 Examples of atypical kinetics involving UGT enzymes

Substrate	Pathway	Enzyme source	Effector	Type of kinetics	References
Morphine	3-Glucuronidation	UGT2B7		Biphasic	Stone et al. (2003)
Morphine	6-Glucuronidation	UGT2B7		Biphasic	Stone et al. (2003)
4-Methylumbelliferone	Glucuronidation	UGT1A9		Weak substrate inhibition	Tsoutsikos et al. (2004)
4-Methylumbelliferone	Glucuronidation	UGT2B7		Autoactivation	Tsoutsikos et al. (2004)
Naproxen	Acyl glucuronidation	HLM		Biphasic	Bowalgaha et al. (2005)
Desmethylnaproxen	Glucuronidation	HLM		Biphasic	Bowalgaha et al. (2005)
Naproxen	Acyl glucuronidation	UGT1A9		Autoactivation	Bowalgaha et al. (2005)
Naproxen	Acyl glucuronidation	UGT1A10		Weak substrate inhibition	Bowalgaha et al. (2005)
Mycophenolic acid	Acyl glucuronidation	UGT1A7		Substrate inhibition	Basu et al. (2004)
Mycophenolic acid	Acyl glucuronidation	UGT1A10		Substrate inhibition	Basu et al. (2004)
trans-Resveratrol	3-Glucuronidation	HLM, HIM		Substrate inhibition	Iwuchukwu and Nagar (2008)
trans-Resveratrol	4′-Glucuronidation	HLM		Biphasic	Iwuchukwu and Nagar (2008)
trans-Resveratrol	4′-Glucuronidation	HIM (intestinal)		Substrate inhibition	Iwuchukwu and Nagar (2008)
trans-Resveratrol	3-Glucuronidation	UGT1A1, UGT1A9, UGT1A10		Substrate inhibition	Iwuchukwu and Nagar (2008)
trans-Resveratrol	4′-Glucuronidation	UGT1A1, UGT1A10		Substrate inhibition	Iwuchukwu and Nagar (2008)
trans-Resveratrol	4′-Glucuronidation	UGT1A9		Autoactivation	Iwuchukwu and Nagar, (2008)
JNJ-10198409	N-amine-glucuronidation	HLM		Biphasic	Yan et al. (2006)
JNJ-10198409	1-N-pyrazole-glucuoronidation	HLM		Biphasic	Yan et al. (2006)

Table 20.2 (continued)

Substrate	Pathway	Enzyme source	Effector	Type of kinetics	References
Estradiol	3-Glucuronidation	pHLM (UGT1A1)		Autoactivation	Fisher et al. (2000)
Acetaminophen	O-Glucuronidation	pHLM (UGT1A6)		Autoactivation	Fisher et al. (2000)
Estradiol	3-Glucuronidation	pHLM (UGT1A1)	17-α-EE, anthraflavic acid, PhIP	Heteroactivation	Williams et al. (2002)
Troglitazone	Glucuronidation	UGT1A10, UGT1A1		Substrate inhibition	Watanabe et al. (2002)
1′-Hydroxyestragole	Glucuronidation	HLM		Autoactivation	Iyer et al. (2003)
Imipramine	N-Glucuronidation	HLM, UGT1A4		Biphasic	Nakajima et al. (2002)
17-β-estradiol	3-Glucuronidation	HLM, UGT1A1	Daidzein, formononetin, eqoul, 3′-OH-daidzein, 6-OH-daidzein	Heteroactivation	Pfeiffer et al. (2005)
1-Napthol	Glucuronidation	HLM		Biphasic	Miners et al. (1988)
Dihydrocodeine	6-Glucuronidation	HLM		Autoactivation	Kirkwood et al. (1998)
1-Napththol	Glucuronidation	UGT2B7		Autoactivation	Uchaipichat et al. (2008)

testosterone by diazepam, quinidine, haloperidol, and azoles by CYP3A4, whereby activation is observed but the testosterone kinetics still retain their autoactivation nature (Houston and Galetin, 2005).

20.4.2 Sulfotransferases (SULTs)

Several types of atypical kinetic profiles have been observed with SULT1A1, 1A2, 2A1, and 1A3, including substrate inhibition, sigmoidal, and possibly biphasic kinetics. For example, substrate inhibition was observed for SULT1A1- and SULT1A2-catalyzed resveratrol sulfation to form *trans*-resveratrol-3-*O*-sulfate. However, formation of *trans*-resveratrol-4′-*O*-sulfate was observed to exhibit sigmoidal and substrate inhibition kinetics with SULT1A1 and SULT1A2, respectively. Finally, the formation of *trans*-resveratrol-3-*O*-4′-*O*-disulfate by SULT1A1 and SULT1A2 demonstrated sigmoidal kinetics (Miksits et al., 2005). With respect to other sulfotransferase substrates, SULT2A1 exhibited substrate inhibition kinetics with dehydroepiandrosterone and 4-hydroxy-2′,3,5-trichlorobiphenyl (Liu et al., 2006). SULT1A1 exhibits substrate inhibition kinetics for the probe substrates 2-aminophenol and 4-nitrophenol, while SULT1A3 appears to demonstrate biphasic kinetics for the same two substrates, although the authors do not make this distinction (Riches et al., 2007). Finally, SULT1A1*1 and 1A1*2 both exhibit substrate inhibition kinetics with the probe substrate 4-nitrophenol (Tabrett and Coughtrie 2003).

20.5 Drug Transporter-Based Allosteric Interactions

20.5.1 P-glycoprotein (ABCB1)

The P-glycoprotein (P-gp) efflux transporter has been demonstrated to play a role in multiple drug resistance by tumor cells, in the secretion of drugs into the bile or kidney, as well as reducing the intestinal and CNS uptake of drugs. Several studies have examined the interaction of multiple substrates with P-gp, providing evidence of multiple binding domains. It has been speculated that these sites may exhibit both a high and a low affinity conformation and can allosterically influence each other's activity (Martin et al., 2000; Safa, 2004). Because of these characteristics, atypical kinetics have been reported for P-gp. For example, hyperbolic kinetics have been observed for verapamil transport by P-gp, and ATPase activity measured as an indication of P-glycoprotein function (Litman et al., 1997). However, when the substrates daunorubicin or gramicidin were added to the system, heteroactivation of ATPase activity was observed resulting in a sigmoidal kinetic profile.

Many of the studies examining the effect of multiple concurrent substrates on P-gp activity have been carried out in the context of drug interaction studies. A recent review summarizes the impact of a second compound (effector) on P-gp transporter activity (Calabrese, 2008). Numerous examples of an effector being stimulatory at low concentrations and inhibitory at high concentrations have been reported (Calabrese, 2008; Garrigues et al., 2002; Romsicki and Sharom, 1999; Sreeramulu et al., 2007; Taub et al., 2005; Wang et al., 2000a). For example, Taub et al. demonstrated that low concentrations of ketoconazole (up to 1 μM) activated the basolateral to apical transport of [^{3}H]vinblastine and [^{3}H]digoxin in MDCK-MDR (Madin Darby canine kidney-multiple drug resistance) cells, but higher concentrations (10–30 μM) of ketoconazole resulted in inhibition of transport (Romsicki and Sharom, 1999; Taub et al., 2005). Similar results were observed for verapamil, thapsigargin, vinblastine, and nigericine which caused activation at low concentrations up to 10 μM but were inhibitory at high concentrations (Garrigues et al., 2002; Romsicki and Sharom, 1999). The presence of two binding sites, one stimulatory and the other inhibitory, has been postulated to explain these "biphasic" kinetics (Calabrese, 2008; Dey et al., 1997; Gottesman et al., 1996). Hence, low concentrations of effectors would bind to the high-affinity stimulatory site, while at higher concentrations the low-affinity inhibitory site would be occupied.

A number of inhibitory studies have also been performed with P-gp, demonstrating competitive inhibition, mixed, and noncompetitive inhibition (Tamai and Safa, 1991; Wang et al., 2000a, b; Litman et al., 1997; Garrigos et al., 1997; Orlowski et al., 1996; Ayesh et al., 1996; Spoelstra et al., 1994; Pascaud et al., 1998; Litman et al., 1997). For example, the apparent binding of compound H33342 to one substrate-binding site appeared to preclude active transport of progesterone, quinidine, and propranolol at a purportedly distinct site (Wang et al., 2000b). Additionally, a negative allosteric effect was observed when nicardipine, nifedipine, and dexniguldipine increased the dissociation of [^{3}H]vinblastine from P-gp by 4-fold (Wang et al., 2000a). These mixed and noncompetitive inhibition events provide further evidence of allosteric interactions between various substrates binding to different regions within the "active" site.

This preponderance of "biphasic" and mixed inhibition data has led to the development of theories of multiple binding regions to explain P-gp's ability to transport a vast number of substrates. The earliest data suggested two unequal binding sites capable of allosteric interactions (Dey et al., 1997; Wang et al., 2001). These positively cooperative sites were termed the H-site and R-site for their ability to accommodate Hoechst 33342 and Rhodamine-123 (Shapiro and Ling, 1997). Recent studies, however, used single and sequential fluorescence titrations to demonstrate that rhodamine-123 and LDS-71 both bind to the putative R-site and reduce each other's binding affinity 5-fold (Lugo and Sharom, 2005). This indicates that both these substrates bind simultaneously to the R-site and form a ternary complex that interacts in a noncompetitive manner (Lugo and Sharom, 2005).

Martin et al. proposed a model with three binding regions in P-gp, each with two spatially independent but interacting substrate-binding sites (Martin et al., 2000). In addition, these binding sites were proposed to exist in a high or low affinity conformation, and that these sites could allosterically communicate with each other to alter binding affinity. These authors proposed that not all the sites were capable of transport, and that some sites were only regulatory in nature. A more recent review suggests that up to seven binding sites may be present in P-gp, including transport and regulatory sites, and these sites may exist in a high- or low-affinity conformation, allosterically interacting (Safa, 2004). Hence, P-gp and its ability to exhibit atypical kinetics may be more complicated than P450s in their kinetics.

20.5.2 Breast Cancer Resistance Protein (ABCG2)

Though much less studied than P-gp with respect to atypical kinetics, the breast cancer resistance protein (BCRP) also appears to exhibit atypical kinetic profiles with various substrates, suggesting the presence of multiple binding sites. Giri et al. demonstrated that co-incubation of nucleoside analogs was able to completely inhibit the transport of GF120918 by BCRP, while prazosin and imatinib were only able to partially inhibit the transport of GF120918, suggesting the existence of multiple binding sites in BCRP (Giri et al., 2009).

20.6 In Vivo Examples of Allosteric Interactions

The role of dapsone in the heteroactivation of flurbiprofen metabolism by CYP2C9 has been well established in microsomes and reconstituted expressed enzymes (Hutzler et al., 2001b). To determine if this effect persisted in vivo, a clinical trial was conducted wherein 12 subjects were administered flurbiprofen alone, or flurbiprofen 2 h after one 100 mg/day dose of dapsone, or flurbiprofen after 7 days of dapsone administration at 100 mg/day (Hutzler et al., 2001a). An 11% increase ($p<0.02$) in the clearance of flurbiprofen was observed when dapsone was administered for 7 days at a dosage of 100 mg/day prior to flurbiprofen administration. Hence, a statistically significant, yet minimal activation of flurbiprofen metabolism was observed in vivo when dapsone was coadministered with flurbiprofen. The level of activation was much lower than that observed in vitro, but the measurable increase suggests that heteroactivation occurs in vivo.

The role of felbamate in the heteroactivation of CBZ metabolism to CBZ-E by CYP3A4 has also been evaluated in in vitro studies (Egnell et al., 2003). Additionally, clinical trials had observed that felbamate coadministration with CBZ monotherapy caused a decrease in CBZ plasma concentration and increased the concentration of CBZ-E in the plasma (Graves et al., 1989). Incidentally, felbamate also inhibits the clearance of CBZ-E (38% by 500 μM felbamate). In vitro–in vivo correlations of CYP3A4 heteroactivation of CBZ by felbamate were combined with a meta-analysis of the clinical studies to predict in vivo the increases in the ratio of steady-state plasma concentrations $C_{SS\text{-}CBZ\text{-}E}/C_{SS\text{-}CBZ}$ due to heteroactivation (Egnell et al., 2003). The estimations predicted a 20–47% increase in the ratio of epoxide to parent drug for a 100–300 μM range of felbamate, while the observed in vivo increases were 25–40% for 85–252 μM felbamate concentrations. The close magnitudes of the percentage increases were suggested to provide evidence of human in vivo heteroactivation of CBZ clearance by felbamate.

Diclofenac (minor pathway) and quinidine are both substrates for CYP3A4 and in vitro studies have shown that while quinidine stimulated diclofenac metabolism, there was no effect vice versa (Ngui et al., 2000). This one-sided heteroactivation suggests that two ligands are present in the CYP3A4 active site simultaneously. The metabolism of diclofenac in the presence and absence of quinidine in monkeys was studied and an ~57% decrease in the plasma concentration of diclofenac was observed when quinidine was coadministered as compared to diclofenac alone (Tang et al., 1999). Since plasma protein binding and plasma to blood ratio were unaffected by the coadministration of quinidine, heteroactivation of CYP3A4 was presumed to be responsible for the increase in diclofenac metabolism. A 2-fold increase in 5-hydroxy-diclofenac (CYP3A metabolite) concentrations was also observed upon the coadministration. This is the most pronounced in vivo effect of heteroactivation reported to date, but must be taken in context that the species were monkeys and not humans.

Substrate-dependent inhibition, a phenomenon attributed to multiple binding regions within the active site, has also been observed in a randomized, double-blind crossover three-phase clinical trial (Backman et al., 1999). Nine subjects were orally administered 50 mg mibefradil or 5 mg isradipine or placebo for 3 days followed by the administration of a single dose of 0.25 mg triazolam; pharmacokinetics and pharmacodynamics were then monitored. Mibefradil coadministration increased the AUC, plasma concentration, and elimination half-life of triazolam 9-fold, 1.8-fold, and 4.9-fold, respectively, compared to placebo. In contrast, isradipine had a minimal effect on triazolam decreasing the AUC and half-life by only 20%. Hence, although mibefradil, isradipine, and triazolam are substrates of CYP3A4, only mibefradil had an effect on triazolam disposition, demonstrating the occurrence of substrate-dependent inhibition in humans.

References

Andersson T, Miners JO, Veronese ME and Birkett DJ (1994) Diazepam metabolism by human liver-microsomes is mediated by both S-mephenytoin hydroxylase and Cyp3A isoforms. *Br J Clin Pharmacol* **38**:131–137.

Atkins WM, Wang RW and Lu AYH (2001) Allosteric behavior in cytochrome P450-dependent in vitro drug-drug interactions: a prospective based on conformational dynamics. *Chem Res Toxicol* **14**:338–347.

Ayesh S, Shao YM and Stein WD (1996) Co-operative, competitive and non-competitive interactions between modulators of P-glycoprotein. *Biochim Biophys Acta* **1316**:8–18.

Baas BJ, Denisov IG and Sliger SG (2004) Homotropic cooperativity of monomeric cytochrome P450 3A4 in a nanoscale native bilayer environment. *Arch Biochem Biophys* **430**:218–228.

Backman JT, Wang JS, Wen X, Kivisto KT and Neuvonen PJ (1999) Mibefradil but not isradipine substantially elevates the plasma concentrations of the CYP3A4 substrate triazolam. *Clin Pharmacol Ther* **66**:401–407.

Basu NK, Kole L, Kubota S and Owens IS (2004) Human UDP-glucuronosyltransferases show atypical metabolism of mycophenolic acid and inhibition by curcumin. *Drug Metab Dispos* **32**:768–773.

Bowalgaha K, Elliot DJ, Mackenzie PI, Knights KM, Swedmark S and Miners JO (2005) S-Naproxen and desmethylnaproxen glucuronidation by human liver microsomes and recombinant human UDP-glucuronosyltransferases (UGT): role of UGT2B7 in the elimination of naproxen. *Br J Clin Pharmacol* **60**:423–433.

Calabrese EJ (2008) P-glycoprotein efflux transporter activity often displays biphasic dose-response relationships. *Crit Rev Toxicol* **38**:473–487.

Davydov DR and Halpert JR (2008) Allosteric P450 mechanisms: multiple binding sites, multiple conformers or both? *Expert Opin Drug Metab Toxicol* **4**:1523–1535.

Dey S, Ramachandra M, Pastan I, Gottesman MM and Ambudkar SV (1997) Evidence for two nonidentical drug-interaction sites in the human P-glycoprotein. *Proc Natl Acad Sci U S A* **94**:10594–10599.

Domanski TL, He YA, Harlow GR and Halpert JR (2000) Dual role of human cytochrome P450 3A4 residue phe-304 in substrate specificity and cooperativity. *J Pharmacol Exp Ther* **293**:585–591.

Domanski TL, He YA, Khan KK, Roussel F, Wang QM and Halpert JR (2001) Phenylalanine and tryptophan scanning mutagenesis of CYP3A4 substrate recognition site residues and effect on substrate oxidation and cooperativity. *Biochemistry* **40**:10150–10160.

Domanski TL, Liu JP, Harlow GR and Halpert JR (1998) Analysis of four residues within substrate recognition site 4 of human cytochrome p450 3A4: role in steroid hydroxylase activity and α-naphthoflavone stimulation. *Arch Biochem Biophys* **350**:223–232.

Egnell AC, Houston B and Boyer S (2003) In vivo CYP3A4 heteroactivation is a possible mechanism for the drug interaction between felbamate and carbamazepine. *J Pharmacol Exp Ther* **305**:1251–1262.

Ekroos M and Sjogren T (2006) Structural basis for ligand promiscuity in cytochrome P450 3A4. *Proc Natl Acad Sci U S A* **103**:13682–13687.

Emoto C, Yamazaki H, Iketaki H, Yamasaki S, Satoh T, Shimizu R, Suzuki S, Shimada N, Nakajima M and Yokoi T (2001) Cooperativity of α -naphthoflavone in cytochrome P450 3A-dependent drug oxidation activities in hepatic and intestinal microsomes from mouse and human. *Xenobiotica* **31**:265–275.

Fisher MB, Campanale K, Ackermann BL, Vandenbranden M and Wrighton SA (2000) In vitro glucuronidation using human liver microsomes and the pore-forming peptide alamethicin. *Drug Metab Dispos* **28**:560–566.

Galetin A, Clarke SE and Houston JB (2002) Quinidine and haloperidol as modifiers of CYP3A4 activity: multisite kinetic model approach. *Drug Metab Dispos* **30**:1512–1522.

Garrigos M, Mir LM and Orlowski S (1997) Competitive and non-competitive inhibition of the multidrug-resistance-associated P-glycoprotein ATPase – Further experimental evidence for a multisite model. *Eur J Biochem* **244**:664–673.

Garrigues A, Nugier J, Orlowski S and Ezan E (2002) A high-throughput screening microplate test for the interaction of drugs with P-glycoprotein. *Anal Biochem* **305**:106–114.

Ghosal A, Satoh H, Thomas PE, Bush E and Moore D (1996) Inhibition and kinetics of cytochrome P4503A activity in microsomes from rat, human, and cDNA-expressed human cytochrome P450. *Drug Metab Dispos* **24**:940–947.

Giri N, Agarwal S, Shaik N, Pan G, Chen Y and Elmquist GF (2009) Substrate-dependent breast cancer resistance protein (Bcrp1/Abcg2)-mediated interactions: consideration of multiple binding sites in in vitro assay design. *Drug Metab Dispos* **37**:560–570.

Gottesman MM, Pastan I and Ambudkar SV (1996) P-glycoprotein and multidrug resistance. *Curr Opin Genet Dev* **6**:610–617.

Graves NM, Holmes GB, Fuerst RH and Leppik IE (1989) Effect of felbamate on phenytoin and carbamazepine serum concentrations. *Epilepsia* **30**:225–229.

Hallifax D, Galetin A and Houston JB (2008) Prediction of metabolic clearance using fresh human hepatocytes: comparison with cryopreserved hepatocytes and hepatic microsomes for five benzodiazepines. *Xenobiotica* **38**:353–367.

Henshall J, Galetin A, Harrison A and Houston JB (2008) Comparative analysis of CYP3A heteroactivation by steroid hormones and flavonoids in different in vitro systems and potential in vivo implications. *Drug Metab Dispos* **36**:1332–1340.

Houston JB and Galetin A (2005) Modelling atypical CYP3A4 kinetics: principles and pragmatism. *Arch Biochem Biophys* **433**:351–360.

Hummel MA, Gannett PM, Aguilar JS and Tracy TS (2004) Effector-mediated alteration of substrate orientation in cytochrorne P4502C9. *Biochemistry* **43**:7207–7214.

Hutzler JM, Frye RF, Korzekwa KR, Branch RA, Huang SM and Tracy TS (2001a) Minimal in vivo activation of CYP2C9-mediated flurbiprofen metabolism by dapsone. *Eur J Pharma Sci* **14**:47–52.

Hutzler JM, Hauer MJ and Tracy TS (2001b) Dapsone activation of CYP2C9-mediated metabolism: Evidence for activation of multiple substrates and a two-site model. *Drug Metab Dispos* **29**:1029–1034.

Hutzler JM and Tracy TS (2002) Atypical kinetic profiles in drug metabolism reactions. *Drug Metab Dispos* **30**:355–362.

Hutzler JM, Wienkers LC, Wahlstrom JL, Carlson TJ and Tracy TS (2003) Activation of cytochrome P4502C9-mediated metabolism: mechanistic evidence in support of kinetic observations. *Arch Biochem Biophys* **410**:16–24.

Inouye K, Mizokawa T, Saito A, Tonomura B and Ohkawa H (2000) Biphasic kinetic behavior of rat cytochrome P-4501A1-dependent monooxygenation in recombinant yeast microsomes. *Biochim Biophys Acta-Protein Struct Mol Enzymol* **1481**:265–272.

Iwuchukwu OF and Nagar S (2008) Resveratrol (trans-resveratrol, 3,5,4 ′-trihydroxy-trans-stilbene) glucuronidation exhibits atypical enzyme kinetics in various protein sources. *Drug Metab Dispos* **36**:322–330.

Iyer LV, Ho MN, Shinn WM, Bradford WW, Tanga MJ, Nath SS and Green CE (2003) Glucuronidation of 1′-hydroxyestragole (1′-HE) by human UDP-glucuronosyltransferases UGT2B7 and UGT1A9. *Toxicol Sci* **73**:36–43.

Jushchyshyn MI, Hutzler JM, Schrag ML and Wienkers LC (2005) Catalytic turnover of pyrene by CYP3A4: evidence that cytochrome b(5) directly induces positive cooperativity. *Arch Biochem Biophys* **438**:21–28.

Kenworthy KE, Bloomer JC, Clarke SE and Houston JB (1999) CYP3A4 drug interactions: correlation of 10 in vitro probe substrates. *Br J Clin Pharmacol* **48**:716–727.

Kenworthy KE, Clarke SE, Andrews J and Houston JB (2001) Multisite kinetic models for CYP3A4: simultaneous activation and inhibition of diazepam and testosterone metabolism. *Drug Metab Dispos* **29**:1644–1651.

Kerlan V, Dreano Y, Bercovici JP, Beaune PH, Floch HH and Berthou F (1992) Nature of cytochromes P450 involved in the 2-/4-hydroxylations of estradiol in human liver-microsomes. *Biochem Pharmacol* **44**:1745–1756.

Kerr BM, Thummel KE, Wurden CJ, Klein SM, Kroetz DL, Gonzalez FJ and Levy RH (1994) Human liver carbamazepine metabolism – role of Cyp3A4 and Cyp2C8 in 10,11-epoxide formation. *Biochem Pharmacol* **47**:1969–1979.

Kirkwood LC, Nation RL and Somogyi AA (1998) Glucuronidation of dihydrocodeine by human liver microsomes and the effect of inhibitors. *Clin Exp Pharmacol Physiol* **25**:266–270.

Korzekwa KR, Krishnamachary N, Shou M, Ogai A, Parise RA, Rettie AE, Gonzalez FJ and Tracy TS (1998) Evaluation of atypical cytochrome P450 kinetics with two-substrate models: evidence that multiple substrates can simultaneously bind to cytochrome P450 active sites. *Biochemistry* **37**:4137–4147.

Kronbach T, Mathys D, Umeno M, Gonzalez FJ and Meyer UA (1989) Oxidation of midazolam and triazolam by human-liver cytochrome P450Iiia4. *Mol Pharmacol* **36**:89–96.

Kumar V, Wahlstrom JL, Rock DA, Warren CJ, Gorman LA and Tracy TS (2006) CYP2C9 inhibition: impact of probe selection and pharmacogenetics on in vitro inhibition profiles. *Drug Metab Dispos* **34**:1966–1975.

Lee CA, Kadwell SH, Kost TA and SerabjitSingh CJ (1995) Cyp3A4 expressed by insect cells infected with a recombinant baculovirus containing both Cyp3A4 and human Nadph-cytochrome P450 reductase is catalytically similar to human liver microsomal Cyp3A4. *Arch Biochem Biophys* **319**:157–167.

Lin Y, Lu P, Tang C, Mei Q, Sandig G, Rodrigues AD, Rushmore TH and Shou M (2001) Substrate inhibition kinetics for cytochrome P450-catalyzed reactions. *Drug Metab Dispos* **29**:368–374.

Litman T, Zeuthen T, Skovsgaard T and Stein WD (1997) Competitive, non-competitive and cooperative interactions between substrates of P-glycoprotein as measured by its ATPase activity. *Biochim Biophys Acta* **1361**:169–176.

Liu Y, Apak TI, Lehmler HJ, Robertson LW and Duffel MW (2006) Hydroxylated polychlorinated biphenyls are substrates and inhibitors of human hydroxysteroid sulfotransferase SULT2A1. *Chemical Research in Toxicology* **19**:1420–1425.

Ludwig E, Schmid J, Beschke K and Ebner T (1999) Activation of human cytochrome P-450 3A4-catalyzed meloxicam 5′-methylhydroxylation by quinidine and hydroquinidine in vitro. *J Pharmacol Exp Ther* **290**:1–8.

Lugo MR and Sharom FJ (2005) Interaction of LDS-751 and rhodamine 123 with P-glycoprotein: evidence for simultaneous binding of both drugs. *Biochemistry* **44**:14020–14029.

Martin C, Berridge G, Higgins CF, Mistry P, Charlton P and Callaghan R (2000) Communication between multiple drug binding sites on P-glycoprotein. *Mol Pharmacol* **58**:624–632.

McMasters DR, Torres RA, Crathern SJ, Dooney DL, Nachbar RB, Sheridan RP and Korzekwa KR (2007) Inhibition of recombinant cytochrome p450 isoforms 2D6 and 2C9 by diverse drug-like molecules. *J Med Chem* **50**:3205–3213.

Miksits M, Maier-Salamon A, Aust S, Thalhammer T, Reznicek G, Kunert O, Haslinger E, Szekeres T and Jaeger W (2005) Sulfation of resveratrol in human liver: Evidence of a major role for the sulfotransferases SULT1A1 and SULT1E1. *Xenobiotica* **35**:1101–1119.

Miller GP and Guengerich FP (2001) Binding and oxidation of alkyl 4-nitrophenyl ethers by rabbit cytochrome P450 1A2: evidence for two binding sites. *Biochemistry* **40**:7262–7272.

Miners JO, Lillywhite KJ, Matthews AP, Jones ME and Birkett DJ (1988) Kinetic and inhibitor studies of 4-methylumbelliferone and 1-naphthol glucuronidation in human-liver microsomes. *Biochem Pharmacol* **37**:665–671.

Nakajima M, Tanaka E, Kobayashi T, Ohashi N, Kume T and Yokoi T (2002) Imipramine N-glucuronidation in human liver microsomes: biphasic kinetics and characterization of UDP-glucuronosyltransferase isoforms. *Drug Metab Dispos* **30**:636–642.

Nakamura H, Torimoto N, Ishii I, Ariyoshi N, Nakasa H, Ohmori S and Kitada M (2003) CYP3A4 and CYP3A7-mediated carbamazepine 10,11-epoxidation are activated by differential endogenous steroids. *Drug Metab Dispos* **31**:432–438.

Ngui JS, Chen Q, Shou MG, Wang RW, Stearns RA, Baillie TA and Tang W (2001) In vitro stimulation of warfarin metabolism by quinidine: Increases in the formation of 4′- and 10-hydroxywarfarin. *Drug Metab Dispos* **29**:877–886.

Ngui JS, Tang W, Stearns RA, Shou MG, Miller RR, Zhang Y, Lin JH and Baillie TA (2000) Cytochrome P450 3A4-mediated interaction of diclofenac and quinidine. *Drug Metab Dispos* **28**:1043–1050.

Niwa T, Murayama N and Yamazaki H (2008a) Heterotropic cooperativity in oxidation mediated by cytochrome P450. *Curr Drug Metab* **9**:453–462.

Niwa T, Okada K, Hiroi T, Imaoka S, Narimatsu S and Funae Y (2008b) Effect of psychotropic drugs on the 21-hydroxylation of neurosteroids, progesterone and allopregnanolone, catalyzed by rat CYP2D4 and human CYP2D6 in the brain. *Biol Pharmaceut Bull* **31**:348–351.

Niwa T, Shiraga T, Yamasaki S, Ishibashi K, Ohno Y and Kagayama A (2003) In vitro activation of 7-benzyloxyresorufin O-debenzylation and nifedipine oxidation in human liver microsomes. *Xenobiotica* **33**:717–729.

Orlowski S, Mir LM, Belehradek J and Garrigos M (1996) Effects of steroids and verapamil on P-glycoprotein ATPase activity: progesterone, desoxycorticosterone, corticosterone and verapamil are mutually non-exclusive modulators. *Biochem J* **317**:515–522.

Pascaud C, Garrigos M and Orlowski S (1998) Multidrug resistance transporter P-glycoprotein has distinct but interacting binding sites for cytotoxic drugs and reversing agents. *Biochem J* **333**:351–358.

Pfeiffer E, Treiling CR, Hoehle SI and Metzler M (2005) Isoflavones modulate the glucuronidation of estradiol in human liver microsomes. *Carcinogenesis* **26**:2172–2178.

Ramamoorthy Y, Tyndale RF and Sellers EM (2001) Cytochrome P450 2D6.1 and cytochrome P450 2D6.10 differ in catalytic activity for multiple substrates. *Pharmacogenetics* **11**:477–487.

Riches Z, Bloomer JC and Coughtrie MWH (2007) Comparison of 2-aminophenol and 4-nitrophenol as in vitro probe substrates for the major human hepatic sulfotransferase, SULT1A1, demonstrates improved selectivity with 2-aminophenol. *Biochemical Pharmacology* **74**:352–358.

Ring BJ, Eckstein JA, Gillespie JS, Binkley SN, Vandenbranden M and Wrighton SA (2001) Identification of the human cytochromes P450 responsible for in vitro formation of R- and S-norfluoxetine. *J Pharmacol Exp Ther* **297**:1044–1050.

Romsicki Y and Sharom FJ (1999) The membrane lipid environment modulates drug interactions with the P-glycoprotein multidrug transporter. *Biochemistry* **38**:6887–6896.

Safa AR (2004) Identification and characterization of the binding sites of P-glycoprotein for multidrug resistance-related drugs and modulators. *Curr Med Chem Anticancer Agents* **4**:1–17.

Schmider J, Greenblatt DJ, VonMoltke LL, Harmatz JS and Shader RI (1995) N-demethylation of amitriptyline in-vitro – role of cytochrome-P-450 3A (Cyp3A) isoforms and effect of metabolic-inhibitors. *J Pharmacol Exp Ther* **275**:592–597.

Schoch GA, Yano JK, Wester MR, Griffin KJ, Stout CD and Johnson EF (2004) Structure of human microsomal cytochrome P4502C8 – evidence for a peripheral fatty acid binding site. *J Biol Chem* **279**:9497–9503.

Schrag ML and Wienkers LC (2001) Triazolam substrate inhibition: evidence of competition for heme-bound reactive oxygen within the CYP3A4 active site. *Drug Metab Dispos* **29**:70–75.

Shapiro AB and Ling V (1997) Positively cooperative sites for drug transport by P-glycoprotein with distinct drug specificities. *Eur J Biochem* **250**:130–137.

Shaw PM, Hosea NA, Thompson DV, Lenius JM and Guengerich FP (1997) Reconstitution premixes for assays using purified recombinant human cytochrome P450, NADPH-cytochrome P450 reductase, and cytochrome b(5). *Arch Biochem Biophys* **348**:107–115.

Shimada T, Yamazaki H, Mimura M, Inui Y and Guengerich FP (1994) Interindividual variations in human liver cytochrome-P-450 enzymes involved in the oxidation of drugs, carcinogens and toxic-chemicals – studies with liver-microsomes of 30 Japanese and 30 Caucasians. *J Pharmacol Exp Ther* **270**:414–423.

Shou M, Grogan J, Mancewicz JA, Krausz KW, Gonzalez FJ, Gelboin HV and Korzekwa KR (1994) Activation of Cyp3A4 – evidence for the simultaneous binding of 2 substrates in a cytochrome-P450 active-site. *Biochemistry* **33**:6450–6455.

Shou M, Lin Y, Lu P, Tang C, Mei Q, Cui D, Tang W, Ngui JS, Lin CC, Singh R, Wong BK, Yergey JA, Lin JH, Pearson PG, Baillie TA, Rodrigues AD and Rushmore TH (2001) Enzyme kinetics of cytochrome P450-mediated reactions. *Curr Drug Metab* **2**:17–36.

Shou MG, Mei Q, Ettore MW, Dai RK, Baillie TA and Rushmore TH (1999) Sigmoidal kinetic model for two co-operative substrate-binding sites in a cytochrome P450 3A4 active site: an example of the metabolism of diazepam and its derivatives. *Biochem J* **340**:845–853.

Spoelstra EC, Westerhoff HV, Pinedo HM, Dekker H and Lankelma J (1994) The multidrug-resistance-reverser verapamil interferes with cellular P-glycoprotein-mediated pumping of daunorubicin as a non-competing substrate. *Eur J Biochem* **221**:363–373.

Sreeramulu K, Liu R and Sharom FJ (2007) Interaction of insecticides with mammalian P-glycoprotein and their effect on its transport function. *Biochim Biophys Acta* **1768**: 1750–1757.

Stevens JC and Wrighton SA (1993) Interaction of the enantiomers of fluoxetine and norfluoxetine with human liver cytochromes-P450. *J Pharmacol Exp Ther* **266**:964–971.

Stone AN, Mackenzie PI, Galetin A, Houston JB and Miners JO (2003) Isoform selectivity and kinetics of morphine 3-and 6-glucuronidation by human UDP-glucuronosyltransferases:

evidence for atypical glucuronidation kinetics by UGT2B7. *Drug Metab Dispos* **31**: 1086–1089.

Stresser DM, Blanchard AP, Turner SD, Erve JCL, Dandeneau AA, Miller VP and Crespi CL (2000) Substrate-dependent modulation of CYP3A4 catalytic activity: analysis of 27 test compounds with four fluorometric substrates. *Drug Metab Dispos* **28**:1440–1448.

Tabrett CA and Coughtrie MWH (2003) Phenol sulfotransferase 1A1 activity in human liver: kinetic properties, interindividual variation and re-evaluation of the suitability of 4-nitrophenol as a probe substrate. *Biochemical Pharmacology* **66**:2089–2097.

Tamai I and Safa AR (1991) Azidopine noncompetitively interacts with vinblastine and cyclosporine-A binding to P-glycoprotein in multidrug resistant cells. *J Biol Chem* **266**:16796–16800.

Tang W, Stearns RA, Kwei GY, Iliff SA, Miller RR, Egan MA, Yu NX, Dean DC, Kumar S, Shou MG, Lin JH and Baillie TA (1999) Interaction of diclofenac and quinidine in monkeys: stimulation of diclofenac metabolism. *J Pharmacol Exp Ther* **291**:1068–1074.

Taub ME, Podila L, Ely D and Almeida I (2005) Functional assessment of multiple P-glycoprotein (P-gp) probe substrates: influence of cell line and modulator concentration on P-GP activity. *Drug Metab Dispos* **33**:1679–1687.

Tracy TS (2003) Atypical enzyme kinetics: their effect on in vitro-in vivo pharmacokinetic predictions and drug interactions. *Curr Drug Metab* **4**:341–346.

Tracy TS (2006) Atypical cytochrome P450 kinetics, implications for drug discovery. *Drugs R D* **7**:349–363.

Tracy TS and Hummel MA (2004) Modeling kinetic data from in vitro drug metabolism enzyme experiments. *Drug Metab Rev* **36**:231–242.

Tsoutsikos P, Miners JO, Stapleton A, Thomas A, Sallustio BC and Knights KM (2004) Evidence that unsaturated fatty acids are potent inhibitors of renal UDP-glucuronosyltransferases (UGT): kinetic studies using human kidney cortical microsomes and recombinant UGT1A9 and UGT2B7. *Biochem Pharmacol* **67**:191–199.

Uchaipichat V, Galetin A, Houston JB, Mackenzie PI, Williams JA and Miners JO (2008) Kinetic modeling of the interactions between 4-methylumbelliferone, 1-naphthol, and zidovudine glucuronidation by UDP-glucuronosyltransferase 2B7 (UGT2B7) provides evidence for multiple substrate binding and effector sites. *Mol Pharmacol* **74**:1152–1162.

Ueng YF, Kuwabara T, Chun YJ and Guengerich FP (1997) Cooperativity in oxidations catalyzed by cytochrome P450 3A4. *Biochemistry* **36**:370–381.

Venkatakrishnan K, von Moltke LL, Duan SX, Fleishaker JC, Shader RI and Greenblatt DJ (1998) Kinetic characterization and identification of the enzymes responsible for the hepatic biotransformation of adinazolam and N-desmethyladinazolam in man. *J Pharm Pharmacol* **50**:265–274.

Wang EJ, Casciano CN, Clement RP and Johnson WW (2000a) Cooperativity in the inhibition of P-glycoprotein-mediated daunorubicin transport: Evidence for half-of-the-sites reactivity. *Arch Biochem Biophys* **383**:91–98.

Wang EJ, Casciano CN, Clement RP and Johnson WW (2000b) Two transport binding sites of P-glycoprotein are unequal yet contingent: initial rate kinetic analysis by ATP hydrolysis demonstrates intersite dependence. *Biochim Biophys Acta-Protein Struct Mol Enzymol* **1481**:63–74.

Wang EJ, Casciano CN, Clement RP and Johnson WW (2001) Active transport of fluorescent P-glycoprotein substrates: evaluation as markers and interaction with inhibitors. *Biochem Biophys Res Comm* **289**:580–585.

Wang RW, Newton DJ, Liu N, Atkins WM and Lu AYH (2000c) Human cytochrome P-450 3A4: in vitro drug-drug interaction patterns are substrate-dependent. *Drug Metab Dispos* **28**:360–366.

Watanabe Y, Nakajima M and Yokoi T (2002) Troglitazone glucuronidation in human liver and intestine microsomes: high catalytic activity of UGT1A8 and UGT1A10. *Drug Metab Dispos* **30**:1462–1469.

Williams JA, Ring BJ, Cantrell VE, Campanale K, Jones DR, Hall SD and Wrighton SA (2002) Differential modulation of UDP-glucuronosyltransferase 1A1 (UGT1A1) catalyzed estradiol-3-glucuronidation by the addition of UGT1A1 substrates and other compounds to human liver microsomes. *Drug Metab Dispos* **30**:1266–1273.

Williams PA, Cosme J, Vinkovic DM, Ward A, Angove HC, Day PJ, Vonrhein C, Tickle IJ and Jhoti H (2004) Crystal structures of human cytochrome P450 3A4 bound to metyrapone and progesterone. *Science* **305**:683–686.

Yan Z, Caldwell GW, Gauthier D, Leo GC, Mei J, Ho CY, Jones WJ, Masucci JA, Tuman RW, Galemmo RA and Johnson DL (2006) N-glucuronidation of the platelet-derived growth factor receptor tyrosine kinase inhibitor 6,7-(dimethoxy-2,4-dihydroindeno[1,2-c]pyrazol-3-yl)-(3-fluoro-phenyl)-amine by human UDP-glucuronosyltransferases. *Drug Metab Dispos* **34**:748–755.

Chapter 21
The Impact and In Vitro to In Vivo Prediction of Transporter-Based Drug–Drug Interactions in Humans

Jashvant D. Unadkat, Brian J. Kirby, Christopher J. Endres, and Joseph K. Zolnerciks

Abstract Drug transport proteins have been recognized as significant contributors to drug absorption, distribution, elimination, toxicity, and efficacy. This chapter will discuss the key concepts of transporter-based drug–drug interactions (DDIs), provide a concise review of DDIs involving the major drug transporters, and describe methodologies used to quantitatively predict the magnitude of transporter-based DDIs in humans.

21.1 Introduction

Our understanding of the role of drug transporters in drug disposition has increased greatly with the completion of the human genome project and the creation of a variety of transgenic and knockout mouse models. With this understanding, transport proteins have been recognized as significant contributors to drug absorption, distribution, elimination, toxicity, and efficacy.

Drug transporters are extensively expressed in the key organs governing drug disposition, the intestine, liver, and kidney, as well as sites of toxicity or therapy as shown in Fig. 21.1 (Choudhuri and Klaassen, 2006; Eyal et al., 2009; Glaeser et al., 2007; Kusuhara and Sugiyama, 2009; Marzolini and Kim, 2005; Otsuka et al., 2005). Based on their cellular localization in the intestine, liver, and kidney, they can act to either remove drug substrates from the body or aid in their entry to or distribution within the body. For example, the efflux transporter P-glycoprotein (P-gp) is expressed on the apical membrane of the intestine, the canalicular membrane of hepatocytes, and the luminal membrane of the proximal renal tubule cells (Fig. 21.1). The localization of P-gp in all three organs acts to remove drug from the body. In contrast, the organic anion-transporting polypeptide (OATP) family of transporters are uptake transporters that are expressed on the apical membrane of the intestine,

J.D. Unadkat (✉)
Department of Pharmaceutics, School of Pharmacy, University of Washington, Seattle, WA, USA
e-mail: jash@u.washington.edu

K.S. Pang et al. (eds.), *Enzyme- and Transporter-Based Drug–Drug Interactions*,
DOI 10.1007/978-1-4419-0840-7_21,

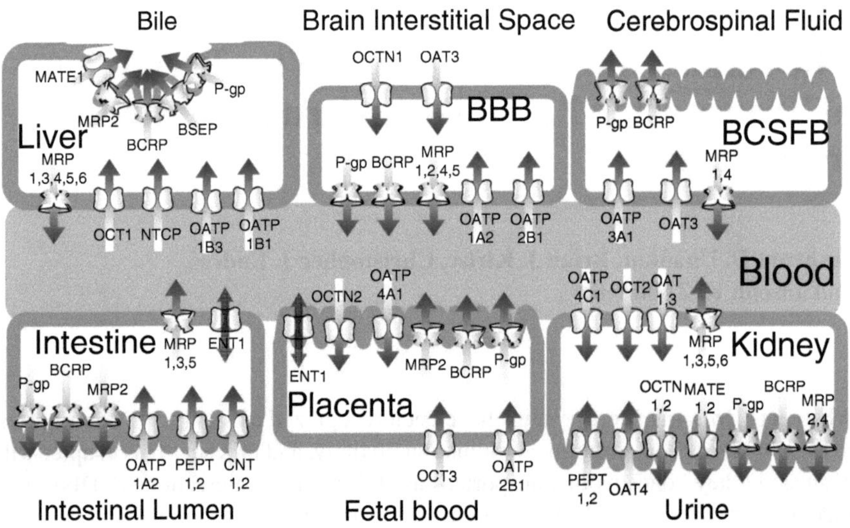

Fig. 21.1 Tissue and membrane localization of transporters in the human intestine, liver, placenta, kidney, blood–brain barrier (BBB), and blood–cerebrospinal fluid barrier (BCSFB)

sinusoidal membrane of hepatocytes, and the basolateral membrane of renal proximal tubule cells (Fig. 21.1). In these organs, OATPs contribute to the uptake of drug substrates into the cells, but the overall effects on drug disposition are different. In the intestine, the OATPs act to increase access of the drug substrate to the body, whereas in the liver and kidney they aid in the removal of substrates from the blood. For this reason, it is imperative to know not only in which tissues drug transporters are located, but also their level of expression, membrane localization, and direction of transport. Only when this is known can the effect of these transporters on drug disposition be hypothesized.

Transporter-based inhibitory or inductive drug–drug interactions (DDIs) may be pharmacokinetic or pharmacodynamic in nature. Pharmacodynamic interactions occur when there is little to no effect on systemic exposure of the drug, but exposure to a tissue driving a therapeutic or toxic response is modulated. The interactions may involve equilibrative (facilitative) influx or efflux transporters. A drug transporter must be the rate-limiting step in the organ clearance (or absorption) of the drug substrate for modulation of drug transporter activity to significantly contribute to the pharmacokinetics (or absorption) of the drug. Additionally, the organ must also have a significant impact on the overall disposition (or absorption) of the drug. In the case where neither of these is true, a drug transporter may still play a crucial role in the pharmacodynamics and toxicity of a drug by controlling drug access to the site of these actions.

All of the in vivo pharmacokinetic parameters of a probe drug can be modulated by inhibition or induction of drug transporters if the conditions stated above are met. The manner in which these parameters are altered will depend on the tissue and membrane localization of the transporter activity being modulated as well as the importance of that transporter and tissue in the disposition of the drug. This is described in more detail in Section 21.2.

Like certain metabolism-based DDIs, transporter-based drug–drug interactions may be used for a therapeutic advantage. In HIV/AIDS treatment, the protease inhibitor ritonavir is extensively used not for its antiviral efficacy but for its ability to inhibit the CYP3A-mediated metabolism of other concomitantly administered anti-HIV drugs thereby increasing their antiviral efficacy (Cooper et al., 2003). A similar deliberate transporter-based DDI, the use of probenecid to inhibit organic anion transporter activity in the renal epithelial cells, is used to reduce the nephrotoxicity of cidofovir (Lalezari et al., 1995; Cundy, 1999).

In recent years, the number of drugs identified as substrates of drug transporters has increased greatly. This, in combination with the understanding that transporters play a significant role in the disposition of many drugs, has lead to the characterization of new chemical entities (NCEs) as substrates, inhibitors, or inducers of drug transporters. In addition, significant effort has been focused on models to predict the contribution of transporters to drug disposition as well as the impact of transporter-based DDIs in vivo. Considerable progress has been made in predicting these types of interactions with respect to cytochrome P450 metabolism. Similar approaches are being applied to transporter-based drug–drug interactions. This chapter is not intended to be a comprehensive review of these interactions in humans. Rather, it is intended to describe the key concepts of transporter-based DDIs and to highlight the varying types of transporter-based DDIs that have been reported. In the next section (Key Concepts of Transporter-Based DDIs) we will discuss some of the important concepts necessary for understanding transporter-based DDIs. This will be followed by a concise review of DDIs involving the major drug transporters and a description of a methodology used to quantitatively predict the magnitude of transporter-based DDIs in man.

21.2 Key Concepts of Transporter-Based DDIs

21.2.1 Effect on PK Parameters

The potential effects on PK parameters of a drug as a result of inhibition or induction of transport activity in eliminating organs (intestine, liver, and kidney) and tissues are listed in Table 21.1. Of course the extent to which inhibition or induction of the transporter will affect the parameters listed is dependent on two conditions: (1) the significance of the transporter in drug permeability across the membrane in the specific organ and (2) the significance of that organ in determination of the specific PK parameter. The potential changes listed in Table 21.1 assume that both these conditions are significant.

Table 21.1 Effect of transporter inhibition (double arrows ⇑, ⇓, or ⇔) or induction (single arrows ↑, ↓, or ↔) on plasma pharmacokinetic parameters of drug substrates

			Pharmacokinetic parameter				
Tissue	Membrane	Transporter	C_{max}	T_{max}	k_a	F_G	
Intestine	Apical (luminal)	Uptake	⇓↑	⇑↓	⇓↑	⇓↑	
		Efflux	⇑↓	⇓↑	⇑↓	⇑↓	
	Basolateral	Efflux	⇓↑	⇑↓	⇓↑	⇓↑	
			C_{max}	T_{max}	Cl_{oral}	Cl_{sys}	F_H
Liver	Canalicular	Efflux	⇑↓	⇑↓	⇓↑	⇓↑	⇑↓
	Sinusoidal	Uptake	⇑↓	⇑↓	⇓↑	⇓↑	⇑↓
		Efflux	⇓↑	⇓↑	⇑↓	⇑↓	⇓↑
			C_{max}	T_{max}	Cl_{oral}	Cl_{sys}	Cl_{renal}
Kidney	Apical (luminal)	Uptake	⇓↑	⇓↑	⇑↓	⇑↓	⇑↓
		Efflux	⇑↓	⇑↓	⇓↑	⇓↑	⇓↑
	Basolateral	Uptake	⇑↓	⇑↓	⇓↑	⇓↑	⇓↑
		Efflux	⇓↑	⇓↑	⇑↓	⇑↓	⇑↓
			C_{max}	T_{max}	Cl_{oral}	Cl_{sys}	V
Tissue (nonclearance, distribution only)		Uptake	⇑↓	⇓↑	⇔↔	⇔↔	⇓↑
		Efflux	⇓↑	⇑↓	⇔↔	⇔↔	⇑↓

All of the changes listed above are made under the base assumption that the particular transporter and organ/tissue play a significant role in the physiological process governing the pharmacokinetic parameters listed.

21.2.2 Synergistic Effect of two Transporters

In the situation where two transporters significantly contribute to an organ clearance or distribution into or out of a tissue an apparent synergistic effect can be seen when the two transporters are simultaneously inhibited or induced. For example, consider the situation described in Fig. 21.2 where the systemic clearance of a drug given by IV administration is 125 mL/min and the biliary clearance of the drug is 100 mL/min ($f_{e,bile}$, fraction of clearance that is biliary = 0.8). If the biliary clearance is equally contributed to by P-gp and BCRP ($Cl_{bile,p\text{-}gp}$ and $Cl_{bile,\ BCRP} = 50$ mL/min), selective inhibition of one of the transporters would result in new systemic clearance of 75 mL/min which would result in a 70% increase in the area under the curve of the plasma concentration–time profile (AUC) of the drug or an AUC ratio ($AUC_{inhibited}/AUC_{control}$) of 1.7. If both of the transporters were inhibited, the new systemic clearance would be 25 mL/min and the AUC ratio would be 5. In this scenario, the biliary clearance must be at least in part an excretory process, that is that entero-hepatic recirculation of the drug excreted in the bile is not complete. Otherwise this would be a purely distributional phenomenon that would

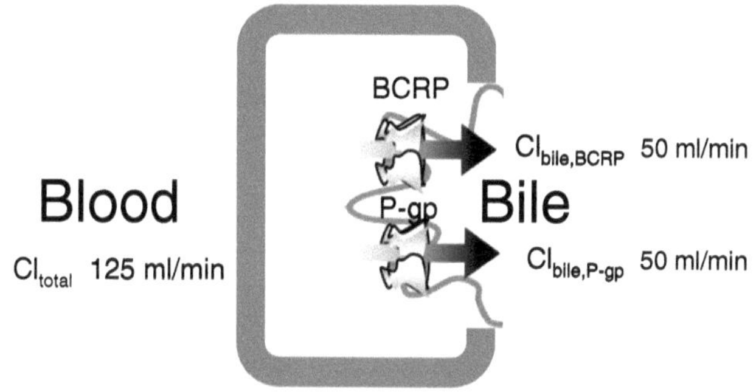

Conditions	Cl_{bile} (ml/min)	Cl_{total} (ml/min)	$f_{e,bile}$	AUC Ratio
Normal	100	125	0.8	
Inhibition of P-gp or BCRP	50	75	0.67	1.67
Inhibition of P-gp and BCRP	0	25	0	5.0

Fig. 21.2 Transport synergism: If an organ clearance is mediated by more than one transport process, inhibition of each process individually may result in a relatively small change compared to inhibition of all processes simultaneously. This example shows how inhibition of BCRP or P-gp would result in an AUC ratio of approximately 1.7 where inhibition of both processes would result in an AUC ratio of 5

change plasma concentration, but not AUC. In addition, by inhibiting one of the biliary efflux processes, the concentration of drug in the hepatocytes relative to plasma concentration will be increased which will appear as an apparent increase in the biliary clearance of the unaffected transporter.

This apparent synergism is similar to a situation in which inhibition of two metabolic clearance process can have a larger than additive effect on the AUC ratio of a substrate drug. This example illustrates where caution must be taken when predicting the effect of inhibiting multiple processes that comprise a clearance process. A similar type of apparent synergism could be observed for intestinal or renal clearance as well as tissue distribution of a drug.

21.2.3 Transporter DDI "Masked" as Metabolism DDI

The endpoint of many DDI studies is the AUC ratio of the parent (also called object or victim) drug in the presence and absence of the precipitant drug. When understanding drug interactions from a mechanistic standpoint often times the AUC ratio of metabolite to parent drug or urinary formation clearance is used to quantify the effect on metabolic intrinsic clearance. Typically the fold change in metabolite to parent AUC ratio or urinary formation clearance is a more sensitive measure than fold change in AUC of the parent as a result of clearance mechanisms of the parent drug other than metabolism. All of these methods for estimating metabolic DDIs operate under the assumption that the interaction has not been a result of altering the intrahepatic to plasma free drug concentration ratio. That is that the interaction was not a result of transporters. The reason for this assumption is that the estimation of clearance is a blood or plasma clearance rather than the true intrinsic clearance (metabolic rate relative to the drug concentration at the enzyme active site) which can only be calculated if the intrahepatic free drug concentration is known. Since this is not measurable we are left with the blood or plasma clearance measurement and assumptions of intrahepatic/plasma concentration ratios. Therefore a change in this ratio is indistinguishable from a change in the true metabolic intrinsic clearance. As we become more aware of the role of transporters in hepatic disposition of drugs, the likelihood of violating this assumption increases. Mechanisms by which transporter interactions can be "masked" as metabolic interactions are describe below and graphically in Fig. 21.3.

DDIs that appear to be a result of inhibition of hepatic metabolism can be the result of at least two different types of interactions with transporters (illustrated in Fig. 21.3 panels B and D). Inhibition of a sinusoidal uptake transporter (Fig. 21.3 panel B) in the liver would result in a decrease of the intrahepatic/plasma free drug concentration ratio (denoted by the difference in text size of the [D] symbol in the plasma and inside the hepatocyte) and may also increase the plasma AUC of the drug. Similarly, induction of an efflux transporter either on the canalicular or sinusoidal membrane (Fig. 21.3 panel D) would result in a similar decrease in the intrahepatic/plasma free drug concentration ratio. As a result of the decrease in intrahepatic drug concentration relative to the plasma, the formation rate of the metabolite would be decreased (denoted in the figure as a change in text size of the [M] symbol). The formation clearance (total amount of metabolite excreted relative to plasma AUC of the drug) would be significantly decreased because the amount excreted would be decreased and the plasma AUC could be increased. This would appear as if the enzyme activity was reduced, but in reality the enzyme was exposed to a lower concentration of drug resulting in an apparent lower metabolic formation clearance.

DDIs that appear to be induction or activation of hepatic metabolism may also be the result of at least two different types of transporter-mediated DDIs (illustrated in Fig. 21.3 panels C and E). Induction of a basolateral uptake transporter (Fig. 21.3 panel C) in the liver would increase the intrahepatic/plasma free drug concentration ratio and potentially decrease the plasma AUC of the drug. Inhibition of an efflux transporter localized either on the sinusoidal or canalicular membrane (Fig. 21.3 panel D) would also result in an increase in the intrahepatic/plasma free drug concentration ratio. As a result of the increased intrahepatic free drug concentration, the formation rate of metabolite would be increased; in addition the plasma AUC may be decreased. When the formation clearance is estimated, the numerator (amount of metabolite excreted) would be increased and the denominator (AUC of the parent) could be decreased resulting in a large increase when the true enzyme activity has not been increased.

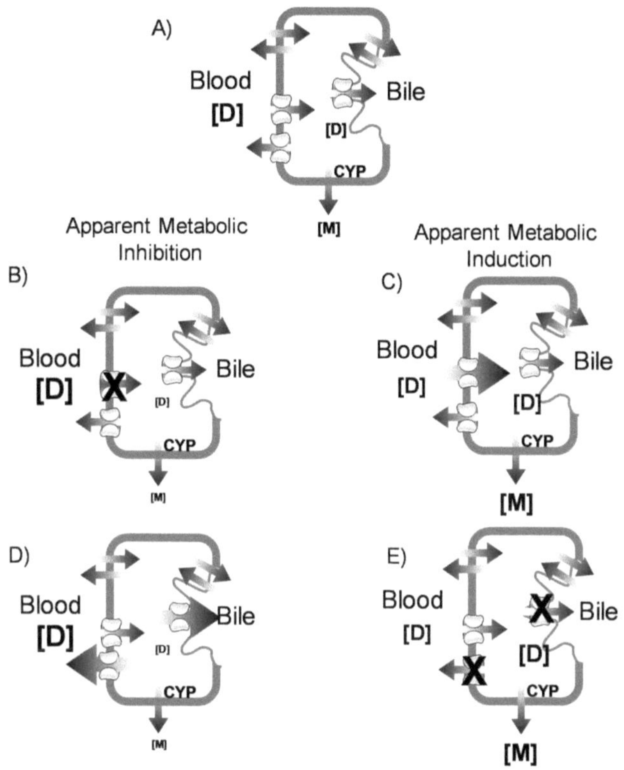

Fig. 21.3 Transporter DDIs masked as metabolism DDIs: Inhibition or induction of a transport process that determines the plasma to intrahepatic free drug concentration ratio can be perceived as an apparent metabolic interaction. The text size of the symbol [D] and [M] relative to the symbol size in panel A represent the relative change in drug ([D]) in the cell or plasma and the metabolite [M] formed in the inhibited or induced state (panels C and D) relative to the uninhibited, uninduced state (panel A)

In both of the cases above where a transporter interaction is manifested as a metabolic interaction, if the true intrahepatic concentration (or ratio to that in the plasma) were known and used to calculate the metabolic formation clearance, there would be no change as a result of transporter interaction.

21.2.4 Vectorial Transport

Two transporters can work in a vectorial manner to move a hydrophilic substrate across a cell. For example, the concentrative nucleoside transporter 2 (CNT2) and equilibrative nucleoside transporter 1 (ENT1) mediate the transport of nucleoside analog drugs across the intestinal epithelium (Fig. 21.4). Assuming that the nucleoside drug is metabolized in the enterocytes, inhibition of either of the two

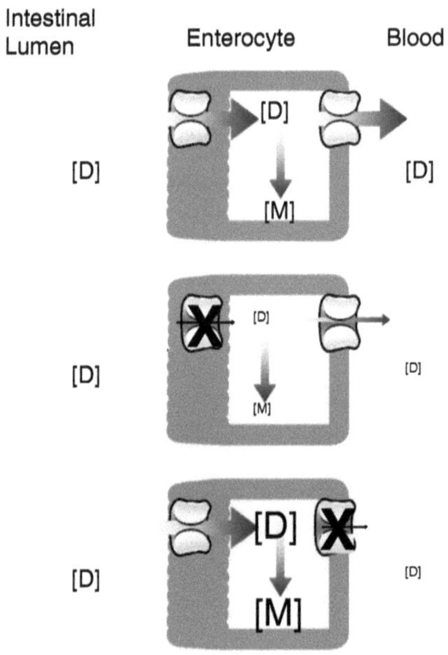

Fig. 21.4 Vectorial transport: When two transporters work in concert to move a hydrophilic drug across a cell, inhibition of either of the transporters will have the same effect on the overall transport of the drug but drastically different effects on the intracellular exposure of the drug. The relative text size of the [D] symbol indicates change in drug concentration relative to the uninhibited conditions (*top panel*). The *middle panel* shows that inhibition of the uptake process into the cell decreases the intracellular and blood concentration of the drug. The *bottom panel* shows that inhibition of the basolateral transporter also results in lower plasma concentration, but the exposure of the enterocyte is drastically increased

transporters will have the same effect systemically, potentially decreasing bioavailability (denoted by the decrease in the text size of the [D] symbol on the blood side), but will have drastically different effects on the exposure of the intestinal enterocytes. In this example, if CNT2 on the apical membrane is inhibited, the enterocyte drug and metabolite exposure as well as systemic exposure of the drug will be decreased significantly. On the other hand, if ENT1 on the basolateral membrane is inhibited, the systemic drug exposure will be reduced, but the enterocyte drug and metabolite exposure will be greatly increased. In this case, inhibition of the equilibrative transporter may have a significant impact on bioavailability of the drug and also on intestinal epithelial toxicity. This type of phenomenon is applicable to many other organs such as the liver and kidney and other tissues. If inhibition of ENT1 was to happen in a distributional type interaction, the acute exposure of the tissue relative to the blood may be decreased, but eventually the same plasma to tissue equilibrium will be reached, it will only take longer to establish the equilibrium. This equilibrium may not be achieved if the drug substrate is rapidly metabolized in the tissue, which may result in a "sink" effect.

21.2.5 Equilibrative Transporter DDI

Drug transporters can be either active or equilibrative. The active transporters use a source of energy to transport drugs unidirectionally, either into or out of a cell, typically against a concentration gradient, whereas the equilibrative transporters (also known as facilitative transporters) transport drugs bidirectionally and are driven by the concentration gradient of a drug aiding the establishment of drug concentration equilibrium across the cell membrane. As a result of these differences, it is apparent that DDIs involving active and equilibrative transporters can have substantially different consequences. Equilibrative transporters augment passive diffusion, therefore inhibition and induction of these transporters would increase and decrease, respectively, the time to reach concentration equilibrium of a drug across a membrane, but eventually the equilibrium would be established irrespective of the transporter. Obviously, the impact of DDIs with equilibrative transporters is substrate dependent. If a substrate has very low passive permeability, the time to equilibrium in the absence of the transporter activity may be consequential with respect to the PK or PD of a drug. On the other hand, for a drug with adequate passive permeability the inhibition or induction of an equilibrative transporter may have an inconsequential effect on the time to reach concentration equilibrium. This phenomenon is illustrated in Fig. 21.5 for a hypothetical drug given by IV bolus with varying passive permeability rate constants (k_{diff} of 1.0 {panel A}, 0.1 {panel B}, and 0.005 min^{-1} {panel C}) in the absence and presence of an equilibrative transporter augmenting transport into and out of a tissue compartment. The tissue was assumed to be purely distributional (no metabolism) and inconsequential in determining the volume of distribution of the drug. Therefore, the same plasma profile was used as the driving force for all simulations. The equilibrative transporter was assumed to be operating at concentrations well below its K_{m} such that the transport rate constant was constant at 0.2 min^{-1}. Figure 21.5 panel A shows that for the high-permeability drug, the absence or presence of the equilibrative transporter has virtually no effect on the tissue concentration profile. For the intermediate permeability, Fig. 21.5 panel B (k_{diff} = 0.1 min^{-1} or ½ of the transport rate constant of 0.2 min^{-1}), the absence of the transporter results in only a very minor transient effect in the very early portion of the tissue profile. In stark contrast is Fig. 21.5 panel C, the case where k_{diff} = 0.005 min^{-1} (1/40th of the transport rate constant). In this case there is a significant change in the tissue profile as a result of inhibition of the equilibrative transporter, a lengthening of the tissue half-life, but no change in the tissue exposure (AUC when extrapolated to infinite time). This simulation was done for a single dose experiment. In the case of multiple doses the drug would accumulate in the tissue to a greater extent than accumulation in the plasma in the absence of the equilibrative transporter resulting in a net increase in the tissue to plasma concentration ratio predicted from the single dose experiment (Fig. 21.5 panel D). In the presence of the equilibrative transporter there is no accumulation in the tissue upon multiple dose administration. When the multiple dose scenario is applied to the intermediate- and high-permeability drugs, there is very little to no effect of loss of the equilibrative

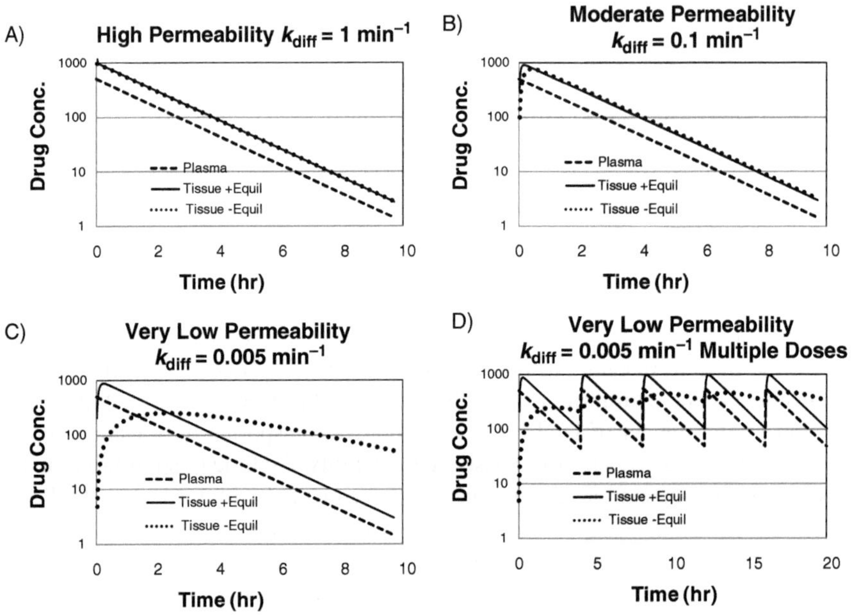

Fig. 21.5 Equilibrative transporter inhibition: Inhibition of an equilibrative transporter will have drastically different effects on tissue concentration depending on the passive permeability of the drug. (**A**) For a drug with high permeability ($k_{\text{diff}} = 1.0$ min^{-1}) inhibition of the equilibrative transporter (Tissue – Equil) has no effect on tissue exposure relative to the presence of the transporter (Tissue + Equil). (**B**) For a moderate permeability drug ($k_{\text{diff}} = 0.1$ min^{-1}) the effect of the transporter is transient and minor. (**C**) For a very low-permeability drug ($k_{\text{diff}} = 0.005$ min^{-1}) the absence of the equilibrative transporter (Tissue – Equil) has a profound effect on the tissue concentration profile. (**D**) Upon multiple dosing in the absence of an equilibrative transporter, tissue accumulation of a very low-permeability drug can occur

transporter on the accumulation of drug in the tissue as expected (simulations not shown).

21.3 ATP-Binding Cassette (ABC) Super Family Transporter Interactions

21.3.1 P-gp (MDR1, ABCB1) Interactions

P-glycoprotein (P-gp; encoded by the *ABCB1* gene) is a 170 kDa efflux transporter that belongs to the ATP-binding cassette (ABC) protein super family. First reported to reduce the permeability of resistant cells to colchicine in the 1970s (Juliano and Ling, 1976), P-gp has since been shown to transport a remarkably diverse range of substrates, typically hydrophobic, amphipathic molecules. The anatomic expression of P-gp strongly suggests that it plays a protective role, shielding important organs from potentially toxic xenobiotics. Expression of P-gp (Fig. 21.1) has been shown on the luminal surface of intestinal epithelia, the bile canalicular membrane of hepatocytes, and the luminal surface of the renal proximal tubule (Thiebaut et al., 1987), as well as the blood–brain barrier (BBB), blood–cerebrospinal fluid barrier (BCSFB) (Cordon-Cardo et al., 1989), placenta, ovaries, and testes (Sugawara et al., 1988). The net effect of this expression profile, combined with the broad substrate specificity, is that P-gp plays a major role in the absorption, distribution, elimination, toxicity, and efficacy of many drugs, and that clinically significant DDIs may occur when a combination of P-gp substrates, inhibitors, or inducers is co-administered.

One of the most important model drugs used to examine P-gp-mediated interactions is the cardiac glycoside, digoxin. Because it is a P-gp substrate that is not metabolized by cytochrome P450 enzymes (Hinderling and Hartmann, 1991; Lacarelle et al., 1991), digoxin is widely used as a marker of P-gp activity in vivo. By administering digoxin both orally and intravenously, the effect of drug on P-gp activity in the gut can be distinguished from the effect on P-gp-mediated systemic clearance (biliary and renal secretion). In one example of this, cardiac arrhythmia patients receiving chronic oral digoxin therapy were administered an intravenous dose of [^{3}H]-digoxin in the presence or absence of the P-gp inhibitor quinidine, also administered orally (Pedersen et al., 1983). It was found that co-administration of quinidine increased the absorption rate constant of unlabeled digoxin by 30% (2.72 $\pm$ 1.04 to 3.53 $\pm$ 1.34 h^{-1}), plasma AUC by 77%, and dose-normalized [^{3}H]-digoxin AUC by 54%. This resulted in a 15% increase in the relative systemic availability of digoxin, primarily due to inhibition of P-gp-mediated efflux in the gut and liver during the first-pass effect. Similarly, it was reported that oral co-administration of digoxin and talinolol, another P-gp substrate, increased the AUC and maximum serum concentrations of digoxin by 23% and 45%, respectively (Westphal et al., 2000). In contrast, there was no change in the pharmacokinetics of orally administered digoxin when talinolol was given intravenously; further demonstrating that modulation of P-gp activity within the small intestine can have a significant impact upon the bioavailability of digoxin.

Digoxin has also been used by many researchers to examine P-gp-mediated drug–drug interactions occurring in the liver and kidneys due to the presence of the transporter on the bile canalicular membrane and the luminal membrane of the renal proximal tubules, respectively. In one such study, healthy volunteers were administered digoxin orally for 8 days in the presence or absence of quinidine before digoxin biliary clearance was measured by duodenal perfusion (Hedman et al., 1990). Co-administration of quinidine was found to decrease digoxin biliary clearance by 35% and renal clearance by 45%. Likewise, administration of itraconazole has been shown to increase the AUC of digoxin by around 50%, while reducing digoxin renal clearance by 20% (Jalava et al., 1997).

Within the placenta, P-gp expression is found on the maternal-facing membrane of the syncytiotrophoblasts (Atkinson et al., 2003; Nagashige et al., 2003; Mathias et al., 2005), where it acts to protect the developing fetus from potentially harmful xenobiotics (Unadkat et al., 2004). In lieu of conducting fetal drug distribution studies in humans, which carries obvious ethical objections, the ex vivo use of perfused human term placenta is commonly employed to determine the clearance associated with maternal-to-fetal or fetal-to-maternal transfer of the drug of interest. These values are routinely normalized to the maternal-to-fetal and fetal-to-maternal clearances of antipyrine, a reference drug which can passively diffuse across the placenta, to give the clearance index of the drug. Inhibition of placental P-gp using the highly potent P-gp inhibitors PSC833 or elacridar has been demonstrated to increase by 3-fold the maternal-to-fetal clearance index of saquinavir, a protease inhibitor and P-gp substrate (Molsa et al., 2005).

P-gp is highly expressed at the BBB (Cordon-Cardo et al., 1989; Thiebaut et al., 1989), an impermeable structure comprised of endothelial cells which form tight junctions, severely limiting the entry of drugs into the central nervous system (CNS) (de Boer et al., 2003; Bernacki et al., 2008). Animal models in which one or both the mouse *ABCB1* homologs, *mdr1a* and *mdr1b*, have been knocked out have been extensively used to demonstrate the importance of P-gp within this structure (Schinkel et al., 1994; 1995; 1997). Exposure of these *mdr1a*$^{-/-}$ mice to P-gp substrate drugs results in a striking increase of the distribution into the brain. Despite the obvious utility of these P-gp knockout mouse models, the study of P-gp-mediated drug interactions at the human BBB remains a far greater technical challenge. Recent developments in noninvasive positron emission tomography (PET) imaging, however, have begun to transform this area of study.

Recently, we developed a novel methodology to image P-gp activity at the human BBB using PET imaging to examine the brain distribution of [^{11}C]-verapamil, a P-gp substrate (Sasongko et al., 2005). We found that the brain:blood AUC of [^{11}C]-radioactivity increased by 88% when the P-gp inhibitor cyclosporin A (CsA) was co-administered at a pseudo steady-state dose of 2.8 μM. When comparing the unbound concentration of CsA to the in vitro IC_{50} (Hsiao et al., 2008) of CsA for P-gp efflux of verapamil, it appears that inhibition of P-gp at the human BBB was not complete. Based on in vitro and rat in vivo data (Hsiao et al., 2008) it appears that upon complete inhibition of P-gp at the human BBB, the increase in brain exposure of verapamil could be greater than 88% exemplifying the important role that P-gp can play in limiting brain exposure of drugs.

In addition to inhibition of P-gp efflux, a number of drug interactions have been demonstrated wherein expression of the transporter is induced. Expression of P-gp is controlled by several regulatory elements, one of which binds the pregnane X receptor (PXR), a nuclear xenobiotic receptor that transcriptionally upregulates expression of both *ABCB1* and *CYP3A4* following binding of certain ligands (Geick et al., 2001). One example of PXR-mediated upregulation of P-gp leading to a drug interaction is demonstrated by measuring the systemic availability of digoxin before and after 10-day administration of the PXR ligand rifampin (RIF) in healthy volunteers (Greiner et al., 1999). It was found that RIF treatment induced a 3.5-fold increase in intestinal P-gp expression, and concomitantly the bioavailability of digoxin decreased from 63 to 44%. Likewise, administration of

St. John's Wort extract (another PXR ligand) for 14 days resulted in a 1.4-fold increase in duodenal P-gp expression and an 18% decrease in digoxin plasma AUC (Durr et al., 2000).

The impact of P-gp on drug absorption, distribution, excretion, and toxicity has become clearer with the development of new tools to study drug interactions, such as knockout mouse models and noninvasive PET imaging modalities. The importance of this work is demonstrated by the efforts to which pharmaceutical companies attempt to screen for inhibition or induction of P-gp at the very earliest step in the drug development process. Although many P-gp-mediated drug interactions are unfavorable within the clinic, a better understanding of this phenomenon may potentially be of use in the treatment of multidrug-resistant tumors using chemotherapy or delivering anti-retroviral drugs to HIV sanctuary sites such as the CNS or the fetus.

21.3.2 BCRP (ABCG2) Interactions

Breast cancer resistance protein (BCRP), encoded by the *ABCG2* gene, is an ABC half-transporter that is abundantly expressed in the intestine, placenta, BBB, and liver (Fig. 21.1). Like P-gp, BCRP also demonstrates wide substrate specificity and has been shown to transport anthracyclins, bisantrene, etoposide, flavopiridol, mitoxantrone, prazosin, and topotecan derivatives (Allen et al., 1999; Brangi et al., 1999; Maliepaard et al., 1999; Litman et al., 2000; Robey et al., 2001). In contrast to P-gp, however, the involvement of BCRP in drug interactions is much less understood.

Elacridar, a potent inhibitor of both BCRP and P-gp, has been shown to increase the bioavailability of topotecan from 40 to over 97% when both drugs were orally co-administered (Kruijtzer et al., 2002). Additionally, intravenous administration of elacridar had a more modest effect, increasing the plasma AUC of topotecan by 10%. Because topotecan has been shown to have a low affinity for P-gp, these data suggest that the interaction between elacridar and topotecan may be caused by modulation of BCRP-mediated intestinal absorption. The potential confounding effect of P-gp can be avoided in the *mdr1a*$^{-/-}$ mouse model, which has been utilized to examine the role of BCRP on the brush border membrane of the placental syncytioblast (Jonker et al., 2000). It was found that administration of elacridar increased fetal distribution of topotecan 2-fold in *mdr1a*$^{-/-}$ mice, suggesting that the presence of BCRP within the placenta helped to protect the fetus from exposure to toxic xenobiotics.

Although the current understanding of BCRP-mediated drug interactions is at an early stage, the diversity of transported substrates combined with tissue localization of the protein strongly indicate that future research will determine an important role for BCRP in drug absorption, distribution, and elimination.

21.3.3 MRPs (ABCC Family) Interactions

Members of the C subfamily of the ABC transporters have been shown to confer multiple drug resistance to cancer cells (Mirski et al., 1987; Cole, 1990; Young et al., 1999). This subfamily contains 13 members, 9 of which have been shown to have the capacity to transport drugs. The tissue expression, subcellular localization, drug substrates, and inhibitors of MRP1 through 9 are listed in Table 21.2 (Kruh et al., 2001; Russel et al., 2002; Kruh and Belinsky, 2003; Haimeur et al., 2004; Dallas et al., 2006) and Fig. 21.1. The exact physiological role of the MRPs is unclear, although dysfunction of MRP2 manifests clinically as Dubin–Johnson syndrome (Paulusma et al., 1997) which is characterized by an increase in conjugated bilirubin without elevation of the liver enzymes ALT and AST. This is believed to be a result of the lack of canalicular membrane MRP2-mediated efflux of conjugated bilirubin into the bile. MRP2 substrates are typically phase II glucuronide or glutathione drug conjugates. The active metabolite of the pro-drug irinotecan, SN-38, and the glucuronide conjugate of SN-38 have been shown to be substrates of MRP2 (Chu et al., 1999). Probenecid, an inhibitor of multiple MRPs, increased the AUC of SN-38 in rats by 50% and decreased the biliary clearance by 60% (Horikawa et al., 2002). Thalidomide was shown to decrease the intestinal toxicity associated with irinotecan in colorectal cancer patients, possibly through inhibition of MRP2 or P-gp biliary efflux of SN-38 (Govindarajan, 2002). This hypothesis was evaluated in rats, rat hepatocytes, and MRP1-, MRP2-, and P-gp-overexpressing MCDK cells, indicating that thalidomide and its hydrolysis products do inhibit MRP1, MRP2, and P-gp and production of pro-inflammatory cytokines (Yang et al., 2006a,b).

Saquinavir, a known P-gp substrate, is also a substrate of MRPs which play a role in distribution of saquinavir into the mouse brain (Eilers et al., 2008). MK571, an MRP inhibitor, increased the brain uptake of saquinavir by 4.4-fold and the P-gp inhibitor GF120918 increased the brain uptake by 7-fold (Park and Sinko, 2005). Recently, in human brain microvesicles, MK571 was shown to increase the accumulation of saquinavir by ~4-fold (Eilers et al., 2008).

A substantial amount of information has been generated on the role of some of the MRPs in drug transport in vitro and in vivo models. Unfortunately, our understanding of the role of the MRPs in drug disposition and interactions clinically is somewhat limited. Further studies are necessary to determine the impact of the MRPs in man and to predict the impact of MRP-mediated drug–drug interactions.

21.3.4 Bile Salt Export Pump (BSEP, ABCB11) Interactions

Bile salt export pump (ABCB11), originally named sister of P-gp, is expressed on the canalicular membrane of human hepatocytes (Gerloff et al., 1998). BSEP's physiological role is the export of conjugated bile salts from the hepatocytes into the bile against an enormous concentration gradient (Oude Elferink et al., 1995). Progressive familial intrahepatic cholestasis (PFIC), characterized by

Table 21.2 ABCC (MRP) transporter localization, substrates, and inhibitors*

Transporter	Alternative name	Tissue localization	Subcellular localization	Substrates	Inhibitors
MRP1 (ABCC1)	MRP GS-X	Ubiquitous (high expression in lungs, kidney, testis)	Basolateral	Methotrexate, estradiol-17-β-D-glucuronide, GSH, doxorubicin, vincristine	MK571, CsA, indomethacin, probenecid, sulfinpyrazone, LY475776, PSC833
MRP2 (ABCC2)	cMOAT cMRP	Liver, intestine, kidney	Apical	Methotrexate, 2,4-dinitrophenyl-S-glutathione, vinblastine, etoposide, vincristine	Probenecid, MK571, furosemide, indomethacin, benzbromarone
MRP3 (ABCC3)	MOAT-D cMOAT2 MLP-2	Liver, intestine, kidney, bile duct, pancreas, prostate, placenta	Basolateral	Methotrexate, estradiol-17-β-D-glucuronide, vincristine	
MRP4 (ABCC4)	MOAT-B	Prostate, lung, muscle, pancreas, bladder	Apical	Estradiol-17-β-D-glucuronide, cyclic nucleotides (cAMP, cGMP), GSH, cholates, 9-(2-phosphonylmethoxyethyl) adenine (PMEA)	Probenecid, dipyridamole

Table 21.2 (continued)

Transporter	Alternative name	Tissue localization	Subcellular localization	Substrates	Inhibitors
MRP5 (ABCC5)	MOAT-C SMRP	Ubiquitous (high expression in skeletal muscle and heart)	Basolateral	Cyclic nucleotides analogs, heavy metals (K and Cd), GSH, 6-mercaptopurine	Probenecid, sulfinpyrazone, zaprinast, trequinsin, sildenafil
MRP6 (ABCC6)	MOAT-E MLP-1	Liver, kidney	Basolateral	Endothelin receptor antagonist BQ-123 (peptide), leukotriene C4, 6-mercaptopurine	
MRP7 (ABCC10)		Colon, skin, testis, spleen		Leukotriene C4, docetaxel, estradiol-17-β-D-glucuronide	MK571, CsA
MRP8 (ABCC11)		Liver, lung, kidney, fetal tissue		Cyclic nucleotides (cAMP, cGMP), 5-fluorouracil	
MRP9 (ABCC12)		Breast, testis, brain, skeletal muscle, ovary			

*Dallas et al. (2006), Russel et al. (2002), Haimeur et al. (2004), Kruh et al. (2001), Kruh and Belin (2003) hepatotoxicity arising from accumulation of bile salts in hepatocytes, can result from mutations in BSEP or other canalicular bile salt-transporting proteins such as MDR3 (ABCB4) (Davit-Spraul et al., 2009). A number of drugs have been shown to be substrates or inhibitors of BSEP including pravastatin (Hirano et al., 2005), flutamide (Hirano et al., 2005; Iwanaga et al., 2007), and bosentan (Fattinger et al., 2001). Clinically significant interactions with BSEP are of the drug–endogenous substrate nature rather than DDIs. The significance of this interaction is demonstrated by the hepatotoxicity caused by troglitazone which led to its removal from the market (Herrine and Choudhary, 1999; Funk et al., 2001). Troglitazone and its sulfate conjugate were shown to inhibit rat BSEP with a K_i of 1.3 and 0.23 μM, respectively, in liver canalicular membrane vesicles (Funk et al., 2001). Plasma troglitazone concentrations of 2–4 μM were achieved at the therapeutic doses of 400–600 mg/day (Loi et al., 1999).

Many drugs that can cause cholestasis such as RIF, glyburide, rifamycin, and CsA have been shown to be inhibitors of either BSEP or the sodium-dependent taurocholate co-transporter (NTCP) in cell lines co-transfected with both transporters (Mita et al., 2006). A qualitative structure–activity relationship (QSAR) study was done to aid in predicting drug-induced intrahepatic cholestasis in the drug development process (Hirano et al., 2006). Further refinement of this QSAR will ultimately aid in reducing the incidence of BSEP inhibition and cholestatic hepatotoxicity in new chemical entities (NCEs) as well as drawing attention to currently marketed drugs with the potential to inhibit BSEP.

21.4 Solute Carrier (SLC) Super Family Transporter Interactions

21.4.1 OATPs (SLC21 Family) Interactions

The organic anion-transporting polypeptides (OATPs) are Na^+-independent transporters that transport a wide variety of xenobiotics as well as endogenous compounds (Hagenbuch and Meier, 2004). OATPs are encoded by the gene family SLC21 and are expressed in organs important in drug disposition (Fig. 21.1) such as the intestine, liver, and kidney (Mikkaichi et al., 2004b). Because of their tissue expression and broad substrate selectivity, inhibition of the OATPs may result in clinically significant drug interactions at the levels of drug absorption, distribution, and excretion.

OATPs are expressed on the apical membrane of the intestinal epithelial cells (Glaeser et al., 2007), on sinusoidal membrane of hepatocytes (Abe et al., 1999; Hsiang et al., 1999; Tamai et al., 2000), and on the basolateral membrane of the renal proximal tubule cells (Tamai et al., 2000). In all three of these locations the OATPs actively transport substrates into the cells, but the effect that these transporters have on the pharmacokinetics of substrate drugs is different. Inhibition of

OATPs in the intestine would cause a decrease in either the extent or rate of absorption which would be manifested by either no change or a decrease in C_{max} or AUC. On the other hand, inhibition of OATPs in the liver or kidney would be manifested by either a decrease in volume of distribution (if the liver/kidney is a significant contributor to the distribution volume) and/or a decrease in clearance (if the OATP is the rate-limiting step in the liver or kidney clearance) resulting in an increased AUC. Because of the dual effect that inhibition of OATPs can have on the AUC of a substrate, a thorough understanding of the OATP isoform tissue expression and contribution of those isoforms to substrate permeability is necessary to predict the outcome of inhibitory OATP DDIs.

Intestinal expression of many OATPs including OATP1A2 and 2B1 as well as the previously believed liver-specific OATPs 1B1 and 1B3 has been shown (Mikkaichi et al., 2004b; Glaeser et al., 2007). Inhibition of OATP1A2 in the intestine is believed to be the cause of the decrease in AUC of the OATP and P-gp substrate fexofenadine when co-administered with grapefruit juice (Glaeser et al., 2007). Grapefruit juice and some of its components inhibit transport of OATP 1A2, 2B1, and P-gp substrates in vitro. Interestingly, grapefruit juice has no effect on the AUC of the P-gp substrate digoxin but does significantly reduce the absorption rate constant of digoxin indicating potential inhibition of an uptake process or activation of P-gp activity (Soldner et al., 1999). Digoxin has also been shown to be a substrate of OATP1B3 (Kullak-Ublick et al., 2001) and 4C1 (Mikkaichi et al., 2004a) indicating that like fexofenadine, digoxin may be a dual substrate of OATP and P-gp in the intestine. It must be noted that there is evidence indicating that P-gp may be activated by grapefruit juice (Soldner et al., 1999). Activation of intestinal P-gp efflux would be difficult to differentiate from inhibition of OATP uptake since digoxin and fexofenadine are substrates of both OATPs and P-gp.

The two major isoforms of OATP expressed on the sinusoidal membrane of hepatocytes are OATP1B1 and 1B3 (Mikkaichi et al., 2004b). Substrates of these two isoforms include digoxin, fexofenadine, RIF, bosentan, glyburide, repaglinide, rosiglitazone, and many of the statins including cerivastatin, pravastatin, rosuvasatin, and the extensively metabolized atorvastatin and simvastatin. Clinically relevant interactions precipitated by gemfibrozil resulted in approximately 2-fold increase in the AUCs of simvastatin (Backman et al., 2000), rosuvastatin (Schneck et al., 2004), pravastatin (Mikkaichi et al., 2004b), and lovastatin (Kyrklund et al., 2001) and a 5.6-fold increase in cerivastatin AUC (Backman et al., 2002). These interactions must be interpreted with caution, as gemfibrozil inhibition of other transporters such as MRP2 (Sasaki et al., 2002), BCRP (Huang, 2009) or metabolism by CYP2C8 (Wang et al., 2002) or 2C9 (Wen et al., 2001) may contribute significantly to these interactions. Drug interactions between CsA and the statins are believed to involve the OATPs (Asberg et al., 2001; Park et al., 2001). It is not easy to see the OATP contribution to these interactions because of the contribution of CYP3A metabolism to many of the statins and the ability of CsA to inhibit CYP3A. The clearest example is the interaction between CsA and rosuvastatin (Simonson et al., 2004). Heart transplant patients receiving an anti-rejection regime of CsA showed a 7-fold increase in rosuvastatin AUC compared to historical controls. Rosuvastatin is eliminated primarily unchanged (~90% fecal and ~10% renal) with a minor metabolic clearance mediated by CYP2C9 and 2C19 (McTaggart et al., 2001). Rosuvastatin is not transported by P-gp or MRP2 (Huang, 2009) and its metabolic clearance is not mediated by CYP3A, therefore it is likely that this interaction is a result of CsA inhibition of OATP1B1 and/or inhibition of BCRP-mediated biliary excretion. The interactions between CsA or RIF and the anti-pulmonary hypertension drug bosentan (TRACLEER®) highlight the differences between single and multiple dose interactions when the object and/or precipitant are inducers. Bosentan is extensively metabolized by CYP3A and 2C9, is a substrate of OATP1B1 and 1B3 (Treiber et al., 2007), and is an autoinducer of its own metabolism, resulting in a 30% decrease in AUC after multiple doses (van Giersbergen et al., 2003). When bosentan is given with ketoconazole (200 mg qd 6 days), bosentan AUC is increased 130% (van Giersbergen et al., 2002). Single dose administration of CsA caused a 30-fold increase in bosentan trough concentrations, where after multiple doses of CsA, the bosentan steady-state plasma concentrations were only 3- to 4-fold higher (TRACLEER® Package Insert). In this case it appears that the inductive effect of bosentan is capable of dampening the inhibitory (CYP and/or OATP) effect of CsA. When bosentan is administered with RIF, the bosentan trough concentration is increased 6.5-fold on the first day of RIF treatment but reduced 60% after extended RIF treatment (van Giersbergen et al., 2007). Since RIF is not expected to inhibit CYP3A activity based on in vitro data (Kajosaari et al., 2005), the 6.5-fold bosentan trough concentration increase (first day dosing of RIF) appears to be the result of OATP and/or CYP2C9 inhibition which upon multiple day RIF dosing is overshadowed by induction (CYP3A, 2C9, and/or OATP) resulting in a net decrease in trough concentrations. RIF, a broad spectrum inducer of drug-metabolizing enzymes and drug transporters, is also a substrate and inhibitor of OATP1B1 and 1B3 in the liver with OATP1B1 playing the predominant uptake role (Niemi et al., 2003; Tirona et al., 2003). Hepatic uptake of RIF by OATP1B1 has been suggested as a necessary component of PXR activation and induction of drug-metabolizing enzymes and transporters (Tirona et al., 2003). This indicates that individuals with deficient RIF uptake as a result of OATP1B1 polymorphisms *1b, *5, or *9 may exhibit a reduced degree of PXR activation. To circumvent the inductive effects of RIF and observe the acute inhibitory activity against OATPs, a single 30-min intravenous infusion of RIF was given in combination with oral atorvastatin (Lau et al., 2007). The AUC of atorvastatin acid and lactone were increased 8- and 2.7-fold, respectively, while only modest inhibition of metabolism (~40% decrease and no change in 4-OH and 2-OH metabolite/parent AUC ratios, respectively) was observed.

The examples listed above illustrate the drug interactions that can be attributed at least in part to inhibition of OATP-mediated transport of substrates in the intestine or liver. Because of the broad tissue expression of the OATPs and the increasing number of drugs identified as substrates, it is possible for OATP-mediated interactions to also occur in organs such as the placenta, kidney, and BBB. In recent years, research into the role of OATPs in drug disposition has intensified and increased our understanding of the importance of the OATP family of transporters.

21.4.2 OAT (SLC22 Family) Interactions

The organic anion transporters (OATs) are also encoded by the SLC22 gene family and primarily expressed in the renal epithelial cells, although there is some evidence of their expression in the liver and the brain (Wright and Dantzler, 2004). OATs are capable of transporting both endogenous (e.g., para-amino hippurate, riboflavin) substrates and a wide variety of weakly acidic drugs. OATs 1–4 have been cloned and are expressed primarily in the kidney and the liver (Srimaroeng et al., 2008). The organic anion/dicarboxylate exchanger hOAT1 (PAH/dicarboxylate exchanger) is a Na^+-independent transporter and relies on the co-transport of a dicarboxylate (e.g., α-ketoglutarate) down its concentration gradient, whereas OAT2 and 3 are Na^+-independent facilitative transporters. OAT1–3 are expressed in the basolateral membrane of the renal epithelia cells and facilitate the uptake and secretion of substrates. OAT4 is expressed on the brush border membrane of renal epithelial cells and facilitates the re-uptake of substrates (Srimaroeng et al., 2008).

For drugs that are actively secreted by OATs and whose secretory renal clearance is a substantial component of the total clearance, drug interactions involving OATs will result in clinically significant increases in AUCs and decreases in clearance. Conversely, for drugs by which re-uptake by OAT4 attenuates the renal clearance, drug interactions involving OAT4 will result in a decrease in AUC and increase in clearance of the drugs.

Classically described OAT interactions include the interactions of penicillins or cephalosporins with the hOAT inhibitor probenecid and have been well characterized since the 1950s. It was believed very early on that inhibition of a transporter involved in active tubular secretion was the mechanism by which the renal elimination of these drugs was decreased by probenecid (Weiner et al., 1960). For example, when the cephalosporin, cephradrine, was administered intravenously with and without oral probenecid, a 2.4-fold increase in AUC (from 24.0 to 57.3 μg·h/mL) and a 1.8-fold decrease in renal clearance (363–198 mL/min) were observed (Roberts et al., 1981). The authors concluded that the change in AUC of cephradrine was the result of probenecid inhibition of the renal secretion of the drug. Recently, probenecid has been shown to increase the AUC of dicloxacillin by 1.9-fold (Beringer et al., 2008). Additionally, in this study, the renal clearance of dicloxacillin was decreased 3.6-fold, and the renal clearance to GFR ratio was decreased 3.1-fold.

Cidofovir is a nucleoside antiviral drug used in the treatment of cytomegalovirus (CMV) retinitis in patients with AIDS (Lalezari et al., 1997). Because of poor bioavailability, cidofovir is administered as an intravenous infusion, typically over 1 h (5 mg/kg). After an intravenous dose, cidofovir is predominantly excreted in the urine as the unchanged drug (90.3% recovered in 24 h) (Cundy, 1999). For a 70-kg man, the total systemic and renal clearance (Cl_{renal}) of cidofovir were found to be 173 and 150 mL/min, respectively, whereas the creatinine clearance (Cl_{GFR}) in the same subjects was 97 mL/min (Cundy, 1999). These data indicate that cidofovir is eliminated unchanged in the urine by active tubular secretion (secretory clearance, $Cl_{sec} = 53$ mL/min, $Cl_{renal}/Cl_{GFR} = 1.54$). The dose-limiting toxicity of cidofovir is nephrotoxicity (Lalezari et al., 1995). This toxicity is believed to be due to high concentrations of the drug in the renal epithelium produced by rapid drug uptake at the basolateral membrane and slower efflux into the urine via the brush border membrane. Oral co-administration of probenecid with cidofovir reduced the total clearance of cidofovir by 15% (to 147 mL/min) and the renal clearance by 27% (to 102 mL/min). The latter was not significantly different from the Cl_{GFR} in the same subjects (109 mL/min) (Cundy, 1999). Co-administration of probenecid significantly reduced the nephrotoxicity of cidofovir. For this reason, these two drugs are co-administered in the clinic. In vitro, both OAT1 and OAT3 have been shown to transport cidofovir, and this transport has been shown to be inhibited by probenecid (Uwai et al., 2007).

This interesting cidofovir–probenecid drug interaction illustrates the role a transporter can have in drug toxicity and the potential therapeutic benefit of a drug interaction. Because the expression of OATs is highest in the kidney, OAT-mediated interactions will clearly have the greatest impact on drugs that are OAT substrates and for which renal secretory clearance is the major clearance mechanism. Given the evidence of OAT expression in other tissues such as the liver, the intestine, and the brain, the OAT family of transporters may also be shown to be important in drug interactions in these other tissues as well.

21.4.3 OCT (SLC22 Family) Interactions

The organic cation transporters (OCTs) are also encoded by the SLC22 gene family and are primarily expressed on the basolateral membrane of epithelial cells in the renal proximal tubule (Wright and Dantzler, 2004). OCTs are typically capable of transporting heterocyclic weak bases, including both endogenous compounds (e.g., dopamine, epinephrine, and choline) and a variety of drugs. OCTs 1–3 have been cloned and have been shown to be expressed in the kidney, liver, placenta, and other tissues (Wright and Dantzler, 2004), although OCT1 and OCT2 appear to be the primary OCTs in the kidney.

A number of OCT-mediated drug interactions have been described. For example, both procainamide and cimetidine undergo substantial net tubular secretion and are predominantly renally cleared (Reidenberg et al., 1980; Somogyi et al., 1980; Drayer et al., 1982). Co-administration of cimetidine increased the AUC of procainamide by 35% and reduced the Cl_{renal} by 43% (from 347 to 196 mL/min) (Somogyi et al., 1983). Metformin also undergoes substantial renal secretion ($Cl_{renal}/Cl_{GFR} \sim 5$) (Somogyi et al., 1987). Co-administration of cimetidine with metformin increased the AUC of metformin by 46% and reduced the Cl_{renal} by 28%

(from 527 to 378 mL/min) (Somogyi et al., 1987). Cimetidine has also been shown to decrease the renal clearance of varenicline by ~25%, most likely mediated by OCT2 (Feng et al., 2008).

Like the OATs, OCT interactions typically have the greatest impact on the level of active tubular secretion in the kidney. When a drug is cleared predominantly by OCTs, inhibition of these transporters can affect both systemic exposure (AUC) and tissue (kidney)-specific exposure of the OCT substrate drug. Additionally, the broad tissue expression of the OCTs suggests that OCT inhibition may also affect drug distribution into other organs (e.g., liver and placenta).

Various nonsynonymous SNPs have been described that effect the activity of the OCTs in vitro (Shikata et al., 2007; Choi and Song, 2008). OCT1 SNPs that decrease the transport of metformin in vitro (Shu et al., 2007) have also been shown to significantly increase AUC and decrease V/F and CL/F of metformin (Shu et al., 2008) clinically. Additionally, OCT2 SNPs that reduce metformin transport in oocytes (Song et al., 2008b) also result in increased metformin AUC and decreased metformin renal clearance clinically (Song et al., 2008a; Wang et al., 2008).

21.4.4 Peptide Transporters (PEPT1-2, SLC15A1-2) Interactions

The oligopeptide transporters PEPT1 and PEPT2 are proton-coupled di- and tripeptide transporters (Liang et al., 1995; Adibi, 1997) that have been shown to transport peptidomimetic drugs including β-lactam antibiotics (Terada et al., 1998) and angiotensin-converting enzyme inhibitors (Boll et al., 1994). PEPT1 is believed to be the major absorptive transport mechanism of the β-lactam antibiotics in intestinal enterocytes (Liang et al., 1995). PEPT1 and PEPT2 are expressed at the brush border membrane of the proximal tubule cells in the kidney where they act to reabsorb peptides and peptide-like drugs from the renal filtrate. In vitro and animal studies have explored the substrates and inhibitors of these two transporters, but there are minimal clinical data suggesting a significant contribution of DDIs mediated through the PEPT transporters. Given their expression profiles, PEPT1 interactions would most likely occur in the intestine and could potentially reduce the absorption rate or bioavailability of substrate drugs. PEPT2-mediated DDIs would most likely occur in the kidney and result in the reduced re-absorption of substrate drugs ultimately increasing their renal clearance.

21.4.5 Multidrug and Toxin Extrusion (MATE, SLC47A) Interactions

The human multidrug and toxin extrusion transporters (MATEs) function as an exchanger of protons and a variety of organic cations such as tetraethylammonium

(TEA) and cimetidine and zwitterionic compounds such as cephalexin. The localization of MATE1 and MATE2-K on the brush border membrane of the renal proximal tubular cells indicates that they may be candidates for efflux of cationic drugs in the kidney (Liang et al., 1995; Otsuka et al., 2005). Cimetidine, the OCT inhibitor, caused a 34% decrease in the renal clearance of fexofenadine (Yasui-Furukori et al., 2005). Cimetidine-unbound plasma concentrations were well below its K_m values for OAT3 and OCT2 expressed on the basolateral membrane of the renal epithelial cells indicating that renal uptake of fexofenadine by these transporters was likely not inhibited significantly. Recently, using MATE1-transfected HEK293 cells, the uptake of fexofenadine was shown to be inhibited 40% by 3 μM cimetidine (Matsushima et al., 2009) which is comparable to the unbound plasma concentration during the clinical study. Inhibition of MATE1 has also recently been implicated in the 45% reduction of metformin renal clearance by cimetidine (Wang et al., 2008; Tsuda et al., 2009). MATE1 is also expressed on the canalicular membrane in hepatocytes, but interestingly cimetidine had only a very minor effect on the systemic clearance of fexofenadine. The MATE interactions characterized to date are significant, yet result in only modest increases in systemic exposure of the affected drug. More investigation is needed to increase our understanding of the role of MATEs in drug disposition and DDIs in man.

21.5 In Vitro to In Vivo DDI Prediction

Due to the vast number of potential transporter-based interactions, it would be nearly impossible to test each of the possible precipitant–object pairs in a clinical setting. Therefore, if we are to predict the magnitude of transporter-based drug interactions, an approach similar to that which is being applied for metabolic-based drug interactions must be employed. This method involves using high throughput in vitro techniques to determine the IC_{50} or K_i values of drug substrates to inhibit specific transporters. An R value is calculated by comparison of the unbound IC_{50} or K_i value to the unbound plasma concentration of the inhibitor ($f_u[I]$). In the simplest case, in which the clearance mechanism of the object drug is solely mediated through the inhibited transporter, Equation (21.1) can be used to estimate the fold change in clearance ratio, or Equation (21.2) for AUC ratio:

$$\frac{Cl_{\text{inhibited}}}{Cl} = \frac{1}{1 + \frac{f_u \bullet [I]}{K_i}} \tag{21.1}$$

$$R = \frac{\text{AUC}_{\text{inhibited}}}{\text{AUC}} = \frac{Cl}{Cl_{\text{inhibited}}} = 1 + \frac{f_u \cdot [I]}{K_i} \tag{21.2}$$

In the case in which the clearance of the object drug is not solely mediated by the inhibited transporter, Equation (21.2) can be modified with the addition of an f_{cl}

term which is the fraction of the total clearance of the object that is mediated by the transporter:

$$R = \frac{\text{AUC}_{\text{inhibited}}}{\text{AUC}} = \frac{1}{\frac{f_{\text{cl}}}{\left(1+f_{\text{u}} \cdot [I]/K_i\right)} + (1 - f_{\text{cl}})} \tag{21.3}$$

Equation (21.3) assumes that the transport mechanism being inhibited is the rate-limiting step in the elimination of the drug. For example, if a drug is a substrate of hepatic uptake by OATPs and biliary efflux by P-gp, the prediction must be made with respect to the transporter which is the rate-limiting step in the clearance process. Conversely, if there are two transporters that contribute to the same process (efflux or uptake) in the same tissue, then the inhibition of both of these transporters must be taken into account. A second assumption of Equations (21.1)–(21.3) is that the transport process is a typical Michaelis–Menten process that is subjected to competitive inhibition and the substrate concentration is well below the transport K_{m} (no saturation of the transport). Another key assumption for this model is that the unbound systemic plasma concentration of the inhibitor is indicative of the concentration at the site of the interaction. This assumption may be violated in the case of active uptake or efflux transport into or out of the site of action, a high rate of tissue elimination (e.g., metabolism), or the case of elevated portal vein inhibitor concentrations during first pass after an oral dose.

In the case where a clearance process (e.g., biliary or renal secretion) is mediated by multiple transport mechanisms, the estimated effect on the clearance as a result of inhibiting one or more of the transporters can be estimated by Equation (21.4) where n is the total number of transporters (t) that account for the clearance process, $f_{\text{cl},t}$ is the fraction of the clearance attributed to the specific transporter t and $K_{\text{i},t}$ is the inhibition constant for the inhibitor against transporter t:

$$\frac{Cl_{\text{inhibited}}}{Cl} = \sum_{t=1}^{n} f_{\text{cl},t} \cdot \frac{1}{1 + \frac{f_{\text{u}} \cdot [I]}{K_{\text{i},t}}} \tag{21.4}$$

Simulations using Equation (21.3) for varying f_{cl} and $f_{\text{u}} * [I]/K_{\text{i}}$ values are shown in Fig. 21.6. The partial surfaces in Fig. 21.6 show conditions under which AUC ratios (R) >2, 5, or 10 would be predicted. These simulations indicate that for an object drug with an f_{cl} value of 0.5, the expected AUC ratio (R) observed when the transport clearance mechanism is completely inhibited is 2.0. This maximum AUC ratio for any f_{cl} can be calculated by Equation (21.5) as

$$R_{\text{max}} = \frac{1}{1 - f_{\text{cl}}} \tag{21.5}$$

This type of prediction is applicable only to the inhibition of a clearance process. In the case where inhibition of transport alters a purely distributional process,

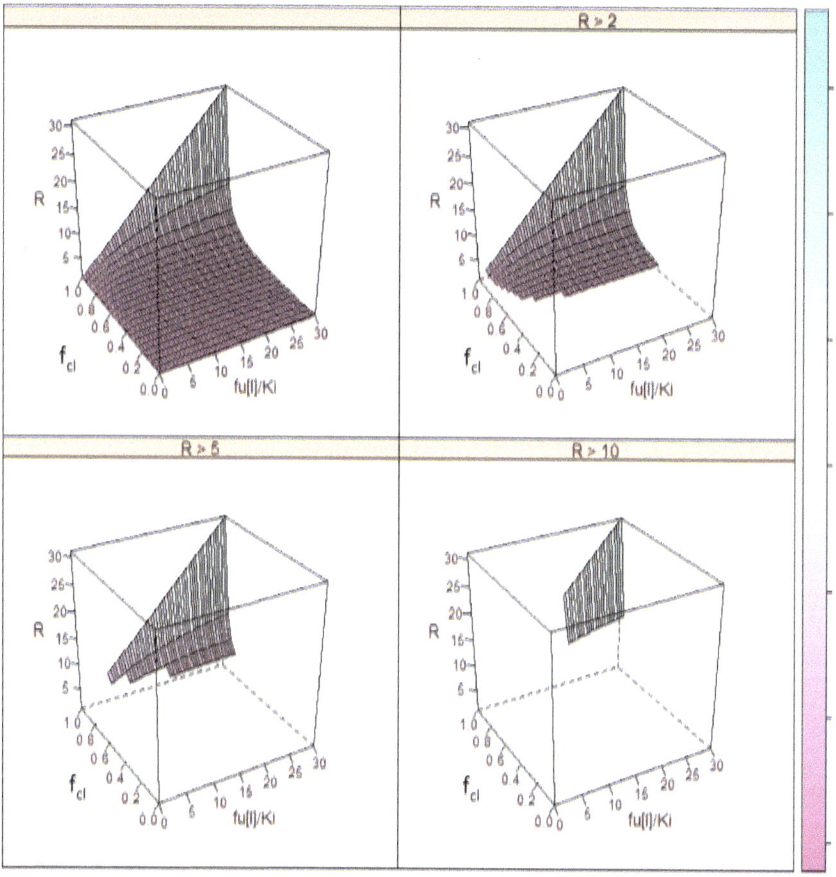

Fig. 21.6 Simulations of predicted AUC ratio using Equation (21.3): Partial surfaces indicate the conditions under which AUC ratios (R) >2, 5, and 10 would be predicted across a range of f_{cl} (fractional clearance due to the inhibited transport process) and $f_{\mathrm{u}}[I]/K_{\mathrm{i}}$ (inhibition potency of the precipitant drug)

the fold increase in drug exposure to the tissue relative to the blood or plasma concentration (R_{dist}, Equation (21.6)) could be predicted if both the diffusion and transport-mediated clearances are known. For example, to predict the change in brain/blood ratio of a P-gp substrate in the presence of an inhibitor, the relative rates of P-gp-mediated efflux clearance and the diffusion clearance across the BBB are necessary. The exposure of the organ relative to the blood is determined by the ratio of the net influx and net efflux clearances. Using a simple two-compartment model to describe this process and assuming no active uptake process and no active efflux process other than P-gp, the clearance into the brain would be the diffusion clearance, but clearance out of the brain would be the sum of diffusion and P-gp efflux clearances:

$$R_{\text{dist}} = \frac{\text{AUC}_{\text{tissue,inhib}}/\text{AUC}_{\text{blood,inhib}}}{\text{AUC}_{\text{tissue}}/\text{AUC}_{\text{blood}}} = \frac{Cl_{\text{efflux}}}{Cl_{\text{efflux,inhib}}} = \frac{Cl_{\text{diff}} + Cl_{\text{pgp}}}{Cl_{\text{diff}} + CL_{\text{pgp,inhib}}}$$
$$= \frac{Cl_{\text{diff}} + Cl_{\text{pgp}}}{Cl_{\text{diff}} + CL_{\text{pgp}}\Big/\left(1 + f_{\text{u}[I]}/K_{\text{i}}\right)} \qquad (21.6)$$

If the P-gp-mediated efflux clearance is equal to the diffusional clearance, then the brain/blood ratio would be expected to increase by 100% ($R_{\text{dist}} = 2$) with complete inhibition ($f_{\text{u}}[I] >>> K_{\text{i}}$) of P-gp efflux activity using Equation (21.6). If there are multiple transporters actively effluxing the drug from the brain (e.g., P-gp and BCRP), the total efflux clearance would be the sum of the diffusional, P-gp-mediated, and BCRP-mediated clearances. In this type of situation where there are two efflux transporters, inhibition of both transporters simultaneously would appear to have a synergistic effect. For example, if the total efflux clearance is 10% diffusional, 45% P-gp, and 45% BCRP, complete inhibition of P-gp or BCRP alone would result in an R_{dist} of 1.8, whereas complete inhibition of both P-gp and BCRP would result in an R_{dist} of 10. Large changes in the brain/blood ratio are expected as a result of the impermeable nature of the BBB and the rich expression of efflux transport proteins such as P-gp.

21.5.1 P-glycoprotein (P-gp, MDR1) Drug–Drug Interaction Prediction

P-gp is arguably the most important drug transporter because of the large number of drugs on the market that are substrates and the localization of P-gp in key organs for drug elimination (intestine, liver, and kidney) as well as privileged compartments in the body including the brain and placenta. Therefore, it is not surprising that a substantial amount of effort has been focused on the ability to predict the role of P-gp in drug disposition and the impact of P-gp-mediated DDIs. A substantial amount of in vitro data indicates that P-gp may have multiple binding sites and therefore evaluating drugs as inhibitors and substrates is more challenging. As a result, inhibitory constants (IC_{50}) may be substrate–inhibitor pair dependent. The magnitudes of the P-gp-mediated DDIs listed in Table 21.3 (Hedman et al., 1990; Jalava et al., 1997; Hedman et al., 1991; Yang and Elmquist, 1996; Simonson et al., 2004; Schneck et al., 2004; Sasongko et al., 2005; Woodland et al., 1998; Wandel et al., 1999; Ekins et al., 2001; Thummel et al., 2006; He and Liu, 2002; Ding et al., 2004; Penzak et al., 2004; Templeton et al., 2008) were predicted using Equation (21.2). One key assumption of Equation (21.2) is that the entire clearance mechanism of the probe drug is mediated through the affected transporter. This assumption, if invalid, will cause an overprediction of the decrease in clearance.

Table 21.3 In vitro to in vivo prediction of transporter-mediated DDIs

				Inhibitor parameters				Predicted $Cl_{inhibit}/Cl$			Predicted AUC ratio	
Transporter	Substrate	Inhibitor	Site of DDI	K_i or IC_{50} (uM)	$[I]C_{max}$ (uM)	f_u	Observed $Cl_{inhibit}/Cl$	Using $[I]C_{max,u}$	Using $[I]C_{max}$	Observed AUC ratio	Using $[I]C_{max,u}$	Using $[I]C_{max}$
P-gp	Digoxin	Itraconazole	Oral Cl	0.5[a]	0.3[b]	0.036[c]				1.68[b]	1.02	**1.68**
		Ritonavir	Oral Cl	3.8[d]	6.2[e]	0.01[f]	0.70[e]	**0.98**	0.38			
		Ritonavir	Oral Cl	3.8[d]	8.2[g]	0.01[f]				1.86[g]	**1.02**	3.2
		Quinidine	Biliary Cl	2.2[h]	4.5[i]	0.13[f]	0.65[i]	**0.79**	0.33			
		Verapamil	Biliary Cl	2.2[h]	0.13[j]	0.1[f]	0.57[j]	0.98	**0.94**			
		Itraconazole	Biliary Cl	0.5[a]	0.3[b]	0.036[c]	0.8[b]	0.98	**0.63**			
		Quinidine	Renal Cl	2.2[h]	4.5[i]	0.13[f]	0.71[i]	**0.79**	0.33			
		Ritonavir	Renal Cl	3.8[d]	6.2[e]	0.01[f]	0.90[e]	**0.98**	0.38			
		Ritonavir	Renal Cl	3.8[d]	8.2[g]	0.01[f]	0.65[g]	**0.98**	**0.32**			
	Verapamil	CsA	Brain	0.46[k]	2.8[l]	0.02[m]	0.56[l]	**0.89**	0.14			
OATP1B1	Rosuvastatin	Gemfibrozil	Oral Cl	4.0[n]	102[n]	0.05[n]				1.88[n]	**2.25**	27
		CsA	Oral Cl	2.2[o]	5[o]	0.02[o]				7.08[o]	1.05	**3.27**

Bold values are the predicted value closest to the observed value. In some cases, *fu*, [*I*]*C*max, or [*I*]*C*max,u were calculated from data provided in the reference.
[a]Woodland et al. (1998), [b]Jalava et al. (1997), [c]Templeton et al. (2008), [d]Ekins et al. (2001), [e]Penzak et al. (2004), [f]Hardamn (2001) [g]Ding et al. (2004), [h]Wandel et al. (1999), [i]Hedman et al. (1990), [j]Hedman et al. (1991), [k]He and Liu (2002), lSasongko et al. (2005), [m]Yang and Elmquist (1996), [n]Schneck et al. (2004), [o]Simonson et al. (2004).

21.5.2 Distributional Drug–Drug Interaction Prediction

Inhibition of P-gp by CsA in vitro has produced IC_{50} (or K_i) values of 0.13–3.8 μM, with the IC_{50} for the CsA–verapamil pair reported at 4.6 μM (Endres et al., 2006). The in vivo-unbound EC_{50} for P-gp transport of verapamil by CsA into the rat brain was calculated as 0.47 μM (Hsiao et al., 2006). The data generated in the rat model were used to predict the fold change in brain:blood ratio that we observed in a human study (Sasongko et al., 2005) using the CsA–verapamil inhibitor–substrate pair (Table 21.3). The change in brain:blood ratio observed in the rat, 75% increase, at blood CsA concentrations of ~3 μM was very similar to the 79% increase observed in the human study (Sasongko et al., 2005) in which blood CsA concentrations were also 3 μM. To attempt to simplify the prediction process and apply it to a high throughput in vitro methodology, a similar experiment was carried out in LLCPK-MDR1 cells using CsA as the P-gp inhibitor and fluorescently labeled verapamil–bodipy as the substrate (Hsiao et al., 2008). This cell model produced an EC_{50} of 0.6 μM which is comparable to the unbound EC_{50} calculated from the rat in vivo study. The in vitro method predicts a 129% increase in the brain:blood ratio for the human study.

21.5.3 Absorption/Elimination Drug–Drug Interaction Prediction

An in vitro to in vivo correlation of the effect of 19 P-gp inhibitors on the AUC ratio of digoxin showed that there is a positive trend between the predicted $[I]/IC_{50}$ ratio and the observed in vivo AUC_i/AUC ratio (Fenner et al., 2009). They showed a 41% false negative rate using the $[I]/IC_{50} > 0.1$ cut-off described in a US Food and Drug Administration (FDA) draft guidance (http://www.fda.gov/cder/guidance/6695dft.htm) on drug interaction studies to the pharmaceutical industry. Using the $[I_2]/IC_{50} < 10$ cutoff recently proposed by Zhang et al. (Zhang et al., 2008), where $[I_2]$ is the hypothetical intestinal concentration of the inhibitor, the incidence of false negatives was reduced substantially, but the incidence of false positives was increased.

21.5.4 OATP Drug–Drug Interaction Prediction

OATP1B1 and 1B3 are expressed on the sinusoidal membrane of hepatocytes (Fig. 21.1). Substrates of OATP 1B1 and 1B3 include many of the lipid-lowering statins, the antidiabetics repaglinide, rosiglitazone and glyburide, the cardiac glycoside digoxin, as well as fexofenadine and RIF. The effects of CsA and gemfibrozil on the disposition of rosuvastatin are the best available opportunities for in vitro to in vivo DDI prediction with respect to OATPs. This is because the contribution of metabolism to the total clearance of rosuvastatin is minimal. Many of the other interactions described earlier are complicated by potential metabolic inhibition or induction. Since these interactions have not been studied extensively enough to confidently assign fractional clearances to the transporter and metabolic components, prediction of these interactions is not possible. The CsA and gemfibrozil interaction predictions with rosuvastatin are listed in Table 21.3. Transport of [^{3}H]-rosuvastatin in *Xenopus laevis* oocytes expressing OATP1B1 was saturable with a K_m of 8.5 μM (Schneck et al., 2004; Simonson et al., 2004). Gemfibrozil inhibited the [^{3}H]-rosuvastatin OATP1B1-mediated uptake into oocytes with an IC_{50} of 4.0 μM (Schneck et al., 2004). In the in vivo interaction study, the average unbound gemfibrozil C_{max} concentration was approximately 5 μM indicating that OATP1B1 transport was inhibited substantially (~≥50%). Under the assumption that the IC_{50} is similar to the K_i, the predicted AUC ratio in Table 21.3 is 2.25 where the observed ratio is 1.88. In this situation using the total drug concentration drastically overpredicts the magnitude of the interactions ($R = 27$). In contrast, the observed AUC ratio of 7.8 for rosuvastatin as a result of CsA administration is drastically underpredicted using the unbound C_{max} ($R = 1.05$). Using the total C_{max}, the AUC ratio is predicted at ~3.3 indicating that either CsA concentrations at the site of inhibition are not well estimated by blood concentrations, the in vitro inhibition potential of CsA (IC_{50} = 2.2 μM) for OATP1B1 is underestimated, or another transporter is involved in the CsA–rosuvastatin interaction.

21.5.5 In Vitro to In Vivo Prediction Summary

A review of the predicted interactions in Table 21.3 shows that using the unbound C_{max} concentration was the better predictor of the interaction for 58% (7 of 12) of the interactions, total C_{max} was the better predictor for 33% (4 of 12) of the interactions, and one interaction was equally well predicted. In addition, for all but one of the predicted interactions, the unbound plasma concentration underpredicted the AUC ratio and overpredicted the Cl_{inhib}/Cl. All but three of the interaction AUC ratios were overpredicted and Cl_{inhib}/CL ratios were underpredicted using the total plasma concentration. This implies that for the most part, the unbound plasma C_{max} concentration underrepresents the concentration of drug at the active site of the transporter or the in vitro inhibition parameters (IC_{50}, K_i) are on average higher than the true in vivo parameters. The converse is true with respect to total C_{max} concentrations.

21.6 Conclusions

Our understanding of the role and impact of drug transporters on drug disposition has been steadily increasing. We have made significant progress in categorizing the location of drug transporters and determining drug substrates/inhibitors of transporters, but there are at least three critical pieces of information that still remain. Those are (1) determining the relationship between the plasma concentration and the concentration of drug the transporter is exposed to, (2) determining if or to what extent the transporter is a limiting factor in the drug permeability across a membrane or in the organ clearance, and (3) quantifying the level of transporter expression in tissues. To address these problems and ultimately increase our ability to predict transporter-based DDIs, it is necessary to quantify the relative expression of transporters in tissues using quantitative methods such as mass spectrometry and to identify probe drugs that can be used in vivo that are minimally metabolized and are selective substrates of individual transporters.

Acknowledgments This work was supported by NIH grants GM032165 and GM54447. Brian Kirby was supported in part by an ARCS fellowship and an NIH Pharmacological Sciences training grant (GM07550).

References

Abe T, Kakyo M, Tokui T, Nakagomi R, Nishio T, Nakai D, Nomura H, Unno M, Suzuki M, Naitoh T, Matsuno S and Yawo H (1999) Identification of a novel gene family encoding human liver-specific organic anion transporter LST-1. *J Biol Chem* **274**:17159–17163.

Adibi SA (1997) The oligopeptide transporter (Pept-1) in human intestine: biology and function. *Gastroenterology* **113**:332–340.

Allen JD, Brinkhuis RF, Wijnholds J and Schinkel AH (1999) The mouse Bcrp1/Mxr/Abcp gene: amplification and overexpression in cell lines selected for resistance to topotecan, mitoxantrone, or doxorubicin. *Cancer Res* **59**:4237–4241.

Asberg A, Hartmann A, Fjeldsa E, Bergan S and Holdaas H (2001) Bilateral pharmacokinetic interaction between cyclosporine A and atorvastatin in renal transplant recipients. *Am J Transplant* **1**:382–386.

Atkinson DE, Greenwood SL, Sibley CP, Glazier JD and Fairbairn LJ (2003) Role of MDR1 and MRP1 in trophoblast cells, elucidated using retroviral gene transfer. *Am J Physiol Cell Physiol* **285**:C584–C591.

Backman JT, Kyrklund C, Kivisto KT, Wang JS and Neuvonen PJ (2000) Plasma concentrations of active simvastatin acid are increased by gemfibrozil. *Clin Pharmacol Ther* **68**:122–129.

Backman JT, Kyrklund C, Neuvonen M and Neuvonen PJ (2002) Gemfibrozil greatly increases plasma concentrations of cerivastatin. *Clin Pharmacol Ther* **72**:685–691.

Beringer PM, Kriengkauykiat J, Zhang X, Hidayat L, Liu S, Louie S, Synold T, Burckart GJ, Rao PA, Shapiro B and Gill M (2008) Lack of effect of P-glycoprotein inhibition on renal clearance of dicloxacillin in patients with cystic fibrosis. *Pharmacotherapy* **28**:883–894.

Bernacki J, Dobrowolska A, Nierwinska K and Malecki A (2008) Physiology and pharmacological role of the blood-brain barrier. *Pharmacol Rep* **60**:600–622.

Boll M, Markovich D, Weber WM, Korte H, Daniel H and Murer H (1994) Expression cloning of a cDNA from rabbit small intestine related to proton-coupled transport of peptides, beta-lactam antibiotics and ACE-inhibitors. *Pflugers Arch* **429**:146–149.

Brangi M, Litman T, Ciotti M, Nishiyama K, Kohlhagen G, Takimoto C, Robey R, Pommier Y, Fojo T and Bates SE (1999) Camptothecin resistance: role of the ATP-binding cassette (ABC), mitoxantrone-resistance half-transporter (MXR), and potential for glucuronidation in MXR-expressing cells. *Cancer Res* **59**:5938–5946.

Choi MK and Song IS (2008) Organic cation transporters and their pharmacokinetic and pharmacodynamic consequences. *Drug Metab Pharmacokinet* **23**:243–253.

Choudhuri S and Klaassen CD (2006) Structure, function, expression, genomic organization, and single nucleotide polymorphisms of human ABCB1 (MDR1), ABCC (MRP), and ABCG2 (BCRP) efflux transporters. *Int J Toxicol* **25**:231–259.

Chu XY, Suzuki H, Ueda K, Kato Y, Akiyama S and Sugiyama Y (1999) Active efflux of CPT-11 and its metabolites in human KB-derived cell lines. *J Pharmacol Exp Ther* **288**: 735–741.

Cole SP (1990) Patterns of cross-resistance in a multidrug-resistant small-cell lung carcinoma cell line. *Cancer Chemother Pharmacol* **26**:250–256.

Cooper CL, van Heeswijk RP, Gallicano K and Cameron DW (2003) A review of low-dose ritonavir in protease inhibitor combination therapy. *Clin Infect Dis* **36**:1585–1592.

Cordon-Cardo C, O'Brien JP, Casals D, Rittman-Grauer L, Biedler JL, Melamed MR and Bertino JR (1989) Multidrug-resistance gene (P-glycoprotein) is expressed by endothelial cells at blood-brain barrier sites. *Proc Natl Acad Sci U S A* **86**:695–698.

Cundy KC (1999) Clinical pharmacokinetics of the antiviral nucleotide analogues cidofovir and adefovir. *Clin Pharmacokinet* **36**:127–143.

Dallas S, Miller DS and Bendayan R (2006) Multidrug resistance-associated proteins: expression and function in the central nervous system. *Pharmacol Rev* **58**:140–161.

Davit-Spraul A, Gonzales E, Baussan C and Jacquemin E (2009) Progressive familial intrahepatic cholestasis. *Orphanet J Rare Dis* **4**:1.

de Boer AG, van der Sandt ICJ and Gaillard PJ (2003) The role of drug transporters at the blood-brain barrier. *Annu Rev Pharmacol Toxicol* **43**:629–656.

Ding R, Tayrouz Y, Riedel KD, Burhenne J, Weiss J, Mikus G and Haefeli WE (2004) Substantial pharmacokinetic interaction between digoxin and ritonavir in healthy volunteers. *Clin Pharmacol Ther* **76**:73–84.

Drayer DE, Romankiewicz J, Lorenzo B and Reidenberg MM (1982) Age and renal clearance of cimetidine. *Clin Pharmacol Ther* **31**:45–50.

Durr D, Stieger B, Kullak-Ublick GA, Rentsch KM, Steinert HC, Meier PJ and Fattinger K (2000) St John's Wort induces intestinal P-glycoprotein/MDR1 and intestinal and hepatic CYP3A4. *Clin Pharmacol Ther* **68**:598–604.

Eilers M, Roy U and Mondal D (2008) MRP (ABCC) transporters-mediated efflux of anti-HIV drugs, saquinavir and zidovudine, from human endothelial cells. *Exp Biol Med (Maywood)* **233**:1149–1160.

Ekins S, Durst GL, Stratford RE, Thorner DA, Lewis R, Loncharich RJ and Wikel JH (2001) Three-dimensional quantitative structure-permeability relationship analysis for a series of inhibitors of rhinovirus replication. *J Chem Inf Comput Sci* **41**:1578–1586.

Endres CJ, Hsiao P, Chung FS and Unadkat JD (2006) The role of transporters in drug interactions. *Eur J Pharm Sci* **27**:501–517.

Eyal S, Hsiao P and Unadkat JD (2009) Drug interactions at the blood-brain barrier: fact or fantasy? *Pharmacol Ther* **123**(1):80-104

Fattinger K, Funk C, Pantze M, Weber C, Reichen J, Stieger B and Meier PJ (2001) The endothelin antagonist bosentan inhibits the canalicular bile salt export pump: a potential mechanism for hepatic adverse reactions. *Clin Pharmacol Ther* **69**:223–231.

Feng B, Obach RS, Burstein AH, Clark DJ, de Morais SM and Faessel HM (2008) Effect of human renal cationic transporter inhibition on the pharmacokinetics of varenicline, a new therapy for smoking cessation: an in vitro-in vivo study. *Clin Pharmacol Ther* **83**: 567–576.

Fenner KS, Troutman MD, Kempshall S, Cook JA, Ware JA, Smith DA and Lee CA (2009) Drug–drug interactions mediated through P-glycoprotein: clinical relevance and in vitro-in vivo correlation using digoxin as a probe drug. *Clin Pharmacol Ther* **85**: 173–181.

Funk C, Ponelle C, Scheuermann G and Pantze M (2001) Cholestatic potential of troglitazone as a possible factor contributing to troglitazone-induced hepatotoxicity: in vivo and in vitro interaction at the canalicular bile salt export pump (Bsep) in the rat. *Mol Pharmacol* **59**: 627–635.

Geick A, Eichelbaum M and Burk O (2001) Nuclear receptor response elements mediate induction of intestinal MDR1 by rifampin. *J Biol Chem* **276**:14581–14587.

Gerloff T, Stieger B, Hagenbuch B, Madon J, Landmann L, Roth J, Hofmann AF and Meier PJ (1998) The sister of P-glycoprotein represents the canalicular bile salt export pump of mammalian liver. *J Biol Chem* **273**:10046–10050.

Glaeser H, Bailey DG, Dresser GK, Gregor JC, Schwarz UI, McGrath JS, Jolicoeur E, Lee W, Leake BF, Tirona RG and Kim RB (2007) Intestinal drug transporter expression and the impact of grapefruit juice in humans. *Clin Pharmacol Ther* **81**:362–370.

Govindarajan R (2002) Irinotecan/thalidomide in metastatic colorectal cancer. *Oncology (Williston Park)* **16**:23–26.

Greiner B, Eichelbaum M, Fritz P, Kreichgauer HP, von RO, Zundler J and Kroemer HK (1999) The role of intestinal P-glycoprotein in the interaction of digoxin and rifampin. *J Clin Invest* **104**:147–153.

Hagenbuch B and Meier PJ (2004) Organic anion transporting polypeptides of the OATP/ SLC21 family: phylogenetic classification as OATP/ SLCO superfamily, new nomenclature and molecular/functional properties. *Pflugers Arch* **447**:653–665.

Haimeur A, Conseil G, Deeley RG and Cole SP (2004) The MRP-related and BCRP/ABCG2 multidrug resistance proteins: biology, substrate specificity and regulation. *Curr Drug Metab* **5**:21–53.

Hardamn JG, Limbird LE and Gilman AG (2001) *Goodman and Gilman's The pharmacological Basis of Therapeutics*. McGraw-Hill.

He L and Liu GQ (2002) Interaction of multidrug resistance reversal agents with P-glycoprotein ATPase activity on blood-brain barrier. *Acta Pharmacol Sin* **23**:423–429.

Hedman A, Angelin B, Arvidsson A, Beck O, Dahlqvist R, Nilsson B, Olsson M and Schenck-Gustafsson K (1991) Digoxin-verapamil interaction: reduction of biliary but not renal digoxin clearance in humans. *Clin Pharmacol Ther* **49**:256–262.

Hedman A, Angelin B, Arvidsson A, Dahlqvist R and Nilsson B (1990) Interactions in the renal and biliary elimination of digoxin: stereoselective difference between quinine and quinidine. *Clin Pharmacol Ther* **47**:20–26.

Herrine SK and Choudhary C (1999) Severe hepatotoxicity associated with troglitazone. *Ann Intern Med* **130**:163–164.

Hinderling PH and Hartmann D (1991) Pharmacokinetics of digoxin and main metabolites/derivatives in healthy humans. *Ther Drug Monit* **13**:381–401.

Hirano H, Kurata A, Onishi Y, Sakurai A, Saito H, Nakagawa H, Nagakura M, Tarui S, Kanamori Y, Kitajima M and Ishikawa T (2006) High-speed screening and QSAR analysis of human ATP-binding cassette transporter ABCB11 (bile salt export pump) to predict drug-induced intrahepatic cholestasis. *Mol Pharm* **3**:252–265.

Hirano M, Maeda K, Hayashi H, Kusuhara H and Sugiyama Y (2005) Bile salt export pump (BSEP/ABCB11) can transport a nonbile acid substrate, pravastatin. *J Pharmacol Exp Ther* **314**:876–882.

Horikawa M, Kato Y, Tyson CA and Sugiyama Y (2002) The potential for an interaction between MRP2 (ABCC2) and various therapeutic agents: probenecid as a candidate inhibitor of the biliary excretion of irinotecan metabolites. *Drug Metab Pharmacokinet* **17**:23–33.

Hsiang B, Zhu Y, Wang Z, Wu Y, Sasseville V, Yang WP and Kirchgessner TG (1999) A novel human hepatic organic anion transporting polypeptide (OATP2). Identification of a liver-specific human organic anion transporting polypeptide and identification of rat and human hydroxymethylglutaryl-CoA reductase inhibitor transporters. *J Biol Chem* **274**:37161–37168.

Hsiao P, Bui T, Ho RJ and Unadkat JD (2008) In vitro-to-in vivo prediction of P-glycoprotein-based drug interactions at the human and rodent blood-brain barrier. *Drug Metab Dispos* **36**:481–484.

Hsiao P, Sasongko L, Link JM, Mankoff DA, Muzi M, Collier AC and Unadkat JD (2006) Verapamil P-glycoprotein transport across the rat blood-brain barrier: cyclosporine, a concentration inhibition analysis, and comparison with human data. *J Pharmacol Exp Ther* **317**:704–710.

Huang L (2009) *Role of drug transporters in the disposition of statins*. AAPS Workshop on Drug Transporters in ADME: From the Bench to the Bedside, Parsippany, NJ. 2005. Ref Type: Conference Proceeding

Iwanaga T, Nakakariya M, Yabuuchi H, Maeda T and Tamai I (2007) Involvement of bile salt export pump in flutamide-induced cholestatic hepatitis. *Biol Pharm Bull* **30**:739–744.

Jalava KM, Partanen J and Neuvonen PJ (1997) Itraconazole decreases renal clearance of digoxin. *Ther Drug Monit* **19**:609–613.

Jonker JW, Smit JW, Brinkhuis RF, Maliepaard M, Beijnen JH, Schellens JH and Schinkel AH (2000) Role of breast cancer resistance protein in the bioavailability and fetal penetration of topotecan. *J Natl Cancer Inst* **92**:1651–1656.

Juliano RL and Ling V (1976) A surface glycoprotein modulating drug permeability in Chinese hamster ovary cell mutants. *Biochim Biophys Acta* **455**:152–162.

Kajosaari LI, Laitila J, Neuvonen PJ and Backman JT (2005) Metabolism of repaglinide by CYP2C8 and CYP3A4 in vitro: effect of fibrates and rifampicin. *Basic Clin Pharmacol Toxicol* **97**:249–256.

Kruh GD and Belinsky MG (2003) The MRP family of drug efflux pumps. *Oncogene* **22**: 7537–7552.

Kruh GD, Zeng H, Rea PA, Liu G, Chen ZS, Lee K and Belinsky MG (2001) MRP subfamily transporters and resistance to anticancer agents. *J Bioenerg Biomembr* **33**:493–501.

Kruijtzer CM, Beijnen JH, Rosing H, ten Bokkel Huinink WW, Schot M, Jewell RC, Paul EM and Schellens JH (2002) Increased oral bioavailability of topotecan in combination with the breast cancer resistance protein and P-glycoprotein inhibitor GF120918. *J Clin Oncol* **20**: 2943–2950.

Kullak-Ublick GA, Ismair MG, Stieger B, Landmann L, Huber R, Pizzagalli F, Fattinger K, Meier PJ and Hagenbuch B (2001) Organic anion-transporting polypeptide B (OATP-B) and its functional comparison with three other OATPs of human liver. *Gastroenterology* **120**:525–533.

Kusuhara H and Sugiyama Y (2009) In vitro-in vivo extrapolation of transporter-mediated clearance in the liver and kidney. *Drug Metab Pharmacokinet* **24**:37–52.

Kyrklund C, Backman JT, Kivisto KT, Neuvonen M, Laitila J and Neuvonen PJ (2001) Plasma concentrations of active lovastatin acid are markedly increased by gemfibrozil but not by bezafibrate. *Clin Pharmacol Ther* **69**:340–345.

Lacarelle B, Rahmani R, de SG, Durand A, Placidi M and Cano JP (1991) Metabolism of digoxin, digoxigenin digitoxosides and digoxigenin in human hepatocytes and liver microsomes. *Fundam Clin Pharmacol* **5**:567–582.

Lalezari JP, Drew WL, Glutzer E, James C, Miner D, Flaherty J, Fisher PE, Cundy K, Hannigan J and Martin JC (1995) (S)-1-[3-hydroxy-2-(phosphonylmethoxy)propyl]cytosine (cidofovir): results of a phase I/II study of a novel antiviral nucleotide analogue. *J Infect Dis* **171**:788–796.

Lalezari JP, Stagg RJ, Kuppermann BD, Holland GN, Kramer F, Ives DV, Youle M, Robinson MR, Drew WL and Jaffe HS (1997) Intravenous cidofovir for peripheral cytomegalovirus retinitis in patients with AIDS. A randomized, controlled trial. *Ann Intern Med* **126**:257–263.

Lau YY, Huang Y, Frassetto L and Benet LZ (2007) Effect of OATP1B transporter inhibition on the pharmacokinetics of atorvastatin in healthy volunteers. *Clin Pharmacol Ther* **81**:194–204.

Liang R, Fei YJ, Prasad PD, Ramamoorthy S, Han H, Yang-Feng TL, Hediger MA, Ganapathy V and Leibach FH (1995) Human intestinal H+/peptide cotransporter. Cloning, functional expression, and chromosomal localization. *J Biol Chem* **270**:6456–6463.

Litman T, Brangi M, Hudson E, Fetsch P, Abati A, Ross DD, Miyake K, Resau JH and Bates SE (2000) The multidrug-resistant phenotype associated with overexpression of the new ABC half-transporter, MXR (ABCG2). *J Cell Sci* **113 (Pt 11)**:2011-2021.

Loi CM, Young M, Randinitis E, Vassos A and Koup JR (1999) Clinical pharmacokinetics of troglitazone. *Clin Pharmacokinet* **37**:91–104.

Maliepaard M, van Gastelen MA, de Jong LA, Pluim D, van Waardenburg RC, Ruevekamp-Helmers MC, Floot BG and Schellens JH (1999) Overexpression of the BCRP/MXR/ABCP gene in a topotecan-selected ovarian tumor cell line. *Cancer Res* **59**:4559–4563.

Marzolini C and Kim RB (2005) Placental transfer of antiretroviral drugs. *Clin Pharmacol Ther* **78**:118–122.

Mathias AA, Hitti J and Unadkat JD (2005) P-glycoprotein and breast cancer resistance protein expression in human placentae of various gestational ages. *Am J Physiol Regul Integr Comp Physiol* **289**:R963–R969.

Matsushima S, Maeda K, Inoue K, Ohta KY, Yuasa H, Kondo T, Nakayama H, Horita S, Kusuhara H and Sugiyama Y (2009) The inhibition of human multidrug and toxin extrusion 1 is involved in the drug–drug interaction caused by cimetidine. *Drug Metab Dispos* **37**:555–559.

McTaggart F, Buckett L, Davidson R, Holdgate G, McCormick A, Schneck D, Smith G and Warwick M (2001) Preclinical and clinical pharmacology of Rosuvastatin, a new 3-hydroxy-3-methylglutaryl coenzyme A reductase inhibitor. *Am J Cardiol* **87**:28B–32B.

Mikkaichi T, Suzuki T, Onogawa T, Tanemoto M, Mizutamari H, Okada M, Chaki T, Masuda S, Tokui T, Eto N, Abe M, Satoh F, Unno M, Hishinuma T, Inui K, Ito S, Goto J and Abe T (2004a) Isolation and characterization of a digoxin transporter and its rat homologue expressed in the kidney. *Proc Natl Acad Sci U S A* **101**:3569–3574.

Mikkaichi T, Suzuki T, Tanemoto M, Ito S and Abe T (2004b) The organic anion transporter (OATP) family. *Drug Metab Pharmacokinet* **19**:171–179.

Mirski SE, Gerlach JH and Cole SP (1987) Multidrug resistance in a human small cell lung cancer cell line selected in adriamycin. *Cancer Res* **47**:2594–2598.

Mita S, Suzuki H, Akita H, Hayashi H, Onuki R, Hofmann AF and Sugiyama Y (2006) Inhibition of bile acid transport across Na^+/taurocholate cotransporting polypeptide (SLC10A1) and bile salt export pump (ABCB 11)-coexpressing LLC-PK1 cells by cholestasis-inducing drugs. *Drug Metab Dispos* **34**:1575–1581.

Molsa M, Heikkinen T, Hakkola J, Hakala K, Wallerman O, Wadelius M, Wadelius C and Laine K (2005) Functional role of P-glycoprotein in the human blood-placental barrier. *Clin Pharmacol Ther* **78**:123–131.

Nagashige M, Ushigome F, Koyabu N, Hirata K, Kawabuchi M, Hirakawa T, Satoh S, Tsukimori K, Nakano H, Uchiumi T, Kuwano M, Ohtani H and Sawada Y (2003) Basal membrane localization of MRP1 in human placental trophoblast. *Placenta* **24**:951–958.

Niemi M, Backman JT, Fromm MF, Neuvonen PJ and Kivisto KT (2003) Pharmacokinetic interactions with rifampicin: clinical relevance. *Clin Pharmacokinet* **42**:819–850.

Otsuka M, Matsumoto T, Morimoto R, Arioka S, Omote H and Moriyama Y (2005) A human transporter protein that mediates the final excretion step for toxic organic cations. *Proc Natl Acad Sci U S A* **102**:17923–17928.

Oude Elferink RP, Meijer DK, Kuipers F, Jansen PL, Groen AK and Groothuis GM (1995) Hepatobiliary secretion of organic compounds; molecular mechanisms of membrane transport. *Biochim Biophys Acta* **1241**:215–268.

Park JW, Siekmeier R, Lattke P, Merz M, Mix C, Schuler S and Jaross W (2001) Pharmacokinetics and pharmacodynamics of fluvastatin in heart transplant recipients taking cyclosporine A. *J Cardiovasc Pharmacol Ther* **6**:351–361.

Park S and Sinko PJ (2005) P-glycoprotein and multidrug resistance-associated proteins limit the brain uptake of saquinavir in mice. *J Pharmacol Exp Ther* **312**:1249–1256.

Paulusma CC, Kool M, Bosma PJ, Scheffer GL, ter BF, Scheper RJ, Tytgat GN, Borst P, Baas F and Oude Elferink RP (1997) A mutation in the human canalicular multispecific organic anion transporter gene causes the Dubin-Johnson syndrome. *Hepatology* **25**: 1539–1542.

Pedersen KE, Christiansen BD, Klitgaard NA and Nielsen-Kudsk F (1983) Effect of quinidine on digoxin bioavailability. *Eur J Clin Pharmacol* **24**:41–47.

Penzak SR, Shen JM, Alfaro RM, Remaley AT, Natarajan V and Falloon J (2004) Ritonavir decreases the nonrenal clearance of digoxin in healthy volunteers with known MDR1 genotypes. *Ther Drug Monit* **26**:322–330.

Reidenberg MM, Camacho M, Kluger J and Drayer DE (1980) Aging and renal clearance of procainamide and acetylprocainamide. *Clin Pharmacol Ther* **28**:732–735.

Roberts DH, Kendall MJ, Jack DB and Welling PG (1981) Pharmacokinetics of cephradine given intravenously with and without probenecid. *Br J Clin Pharmacol* **11**:561–564.

Robey RW, Medina-Perez WY, Nishiyama K, Lahusen T, Miyake K, Litman T, Senderowicz AM, Ross DD and Bates SE (2001) Overexpression of the ATP-binding cassette half-transporter,

ABCG2 (Mxr/BCrp/ABCP1), in flavopiridol-resistant human breast cancer cells. *Clin Cancer Res* **7**:145–152.

Russel FG, Masereeuw R and van Aubel RA (2002) Molecular aspects of renal anionic drug transport. *Annu Rev Physiol* **64**:563–594.

Sasaki M, Suzuki H, Ito K, Abe T and Sugiyama Y (2002) Transcellular transport of organic anions across a double-transfected Madin-Darby canine kidney II cell monolayer expressing both human organic anion-transporting polypeptide (OATP2/SLC21A6) and Multidrug resistance-associated protein 2 (MRP2/ABCC2). *J Biol Chem* **277**:6497–6503.

Sasongko L, Link JM, Muzi M, Mankoff DA, Yang X, Collier AC, Shoner SC and Unadkat JD (2005) Imaging P-glycoprotein transport activity at the human blood-brain barrier with positron emission tomography. *Clin Pharmacol Ther* **77**:503–514.

Schinkel AH, Mayer U, Wagenaar E, Mol CA, van Deemter L, Smit JJ, van der Walk MA, Voordouw AC, Spits H, van Tellingen O, Zijlmans JM, Fibbe WE and Borst P (1997) Normal viability and altered pharmacokinetics in mice lacking mdr1-type (drug-transporting) P-glycoproteins. *Proc Natl Acad Sci U S A* **94**:4028–4033.

Schinkel AH, Mol CA, Wagenaar E, van DL, Smit JJ and Borst P (1995) Multidrug resistance and the role of P-glycoprotein knockout mice. *Eur J Cancer* **31A**:1295–1298.

Schinkel AH, Smit JJ, Van TO, Beijnen JH, Wagenaar E, van DL, Mol CA, van **der** valk MA, Robanus-Maandag EC, Teriele HP et al. (1994) Disruption of the mouse mdr1a P-glycoprotein gene leads to a deficiency in the blood-brain barrier and to increased sensitivity to drugs. *Cell* **77**:491–502.

Schneck DW, Birmingham BK, Zalikowski JA, Mitchell PD, Wang Y, Martin PD, Lasseter KC, Brown CD, Windass AS and Raza A (2004) The effect of gemfibrozil on the pharmacokinetics of rosuvastatin. *Clin Pharmacol Ther* **75**:455–463.

Shikata E, Yamamoto R, Takane H, Shigemasa C, Ikeda T, Otsubo K and Ieiri I (2007) Human organic cation transporter (OCT1 and OCT2) gene polymorphisms and therapeutic effects of metformin. *J Hum Genet* **52**:117–122.

Shu Y, Brown C, Castro RA, Shi RJ, Lin ET, Owen RP, Sheardown SA, Yue L, Burchard EG, Brett CM and Giacomini KM (2008) Effect of genetic variation in the organic cation transporter 1, OCT1, on metformin pharmacokinetics. *Clin Pharmacol Ther* **83**:273–280.

Shu Y, Sheardown SA, Brown C, Owen RP, Zhang S, Castro RA, Ianculescu AG, Yue L, Lo JC, Burchard EG, Brett CM and Giacomini KM (2007) Effect of genetic variation in the organic cation transporter 1 (OCT1) on metformin action. *J Clin Invest* **117**:1422–1431.

Simonson SG, Raza A, Martin PD, Mitchell PD, Jarcho JA, Brown CD, Windass AS and Schneck DW (2004) Rosuvastatin pharmacokinetics in heart transplant recipients administered an antirejection regimen including cyclosporine. *Clin Pharmacol Ther* **76**:167–177.

Soldner A, Christians U, Susanto M, Wacher VJ, Silverman JA and Benet LZ (1999) Grapefruit juice activates P-glycoprotein-mediated drug transport. *Pharm Res* **16**:478–485.

Somogyi A, McLean A and Heinzow B (1983) Cimetidine-procainamide pharmacokinetic interaction in man: evidence of competition for tubular secretion of basic drugs. *Eur J Clin Pharmacol* **25**:339–345.

Somogyi A, Rohner HG and Gugler R (1980) Pharmacokinetics and bioavailability of cimetidine in gastric and duodenal ulcer patients. *Clin Pharmacokinet* **5**:84–94.

Somogyi A, Stockley C, Keal J, Rolan P and Bochner F (1987) Reduction of metformin renal tubular secretion by cimetidine in man. *Br J Clin Pharmacol* **23**:545–551.

Song IS, Shin HJ, Shim EJ, Jung IS, Kim WY, Shon JH and Shin JG (2008a) Genetic variants of the organic cation transporter 2 influence the disposition of metformin. *Clin Pharmacol Ther* **84**:559–562.

Song IS, Shin HJ and Shin JG (2008b) Genetic variants of organic cation transporter 2 (OCT2) significantly reduce metformin uptake in oocytes. *Xenobiotica* **38**:1252–1262.

Srimaroeng C, Perry JL and Pritchard JB (2008) Physiology, structure, and regulation of the cloned organic anion transporters. *Xenobiotica* **38**:889–935.

Sugawara I, Kataoka I, Morishita Y, Hamada H, Tsuruo T, Itoyama S and Mori S (1988) Tissue distribution of P-glycoprotein encoded by a multidrug-resistant gene as revealed by a monoclonal antibody, MRK 16. *Cancer Res* **48**:1926–1929.

Tamai I, Nezu J, Uchino H, Sai Y, Oku A, Shimane M and Tsuji A (2000) Molecular identification and characterization of novel members of the human organic anion transporter (OATP) family. *Biochem Biophys Res Commun* **273**:251–260.

Templeton IE, Thummel KE, Kharasch ED, Kunze KL, Hoffer C, Nelson WL and Isoherranen N (2008) Contribution of itraconazole metabolites to inhibition of CYP3A4 in vivo. *Clin Pharmacol Ther* **83**:77–85.

Terada T, Saito H and Inui K (1998) Interaction of beta-lactam antibiotics with histidine residue of rat H+/peptide cotransporters, PEPT1 and PEPT2. *J Biol Chem* **273**:5582–5585.

Thiebaut F, Tsuruo T, Hamada H, Gottesman MM, Pastan I and Willingham MC (1987) Cellular localization of the multidrug-resistance gene product P-glycoprotein in normal human tissues. *Proc Natl Acad Sci U S A* **84**:7735–7738.

Thiebaut F, Tsuruo T, Hamada H, Gottesman MM, Pastan I and Willingham MC (1989) Immunohistochemical localization in normal tissues of different epitopes in the multidrug transport protein P170: evidence for localization in brain capillaries and crossreactivity of one antibody with a muscle protein. *J Histochem Cytochem* **37**:159–164.

Thummel KE, Shen DD, Soherranen N and Mith HE (2006) Design and optimization of dosage regimens; pharmacokinetic data, in *Goodman & Gilman's the Pharmacological Basis of Therapeutics* (Goodman LS, Gilman A, Brunton LL, Lazo JS and Parker KL eds) pp 1787–1887, McGraw-Hill, New York.

Tirona RG, Leake BF, Wolkoff AW and Kim RB (2003) Human organic anion transporting polypeptide-C (SLC21A6) is a major determinant of rifampin-mediated pregnane X receptor activation. *J Pharmacol Exp Ther* **304**:223–228.

Treiber A, Schneiter R, Hausler S and Stieger B (2007) Bosentan is a substrate of human OATP1B1 and OATP1B3: inhibition of hepatic uptake as the common mechanism of its interactions with cyclosporin A, rifampicin, and sildenafil. *Drug Metab Dispos* **35**:1400–1407.

Tsuda M, Terada T, Ueba M, Sato T, Masuda S, Katsura T and Inui K (2009) Involvement of human multidrug and toxin extrusion 1 in the drug interaction between cimetidine and metformin in renal epithelial cells. *J Pharmacol Exp Ther* **329**:185–191.

Unadkat JD, Dahlin A and Vijay S (2004) Placental drug transporters. *Curr Drug Metab* **5**: 125–131.

Uwai Y, Ida H, Tsuji Y, Katsura T and Inui K (2007) Renal transport of adefovir, cidofovir, and tenofovir by SLC22A family members (hOAT1, hOAT3, and hOCT2). *Pharm Res* **24**: 811–815.

van Giersbergen PL, Halabi A and Dingemanse J (2002) Single- and multiple-dose pharmacokinetics of bosentan and its interaction with ketoconazole. *Br J Clin Pharmacol* **53**: 589–595.

van Giersbergen PL, Popescu G, Bodin F and Dingemanse J (2003) Influence of mild liver impairment on the pharmacokinetics and metabolism of bosentan, a dual endothelin receptor antagonist. *J Clin Pharmacol* **43**:15–22.

van Giersbergen PL, Treiber A, Schneiter R, Dietrich H and Dingemanse J (2007) Inhibitory and inductive effects of rifampin on the pharmacokinetics of bosentan in healthy subjects. *Clin Pharmacol Ther* **81**:414–419.

Wandel C, Kim RB, Kajiji S, Guengerich FP, Wilkinson GR and Wood AJ (1999) P-glycoprotein and cytochrome P-450 3A inhibition: dissociation of inhibitory potencies. *Cancer Res* **59**:3944–3948.

Wang JS, Neuvonen M, Wen X, Backman JT and Neuvonen PJ (2002) Gemfibrozil inhibits CYP2C8-mediated cerivastatin metabolism in human liver microsomes. *Drug Metab Dispos* **30**:1352–1356.

Wang ZJ, Yin OQ, Tomlinson B and Chow MS (2008) OCT2 polymorphisms and in-vivo renal functional consequence: studies with metformin and cimetidine. *Pharmacogenet Genom* **18**:637–645.

Weiner IM, Washington JA and Mudge GH (1960) On the mechanism of action of probenecid on renal tubular secretion. *Bull Johns Hopkins Hosp* **106**:333–346.

Wen X, Wang JS, Backman JT, Kivisto KT and Neuvonen PJ (2001) Gemfibrozil is a potent inhibitor of human cytochrome P450 2C9. *Drug Metab Dispos* **29**:1359–1361.

Westphal K, Weinbrenner A, Giessmann T, Stuhr M, Franke G, Zschiesche M, Oertel R, Terhaag B, Kroemer HK and Siegmund W (2000) Oral bioavailability of digoxin is enhanced by talinolol: evidence for involvement of intestinal P-glycoprotein. *Clin Pharmacol Ther* **68**:6–12.

Woodland C, Ito S and Koren G (1998) A model for the prediction of digoxin-drug interactions at the renal tubular cell level. *Ther Drug Monit* **20**:134–138.

Wright SH and Dantzler WH (2004) Molecular and cellular physiology of renal organic cation and anion transport. *Physiol Rev* **84**:987–1049.

Yang H and Elmquist WF (1996) The binding of cyclosporin A to human plasma: an in vitro microdialysis study. *Pharm Res* **13**:622–627.

Yang XX, Hu ZP, Chan SY, Duan W, Ho PC, Boelsterli UA, Ng KY, Chan E, Bian JS, Chen YZ, Huang M and Zhou SF (2006a) Pharmacokinetic mechanisms for reduced toxicity of irinotecan by coadministered thalidomide. *Curr Drug Metab* **7**:431–455.

Yang XX, Hu ZP, Xu AL, Duan W, Zhu YZ, Huang M, Sheu FS, Zhang Q, Bian JS, Chan E, Li X, Wang JC and Zhou SF (2006b) A mechanistic study on reduced toxicity of irinotecan by coadministered thalidomide, a tumor necrosis factor-alpha inhibitor. *J Pharmacol Exp Ther* **319**:82–104.

Yasui-Furukori N, Uno T, Sugawara K and Tateishi T (2005) Different effects of three transporting inhibitors, verapamil, cimetidine, and probenecid, on fexofenadine pharmacokinetics. *Clin Pharmacol Ther* **77**:17–23.

Young LC, Campling BG, Voskoglou-Nomikos T, Cole SP, Deeley RG and Gerlach JH (1999) Expression of multidrug resistance protein-related genes in lung cancer: correlation with drug response. *Clin Cancer Res* **5**:673–680.

Zhang L, Zhang YD, Strong JM, Reynolds KS and Huang SM (2008) A regulatory viewpoint on transporter-based drug interactions. *Xenobiotica* **38**:709–724.

Chapter 22
Herbal Supplement-Based Interactions

Guohua An and Marilyn E. Morris

Abstract As herbal supplements are purchased and used by more than 25% of the general population, herbal supplement-based interactions have become increasingly reported. The primary mechanism of herb–drug interactions is modulation of metabolizing enzymes and/or transporters. This chapter highlights the modulation effects of several top selling herbal supplements (St. John's wort, garlic, ginseng, milk thistle, ginkgo, and others) and important herbal constituents (flavonoids and isothiocyanates) on metabolizing enzymes/transporters and the subsequent herb–drug interactions. Literature reports indicate that these herbal supplements/constituents modulate various phase I/II enzymes and/or transporters at both functional and expression levels and have an impact on the pharmacokinetics of co-administered drugs, which are mainly metabolized/transported by these enzymes/transporters. The alteration in enzyme/transporter function by an herbal product may vary markedly due to different manufacturers or lots of herbal products, doses, drug substrates, and experimental systems used to evaluate potential interactions. Many effects of herbal supplements/constituents on enzymes/transporters have been investigated only in vitro. Further in vivo animal studies and clinical trials need to be carried out to confirm these in vitro observations and determine the clinical relevance of herbal–drug interactions.

Abbreviations

AhR	aryl hydrocarbon receptor
BITC	benzyl isothiocyanate
BCRP	breast cancer resistance protein
CAR	constitutive androstane receptor
CYP	cytochrome P450
DNM	daunomycin

M.E. Morris (✉)
Department of Pharmaceutical Sciences, School of Pharmacy and Pharmaceutical Sciences, University at Buffalo, State University of New York, Amherst, NY, USA
e-mail: memorris@buffalo.edu

K.S. Pang et al. (eds.), *Enzyme- and Transporter-Based Drug–Drug Interactions*,
DOI 10.1007/978-1-4419-0840-7_22,

ECG	epicatechin gallate
EGCG	epigallocatechin gallate
GBE	*Ginkgo biloba* extract
GSH	glutathione
GST	glutathione *S*-transferase
INR	International Normalized Ratio
ITC	isothiocyanates
MCT	monocarboxylic transporter
MRP	multidrug resistance protein
Nrf2	E2-related factor2
OATP	organic anion-transporting polypeptide
OSCs	organosulfur compounds
PEITC	phenethyl isothiocyanate
P-gp	P-glycoprotein
PXR	pregnane X receptor
QR	quinone reductase
SFN	sulforaphane
SJW	St. John's wort
SULT	sulfotransferase
UGT	UDP-glucuronosyltransferase

22.1 Introduction

The consumption of herbal supplements has increased rapidly over the last decade and herbal supplements are often administered in combination with conventional therapeutic drugs. Approximately 25% of Americans taking prescription medications also take a dietary supplement (Kaufman et al., 2002). Potential herb–drug interactions may occur and the incidence of herb–drug interactions may even be higher than drug–drug interactions because each herbal supplement, unlike marketed drug products, usually is a mixture of a large number of individual active components. Clinically significant herb–drug interactions are not uncommon. For example, St. John's wort (SJW) has been observed to interfere with numerous therapeutic compounds (such as digoxin, cyclosporine, indinavir, irinotecan), resulting in therapeutic failure (Johne et al., 1999; Piscitelli et al., 2000; Ruschitzka et al., 2000; Hu et al., 2005). Additional investigations have revealed that SJW influenced the pharmacokinetics (PK) of these drugs mainly through CYP3A4 and/or P-gp induction (Durr et al., 2000). To inform the public of the risk of SJW–drug interactions, the FDA issued a public health advisory in 2000. In addition to SJW, herb–drug interactions have also been observed for various other herbal products, including garlic, kava, and ginkgo (Izzo and Ernst, 2001).

The primary mechanism of herb–drug interactions is modulation of metabolizing enzymes and/or transporters. To elucidate the underlying mechanisms and prevent unwanted herb–drug interactions, numerous studies have been conducted recently to investigate the effects of various herbal products and/or their active ingredients on phase I/II enzymes and transporters. The information on interactions of herbal supplements with metabolizing enzymes/transporters indicates that herbal supplements may be inhibitors, inducers, or substrates of metabolizing enzymes and/or transporters and have an impact on the PK of co-administered drugs which are mainly metabolized /transported by these enzymes/transporters. This chapter will highlight the modulation effects of several top selling herbal supplements (SJW, garlic, ginseng, milk thistle, ginkgo, and others) on metabolizing enzymes/transporters and the subsequent herb–drug interactions. Flavonoids and organic isothiocyanates, which are not only widely present in human diet but important components of various herbal products, also will be discussed in detail. The pharmacodynamic interactions between herbal supplements and therapeutic compounds, another type of herb–drug interaction observed for a few herbal products such as ginseng and kava, are beyond the scope of this chapter.

22.2 Metabolic Enzyme–Transporter Interactions

22.2.1 St. John's Wort

St. John's wort (SJW), an extract of the medicinal plant *Hypericum perforatum*, is the most prescribed herbal drug for the treatment of mild to moderate depression. SJW contains more than two dozen constituents among which hypericin and hyperforin have been identified as the possible active constituents for its antidepressive activities. With SJW's widespread use in the United States and Europe, there have been increasing numbers of case reports and clinical trials which have documented clinically relevant drug interactions between SJW and co-administered drugs. Concomitant intake of hypericum extract 900 mg/day for 2 weeks significantly increased the oral clearance of midazolam by 108.9% with >50% decrease in the AUC (Wang et al., 2001). Long-term administration of hypericum also decreased the AUC of amitriptyline by 23% and nortriptyline by 41% (Johne et al., 2002). In a study carried out by Breidenbach et al. (Breidenbach et al., 2000), the blood concentration of cyclosporine dropped by 49% with the ingestion of hypericum. Transplant graft rejection due to reduced cyclosporine exposure was observed in several case reports when those patients co-administered SJW (Breidenbach et al., 2000; Ruschitzka et al., 2000). Similar interactions have also been documented for SJW and imatinib, indinavir, irinotecan, simvastatin, quazepam, and oral contraceptives (Piscitelli et al., 2000; Ruschitzka et al., 2000; Sugimoto et al., 2001; Kawaguchi et al., 2004; Hu et al., 2005). Several similarities were noticed between these reports. In most cases plasma concentration of the concomitant medication was reduced after long-term SJW usage and most of these co-administered drugs are metabolized mainly, or at least partly, by the cytochrome P450 (CYP) 3A4 isoform. These findings indicated that SJW might interact with those drugs through CYP3A4 induction.

The direct evidence of CYP3A induction by SJW has been provided by several in vitro and in vivo studies. Following 3 weeks of SJW treatment at moderate doses (140 and 280 mg/kg) to male mice, SJW at both doses resulted in a 2-fold increase in CYP3A catalytic activity and a 6-fold increase in protein levels (Bray et al., 2002). After chronic exposure of human hepatocytes to hyperforin or hypericin, hyperforin but not hypericin treatment resulted in significant increases in mRNA, protein, and activity of CYP3A4 (Komoroski et al., 2004). Significant induction of CYP3A by SJW was also observed after treatment with SJW extract (900 mg/day) for 14 days in 16 healthy volunteers (Wenk et al., 2004). A trend for a sex difference in CYP3A inducibility was also found in the same study, although this was not statistically significant. In contrast to the significant CYP3A induction observed after long-term dosing (>2 weeks) of SJW, there is no evidence that SJW alters CYP3A activity after short-term administration. SJW did not influence midazolam pharmacokinetics when administered to healthy volunteers (900 mg/day for 2 days) (Wang et al., 2001) and it also failed to change the metabolism of alprazolam (CYP3A4 probe drug) when 900 mg/day was administered for 3 days (Markowitz et al., 2000). It appears that repeated dosing for extended periods is necessary before changes in CYP3A activity and expression are observed. The acute effect of SJW on CYP3A4 activity was also investigated in a number of in vitro studies. The study conducted by Obach using rCYP3A4 microsomes found that SJW can potently inhibit CYP3A4 after a 10-min incubation period (Obach, 2000). Another study also showed that after acute administration of hyperforin at 5 and 10 μM for 1 h along with the CYP3A4 probe substrate testosterone, the formation rate of the metabolite 6β-hydroxytestosterone was significantly decreased (Komoroski et al., 2004). Since the inhibitory effects of SJW on CYP3A4 were all observed in vitro using very short incubation times (10 min to 1 h), extrapolation to in vivo is not possible at this time. The pregnane X receptor (PXR, also known as SXR or PAR) serves as a key regulator of CYP3A4 transcription, and thus SJW might increase CYP3A4 expression via PXR activation. This hypothesis was confirmed by studies conducted by Moore et al. (2000) and Wentworth et al. (2000). In Moore's study, treatment of primary human hepatocytes with hyperforin resulted in a marked induction of CYP3A4 expression and the results from a PXR transfection assay clearly showed that hyperforin is a potent ligand for the PXR. The concentrations of hyperforin required to activate PXR are well below those that are achieved in human plasma (~380 nM) in individuals taking the standard regimen of SJW (900 mg/day). Wentworth et al. showed that hyperforin, but not hypericin, mediates PXR activation and that the induction was dose dependent.

In addition to CYP3A4, the effect of SJW on the other CYP isoforms has also been investigated. In a case report a substantially decreased theophylline plasma concentration was observed in a female patient after ingestion of SJW (Nebel et al., 1999). Since theophylline is metabolized mainly by CYP1A2, it is plausible that SJW might interact with theophylline through CYP1A2 induction. The expression of CYP1A2 in the liver was profoundly increased by 357% after 10 days of administration of SJW in an animal study (Shibayama et al., 2004). Elevation of CYP1A2 protein expression was also observed when incubated with SJW in human intestinal LS180 cells, in both a time- and concentration-dependent manner (Karyekar et al., 2002). On the other hand, among three clinical trials which have similar study designs and all used caffeine as the probe drug, a slight but significant increase in CYP1A2 activity after long-term usage of SJW was found only for females in one study (Wenk et al., 2004) and no change of CYP1A2 activity was observed in the other two studies (Wang et al., 2001; Gurley et al., 2005a). CYP1A2 contributes to ~15% of hepatic CYP content in humans and is involved in the metabolism of many therapeutic compounds, including clozapine, olanzapine, propranolol, and clomipramine. Therefore, although CYP1A2 induction by SJW has been less consistently reported than for CYP3A4, there is a need for caution with the concomitant use of SJW with CYP1A2 substrates.

Other CYP enzymes have been reported to be inhibited or induced by SJW. Case reports of increased clearance of warfarin (partially metabolized by CYP2C9) following hypericum treatment suggested that CYP2C9 might be involved in the drug interactions (Yue et al., 2000). However, the effect of SJW on CYP2C9 has not been consistent and different results were observed in different studies. Hyperforin treatment resulted in significant increase in mRNA, protein, and activity of CYP2C9 in a human hepatocytes (Komoroski et al., 2004). Potent inhibition was observed when hyperforin was added to rCYP2C9 microsomes (Obach, 2000). In a clinical trial, neither short-term nor long-term administration of SJW altered the CYP2C9 activities in 12 volunteers (Wang et al., 2001). The effect of SJW on CYP2D6 was also inconsistent in that certain data suggested hyperforin can inhibit CYP2D6 (Obach, 2000) while other studies indicated that it has little impact on CYP2D6 (Markowitz et al., 2000; Wang et al., 2001; Wenk et al., 2004). The different probes used to access CYP2D6 activity may contribute to this difference: studies with a significant interaction used the substrate bufuralol while dextromethorphan was used in all studies with no effect. SJW was reported to cause an increase in the catalytic activity and protein expression of CYP2E1 after 3 weeks of administration in mice (Bray et al., 2002). Significant induction of CYP2E1 activity (~28%) was also found in elderly subjects following 4 weeks of SJW administration (Gurley et al., 2005a). Unlike CYP3A4, CYP2E1 induction is not mediated by PXR, but may occur through other mechanisms.

The modulation effect of SJW on P-glycoprotein (ABCB1, P-gp) has been reported in a number of publications. Ten days of treatment with SJW decreased digoxin exposure in healthy volunteers by 25% (Johne et al., 1999). Long-term administration of SJW caused a 35% decrease in the $C_{\max}$ of fexofenadine in a clinical trial (Wang et al., 2002). These drug interactions cannot be explained by CYP3A4 induction because both digoxin and fexofenadine are good substrates of P-gp and oxidative hepatic metabolism plays only a minor role in their elimination. CYP3A4 induction also may not represent the only mechanism for the decrease in cyclosporine and indinavir concentrations when co-administered with SJW, since these two compounds are also P-gp substrates. It is possible that some of the previous interactions involve, at least in part, P-gp induction. Strongly supportive evidence of P-gp induction by SJW was provided by both in vitro and in vivo studies. SJW and hyperforin, but not hypericin, greatly increased the P-gp protein expression in LS180 cells (Tian et al., 2005). Chronic treatment with SJW produced a 4.2-fold increase in P-gp expression in peripheral blood lymphocytes of healthy volunteers (Hennessy et al., 2002). A direct inducing effect of SJW on P-gp protein expression was also observed in both rats and humans (Durr et al., 2000).

There have been fewer studies investigating the effect of SJW on phase II enzymes and on other transporters besides P-gp. When SJW was administered at a dose of 400 mg/kg/day in rats for 10 days, the multidrug resistance protein 2 (MRP2) and glutathione-*S*-transferase-P (GST-P) in the liver were increased by 304 and 252%, respectively (Shibayama et al., 2004). SJW caused a significant decrease of MRP1 and MRP2 in the fetuses and an increase of MRP2 in the mothers when SJW was administered prenatally and during breastfeeding in rats (Garrovo et al., 2006). The effect of SJW on UDP-glucuronosyltransferase (UGT), *N*-acetyltransferase 2, and xanthine oxidase were also investigated in a few studies, but none of these enzymes were significantly altered by SJW (Bray et al., 2002; Wenk et al., 2004).

Collectively, SJW can modulate several CYP isoforms and transporters. Strong induction of CYP3A4 and P-gp after long-term usage of SJW was consistently observed both in vitro and in vivo, resulting in clinically relevant SJW–drug interactions. Since more than 50% of all prescription drugs are metabolized by CYP3A4 and a large number of compounds are substrates of P-gp, caution is needed when SJW is co-administered with compounds, which are substrates of CYP3A4 and/or P-gp.

22.2.2 Garlic

Garlic (*Allium sativum*) has gained a worldwide reputation because of its beneficial effects for improving health. It is reported to have antioxidative, antimicrobial, cardioprotective, chemopreventive, and anticancer effects. The garlic-rich organosulfur compounds (OSCs) are believed to play key roles in these biological effects (Wu et al., 2002). Of the OSCs naturally occurring in garlic, diallyl sulfide, diallyl disulfide, and diallyl trisulfide, which differ in their number of sulfur atoms, are the most abundant.

Since OSCs have been shown to inhibit several types of chemically induced cancer in rodents, the effects of OSCs on phase I and II enzymes (especially CYP1A, 2B and 2E1, and GST) have been extensively investigated. CYP2E1 activates a wide variety of low molecular weight compounds including many important chemical carcinogens and environmental toxins (Wargovich, 2006). Diallyl disulfide was reported to strongly inhibit the mutagenicity of dimethylnitrosamine, a nitrosamine selectively activated by CYP2E1 (Guyonnet et al., 2000). After orally administering garlic oil, diallyl sulfide, diallyl disulfide, or diallyl trisulfide to rats three times a week for 6 weeks, CYP2E1 protein expression was greatly decreased by garlic oil and by each of the three allyl sulfides (Wu et al., 2002). The inhibition of CYP2E1 by OSCs was also consistently observed in studies (Le Bon et al., 2003; Wargovich, 2006). In contrast to CYP2E1 inhibition, CYP1A1/2 and CYP2B1/2 were usually increased by garlic consumption. A single dose of 200 mg/kg diallyl sulfide to rats increased hepatic CYP1A2 protein by 282% and CYP1A1 protein by 684% (Davenport and Wargovich, 2005). The stimulation of CYP2B1/2 activity in rats by diallyl disulfide (1 mmol/kg for 4 days) was accompanied by a marked elevation of CYP 2B1 and 2B2 apoproteins (Siess et al., 1997). OSCs, particularly diallyl sulfide, was found to induce CYP2B1/2 via constitutive androstane receptor (CAR) and nuclear factor E2-related factor2 (Nrf2) activation (Fisher et al., 2007). The effect of OSCs on GST is also well documented. Through the glutathione conjugation of many electrophile agents, GST plays a major role in the detoxification of carcinogens. Numerous studies have shown that garlic, as well as OSCs, strongly induced GST activity (Haber et al., 1994; Guyonnet et al., 1999; Fukao et al., 2004). Garlic was also reported to inhibit CYP2C9 and CYP2C19 (Zou et al., 2002) and induce CYP3A1/2 (Haber et al., 1994; Wu et al., 2002; Davenport and Wargovich, 2005), quinone reductase (QR), UGT, and microsomal epoxide hydrolase (Guyonnet et al., 1999). The effect of garlic on transporters is less well documented. Diallyl disulfide (200 mg/kg/day) induced Mrp2 expression by 7-fold after oral administration for 3 days in rats, which correlated with an increase in GST activity and GSH levels (Demeule et al., 2004). Also, the results of an in vitro study showed that garlic is a weak to moderate P-gp modulator compared with the positive control verapamil (Foster et al., 2001).

Compared to SJW, information on enzyme/transporter-based drug–garlic interactions is very limited. Garlic did not affect the activity of CYP2D6 and 3A4 after 14 days of exposure in healthy volunteers (Markowitz et al., 2003). This indicated that garlic was unlikely to influence the disposition of co-administered drugs, which are primarily eliminated through CYP2D6 or CYP3A4. Garlic supplements reduced saquinavir plasma concentration by 51% and decreased the AUC of ritonavir by 17% in healthy volunteers (James, 2001; Gallicano et al., 2003). Although CYP3A4 and P-gp are involved in the disposition of these compounds, the mechanism of these interactions is not clear since garlic did not significantly influence CYP3A4 and P-gp activities. Garlic decreased the plasma concentration of oxidative acetaminophen metabolites in an animal study and caused a slight increase in acetaminophen sulfation in healthy volunteers (Gwilt et al., 1994; Lin et al., 1996). Since CYP2E1 is the major enzyme responsible for bioactivation of acetaminophen, CYP2E1 inhibition, a well-documented effect of garlic, maybe involved in the interaction between garlic and acetaminophen.

Overall, effects of garlic on CYP2E1, CYP1A1/2, CYP2B1/2, UGT, and GST are consistently reported, and garlic should be used with caution when co-administered with drugs that are mainly metabolized by these enzymes.

22.2.3 Ginseng

Ginseng is one of the most common herbal products consumed today. There are several species of ginseng including panax ginseng (or Asian ginseng), panax quinquefolius (or American ginseng), and Siberian ginseng. Ginseng is widely used as a general body tonic because of its antifatigue, antioxidation, neuroprotective, cognition-enhancing, and immuno-enhancing effects. Seven ginsenosides (Rb1, Rb2, Rc, Rd, Re, Rf, Rg1) are thought to be the major pharmacologically active constituents of Asian and American ginseng and two eleutherosides (B and E) are the most active constituents of Siberian ginseng.

The effects of ginseng extracts on CYP isoforms have been investigated. Seven ginsenosides and two eleutherosides were tested for their effects on c-DNA-expressed CYP isoforms (1A2, 2C9, 2C19, 2D6, and 3A4) in a study conducted by Henderson et al. (Henderson et al., 1999). The results showed that ginsenoside Rd displayed weak inhibitory activity on CYP3A4 (at 74 μM) and 2D6 (at 76 μM). Ginsenoside Rd and Rf (200 μM) produced a 70 and 54% increase in the activity of CYP3A4, respectively. These effects are not likely to be clinically significant since the concentrations necessary to produce these effects are very high. In another in vitro study (Chang et al., 2002), the ginseng extracts were found to inhibit CYP1A1/2 and CYP1B1 catalytic activity in an enzyme-selective and extract-specific manner and these effects were not due to those individual seven ginsenosides. Other chemical constituents in ginseng might be responsible for this CYP1 enzyme inhibition. Asian ginseng inhibition of CYP2D6 was statistically significant in a clinical trial (Gurley et al., 2005a). However, the magnitude of the effect (~7%) was small. The inhibitory effect of ginseng on P-gp has been confirmed in several in vitro studies. Ginsenoside Rg1, Rc, Rd, and Re were found to have a moderate inhibitory effect on P-gp (Molnar et al., 2000). Rg3, found only in red ginseng, was shown to inhibit vinblastine efflux and reverse multidrug resistance to doxorubicin and VP-16 in KBV20C cells. Further experiments revealed that Rg3 blocked drug efflux through direct binding to P-gp (Kim et al., 2003).

Clinical trials or case reports on ginseng–drug interactions are limited and the mechanism remains unclear in most cases. Potential interactions between ginseng and phenelzine have been documented in several case reports, but without a clear explanation of the mechanism. When alcohol was co-administered with ginseng in 14 healthy volunteers, blood alcohol levels were 30% lower than those following alcohol ingestion alone (Lee et al., 1987). Delayed gastric emptying by ginseng and/or induction of alcohol-oxidizing systems such as CYP2E1 might be the possible reason. Treatment with American ginseng for 3 weeks significantly reduced the INR, C_{max}, and AUC of warfarin in healthy volunteers (Yuan et al., 2004). This potential interaction was also supported by a case report (Janetzky and Morreale, 1997). However, co-administration of warfarin with ginseng did not affect the PK and pharmacodynamics of warfarin in a rat model (Zhu et al., 1999). Although the mechanism of ginseng-warfarin interaction is not clear, modulation of CYP isoforms might be one of the possible reasons. Since the investigations of ginseng on enzymes or transporters have been mostly limited to in vitro studies, the clinical relevance of ginseng–drug interactions remains unknown and further in vivo studies are warranted.

22.2.4 Milk Thistle

Milk thistle (*Silybum marianum*) extracts have been used for almost 2000 years as medical remedies. The principal active component of milk thistle is a mixture of flavonolignans, termed silymarin, which is therapeutically used for the treatment of acute and chronic liver diseases. Silymarin is composed mainly of silybin (about 50~70%), silychristin, silydianin, and other related flavonolignans.

Since hepatotoxins such as ethanol, halothane, and aflatoxin B1 can be activated by CYP enzymes (Beckmann-Knopp et al., 2000), several in vitro studies have been conducted to test whether hepatoprotective effects of silymarin are due to CYP inhibition. The inhibitory effect of silymarin or silybin on CYP3A4 or CYP2C9 was consistently observed in many studies, with the K_i values ranging from 4.9 to 160 μM for CYP3A4 (Beckmann-Knopp et al., 2000; Zuber et al., 2002; Sridar et al., 2004; Doehmer et al., 2008) and 5 to 19 μM for CYP2C9 (Beckmann-Knopp et al., 2000; Sridar et al., 2004) in different experiments. The lack of inhibitory silibin effects on CYP2E1 was confirmed by several studies (Beckmann-Knopp et al., 2000; Sridar et al., 2004; Doehmer et al., 2008). Either weak or no inhibitory effect of silymarin or silybin was found on other CYP enzymes such as CYP1A2, 2B6, 2C8, 2D6, and 2C19 (Beckmann-Knopp et al., 2000; Zuber et al., 2002; Doehmer et al., 2008). The effects of silymarin on UGT have also been investigated. Silybin was shown to be a potent inhibitor of UGT1A1 (IC_{50} of 1.4 μM) and was 14- and 20-fold more selective for UGT1A1 than for UGT1A9 and UGT1A6, respectively (Sridar et al., 2004). The inhibitory effects of silymarin on transporters were also reported recently. In two studies conducted by Zhang et al. (Zhang and Morris, 2003; Zhang et al., 2004b), 50 μM silymarin significantly increased the cellular accumulation of [^{3}H]daunomycin (DNM) (P-gp substrate) and mitoxantrone (BCRP substrate) in P-gp and BCRP overexpressing cells, respectively. A significant decrease in [^{3}H]dehydroepiandrosterone (DHEA) uptake was observed when co-incubated with 50 μM silymarin in OATP1B1-expressing cells, indicating that silymarin is also a OATP1B1 inhibitor (Wang et al., 2005).

Since inhibitory effects of silymarin on phase I (CYP3A4, 2C9), phase II (UGT1) enzymes and transporters (P-gp, BCRP, and OATP1B1) were observed in vitro, the effects of silymarin on the PK of various drugs, which are substrates of the enzymes and/or transporters mentioned above, were investigated recently. However, most results from these clinical trials were not consistent with the in vitro observations. Pretreatment with silymarin for 6 days increased the clearance of metronidazole (a dual P-gp/CYP3A4 substrate) by 29.5% (Rajnarayana et al., 2004). P-gp and/or CYP3A4 induction with long-term treatment might be one of the possible reasons. Neither short-term (4 days) nor long-term (12 days) intake of milk thistle (600 mg/day) significantly influenced the PK of irinotecan, a substrate of CYP3A4, in patients with cancer (van Erp et al., 2005). No significant changes in PK parameters of rosuvastatin were observed in human subjects following 3 days of silymarin pretreatment with a dosage of 520 mg/day, even though rosuvastatin is a good substrate for both OATP1B1 and BCRP and does not undergo extensive metabolism (Deng et al., 2008). Silymarin also failed to influence the PK of many other drugs (Wu et al., 2009) such as nifedipine (CYP3A4 substrate), indinavir (dual P-gp/CYP3A4 substrate), digoxin (P-gp substrate) (Gurley et al., 2006), ranitidine, and midazolam (CYP3A4 substrates) (Gurley et al., 2004). Low bioavailability of silymarin (~0.95%) was believed to be one of the reasons contributing to this in vitro–in vivo disconnect. The plasma concentrations of silymarin in clinical studies are likely to be too low for enzyme/transporter inhibition. For example, the C_{max} of total silybin B was only 131 ng/mL after the ingestion of 600 mg standardized milk thistle extract in healthy volunteers, which is much lower than the inhibitory concentrations used in vitro (Wen et al., 2008). Collectively, the modulation effect of milk thistle on several enzymes and transporters was observed in vitro but not in vivo. The discrepancy may be caused by the low bioavailability of silymarin.

22.2.5 Ginkgo

Ginkgo, a leaf extract of *Ginkgo biloba*, is one of the most popular herbal products in the world due to its effectiveness in Alzheimer's disease, poor peripheral circulation, and age-related dementia. Flavonols (mainly quercetin, kaempferol, isorhamnetin) and terpene lactones (ginkgolide and bilobalide) are the primary active constituents of *Ginkgo biloba* extract (GBE) and they account for 26 and 7% of GBE, respectively (Gurley et al., 2005a).

The effects of GBE on CYPs have been well documented both in vitro and in vivo. In a study conducted by Hellum et al. (Hellum et al., 2007), GBE significantly increased the activity of CYP1A2 by 140% at a concentration of 2.19 μg/mL in hepatocytes. The inductive effect of GBE on CYP1A2 was also consistently observed in two animal studies. Pretreatment of GBE at 100 mg/kg for 10 days significantly reduced the AUC and C_{max} of propranolol, a typical CYP1A2 substrate, in rats (Zhao et al., 2006). In another animal study, the AUC of theophylline, which is also predominantly metabolized by CYP1A2, was reduced by 40% following pretreatment with GBE 100 mg/kg for 5 days (Tang et al., 2007). Compared to CYP1A2, the effects of GBE on CYP3A are less consistent. Feeding GBE (0.5%, w/w) for 4 weeks to rats increased the mRNA of CYP3A1 and 3A2 in the liver and reduced the hypotensive effect of nicardipine in rats (Shinozuka et al., 2002). On the other hand, GBE inhibited CYP3A activities in a dose-dependent manner in small intestine and liver microsomes in a study conducted by Ohnishi et al. (Ohnishi et al., 2003). A single oral treatment of GBE (20 mg/kg) in rats profoundly increased the AUC and bioavailability of diltiazem and nifedipine, both of which are good substrates of CYP3A (Ohnishi et al., 2003; Yoshioka et al., 2004). A CYP3A4-mediated GBE–midazolam interaction was also observed in a clinical trial. The AUC of midazolam was significantly increased by 25% after 28 days intake of GBE in 10 healthy volunteers (Uchida et al., 2006). A potential interaction between GBE and CYP2C9 substrates remains controversial. The inhibitory effect of GBE on CYP2C9 was shown in human liver microsomes with a K_i value of 14.8 μg/mL (Mohutsky et al., 2006). However, in contrast to the in vitro inhibition of CYP2C9, treatment of GBE at 240 mg/day for 1 week failed to influence the PK of two CYP2C9 probe substrates, tolbutamide and diclofenac, in healthy subjects (Mohutsky et al., 2006). Another clinical trial also showed that GBE did not significantly affect the PK of warfarin in healthy subjects at recommended doses (Jiang et al., 2005). On the other hand, after intake of GBE at an oral dose of 360 mg/day for 28 days, a higher dose and longer period than previous studies, the AUC of tolbutamide was slightly but significantly (16%) lower than that without GBE intake, indicating CYP2C9 induction (Uchida et al., 2006). The effect of GBE on CYP2C19 was also investigated in clinical trials. In a study carried out by Yin et al. (Yin et al., 2004), the AUC ratio of omeprazole to 5-hydroxyomeprazole was decreased by 24.5% in poor metabolizers and 57.5% in homozygous extensive metabolizers after treatment of GBE, suggesting that GBE is an CYP2C19 inducer and manifests its inductive effect in a CYP2C19 genotype-dependent manner. GBE demonstrated a biphasic effect on CYP2D6 in vitro with an inhibitory effect at low concentration (2.19 μg/mL) and an inductive effect at high concentration (219 μg/mL) (Hellum et al., 2007). GBE did not affect CYP2D6 activity in a clinical trial using debrisoquine as the probe substrate (Gurley et al., 2005a).

Compared to the CYP enzymes mentioned above, information on the effect of GBE on phase II enzymes and transporters is limited and involves predominantly in vitro evaluations. GST-P1 was increased at both the expression and activity levels after exposure to GBE in Hep G2 and Hep1c1c7 cell lines (Liu et al., 2009). Feeding GBE for 1 week markedly increased GST activity in the liver in an animal study (Sugiyama et al., 2004). GBE potently inhibited estrone-3-sulfate uptake by 85.4% in OATP-B (SLCO2B1)-transfected HEK 293 cells, suggesting the OATP-B inhibition of GBE. A further experiment revealed that flavonols, but not terpenoids, contribute to this inhibitory effect (Fuchikami et al., 2006).

In conclusion, GBE modulates several CYPs: CYP1A2, 3A4, 2C9, and 2C19. GBE–drug interactions were demonstrated in many animal studies and clinical trials, and compounds, which are predominantly metabolized by those enzymes, should be used with caution with concomitant intake of GBE.

22.2.6 Other Herbal Products

Many other herbal products are also able to modulate enzymes and transporters, raising the potential of herb–drug interactions. The enzyme- and transporter-mediated interactions between these herbs and drugs, either in animal studies or clinical trials, are listed in Table 22.1.

22.3 Interactions of Flavonoids with Metabolizing Enzymes/Transporters

With more than 8000 different flavonoids identified (Galli, 2007), flavonoids comprise the most prevalent group of plant polyphenols in fruits and vegetables and

Table 22.1 In vivo herb–drug interactions

Herb	Probe drug	Interaction outcomes	Possible mechanism	Study type	References
Kava	Chlorzoxazone	Metabolic ratio ↑	CYP2E1↓	Healthy volunteers	Gurley et al. (2005b)
Kava	Caffeine	Metabolic ratio ↑	CYP1A2↓	Healthy volunteers	Russmann et al. (2005)
Curcumin	Midazolam	AUC↑ CL_{oral} ↓	CYP3A↓	Rats	Zhang et al. (2007)
Curcumin	Celiprolol	AUC↑ C_{max} ↓ CL_{oral} ↓	P-gp↓	Rats	Zhang et al. (2007)
Piperine	Nevirapine	AUC↑ C_{max} ↑	CYP3A4↓	Healthy volunteers	Kasibhatta and Naidu (2007)
Piperine	Phenytoin	AUC↑ C_{max} ↑ k_a ↑	P-gp↓	Patients	Pattanaik et al. (2006)
Piperine	Propranolol	AUC↑ C_{max} ↑	CYP3A4↓	Healthy volunteers	Bano et al. (1991)
Piperine	Theophylline	AUC↑ C_{max} ↑ $t_{1/2}$ ↑	CYP1A2↓	Healthy volunteers	Bano et al. (1991)
Piperine	Rifampin	C_{max} ↓	P-gp↓	Patients	Zutshi et al. (1985)

are the main components of many herbal supplements such as *Sophora japonica, Citrus grandis,* and *Scutellaria radix* (Jang et al., 2003). The basic structure of flavonoids consists of two aromatic rings (A and B) linked through three carbons that usually form a heterocyclic ring (C). Based on the variations in the patterns of hydroxylation and substitutions in ring C, flavonoids are divided into seven subclasses: flavones, flavonols, flavanones, flavanols, anthocyanidines, chalcones, and isoflavones. Despite structural similarities, the biological and biochemical properties of flavonoids can vary considerably.

Flavonoids have been reported to have a wide variety of beneficial pharmacological effects including antioxidative, anti-inflammatory, antiviral, anticancer, and chemopreventive properties. One of the mechanisms that might contribute to anticancer effects of flavonoids is through interaction with phase I and/or phase II enzymes, especially those involved in carcinogen activation/detoxification, leading to reduced carcinogen formation and increased elimination. The effects of flavonoids on phase I/II enzymes have been extensively investigated. Since many enzymes, which are modulated by flavonoids, play an important role in not only carcinogenesis but also metabolism of clinically important compounds, the interactions of flavonoids with these metabolizing enzymes will be discussed in Section 22.3.1. In addition to enzymes, many efflux and uptake transporters, which are widely distributed in normal tissues and involved in the disposition of various xenobiotics, are also modulated by flavonoids. The interactions of flavonoids with transporters will be discussed in 22.3.2. In vivo enzyme- and/or transporter-mediated flavonoid–drug interactions are summarized in Table 22.2.

Table 22.2 In vivo enzyme- and/or transporter-mediated flavonoid–drug interactions

Flavonoid	Probe drug	Study outcome	Possible mechanism	Study type	References
Biochanin A	Paclitaxel	AUC↑ C_{max}↑ k_a↑ F↑	CYP3A and/or P-gp ↓	Rats	Peng et al. (2006)
Biochanin A	Digoxin	AUC↑ C_{max} ↑ F↑	CYP3A and/or P-gp ↓	Rats	Peng et al. (2006)
Diosmin	Metronidazole	AUC↑ C_{max} ↑	CYP3A4 and/or CYP2C9 ↓	Human	Rajnarayana et al. (2003)
Morin	Tamoxifen	AUC↑ C_{max} ↑ F↑	CYP3A and/or P-gp ↓	Rats	Shin et al. (2008)
Flavone	Paclitaxel	AUC↑ C_{max} ↑ k_a↑ F↑ $t_{1/2}$↑	CYP3A and/or P-gp ↓	Rats	Choi et al. (2004)
Morin	Etoposide	AUC↑ C_{max} ↑ F↑	CYP3A and/or P-gp ↓	Rats	Li et al. (2007)
Morin	Nicardipine	AUC↑ C_{max} ↑ F↑ CL↑	CYP3A and/or P-gp ↓	Rats	Piao and Choi (2008)
Quercetin	Diltiazem	AUC↑ C_{max} ↑ F↑	CYP3A and/or P-gp ↓	Rabbits	Choi and Li (2005)
Quercetin	Verapamil	AUC↑ C_{max} ↑ F↑	CYP3A and/or P-gp ↓	Rabbits	Choi and Han (2004)
Quercetin	Moxidectin	AUC↑	CYP3A and/or P-gp ↓	Lambs	Dupuy et al. (2003)
Naringin	Paclitaxel	AUC↑ C_{max} ↑ F↑ $t_{1/2}$↑	CYP3A and/or P-gp ↓	Rats	Choi and Shin (2005)
Quercetin	Digoxin	AUC↑ C_{max} ↑	P-gp ↓	Pigs	Wang et al. (2004)
Chrysin	Nitrofurantoin	AUC↑ C_{max} ↑ CL↑	BCRP ↓	Rats	Wang and Morris, (2007b)
Luteolin	GHB	AUC↑ CL↑	MCT1 ↓	Rats	Wang et al. (2008)

22.3.1 Interactions of Flavonoids with Metabolizing Enzymes

The CYP1 family is comprised of 1A1, 1A2, and 1B1, which are expressed in a tissue-specific manner and activate a large number of procarcinogens to reactive intermediates. The effects of flavonoids on the CYP1 family have been extensively explored due to its important role in carcinogenesis. Acacetin and diosmetin were shown to be potent inhibitors of CYP1A1/2 and CYP1B1 and their potency is enzyme specific. Eriodictyol and naringin were poor inhibitors of CYP1A and homoeriodictyol was a selective inhibitor of CYP1B1 (Doostdar et al., 2000). Flavone was reported to be a less potent inhibitor of CYP1A1 ($IC_{50} = 0.14$ μM) than CYP1A2 ($IC_{50} = 0.066$ μM). Galangin showed a 5-fold selectivity in its inhibition of CYP1A2 over CYP1A1 (Zhai et al., 1998a). Compared to the potent inhibition of tangeretin ($IC_{50} = 0.8$ μM) on CYP1A, green tea flavonoids, such as epicatechin gallate (ECG) and epigallocatechin gallate (EGCG), were much less potent with IC_{50} values ranging from 75 to 400 μM (Obermeier et al., 1995). Inhibitory effects of genistein, equol, quercetin, chrysin, flavanone, 7,8-benzoflavone, and 5,7-dimethoxyflavone on CYP1A have also been reported (Siess et al., 1995; Zhai et al., 1998b; Wen et al., 2005). In contrast to the inhibitory effect of flavonoids on CYP1A following a short incubation (within 30 min), increased CYP1A activity and/or expression was observed in many studies following long-term incubation (in vitro) or administration (in vivo). Feeding a diet containing 0.3% (w/w) flavone or tangeretin to rats for 2 weeks increased the activities and protein expression of CYP1A1 and CYP1A2 (Canivenc-Lavier et al., 1996). A number of flavonoids (apigenin, chrysin, galangin, isorhamnetin, diosmin, diosmetin, and luteolin) have been reported to increase CYP1A1 activity after a 72-h incubation in Hep G2 cells (Wen et al., 2005). Transcriptional activation of CYP1A1 is regulated by aryl hydrocarbon receptor (AhR) and several flavonoids such as quercetin, diosmin, diosmetin, and galangin were verified to be AhR ligands and could increase CYP1A1 mRNA through AhR (Ciolino et al., 1998, 1999; Ciolino and Yeh, 1999).

In addition to the CYP1 family, the effects of flavonoids on several other CYP isoforms were also investigated. Treatment of mice with 300 mg/kg baicalin orally decreased both CYP2E1 activity and protein expression (Breinholt et al., 2002). Quercetin caused a ~20% inhibition of CYP2E1-dependent acetaminophen oxidation (Li et al., 1994). Genistein and equol inhibited *p*-nitrophenol (CYP2E1 substrate) metabolism in liver microsomes with IC_{50} values of 10 mM and 560 μM, respectively (Helsby et al., 1998). Inhibition of CYP2C9 by a series of flavonoids was observed in an in vitro study. Among the 14 flavonoids screened (9 flavones and 5 flavonols), all flavonoids, except flavone, were potent inhibitors (K_i of 0.15–2.2 μM) of CYP2C9-mediated diclofenac 4′-hydroxylation (Si et al., 2008). Quercetin (10 μM) enhanced the expression of CYP3A4 mRNA in hepatocyte cultures after 48 h of exposure and a further study indicated that this induction was not mediated by PXR (Raucy, 2003). In contrast to the inductive effect, quercetin inhibited CYP3A4 in human liver microsomes after short incubation times (Li et al., 1994). Several kaempferol glycosides from strawberries showed CYP3A4 inhibition even at nM concentrations (Tsukamoto et al., 2004).

Many studies have explored the effect of flavonoids on UGTs, a group of important detoxification and drug-metabolizing enzymes. After pretreating Hep G2 cells with 25 μM chrysin for 3 days, UGT1A1 activity, mRNA, and protein expression were greatly increased (Walle et al., 2000). Using the same experimental conditions, 22 flavonoids were screened and 4 flavonoids (acacetin, apigenin, luteolin, and diosmetin) also induced UGT1A1 similarly to that of chrysin (Walle and Walle, 2002). A further investigation performed by another group indicated that UGT1A1 induction by chrysin occurs after activation of AhR (Bonzo et al., 2007). UGT activation by flavonoids was also demonstrated in vivo. Exposure of rats to green tea (2.5%, w/v) for 4 weeks increased UGT activity by 100%, whereas no effect was observed on SULT, epoxide hydrolase, and GSTs (Bu-Abbas et al., 1995). Both rat small and large intestinal UGT enzyme activities were increased after 2 weeks of feeding quercetin (1%, w/w) or flavone (0.5%, w/w) (van der Logt et al., 2003).

Compared to UGTs, reports of flavonoid effects on GSTs, another major phase II enzyme, have been less consistent. Catechins like EGCG were reported to induce GST activity in human liver cells (Steele et al., 2000), whereas the same flavonoids failed to change the GST activity after 4 weeks of administration in vivo (Bu-Abbas et al., 1995). No effect on GST1A was observed after a 2-h incubation of quercetin in Caco-2 cells (Petri et al., 2003). However, a time- and concentration-dependent inhibition of GST P1-1 by quercetin was clearly shown in another in vitro study (van Zanden et al., 2003). Feeding rats with flavone, tangeretin, or flavanone for 2 weeks (0.3%, w/w) increased the activities of GST in an in vivo study (Canivenc-Lavier et al., 1996).

Many flavonoids have been reported to be potent inhibitors of sulfotransferase (SULT). EGCG was found to be the most potent inhibitor of SULT1A1 among the catechins tested with an IC_{50} of 0.93 μM in an in vitro study (Tamura and Matsui, 2000). Quercetin showed a very potent inhibitory effect on SULT1A1 and the IC_{50} values in adult and fetal livers were 13 and 12 nM, respectively (De Santi et al., 2002). Similarly, a number of other flavonoids such as fisetin, galangin, myricetin, kaempferol, chrysin, and apigenin were also reported to be potent SULT1A1 inhibitors ($EC_{50} < 1$ μM) in human liver cytosols (Eaton et al., 1996).

Activation of QR by flavonoids was also demonstrated in many studies. A significant increase of QR activity and mRNA was observed in human Colo205 cells in the presence of genistein and biochanin A at various concentrations (0.1–10 μM) (Wang et al., 1998). In addition to isoflavonoids, other subclasses of flavonoids such as flavonols (kaempferide, quercetin, kaempferol, and galangin), flavones (apigenin, flavone) were also found to be QR inhibitors (Uda et al., 1997; Yannai et al., 1998). Flavonoids also showed inhibitory effect on many other enzymes, including *N*-acetyltransferase(Mizoyama et al., 2004), carbonyl reductase 1(Carlquist et al., 2008), xanthine oxidase(Nagao et al., 1999), and aromatase (CYP19) (Kao et al., 1998).

22.3.2 Interactions of Flavonoids with Transporters

P-gp plays an important role in not only multidrug resistance but also drug disposition due to its expression in normal tissues. Various widely used drugs, such as anthracyclines, paclitaxel, vinblastine, digoxin, indinavir, and cyclosporine, are substrates of P-gp and their absorption, distribution, or elimination can be greatly influenced by co-administration of P-gp modulators. Results from numerous studies showed that P-gp can be inhibited, activated, or induced by flavonoids. EGCG (100 μM) significantly increased rhodamine 123 (R-123) accumulation in P-gp overexpressing CH^RC5 cells (Jodoin et al., 2002) and KB-C_2 cells (Kitagawa, 2006). An inhibitory effect of EGCG on P-gp was also observed using vinblastine as a substrate in Caco-2 cells (Jodoin et al., 2002). Among 20 naturally occurring flavonoids screened, 50 μM biochanin A, morin, phloretin, and silymarin can substantially increase the [^{3}H]DNM accumulation in MCF7/ADR cells, but not in MCF7-sensitive cells (Zhang and Morris, 2003). When co-incubated with

50 μM diosmin, the digoxin efflux ratio significantly decreased in Caco-2 cells due to effective P-gp inhibition (Yoo et al., 2007). In the presence of quercetin, kaempferol, and galangin, stimulation of P-gp was observed when using 7,12-dimethylbenz(a)anthracene (Phang et al., 1993) and doxorubicin (Chieli et al., 1995), (Rajnarayana et al., 2003) as the P-gp substrates and inhibition of P-gp was found when using R-123 as substrate (Chieli et al., 1995). There is no clear explanation for this discrepancy. Quercetin and kaempferol also demonstrated a biphasic effect on vincristine efflux by activating P-gp at low concentration (10 μM) and inhibiting P-gp at high concentration (50 μM) (Mitsunaga et al., 2000). In addition to modulating P-gp through inhibition or activation, flavonoids can induce P-gp expression both in vitro and in vivo. Among the 14 flavonoids tested, all flavonoids at 10 μM concentrations, except apigenin, greatly increased P-gp protein expression in Caco-2 cells after 4 weeks of exposure (Lohner et al., 2007). Activation of PXR, a key regulator of P-gp, might be one possible mechanism (Lohner et al., 2007).

With a similar tissue distribution as P-gp, breast cancer resistance protein (ABCG2, BCRP) also plays an important role in drug disposition and is modulated by flavonoids. Among the 20 dietary flavonoids covering 7 different subclasses that were tested for inhibition (Zhang et al., 2004b), 13 flavonoids significantly increased mitoxantrone accumulation in BCRP-overexpressing MCF7/MX100 cells, with no effect on the MCF7-sensitive cells, indicating that these flavonoids are BCRP inhibitors. The potent inhibition of BCRP by a large number of flavonoids was also consistently observed by another group using a different substrate (SN-38) and a different cell line (K562 cells) (Imai et al., 2004). Because food or herbal products usually contain multiple flavonoids, instead of one, the combined effects of multiple flavonoids on BCRP have been investigated and the tested flavonoids were found to inhibit BCRP in an additive fashion when given as 2-, 3-, 5-, or 8-flavonoid combinations (Zhang et al., 2004a; Ebert et al., 2007). Flavonoids modulate BCRP not only on a functional level but also on a transcriptional level. Quercetin and flavone greatly induced BCRP mRNA and protein expression after a 72-h incubation in Caco-2 cells (Ebert et al., 2007). Quercetin (25 μM) substantially increased the expression of BCRP in MCF7/WT cells, whereas BCRP in MCF7/AHR$_{200}$ cells was almost nondetectable, indicating that quercetin induces BCRP via AhR (Ebert et al., 2007). On the other hand, downregulation of BCRP was observed after treatment of Caco-2 cells with epicatechin, suggesting an AhR-antagonistic effect (Ebert et al., 2007).

In addition to P-gp and BCRP, flavonoids can also modulate other efflux transporters including multidrug resistance-associated proteins, MRP1 and MRP2. From screening 22 flavonoids, 8 flavonoids (biochanin A, genistein, quercetin, chalcone, silymarin, phloretin, morin, and kaemperol) were found to significantly increase the accumulation of both DNM and vinblastine in Panc-1 cells (Nguyen et al., 2003). Competitive inhibition of MRP1 by flavonoids was demonstrated in one study with the following rank order of potency: kaempferol > apigenin > quercetin > myricetin > naringenin (Leslie et al., 2001). Although MRP2 shares many similarities with MRP1, fewer flavonoids inhibit MRP2 and K_i values are higher. In a study conducted by Van Zanden et al. (van Zanden et al., 2005), the inhibitory effects of

29 flavonoids on both MRP1- and MRP2-mediated calcein efflux in MDCKII cells were evaluated. Fifteen flavonoids inhibited MRP1 with IC_{50} values below 50 μM, whereas only two flavonoids (myricetin and robinetin) inhibited MRP2 with IC_{50} values< 50 μM.

Recently, the effects of flavonoids on influx transporters were also investigated. Quercetin and kaempferol inhibited organic cation transporter 2 (OCT2)-mediated [^{14}C]tetraethylammonium uptake in LLC-PK1 cells with IC_{50} values of 32 and 38 μM, respectively (Ofer et al., 2005). In OATP1B1-expressing Hela cells, 50 μM of the flavonoids biochanin A, fisetin, sillibin, and silymarin produced significant decreases in [^{3}H]DHEA uptake, whereas only negligible effects were observed in OATP1B1 negative Hela cells, indicating OATP1B1 inhibition (Wang et al., 2005). In addition to OATP1B1, the inhibitory effect of flavonoids on other OATPs, such as Oatp1a5 and OATP1A2, have been reported (Dresser et al., 2002). Inhibition of monocarboxylate transporter 1 (MCT1) by flavonoids was observed both in vitro and in vivo. The uptake of γ-hydroxybutyrate (GHB) in MCT1-transfected MDA-MB231 cells was significantly decreased in the presence of 13 dietary flavonoids. Luteolin, the most potent MCT1 inhibitor with an IC_{50} of 0.41 μM, significantly decreased GHB plasma concentration and increased GHB renal and total clearance in rats (Wang and Morris, 2007a).

Many in vivo studies have been carried out to investigate the effect of flavonoids on the PK of therapeutic compounds. The enzyme- and/or transporter-mediated flavonoid–drug interactions in vivo are summarized in Table 22.2. Based on the data, flavonoids can interact with various drugs by inhibiting enzymes or transporters. Caution is needed, particularly for drugs with a narrow therapeutic index, when these drugs are administered with herbal supplements enriched in flavonoids.

22.4 Interactions of Organic Isothiocyanates with Metabolizing Enzymes/Transporters

Isothiocyanates (ITCs), one of the most extensively studied classes of phytochemicals, are widely distributed in cruciferous vegetables such as cabbage, broccoli, brussels sprouts, cauliflower, horseradish, watercress, mustard, and turnip (Hecht, 2000). ITCs occur in plants as thioglucoside conjugates called glucosinolates and over 100 glucosinolates have been identified. Glucosinolates are hydrolyzed into ITCs by myrosinase when the plant is macerated or chewed. Among a wide variety of ITCs that are naturally occurring, the most studied and best characterized are phenethyl isothiocyanate (PEITC), benzyl isothiocyanate (BITC), and 4-methylsulfinylbutyl isothiocyanate (sulforaphane, SFN). ITCs have sparked worldwide interest because of their remarkable ability to prevent cancer, which has been confirmed using various tumor models in different target sites and different species of rodents (Talalay and Fahey, 2001). Modulation of phase I and/or phase II enzymes, which are involved in the activation and detoxification of carcinogens, represents one mechanism important for the chemopreventive effect of ITCs.

Due to the important role of CYP1A and 2B in carcinogenesis, the effects of ITCs on CYP1A and 2B have been extensively studied. Competitive inhibition of CYP1A2 activity by PEITC was observed in two studies using microsomes (K_i of 4.5 μM) and a reconstituted system (K_i of 0.18 μM) (Smith et al., 1996; Nakajima et al., 2001). In a study conducted by Conaway et al.(Conaway et al., 1996), 21 naturally occurring ITCs were screened to test their effects on CYP1A1/2 and 2B1 and most ITCs showed inhibitory effect on these enzymes with inhibition being greater for CYP2B1 than CYP1A1/2 in liver microsomes. For example, the IC_{50} values of BITC and PEITC on CYP2B1 were 5 and 1.8 μM, respectively, compared with 54 and 47 μM for CYP1A1/2. In contrast to the potent inhibition by ITCs observed in vitro, the expression levels of CYP1A1, 1A2, and 2B1 were usually elevated when animals were exposed to ITCs for several days. Feeding rats with glucoraphanin (SFN precursor) at 120 or 240 mg/kg increased the expression levels of CYP1A1 (up to 14-fold), 1A2 (up to 4.7-fold), and 2B1/2 (up to 7.5-fold) (Paolini et al., 2004). Following 14 days of PEITC treatment (0.1%, w/w in diet) to rats, the protein expression of CYP1A1, 1A2, and 2B1/2 in livers was increased modestly(Manson et al., 1997). In a recent study, rats were fed diets supplemented with different doses of PEITC (range from 0.06 to 6 μM/g) and immunoblots revealed a small decrease in CYP1A1 and 1A2 apoprotein levels following a low dose and marked elevation in CYP1A1, 1A2, and 2B1 after a high dose (Konsue and Ioannides, 2008).

In addition to CYP1A and 2B, the effects of ITCs on other CYPs were also investigated. BITC was found to inactivate CYP2E1 in a mechanism-based manner, with a K_i of 13 μM (Moreno et al., 1999). PEITC inactivated both wild-type and mutant CYP2E1, with K_i values of 2.7 and 1.6 μM, respectively (Moreno et al., 2001). Several other ITCs were also found to be CYP2E1 inhibitors, including SFN and *t*-butyl isothiocyanate (Kent et al., 1999; Fimognari et al., 2008). CYP3A4 inhibition was observed with PEITC, and characterized as mixed competitive (K_i of 34 μM) and noncompetitive (K_i of 63.8 μm) in nature (Nakajima et al., 2001). A concentration-dependent inhibitory effect of SFN on CYP3A4 was also reported and this inhibition was through PXR antagonism (Zhou et al., 2007). Human CYP2C9 (K_i of 6.5 μM), CYP2C19 (K_i of 12 μM), and CYP2D6 (K_i of 28 μM) were reported to be noncompetitively inhibited by PEITC (Nakajima et al., 2001). On the other hand, in an in vivo study, the consumption of 50 g of watercress did not influence the PK of debrisoquine (a probe drug of CYP2D6) in humans (Caporaso et al., 1994).

Numerous studies have shown that ITCs can profoundly induce phase II enzymes, especially QR and GST. QR and GST were markedly induced in the liver after 14 days treatment with PEITC (0.06–6 μM) in rats, at both the activity and protein level (Konsue and Ioannides, 2008). After exposure of RL34 cells to 10 μM BITC for 24 h, GST activity was increased 2.3-fold and both mRNA and protein level of GSTP1 were also significantly increased (Nakamura et al., 2000). SFN, even at dietary doses, stimulated QR in a dose-dependent fashion (Yoxall et al., 2005). Zhang and Talalay (Zhang and Talalay, 1998) reported that the intracellular concentrations of ITCs correlated closely to their potencies as inducers of phase II enzymes. Induction of several phase II enzymes is under the transcriptional control of antioxidant responsive element (ARE) and Nrf2 is a key transcription factor of ARE activation. PEITC and SFN have been reported to enhance phase II enzyme expression via Nrf2 activation (Keum et al., 2003; McWalter et al., 2004). Compared to QR and GST, the effects of ITCs on UGT have received less attention. SFN failed to influence UGT activities after exposure to rats at 3 and 12 mg/kg for 10 days (Yoxall et al., 2005). On the other hand, watercress consumption increased the glucuronidation of nicotine, cotinine, and 3′-hydroxycotinine in smokers, suggesting that PEITC might induce glucuronidation (Hecht et al., 1999).

Similar to flavonoids, ITCs can modulate not only phase I and II enzymes but also efflux transporters. ITCs were reported to be weak P-gp and MRP1 inhibitors in a study conducted by Tseng et al. (Tseng et al., 2002). Among a number of ITCs screened, 100 μM naphthyl ITC significantly increased the 2-h accumulation of DNM and vinblastine in both P-gp-overexpressing MCF7/Adr cells and MRP1-expressing PANC-1 cells, with no effect in sensitive MCF7 cells (Tseng et al., 2002). In another in vitro study, the effects of ITCs on BCRP were also determined and significant increases in mitoxantrone accumulation were observed for several ITCs (BITC, hexyl ITC, PEITC, naphthyl ITC, phenylhexyl ITC, phenylpropyl ITC, phenylbutyl ITC) at 50 μM in BCRP overexpressing MCF7/MX100 cells and NCI-H460/MX20 cells, indicating their inhibitory effect on BCRP (Ji and Morris, 2004). Recently, the expression levels of P-gp, MRP1, and MRP2 in different cells lines were examined after treatment of SFN. Elevated protein expression of MRP2 in both HepG2 and Caco-2 cells and MRP1 in HepG2 cells was observed, whereas P-gp expression was not changed by SFN (Fimognari et al., 2008).

The speculation that ITCs are not only modulators but also substrates of transporters is based on the phenomenon that intracellular ITCs are rapidly effluxed from cells. In a study carried out by Callway et al. (Callaway et al., 2004), the accumulation of ITCs (allyl ITC, BITC, and PEITC) in MRP-1 overexpressing HL60/AR cells and P-gp overexpressing 8226/DDX40 cells were significantly higher in the presence of transporter inhibitors, compared to that in the absence of inhibitors, suggesting the involvement of MRP-1 and P-gp in the transport of ITCs. In a second in vitro study, MRP2, but not P-gp, was also found to mediate the transport of PEITC (Ji and Morris, 2005b). Differences in experimental design and confounding effects of metabolites may contribute to this discrepancy regarding P-gp. In addition to MRP1 and MRP2, BCRP was also found to be involved in the transport of PEITC. The transport of [^{14}C]PEITC into BCRP-overexpressing cell membrane vesicles was ATP dependent and inhibited by fumitremorgin C, a specific BCRP inhibitor (Ji and Morris, 2005a).

The effects of ITCs on the PK of co-administered drugs have also been investigated. After consumption of 50 g watercress, which is equivalent to a total intake of 13 mg of PEITC, the levels of oxidative metabolites of acetaminophen in human volunteers were significantly decreased, suggesting CYP2E1 inhibition (Chen et al., 1996). In another clinical study, the AUC of chlorzoxazone, a clinical probe for CYP2E1, was significantly increased by 56% after 50 g of watercress ingestion

(Leclercq et al., 1998). Information on other enzyme- or transporter-mediated ITC–drug interactions in vivo is very limited. Collectively, numerous in vitro studies have shown that ITCs interact with various enzymes/transporters as inhibitors, inducers, or substrates. Further in vivo studies are needed to determine the clinical relevance of potential ITC–drug interactions.

22.5 Conclusions and Future Directions

In vitro and in vivo studies have demonstrated that many herbal supplements/constituents modulate phase I/II enzyme and/or transporter function, suggesting the potential for herbal–drug interactions. The modulation of enzymes/transporters by an herbal product may vary markedly due to product content, dose, short- or long-term administration of the herbal product, and the probe drugs evaluated. After short-term incubation/administration, herbal supplements usually interact with enzymes/transporters as inhibitors. However, after longer exposures, herbal products or their constituents often induce enzyme/transporter protein, through nuclear receptor activation. For example, after long-term incubation, SJW induces CYP3A4 through PXR and several flavonoids induce CYP1A1 and BCRP via AhR activation.

Many herbal constituents especially flavonoids undergo extensive phase II metabolism to glucuronide and/or sulfate conjugates in the intestine and liver. This limits the bioavailability of these flavonoids, resulting in low systemic concentrations of these herbal components. This represents one reason for the disconnect between in vitro and in vivo effects. A second reason is that the metabolites formed in vivo may contribute to the herbal–drug interaction. The same metabolites may not be formed in the in vitro experimental systems used, or the same extent of metabolism may not occur in experimental systems in vitro as observed in vivo. Recently, Van Zanden and coworkers (van Zanden et al., 2007) investigated the inhibitory effect of the conjugative metabolism of quercetin on MRP1 and MRP2 in Sf9 inside-out vesicles. The results revealed that several major phase II metabolites of quercetin exert equal or even more potent inhibition on human MRP1 and MRP2 than quercetin itself.

In contrast to the widespread consumption of herbal supplements, the knowledge of PK herbal–drug interactions is still incomplete at present. Many effects of herbal supplements/constituents on enzymes/transporters have been investigated only in vitro. Since herbal supplements contain multiple components, often at low doses, an understanding of potential additive, synergistic, or antagonistic actions of these multiple components on drug-metabolizing enzymes and transporters is important. Further in vivo animal studies and clinical trials need to be carried out to evaluate reported in vitro observations and determine the potential clinical significance of herbal–drug interactions.

Acknowledgments Funding by the Susan G. Komen Foundation BCTR138506 and NIH grants R01DA023223 and R03CA121404 is acknowledged.

References

Bano G, Raina RK, Zutshi U, Bedi KL, Johri RK and Sharma SC (1991) Effect of piperine on bioavailability and pharmacokinetics of propranolol and theophylline in healthy volunteers. *Eur J Clin Pharmacol* **41**:615–617.

Beckmann-Knopp S, Rietbrock S, Weyhenmeyer R, Bocker RH, Beckurts KT, Lang W, Hunz M and Fuhr U (2000) Inhibitory effects of silibinin on cytochrome P-450 enzymes in human liver microsomes. *Pharmacol Toxicol* **86**:250–256.

Bonzo JA, Belanger A and Tukey RH (2007) The role of chrysin and the ah receptor in induction of the human UGT1A1 gene in vitro and in transgenic UGT1 mice. *Hepatology* **45**:349–360.

Bray BJ, Perry NB, Menkes DB and Rosengren RJ (2002) St. John's wort extract induces CYP3A and CYP2E1 in the Swiss Webster mouse. *Toxicol Sci* **66**:27–33.

Breidenbach T, Hoffmann MW, Becker T, Schlitt H and Klempnauer J (2000) Drug interaction of St John's wort with cyclosporin. *Lancet* **355**:1912.

Breinholt VM, Offord EA, Brouwer C, Nielsen SE, Brosen K and Friedberg T (2002) In vitro investigation of cytochrome P450-mediated metabolism of dietary flavonoids. *Food Chem Toxicol* **40**:609–616.

Bu-Abbas A, Clifford MN, Ioannides C and Walker R (1995) Stimulation of rat hepatic UDP-glucuronosyl transferase activity following treatment with green tea. *Food Chem Toxicol* **33**: 27–30.

Callaway EC, Zhang Y, Chew W and Chow HH (2004) Cellular accumulation of dietary anti-carcinogenic isothiocyanates is followed by transporter-mediated export as dithiocarbamates. *Cancer Lett* **204**:23–31.

Canivenc-Lavier MC, Vernevaut MF, Totis M, Siess MH, Magdalou J and Suschetet M (1996) Comparative effects of flavonoids and model inducers on drug-metabolizing enzymes in rat liver. *Toxicology* **114**:19–27.

Caporaso N, Whitehouse J, Monkman S, Boustead C, Issaq H, Fox S, Morse MA, Idle JR and Chung FL (1994) In vitro but not in vivo inhibition of CYP2D6 by phenethyl isothiocyanate (PEITC), a constituent of watercress. *Pharmacogenetics* **4**:275–280.

Carlquist M, Frejd T and Gorwa-Grauslund MF (2008) Flavonoids as inhibitors of human carbonyl reductase 1. *Chem Biol Interact* **174**:98–108.

Chang TK, Chen J and Benetton SA (2002) In vitro effect of standardized ginseng extracts and individual ginsenosides on the catalytic activity of human CYP1A1, CYP1A2, and CYP1B1. *Drug Metab Dispos* **30**:378–384.

Chen L, Mohr SN and Yang CS (1996) Decrease of plasma and urinary oxidative metabolites of acetaminophen after consumption of watercress by human volunteers. *Clin Pharmacol Ther* **60**:651–660.

Chieli E, Romiti N, Cervelli F and Tongiani R (1995) Effects of flavonols on P-glycoprotein activity in cultured rat hepatocytes. *Life Sci* **57**:1741–1751.

Choi JS, Choi HK and Shin SC (2004) Enhanced bioavailability of paclitaxel after oral coadministration with flavone in rats. *Int J Pharm* **275**:165–170.

Choi JS and Han HK (2004) The effect of quercetin on the pharmacokinetics of verapamil and its major metabolite, norverapamil, in rabbits. *J Pharm Pharmacol* **56**:1537–1542.

Choi JS and Li X (2005) Enhanced diltiazem bioavailability after oral administration of diltiazem with quercetin to rabbits. *Int J Pharm* **297**:1–8.

Choi JS and Shin SC (2005) Enhanced paclitaxel bioavailability after oral coadministration of paclitaxel prodrug with naringin to rats. *Int J Pharm* **292**:149–156.

Ciolino HP, Daschner PJ and Yeh GC (1999) Dietary flavonols quercetin and kaempferol are ligands of the aryl hydrocarbon receptor that affect CYP1A1 transcription differentially. *Biochem J* **340 (Pt 3)**:715–722.

Ciolino HP, Wang TT and Yeh GC (1998) Diosmin and diosmetin are agonists of the aryl hydrocarbon receptor that differentially affect cytochrome P450 1A1 activity. *Cancer Res* **58**:2754–2760.

Ciolino HP and Yeh GC (1999) The flavonoid galangin is an inhibitor of CYP1A1 activity and an agonist/antagonist of the aryl hydrocarbon receptor. *Br J Cancer* **79**:1340–1346.

Conaway CC, Jiao D and Chung FL (1996) Inhibition of rat liver cytochrome P450 isozymes by isothiocyanates and their conjugates: a structure-activity relationship study. *Carcinogenesis* **17**:2423–2427.

Davenport DM and Wargovich MJ (2005) Modulation of cytochrome P450 enzymes by organosulfur compounds from garlic. *Food Chem Toxicol* **43**:1753–1762.

De Santi C, Pietrabissa A, Mosca F, Rane A and Pacifici GM (2002) Inhibition of phenol sulfotransferase (SULT1A1) by quercetin in human adult and foetal livers. *Xenobiotica* **32**:363–368.

Demeule M, Brossard M, Turcotte S, Regina A, Jodoin J and Beliveau R (2004) Diallyl disulfide, a chemopreventive agent in garlic, induces multidrug resistance-associated protein 2 expression. *Biochem Biophys Res Commun* **324**:937–945.

Deng JW, Shon JH, Shin HJ, Park SJ, Yeo CW, Zhou HH, Song IS and Shin JG (2008) Effect of silymarin supplement on the pharmacokinetics of rosuvastatin. *Pharm Res* **25**:1807–1814.

Doehmer J, Tewes B, Klein KU, Gritzko K, Muschick H and Mengs U (2008) Assessment of drug–drug interaction for silymarin. *Toxicol In Vitro* **22**:610–617.

Doostdar H, Burke MD and Mayer RT (2000) Bioflavonoids: selective substrates and inhibitors for cytochrome P450 CYP1A and CYP1B1. *Toxicology* **144**:31–38.

Dresser GK, Bailey DG, Leake BF, Schwarz UI, Dawson PA, Freeman DJ and Kim RB (2002) Fruit juices inhibit organic anion transporting polypeptide-mediated drug uptake to decrease the oral availability of fexofenadine. *Clin Pharmacol Ther* **71**:11–20.

Dupuy J, Larrieu G, Sutra JF, Lespine A and Alvinerie M (2003) Enhancement of moxidectin bioavailability in lamb by a natural flavonoid: quercetin. *Vet Parasitol* **112**:337–347.

Durr D, Stieger B, Kullak-Ublick GA, Rentsch KM, Steinert HC, Meier PJ and Fattinger K (2000) St John's Wort induces intestinal P-glycoprotein/MDR1 and intestinal and hepatic CYP3A4. *Clin Pharmacol Ther* **68**:598–604.

Eaton EA, Walle UK, Lewis AJ, Hudson T, Wilson AA and Walle T (1996) Flavonoids, potent inhibitors of the human P-form phenolsulfotransferase. Potential role in drug metabolism and chemoprevention. *Drug Metab Dispos* **24**:232–237.

Ebert B, Seidel A and Lampen A (2007) Phytochemicals induce breast cancer resistance protein in Caco-2 cells and enhance the transport of benzo[a]pyrene-3-sulfate. *Toxicol Sci* **96**:227–236.

Fimognari C, Lenzi M and Hrelia P (2008) Interaction of the isothiocyanate sulforaphane with drug disposition and metabolism: pharmacological and toxicological implications. *Curr Drug Metab* **9**:668–678.

Fisher CD, Augustine LM, Maher JM, Nelson DM, Slitt AL, Klaassen CD, Lehman-McKeeman LD and Cherrington NJ (2007) Induction of drug-metabolizing enzymes by garlic and allyl sulfide compounds via activation of constitutive androstane receptor and nuclear factor E2-related factor 2. *Drug Metab Dispos* **35**:995–1000.

Foster BC, Foster MS, Vandenhoek S, Krantis A, Budzinski JW, Arnason JT, Gallicano KD and Choudri S (2001) An in vitro evaluation of human cytochrome P450 3A4 and P-glycoprotein inhibition by garlic. *J Pharm Pharm Sci* **4**:176–184.

Fuchikami H, Satoh H, Tsujimoto M, Ohdo S, Ohtani H and Sawada Y (2006) Effects of herbal extracts on the function of human organic anion-transporting polypeptide OATP-B. *Drug Metab Dispos* **34**:577–582.

Fukao T, Hosono T, Misawa S, Seki T and Ariga T (2004) The effects of allyl sulfides on the induction of phase II detoxification enzymes and liver injury by carbon tetrachloride. *Food Chem Toxicol* **42**:743–749.

Galli F (2007) Interactions of polyphenolic compounds with drug disposition and metabolism. *Curr Drug Metab* **8**:830–838.

Gallicano K, Foster B and Choudhri S (2003) Effect of short-term administration of garlic supplements on single-dose ritonavir pharmacokinetics in healthy volunteers. *Br J Clin Pharmacol* **55**:199–202.

Garrovo C, Rosati A, Bartoli F and Decorti G (2006) St John's wort modulation and developmental expression of multidrug transporters in the rat. *Phytother Res* **20**:468–473.

Gurley BJ, Barone GW, Williams DK, Carrier J, Breen P, Yates CR, Song PF, Hubbard MA, Tong Y and Cheboyina S (2006) Effect of milk thistle (Silybum marianum) and black cohosh (Cimicifuga racemosa) supplementation on digoxin pharmacokinetics in humans. *Drug Metab Dispos* **34**:69–74.

Gurley BJ, Gardner SF, Hubbard MA, Williams DK, Gentry WB, Carrier J, Khan IA, Edwards DJ and Shah A (2004) In vivo assessment of botanical supplementation on human cytochrome P450 phenotypes: Citrus aurantium, Echinacea purpurea, milk thistle, and saw palmetto. *Clin Pharmacol Ther* **76**:428–440.

Gurley BJ, Gardner SF, Hubbard MA, Williams DK, Gentry WB, Cui Y and Ang CY (2005a) Clinical assessment of effects of botanical supplementation on cytochrome P450 phenotypes in the elderly: St John's wort, garlic oil, Panax ginseng and Ginkgo biloba. *Drugs Aging* **22**: 525–539.

Gurley BJ, Gardner SF, Hubbard MA, Williams DK, Gentry WB, Khan IA and Shah A (2005b) In vivo effects of goldenseal, kava kava, black cohosh, and valerian on human cytochrome P450 1A2, 2D6, 2E1, and 3A4/5 phenotypes. *Clin Pharmacol Ther* **77**:415–426.

Guyonnet D, Belloir C, Suschetet M, Siess MH and Le Bon AM (2000) Liver subcellular fractions from rats treated by organosulfur compounds from Allium modulate mutagen activation. *Mutat Res* **466**:17–26.

Guyonnet D, Siess MH, Le Bon AM and Suschetet M (1999) Modulation of phase II enzymes by organosulfur compounds from allium vegetables in rat tissues. *Toxicol Appl Pharmacol* **154**:50–58.

Gwilt PR, Lear CL, Tempero MA, Birt DD, Grandjean AC, Ruddon RW and Nagel DL (1994) The effect of garlic extract on human metabolism of acetaminophen. *Cancer Epidemiol Biomarkers Prev* **3**:155–160.

Haber D, Siess MH, de Waziers I, Beaune P and Suschetet M (1994) Modification of hepatic drug-metabolizing enzymes in rat fed naturally occurring allyl sulphides. *Xenobiotica* **24**:169–182.

Hecht SS (2000) Inhibition of carcinogenesis by isothiocyanates. *Drug Metab Rev* **32**:395–411.

Hecht SS, Carmella SG and Murphy SE (1999) Effects of watercress consumption on urinary metabolites of nicotine in smokers. *Cancer Epidemiol Biomarkers Prev* **8**:907–913.

Hellum BH, Hu Z and Nilsen OG (2007) The induction of CYP1A2, CYP2D6 and CYP3A4 by six trade herbal products in cultured primary human hepatocytes. *Basic Clin Pharmacol Toxicol* **100**:23–30.

Helsby NA, Chipman JK, Gescher A and Kerr D (1998) Inhibition of mouse and human CYP 1A- and 2E1-dependent substrate metabolism by the isoflavonoids genistein and equol. *Food Chem Toxicol* **36**:375–382.

Henderson GL, Harkey MR, Gershwin ME, Hackman RM, Stern JS and Stresser DM (1999) Effects of ginseng components on c-DNA-expressed cytochrome P450 enzyme catalytic activity. *Life Sci* **65**:PL209–214.

Hennessy M, Kelleher D, Spiers JP, Barry M, Kavanagh P, Back D, Mulcahy F and Feely J (2002) St Johns wort increases expression of P-glycoprotein: implications for drug interactions. *Br J Clin Pharmacol* **53**:75–82.

Hu Z, Yang X, Ho PC, Chan SY, Heng PW, Chan E, Duan W, Koh HL and Zhou S (2005) Herb-drug interactions: a literature review. *Drugs* **65**:1239–1282.

Imai Y, Tsukahara S, Asada S and Sugimoto Y (2004) Phytoestrogens/flavonoids reverse breast cancer resistance protein/ABCG2-mediated multidrug resistance. *Cancer Res* **64**:4346–4352.

Izzo AA and Ernst E (2001) Interactions between herbal medicines and prescribed drugs: a systematic review. *Drugs* **61**:2163–2175.

James JS (2001) Garlic reduces saquinavir blood levels 50%; may affect other drugs. *AIDS Treat News*:2–3.

Janetzky K and Morreale AP (1997) Probable interaction between warfarin and ginseng. *Am J Health Syst Pharm* **54**:692–693.

Jang SI, Kim HJ, Hwang KM, Jekal SJ, Pae HO, Choi BM, Yun YG, Kwon TO, Chung HT and Kim YC (2003) Hepatoprotective effect of baicalin, a major flavone from Scutellaria radix, on acetaminophen-induced liver injury in mice. *Immunopharmacol Immunotoxicol* **25**:585–594.

Ji Y and Morris ME (2004) Effect of organic isothiocyanates on breast cancer resistance protein (ABCG2)-mediated transport. *Pharm Res* **21**:2261–2269.

Ji Y and Morris ME (2005a) Membrane transport of dietary phenethyl isothiocyanate by ABCG2 (breast cancer resistance protein). *Mol Pharm* **2**:414–419.

Ji Y and Morris ME (2005b) Transport of dietary phenethyl isothiocyanate is mediated by multidrug resistance protein 2 but not P-glycoprotein. *Biochem Pharmacol* **70**: 640–647.

Jiang X, Williams KM, Liauw WS, Ammit AJ, Roufogalis BD, Duke CC, Day RO and McLachlan AJ (2005) Effect of ginkgo and ginger on the pharmacokinetics and pharmacodynamics of warfarin in healthy subjects. *Br J Clin Pharmacol* **59**:425–432.

Jodoin J, Demeule M and Beliveau R (2002) Inhibition of the multidrug resistance P-glycoprotein activity by green tea polyphenols. *Biochim Biophys Acta* **1542**:149–159.

Johne A, Brockmoller J, Bauer S, Maurer A, Langheinrich M and Roots I (1999) Pharmacokinetic interaction of digoxin with an herbal extract from St John's wort (Hypericum perforatum). *Clin Pharmacol Ther* **66**:338–345.

Johne A, Schmider J, Brockmoller J, Stadelmann AM, Stormer E, Bauer S, Scholler G, Langheinrich M and Roots I (2002) Decreased plasma levels of amitriptyline and its metabolites on comedication with an extract from St. John's wort (Hypericum perforatum). *J Clin Psychopharmacol* **22**:46–54.

Kao YC, Zhou C, Sherman M, Laughton CA and Chen S (1998) Molecular basis of the inhibition of human aromatase (estrogen synthetase) by flavone and isoflavone phytoestrogens: A site-directed mutagenesis study. *Environ Health Perspect* **106**:85–92.

Karyekar CS, Eddington ND and Dowling TC (2002) Effect of St. John's Wort extract on intestinal expression of cytochrome P4501A2: studies in LS180 cells. *J Postgrad Med* **48**:97–100.

Kasibhatta R and Naidu MU (2007) Influence of piperine on the pharmacokinetics of nevirapine under fasting conditions: a randomised, crossover, placebo-controlled study. *Drugs R D* **8**: 383–391.

Kaufman DW, Kelly JP, Rosenberg L, Anderson TE and Mitchell AA (2002) Recent patterns of medication use in the ambulatory adult population of the United States: the Slone survey. *JAMA* **287**:337–344.

Kawaguchi A, Ohmori M, Tsuruoka S, Nishiki K, Harada K, Miyamori I, Yano R, Nakamura T, Masada M and Fujimura A (2004) Drug interaction between St John's Wort and quazepam. *Br J Clin Pharmacol* **58**:403–410.

Kent UM, Yanev S and Hollenberg PF (1999) Mechanism-based inactivation of cytochromes P450 2B1 and P450 2B6 by n-propylxanthate. *Chem Res Toxicol* **12**:317–322.

Keum YS, Owuor ED, Kim BR, Hu R and Kong AN (2003) Involvement of Nrf2 and JNK1 in the activation of antioxidant responsive element (ARE) by chemopreventive agent phenethyl isothiocyanate (PEITC). *Pharm Res* **20**:1351–1356.

Kim SW, Kwon HY, Chi DW, Shim JH, Park JD, Lee YH, Pyo S and Rhee DK (2003) Reversal of P-glycoprotein-mediated multidrug resistance by ginsenoside Rg(3). *Biochem Pharmacol* **65**:75–82.

Kitagawa S (2006) Inhibitory effects of polyphenols on p-glycoprotein-mediated transport. *Biol Pharm Bull* **29**:1–6.

Komoroski BJ, Zhang S, Cai H, Hutzler JM, Frye R, Tracy TS, Strom SC, Lehmann T, Ang CY, Cui YY and Venkataramanan R (2004) Induction and inhibition of cytochromes P450 by the St. John's wort constituent hyperforin in human hepatocyte cultures. *Drug Metab Dispos* **32**: 512–518.

Konsue N and Ioannides C (2008) Tissue differences in the modulation of rat cytochromes P450 and phase II conjugation systems by dietary doses of phenethyl isothiocyanate. *Food Chem Toxicol* **46**:3677–3683.

Le Bon AM, Vernevaut MF, Guenot L, Kahane R, Auger J, Arnault I, Haffner T and Siess MH (2003) Effects of garlic powders with varying alliin contents on hepatic drug metabolizing enzymes in rats. *J Agric Food Chem* **51**:7617–7623.

Leclercq I, Desager JP and Horsmans Y (1998) Inhibition of chlorzoxazone metabolism, a clinical probe for CYP2E1, by a single ingestion of watercress. *Clin Pharmacol Ther* **64**: 144–149.

Lee FC, Ko JH, Park JK and Lee JS (1987) Effects of Panax ginseng on blood alcohol clearance in man. *Clin Exp Pharmacol Physiol* **14**:543–546.

Leslie EM, Mao Q, Oleschuk CJ, Deeley RG and Cole SP (2001) Modulation of multidrug resistance protein 1 (MRP1/ABCC1) transport and atpase activities by interaction with dietary flavonoids. *Mol Pharmacol* **59**:1171–1180.

Li X, Yun JK and Choi JS (2007) Effects of morin on the pharmacokinetics of etoposide in rats. *Biopharm Drug Dispos* **28**:151–156.

Li Y, Wang E, Patten CJ, Chen L and Yang CS (1994) Effects of flavonoids on cytochrome P450-dependent acetaminophen metabolism in rats and human liver microsomes. *Drug Metab Dispos* **22**:566–571.

Lin MC, Wang EJ, Patten C, Lee MJ, Xiao F, Reuhl KR and Yang CS (1996) Protective effect of diallyl sulfone against acetaminophen-induced hepatotoxicity in mice. *J Biochem Toxicol* **11**:11–20.

Liu XP, Goldring CE, Wang HY, Copple IM, Kitteringham NR and Park BK (2009) Extract of Ginkgo biloba induces glutathione-S-transferase subunit-P1 in vitro. *Phytomedicine* 16(5): 451–455.

Lohner K, Schnabele K, Daniel H, Oesterle D, Rechkemmer G, Gottlicher M and Wenzel U (2007) Flavonoids alter P-gp expression in intestinal epithelial cells in vitro and in vivo. *Mol Nutr Food Res* **51**:293–300.

Manson MM, Ball HW, Barrett MC, Clark HL, Judah DJ, Williamson G and Neal GE (1997) Mechanism of action of dietary chemoprotective agents in rat liver: induction of phase I and II drug metabolizing enzymes and aflatoxin B1 metabolism. *Carcinogenesis* **18**: 1729–1738.

Markowitz JS, DeVane CL, Boulton DW, Carson SW, Nahas Z and Risch SC (2000) Effect of St. John's wort (Hypericum perforatum) on cytochrome P-450 2D6 and 3A4 activity in healthy volunteers. *Life Sci* **66**:PL133–139.

Markowitz JS, Devane CL, Chavin KD, Taylor RM, Ruan Y and Donovan JL (2003) Effects of garlic (Allium sativum L.) supplementation on cytochrome P450 2D6 and 3A4 activity in healthy volunteers. *Clin Pharmacol Ther* **74**:170–177.

McWalter GK, Higgins LG, McLellan LI, Henderson CJ, Song L, Thornalley PJ, Itoh K, Yamamoto M and Hayes JD (2004) Transcription factor Nrf2 is essential for induction of NAD(P)H:quinone oxidoreductase 1, glutathione S-transferases, and glutamate cysteine ligase by broccoli seeds and isothiocyanates. *J Nutr* **134**:3499S–3506S.

Mitsunaga Y, Takanaga H, Matsuo H, Naito M, Tsuruo T, Ohtani H and Sawada Y (2000) Effect of bioflavonoids on vincristine transport across blood-brain barrier. *Eur J Pharmacol* **395**: 193–201.

Mizoyama Y, Takaki H, Sugihara N and Furuno K (2004) Inhibitory effect of flavonoids on N-acetylation of 5-aminosalicylic acid in cultured rat hepatocytes. *Biol Pharm Bull* **27**: 1455–1458.

Mohutsky MA, Anderson GD, Miller JW and Elmer GW (2006) Ginkgo biloba: evaluation of CYP2C9 drug interactions in vitro and in vivo. *Am J Ther* **13**:24–31.

Molnar J, Szabo D, Pusztai R, Mucsi I, Berek L, Ocsovszki I, Kawata E and Shoyama Y (2000) Membrane associated antitumor effects of crocine-, ginsenoside- and cannabinoid derivates. *Anticancer Res* **20**:861–867.

Moore LB, Goodwin B, Jones SA, Wisely GB, Serabjit-Singh CJ, Willson TM, Collins JL and Kliewer SA (2000) St. John's wort induces hepatic drug metabolism through activation of the pregnane X receptor. *Proc Natl Acad Sci U S A* **97**:7500–7502.

Moreno RL, Goosen T, Kent UM, Chung FL and Hollenberg PF (2001) Differential effects of naturally occurring isothiocyanates on the activities of cytochrome P450 2E1 and the mutant P450 2E1 T303A. *Arch Biochem Biophys* **391**:99–110.

Moreno RL, Kent UM, Hodge K and Hollenberg PF (1999) Inactivation of cytochrome P450 2E1 by benzyl isothiocyanate. *Chem Res Toxicol* **12**:582–587.

Nagao A, Seki M and Kobayashi H (1999) Inhibition of xanthine oxidase by flavonoids. *Biosci Biotechnol Biochem* **63**:1787–1790.

Nakajima M, Yoshida R, Shimada N, Yamazaki H and Yokoi T (2001) Inhibition and inactivation of human cytochrome P450 isoforms by phenethyl isothiocyanate. *Drug Metab Dispos* **29**:1110–1113.

Nakamura Y, Ohigashi H, Masuda S, Murakami A, Morimitsu Y, Kawamoto Y, Osawa T, Imagawa M and Uchida K (2000) Redox regulation of glutathione S-transferase induction by benzyl isothiocyanate: correlation of enzyme induction with the formation of reactive oxygen intermediates. *Cancer Res* **60**:219–225.

Nebel A, Schneider BJ, Baker RK and Kroll DJ (1999) Potential metabolic interaction between St. John's wort and theophylline. *Ann Pharmacother* **33**:502.

Nguyen H, Zhang S and Morris ME (2003) Effect of flavonoids on MRP1-mediated transport in Panc-1 cells. *J Pharm Sci* **92**:250–257.

Obach RS (2000) Inhibition of human cytochrome P450 enzymes by constituents of St. John's Wort, an herbal preparation used in the treatment of depression. *J Pharmacol Exp Ther* **294**: 88–95.

Obermeier MT, White RE and Yang CS (1995) Effects of bioflavonoids on hepatic P450 activities. *Xenobiotica* **25**:575–584.

Ofer M, Wolffram S, Koggel A, Spahn-Langguth H and Langguth P (2005) Modulation of drug transport by selected flavonoids: Involvement of P-gp and OCT? *Eur J Pharm Sci* **25**:263–271.

Ohnishi N, Kusuhara M, Yoshioka M, Kuroda K, Soga A, Nishikawa F, Koishi T, Nakagawa M, Hori S, Matsumoto T, Yamashita M, Ohta S, Takara K and Yokoyama T (2003) Studies on interactions between functional foods or dietary supplements and medicines. I. Effects of Ginkgo biloba leaf extract on the pharmacokinetics of diltiazem in rats. *Biol Pharm Bull* **26**:1315–1320.

Paolini M, Perocco P, Canistro D, Valgimigli L, Pedulli GF, Iori R, Croce CD, Cantelli-Forti G, Legator MS and Abdel-Rahman SZ (2004) Induction of cytochrome P450, generation of oxidative stress and in vitro cell-transforming and DNA-damaging activities by glucoraphanin, the bioprecursor of the chemopreventive agent sulforaphane found in broccoli. *Carcinogenesis* **25**:61–67.

Pattanaik S, Hota D, Prabhakar S, Kharbanda P and Pandhi P (2006) Effect of piperine on the steady-state pharmacokinetics of phenytoin in patients with epilepsy. *Phytother Res* **20**: 683–686.

Peng SX, Ritchie DM, Cousineau M, Danser E, Dewire R and Floden J (2006) Altered oral bioavailability and pharmacokinetics of P-glycoprotein substrates by coadministration of biochanin A. *J Pharm Sci* **95**:1984–1993.

Petri N, Tannergren C, Holst B, Mellon FA, Bao Y, Plumb GW, Bacon J, O'Leary KA, Kroon PA, Knutson L, Forsell P, Eriksson T, Lennernas H and Williamson G (2003) Absorption/metabolism of sulforaphane and quercetin, and regulation of phase II enzymes, in human jejunum in vivo. *Drug Metab Dispos* **31**:805–813.

Phang JM, Poore CM, Lopaczynska J and Yeh GC (1993) Flavonol-stimulated efflux of 7,12-dimethylbenz(a)anthracene in multidrug-resistant breast cancer cells. *Cancer Res* **53**: 5977–5981.

Piao YJ and Choi JS (2008) Effects of morin on the pharmacokinetics of nicardipine after oral and intravenous administration of nicardipine in rats. *J Pharm Pharmacol* **60**:625–629.

Piscitelli SC, Burstein AH, Chaitt D, Alfaro RM and Falloon J (2000) Indinavir concentrations and St John's wort. *Lancet* **355**:547–548.

Rajnarayana K, Reddy MS and Krishna DR (2003) Diosmin pretreatment affects bioavailability of metronidazole. *Eur J Clin Pharmacol* **58**:803–807.

Rajnarayana K, Reddy MS, Vidyasagar J and Krishna DR (2004) Study on the influence of silymarin pretreatment on metabolism and disposition of metronidazole. *Arzneimittelforschung* **54**:109–113.

Raucy JL (2003) Regulation of CYP3A4 expression in human hepatocytes by pharmaceuticals and natural products. *Drug Metab Dispos* **31**:533–539.

Ruschitzka F, Meier PJ, Turina M, Luscher TF and Noll G (2000) Acute heart transplant rejection due to Saint John's wort. *Lancet* **355**:548–549.

Russmann S, Lauterburg BH, Barguil Y, Choblet E, Cabalion P, Rentsch K and Wenk M (2005) Traditional aqueous kava extracts inhibit cytochrome P450 1A2 in humans: protective effect against environmental carcinogens? *Clin Pharmacol Ther* **77**:453–454.

Shibayama Y, Ikeda R, Motoya T and Yamada K (2004) St. John's Wort (Hypericum perforatum) induces overexpression of multidrug resistance protein 2 (MRP2) in rats: a 30-day ingestion study. *Food Chem Toxicol* **42**:995–1002.

Shin SC, Piao YJ and Choi JS (2008) Effects of morin on the bioavailability of tamoxifen and its main metabolite, 4-hydroxytamoxifen, in rats. *In Vivo* **22**:391–395.

Shinozuka K, Umegaki K, Kubota Y, Tanaka N, Mizuno H, Yamauchi J, Nakamura K and Kunitomo M (2002) Feeding of Ginkgo biloba extract (GBE) enhances gene expression of hepatic cytochrome P-450 and attenuates the hypotensive effect of nicardipine in rats. *Life Sci* **70**:2783–2792.

Si D, Wang Y, Guo Y, Wang J, Zhou H, Zhou YH, Li ZS and Fawcett JP (2008) Mechanism of CYP2C9 inhibition by flavones and flavonols. *Drug Metab Dispos* 36:2316–2323.

Siess MH, Le Bon AM, Canivenc-Lavier MC and Suschetet M (1997) Modification of hepatic drug-metabolizing enzymes in rats treated with alkyl sulfides. *Cancer Lett* **120**:195–201.

Siess MH, Leclerc J, Canivenc-Lavier MC, Rat P and Suschetet M (1995) Heterogenous effects of natural flavonoids on monooxygenase activities in human and rat liver microsomes. *Toxicol Appl Pharmacol* **130**:73–78.

Smith TJ, Guo Z, Guengerich FP and Yang CS (1996) Metabolism of 4-(methylnitrosamino)-1-(3-pyridyl)-1-butanone (NNK) by human cytochrome P450 1A2 and its inhibition by phenethyl isothiocyanate. *Carcinogenesis* **17**:809–813.

Sridar C, Goosen TC, Kent UM, Williams JA and Hollenberg PF (2004) Silybin inactivates cytochromes P450 3A4 and 2C9 and inhibits major hepatic glucuronosyltransferases. *Drug Metab Dispos* **32**:587–594.

Steele VE, Kelloff GJ, Balentine D, Boone CW, Mehta R, Bagheri D, Sigman CC, Zhu S and Sharma S (2000) Comparative chemopreventive mechanisms of green tea, black tea and selected polyphenol extracts measured by in vitro bioassays. *Carcinogenesis* **21**:63–67.

Sugimoto K, Ohmori M, Tsuruoka S, Nishiki K, Kawaguchi A, Harada K, Arakawa M, Sakamoto K, Masada M, Miyamori I and Fujimura A (2001) Different effects of St John's wort on the pharmacokinetics of simvastatin and pravastatin. *Clin Pharmacol Ther* **70**:518–524.

Sugiyama T, Kubota Y, Shinozuka K, Yamada S, Yamada K and Umegaki K (2004) Induction and recovery of hepatic drug metabolizing enzymes in rats treated with Ginkgo biloba extract. *Food Chem Toxicol* **42**:953–957.

Talalay P and Fahey JW (2001) Phytochemicals from cruciferous plants protect against cancer by modulating carcinogen metabolism. *J Nutr* **131**:3027S–3033S.

Tamura H and Matsui M (2000) Inhibitory effects of green tea and grape juice on the phenol sulfotransferase activity of mouse intestines and human colon carcinoma cell line, Caco-2. *Biol Pharm Bull* **23**:695–699.

Tang J, Sun J, Zhang Y, Li L, Cui F and He Z (2007) Herb-drug interactions: effect of Ginkgo biloba extract on the pharmacokinetics of theophylline in rats. *Food Chem Toxicol* **45**:2441–2445.

Tian R, Koyabu N, Morimoto S, Shoyama Y, Ohtani H and Sawada Y (2005) Functional induction and de-induction of P-glycoprotein by St. John's wort and its ingredients in a human colon adenocarcinoma cell line. *Drug Metab Dispos* **33**:547–554.

Tseng E, Kamath A and Morris ME (2002) Effect of organic isothiocyanates on the P-glycoprotein- and MRP1-mediated transport of daunomycin and vinblastine. *Pharm Res* **19**:1509–1515.

Tsukamoto S, Tomise K, Aburatani M, Onuki H, Hirorta H, Ishiharajima E and Ohta T (2004) Isolation of cytochrome P450 inhibitors from strawberry fruit, Fragaria ananassa. *J Nat Prod* **67**:1839–1841.

Uchida S, Yamada H, Li XD, Maruyama S, Ohmori Y, Oki T, Watanabe H, Umegaki K, Ohashi K and Yamada S (2006) Effects of Ginkgo biloba extract on pharmacokinetics and pharmacodynamics of tolbutamide and midazolam in healthy volunteers. *J Clin Pharmacol* **46**:1290–1298.

Uda Y, Price KR, Williamson G and Rhodes MJ (1997) Induction of the anticarcinogenic marker enzyme, quinone reductase, in murine hepatoma cells in vitro by flavonoids. *Cancer Lett* **120**:213–216.

van der Logt EM, Roelofs HM, Nagengast FM and Peters WH (2003) Induction of rat hepatic and intestinal UDP-glucuronosyltransferases by naturally occurring dietary anticarcinogens. *Carcinogenesis* **24**:1651–1656.

van Erp NP, Baker SD, Zhao M, Rudek MA, Guchelaar HJ, Nortier JW, Sparreboom A and Gelderblom H (2005) Effect of milk thistle (Silybum marianum) on the pharmacokinetics of irinotecan. *Clin Cancer Res* **11**:7800–7806.

van Zanden JJ, Ben Hamman O, van Iersel ML, Boeren S, Cnubben NH, Lo Bello M, Vervoort J, van Bladeren PJ and Rietjens IM (2003) Inhibition of human glutathione S-transferase P1-1 by the flavonoid quercetin. *Chem Biol Interact* **145**:139–148.

van Zanden JJ, van der Woude H, Vaessen J, Usta M, Wortelboer HM, Cnubben NH and Rietjens IM (2007) The effect of quercetin phase II metabolism on its MRP1 and MRP2 inhibiting potential. *Biochem Pharmacol* **74**:345–351.

van Zanden JJ, Wortelboer HM, Bijlsma S, Punt A, Usta M, Bladeren PJ, Rietjens IM and Cnubben NH (2005) Quantitative structure activity relationship studies on the flavonoid mediated inhibition of multidrug resistance proteins 1 and 2. *Biochem Pharmacol* **69**:699–708.

Walle T, Otake Y, Galijatovic A, Ritter JK and Walle UK (2000) Induction of UDP-glucuronosyltransferase UGT1A1 by the flavonoid chrysin in the human hepatoma cell line hep G2. *Drug Metab Dispos* **28**:1077–1082.

Walle UK and Walle T (2002) Induction of human UDP-glucuronosyltransferase UGT1A1 by flavonoids-structural requirements. *Drug Metab Dispos* **30**:564–569.

Wang Q and Morris ME (2007a) Flavonoids modulate monocarboxylate transporter-1-mediated transport of gamma-hydroxybutyrate in vitro and in vivo. *Drug Metab Dispos* **35**:201–208.

Wang W, Liu LQ, Higuchi CM and Chen H (1998) Induction of NADPH:quinone reductase by dietary phytoestrogens in colonic Colo205 cells. *Biochem Pharmacol* **56**:189–195.

Wang X and Morris ME (2007b) Effects of the flavonoid chrysin on nitrofurantoin pharmacokinetics in rats: potential involvement of ABCG2. *Drug Metab Dispos* **35**:268–274.

Wang X, Wang Q and Morris ME (2008) Pharmacokinetic interaction between the flavonoid luteolin and gamma-hydroxybutyrate in rats: potential involvement of monocarboxylate transporters. *AAPS J* **10**:47–55.

Wang X, Wolkoff AW and Morris ME (2005) Flavonoids as a novel class of human organic anion-transporting polypeptide OATP1B1 (OATP-C) modulators. *Drug Metab Dispos* **33**:1666–1672.

Wang YH, Chao PD, Hsiu SL, Wen KC and Hou YC (2004) Lethal quercetin-digoxin interaction in pigs. *Life Sci* **74**:1191–1197.

Wang Z, Gorski JC, Hamman MA, Huang SM, Lesko LJ and Hall SD (2001) The effects of St John's wort (Hypericum perforatum) on human cytochrome P450 activity. *Clin Pharmacol Ther* **70**:317–326.

Wang Z, Hamman MA, Huang SM, Lesko LJ and Hall SD (2002) Effect of St John's wort on the pharmacokinetics of fexofenadine. *Clin Pharmacol Ther* **71**:414–420.

Wargovich MJ (2006) Diallylsulfide and allylmethylsulfide are uniquely effective among organosulfur compounds in inhibiting CYP2E1 protein in animal models. *J Nutr* **136**: 832S–834S.

Wen X, Walle UK and Walle T (2005) 5,7-dimethoxyflavone downregulates CYP1A1 expression and benzo[a]pyrene-induced DNA binding in Hep G2 cells. *Carcinogenesis* **26**:803–809.

Wen Z, Dumas TE, Schrieber SJ, Hawke RL, Fried MW and Smith PC (2008) Pharmacokinetics and metabolic profile of free, conjugated, and total silymarin flavonolignans in human plasma after oral administration of milk thistle extract. *Drug Metab Dispos* **36**:65–72.

Wenk M, Todesco L and Krahenbuhl S (2004) Effect of St John's wort on the activities of CYP1A2, CYP3A4, CYP2D6, N-acetyltransferase 2, and xanthine oxidase in healthy males and females. *Br J Clin Pharmacol* **57**:495–499.

Wentworth JM, Agostini M, Love J, Schwabe JW and Chatterjee VK (2000) St John's wort, a herbal antidepressant, activates the steroid X receptor. *J Endocrinol* **166**:R11–16.

Wu CC, Sheen LY, Chen HW, Kuo WW, Tsai SJ and Lii CK (2002) Differential effects of garlic oil and its three major organosulfur components on the hepatic detoxification system in rats. *J Agric Food Chem* **50**:378–383.

Wu JW, Lin LC and Tsai TH (2009) Drug–drug interactions of silymarin on the perspective of pharmacokinetics. *J Ethnopharmacol* **121**:185–193.

Yannai S, Day AJ, Williamson G and Rhodes MJ (1998) Characterization of flavonoids as monofunctional or bifunctional inducers of quinone reductase in murine hepatoma cell lines. *Food Chem Toxicol* **36**:623–630.

Yin OQ, Tomlinson B, Waye MM, Chow AH and Chow MS (2004) Pharmacogenetics and herb-drug interactions: experience with Ginkgo biloba and omeprazole. *Pharmacogenetics* **14**: 841–850.

Yoo HH, Lee M, Chung HJ, Lee SK and Kim DH (2007) Effects of diosmin, a flavonoid glycoside in citrus fruits, on P-glycoprotein-mediated drug efflux in human intestinal Caco-2 cells. *J Agric Food Chem* **55**:7620–7625.

Yoshioka M, Ohnishi N, Sone N, Egami S, Takara K, Yokoyama T and Kuroda K (2004) Studies on interactions between functional foods or dietary supplements and medicines. III. Effects of ginkgo biloba leaf extract on the pharmacokinetics of nifedipine in rats. *Biol Pharm Bull* **27**:2042–2045.

Yoxall V, Kentish P, Coldham N, Kuhnert N, Sauer MJ and Ioannides C (2005) Modulation of hepatic cytochromes P450 and phase II enzymes by dietary doses of sulforaphane in rats: implications for its chemopreventive activity. *Int J Cancer* **117**:356–362.

Yuan CS, Wei G, Dey L, Karrison T, Nahlik L, Maleckar S, Kasza K, Ang-Lee M and Moss J (2004) Brief communication: American ginseng reduces warfarin's effect in healthy patients: a randomized, controlled trial. *Ann Intern Med* **141**:23–27.

Yue QY, Bergquist C and Gerden B (2000) Safety of St John's wort (Hypericum perforatum). *Lancet* **355**:576–577.

Zhai S, Dai R, Friedman FK and Vestal RE (1998a) Comparative inhibition of human cytochromes P450 1A1 and 1A2 by flavonoids. *Drug Metab Dispos* **26**:989–992.

Zhai S, Dai R, Wei X, Friedman FK and Vestal RE (1998b) Inhibition of methoxyresorufin demethylase activity by flavonoids in human liver microsomes. *Life Sci* **63**:PL119–123.

Zhang S and Morris ME (2003) Effects of the flavonoids biochanin A, morin, phloretin, and silymarin on P-glycoprotein-mediated transport. *J Pharmacol Exp Ther* **304**: 1258–1267.

Zhang S, Yang X and Morris ME (2004a) Combined effects of multiple flavonoids on breast cancer resistance protein (ABCG2)-mediated transport. *Pharm Res* **21**:1263–1273.

Zhang S, Yang X and Morris ME (2004b) Flavonoids are inhibitors of breast cancer resistance protein (ABCG2)-mediated transport. *Mol Pharmacol* **65**:1208–1216.

Zhang W, Tan TM and Lim LY (2007) Impact of curcumin-induced changes in P-glycoprotein and CYP3A expression on the pharmacokinetics of peroral celiprolol and midazolam in rats. *Drug Metab Dispos* **35**:110–115.

Zhang Y and Talalay P (1998) Mechanism of differential potencies of isothiocyanates as inducers of anticarcinogenic phase 2 enzymes. *Cancer Res* **58**:4632–4639.

Zhao LZ, Huang M, Chen J, Ee PL, Chan E, Duan W, Guan YY, Hong YH, Chen X and Zhou S (2006) Induction of propranolol metabolism by Ginkgo biloba extract EGb 761 in rats. *Curr Drug Metab* **7**:577–587.

Zhou C, Poulton EJ, Grun F, Bammler TK, Blumberg B, Thummel KE and Eaton DL (2007) The dietary isothiocyanate sulforaphane is an antagonist of the human steroid and xenobiotic nuclear receptor. *Mol Pharmacol* **71**:220–229.

Zhu M, Chan KW, Ng LS, Chang Q, Chang S and Li RC (1999) Possible influences of ginseng on the pharmacokinetics and pharmacodynamics of warfarin in rats. *J Pharm Pharmacol* **51**: 175–180.

Zou L, Harkey MR and Henderson GL (2002) Effects of herbal components on cDNA-expressed cytochrome P450 enzyme catalytic activity. *Life Sci* **71**:1579–1589.

Zuber R, Modriansky M, Dvorak Z, Rohovsky P, Ulrichova J, Simanek V and Anzenbacher P (2002) Effect of silybin and its congeners on human liver microsomal cytochrome P450 activities. *Phytother Res* **16**:632–638.

Zutshi RK, Singh R, Zutshi U, Johri RK and Atal CK (1985) Influence of piperine on rifampicin blood levels in patients of pulmonary tuberculosis. *J Assoc Physicians India* **33**:223–224.

Chapter 23
Anticipating and Minimizing Drug Interactions in a Drug Discovery and Development Setting: An Industrial Perspective

Ragini Vuppugalla, Sean Kim, Tatyana Zvyaga, Yong-hae Han, Praveen Balimane, Punit Marathe, and A. David Rodrigues

Abstract In the current age of polypharmacy, it is increasingly likely that a new chemical entity (NCE) will be prescribed with a second drug that demonstrates a narrow therapeutic index. As a result, one has to consider interactions involving drug-metabolizing enzymes and transporters. NCEs with drug–drug interaction (DDI) liabilities may have limited marketing potential, as they may alter the pharmacokinetic profile of a co-administered drug resulting in either unwanted side effects or loss of pharmacological activity. Within the current competitive landscape, therefore, it is highly desirable to select candidates with reduced potential for DDIs and most pharmaceutical companies spend considerable resources screening and triaging NCEs for induction and inhibition of drug-metabolizing enzymes (e.g., cytochromes P450) and transporters. Thus, the purpose of the present chapter is to provide an industrial perspective on how the existing strategies are utilized to enable the selection of suitable candidates with reduced DDI risk. Additional emphasis will be placed on in vitro tools and the challenges associated with the prediction of DDIs prior to first in man.

Abbreviations

NCE	new chemical entity
DDI	drug–drug interaction
P450	cytochrome P450
PK	pharmacokinetics
SAR	structure–activity relationship
AUC	area under the curve
IVIVC	in vitro–in vivo correlation
P-gp	P-glycoprotein
rCYP	recombinant cytochrome P450

A.D. Rodrigues (✉)
Department of Pharmaceutical Candidate Optimization, Bristol-Myers Squibb Co., Princeton, NJ 08543, USA
e-mail: david.rodrigues@bms.com

K.S. Pang et al. (eds.), *Enzyme- and Transporter-Based Drug–Drug Interactions*, DOI 10.1007/978-1-4419-0840-7_23,

HLM	human liver microsomes
LC	liquid chromatography
MS	mass spectrometry
hPXR	human pregnane X receptor
NHR	nuclear hormone receptor
MRP2	multidrug resistance gene-associated protein 2
OATP	organic anion transporter protein
ADME	absorption, distribution, metabolism, and elimination
NTCP	Na^+-taurocholate co-transporting polypeptides
OCT	organic cation transporter
SLC	solute carrier
SLCO	solute carrier organic anion
CAR	constitutive androstane receptor
ABC	ATP-binding cassette
PPAR	peroxisome proliferator activated receptor
TEER	trans epithelial electrical resistance
AhR	aromatic hydrocarbon receptor
FDA	Food and Drug administration
K_m	apparent affinity constant
CL_{int}	intrinsic clearance
$CL_{int,i}$	intrinsic clearance in the presence of inhibitor
AUC,i	area under the curve in the presence of inhibitor
$[I]$	inhibitor concentration
k_i	inhibition constant
E_{max}	maximum induction
EC_{50}	concentration of inducer associated with half maximum induction
IC_{50}	concentration of inhibitor associated with half maximal inhibition
C_{max}	maximum plasma concentration
k_{inact}	maximal rate of enzyme inactivation at saturating concentrations of the inhibitor
K_I	concentration of inhibitor that produces half maximal inactivation
K_{deg}	turnover rate of the inhibited enzyme
f_m	fraction metabolized by all P450s
$f_{m,CYP}$	fraction metabolized by specific P450
UGTs	UDP-glucuronosyltransferases
SULTs	sulfotransferases
hERG	human ether-a-go-go related gene

23.1 Introduction

In an industrial setting, where compound attrition is a significant concern, the development of new chemical entities (NCEs) with improved pharmacokinetic (PK), potency, and liability profiles is important. At the same time, many diseases are treated with multiple drugs, drug cocktails, or co-formulated drugs, and so the risk of clinically significant drug–drug interactions (DDIs) has increased. Such DDIs can bring about unwanted side effects or loss of pharmacological activity. Consequently, DDIs are considered one of the major liabilities for any NCE being considered for full development.

The concept of DDIs is not new and was recognized as early as the late 1960s (Surveillance, 1972). In recent years, however, DDIs have garnered even more attention, because of well-documented cases involving market withdrawals, the impact on market share, and increased regulatory burden (Table 23.1). In light of such cases, the pharmaceutical industry has stepped up screening efforts to enable the selection of NCEs with reduced DDI potential, specifically, identifying drug candidates that not only have the least potential to modulate drug-metabolizing enzymes and transporters (act as perpetrator) but also bring forward molecules that are least susceptible to enzyme inhibition or induction (act as victim). Because the cytochromes P450 (P450) superfamily of proteins plays a major role in the metabolism of drugs, considerable attention is focused on the screening of compounds for inhibition and induction of these enzymes. In particular, the pharmaceutical industry and regulatory agencies have taken a prospective approach toward mitigating and minimizing P450-based DDIs using in vitro data at the early stages of drug development (Bjornsson et al., 2003). Such efforts are possible because of advances in P450 molecular biology, in silico approaches, and access to readily available in vitro models (human liver microsomes and recombinant proteins). With the advent of automation, liquid-handling instrumentation, and increased computing firepower, it has also been possible to develop high-throughput in vitro P450 assays and drive lead optimization campaigns by supporting the generation of structure–activity relationships (SARs) (Lin and Lu, 1998; White, 2000; Riley,

Table 23.1 Documented drug–drug interactions in the clinic

Victim	Perpetrator	Likely mechanism
Terfenadine**	Ketoconazole	CYP3A4 inhibition
Astemizole**	Ketoconazole	CYP3A4 inhibition
Cisapride**	Ketoconazole	CYP3A4 inhibition
Cerivastatin**	Gemfibrozil	CYP2C8 inhibition (OATP inhibition)
Statins	Mibefridil**	CYP3A4 inhibition
Ethinyl Estradiol	Pleconaril*	CYP3A induction
Repaglinide***	Gemfibrozil (+ Itraconazole)	CYP2C8/OATP inhibition (CYP3A4 inhibition)
Tizanidine***	Fluvoxamine (Rofecoxib)	CYP1A2 inhibition
Fluticasone***	Ritonavir	CYP3A4 inhibition
Ramelteon****	Fluvoxamine	CYP1A2 inhibition

*, **, ***, and **** refers to drugs that were not approved, withdrawn from the market, received regulatory feedback for label revisions, or FDA warning letter, respectively.

2001). Slowly, the repertoire of tools has extended to various drug transporters and other drug-metabolizing enzymes (e.g., UDP-glucuronosyltransferases) (Hidalgo, 2001; Miners et al., 2004; Hsiao et al., 2007). Beyond screening, it has been recognized that high-quality in vitro data can be leveraged to support the design, planning, and prioritization of the most relevant clinical DDI studies, in addition to those involving frequently evaluated drugs (e.g., cimetidine, theophylline, warfarin, digoxin, and oral contraceptives). Despite the utility of in vitro data, however, challenges remain and further progress is needed (Rodrigues AD, 2007).

Although our ability to qualitatively predict DDIs using in vitro data has increased, given the complex nature of most interactions, quantitative predictions remain challenging (Huang et al., 1999; Bjornsson et al., 2003). For example, a drug can potentially modulate enzyme activity by different mechanisms (e.g., reversible inhibition, mechanism-based inactivation, and induction) and failure to take into account all mechanisms may result in misleading predictions. Transporters and secondary drug-metabolizing enzymes can further complicate predictions and so more complex and fully integrated models are needed (Yu, 1999; Chiou et al., 2000; Suzuki and Sugiyama, 2000). Second, although numerous mathematical and kinetic models have been proposed as guides to predict DDIs, many are too simplistic. This is particularly evident when attempting to prospectively model the concentration of the inhibitor (perpetrator) in vivo, which is challenging in the absence of clinical data (Rodrigues et al., 2001; Yao and Levy, 2002; Bachmann, 2006).

The purpose of the present chapter is to provide an industrial perspective on how existing DDI screening and prediction strategies are utilized within a drug discovery and development setting. Additionally, some emphasis will be placed on existing strategies, their pitfalls and refinements, in vitro tools, and the challenges associated with the prediction of DDIs.

23.2 Pharmacokinetic Framework

23.2.1 Existing Strategies for Prospective Prediction of DDIs and the Framework for Support of Discovery

As described above, a perpetrator drug can potentially alter the PK of a concomitantly administered (victim) drug. Such an interaction can involve the inhibition or induction of one or more drug-metabolizing enzymes. The former can be manifested as reversible inhibition (e.g., competitive, noncompetitive, and uncompetitive) or mechanism-based inhibition (Lin and Pearson, 2002). Importantly, because the outcome can depend on the type of inhibition, it is advisable to identify the mechanism of inhibition as soon as possible. Fortunately, it is possible to assess whether an NCE is a reversible or mechanism-based inhibitor before attempting the forecasting of DDIs (Madan et al., 2002; Bjornsson et al., 2003; Venkatakrishnan et al., 2007).

Of the mechanisms described above, reversible (competitive) inhibition is the most frequently encountered and well studied. One approach that is commonly used for the prediction of the fold change in the victim area under the plasma concentration versus time curve (AUC) is described below in Equation (23.1).

$$\frac{CL_{\text{int}}}{CL_{\text{int,i}}} = \frac{\text{AUC,i}}{\text{AUC}} = 1 + \frac{[I]}{k_{\text{i}}} \tag{23.1}$$

The equation shows that one is able to assess the change in victim AUC based on knowledge of $[I]$ (inhibitor concentration) and k_{i} (inhibition constant). In this instance, linear PK and first-order enzyme kinetics for the substrate (concentrations are assumed to be lower than K_{m}) are assumed (Ito et al., 1998). CL_{int}, $CL_{\text{int,i}}$, AUC, and AUC,i represent the intrinsic clearance and area under the curve of the substrate in the absence and presence of the inhibitor. Thus, according to Equation (23.1), interactions can be regarded as low risk if the estimated $[I]/k_{\text{i}}$ ratio is less than 0.1, moderate risk if it is between 0.1 and 1.0, and high risk if the ratio is greater than 1.0 (Tucker et al., 2001).

Equation (23.1), although useful, assumes that the substrate is eliminated via a single metabolic pathway. However, this assumption may result in over-prediction of the interaction in cases where multiple pathways are involved in the elimination of the substrate (Brown et al., 2005). Additionally, the above equation does not account for other factors like P-glycoprotein (P-gp), gut first-pass metabolism, and does not describe the simultaneous influences of induction, inhibition, and mechanism-based inhibition. Thus, given its limitations, further refinements have been made in order to improve the predictive performance of this simplistic approach. This is evident in reports that account for the role of the fraction metabolized by the liver (f_{m}) and the fraction metabolized by the specific enzyme (e.g., P450 isoenzyme) such as CYP3A4 ($f_{\text{m,CYP3A4}}$) or CYP2C9 ($f_{\text{m,CYP2C9}}$) (Equation 23.2). Although originally reported by Rowland and Matin, (1973), the relationship has been applied increasingly by later researchers in the field (Ito et al., 1998; Rodrigues et al., 2001; Brown et al., 2005).

$$\frac{\text{AUC,i}}{\text{AUC}} = \frac{1}{\left[\frac{f_{\text{m}} \times f_{\text{m,CYP}}}{R}\right] + \left[1 - (f_{\text{m}} \times f_{\text{m,CYP}})\right]} \tag{23.2}$$

where f_{m}, $f_{\text{m,CYP}}$, AUC, AUC,i represent fraction of the dose metabolized (e.g., by all P450 isoforms), fraction of the dose metabolized by specific P450 enzyme, and area under the curve in the absence and presence of the inhibitor, respectively. Depending upon the mechanism of interaction (competitive inhibition, mechanism-based inhibition), the R term in the equation can be either one of the following:

For competitive and noncompetitive inhibition:

$$R = 1 + \frac{[I]}{k_{\text{i}}} \tag{23.3}$$

For mechanism-based inhibition:

$$R = 1 + \frac{k_{\text{inact}} \times [I]}{K_{\text{I}} \times K_{\text{deg}}} \tag{23.4}$$

where k_{inact}, K_{I}, and K_{deg} represent the maximal rate of enzyme inactivation at saturating concentrations of the inhibitor, concentration of inhibitor that produces half maximal inactivation, and turnover rate of the inhibited enzyme, respectively. Whereas values of k_{inact} and K_{I} are derived experimentally, the K_{deg} values are typically obtained from the literature (Venkatakrishnan et al., 2007).

Equation (23.2) could be applied for low-clearance drugs administered either orally or intravenously and for high-clearance drugs administered orally. This equation assumes that the metabolism of the substrate is restricted to liver, that the well-stirred model of hepatic clearance applies, and that the absorption of the drug is complete and not affected by first-pass gut metabolism (Yao and Levy, 2002). Consequently, Equations (23.5) and (23.6) have been proposed as a means to model DDIs with compounds that are subjected to both liver and gut first-pass metabolism (Wang et al., 2004). However, one should realize that these equations assume that (a) metabolism in the intestine is catalyzed solely by a single enzyme (e.g., CYP3A4); (b) intestinal enzymes do not contribute to the systemic clearance of the substrate; and (c) absorption is complete and is not influenced by the transporters expressed on the intestine (Wang et al., 2004).

$$\frac{\text{AUC,i}}{\text{AUC}} = \frac{1}{F_{\text{g}} + \left((1 - F_{\text{g}}) \times \frac{1}{1 + \frac{[I]_{\text{g}}}{k_{\text{i,g}}}} \right)} \times \frac{1}{\left(\frac{f_{\text{m}} \times f_{\text{m,CYP}}}{1 + \left(\frac{[I]_{\text{h}}}{k_{\text{i,h}}} \right)} \right) + (1 - f_{\text{m}} \times f_{\text{m,CYP}})} \tag{23.5}$$

$$\frac{\text{AUC,i}}{\text{AUC}} = \frac{1}{F_{\text{g}} + \left((1 - F_{\text{g}}) \times \frac{1}{\frac{k_{\text{inact,g}} \times [I]_{\text{g}}}{K_{\text{deg,g}} \times ([I]_{\text{g}} + K_{\text{I,g}})}} \right)} \times \frac{1}{\left(\frac{f_{\text{m}} \times f_{\text{m,CYP}}}{1 + \left(\frac{k_{\text{inact,h}} \times [I]_{\text{h}}}{K_{\text{I,h}} \times K_{\text{deg,h}}} \right)} \right) + (1 - f_{\text{m}} \times f_{\text{m,CYP}})} \tag{23.6}$$

where F_{g} is the fraction of the dose of victim drug that escapes the gut unchanged, $[I]_{\text{g}}$ and $[I]_{\text{h}}$ represent the concentrations of the inhibitor in the intestine and liver, respectively, $k_{\text{i,g}}$, $k_{\text{i,h}}$, $k_{\text{inact,g}}$, $k_{\text{inact,h}}$, $K_{\text{deg,g}}$, $K_{\text{deg,h}}$, $K_{\text{I,g}}$, and $K_{\text{I,h}}$ are the inhibitory constant, maximum rates of enzyme inactivation, turnover rate of the enzyme, and the concentration of inhibitor that produces half maximum inactivation in the gut and liver, respectively.

The concentration of the inhibitor in the intestinal wall ($[I]_g$) is estimated using the below equation:

$$[I]_g = \frac{\text{Dose} \times k_a \times F_a}{Q_g} \tag{23.7}$$

k_a and F_a refer to the absorption rate constant of the inhibitor and the fraction of the drug that reaches the intestine, respectively. A value of 248 mL/min was assumed as the intestinal blood flow (Q_g) (Rostami-Hodjegan and Tucker, 2004).

Although the approaches described above have been the basis of increasing numbers of DDI predictions and in vitro–in vivo correlations (IVIVCs), they do not account for the simultaneous influences of competitive inhibition, mechanism-based inactivation, and induction. This can be problematic for agents such as ritonavir, which serve as both inducers and inhibitors (Greenblatt et al., 2000). To address this concern, a comprehensive mathematical model has been proposed recently that allows one to estimate the net effects of competitive inhibition, mechanism-based inhibition, and induction in both gut and liver and is shown in Equation (23.8) (Fahmi et al., 2008)

$$\frac{\text{AUC,i}}{\text{AUC}} = \left(\frac{1}{\frac{K_{\text{deg,g}}}{K_{\text{deg,g}} + \frac{[I]_g \times k_{\text{inact,g}}}{[I]_g + K_{\text{I,g}}}} + 1 + \frac{E_{\text{max,g}} \times [I]_g}{[I]_g + \text{EC}_{50,\text{g}}} + \frac{1}{1 + \frac{[I]_g}{K_{\text{i,g}}}} \times (1 - F_g) + F_g} \right)$$
$$\times \left(\frac{1}{\frac{K_{\text{deg,h}}}{K_{\text{deg,h}} + \frac{[I]_h \times k_{\text{inact,h}}}{[I]_h + K_{\text{I,h}}}} + 1 + \frac{E_{\text{max,h}} \times [I]_h}{[I]_h + \text{EC}_{50,\text{h}}} + \frac{1}{1 + \frac{[I]_h}{K_{\text{i,h}}}} \times f_{\text{m,CYP}} + (1 - f_{\text{m,CYP}})} \right) \tag{23.8}$$

where $E_{\text{max,g}}$, $E_{\text{max,h}}$, $\text{EC}_{50,\text{g}}$, and $\text{EC}_{50,\text{h}}$ represent the maximum induction and the concentration of inducer associated with half maximum induction in gut and liver, respectively. The rest of the terms have been described above. Despite its ability to integrate different mechanisms of interaction, most of the assumptions of Equation (23.5) and (23.6) are still applicable for Equation (23.8).

The above models can serve as a useful framework to guide screening efforts and for the support of SAR. For a particular chemotype, it is advantageous to minimize DDI risk by screening out potent inhibitors (e.g., $\text{IC}_{50} < 1\ \mu\text{M}$), advancing only weak inhibitors (e.g., $\text{IC}_{50} > 20\ \mu\text{M}$), reduce the k_{inact}/K_I and $E_{\text{max}}/\text{EC}_{50}$ ratios, and have metabolism via multiple P450 isoforms ($f_{\text{m,CYP}} < 0.5$) (Equations 23.2, 23.5, 23.6, and 23.7). At the same time, potent (low dose) compounds are sought after, which further minimizes the potential for DDIs. By extension, this means that effective human PK and dose projections are needed at the NCE nomination stage, so that in vitro data are assessed in context (Huang et al., 2008).

23.2.2 The Impact of Inhibitor Concentration [I], In Vitro Parameters (k_i, k_{inact}, K_I, E_{max}, EC_{50}), and $f_m \times f_{m,CYP}$ on Predictions

Although the importance of [*I*] in DDI predictions has been known for some time, a consensus has not been reached as to which concentration of the inhibitor to use (Yao and Levy, 2002; Bachmann, 2006). The most ideal concentration would be the one at the active site of the enzyme in question. However, since it is not practically possible to measure the inhibitor concentrations surrounding the hepatic enzymes, predictions can be attempted using several surrogates for [*I*] (or $[I]_h$) including plasma total C_{max}, plasma free C_{max}, portal vein C_{max}, or portal vein free C_{max}, among others (Ito et al., 1998; Ito et al., 2004).

The maximum concentration of the inhibitor in the portal vein ($C_{max,\,pv}$) can be estimated using the equation below (Equation 23.9):

$$C_{max,\,pv} = C_{max} + \frac{\text{Dose} \times k_a \times F_a}{Q_h} \tag{23.9}$$

C_{max}, k_a, and F_a refer to the maximum plasma concentration of the inhibitor, absorption rate constant of the inhibitor, and the fraction of the drug that reaches the intestine, respectively. Q_h represents the hepatic blood flow (Obach et al., 2007).

Based on the free drug hypothesis, one would assume that the unbound concentration in plasma is the most meaningful parameter, as it is believed that only the unbound drug would diffuse into the hepatocytes. Although in some cases use of unbound plasma concentrations can improve predictions and IVIVCs (Komatsu et al., 2000), there are several studies that contradict the unbound drug hypothesis (von Moltke et al., 1994; Hsu et al., 1998). For example, Ito et al. (2004) have investigated the DDI potential of nearly 200 compounds using different inhibitor concentrations (total plasma C_{max}, average total plasma concentration, average unbound plasma concentration, and maximum hepatic input concentration) and concluded that total rather than the unbound plasma concentrations yielded better predictions. Conversely, there are several cases where even the total plasma concentrations under-predicted the interaction and additional correction factors (e.g., active uptake leading to higher liver-to-plasma ratio) had to be applied (Preskorn and Magnus, 1994; von Moltke et al., 1996).

In all the above cases, the inhibitor concentrations were assumed as constant. However, in reality, the concentration of the inhibitor changes with time, and therefore, predictions using a fixed value of [*I*] may result in substantial overestimation (or underestimation) of the in vivo interactions. Given this limitation with the existing models, physiologically-based pharmacokinetic (PBPK) models are becoming increasingly accepted due to their ability to take into consideration the temporal changes in concentration of inhibitors for the prediction of DDIs (Ito et al., 1998; Kanamitsu et al., 2000). Toward this end, numerous groups are now using commercially available software such as the Simcyp ADME (absorption, metabolism, distribution, elimination) simulator (Simcyp Ltd., Sheffield, UK). Such software packages allow one to simulate the substrate and inhibitor concentrations over time in a virtual population and thus provide a population-based estimate of DDI risk (Rostami-Hodjegan and Tucker, 2004). Increasingly, pharmaceutical companies are exploring the utility of such software packages with the hope that these tools will enhance our ability to more accurately predict DDIs, enable risk assessment, and support decision making. The merits and drawbacks of suck PBPK approaches will be more apparent in the coming years.

In addition to the problems of estimating the in vivo inhibitor concentrations, the accuracy of the prediction also relies on one's ability to accurately estimate in vitro parameters such as k_i, k_{inact}, and K_I. Depending upon the choice of the in vitro system used (e.g., liver microsomes, recombinant P450s, liver slices, and hepatocytes), the protein concentration in the incubation media, the time of incubation, and the choice of probe substrates and their concentrations, in vitro estimates of IC_{50}, k_i, k_{inact}, and K_I can differ and errors can significantly impact DDI projections. Additional factors such as non-specific binding in the incubations should also be taken into consideration (Grime and Riley, 2006). Thus, in vitro experiments should be performed under optimized conditions in order to reduce the influence of confounding factors on the accuracy of the in vitro estimates (Bjornsson et al., 2003).

Besides $[I]$ and in vitro inhibitory constants, the third factor that is critical for the prediction of in vivo DDIs is the product of f_m and $f_{m,CYP}$ (relative contribution of the affected metabolic pathway to the overall elimination of the drug). The results of some simulations highlighting the importance of $f_m \times f_{m,CYP}$ on the magnitude of an in vivo interaction are presented in Fig. 23.1. These simulations suggest that irrespective of the degree of inhibition and the potency of the inhibitor, a perceivable interaction is possible only when the affected elimination pathway accounts for greater than 50% of the total elimination. The simulation exercise also highlights the importance of multiple clearance routes and suggests that the risk of DDIs is minimal when the $f_m \times f_{m,CYP}$ values are less than 0.5. Recent studies by Brown et al. (2005) investigating the influence of parallel pathways on the accuracy of the

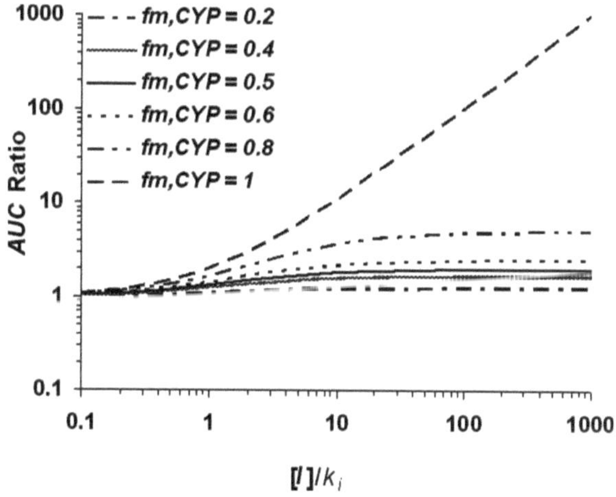

Fig. 23.1 Relationship between the fold change in AUC and $[I]/k_i$ ratio. The f_m (fraction of dose metabolized via P450 catalyzed oxidation) in these simulations was assumed as 1 and the $f_{m,CYP}$ (fraction of dose metabolized by specific P450) values were assumed as 0.2, 0.4, 0.5, 0.6, 0.8, and 1

in vivo predictions corroborated the results of the above simulations and suggested that the prediction accuracy can be improved significantly by accounting for the contribution of the parallel pathways. This means that one needs to obtain high quality metabolic profiling (in vitro and in vivo) and reaction phenotyping data (Zhang et al., 2007a).

23.2.3 Assessment of Drug Interaction Potential in Early Drug Discovery

As indicated above, the parameters that are critical for the prospective prediction of DDIs in the clinic include inhibitor concentration [I], in vitro inhibition or induction parameters (k_i, k_{inact}, K_I, E_{max}, and EC_{50}), and $f_m \times f_{m,CYP}$ values. Unfortunately, neither the human plasma concentration data nor the $f_m \times f_{m,CYP}$ values for NCEs are available at the early stages of drug discovery. Consequently, quantitative assessment of the inhibitory or induction potential of an NCE is not possible at the early stages of drug discovery and the kinetic models discussed above are rarely used for driving decisions. During these early stages, discovery teams solely rely on the in vitro data for the optimization of the P450 inhibition and induction liabilities of the chemotypes by developing SAR for these liabilities. Generally, arbitrary cutoff values are used for categorizing the compounds as strong, moderate, and weak inhibitors/inducers of CYP enzymes. For instance, a compound having an IC_{50} (EC_{50}) value of greater than 10 μM is normally viewed as a weak inhibitor or inducer; a potent inhibitor or inducer is defined as having an IC_{50} (EC_{50}) of less than 1 μM. When screening for time-dependent inhibition, the fold shift in IC_{50} (0 versus 30 min pre-incubation) is commonly used to assess the risk and compounds with greater than 2- to 3-fold shift in IC_{50} are flagged. Despite their arbitrary nature, these approaches serve as practical tools for triaging compounds. However, since these screening strategies are purely empirical and rarely take in vivo inhibitor concentrations, or $f_m \times f_{m,CYP}$ values, into consideration while interpreting the in vitro data, caution should be exercised to prevent the premature termination of some promising candidates. For example, it should be remembered that an IC_{50} (EC_{50}) value of lower than 1 μM will not result in a clinically significant interaction if the in vivo concentrations of the inhibitor are below 1 μM, or if the $f_m \times f_{m,CYP}$ value is less than 0.5. Thus, careful integration of PK and drug metabolism data, along with the projected efficacious dose and time–concentration profile, is critical to enable rational drug discovery decisions and to ensure the advancement of safe and effective therapeutics with a minimal side effect profile.

23.3 DDI Suites at Bristol-Myers Squibb

23.3.1 P450 Inhibition

In order to anticipate and predict the extent of potential DDI liabilities and to make an informed decision about the fate of a particular chemotype or individual compound, it is important to have access to a variety of assay suites ranging from fairly basic to rather complex P450 inhibition assays that can be used in parallel, or in a sequential (tier-like) fashion, at various stages of lead optimization (FDA, 2006).

During lead optimization, a number of high-capacity (high-throughput) P450 inhibition assays are employed that utilize heterologously expressed recombinant P450 enzymes (rCYP). Fluorogenic probe substrates, such as 7-ethoxy-4-trifluoromethylcoumarin, 7-benzyloxyquinoline, 7-methoxy-4-trifluoro methylcoumarin, 3-cyano-7-ethoxycoumarin, 3-[2-(*N,N*-diethyl-*N*-methylamino) ethyl]-7-methoxy-4-methylcoumarin, 7-benzyloxyresorufin, and 7-benzyloxy-4-trifluoromethylcoumarin, are used because they circumvent the need for chromatographic separation and enable the assays to be fully automated (Crespi et al., 1997; Crespi and Miller, 1999).

Recent advances in nanodispensing technologies, sophisticated robotics, automated liquid-handling devices, ultra-fast plate readers, and highly integrated information technology solutions have greatly facilitated miniaturization of these assays. It is now possible to run the assays in 384- and 1536-well formats, with very low reaction volumes (<30 μL) (Miller et al., 2000; Kariv et al., 2001; Yan et al., 2002; Yamamoto et al., 2004; Trubetskoy et al., 2005; Youdim et al., 2008). Such high-capacity, high-throughput, and cost-effective assays permit large numbers of compounds to be screened for P450 inhibition and support the needs of discovery teams.

Operationally, the first tier suite utilizes a panel of rCYPs (CYP1A2, 2B6, 2C8, 2C9, 2C19, 2D6, and 3A4) and is designed to generate reversible IC_{50} values quickly. The same panel can be adapted to assess time-dependent changes in IC_{50} values at multiple time points, from 5 to 60 min (Tier II TD rCYP Panel) (Yan et al., 2002; Naritomi et al., 2004; Yamamoto et al., 2004). For a few P450 enzymes of particular interest, such as CYP3A4, the IC_{50} values can be determined at two time points (5 and 30 min), which can be easily done as part of a first tier screening panel and support early detection of time-dependent inhibition. Typically, such rCYP-fluorogenic assays generate data that are not used to predict inhibitory P450 DDIs per se. However, the data are used to track SAR, rank-order, and bin different compounds within a chemotype or structural series. Compounds can also be flagged for additional follow-up as needed.

Tier II P450 inhibition assay panels utilize human liver microsomes (HLM) as the enzyme source and traditional drugs as substrate probes. The assay results generated in the HLMs are usually more predictive of the drug behavior in vivo and are widely accepted as a gold-standard throughout the pharmaceutical industry and recommended by regulators (FDA, 2006). More often than not, such assays are employed at later stages of lead optimization campaigns and in early preclinical development. In this instance, IC_{50}, k_i, k_{inact}, and K_I data are the basis of DDI predictions and IVIVCs (Silverman, 1995; Fontana et al., 2005). Unfortunately, HLM-based assays necessitate the use of highly selective P450 probe substrates. This means that time-consuming sample processing steps, liquid chromatography (LC) separation methods, and mass spectrometric (MS) detection are needed. In turn, this significantly limits the capacity and possibility of utilizing such assays in the very early stages of lead optimization. Despite such drawbacks, recent advances in analytical technologies have dramatically reduced the sample analysis time (e.g., from minutes to seconds per sample) (Shiau et al., 2008; Youdim et al., 2008; Zientek et al., 2008). For example, RapidFire^{TM} technology (BioTrove Inc. Woburn, MA) enables ultra-fast (in-line) solid phase extraction, coupled with the flow injection MS/MS analysis of the reaction components, within 4–10 sec (per sample). This allows one to analyze a single plate in about 7–16 min (96-well) or 40–60 min (384-well) (Ozbal et al., 2004). Although still slower than fluorescence detection, this ultra-fast analytical read-out can realistically support a relatively high-throughput HLM-P450 inhibition assay panel. Accessibility of these assays and their broader utilization at earlier steps of lead optimization process is beneficial for successful anticipation, proper assessment, and prediction of potential DDI liabilities.

23.3.2 P450 Induction

Our understanding of the mechanisms leading to enzyme induction has significantly improved in the last decade. Human pregnane X receptor (hPXR, SXR, NR1I2) has been identified as a nuclear hormone receptor (NHR) largely responsible for the induction of CYP3A4 and other genes such as CYP2B6, CYP2C8, CYP2C9, P-gp, multidrug resistance gene-associated protein 2 (MRP2), and organic anion transporting polypeptides (e.g., OATP-2) (Kliewer et al., 2002). Besides hPXR, other transcriptional factors play a role in the induction of many other drug-metabolizing enzymes. These include constitutive androstane receptor (CAR), peroxisome proliferator activated receptor (PPAR), and aromatic hydrocarbon receptor (AhR) (Honkakoski et al., 2003; Mandal, 2005; Tirona and Kim, 2005). This advancement in knowledge, and availability of various tools and reagents, has benefited the industry, which is now able to screen for induction of enzymes and transporters, as well as NHR transactivation (Vignati et al., 2004; Zhu et al., 2004; Stanley et al., 2006).

23.3.2.1 Receptor Transactivation and Ligand-Binding Assays

In an industrial setting, receptor-binding or transactivation assays are used as Tier I assays to screen compounds for enzyme induction potential in early drug discovery due to their throughput (96- or 384-well formats) coupled with a low cost. Among them, PXR-derived assays are the most common high-throughput assays due largely to the importance of its target genes in DDIs. PXR ligand-binding assays are usually performed as a scintillation proximity assay, which consists of an expressed receptor ligand-binding domain (sometimes supplemented with a coactivator, SRC-1) and a high-affinity radiolabeled ligand (e.g., SR-12813). The competition for the ligand-binding pocket between the radiolabeled ligand and the test compound usually results in the displacement of the radioligand upon co-incubation. The extent of reduction of the radioactivity reflects the affinity of the test compound. Recently, a fluorescence-based PXR ligand-binding assay (PXR lanthascreen^{TM})

has been developed by the Invitrogen Corporation (Carlsbard, CA). Such an assay will obviate the need for radiolabeled ligand.

Beyond relatively simple binding assays, transactivation (or reporter) assays typically employ an expression vector containing the complete (or partial) promoter region of the target gene. Such a construct minimally requires both proximal response elements and distal enhancer sequences for a maximal induction response. The promoter is coupled to a reporter (e.g., luciferase, green-fluorescent protein, or alkaline phosphatase) and the induction response can be measured as an increase in the production of a luminescent (or fluorescent) metabolite. Such assay formats are compatible with low volumes, use of 384-well plates, and full or semi-automation.

PXR transactivation assays are frequently augmented with co-expression of a full-length hPXR due to a generally low expression of functional PXR in host cell lines (Goodwin et al., 1999; El-Sankary et al., 2001). In concert with the selection of appropriate host cell line, the over-expression of PXR significantly enhances the magnitude of induction responses. HepG2 and CV-1 cells are widely used, because they are thought to possess all of the regulatory partners (co-activators and co-suppressors) that mediate the complex ligand–PXR interaction.

In these assays, test compounds are added at increasing concentrations and thus the subsequent concentration-response Profiles can be tracked. Analysis of the data enables the generation of IC_{50} (ligand binding), E_{max}, and EC_{50} (transactivation) parameters in support of SAR generation. Overall, there is a good correlation between IC_{50} and EC_{50} values as the binding of ligands generally leads to the activation of the receptor (Zhu et al., 2004). However, some compounds can behave as strong PXR binders but elicit no (or weak) PXR transactivation (Synold et al., 2001; Zhou et al., 2007; Healan-Greenberg et al., 2008). Compounds can also serve as ligands that bind to the receptor but do not elicit the appropriate displacement of co-repressors or recruitment of co-activators (Harmsen et al., 2007). Recently, several azole compounds including ketoconazole have also been shown to bind to the outer surface of the PXR-AF2 binding pocket and interfere with the binding of a co-activator, SRC-1 (Wang et al., 2007). Therefore, it is quite possible to generate false positives in PXR-binding assays and thus most discovery teams turn to PXR transactivation assays. Fortunately, for most compounds the correlation of PXR transactivation with induction of CYP3A4 in primary hepatocytes is good (Luo et al., 2002). This is important, because discovery teams do not have to rely on primary hepatocytes to guide their SAR campaigns, which would limit the number of compounds for screening due to low throughput.

At BMS, luciferase is used as the reporter in a HepG2-hPXR transactivation assay. Luciferase activities are normalized against the signal from rifampicin (10 μM) and are expressed as a percent activation (%Act) at each concentration of test article. From the concentration versus %Act plot, E_{max} and EC_{50} values are calculated for each compound. In turn, the parameters are evaluated in light of the observed (or projected) therapeutic plasma concentrations. A risk assessment is made, compounds are flagged and in vitro or in vivo (e.g., clinical) follow-up studies are initiated as appropriate. For example, it has been shown that compounds eliciting a >40% transactivation of hPXR (versus rifampicin), at their corresponding plasma C_{max}, are "likely" to behave as inducers in a clinical setting. Compounds eliciting

15–40% transactivation "may" elicit drug interactions, whereas weak transactivators (<15%) do "not" cause drug interactions (Sinz et al., 2006).

23.3.2.2 Immortalized Cell Lines

Although considered the "gold standard" model for the in vitro assessment of induction, cultures of freshly prepared human primary hepatocytes can be difficult to obtain commercially. At the same time, one has to deal with considerable inter-donor variability and poor cell viability in micro-titer plates (Soars et al., 2007). Up until very recently, this has limited their utility in a discovery setting. Therefore, numerous attempts have been made to generate human hepatocyte-like cells that can ensure a continuous supply and support high-throughput screens. Broadly speaking, three approaches have been described to date. The first involves the refinement hepatocarcinoma-derived cells. The second requires a source of stem cells that can be differentiated into hepatocyte-like cells, whereas the third strategy is based on the immortalization of human primary hepatocytes (Sinz and Kim, 2006). For example, the HepG2 cell line was established in 1979 from a hepatoma tissue. While retaining some induction responses to known CYP1A and CYP3A4 inducers such as β-naphthoflavone, 3-methylcolanthrene, *o, p′*-dichloro-diphenyl-trichloroethane (main isomer of DDT), phenobarbital, and dexamethasone (Hewitt and Hewitt, 2004; Medina-Diaz and Elizondo, 2005; Medina-Diaz et al., 2007), HepG2 cells are not considered an appropriate in vitro model to study induction of P450s due to the low basal expression of enzymes and induction responses (Vermeir et al., 2005). Recently, several HepG2 subtypes have been established by enriching only the type of cells that have better induction responses through means of altered nutritional requirement. Amphioxus' ACTIVTox® (Stem Cell Innovations, Houston, TX) and WGA cells are examples of highly differentiated subclones of HepG2 cells (Rencurel et al., 2005) and both cell lines have been shown to be responsive to known inducers of CYP1A, CYP2B, and CYP3A4.

In comparison to HepG2 cells, immortalized hepatic cells can be used to support induction screening and serve as a surrogate for primary hepatocytes (Hahn, 2002). Immortalization of cells, such as HBG and HepaRG cells, can occur naturally as exemplified by hepatic cells of tumor origin. In fact, both cell lines have been isolated from surgically removed liver tumors (Gripon et al., 2002). HepaRG cells, in particular, express major NHRs (e.g., PXR, AhR, PPARα and CAR) at the levels comparable to those in primary human hepatocytes (aided by the culture of the cells in DMSO) and may be of special interest in studying CAR-mediated induction as no other tumor cells express high levels of CAR (Aninat et al., 2006). In addition to the induction of P450 enzymes, HepaRG cells were shown to respond to the inducers of transporters and increased the expression of multidrug resistance protein 1 (MDR-1), MRP-2, and bile salt export pump (BSEP) (Guillouzo et al., 2007).

As described above, embryonic stem cells provide another potential source of hepatocyte-like cells, which are shown to express CYP1A2, CYP3A4/7, and low levels of CYP1A1 and CYP2C proteins (Ek et al., 2007). When treated with a cocktail of known P450 inducers, CYP1A2 and CYP3A4 protein expression is increased, suggesting the induction of these genes is also retained. It is likely that the application of stem cell technology will increase in the future, once ideal methods of deriving hepatocyte-like cells are identified (Sinz and Kim, 2006).

A considerable effort is also being made to convert primary cells into non-tumorigenic immortalized cells. This is exemplified by the efforts of researchers at Multicell Technologies (Lincoln, RI, USA), who have employed a clonal selection process following immortalization of human primary hepatocyte, to generate the Fa2N-4 cell line. While viral transformation normally results in genetically unstable cells and loss of phenotypic characteristics (Cascio, 2001), Fa2N-4 cells retain normal hepatocellular morphology, as well as expression and induction of P450s and transporters. Concentration-dependent increases in CYP3A4, CYP2C9, CYP1A2, UGT1A, and MDR-1 expression (transcript levels and activity) have been reported following the addition of rifampicin to the cells (Mills et al., 2004). However, there is no (or very low) expression of CAR in these cells and thus the evaluation of CYP2B6 induction is not recommended (Lyon et al., 2005). The procedure for assessing enzyme induction in Fa2N-4 cells is well defined and is similar to the protocol using primary hepatocytes. Perhaps the biggest difference between human primary hepatocytes and Fa2N-4 cells is that the latter cells continuously proliferate in culture. Hence the number of Fa2N-4 cell passages has to be carefully monitored to prevent the deterioration of the induction response. If cryopreserved, however, Fa2N-4 cells can maintain their induction response even in higher-density well formats (e.g., 96-well plate) (Youdim et al., 2007). In general, immortalized hepatocytes are being accepted as a suitable higher-throughput alternative to primary hepatocytes. At the present time, however, the utility of Fa2N-4 cells is limited to CYP1A and CYP3A4 induction screening.

23.3.2.3 Primary Human Hepatocytes

From the standpoint of in vitro induction assays, it is accepted widely that cultures of primary human hepatocytes are the best model, and regulatory agencies have increasingly accepted DDI predictions based on their use (LeCluyse et al., 2000). Assessment of induction in primary human hepatocytes is possible because cells can be plated and sustain their viability on a substratum such as collagen or Matrigel® (LeCluyse et al., 2000; Hamilton et al., 2001; LeCluyse et al., 2005). In the "sandwich" culture format, where Matrigel is also layered on top of the plated cells, dedifferentiation of the hepatocytes is further slowed down and liver-specific phenotypes are maintained (e.g., expression of drug-metabolizing enzymes and transporters, as well as the formation of bile canaliculi) (Runge et al., 2000; Bi et al., 2006; Marion et al., 2007). While most of the primary hepatocytes lose their ability to reattach to a culture plate following cryopreservation, "inducible" or "attachable" cryopreserved hepatocytes retain their ability to attach and respond to known CYP1A2, CYP2B6, CYP2C9, CYP2E1, and CYP3A4 inducers (Schehrer et al., 2000; Roymans et al., 2005; Hewitt et al., 2007a). At BMS, it has been possible to obtain commercially available cryopreserved human hepatocytes and plate them in a 96-well format. After the addition of test articles and positive controls, induction of CYP1A2, CYP3A4, and CYP2B6 is assessed via RT-PCR (real-time polymerase chain reaction). The three enzymes are representative of AhR, PXR, and CAR transactivation, respectively. Such a tier II assay is used coordinately with the hPXR transactivation assay in support of various discovery programs.

In development, hepatocyte induction data are typically collected from a minimum of three different donors due to known individual variations in induction response. The common endpoints for assessment of enzyme induction are RNA expression (RT-PCR), protein level, and enzyme activity. Enzyme activity can be measured by the rate of metabolite formation, or by the rate of parent drug disappearance, using proven substrates that are specific to the enzyme of interest. While enzyme activity is accepted as an end-point by regulatory agencies, quantitation of RNA expression or protein level is also required as supplemental information. A more complete data package is useful when a compound concurrently induces and inhibits the same enzyme. Ritonavir, troleandomycin, and DPC-681 are examples of such compounds that generally exhibit normal or reduced enzyme activity but result in increased CYP3A4 gene transcription (Luo et al., 2002; Luo et al., 2003). It also emphasizes the importance of incorporating PXR transactivation and P450 inhibition data when evaluating hepatocyte results to predict enzyme induction. A recently published draft FDA guidance for industry on drug interactions proposed two numerical endpoints to be used in the prediction; percent increase relative to a positive control and EC_{50} (FDA, 2006). Using an approach similar to that described for the human PXR assay, the guidance indicates that a drug that produces a change (enzyme activity) equal to or greater than 40% of the positive control is a potential inducer and requires further in vivo evaluation. As for the EC_{50}, the guidance suggests that this parameter can be used to compare the potency of different compounds. While simple rank ordering of EC_{50} values and comparisons to known CYP3A4 inducers, such as rifampicin, can provide useful information, more accurate predictions of drug interactions require incorporation of clinically relevant drug concentrations (Ripp et al., 2006; Sinz et al., 2006; Hewitt et al., 2007a).

23.3.3 P-gp Substrate and Inhibition Assays

P-glycoprotein (P-gp, MDR1) is a member of the ATP-binding cassette (ABC) efflux transporter family that is the best understood among all drug transporters. It is an ubiquitous transporter, which is present on the apical surface of enterocytes, the canalicular membrane of hepatocytes, apical surface of the kidneys, placenta, and endothelial cells of the brain. Because of its strategic location, it is widely recognized that P-gp is a major determinant of drug ADME (absorption, distribution, metabolism, and excretion) in humans (Polli et al., 1999; Matheny et al., 2001; Lin and Yamazaki, 2003). For example, it is known to limit the oral absorption of drugs such as cyclosporin and taxol; limit the entry of drugs (e.g., HIV protease inhibitors) into the CNS; and facilitate the excretion of drugs like digoxin, verapamil, and vincristine via biliary and urinary routes. Therefore, increasing efforts are being made in early discovery and development to identify compounds that can potentially interact with P-gp. Regulatory agencies have also acknowledged P-gp as a key drug transporter (FDA, 2006).

There are literature reports of various in vitro and in vivo models used for assessing P-gp interactions (Hidalgo, 2001; Polli et al., 2001; Balimane et al., 2006). In vitro assays such as ATPase activity, rhoadmine-123 uptake, calcein AM uptake, cell-based bi-directional permeability, radio-ligand binding along with in vivo models such as transgenic (knockout mice) animal models are among the commonly used methods in a discovery setting. However, the cell-based bi-directional permeability assay is the most popular method for identification of P-gp substrate in drug discovery labs (Polli et al., 2001; Kerns et al., 2004; Balimane et al., 2006). Similarly, the cell-based bi-directional permeability assay using digoxin as a probe is currently accepted as the "method of choice" for determining the P-gp inhibition potential of test compounds (Polli et al., 2001; Keogh and Kunta, 2006). The FDA also recommends conducting bi-directional permeability studies using Caco-2 cells or other wild-type or transfected cell lines (e.g., MDCK, LLC-PK1) to determine the P-gp interaction potential of test compounds (FDA, 2006).

23.3.3.1 P-gp Assay Methodology

Caco-2 cell-based methodology is currently the "industry standard" for assessing the in vitro P-gp substrate or inhibition potential of test compounds. Caco-2 cells are typically grown in culture medium (maintained at 37°C, 95% relative humidity) consisting of Dulbecco's modified Eagle's medium supplemented with 10% fetal bovine serum, 1% nonessential amino acids, 1% L-glutamine, penicillin-G (100 U/mL), and streptomycin (100 μg/mL). P-gp interaction studies are conducted with physiologically and morphologically well developed Caco-2 cell monolayers that have been cultured for approximately 21 days and demonstrate adequate trans epithelial electrical resistance (TEER) value, as well as low permeability for mannitol (paracellular probe) (Balimane et al., 2006).

23.3.3.2 P-gp Substrate Assay

At BMS, Caco-2 cells are used to identify compounds that are potential P-gp transporter substrates. HTC-24-transwell® inserts (surface area, 0.33 cm^2) consisting of a 0.4-μm pore size polycarbonate membrane are used for these studies. The transport medium used for the bi-directional studies is modified HBSS buffer containing 10 mM HEPES. The pH of both the apical and basolateral compartments is 7.4. Prior to these experiments, each monolayer is washed twice with buffer and TEER is measured to ensure the integrity of the monolayers. The test compound is studied either at a single low concentration (3 μM) for screening purposes or at multiple concentrations (1–100 μM) to assess the K_m and concentration effects. The bi-directional permeability studies are initiated by adding an appropriate volume of buffer containing test compound to either the apical (for apical-to-basolateral transport; A-to-B) or basolateral (for basolateral-to-apical transport; B-to-A) side of the monolayer. The monolayers are then placed in an incubator for 2 h at 37°C. Samples are taken from both the apical and basolateral compartment at the end of the 2 h period and the concentrations of test compound are analyzed by a sensitive LC-MS/MS method. The permeability coefficient (*Pc*) is calculated using Equation 23.10

$$Pc(nm/s) = \frac{\mathrm{d}A}{\mathrm{d}t^{*}S^{*}C_o} \qquad (23.10)$$

where d*A*/d*t* is the flux of the test compound across the monolayer (nmol/sec), *S* is the surface area of the cell monolayer, and C_o is the initial concentration (1–100 μM) in the donor compartment.

Test compounds with an efflux ratio (B-to-A/A-to-B ratio) greater than 2 are considered as P-gp substrates. Non-substrates have an efflux ratio of unity at all concentrations tested. For substrates, a follow-up study is conducted and involves co-incubation with a potent P-gp inhibitor (e.g., 50 μM ketoconazole). In this instance, if the efflux ratio collapses to unity, then the involvement of P-gp is confirmed.

23.3.3.3 P-gp Inhibition Assay

The transport medium used for the Caco-2 cell P-gp inhibition studies is also modified HBSS buffer containing 10 mM HEPES (pH of both the apical and basolateral compartments is 7.4). Prior to these experiments, each monolayer is washed twice with buffer and TEER is measured to ensure the integrity of the monolayers. Both the apical-to-basolateral (A to B) transport and the basolateral-to-apical (B to A) transport of[^{3}H]digoxin (5 μM) is measured in the absence and presence of various concentrations (0.1–100 μM) of test compound. The studies are initiated by adding an appropriate volume of buffer containing digoxin to either the apical (apical-to-basolateral transport) or basolateral (basolateral-to-apical transport) side of the Caco-2 cell monolayer. Test article is added to both sides of the monolayer at various concentrations. The monolayers are then incubated for 2 h at 37°C. An aliquot is removed from both apical and basolateral compartments at the end of the 2 h period and analyzed for [^{3}H]digoxin using a liquid scintillation counter.

The apical-to-basolateral and the basolateral-to-apical permeability coefficient (*Pc*) values for digoxin are calculated in the presence and absence of the test compound. Results are then reported as "% of inhibition" of digoxin transport (Equation 23.11).

$$\%\ \text{of inhibition} = \left[1 - \frac{(iBA - iAB)}{(dBA - dAB)}\right]^{*} 100 \qquad (23.11)$$

where *dBA* and *dAB* are the B-to-A and A-to-B permeability of digoxin alone. *iBA* and *iAB* are the B-to-A and A-to-B permeability of digoxin in presence of the test compound. An IC_{50} value is calculated by using the % of inhibition and the corresponding concentration of test compound. The data are fitted using the Hill equation (Equation 23.12).

$$E = \frac{E^{*}_{\max} C^n}{C^n + \mathrm{IC}_{50}{}^n} \tag{23.12}$$

where E = % inhibition of P-gp at a particular concentration of inhibitor, C = concentration of inhibitor, $E_{\max}$ = maximum inhibition of P-gp, n = Hill coefficient. For test compounds that inhibit P-gp, their IC_{50} value is compared to plasma concentrations in human (known or estimated) to determine P-gp inhibition potential.

23.3.4 P450 Substrate

As discussed above, besides [*I*] and in vitro inhibitory constants, the third factor that has been shown to be critical for the accurate assessment of in vivo DDIs is the $f_m \times f_{m,\mathrm{CYP}}$ (Fig. 23.1). Consequently, it is important to conduct reaction phenotyping studies to assess the role of different drug-metabolizing enzymes in clearance. In this regard, it is advantageous to develop NCEs that are cleared by numerous enzymes and pathways. Moreover, compounds with low rates of metabolic turnover are sought after, because this minimizes the potential for marked gut and liver first pass. For example, at least two (non-polymorphic) P450s are preferred, with a low intrinsic clearance (<10% of hepatic blood flow). Conversely, compounds exhibiting high (to very high) turnover via one major P450 should be screened out. Therefore, this necessitates relatively robust reaction phenotyping screens in a discovery setting and fully integrated data set later in development (Zhang et al., 2007a).

At BMS, a variety of in vitro tools are employed for assessing the role of metabolism in the overall elimination of an NCE. At the very early stages of drug discovery, various chemotypes and compounds are subject to metabolic stability (metstab) screens. Typically, this involves incubation of test article at a low concentration (0.5 μM) with NADPH-fortified liver microsomes. Such incubations can be carried out with 96-well microtiter plates and require considerable LC-MS infrastructure. The in vitro hepatic clearance is then calculated from the first-order rate constant estimated from substrate depletion and compared with the in vivo intravenous clearance of preclinical species (Obach, 2001). In case of a reasonable correlation between the in vitro and in vivo clearance in the preclinical species, the same construct is used for predicting the in vivo clearance in humans from the in vitro microsomal data. A reasonable correlation further implies oxidative metabolism via liver enzymes as the major route of elimination and calls for more detailed in vitro characterization studies to identify the enzymes involved and their contribution to the total elimination of the NCE in question. Toward this end, the FDA has recommended three well-characterized in vitro approaches: (1) use of specific chemicals or antibodies as enzyme inhibitors; (2) use of individual recombinant P450 enzymes; or (3) a bank of HLM characterized for P450 activity prepared from individual donor livers (FDA, 2006). Each of the above approaches have their own advantages and disadvantages and the FDA recommends that at least two out of the three approaches be performed to identify the specific enzyme(s) responsible for drug's metabolism. At BMS, P450 form-selective inhibitors (chemicals and antibodies) and recombinant P450 enzymes are used in combination for assessing the potential role of P450 enzymes in the metabolism of an NCE (Zhang et al., 2007a). In addition, a bile duct-cannulated (BDC) rat (or monkey) study may also be conducted to confirm the role of the oxidative metabolism in vivo versus biliary and renal elimination.

However, in cases where a disconnect is observed between the in vivo clearance and the clearance predicted from the in vitro microsomal data, the metabolism of the compound may be studied using other in vitro systems like 9000 g supernatant fraction (S9) fractions, UDPGA-supplemented microsomes, and fresh or cryopreserved hepatocytes. In this instance, it is important to evaluate the involvement of the non-P450 enzymes (non-microsomal or Phase II enzymes) in the elimination of an NCE. Furthermore, a BDC rat or monkey study will also be conducted to understand the potential roles of biliary and renal routes in the overall elimination of an NCE. Subsequently, all the aforementioned information gathered from the in vitro and in vivo data is carefully integrated to get a reasonable assessment of the $f_m \times f_{m,CYP}$ in humans. Finally, as evidenced in the recent guidance document issued by the FDA, although the regulatory agencies acknowledge the usefulness of such data (FDA, 2006), one should exercise caution while using this information due to the uncertainty associated with the estimation of $f_m \times f_{m,CYP}$ in humans at the early stages of drug discovery. Once in the clinic, radiolabel ADME information can be used to supplement the data set and further refine estimates of $f_m \times f_{m,CYP}$ (Zhang et al., 2007a)

23.3.5 OATP Substrate and Inhibition Assays

In recent years, significant progress has been made in the molecular and functional characterization of pharmaceutically relevant drug uptake transporters. As a result, our understanding of their role in the disposition of drugs has increased tremendously in the past decade. Like P-gp, the various uptake transporters are expressed in several key organs and also govern the ADME properties of drugs. Uptake transporter proteins, or solute carriers (SLCs), are represented by proteins such as Na+-taurocholate co-transporting polypeptides (NTCP), OATPs, organic anion transporters (OATs), and organic cation transporters (OCTs). Among the SLCs, OATPs are the most popular transporters and are often taken into consideration within an industrial setting.

23.3.5.1 OATP Substrate Study

OATP superfamily members are encoded by genes of the solute carrier organic anion transporter (SLCO) and consist of 11 members in human (Hagenbuch and

Meier, 2004). Among OATP members, four OATPs (OATP1B1, 1B3, 1A2, and 2B1) are considered as the major uptake transporters in drug absorption and hepatic entry of various endogenous compounds (e.g., bilirubin, bile acids, steroid conjugates, and thyroid hormone) and drugs (e.g., statins, digoxin, and fexofenadine) (Ayrton and Morgan, 2008). OATP1B1 and OATP1B3 are mainly expressed in human liver (Smith et al., 2005), whereas, OATP1A2 and OATP2B1 are more broadly distributed in a variety of tissues such as brain and intestine (Hagenbuch and Meier, 2003). Therefore, OATP1B1 and OATP1B3 play a major role in hepatic uptake and elimination of organic anion drugs (Ayrton and Morgan, 2008).

Despite the availability of recombinant OATPs, preliminary hepatic uptake studies are performed with suspensions of primary hepatocytes. Since fresh hepatocytes are equipped with all the hepatic transporters, as well as Phase I and Phase II enzymes, hepatocytes are an appropriate model for evaluating transporter (versus passive permeability) and enzyme-mediated hepatic clearance. Generally, the uptake into hepatocytes is measured in suspension by a centrifugation method using oil-layered tubes to separate hepatocytes from the incubation medium (Schwenk, 1980). Alternatively, the assay can be run in a semi-high-throughput fashion with plated hepatocytes. Recently, cryopreserved hepatocytes have become popular in the drug discovery environment due to their increased availability and resource consistency (Hewitt et al., 2007b).

If there is evidence for saturable, energy-dependent uptake into primary hepatocytes, then follow-up studies are conducted with cell lines containing individually expressed (recombinant) OATPs. The most widely utilized host cells for transfection are HEK-293, COS-7, HeLa, CHO, MDCK cells, and *Xenopus laevis* oocytes since they express low levels of endogenous transporter proteins. The use of OATP-expressing systems allow a mechanistic insight into the particular transporter(s) involved in the transport of a test drug.

To ascertain whether a test compound is a substrate of a particular transporter, test compounds are incubated with transfected cells in parallel with their respective mock line. The incubation times depend on the cell type; shorter time for cell lines (e.g., 1–10 min), longer time for *Xenopus laevis* oocytes (e.g., 1 h). When the uptake of test compound into transfected cells is higher relative to that of the mock cells (empirically, 50% higher), the compound is considered to be an OATP substrate. Since the expression levels of transporters in the transfected cells can vary from batch to batch, and passage to passage, the uptake of the positive control compounds needs to be measured for quality control (e.g., estradiol-17β-D-glucuronide, dehydroepiandrosterone-3-sulfate, estrone-3-sulfate, pravastatin, cholecystokinin octapeptide, BQ-123, digoxin) (Hagenbuch and Meier, 2003).

23.3.5.2 OATP Inhibition Study

Drug–drug interactions can occur when an OATP substrate is co-administered with an OATP inhibitor or inducer. Clinically significant DDIs leading to severe side effects have been reported for several statins when they are co-medicated with cyclosporin A or gemfibrozil (Poirier et al., 2007). Other OATP-mediated DDIs include fexofenadine–grapefruit juice, atrasentan–rifampin, bosentan–rifampin, repaglinide–cyclosporin A, and ticlopidine–ergoloid mesylates (Niemi, 2007). Consequently, it is important to assess OATP inhibition potential prior to first in man. In this instance, the uptake of an OATP substrate is measured in the presence and absence of the test article.

For a reliable inhibition study, appropriate probe compounds are carefully selected among OATP substrates, because each OATP substrate has a different dynamic inhibition range. Generally OATP substrates show broad substrate specificity across OATP members. As shown in Fig. 23.2, estrone-3-sulfate has high affinity to OATP2B1. Estradiol-17β-D-glucronide and cholecystokinin octapeptide are high-affinity OATP1B1 and OATP1B3 substrates, respectively. Therefore, these compounds are widely used as probe compounds in each OATP assay and their uptake is measured in the presence and absence of the test article to determine an IC_{50} or k_i.

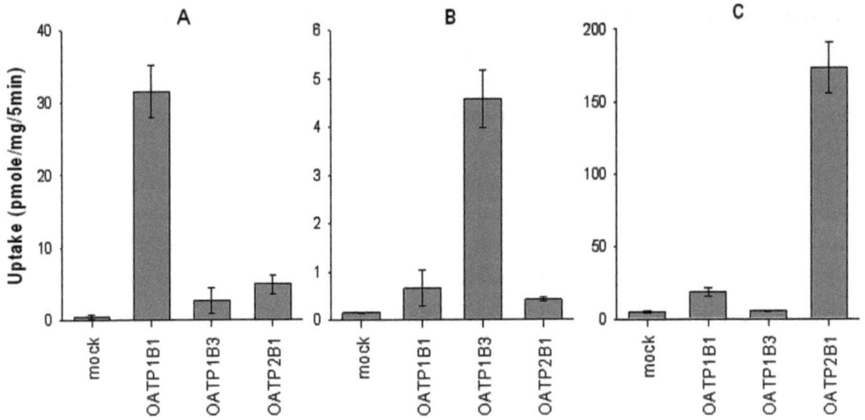

Fig. 23.2 Difference in substrate specificity of OATP substrates between OATP1B1, OATP1B3, and OATP2B1; **A** (estradiol 17-β-D-glucronide), **B** (cholecystokinin octapeptide), **C** (estrone-3-sulfate). Cellular uptake of estradiol 17-β-D-glucronide (1 μM), cholecystokinin octapeptide (0.1 μM), and estrone-3-sulfate (1 μM), into mock-, OATP2B1-, 1B1, and 1B3-HEK293 cells was determined at 5 min. Data are shown as the mean ± S.D. ($n = 4$)

23.3.6 *Animal Models of CYP Induction and Inhibition*

Testing potential enzyme inducers/inhibitors in animal models, at doses that target the therapeutically relevant concentrations in the clinic, could significantly improve one's ability to quantitatively predict DDIs in the clinic. Assuming the ADME characteristics, in vitro inhibitory and induction potencies, and the plasma protein binding for the NCEs are similar between animals and human, one can potentially recapitulate the induction and inhibitory actions of NCEs in a more physiologically relevant environment (Marathe and Rodrigues, 2006). However, the use of animal models has been hampered due to either the species-specific differences in drug metabolism pathways or significant species differences in ligand affinities of various NHRs or transcription factors. For example, human and rodent PXR share only ~75% sequence homology in their ligand-binding domains. Thus the known human CYP3A4 inducer rifampicin preferentially activates human PXR but not rodent PXR, whereas PCN (pregnenolone-16α-carbonitrile) has been shown to be a weak activator of human PXR and a potent activator of rodent PXR (Gonzalez and Yu, 2006). Recently, the humanization of mouse PXR by replacing mouse PXR with human PXR (SXR) has opened the possibility for a new in vivo animal model that may overcome the limitation of species differences (Xie and Evans, 2002; Down et al., 2007; Ma et al., 2007). Another humanized mouse has also been created by transplanting human hepatocytes into a SCID nude mouse liver (Tateno et al., 2004; Nishimura et al., 2007). Of note, the responses in both of these models, when treated with known human inducers, were similar to those of primary human hepatocytes.

At the same time, increased efforts have been made to validate non-human primates (e.g., rhesus or cynomolgus monkey) as animal models for the prediction of CYP3A-mediated induction and inhibitory interactions in humans. Such interest in monkeys has been partly due to similarities in the CYP3A expression and high sequence homology between members of humans and monkey CYP3A subfamilies. Moreover, the K_m and the primary metabolic pathway (1-OH midazolam) of midazolam in cynomolgus monkey liver microsomes, a commonly used probe substrate for CYP3A, are both comparable to that of HLM preparations. Also, a good concordance can be observed between the in vitro potencies (IC_{50}) and the plasma protein binding of common CYP3A inhibitors such as ketoconazole, fluconazole, diltiazem, and cimetidine in both species. These similarities have led to the use of the monkey as an in vivo model for the investigation of CYP3A-mediated DDIs using midazolam as the probe substrate (Kanazu et al., 2004; Prueksaritanont et al., 2006; Ogasawara et al., 2007; Zhang et al., 2007b). In these studies, co-administration of known and unknown inhibitors of CYP3A in monkey resulted in interactions that are comparable to those reported in humans, suggesting that in vivo models could be quite valuable for assessing the DDI risk of NCEs. Furthermore, unlike mouse, rat, or dog PXR, monkey PXR shares high sequence homology (~96%) with its human counterpart. In agreement, the results of monkey primary hepatocyte induction studies and in vivo studies have pointed to human-like induction responses (Prueksaritanont et al., 2006; Nishimura et al., 2007). Although promising, several compounds do exhibit species-specificity (monkey vs. human) in their induction response (not published). Therefore, before a costly in vivo endeavor, one should examine the results from in vitro assays (PXR transactivation and hepatocytes) and ensure consistent induction responses in both human and monkey. In summary, humanized mice and monkeys have emerged as useful in vivo models for the quantitative assessment of human induction and inhibition but will require further characterization before being widely adopted by the pharmaceutical industry.

23.4 Lead Optimization

For most drug discovery programs, the lead optimization process is often arduous and most discovery teams strive to triage multiple compounds belonging to one or more chemotypes. Typically, anywhere from 700 to 4000 different compounds are synthesized and screened en route to the final development candidate. Teams leverage existing assay suites and establish a screening "funnel" that enables the assessment of compounds via "tier I" (high-throughput in vitro) and "tier II" (more complex in vitro, less throughput) assays. Later in the funnel, compounds are dosed to animals in order to screen for PK, efficacy, and establish a pharmacokinetic–pharmacodynamic (PK–PD) relationship. Therefore, the DDI screening suites described previously are only part of a larger repertoire of models, assays, and reagents.

Where possible, a SAR is established for a given set of parameters and Table 23.2 shows the type of data that are generated. It is evident that up to 30 different SARs can be tracked, at one time or another, as teams attempt to obtain

Table 23.2 Examples of various screening end-points during lead optimization

Solubility	Aqueous (buffer)
	Simulated gastrointestinal fluids
Permeability	PAMPA
	Caco-2 cells (apical to basolateral)
	Caco-2 cells (bidirectional)
Inhibition (recombinant P450s)	CYP1A2, CYP3A4, CYP2C9, CYP2D6, CYP2C8, CYP2C19, CYP2B6
Inhibition (liver microsomes)	CYP1A2, CYP3A4, CYP2C9, CYP2D6, CYP2C8, CYP2C19, CYP2B6
Inhibition (Caco-2 cells)	P-glycoprotein
Induction	Human pregnane X receptor (hPXR) transactivation
	Cryopreserved primary human hepatocytes (CYP3A4, CYP2B6, CYP1A2)
Metabolic stability in vitro (predicted CL_{int})	Rate of consumption in the presence of liver microsomes (multiple species)
	Rate of consumption in the presence of suspensions of primary hepatocytes (multiple species)
	Metabolic soft spot determination
	Reaction phenotyping
Protein binding in vitro	Plasma (multiple species)
Hepatotoxicity	Primary human hepatocytes (various assay types)
Cardiotoxicity	hERG assay (in vitro)
	Telemetry animals
Channel effects	Sodium channel, potassium channel assays (in vitro)
Pharmacokinetics (animals)	Clearance, volume of distribution, half life
	Absorption, oral bioavailability
Target pharmacology	Single protein (recombinant, purified) and cell-based assay (multiple species)
Pharmacodynamics (in vivo)	Efficacy model (one or more different models or species)
Pharmacology counter screening	In vitro assay panels (various proteins)

lead compounds with the best balance of parameters. This "balancing act" is illustrated in Fig. 23.3, where data are shown for the inhibition of a target protein and CYP3A4. Subtle changes in structure can give rise to marked shifts in CYP3A4 IC_{50}, with relatively minimal impact on target potency. The reverse is also possible. Results shown in Table 23.3 further illustrate the types of SAR that can be generated for different chemotypes. Although the transactivation of human PXR is exemplified, similar SARs are possible for other end-points such as metabolic stability in vitro, plasma protein binding, permeability, solubility, hERG inhibition (Table 23.2).

From the standpoint of drug interactions, it is advantageous to increase IC_{50} for all P450s and decrease the k_{inact}/K_I and E_{max}/EC_{50} ratio as much as possible. For P450 substrates, metabolism via multiple forms ($f_{m,CYP} < 0.5$) is beneficial

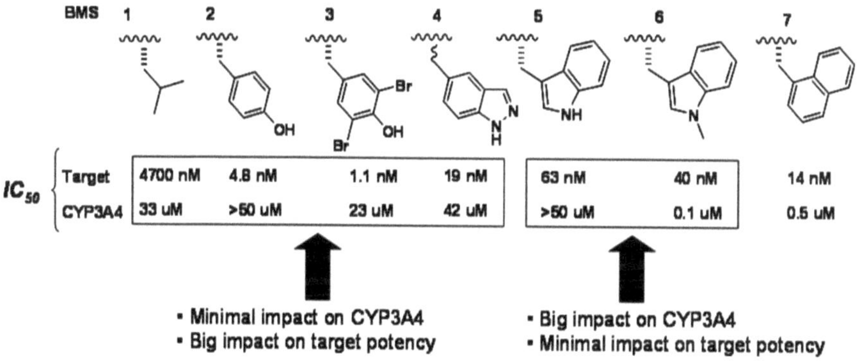

Fig. 23.3 Example of inhibition SAR (pharmacological target and CYP3A4) for a number of closely related compounds (partial structures shown)

Table 23.3 Examples of SAR involving the transactivation of the human pregnane X receptor (PXR)

Chemotype	Compound	Substitution at R_1	Substitution at R_2	Substitution at R_3	PXR EC_{50} (μM)
A	1	Cl	Cl	H	>50
A	2	Cl	H	CH_3	0.79
A	3	CF_3	H	CH_3	0.08
Chemotype	Compound	No fluorine	Fluorine (*ortho*)	Fluorine (*meta*)	PXR EC_{50} (μM)
B	1	X			4.3
B	2		X		17
B	3			X	>50
Chemotype	Compound	Hydroxy	Carbonyl	Substitution at R_1	PXR EC_{50} (μM)
C	1	X		H	>50
C	2		X	H	6.2

(Equations 23.5, 23.6, and 23.7). At the same time, compounds that are pharmacologically more potent usually render lower efficacious doses and lower organ and plasma concentrations during first pass. If weak inhibitors and inducers, such compounds will have a potentially improved DDI profile in a clinical setting. The plot shown in Fig. 23.4 shows in vitro P450 IC_{50} data for 30 different nominated compounds. All of the compounds belonged to different programs and were the products of extensive SAR campaigns. The majority (84%) of the IC_{50}s generated in vitro were high (>40 μM), well above the plasma C_{max} at a projected efficacious dose, and the potential for DDIs was deemed low.

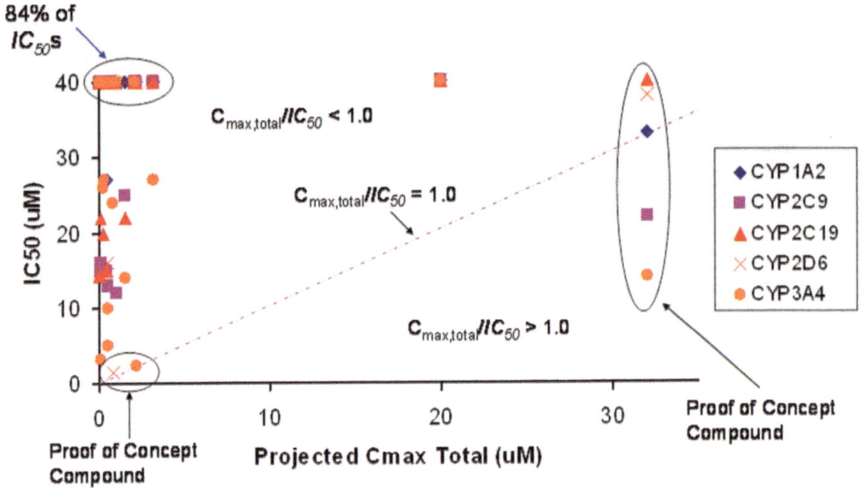

Fig. 23.4 In vitro inhibition of 5 different human cytochromes P450 in human liver microsomes by 30 development candidates (150 different IC_{50} values are plotted). In each case, the IC_{50} value (*Y*-axis) is plotted against the target (total drug) plasma C_{max} concentration (based on projection of human pharmacokinetics). The *dash line* represents a C_{max}-to-IC_{50} ratio of unity

23.5 Lead Characterization and Beyond

23.5.1 Contexting of In Vitro and In Vivo DDI Data

As described above, several cycles of SAR are required to optimize multiple properties in a lead molecule and a number of them pertain to properties associated with DDI potential. An optimized lead may still have, as an example, an IC_{50} (for inhibition) or EC_{50} (for induction relative to a positive control) for CYP3A4 in the low micromolar range. Traditionally, these numbers will have been frowned upon as the lead enters the development stage. However, due to recent advances in the understanding of drug-metabolizing enzymes and their roles in the metabolism of specific drugs, several predictive tools are available for prospective prediction of human PK (Huang et al., 2008). Likewise, advances in various therapeutic areas have led to animal models of disease that incorporate pathophysiological features of human disease and allow the scientists to assess efficacious concentrations of lead molecules (Jones-Bolin et al., 2006; Kai et al., 2006; Lee and Farde, 2006). With the understanding of the projected human PK profile, and efficacious concentrations in pharmacology models, one can now consider likely human efficacious doses and project the range of plasma concentrations of the lead molecule that subjects in clinical trials will be exposed to (Fig. 23.4). Such projected plasma concentration–time profiles form the basis of further studies required to characterize the drug interaction potential in an industrial setting. For example, when a CYP3A4 inhibitor is co-administered, increase in the plasma concentrations of a victim drug may be predicted with consideration of the perpetrator's IC_{50} against midazolam hydroxylation in HLMs and its plasma concentrations at a therapeutic dose. For a CYP3A4 substrate as a victim, one needs to consider its overall intrinsic clearance (CL_{int}), plasma concentrations relative to its K_m for CYP3A4 along with its $f_m \times f_{m,CYP}$ value. The likelihood of gut first pass has to be considered also.

23.5.2 Prioritization of Clinical Studies

Prospective prediction of DDI potential (albeit qualitative) is critical in an industrial setting, because it allows prioritization of studies and enables clinical plans. According to the FDA guidance, although quantitative predictions of in vivo drug–drug interactions from in vitro studies are not possible, rank ordering of in vitro k_is (or IC_{50}s) across different P450s for the same drug may help prioritize in vivo DDI evaluations (FDA, 2006). In the absence of a proven method by which to quantitatively predict DDI, in vitro inhibition data should be able to indicate which enzymes will be more affected than others. This assumes that any in vivo phenomenon that potentially may confound an accurate quantitative prediction should be equally accounted for irrespective of the enzyme that is affected. When the projected $[I]/k_i$ is in the gray zone (~0.1–1.0) for multiple P450s, prioritization of DDI studies can be done using a rank-order approach (Obach et al., 2005). In this approach, a clinical interaction study with the probe substrate of the P450 with the highest projected $[I]/k_i$ is conducted first (e.g., midazolam (CYP3A4), theophylline (CYP1A2), (*S*)-warfarin (CYP2C9), or desipramine (CYP2D6)). The outcome of the clinical study governs the next steps. For example, the NCE in question can be classified as "none" (<1.25-fold increase in probe substrate AUC), "weak" (>1.25 to <2.0-fold increase in AUC), "moderate" (≥2.0 to 4.9-fold increase in AUC), or "strong" (≥5.0-fold increase in AUC) inhibitor (Rodrigues, 2007). Based on this classification, further studies and their timing may be decided. If the results demonstrate a clinically significant increase in the probe substrate AUC, the following interaction study is based on the next highest projected $[I]/k_i$ ratio and the process is then repeated. In so doing, one works through the different P450s until no drug interaction is observed. If no DDI is observed in the first study, then no additional studies are needed and the compound does not behave as a P450 inhibitor. On the other hand, if the probe DDI studies indicate inhibition of multiple P450s, then further development of the NCE has to be questioned and its development has to be viewed in light of medical need. With further refinements, the rank order approach could also be extended to include mechanism-based inhibitors, P450 substrates and inducers (Rodrigues, 2007).

With the emerging novel mechanisms of action to treat diseases for unmet medical need, the pharmaceutical industry is faced with challenges of developing NCEs where clinical proof of concept does not exist. At times, a sponsor may be interested in getting to the clinic quickly in order to assess proof of target modulation with biomarkers. In such scenarios, an NCE with potential for potent P450 inhibition or induction may be put forth for clinical testing if concomitant victim drugs are avoided. Once a clinical proof of concept is achieved, the sponsor can conduct DDI studies with a rank order approach or put forth a different molecule with reduced DDI liability. Such a scenario is represented in Fig. 23.4, where some of the compounds were nominated for development despite the potential for inhibitory DDIs.

23.5.3 Final Integration of Drug–Drug Interaction Data Set

For regulatory filings, the sponsor will need to integrate in vitro IC_{50} (or k_i) values for inhibition in HLMs, EC_{50} (and E_{max}) values (relative to a positive control) in primary human hepatocytes for induction potential, and clinical exposure data of probe substrates when co-administered with highest recommended doses of the NCE as the perpetrator. For an NCE as a victim, a DDI data package will consist of in vitro reaction phenotyping, clinical ADME data supporting $f_m \times f_{m,CYP}$, and exposure data when co-administered with a potent (FDA-recommended) inhibitor of the P450 of interest. If necessary, data describing the impact of P450 genotype (e.g., *CYP2C9*3*, *CYP2C19*2*, *CYP2D6*4*) on PK, or the combined impact of genotype and inhibitor on PK, should be included (Zhang et al., 2007a). The NCE as the victim drug should be dosed at a sufficiently low therapeutic dose such that the risk of achieving high exposures in presence of the inhibitor can be minimized especially in a healthy volunteer setting. In all cases, the perpetrator should be dosed to steady state in order to maximize the potential for interaction.

23.6 Challenges of Predicting DDIs

23.6.1 Prospective vs. Retrospective In Vitro Studies

Although prospective in vitro studies, accompanied by accurate DDI predictions, are sought after, such approaches are difficult and any predictions prior to first in man have to be viewed as semi-quantitative at best (Prueksaritanont et al., 2002). This is particularly true in an industrial setting when one is dealing with complex chemistries, NCEs with unique PK-ADME properties, and no clinical data at the time of nomination for entry into development. After all, NCEs very rarely behave like classical drug-metabolizing enzyme inhibitors and substrates. As a result, various strategies have been employed (e.g., the rank order approach described previously) and in vitro data are regarded only as a means to prioritize (or de-prioritize) DDI studies.

It is worth considering three possible scenarios. In the first, no DDI is observed clinically, which is consistent with preclinical predictions. At this point, the data are integrated and no additional studies are needed. For the second scenario, where a DDI is forecast, but none is observed clinically, the clinical data trump the preclinical data. In the third instance, if a clinical DDI is observed, and predictions forecast no DDI, then the results need to be rationalized. More often than not, this is where in vitro studies are particularly useful. For both the second and third scenarios above, in vitro studies can be conducted retrospectively to explain and rationalize clinical DDIs mechanistically (with relevant parameters; $f_{\mathrm{m}} \times f_{\mathrm{m,CYP}}$, $[I]$, k_i, EC_{50}, E_{max}, fraction unbound, etc.). For example, pioglitazone with its in vitro k_i of 1.7 μM for CYP2C8-mediated paclitaxel 6α-hydroxylation will be expected to cause ~1.5-fold increase in the AUC of a drug with a $f_{\mathrm{m,CYP2C8}}$ ~0.5 (such as repaglinide). Pioglitazone, however, does not increase the plasma concentrations of repaglinide in vivo (Kajosaari et al., 2006), a finding rationalized based on its extensive plasma protein binding. Similar examples exist for RG-12525 with CYP3A4, and montelukast with rosiglitazone and pioglitazone (Fayer et al., 2001; Jaakkola et al., 2006; Kajosaari et al., 2006; Kim et al., 2007).

23.6.2 *What [I] to Use Prior to First in Man?*

Prospective prediction of DDIs is difficult even with the in vitro tools and mathematical models developed in recent years. A fundamental question for an NCE as a perpetrator is which concentrations to use to predict the DDI. As described previously, there is precedence for several concentrations such as total maximum or average plasma concentrations, plasma concentrations corrected by plasma protein binding, liver concentrations accounting for concentrative uptake in the liver or plasma concentrations at the inlet of liver during first pass (Kanamitsu et al., 2000). Other challenges for prospective predictions include non-specific binding and its impact on k_{i}, enzyme variability (e.g., impact of CYP3A5), non-Michaelis–Menten kinetics (co-operativity, autoactivation), substrate-dependent IC_{50} values (e.g., CYP3A4), the contribution of intestinal metabolism, the absorption rate constant (Brown et al., 2005; Bachmann, 2006; Brown et al., 2006) and inhibitory metabolites (Ogilvie et al., 2006). With a set of 193 in vivo DDI studies, the incidence of false negative predictions was largest using the average unbound plasma concentrations and smallest using the hepatic input total plasma concentrations of the inhibitor (Ito et al., 2004). Another approach for quantitative prediction of CYP3A4-mediated oral drug interactions was recently proposed based on the AUC increase of standard CYP3A4 substrates by co-administration of standard CYP3A4 inhibitors (Ohno et al., 2007).

While one realizes the challenges involved in accurate quantitative predictions of reversible interactions, bigger challenges are posed when the interaction involves a time-dependent component where enzyme activity can be restored only by synthesis of new apoprotein. For a time-dependent interaction, additional factors such k_{inact}, K_I, and K_{deg} need to be considered. Accuracy of DDI prediction may improve with consideration of parallel elimination pathways and inhibition of intestinal metabolism by the mechanism-based inhibitor (Galetin et al., 2006; Galetin et al., 2007).

23.6.3 Going Beyond P450s to Other Drug-Metabolizing Enzymes and Transporters

Models describing P450 interactions are more common for P450 substrates (as victim drugs). Prospective quantitative predictions become progressively more complex for perpetrator drugs as P450 inhibitors (combination of reversible and metabolism dependent), P450 inducers and combined P450 inhibitors and inducers. Although the main focus of DDI is on P450-mediated metabolism, there is increasing awareness of the role of other metabolic enzymes and transporters in drug disposition. Although there are reported examples of clinically relevant DDI for UGT substrates, exposure increases of the aglycone are rarely greater than 2-fold in the presence of a UGT inhibitor (Kiang et al., 2005). This is rationalized based on the low affinity for UGT inhibitors and multiple UGTs involved in metabolism (Williams et al., 2004). Likewise, DDIs with sulfotransferases (SULTs) should be considered also. Many SULTs are expressed in the gut and liver and play a role in first pass. This is evident with ethinyl estradiol, which undergoes extensive first-pass sulfation in the gut. Therefore, DDIs are possible because of SULT inhibition, induction or depletion of its cofactor (3'-phosphoadenosine 5'-phosphosulfate (PAPS)) (Schrag et al., 2004). Likewise, carboxylesterases are expressed in the gut and liver, and there are reports of possible DDIs involving these enzymes also (Li et al., 2007).

Efflux transporters such as P-gp have been long recognized in mediating clinical DDI (Lin, 2007). Focus is now shifted toward the interplay between drug-metabolizing enzymes and transporters (Lam et al., 2006; Zheng et al., 2009). Because the topographical relationship between efflux transporters and drug-metabolizing enzymes is different in the enterocyte (vs. hepatocyte), the combined inhibition of efflux transporters (e.g., P-gp) and P450 activity may lead to different DDI outcomes depending on which of the tissues is the site of the inhibitory interaction. In the intestine, CYP3A4 and P-gp act in concert to lead to greater first-pass elimination, their combined inhibition is predicted to cause a substantial increase in the in vivo AUC of substrate. The same inhibition of both processes in the liver may increase the substrate AUC, decrease the AUC or cause no net change in the AUC

(Benet et al., 2003; Wu and Benet, 2005). Other efflux transporters such as MRP2 (Luo et al., 2007) and influx transporters such as OATP1B1 (Lau et al., 2007) are also implicated in DDIs. For example, inhibition of hepatic uptake by an intravenous dose of rifampin has been shown to cause a significant increase in the AUC of atorvastatin and its active metabolites (Backman et al., 2005). There is also an increased focus on genotype and its impact on drug interactions (Mikus et al., 2006; Niemi et al., 2006). Prediction of DDI based on the Biopharmaceutics Drug Disposition Classification System (BDDCS) has been recently proposed (Benet et al., 2008). According to this proposal, Class 1 drugs are not substrates for transporters in the intestine and liver and are unlikely to be subject to transporter-mediated DDI. This theory also predicts that for Class 2 drugs, efflux transporter effects in the intestine and liver (along with transporter-enzyme interplay) will be important; while for Class 3 drugs, uptake transporters (along with uptake-efflux transporter interplay) will be important for intestinal absorption and liver entry. When a drug is transported by several transporters, the contribution of each transporter needs to be estimated to predict the degree of overall DDI (similar to the contribution of $f_m \times f_{mCYP}$ in determining the extent of P450-mediated DDI) (Hirano et al., 2004; Hinton et al., 2008).

Ultimately the pharmaceutical industry is responsible for integrating all the in vitro and in vivo data relating to disposition of NCE (with potential role for any drug-metabolizing and transporters) and establishing a safe dosing paradigm in a therapeutic setting. Unfortunately, many of the factors described above serve to further complicate DDI predictions, necessitate even more data integration and complex models. Furthermore, many of the preclinical and clinical tools to assess DDIs involving Phase II enzymes and transporters are either lacking or require considerably more validation. This is a far cry from P450s such as CYP2D6, CYP2C9, CYP1A2, CYP3A4, and CYP2C19, where in vitro reagents are plentiful, their use has been standardized, and clinical strategies (involving probe inhibitors, substrates, and genotyping) are more established (Bjornsson et al., 2003; FDA, 2006).

23.7 Conclusions

In recent years, there have been numerous examples of marketing failures because of clinically significant DDIs. As a result, the industry has spent considerable resources establishing DDI screens to support lead optimization and characterization. This has been possible because of automation, the increased availability of in vitro reagents, and commercial supplies of human tissue. It is now possible to screen large libraries of compounds and select only those compounds that are weak inhibitors of P450s, P-gp and OATPs, behave as weak inducers of P450s, and serve as low-turnover substrates of multiple P450s and other enzymes. Later in development, it is possible to attempt IVIVCs and forecast (or predict) DDIs by leveraging the in vitro data and predictions of human PK and dose. It is also possible to simply rank order the in vitro data and prioritize, or de-prioritize, clinical DDI and radiolabel (ADME) studies.

Once the clinical data are available, further in vitro (mechanistic) studies can be conducted if an unanticipated DDI is observed. Finally, all clinical and preclinical data are integrated in preparation for filing and labeling.

Despite the advances, it is clear that challenges remain. Prior to first in man, for example, quantitative predictions of DDIs are difficult. At the same time, it is recognized that one has to consider reversible and mechanism-based inhibition, along with induction, of drug-metabolizing enzymes and transporters in an integrated fashion. More complex models such as PBPK approaches may prove useful in the long run. The development of animal models also requires further effort, so that they can complement the growing arsenal of in vitro tools and reagents. From an in vitro standpoint, the development of additional reagents to enable a more complete evaluation of non-P450 enzymes (e.g., glucuronosyltransferases and SULTs) and transporters is needed also. At the end of the day, only time will tell if the industry's efforts have paid off. If successful, market withdrawals like those involving terfenadine and mibefradil (Table 23.1) will be a thing of the past.

References

Aninat C, Piton A, Glaise D, Le Charpentier T, Langouet S, Morel F, Guguen-Guillouzo C and Guillouzo A (2006) Expression of cytochromes P450, conjugating enzymes and nuclear receptors in human hepatoma HepaRG cells. *Drug Metab Dispos* **34**:75–83

Ayrton A and Morgan P (2008) Role of transport proteins in drug discovery and development: a pharmaceutical perspective. *Xenobiotica* **38**:676–708

Bachmann KA (2006) Inhibition constants, inhibitor concentrations and the prediction of inhibitory drug drug interactions: pitfalls, progress and promise. *Curr Drug Metab* **7**:1–14

Backman JT, Luurila H, Neuvonen M and Neuvonen PJ (2005) Rifampin markedly decreases and gemfibrozil increases the plasma concentrations of atorvastatin and its metabolites. *Clin Pharmacol Ther* **78**:154–167

Balimane PV, Han YH and Chong S (2006) Current industrial practices of assessing permeability and P-glycoprotein interaction. *Aaps J* **8**:E1–E13

Benet LZ, Amidon GL, Barends DM, Lennernas H, Polli JE, Shah VP, Stavchansky SA and Yu LX (2008) The use of BDDCS in classifying the permeability of marketed drugs. *Pharm Res* **25**:483–488

Benet LZ, Cummins CL and Wu CY (2003) Transporter-enzyme interactions: implications for predicting drug-drug interactions from in vitro data. *Curr Drug Metab* **4**:393–398

Bi YA, Kazolias D and Duignan DB (2006) Use of cryopreserved human hepatocytes in sandwich culture to measure hepatobiliary transport. *Drug Metab Dispos* **34**:1658–1665

Bjornsson TD, Callaghan JT, Einolf HJ, Fischer V, Gan L, Grimm S, Kao J, King SP, Miwa G, Ni L, Kumar G, McLeod J, Obach RS, Roberts S, Roe A, Shah A, Snikeris F, Sullivan JT, Tweedie D, Vega JM, Walsh J and Wrighton SA (2003) The conduct of in vitro and in vivo drug-drug interaction studies: a Pharmaceutical Research and Manufacturers of America (PhRMA) perspective. *Drug Metab Dispos* **31**:815–832

Brown HS, Galetin A, Hallifax D and Houston JB (2006) Prediction of in vivo drug-drug interactions from in vitro data: factors affecting prototypic drug-drug interactions involving CYP2C9, CYP2D6 and CYP3A4. *Clin Pharmacokinet* **45**:1035–1050

Brown HS, Ito K, Galetin A and Houston JB (2005) Prediction of in vivo drug-drug interactions from in vitro data: impact of incorporating parallel pathways of drug elimination and inhibitor absorption rate constant. *Br J Clin Pharmacol* **60**:508–518

Cascio SM (2001) Novel strategies for immortalization of human hepatocytes. *Artif Organs* **25**:529–538

Chiou WL, Chung SM and Wu TC (2000) Potential role of P-glycoprotein in affecting hepatic metabolism of drugs. *Pharm Res* **17**:903–905

Crespi CL and Miller VP (1999) The use of heterologously expressed drug metabolizing enzymes–state of the art and prospects for the future. *Pharmacol Ther* **84**:121–131

Crespi CL, Miller VP and Penman BW (1997) Microtiter plate assays for inhibition of human, drug-metabolizing cytochromes P450. *Anal Biochem* **248**:188–190

Down MJ, Arkle S and Mills JJ (2007) Regulation and induction of CYP3A11, CYP3A13 and CYP3A25 in C57BL/6 J mouse liver. *Arch Biochem Biophys* **457**:105–110

Ek M, Soderdahl T, Kuppers-Munther B, Edsbagge J, Andersson TB, Bjorquist P, Cotgreave I, Jernstrom B, Ingelman-Sundberg M and Johansson I (2007) Expression of drug metabolizing enzymes in hepatocyte-like cells derived from human embryonic stem cells. *Biochem Pharmacol* **74**:496–503

El-Sankary W, Gibson GG, Ayrton A and Plant N (2001) Use of a reporter gene assay to predict and rank the potency and efficacy of CYP3A4 inducers. *Drug Metab Dispos* **29**: 1499–1504

Fahmi OA, Maurer TS, Kish M, Cardenas E, Boldt S and Nettleton D (2008) A combined model for predicting CYP3A4 clinical net drug-drug interaction based on CYP3A4 inhibition, inactivation, and induction determined in vitro. *Drug Metab Dispos* **36**:1698–1708

Fayer JL, Zannikos PN, Stevens JC, Luo Y, Sidhu R and Kirkesseli S (2001) Lack of correlation between in vitro inhibition of CYP3A-mediated metabolism by a PPAR-gamma agonist and its effect on the clinical pharmacokinetics of midazolam, an in vivo probe of CYP3A activity. *J Clin Pharmacol* **41**:305–316

FDA (2006) FDA Draft Guidance for industry: drug interaction studies - study design, data analysis, and implications for dosing and labeling, in.

Fontana E, Dansette PM and Poli SM (2005) Cytochrome p450 enzymes mechanism based inhibitors: common sub-structures and reactivity. *Curr Drug Metab* **6**:413–454

Galetin A, Burt H, Gibbons L and Houston JB (2006) Prediction of time-dependent CYP3A4 drug-drug interactions: impact of enzyme degradation, parallel elimination pathways, and intestinal inhibition. *Drug Metab Dispos* **34**:166–175

Galetin A, Hinton LK, Burt H, Obach RS and Houston JB (2007) Maximal inhibition of intestinal first-pass metabolism as a pragmatic indicator of intestinal contribution to the drug-drug interactions for CYP3A4 cleared drugs. *Curr Drug Metab* **8**:685–693

Gonzalez FJ and Yu AM (2006) Cytochrome P450 and xenobiotic receptor humanized mice. *Annu Rev Pharmacol Toxicol* **46**:41–64

Goodwin B, Hodgson E and Liddle C (1999) The orphan human pregnane X receptor mediates the transcriptional activation of CYP3A4 by rifampicin through a distal enhancer module. *Mol Pharmacol* **56**:1329–1339

Greenblatt DJ, von Moltke LL, Harmatz JS, Durol AL, Daily JP, Graf JA, Mertzanis P, Hoffman JL and Shader RI (2000) Differential impairment of triazolam and zolpidem clearance by ritonavir. *J Acquir Immune Defic Syndr* **24**:129–136

Grime K and Riley RJ (2006) The impact of in vitro binding on in vitro-in vivo extrapolations, projections of metabolic clearance and clinical drug-drug interactions. *Curr Drug Metab* **7**:251–264

Gripon P, Rumin S, Urban S, Le Seyec J, Glaise D, Cannie I, Guyomard C, Lucas J, Trepo C and Guguen-Guillouzo C (2002) Infection of a human hepatoma cell line by hepatitis B virus. *Proc Natl Acad Sci U S A* **99**:15655–15660

Guillouzo A, Corlu A, Aninat C, Glaise D, Morel F and Guguen-Guillouzo C (2007) The human hepatoma HepaRG cells: a highly differentiated model for studies of liver metabolism and toxicity of xenobiotics. *Chem Biol Interact* **168**:66–73

Hagenbuch B and Meier PJ (2003) The superfamily of organic anion transporting polypeptides. *Biochim Biophys Acta* **1609**:1–18

Hagenbuch B and Meier PJ (2004) Organic anion transporting polypeptides of the OATP/ SLC21 family: phylogenetic classification as OATP/ SLCO superfamily, new nomenclature and molecular/functional properties. *Pflugers Arch* **447**:653–665

Hahn WC (2002) Immortalization and transformation of human cells. *Mol Cells* **13**:351–361

Hamilton GA, Jolley SL, Gilbert D, Coon DJ, Barros S and LeCluyse EL (2001) Regulation of cell morphology and cytochrome P450 expression in human hepatocytes by extracellular matrix and cell-cell interactions. *Cell Tissue Res* **306**:85–99

Harmsen S, Meijerman I, Beijnen JH and Schellens JH (2007) The role of nuclear receptors in pharmacokinetic drug-drug interactions in oncology. *Cancer Treat Rev* **33**:369–380

Healan-Greenberg C, Waring JF, Kempf DJ, Blomme EA, Tirona RG and Kim RB (2008) A human immunodeficiency virus protease inhibitor is a novel functional inhibitor of human pregnane X receptor. *Drug Metab Dispos* **36**:500–507

Hewitt NJ, de Kanter R and LeCluyse E (2007a) Induction of drug metabolizing enzymes: a survey of in vitro methodologies and interpretations used in the pharmaceutical industry—do they comply with FDA recommendations?. *Chem Biol Interact* **168**:51–65

Hewitt NJ and Hewitt P (2004) Phase I and II enzyme characterization of two sources of HepG2 cell lines. *Xenobiotica* **34**:243–256

Hewitt NJ, Lechon MJ, Houston JB, Hallifax D, Brown HS, Maurel P, Kenna JG, Gustavsson L, Lohmann C, Skonberg C, Guillouzo A, Tuschl G, Li AP, LeCluyse E, Groothuis GM and Hengstler JG (2007b) Primary hepatocytes: current understanding of the regulation of metabolic enzymes and transporter proteins, and pharmaceutical practice for the use of hepatocytes in metabolism, enzyme induction, transporter, clearance, and hepatotoxicity studies. *Drug Metab Rev* **39**:159–234

Hidalgo I (2001) Assessing the absorption of new pharmaceuticals. *Curr Topics Med Chem* **1**: 385–401

Hinton LK, Galetin A and Houston JB (2008) Multiple inhibition mechanisms and prediction of drug-drug interactions: status of metabolism and transporter models as exemplified by gemfibrozil-drug interactions. *Pharm Res* **25**:1063–1074

Hirano M, Maeda K, Shitara Y and Sugiyama Y (2004) Contribution of OATP2 (OATP1B1) and OATP8 (OATP1B3) to the hepatic uptake of pitavastatin in humans. *J Pharmacol Exp Ther* **311**:139–146

Honkakoski P, Sueyoshi T and Negishi M (2003) Drug-activated nuclear receptors CAR and PXR. *Ann Med* **35**:172–182

Hsiao P, Bui T, Ho RJ and Unadkat JD (2007) In vitro to in vivo prediction of p-glycoprotein based drug interactions at the human and rodent blood-brain barrier. *Drug Metab Dispos* **34**: 786–792.

Hsu A, Granneman GR, Cao G, Carothers L, el-Shourbagy T, Baroldi P, Erdman K, Brown F, Sun E and Leonard JM (1998) Pharmacokinetic interactions between two human immunodeficiency virus protease inhibitors, ritonavir and saquinavir. *Clin Pharmacol Ther* **63**:453–464

Huang C, Zheng M, Yang Z, Rodrigues AD and Marathe P (2008) Projection of exposure and efficacious dose prior to first-in-human studies: how successful have we been?. *Pharm Res* **25**:713–726

Huang SM, Lesko LJ and Williams RL (1999) Assessment of the quality and quantity of drug-drug interaction studies in recent NDA submissions: study design and data analysis issues. *J Clin Pharmacol* **39**:1006–1014

Ito K, Brown HS and Houston JB (2004) Database analyses for the prediction of in vivo drug-drug interactions from in vitro data. *Br J Clin Pharmacol* **57**:473–486

Ito K, Iwatsubo T, Kanamitsu S, Ueda K, Suzuki H and Sugiyama Y (1998) Prediction of pharmacokinetic alterations caused by drug-drug interactions: metabolic interaction in the liver. *Pharmacol Rev* **50**:387–412

Jaakkola T, Backman JT, Neuvonen M, Niemi M and Neuvonen PJ (2006) Montelukast and zafirlukast do not affect the pharmacokinetics of the CYP2C8 substrate pioglitazone. *Eur J Clin Pharmacol* **62**:503–509

Jones-Bolin S, Zhao H, Hunter K, Klein-Szanto A and Ruggeri B (2006) The effects of the oral, pan-VEGF-R kinase inhibitor CEP-7055 and chemotherapy in orthotopic models of glioblastoma and colon carcinoma in mice. *Mol Cancer Ther* **5**:1744–1753

Kai H, Shibuya K, Wang Y, Kameta H, Kameyama T, Tahara-Hanaoka S, Miyamoto A, Honda S, Matsumoto I, Koyama A, Sumida T and Shibuya A (2006) Critical role of M. tuberculosis for dendritic cell maturation to induce collagen-induced arthritis in H-2b background of C57BL/6 mice. *Immunology* **118**:233–239

Kajosaari LI, Jaakkola T, Neuvonen PJ and Backman JT (2006) Pioglitazone, an in vitro inhibitor of CYP2C8 and CYP3A4, does not increase the plasma concentrations of the CYP2C8 and CYP3A4 substrate repaglinide. *Eur J Clin Pharmacol* **62**:217–223

Kanamitsu S, Ito K and Sugiyama Y (2000) Quantitative prediction of in vivo drug-drug interactions from in vitro data based on physiological pharmacokinetics: use of maximum unbound concentration of inhibitor at the inlet to the liver. *Pharm Res* **17**:336–343

Kanazu T, Yamaguchi Y, Okamura N, Baba T and Koike M (2004) Model for the drug-drug interaction responsible for CYP3A enzyme inhibition. I: evaluation of cynomolgus monkeys as surrogates for humans. *Xenobiotica* **34**:391–402

Kariv I, Fereshteh MP and Oldenburg KR (2001) Development of a miniaturized 384-well high throughput screen for the detection of substrates of cytochrome P450 2D6 and 3A4 metabolism. *J Biomol Screen* **6**:91–99

Keogh JP and Kunta JR (2006) Development, validation and utility of an in vitro technique for assessment of potential clinical drug-drug interactions involving P-glycoprotein. *Eur J Pharm Sci* **27**:543–554

Kerns E, Di L, Petusky S, Farris M, Ley R and Jupp P (2004) Combined application of parallel artificial membrane permeability assay and Caco-2 permeability assays in drug discovery. *J Pharma Sci* **93**:1440–1453

Kiang TK, Ensom MH and Chang TK (2005) UDP-glucuronosyltransferases and clinical drug-drug interactions. *Pharmacol Ther* **106**:97–132

Kim KA, Park PW, Kim KR and Park JY (2007) Effect of multiple doses of montelukast on the pharmacokinetics of rosiglitazone, a CYP2C8 substrate, in humans. *Br J Clin Pharmacol* **63**:339–345

Kliewer SA, Goodwin B and Willson TM (2002) The nuclear pregnane X receptor: a key regulator of xenobiotic metabolism. *Endocr Rev* **23**:687–702

Komatsu K, Ito K, Nakajima Y, Kanamitsu S, Imaoka S, Funae Y, Green CE, Tyson CA, Shimada N and Sugiyama Y (2000) Prediction of in vivo drug-drug interactions between tolbutamide and various sulfonamides in humans based on in vitro experiments. *Drug Metab Dispos* **28**:475–481

Lam JL, Okochi H, Huang Y and Benet LZ (2006) In vitro and in vivo correlation of hepatic transporter effects on erythromycin metabolism: characterizing the importance of transporter-enzyme interplay. *Drug Metab Dispos* **34**:1336–1344

Lau YY, Huang Y, Frassetto L and Benet LZ (2007) Effect of OATP1B transporter inhibition on the pharmacokinetics of atorvastatin in healthy volunteers. *Clin Pharmacol Ther* **81**:194–204

LeCluyse E, Madan A, Hamilton G, Carroll K, DeHaan R and Parkinson A (2000) Expression and regulation of cytochrome P450 enzymes in primary cultures of human hepatocytes. *J Biochem Mol Toxicol* **14**:177–188

LeCluyse EL, Alexandre E, Hamilton GA, Viollon-Abadie C, Coon DJ, Jolley S and Richert L (2005) Isolation and culture of primary human hepatocytes. *Methods Mol Biol* **290**:207–229

Lee CM and Farde L (2006) Using positron emission tomography to facilitate CNS drug development. *Trends Pharmacol Sci* **27**:310–316

Li P, Callery PS, Gan LS and Balani SK (2007) Esterase inhibition attribute of grapefruit juice leading to a new drug interaction. *Drug Metab Dispos* **35**:1023–1031

Lin J and Yamazaki M (2003) Role of P-glycoprotein in pharmacokinetics. *Clin Pharmacokinet* **42**:59–98

Lin JH (2007) Transporter-mediated drug interactions: clinical implications and in vitro assessment. *Expert Opin Drug Metab Toxicol* **3**:81–92

Lin JH and Lu AY (1998) Inhibition and induction of cytochrome P450 and the clinical implications. *Clin Pharmacokinet* **35**:361–390

Lin JH and Pearson PG (2002) Prediction of metabolic drug interactions: quantitative or qualitative?, in *Drug-Drug Interactions* (Rodrigues AD ed) pp 415-438, Marcel Dekker, Basel, Switzerland.

Luo G, Cunningham M, Kim S, Burn T, Lin J, Sinz M, Hamilton G, Rizzo C, Jolley S, Gilbert D, Downey A, Mudra D, Graham R, Carroll K, Xie J, Madan A, Parkinson A, Christ D, Selling B, LeCluyse E and Gan LS (2002) CYP3A4 induction by drugs: correlation between a pregnane X receptor reporter gene assay and CYP3A4 expression in human hepatocytes. *Drug Metab Dispos* **30**:795–804

Luo G, Garner CE, Xiong H, Hu H, Richards LE, Brouwer KL, Duan J, Decicco CP, Maduskuie T, Shen H, Lee FW and Gan LS (2007) Effect of DPC 333 [(2R)-2-{(3R)-3-amino-3-[4-(2-methylquinolin-4-ylmethoxy)phenyl]-2-oxopyrr olidin-1-yl}-N-hydroxy-4-methylpentanamide], a human tumor necrosis factor alpha-converting enzyme inhibitor, on the disposition of methotrexate: a transporter-based drug-drug interaction case study. *Drug Metab Dispos* **35**:835–840

Luo G, Lin J, Fiske WD, Dai R, Yang TJ, Kim S, Sinz M, LeCluyse E, Solon E, Brennan JM, Benedek IH, Jolley S, Gilbert D, Wang L, Lee FW and Gan LS (2003) Concurrent induction and mechanism-based inactivation of CYP3A4 by an L-valinamide derivative. *Drug Metab Dispos* **31**:1170–1175

Lyon K, Faris R, Liu J, Snyder J, Brobst D, Lichti K, Staudinger J, Czerwinski M, Toren P and Parkinson A (2005) Improving lot-to-lot variability and dynamic range of CYP3A4 indution in immortalized human hepatocytes (Fa2N-4 cells), in *Internation Society for the Study of Xenobiotics*, Maui, HI.

Ma X, Shah Y, Cheung C, Guo GL, Feigenbaum L, Krausz KW, Idle JR and Gonzalez FJ (2007) The PREgnane X receptor gene-humanized mouse: a model for investigating drug-drug interactions mediated by cytochromes P450 3A. *Drug Metab Dispos* **35**:194–200

Madan A, Usuki E, Burton LA, Ogilvie BW and Parkinson A (2002) In vitro approaches for studying the inhibition of drug-metabolizing enzymes and identifying the drug-metabolizing enzymes responsible for the metabolism of drugs, in *Drug-drug interactions* (Rodrigues AD ed) pp 217-294, Marcel Dekker, Inc, Basel-Switzerland.

Mandal PK (2005) Dioxin: a review of its environmental effects and its aryl hydrocarbon receptor biology. *J Comp Physiol [B]* **175**:221–230

Marathe PH and Rodrigues AD (2006) In vivo animal models for investigating potential CYP3A- and Pgp-mediated drug-drug interactions. *Curr Drug Metab* **7**:687–704

Marion TL, Leslie EM and Brouwer KL (2007) Use of sandwich-cultured hepatocytes to evaluate impaired bile acid transport as a mechanism of drug-induced hepatotoxicity. *Mol Pharm* **4**: 911–918

Matheny C, Lamb M, Brouwer K and Pollak G (2001) Pharmacokinetic and pharmacodynamic implications of P-gp modulation. *Pharmacotherapy* **21**:778–796

Medina-Diaz IM, Arteaga-Illan G, de Leon MB, Cisneros B, Sierra-Santoyo A, Vega L, Gonzalez FJ and Elizondo G (2007) Pregnane X receptor-dependent induction of the CYP3A4 gene by o,p′-1,1,1,-trichloro-2,2-bis (p-chlorophenyl)ethane. *Drug Metab Dispos* **35**: 95–102

Medina-Diaz IM and Elizondo G (2005) Transcriptional induction of CYP3A4 by o,p′-DDT in HepG2 cells. *Toxicol Lett* **157**:41–47

Mikus G, Schowel V, Drzewinska M, Rengelshausen J, Ding R, Riedel KD, Burhenne J, Weiss J, Thomsen T and Haefeli WE (2006) Potent cytochrome P450 2C19 genotype-related interaction between voriconazole and the cytochrome P450 3A4 inhibitor ritonavir. *Clin Pharmacol Ther* **80**:126–135

Miller VP, Stresser DM, Blanchard AP, Turner S and Crespi CL (2000) Fluorometric high-throughput screening for inhibitors of cytochrome P450. *Ann N Y Acad Sci* **919**:26–32

Mills JB, Rose KA, Sadagopan N, Sahi J and de Morais SM (2004) Induction of drug metabolism enzymes and MDR1 using a novel human hepatocyte cell line. *J Pharmacol Exp Ther* **309**: 303–309

Miners JO, Smith PA, Sorich MJ, McKinnon RA and Mackenzie PI (2004) Predicting human drug glucuronidation parameters: application of in vitro and in silico modeling approaches. *Annu Rev Pharmacol Toxicol* **44**:1–25

Naritomi Y, Teramura Y, Terashita S and Kagayama A (2004) Utility of microtiter plate assays for human cytochrome P450 inhibition studies in drug discovery: application of simple method for detecting quasi-irreversible and irreversible inhibitors. *Drug Metab Pharmacokinet* **19**:55–61

Niemi M (2007) Role of OATP transporters in the disposition of drugs. *Pharmacogenomics* **8**: 787–802

Niemi M, Kivisto KT, Diczfalusy U, Bodin K, Bertilsson L, Fromm MF and Eichelbaum M (2006) Effect of SLCO1B1 polymorphism on induction of CYP3A4 by rifampicin. *Pharmacogenet Genom* **16**:565–568

Nishimura M, Koeda A, Suganuma Y, Suzuki E, Shimizu T, Nakayama M, Satoh T, Narimatsu S and Naito S (2007) Comparison of inducibility of CYP1A and CYP3A mRNAs by prototypical inducers in primary cultures of human, cynomolgus monkey, and rat hepatocytes. *Drug Metab Pharmacokinet* **22**:178–186

Obach RS (2001) The prediction of human clearance from hepatic microsomal metabolism data. *Curr Opin Drug Discovery Dev* **4**:36–44

Obach RS, Walsky RL and Venkatakrishnan K (2007) Mechanism-based inactivation of human cytochrome p450 enzymes and the prediction of drug-drug interactions. *Drug Metab Dispos* **35**:246–255

Obach RS, Walsky RL, Venkatakrishnan K, Houston JB and Tremaine LM (2005) In vitro cytochrome P450 inhibition data and the prediction of drug-drug interactions: qualitative relationships, quantitative predictions, and the rank-order approach. *Clin Pharmacol Ther* **78**:582–592

Ogasawara A, Kume T and Kazama E (2007) Effect of oral ketoconazole on intestinal first-pass effect of midazolam and fexofenadine in cynomolgus monkeys. *Drug Metab Dispos* **35**: 410–418

Ogilvie BW, Zhang D, Li W, Rodrigues AD, Gipson AE, Holsapple J, Toren P and Parkinson A (2006) Glucuronidation converts gemfibrozil to a potent, metabolism-dependent inhibitor of CYP2C8: implications for drug-drug interactions. *Drug Metab Dispos* **34**:191–197

Ohno Y, Hisaka A and Suzuki H (2007) General framework for the quantitative prediction of CYP3A4-mediated oral drug interactions based on the AUC increase by coadministration of standard drugs. *Clin Pharmacokinet* **46**:681–696

Ozbal CC, LaMarr WA, Linton JR, Green DF, Katz A, Morrison TB and Brenan CJ (2004) High throughput screening via mass spectrometry: a case study using acetylcholinesterase. *Assay Drug Dev Technol* **2**:373–381

Poirier A, Funk C, Lave T and Noe J (2007) New strategies to address drug-drug interactions involving OATPs. *Curr Opin Drug Discovery Dev* **10**:74–83

Polli J, Jerrett J, Studenberg J, Humphreys J, Dennis S, Brower K and Wooley J (1999) Role of P-gp on CNS disposition of amprenavir, an HIV protease inhibitor. *Pharma Res* **16**:1206–1212

Polli J, Wring S, Humphreys J, Huang L, Morgan J, Webster L and Serabjit-Singh C (2001) Rational use of in vitro P-gp assays in drug discovery. *J Pharmacol Exp Ther* **299**:620–628

Preskorn SH and Magnus RD (1994) Inhibition of hepatic P-450 isoenzymes by serotonin selective reuptake inhibitors: in vitro and in vivo findings and their implications for patient care. *Psychopharmacol Bull* **30**:251–259

Prueksaritanont T, Kuo Y, Tang C, Li C, Qiu Y, Lu B, Strong-Basalyga K, Richards K, Carr B and Lin JH (2006) In vitro and in vivo CYP3A64 induction and inhibition studies in rhesus monkeys: a preclinical approach for CYP3A-mediated drug interaction studies. *Drug Metab Dispos* **34**:1546–1555

Prueksaritanont T, Zhao JJ, Ma B, Roadcap BA, Tang C, Qiu Y, Liu L, Lin JH, Pearson PG and Baillie TA (2002) Mechanistic studies on metabolic interactions between gemfibrozil and statins. *J Pharmacol Exp Ther* **301**:1042–1051

Rencurel F, Stenhouse A, Hawley SA, Friedberg T, Hardie DG, Sutherland C and Wolf CR (2005) AMP-activated protein kinase mediates phenobarbital induction of CYP2B gene expression in hepatocytes and a newly derived human hepatoma cell line. *J Biol Chem* **280**:4367–4373

Riley RJ (2001) The potential pharmacological and toxicological impact of P450 screening. *Curr Opin Drug Discov Dev* **4**:45–54

Ripp SL, Mills JB, Fahmi OA, Trevena KA, Liras JL, Maurer TS and de Morais SM (2006) Use of immortalized human hepatocytes to predict the magnitude of clinical drug-drug interactions caused by CYP3A4 induction. *Drug Metab Dispos* **34**:1742–1748

Rodrigues AD (2007) Prioritization of clinical drug inteaction studies using in vitro cytochrome P450 data: proposed refinement and expansion of the "rank order" approach. *Drug Metab Lett* **1**:31–35

Rodrigues AD, Winchell GA and Dobrinska MR (2001) Use of in vitro drug metabolism data to evaluate metabolic drug-drug interactions in man: the need for quantitative databases. *J Clin Pharmacol* **41**:368–373

Rostami-Hodjegan A and Tucker G (2004) "In silico" simulations to assess the "in vivo" consequences of "in vitro" metabolic drug-drug interactions. *Drug Discovery Today* **1**:441–448

Rowland M and Matin SB (1973) Kinetics of drug-drug interactions. *J Pharmacokinet Biopharma* **1**:553–567

Roymans D, Annaert P, Van Houdt J, Weygers A, Noukens J, Sensenhauser C, Silva J, Van Looveren C, Hendrickx J, Mannens G and Meuldermans W (2005) Expression and induction potential of cytochromes P450 in human cryopreserved hepatocytes. *Drug Metab Dispos* **33**:1004–1016

Runge D, Michalopoulos GK, Strom SC and Runge DM (2000) Recent advances in human hepatocyte culture systems. *Biochem Biophys Res Commun* **274**:1–3

Schehrer L, Regan JD and Westendorf J (2000) UDS induction by an array of standard carcinogens in human and rodent hepatocytes: effect of cryopreservation. *Toxicology* **147**:177–191

Schrag ML, Cui D, Rushmore TH, Shou M, Ma B and Rodrigues AD (2004) Sulfotransferase 1E1 is a low km isoform mediating the 3-O-sulfation of ethinyl estradiol. *Drug Metab Dispos* **32**:1299–1303

Schwenk M (1980) Transport systems of isolated hepatocytes. Studies on the transport of biliary compounds. *Arch Toxicol* **44**:113–126

Shiau AK, Massari ME and Ozbal CC (2008) Back to basics: label-free technologies for small molecule screening. *Comb Chem High Throughput Screen* **11**:231–237

Silverman RB (1995) Mechanism-based enzyme inactivators. *Methods Enzymol* **249**:240–283

Sinz M, Kim S, Zhu Z, Chen T, Anthony M, Dickinson K and Rodrigues AD (2006) Evaluation of 170 xenobiotics as transactivators of human pregnane X receptor (hPXR) and correlation to known CYP3A4 drug interactions. *Curr Drug Metab* **7**:375–388

Sinz MW and Kim S (2006) Stem cells, immortalized cells and primary cells in ADMET assays. *Drug Discovery Today* **3**:79–85

Smith NF, Figg WD and Sparreboom A (2005) Role of the liver-specific transporters OATP1B1 and OATP1B3 in governing drug elimination. *Expert Opin Drug Metab Toxicol* **1**: 429–445

Soars MG, McGinnity DF, Grime K and Riley RJ (2007) The pivotal role of hepatocytes in drug discovery. *Chem Biol Interact* **168**:2–15

Stanley LA, Horsburgh BC, Ross J, Scheer N and Wolf CR (2006) PXR and CAR: nuclear receptors which play a pivotal role in drug disposition and chemical toxicity. *Drug Metab Rev* **38**:515–597

Surveillance BCD (1972) Adverse drug interactions. *J Am Med Assoc* **220**:1238–1239

Suzuki H and Sugiyama Y (2000) Role of metabolic enzymes and efflux transporters in the absorption of drugs from the small intestine. *Eur J Pharm Sci* **12**:3–12

Synold TW, Dussault I and Forman BM (2001) The orphan nuclear receptor SXR coordinately regulates drug metabolism and efflux. *Nat Med* **7**:584–590

Tateno C, Yoshizane Y, Saito N, Kataoka M, Utoh R, Yamasaki C, Tachibana A, Soeno Y, Asahina K, Hino H, Asahara T, Yokoi T, Furukawa T and Yoshizato K (2004) Near completely humanized liver in mice shows human-type metabolic responses to drugs. *Am J Pathol* **165**:901–912

Tirona RG and Kim RB (2005) Nuclear receptors and drug disposition gene regulation. *J Pharm Sci* **94**:1169–1186

Trubetskoy OV, Gibson JR and Marks BD (2005) Highly miniaturized formats for in vitro drug metabolism assays using vivid fluorescent substrates and recombinant human cytochrome P450 enzymes. *J Biomol Screen* **10**:56–66

Tucker GT, Houston JB and Huang SM (2001) Optimizing drug development: strategies to assess drug metabolism/transporter interaction potential-toward a consensus. *Clin Pharmacol Ther* **70**:103–114

Venkatakrishnan K, Obach RS and Rostami-Hodjegan A (2007) Mechanism-based inactivation of human cytochrome P450 enzymes: strategies for diagnosis and drug-drug interaction risk assessment. *Xenobiotica* **37**:1225–1256

Vermeir M, Annaert P, Mamidi RN, Roymans D, Meuldermans W and Mannens G (2005) Cell-based models to study hepatic drug metabolism and enzyme induction in humans. *Expert Opin Drug Metab Toxicol* **1**:75–90

Vignati LA, Bogni A, Grossi P and Monshouwer M (2004) A human and mouse pregnane X receptor reporter gene assay in combination with cytotoxicity measurements as a tool to evaluate species-specific CYP3A induction. *Toxicology* **199**:23–33

von Moltke LL, Greenblatt DJ, Duan SX, Harmatz JS and Shader RI (1994) In vitro prediction of the terfenadine-ketoconazole pharmacokinetic interaction. *J Clin Pharmacol* **34**:1222–1227

von Moltke LL, Greenblatt DJ, Harmatz JS, Duan SX, Harrel LM, Cotreau-Bibbo MM, Pritchard GA, Wright CE and Shader RI (1996) Triazolam biotransformation by human liver microsomes in vitro: effects of metabolic inhibitors and clinical confirmation of a predicted interaction with ketoconazole. *J Pharmacol Exp Ther* **276**:370–379

Wang H, Huang H, Li H, Teotico DG, Sinz M, Baker SD, Staudinger J, Kalpana G, Redinbo MR and Mani S (2007) Activated pregnenolone X-receptor is a target for ketoconazole and its analogs. *Clin Cancer Res* **13**:2488–2495

Wang YH, Jones DR and Hall SD (2004) Prediction of cytochrome P450 3A inhibition by verapamil enantiomers and their metabolites. *Drug Metab Dispos* **32**:259–266

White RE (2000) High-throughput screening in drug metabolism and pharmacokinetic support of drug discovery. *Annu Rev Pharmacol Toxicol* **40**:133–157

Williams JA, Hyland R, Jones BC, Smith DA, Hurst S, Goosen TC, Peterkin V, Koup JR and Ball SE (2004) Drug-drug interactions for UDP-glucuronosyltransferase substrates: a pharmacokinetic explanation for typically observed low exposure (AUCi/AUC) ratios. *Drug Metab Dispos* **32**:1201–1208

Wu CY and Benet LZ (2005) Predicting drug disposition via application of BCS: transport/absorption/ elimination interplay and development of a biopharmaceutics drug disposition classification system. *Pharm Res* **22**:11–23

Xie W and Evans RM (2002) Pharmaceutical use of mouse models humanized for the xenobiotic receptor. *Drug Discovery Today* **7**:509–515

Yamamoto T, Suzuki A and Kohno Y (2004) High-throughput screening for the assessment of time-dependent inhibitions of new drug candidates on recombinant CYP2D6 and CYP3A4 using a single concentration method. *Xenobiotica* **34**:87–101

Yan Z, Rafferty B, Caldwell GW and Masucci JA (2002) Rapidly distinguishing reversible and irreversible CYP450 inhibitors by using fluorometric kinetic analyses. *Eur J Drug Metab Pharmacokinet* **27**:281–287

Yao C and Levy RH (2002) Inhibition-based metabolic drug-drug interactions: predictions from in vitro data. *J Pharm Sci* **91**:1923–1935

Youdim KA, Lyons R, Payne L, Jones BC and Saunders K (2008) An automated, high-throughput, 384 well cytochrome P450 cocktail IC50 assay using a rapid resolution LC-MS/MS end-point. *J Pharm Biomed Anal* **48**:92–99

Youdim KA, Tyman CA, Jones BC and Hyland R (2007) Induction of cytochrome P450: assessment in an immortalized human hepatocyte cell line (Fa_2N_4) using a novel higher throughput cocktail assay. *Drug Metab Dispos* **35**:275–282

Yu DK (1999) The contribution of P-glycoprotein to pharmacokinetic drug-drug interactions. *J Clin Pharmacol* **39**:1203–1211

Zhang H, Davis CD, Sinz MW and Rodrigues AD (2007a) Cytochrome P450 reaction-phenotyping: an industrial perspective. *Expert Opin Drug Metab Toxicol* **3**:667–687

Zhang H, Zhang D, Li W, Yao M, D'Arienzo C, Li YX, Ewing WR, Gu Z, Zhu Y, Murugesan N, Shyu WC and Humphreys WG (2007b) Reduction of site-specific CYP3A-mediated metabolism for dual angiotensin and endothelin receptor antagonists in various in vitro systems and in cynomolgus monkeys. *Drug Metab Dispos* **35**:795–805

Zheng HX, Huang Y, Frassetto LA and Benet LZ (2009) Elucidating rifampin's inducing and inhibiting effects on glyburide pharmacokinetics and blood glucose in healthy volunteers: unmasking the differential effects of enzyme induction and transporter inhibition for a drug and its primary metabolite. *Clin Pharmacol Ther* **85**:78–85

Zhou C, Poulton EJ, Grun F, Bammler TK, Blumberg B, Thummel KE and Eaton DL (2007) The dietary isothiocyanate sulforaphane is an antagonist of the human steroid and xenobiotic nuclear receptor. *Mol Pharmacol* **71**:220–229

Zhu Z, Kim S, Chen T, Lin JH, Bell A, Bryson J, Dubaquie Y, Yan N, Yanchunas J, Xie D, Stoffel R, Sinz M and Dickinson K (2004) Correlation of high-throughput pregnane X receptor (PXR) transactivation and binding assays. *J Biomol Screen* **9**:533–540

Zientek M, Miller H, Smith D, Dunklee MB, Heinle L, Thurston A, Lee C, Hyland R, Fahmi O and Burdette D (2008) Development of an in vitro drug-drug interaction assay to simultaneously monitor five cytochrome P450 isoforms and performance assessment using drug library compounds. *J Pharmacol Toxicol Methods* **58**:206–214

Chapter 24
Clinical Studies of Drug–Drug Interactions: Design and Interpretation

David J. Greenblatt and Lisa L. von Moltke

Abstract The potential importance of drug-drug interaction (DDIs) is increasing as polypharmacy becomes more and more prevalent. In vitro data cannot directly predict clinical DDIs, but may provide a rationale for initiation of human studies to confirm or exclude possible interactions. Clinical DDI studies are designed to determine whether there is a real drug interaction not due to chance, how big the interaction is, and whether the DDI is of clinical importance. Statistical significance is not equivalent to clinical significance, and supplemental pharmacodynamic or clinical outcome information is needed to address the importance of a pharmacokinetic DDI.

24.1 Introduction

Drug–drug interactions (DDIs) have become a topic of substantial scientific and public health concern over the last 20 years. While the clinical phenomenon of DDIs had been recognized for a number of decades, several events in and around the years 1988–1993 brought the topic of DDIs to a position of high attention and priority in the scientific community, as well as in the public arena. During this period, multiple human cytochrome P450 (CYP) isoforms became identified, along with increasing understanding of their substrate and inhibitor specificities, relative quantitative importance in human drug metabolism, and mechanisms of genetic regulation (Clarke, 1998; Smith et al., 1998; b; Venkatakrishnan et al., 2001; Venkatakrishnan et al., 2003). Of particular importance in this context was CYP3A, with its unique hepatic and enteric distribution, and its major contribution to clearance of many clinically relevant drugs as well as naturally occurring chemicals (Venkatakrishnan et al., 2001; Venkatakrishnan et al., 2003; Guengerich,

D.J. Greenblatt (✉)
Department of Pharmacology and Experimental Therapeutics, Tufts University School of Medicine and Tufts Medical Center, Boston, MA, USA
e-mail: dj.greenblatt@tufts.edu

K.S. Pang et al. (eds.), *Enzyme- and Transporter-Based Drug–Drug Interactions*,
DOI 10.1007/978-1-4419-0840-7_24,

1999; Greenblatt et al., 2008). At the same time, in vitro techniques for studying human drug metabolism became increasingly developed and refined, including predictive models for in vitro–in vivo scaling, and the availability of heterologously expressed individual human CYPs. At a clinical level, polypharmacy was becoming increasingly prevalent, as the population aged, the number of patients with multiple illnesses increased, and our capacity to provide pharmacologic treatments for serious disorders became more and more effective. Some newly introduced classes of medications – such as the azole antifungal agents and the selective serotonin reuptake inhibitor (SSRI) antidepressants – offered unique therapeutic options, but also had the secondary property of inhibiting certain human CYPs, thereby elevating the risk of DDIs (Greenblatt et al., 1999; Hemeryck and Belpaire, 2002; Venkatakrishnan et al., 2000). A dramatic and widely publicized event was the interaction of the nonsedating antihistamine terfenadine with potent CYP3A inhibitors such as ketoconazole and erythromycin (Honig et al., 1993b; Honig et al., 1992; Honig et al., 1994; Honig et al., 1993a). Under usual circumstances, terfenadine itself served only as a prodrug, being essentially completely transformed via hepatic and enteric CYP3A into fexofenadine, which was the entity having antihistaminic properties. Although terfenadine had effects on the cardiac QT_c interval (Rampe et al., 1993; Crumb et al., 1995), this was of minimal concern since intact terfenadine does not ordinarily reach the systemic circulation. However, during co-treatment with CYP3A inhibitors, conversion of terfenadine to fexofenadine is blocked, and potentially hazardous levels of the parent drug reach the circulation (Honig et al.,

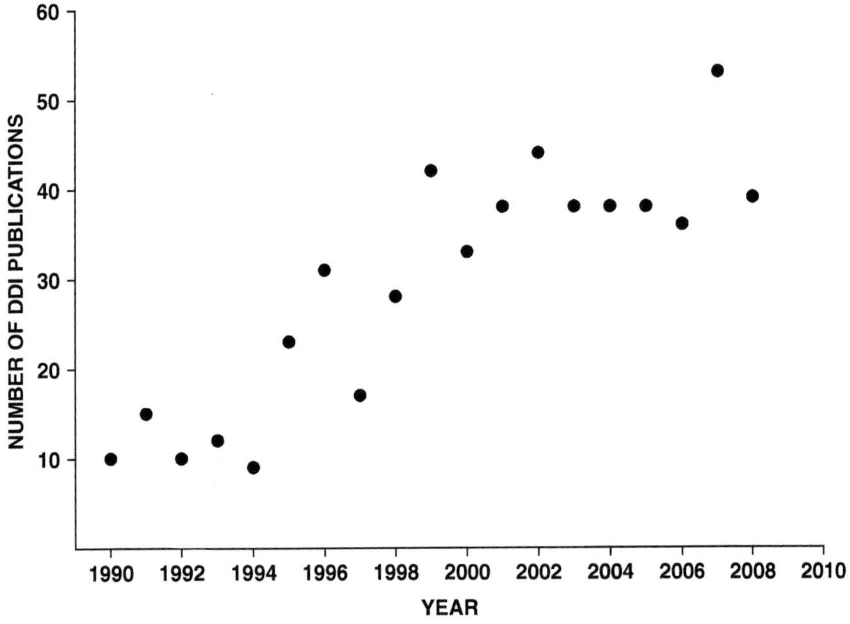

Fig. 24.1 Number of articles indexed as DDI studies published per year in the Journal of Clinical Pharmacology, 1990–2008

1994; von Moltke et al., 1994b). A few cases of serious and even fatal cardiac arrhythmias were reported as a consequence (Monahan et al., 1990; Woosley et al., 1993). The "terfenadine affair" led to an acutely increased awareness of the potential importance of DDIs. Terfenadine was withdrawn from clinical practice, and a number of regulatory reforms increased the requirements for DDI assessments as a component of drug development. The overall shift in focus of the scientific and drug development community is clearly evident from the prevalence of DDI studies among scientific publications (Fig. 24.1).

24.2 Epidemiology of Drug–Drug Interactions

Given the prevalence of polypharmacy in contemporary clinical practice, the number of *possible* DDIs can become very large. If an individual patient is taking n drugs concurrently, the number of pairwise combinations of these two drugs can be calculated as follows:

$$\frac{n!}{(n-2)!2!} \tag{24.1}$$

The larger the value of n, the greater the number of different drug combination pairs, and potential pairwise DDIs (Table 24.1). A patient with diabetes, hypertension, ischemic heart disease, and depression might well be taking 10 drugs concurrently, in which case the number of possible drug interactions is 45. Considering this large "denominator" of possibilities, the number of clinically important DDIs encountered in contemporary therapeutics actually is relatively small.

Table 24.1 Relation of number of drugs concurrently administered to the number of possible pairwise drug–drug interactions

Number of drugs	Possible pairwise drug interactions
2	1
3	3
4	6
5	10
6	15
7	21
8	28
9	36
10	45
11	55
12	66

The outcome options following concurrent administration of two drugs can be constructed based on a probability hierarchy (Fig. 24.2). The most probable outcome is that the two drugs act independently, with no evidence of any interaction. Less probable is a DDI which can be demonstrated in a controlled laboratory setting, but is not detectable in clinical practice either because the magnitude of the change

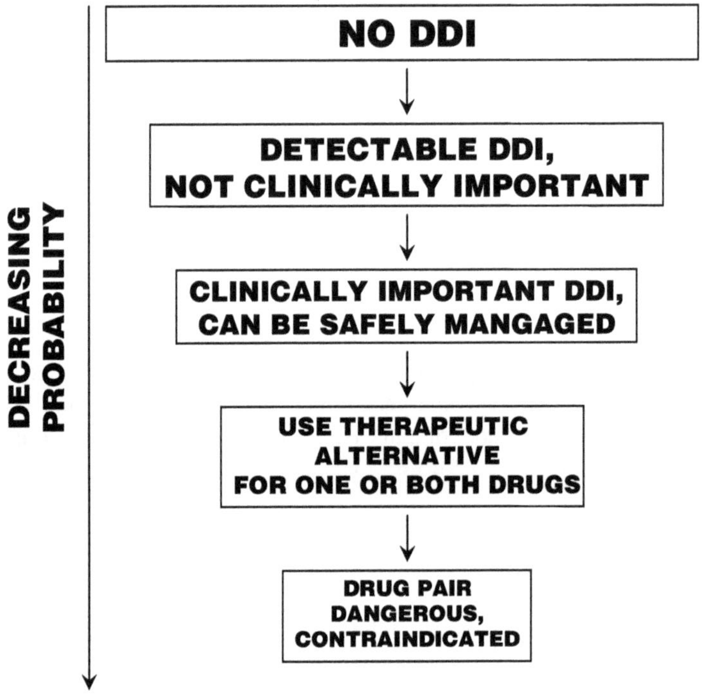

Fig. 24.2 Outcome possibilities in terms of DDIs when two drugs are coadministered, in order of decreasing probability from *top* to *bottom*

in plasma levels of the "victim" drug is so small as to be unimportant, or that the therapeutic index of the victim drug is very large. Still less probable is a DDI that is clinically important, but can be managed, for example, by reducing the dosage of the victim drug or by instituting closer monitoring of plasma levels and/or clinical outcome. Even lower in probability ranking is a DDI that is difficult to manage, such that co-treatment should be avoided if possible, and an alternative choice should be made for one or both drugs in the pair. The very least likely outcome – in fact, quite rare – is that the DDI potential carries an unacceptably serious risk, and the drug pair is contraindicated.

This probability hierarchy has been confirmed in studies of the epidemiology of DDIs. In a study of 9481 ambulatory patients in Germany, 13,672 actual drug combination pairs were identified (Bergk et al., 2004). Of these pairs, only 6.4% were known to cause DDIs with predicted outcome of moderate or major severity, and 0.5% were unmanageable DDIs such that the pair should be avoided. Findings were similar in a study of hospitalized patients in Denmark (Glintborg et al., 2005). The authors conclude that "although potential drug–drug interactions are highly prevalent, serious and clinically significant interactions are rare among recently hospitalized patients." In the specialty area of clinical psychopharmacology, there is extensive literature on the capacity of fluoxetine and paroxetine to inhibit the activity of human CYP2D6, leading to large inhibitory pharmacokinetic interactions with CYP2D6 substrate drugs such as desipramine (Hemeryck and Belpaire, 2002; von Moltke et al., 1994a; von Moltke et al., 1995; Preskorn et al., 1994; Alderman et al., 1997). Yet clinically important drug interactions are rarely reported in actual practice (Davies et al., 2004; deVane, 2006; Molden et al., 2005). One possible explanation is that the therapeutic index of the victim drug or drugs is large enough that even a substantial change in plasma levels is not clinically relevant. Another explanation is that clinicians recognize the potential DDI, and make a pre-emptive downward adjustment in the dose of the victim to prevent the DDI.

24.3 Drug Interaction Mechanisms and Terminology

We have used the term "perpetrator" to indicate the drug that is causing the DDI, while "victim" or "substrate" is the drug that is being interacted with (Greenblatt and von Moltke, 2008). In a pure pharmacodynamic DDI, the perpetrator does not alter the plasma concentrations or systemic pharmacokinetics of the victim. Instead, the two drugs produce either additive or antagonistic pharmacodynamic effects. The interaction may occur via additive or opposite actions on the same receptor systems that yield additive or opposite clinical actions. Ethyl alcohol and benzodiazepines produce additive sedation through their actions on the gamma-aminobutyric acid (GABA) receptor system (Chan, 1984; Greenblatt and von Moltke, 2008); the pharmacokinetic interaction between alcohol and benzodiazepines, if any, is small, and does not account for the additive sedative effects (Greenblatt et al., 1978; Greenblatt and von Moltke, 2008; Ochs et al., 1984;). Benzodiazepine agonists and caffeine have antagonistic pharmacodynamic actions. Benzodiazepines produce sedation via the GABA–benzodiazepine receptor system, whereas caffeine produces alertness due to its action as an adenosine receptor antagonist (Biaggioni et al., 1991; Kaplan et al., 1992a, b; Kaplan et al., 1993). When caffeine is given together with a benzodiazepine agonist such as zolpidem, the sedative effects of zolpidem are partially reversed (Cysneiros et al., 2007). However, there is minimal, if any, pharmacokinetic interaction between these two agents.

A pure pharmacokinetic interaction involves only the effect of the perpetrator on the systemic clearance of the victim drug, causing plasma levels of the victim to increase or decrease. The clinical actions of the victim may be correspondingly increased or decreased, but only because of the indirect effects of the perpetrator on systemic clearance, rather than a direct effect of the perpetrator on the target receptor mediating clinical action.

Pharmacokinetic DDIs involving drug-metabolizing enzyme systems (such as the CYPs) are generally classified as inhibition or induction. With metabolic inhibition, the perpetrator impairs the clearance of the victim drug, systemic exposure increases, and the clinical concern is toxicity. With induction, clearance of the victim increases, systemic exposure decreases, and the clinical concern is lack of efficacy (Table 24.2). However, inhibition and induction are not simply the same process in opposite directions – they involve fundamentally different mechanisms.

Inhibition of victim drug clearance happens rapidly upon exposure to the perpetrator, and represents a direct effect of the perpetrator on the drug-metabolizing enzyme. Metabolic inhibition can be studied in vitro using cell homogenates from human liver, or cells expressing human metabolic enzymes (Venkatakrishnan et al., 2001; Venkatakrishnan et al., 2003). From these in vitro systems it is straightforward to derive metrics of inhibitory potency such as the inhibition constant (K_i) or the 50% inhibitory concentration (IC_{50}). In contrast, induction is an indirect process – the perpetrator (inducer) initiates a signal for the cell to produce more metabolic protein. This is slower than inhibition, and requires cultures of intact cells to study in vitro. The metric of induction potency is not so straightforward. Generally, the inductive effect of a candidate inducer is expressed as the fractional degree of induction relative to the hypothetical "maximum" induction by an index inducer such as rifampin.

Table 24.2 Comparison of metabolic inhibition and induction

	Inhibition	Induction
Effect on victim drug		
Clearance	Decreased	Increased
Plasma levels	Increased	Decreased
Principal clinical concern	Toxicity	Loss of efficacy
Onset (after exposure to perpetrator)	Rapid	Slow
Offset (after perpetrator is discontinued)	Rapid	Slow
Mechanism	Direct chemical effect	Indirect signal to increase protein synthesis
In vitro system	Cell homogenates	Cell culture
Metric of potency in vitro	K_i or IC_{50}	Induction relative to maximum

24.4 The Design of Clinical Drug Interaction Studies

The general objective of DDI studies is to answer the following scientific questions:

1. Given candidate "victim" and "perpetrator" drugs, is there a pharmacokinetic interaction between these two drugs that is not a chance event?
2. What is the magnitude of the pharmacokinetic interaction?
3. Is the interaction likely to be of clinical importance?

Answers to the first two questions are largely objective and numerical, with little need for subjective interpretation or supplemental information. The third question is different – unless the DDI study incorporates measures of pharmacodynamic effect that are applicable to the target patient population, some supplemental information on the exposure–response relationship for the victim substrate drug is needed before a judgment can be made.

24.4.1 Study Rationale

The majority of clinical DDI studies involve healthy volunteers who do not have a medical need for the drugs under study. As such, study participation for these individuals is of no clinical benefit, but does entail some risk (though presumably low, and acceptable to an Institutional Review Board). There is also a dollar cost involved in the conduct of DDI studies. The cost is borne by the pharmaceutical sponsor in the case of an investigational drug, by the general public in the case of an NIH-supported study, or by some other entity. The core assumption is that the risk and cost of the DDI study are justified based on the potential public health benefit of the information to be acquired.

Clinical observations raising the possibility of a DDI may form the basis for initiating a formal study to either confirm or rule out a DDI. In the course of drug development, in vitro data is commonly used to identify drug pairs for which a DDI needs to be evaluated in a clinical study. "Drug X" may be identified as an inhibitor of a certain human CYP isoform in vitro, with a quantitative potency metric of K_i or IC_{50}. If $[I]$ is a typical plasma concentration of Drug X encountered during treatment with the highest approved dosage, then the ratio $[I]/K_i$ or $[I]/IC_{50}$ is used to judge whether a clinical DDI is unlikely, possible, or probable, based on FDA guidelines. A DDI is termed "possible" if

$$[I]/K_i > 0.1 \tag{24.2}$$

This boundary is arguably too conservative on scientific grounds and triggers a large number of clinical DDI studies which turn out to be negative. Nonetheless that boundary reflects the current regulatory outlook, and sponsors often will initiate a DDI study on that basis.

A second category of rationale for DDI studies is not directly scientific, but rather epidemiologic, based on a high probability of concurrent drug therapy. Drug X may be under development for a medical condition (such as diabetes, hypertension, or hyperlipidemia) that has high co-morbidity with ischemic heart disease. The sponsor may choose to initiate DDI studies of Drug X with digoxin or with warfarin because the probability of concurrent therapy is high, and because digoxin and warfarin (as potential victim drugs) have a narrow therapeutic index. Even if there is no direct scientific rationale raising the possibility of a DDI, it could be argued that clinical data excluding DDIs with digoxin or warfarin is needed to assure safe co-treatment of Drug X with these potentially hazardous medications.

Finally, an inevitable consequence of the industry-based system of drug development is that research may be initiated solely for business reasons. Within a given drug class, a number of therapeutic options may be available, for which differences in therapeutic efficacy or toxicity may be subtle at most (Table 24.3). Competitive advantage then may turn on pharmacokinetic properties, such as mechanism of clearance, elimination half-life, and the risk of DDIs. A clinical study may be initiated to show that the sponsor's Drug X is not an inhibitor of a specific CYP isoform, whereas a competitor drug within the same class is in fact a significant inhibitor of the same CYP. These properties can be included in a product label, and used by pharmaceutical representatives or advertising materials for competitive advantage. An example is the interaction of macrolide antimicrobials with human CYP3A. Erythromycin, clarithromycin, and telithromycin are significant CYP3A inhibitors, whereas azithromycin is not (Greenblatt et al., 1998a).

Table 24.3 Examples of drug classes for which individual drugs can be distinguished based on pharmacokinetic properties or drug interaction potential

Newer antidepressants
Fluoxetine
Sertraline
Paroxetine
Fluvoxamine
Citalopram
Venlafaxine
Drugs to treat erectile dysfunction
Sildenafil
Tadalafil
Vardenafil
Macrolide antimicrobials
Erythromycin
Clarithromycin
Azithromycin
Telithromycin
Hypnotics
Triazolam
Zolpidem
Eszopiclone
Temazepam

24.4.2 Protocol Construction

The customary design is a typical DDI protocol that involves a randomized, two-way crossover study in a series of healthy volunteers. On one occasion, the victim substrate is administered in the control or baseline condition, without coadministration of the perpetrator. Total area under the plasma concentration curve from zero to infinity is calculated (AUC_0). On a separate occasion, area under the curve is determined in the same subjects during coadministration of inhibitor (AUC_I). The AUC ratio (R_AUC) is calculated as

$$R_\text{AUC} = \frac{\text{AUC}_\text{I}}{\text{AUC}_0} \qquad (24.3)$$

This represents the fractional increase in substrate AUC attributable to coadministration of the perpetrator. The reciprocal of R_{AUC} is the fractional change in clearance of the substrate.

A key requirement is that the exposure of the volunteer subject to the perpetrator has to span the duration of blood sampling for plasma concentrations of the substrate. The shorter the half-life of the substrate, the shorter the duration of sampling, and the lower the cost and risk of the DDI study. If the perpetrator drug is a metabolic inhibitor, this will prolong the necessary exposure duration and sampling time, but short half-life victim drugs, nonetheless, are easier to deal with in DDI studies. If the perpetrator is an inducer, this if anything decreases the necessary sampling duration, but this advantage may be offset by the need for a period of pretreatment with the inducer due to the time required to attain maximum induction.

An alternative design is to study the kinetics of the victim drug at steady state. With this design, Equation (24.3) is modified to represent the ratio of substrate AUC values over a dosage interval segment at steady state. If the intrinsic kinetics of the victim drug are nonlinear, this may constitute support for the steady-state DDI study design. Beyond that, the steady-state design only has a "showcase" advantage in that it more closely mimics the usual therapeutic situation in which the substrate is given on an extended basis. However, if the kinetics of the substrate victim are linear (dose independent), single-dose kinetics are predictive of what will happen at steady state, and the single-dose design provides DDI data of equivalent quality. An obvious drawback of the steady-state design is that duration, cost, and risk of the study are substantially increased, since the substrate drug must be dosed to steady state both in the control condition and during coadministration of the perpetrator.

24.4.3 Studies of Specific Drug Pairs

The initiator of a DDI study may have a clinical or research question that applies only to a specific drug pair, without the objective of information that is more generalizable. With the limited research objective, the study design involves administration of the substrate victim on two occasions, with and without perpetrator, as described above. The forthcoming research outcome applies to that drug pair, but not necessarily applicable to any other pair. An example is the pharmacokinetic interaction of diazepam and fluvoxamine (Perucca et al., 1994), applicable to that particular combination of substrate and perpetrator, but with no obvious connection to other drug combinations.

24.4.4 Candidate Drug as Victim

"Drug X" may be identified as a potential DDI victim either through in vitro data, clinical observations, or both. The in vitro model may have identified the one or more CYP isoforms responsible for clearance of Drug X. The commercial entity developing Drug X, or a group of academic investigators, then pose the question: what happens to the in vivo clearance of Drug X if one or more of the CYP isoforms responsible for clearance is either induced or inhibited by a perpetrator. This question may represent a critical point in drug development. The outcome could influence the drug's clinical safety profile, and the degree of restrictiveness of the product label if the drug is eventually approved. The DDI study outcome could even lead to discontinuation of the drug as a development candidate.

The choice of perpetrator in the DDI study usually will be whatever produces the "worst case" – that is, the interaction of largest possible magnitude. The scientific community and the FDA largely agree on what those specific perpetrators are, sometimes termed "index inhibitors" or "index inducers" (Table 24.4). Whatever the degree of inhibition or induction produced by the index compound, no other perpetrator will be any worse. Ketoconazole and ritonavir are typical choices of index inhibitor for studies of substrate victims metabolized by CYP3A (Lee et al., 2002; Tsunoda et al., 1999; Knox et al., 2008; Greenblatt et al., 2000). The sponsor or investigator may also wish to concurrently study a less potent perpetrator, in which case the DDI trial design would be modified to become a three-way crossover. For example, a candidate drug that is a CYP3A substrate may be studied with ketoconazole and erythromycin as perpetrators, representing strong and moderate CYP3A inhibitors, respectively.

The impact of a DDI on the clearance of a victim drug is greatest when that drug is extensively metabolized, and a single CYP isoform mediates clearance. Candidate victim drugs metabolized mainly by CYP3A isoforms are a target of concern, since inhibition of CYP3A by a strong inhibitor such as ketoconazole or ritonavir may produce large values of R_{AUC} (Equation (24.3)) (Lee et al., 2002; Tsunoda et al., 1999; Knox et al., 2008; Greenblatt et al., 2000). Concern is augmented when the substrate victim has high clearance, and undergoes significant presystemic extraction after oral dosage (Fig. 24.3).

An important feature of study design is the optimal duration of pre-exposure to the perpetrator drug prior to administration of the substrate victim. To minimize study cost and risk, exposure duration should be the minimum necessary to produce maximum inhibition or induction. In the case of CYP3A inhibition studies, there is strong data to indicate that 24 h of pre-exposure to ketoconazole or ritonavir is sufficient to produce maximal inhibition (Fig. 24.4) (Stoch et al., 2009). For a time-dependent (mechanism-based) CYP3A inhibitor such as erythromycin or clarithromycin, 48 h of pre-exposure is sufficient (Okudaira et al., 2007). On the other hand, if the perpetrator is an inducer (rifampin), a pre-treatment period of 5–7 days is needed for induction to become maximal (Ohnhaus et al., 1989; Lin, 2006).

24.4.5 Candidate Drug as Perpetrator

If the candidate drug is being evaluated as a possible perpetrator of DDIs, the study design requires selection of an index substrate – that is, a substrate victim

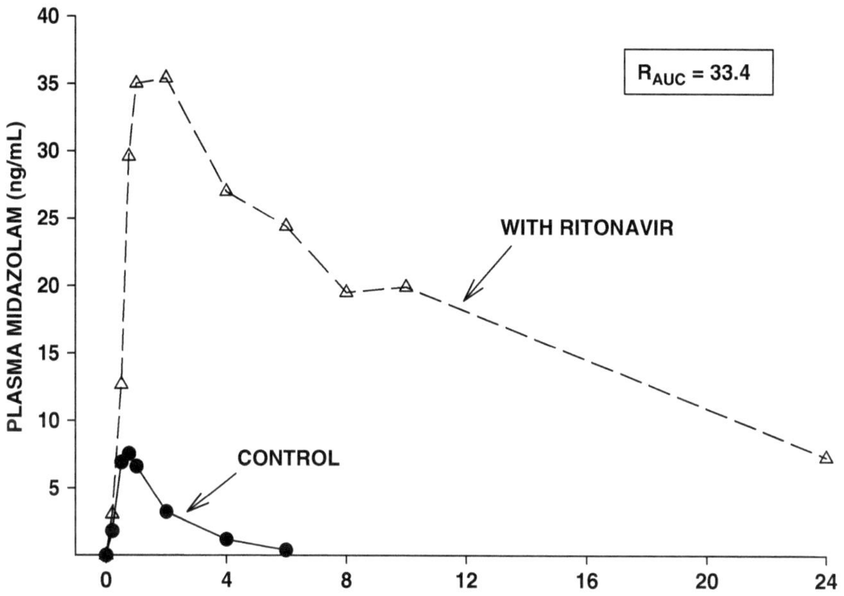

Fig. 24.3 Plasma concentrations of midazolam, an index substrate used to study DDIs involving CYP3A, after a single 3-mg oral dose administered to a healthy volunteer on two occasions: once in the control condition (with no perpetrator coadministered), and again during coadministration of ritonavir (3 doses of 100 mg over 24 h), a strong inhibitor of CYP3A. R_{AUC} is defined in Equation (24.3)

whose clearance is established as mediated largely or entirely by a specific CYP isoform. Representative index substrates are shown in Table 24.4. As an example, if the candidate perpetrator is a potential inhibitor of CYP3A, a DDI protocol could be structured as a two-way crossover study, with buspirone administered in the control condition, and again during coadministration of the candidate inhibitor. Investigators may wish to modify the study to become a three-way crossover, with an additional buspirone trial during co-treatment with the index inhibitor (ritonavir or ketoconazole) as a "positive control" to demonstrate maximum possible inhibition. This provides valuable additional information on the degree of inhibition by the candidate perpetrator in relation to the maximal inhibition achievable in the experimental setting.

A limitation of a DDI study using an index substrate is that the outcome cannot be directly extrapolated to other substrates cleared by the same CYP isoform. In general, strong inhibitors are "strong" and weak inhibitors are "weak" regardless of the substrate. However, the actual numerical R_{AUC} value for a given inhibitor will vary among victim substrates cleared by the same CYP isoform (Venkatakrishnan et al., 2001; Venkatakrishnan et al., 2003; Greenblatt et al., 2008; Ragueneau-Majlessi et al., 2007; Galetin et al., 2005; Obach et al., 2005; Brown et al., 2006; Obach et al., 2006). Factors contributing to variability among substrates include

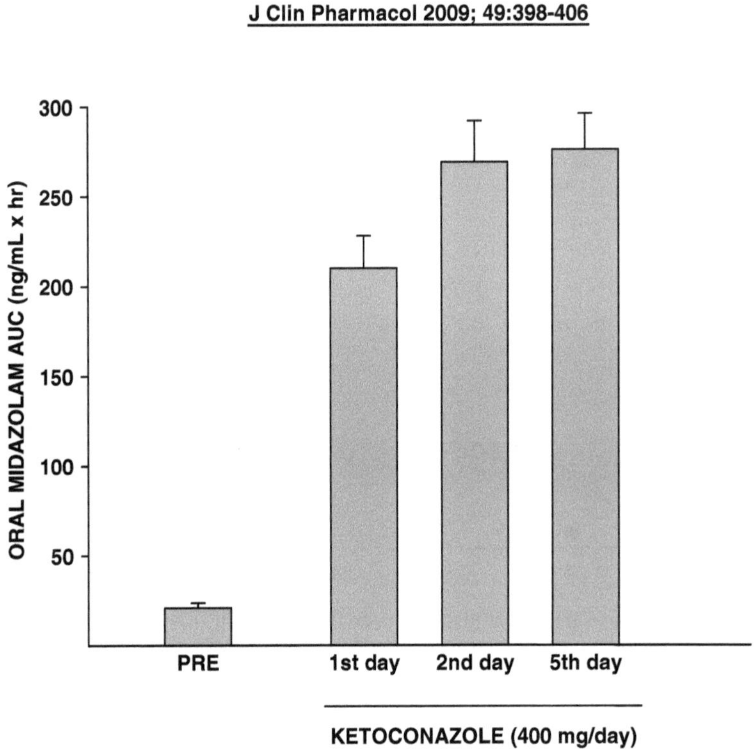

Fig. 24.4 Mean (±SE) total AUC for midazolam after oral administration of midazolam on four occasions (Stoch et al., 2009). PRE represents the control condition, prior to co-treatment with ketoconazole. Midazolam was administered again at the beginning of days 1, 2, and 5 of co-treatment with ketoconazole. For the day 1 study, midazolam was given with the first dose of ketoconazole; for the day 2 study, midazolam was given at the beginning of the second day, with 24 h of midazolam pre-treatment. The results indicate that maximal inhibition – equivalent to that observed at steady state on day 5 – is attained on day 2, after 24 h of pre-treatment with ketoconazole

the intrinsic clearance of the substrate, the fraction of total clearance attributed to the specific CYP isoform, the extent of presystemic extraction after oral dosage, and the relative contribution of enteric and hepatic metabolism to net presystemic extraction. A typical example is the substantially different effect of ketoconazole coadministration on the kinetics of triazolam and alprazolam (Greenblatt et al., 1998c). Both drugs are structurally similar CYP3A substrates, but differ in the other features described above (Greenblatt et al., 2008; Greenblatt et al., 2002).

A research design dilemma arises when the candidate drug is suspected as being an inhibitor of more than one CYP isoform. Individual DDI studies could be conducted, each with a separate cohort of volunteer subjects, and each utilizing an index substrate corresponding to the specific CYP isoform. A second approach is to conduct a single DDI study using a substrate drug "cocktail" (Fuhr et al., 2007; Tanaka

Table 24.4 Representative index substrates, inhibitors, and inducers applicable to the design of drug interaction studies*

CYP isoform	Index substrates	Index inhibitors	Index inducers
CYP1A2	Caffeine	Fluvoxamine	[Cigarette smoking]
CYP2B6	Bupropion, efavirenz	Clopidogrel	Rifampin
CYP2C9	Flurbiprofen	Fluconazole	Rifampin
CYP2C19	Omeprazole	Fluvoxamine	Rifampin
CYP2D6	Desipramine, dextromethorphan	Quinidine, paroxetine	[None known]
CYP3A	Midazolam, triazolam, buspirone	Ritonavir, ketoconazole	Rifampin

*Table entries are intended to be representative, not inclusive.

et al., 2003; Zhu et al., 2001; Chainuvati et al., 2003; Blakey et al., 2004; Gurley et al., 2002; Chow et al., 2006; Zhou et al., 2004; Christensen et al., 2003). Instead of separate studies, volunteer subjects receive a mix of substrates concurrently, or in close temporal proximity, in a single study. Many possible substrate combinations have been proposed and utilized in cocktail DDI studies. A key piece of preliminary information is an unequivocal demonstration that each pairwise combination of substrates in the cocktail does not itself create DDIs with each other.

24.4.6 Approach to Analysis of Data

If the clearance of a substrate drug is not dependent on a polymorphically regulated process, the population distribution of AUC values following a fixed dose of that substrate will not be consistent with a normal distribution, but rather will have a positive skew. Generally the skewed pattern is consistent with a log-normal distribution (Fig. 24.5) (Greenblatt, 2008; Friedman et al., 1986; Lacey et al., 1997; Greenblatt et al., 1989). In any given DDI study, the number of AUC values is usually not sufficient for a stable characterization of the underlying statistical distribution. Nonetheless a log-normal distribution is generally assumed, based on experience with larger population studies.

Going "by the book," a calculation of arithmetic mean and standard deviation (SD) of AUC is theoretically precluded if the underlying distribution is non-normal. In practice, however, the values of mean and SD calculated based on the assumption of underlying normal or log-normal distributions are nearly identical. Statistical theory reportedly justifies calculation of geometric mean AUC, along with a 90% confidence interval. However, the geometric mean value will underestimate the "real" mean value calculated using the assumption of a normal or log-normal distribution (Fig. 24.5).

Statistical analysis of the significance of the DDI – that is, whether the aggregate value of AUC_0 differs from AUC_I – is most straightforwardly done via a

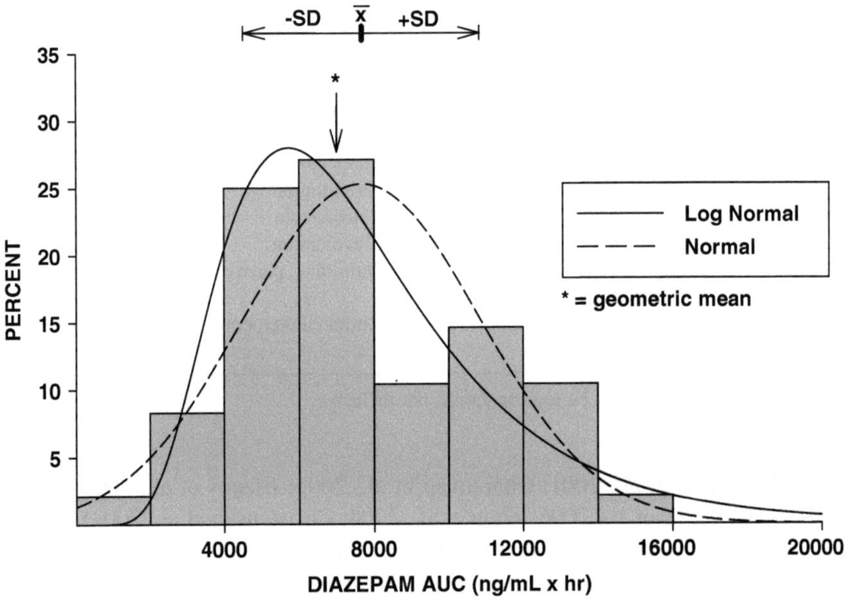

Fig. 24.5 A series of 48 healthy male volunteers received a single 10-mg oral dose of diazepam (Greenblatt et al. 1989). The picture shows the frequency distribution of total AUC among the study population, along with the *fitted curves* consistent with either a normal or log-normal distribution. Statistical analysis indicated that the actual pattern was consistent with the log-normal distribution, but not with the normal distribution. However, the mean and standard deviation (shown as $\bar{x}$ and SD) were essentially identical regardless of which underlying distribution was assumed. The geometric mean (*arrow with asterisk above*) underestimated the true mean

nonparametric equivalent of Student's paired *t*-test, such as with rank transformation of the individual values. With this approach, the underlying distribution of the AUC values is not relevant. Since the underlying distribution of R_{AUC} values is not established as being non-normal, R_{AUC} can be aggregated as an arithmetic mean and SD, which is then tested in comparison with 1.00 using Student's *t*-test. FDA guidance requires that R_{AUC} be aggregated as the geometric mean and 90% CI, with statistical inference based on whether 1.00 falls within the boundaries of the 90% CI.

To summarize a recommended approach that satisfies research and regulatory needs:

for AUC_0 and AUC_I individually:
- Arithmetic mean and SD
- Geometric mean and 90% CI

for R_{AUC}:
- Arithmetic mean and SD
- Geometric mean and 90% CI

statistical analysis:

Compare mean AUC_0 and AUC_I using nonparametric test
Compare arithmetic mean R_{AUC} vs. 1.00 using Student's *t*-test
Determine whether 1.00 falls within 90% CI around geometric mean of R_{AUC}

In our experience, the different statistical options yield conclusions that are essentially identical.

24.5 Is a Drug Interaction of Clinical Importance?

A key and critical point is that *none of these options for data aggregation and statistical analysis described above provides insight on whether a DDI is clinically important.* That judgment must be based on supplemental knowledge of the exposure–response relationship for the victim drug. A quantitatively large interaction is more likely to be clinically significant, but this is not necessarily so (Culm-Merdek et al., 2005). Also, statistical significance is not equivalent to clinical significance.

If a perpetrator drug, acting as a CYP inhibitor, causes a 50% increase in exposure to the substrate victim, this would be numerically evident as

$$\frac{\text{mean } AUC_I - \text{mean } AUC_0}{\text{mean } AUC_0} = 0.5$$

or

$$\text{mean } R_{AUC} = 1.5$$

To judge clinical importance of the interaction, one would need another data source to determine whether a 50% increase in mean plasma levels of the substrate drug was sufficient to cause a meaningful change in efficacy or the occurrence of toxicity. The most convincing data on this question comes from clinical trials evaluating efficacy and adverse events in relation to dosage among patients taking the substrate drug for clinical purposes. If the daily dosage is increased by a factor of 1.5 – on average, the equivalent consequence of the DDI – is there greater efficacy and/or an increased frequency of adverse events? In some DDI studies, pharmacodynamic endpoints are included as part of the study design, in which case the DDI study itself may provide data on the clinical consequences of the interaction (Cysneiros et al., 2007; Greenblatt et al., 1998a; Greenblatt et al., 2000; Culm-Merdek et al., 2005; Greenblatt et al., 1984; Greenblatt et al., 1998b; Greenblatt et al., 2003; Culm-Merdek et al., 2006; von Moltke et al., 1996). The limitation of kinetic–dynamic studies in healthy volunteers is that extrapolation to patient populations taking the substrate drug on an extended basis is not necessarily straightforward.

Investigators with a pre-existing understanding of dose/concentration/response relationships for the substrate victim may have the option of incorporating this information into the statistical inference plan for the DDI study. Suppose it is established that R_{AUC} can range from 0.7 to 1.4 with no evident clinical consequence. The DDI protocol could then adopt this range as pre-determined "no-effect boundaries." If the 90% CI around the mean R_{AUC} falls entirely within the 0.7–1.4 range, then the DDI – even if statistically significant – can be deemed as unlikely to be clinically important.

24.6 Are Clinical Drug Interactions Predictable from In Vitro Models?

In vitro data now is commonly used in the course of drug development to identify potential DDIs that may require clinical studies to rule out or confirm, and to quantitate the magnitude of the DDI if there is one. Current FDA guidelines deem that a DDI is possible if $[I]/K_i$ is greater than 0.1, as discussed above (Huang et al., 2008; Huang et al., 2007; Tucker et al., 2001). These guidelines do allow more informed targeting of the extensive resources needed to conduct clinical studies, but the guidelines nonetheless are very conservative and minimally quantitative. For more than a decade, there has been substantial investment of research energy into the question of whether in vitro models can provide more specific quantitative information that can forecast not only whether or not a clinical DDI will happen, but also how small or big the interaction will be (Venkatakrishnan et al., 2001, 2003; Greenblatt et al., 2008; von Moltke et al., 1994a; Brown et al., 2006; Galetin et al., 2005; Lin, 2006; Obach et al., 2005, 2006; Ragueneau-Majlessi et al., 2007; von Moltke et al., 1998; Youdim et al., 2008; Zhou, 2008; Williams et al., 2004; Ito et al., 2003; Rostami-Hodjegan and Tucker, 2007; Galetin and Houston, 2006; Galetin et al., 2006, 2007; Ito et al., 2004; Ohno et al., 2007; Kanamitsu et al., 2000; Brown et al., 2005; Bachmann and Lewis, 2005; Bachmann, 2006; Lin, 2000; Volak et al., 2007; Farkas et al., 2008).

The basis for all predictive in vitro–in vivo scaling models is a link between what is observed in a clinical DDI study (as in Equation (24.3)), what is available from the in vitro model (such as K_i or IC_{50}), and some measured or assumed in vivo exposure to the inhibitor ($[I]$). The most straightforward linkage relationship is

$$R_{AUC} = \frac{AUC_I}{AUC_0} = 1 + \frac{[I]}{K_i} \tag{24.4}$$

R_{AUC} is objectively determined in a clinically DDI study, and K_i is determined in vitro based on generally accepted procedures. A principal uncertainty is the inhibitor exposure. Theoretical model validity requires that $[I]$ be the inhibitor concentration to which the enteric or hepatic CYP enzyme is exposed. This is not ordinarily available to measurement, and many surrogates have been evaluated in Equation (24.4). These include the minimum, maximum, and mean values of total systemic plasma concentration, unbound systemic plasma concentration, total hepatic inlet (portal) concentration, and unbound hepatic inlet (portal) concentration. Based on extensive analyses of available data, it is evident that none of these schemes allows a reliable prediction of R_{AUC} based on in vitro data (Fig. 24.6). This probably explains

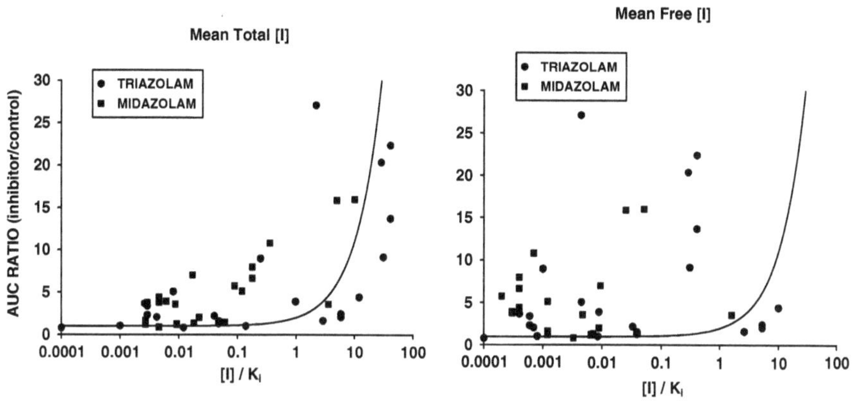

Fig. 24.6 For a series of studies on midazolam and triazolam (two index substrates used in studies of DDIs involving CYP3A), a plot of the predicted value of R_{AUC} based on in vitro data (x-axis) and the actual R_{AUC} value (y-axis) observed in a clinical study, using data from a number of clinical studies of various inhibitors as perpetrators. The *solid line* is the theoretical function in Equation (24.4). Two different assumptions were made. *Left*: Inhibitor exposure $[I]$ is equal to mean total (free plus bound) plasma concentration of inhibitor; *Right*: $[I]$ equal to mean unbound (free) inhibitor concentration. Either way, the predictive model is poor (data taken in part from Ito et al. (2004), with correction of some errors and addition of other data from recently published studies)

Table 24.5 Sources of bias and uncertainty in prediction of clinical drug interactions from in vitro data

Estimation of metabolic enzyme exposure to inhibitor
- Restriction of hepatic availability due to plasma protein binding
- Concentrative uptake into hepatic tissue
- Hepatic uptake or efflux transport of inhibitor
- Contribution of inhibitor metabolites to overall extent of inhibition
- Fluctuation in inhibitor exposure over a dosage interval

Bias in estimation of K_i or IC_{50} in vitro
- Time-dependent (mechanism-based) inhibition
- Nonspecific binding by microsomal protein or other microsomal components
- Metabolic consumption of inhibitor
- Solvent effects on reaction kinetics and/or inhibitor potency

Physiologic factors modulating substrate or inhibitor disposition in vivo
- Route of administration (oral vs. parenteral)
- Intravenous clearance dependent on hepatic flow
- Clearance mediated in part by a non-inhibited CYP or other enzyme
- Extrahepatic routes of clearance

why FDA guidelines for in vitro–in vivo DDI prediction continue to be conservative and broad. Reasons for poor predictive performance are summarized in Table 24.5. In addition to the uncertainty in estimation of $[I]$, these include experimental bias in the determination of K_i, as well as physiologic aspects of drug disposition not accounted for in the available in vitro models.

If in vitro data "over-predicts" the extent of a DDI – that is, the model predicts a DDI that is much larger than what actually happens clinically – research resources may be expended on a clinical study that proves to be negative, but there is no public health hazard. On the other hand, "under-prediction" is of concern, since a clinically important interaction may happen when a much smaller DDI, or none at all, is predicted. One example is bupropion and CYP2D6. Based on in vitro data and the known total systemic exposure to bupropion and its principal metabolite (hydroxybupropion), $[I]/K_i$ is less than 0.1, and no clinical DDI is predicted (Hesse et al., 2000). Yet bupropion is a significant inhibitor of CYP2D6 in vivo (Kotlyar et al., 2005; Guzey et al., 2002; Reese et al., 2008). Another example is fluvoxamine, a potent inhibitor of CYP1A2 and CYP2C19 in vitro and in vivo (Christensen et al., 2002; Jeppesen et al., 1996a, b; Granfors et al., 2004b; Becquemont et al., 1997; Granfors et al., 2004a). The in vitro scaling model (Equation (24.4)) greatly underpredicts the actual extent of AUC increase caused by fluvoxamine in clinical DDI studies involving substrate victim drugs metabolized by CYP1A2 or CYP2C19 (Culm-Merdek et al., 2005; Yao et al., 2001; Yao et al., 2003). The discrepancy in the case of ramelteon is especially glaring. The in vitro model predicts an R_{AUC} value of approximately 3.0, but the actual R_{AUC} in a clinical DDI study was 190 (Fig. 24.7).

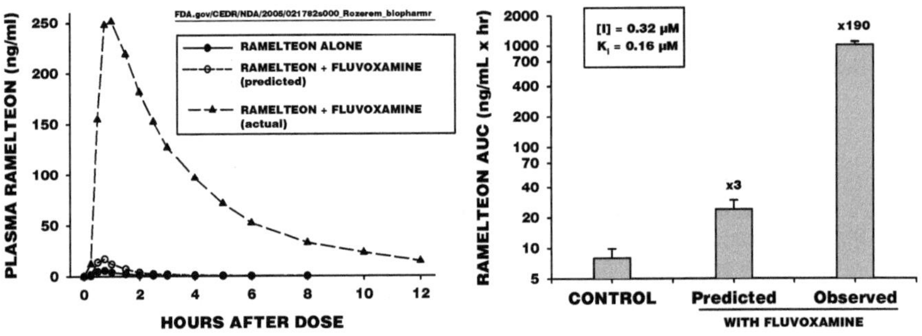

Fig. 24.7 Analysis of the fluvoxamine–ramelteon DDI, based on data available from public documents. *Left*: Mean plasma concentrations in healthy volunteers after administration of ramelteon alone (*closed circles*) or ramelteon plus fluvoxamine (*closed triangles*). The *open circles* represent the predicted plasma ramelteon concentration based on Equation (24.4), assuming an in vitro K_i of 0.16 μM for fluvoxamine as an inhibitor of ramelteon biotransformation, and exposure to fluvoxamine ($[I]$) as the total systemic concentration equal to 0.32 μM. *Right*: Mean (±SE) ramelteon AUC (logarithmic scale) in the control condition, as predicted from Equation (24.4), and as observed in the actual clinical study. The predicted R_{AUC} is 3.0, while the actual value of R_{AUC} was 190

24.7 Boosting or Augmentation

Implicit in everything said so far is that DDIs are a clinical hazard, and should be avoided if possible. In some cases, however, DDIs are desirable, and are deliberately produced for therapeutic benefit. Different terminology is then applied: the DDI is termed "augmentation" or "boosting," and the perpetrator drug (usually an inhibitor) is coadministered to impair clearance of the substrate and augment its plasma levels. Anti-rejection therapy with cyclosporine in transplant patients is very costly. With deliberate coadministration of a CYP3A inhibitor such as ketoconazole or itraconazole, therapeutically effective plasma concentrations of cyclosporine can be sustained with lower daily doses of cyclosporine and lower cost to the medical system (Jones, 1997). In antiretroviral therapy, Kaletra is a fixed-dose combination of the viral protease inhibitors lopinavir and ritonavir. Lopinavir is the principal therapeutic agent. When administered alone, lopinavir undergoes extensive presystemic extraction by hepatic and enteric CYP3A, and probably also efflux transport by P-glycoprotein. It is difficult with tolerable doses to sustain adequate antiretroviral concentrations of lopinavir given as a sale agent. When coadministered with ritonavir (a strong inhibitor of CYP3A and P-glycoprotein), clearance of lopinavir is greatly reduced, and effective plasma concentrations can be sustained with tolerable doses (Sham et al., 1998). It is possible that the next decade of clinical therapeutics will have an increasing focus on the deliberate use of DDIs for therapeutic benefit.

References

Alderman J, Preskorn SH, Greenblatt DJ, Harrison W, Penenberg D, Allison J and Chung M (1997) Desipramine pharmacokinetics when coadministered with paroxetine or sertraline in extensive metabolizers. *J Clin Psychopharmacol* **17**:284–291.

Bachmann KA (2006) Inhibition constants, inhibitor concentrations and the prediction of inhibitory drug drug interactions: pitfalls, progress and promise. *Curr Drug Metab* **7**:1–14.

Bachmann KA and Lewis JD (2005) Predicting inhibitory drug–drug interactions and evaluating drug interaction reports using inhibition constants. *Ann Pharmacother* **39**:1064–1072.

Becquemont L, Ragueneau I, Le Bot MA, Riche C, Funck-Brentano C and Jaillon P (1997) Influence of the CYP1A2 inhibitor fluvoxamine on tacrine pharmacokinetics in humans. *Clin Pharmacol Ther* **61**:619–627.

Bergk V, Gasse C, Rothenbacher D, Loew M, Brenner H and Haefeli WE (2004) Drug interactions in primary care: impact of a new algorithm on risk determination. *Clin Pharmacol Ther* **76**: 85–96.

Biaggioni I, Paul S, Puckett A and Arzubiaga C (1991) Caffeine and theophylline as adenosine receptor antagonists in humans. *J Pharmacol Exp Ther* **258**:588–593.

Blakey GE, Lockton JA, Perrett J, Norwood P, Russell M, Aherne Z and Plume J (2004) Pharmacokinetic and pharmacodynamic assessment of a five-probe metabolic cocktail for CYPs 1A2, 3A4, 2C9, 2D6 and 2E1. *Br J Clin Pharmacol* **57**:162–169.

Brown HS, Galetin A, Hallifax D and Houston JB (2006) Prediction of in vivo drug–drug interactions from in vitro data: factors affecting prototypic drug–drug interactions involving CYP2C9, CYP2D6 and CYP3A4. *Clin Pharmacokinet* **45**:1035–1050.

Brown HS, Ito K, Galetin A and Houston JB (2005) Prediction of in vivo drug–drug interactions from in vitro data: impact of incorporating parallel pathways of drug elimination and inhibitor absorption rate constant. *Br J Clin Pharmacol* **60**:508–518.

Chainuvati S, Nafziger AN, Leeder JS, Gaedigk A, Kearns GL, Sellers E, Zhang Y, Kashuba AD, Rowland E and Bertino JS, Jr. (2003) Combined phenotypic assessment of cytochrome p450 1A2, 2C9, 2C19, 2D6, and 3A, N-acetyltransferase-2, and xanthine oxidase activities with the "Cooperstown 5+1 cocktail". *Clin Pharmacol Ther* **74**:437–447.

Chan AW (1984) Effects of combined alcohol and benzodiazepine: a review. *Drug Alcohol Depend* **13**:315–341.

Chow HH, Hakim IA, Vining DR, Crowell JA, Cordova CA, Chew WM, Xu MJ, Hsu CH, Ranger-Moore J and Alberts DS (2006) Effects of repeated green tea catechin administration on human cytochrome P450 activity. *Cancer Epidemiol Biomarkers Prev* **15**:2473–2476.

Christensen M, Andersson K, Dalen P, Mirghani RA, Muirhead GJ, Nordmark A, Tybring G, Wahlberg A, Yasar U and Bertilsson L (2003) The Karolinska cocktail for phenotyping of five human cytochrome P450 enzymes. *Clin Pharmacol Ther* **73**:517–528.

Christensen M, Tybring G, Mihara K, Yasui-Furokori N, Carrillo JA, Ramos SI, Andersson K, Dahl ML and Bertilsson L (2002) Low daily 10-mg and 20-mg doses of fluvoxamine inhibit the metabolism of both caffeine (cytochrome P4501A2) and omeprazole (cytochrome P4502C19). *Clin Pharmacol Ther* **71**:141–152.

Clarke SE (1998) In vitro assessment of human cytochrome P450. *Xenobiotica* **28**:1167–1202.

Crumb WJ, Jr., Wible B, Arnold DJ, Payne JP and Brown AM (1995) Blockade of multiple human cardiac potassium currents by the antihistamine terfenadine: possible mechanism for terfenadine-associated cardiotoxicity. *Mol Pharmacol* **47**:181–190.

Culm-Merdek KE, von Moltke LL, Gan L, Horan KA, Reynolds R, Harmatz JS, Court MH and Greenblatt DJ (2006) Effect of extended exposure to grapefruit juice on cytochrome P450 3A activity in humans: comparison with ritonavir. *Clin Pharmacol Ther* **79**:243–254.

Culm-Merdek KE, von Moltke LL, Harmatz JS and Greenblatt DJ (2005) Fluvoxamine impairs single-dose caffeine clearance without altering caffeine pharmacodynamics. *Br J Clin Pharmacol* **60**:486–493.

Cysneiros RM, Farkas D, Harmatz JS, von Moltke LL and Greenblatt DJ (2007) Pharmacokinetic and pharmacodynamic interactions between zolpidem and caffeine. *Clin Pharmacol Ther* **82**:54–62.

Davies SJ, Eayrs S, Pratt P and Lennard MS (2004) Potential for drug interactions involving cytochromes P450 2D6 and 3A4 on general adult psychiatric and functional elderly psychiatric wards. *Br J Clin Pharmacol* **57**:464–472.

deVane CL (2006) Antidepressant-drug interactions are potentially but rarely clinically significant. *Neuropsychopharmacol* **31**:1594–1604.

Farkas D, Shader RI, von Moltke LL and Greenblatt DJ (2008) Mechanisms and consequences of drug–drug interactions, in *Preclinical Development Handbook* (Gad SC ed) pp 879–917, Wiley-Interscience, Hoboken, NJ.

Friedman H, Greenblatt DJ, Burstein ES, Harmatz JS and Shader RI (1986) Population study of triazolam pharmacokinetics. *Br J Clin Pharmacol* **22**:639–642.

Fuhr U, Jetter A and Kirchheiner J (2007) Appropriate phenotyping procedures for drug metabolizing enzymes and transporters in humans and their simultaneous use in the "cocktail" approach. *Clin Pharmacol Ther* **81**:270–283.

Galetin A, Burt H, Gibbons L and Houston JB (2006) Prediction of time-dependent CYP3A4 drug–drug interactions: impact of enzyme degradation, parallel elimination pathways, and intestinal inhibition. *Drug Metab Dispos* **34**:166–175.

Galetin A, Hinton LK, Burt H, Obach RS and Houston JB (2007) Maximal inhibition of intestinal first-pass metabolism as a pragmatic indicator of intestinal contribution to the drug–drug interactions for CYP3A4 cleared drugs. *Curr Drug Metab* **8**:685–693.

Galetin A and Houston JB (2006) Intestinal and hepatic metabolic activity of five cytochrome P450 enzymes: impact on prediction of first-pass metabolism. *J Pharmacol Exp Ther* **318**: 1220–1229.

Galetin A, Ito K, Hallifax D and Houston JB (2005) CYP3A4 substrate selection and substitution in the prediction of potential drug–drug interactions. *J Pharmacol Exp Ther* **314**:180–190.

Glintborg B, Andersen SE and Dalhoff K (2005) Drug–drug interactions among recently hospitalised patients – frequent but mostly clinically insignificant. *Eur J Clin Pharmacol* **61**:675–681.

Granfors MT, Backman JT, Laitila J and Neuvonen PJ (2004a) Tizanidine is mainly metabolized by cytochrome p450 1A2 in vitro. *Br J Clin Pharmacol* **57**:349–353.

Granfors MT, Backman JT, Neuvonen M, Ahonen J and Neuvonen PJ (2004b) Fluvoxamine drastically increases concentrations and effects of tizanidine: a potentially hazardous interaction. *Clin Pharmacol Ther* **75**:331–341.

Greenblatt DJ (2008) Preparation of scientific reports on pharmacokinetic drug interaction studies. *J Clin Psychopharmacol* **28**:369–373.

Greenblatt DJ, Abernethy DR, Morse DS, Harmatz JS and Shader RI (1984) Clinical importance of the interaction of diazepam and cimetidine. *N Engl J Med* **310**:1639–1643.

Greenblatt DJ, Harmatz JS, Friedman H, Locniskar A and Shader RI (1989) A large-sample study of diazepam pharmacokinetics. *Ther Drug Monit* **11**:652–657.

Greenblatt DJ, He P, von Moltke LL and Court MH (2008) The CYP3 family, in *Cytochromes P450: Role in the Metabolism and Toxicity of Drugs and Other Xenobiotics* (Ioannides C ed) pp 354–383, Royal Society of Chemistry (Great Britain) RSC Pub; Springer, Cambridge; New York.

Greenblatt DJ, Shader RI, Weinberger DR, Allen MD and MacLaughlin DS (1978) Effect of a cocktail on diazepam absorption. *Psychopharmacology (Berl)* **57**:199–203.

Greenblatt DJ and von Moltke LL (2008) Drug–drug interactions: clinical perspectives, in *Drug–Drug Interactions* (Rodrigues AD ed) pp 643–664, Informa Healthcare, New York.

Greenblatt DJ, von Moltke LL, Harmatz JS, Counihan M, Graf JA, Durol AL, Mertzanis P, Duan SX, Wright CE and Shader RI (1998a) Inhibition of triazolam clearance by macrolide antimicrobial agents: in vitro correlates and dynamic consequences. *Clin Pharmacol Ther* **64**:278–285.

Greenblatt DJ, von Moltke LL, Harmatz JS, Durol AL, Daily JP, Graf JA, Mertzanis P, Hoffman JL and Shader RI (2000) Differential impairment of triazolam and zolpidem clearance by ritonavir. *J Acquir Immune Defic Syndr* **24**:129–136.

Greenblatt DJ, von Moltke LL, Harmatz JS, Fogelman SM, Chen G, Graf JA, Mertzanis P, Byron S, Culm KE, Granda BW, Daily JP and Shader RI (2003) Short-term exposure to low-dose ritonavir impairs clearance and enhances adverse effects of trazodone. *J Clin Pharmacol* **43**:414–422.

Greenblatt DJ, von Moltke LL, Harmatz JS, Mertzanis P, Graf JA, Durol AL, Counihan M, Roth-Schechter B and Shader RI (1998b) Kinetic and dynamic interaction study of zolpidem with ketoconazole, itraconazole, and fluconazole. *Clin Pharmacol Ther* **64**:661–671.

Greenblatt DJ, von Moltke LL, Harmatz JS and Shader RI (2002) Pharmacokinetics, pharmacodynamics, and drug disposition, in *Neuropsychopharmacology the Fifth Generation of Progress* (Davis KL and American College of Neuropsychopharmacology eds) pp 507–524, Lippincott Williams & Wilkins, Philadelphia.

Greenblatt DJ, von Moltke LL, Harmatz JS and Shader RI (1999) Human cytochromes and some newer antidepressants: kinetics, metabolism, and drug interactions. *J Clin Psychopharmacol* **19**(supp.1):23S–35S.

Greenblatt DJ, Wright CE, von Moltke LL, Harmatz JS, Ehrenberg BL, Harrel LM, Corbett K, Counihan M, Tobias S and Shader RI (1998c) Ketoconazole inhibition of triazolam and alprazolam clearance: differential kinetic and dynamic consequences. *Clin Pharmacol Ther* **64**:237–247.

Guengerich FP (1999) Cytochrome P-450 3A4: regulation and role in drug metabolism. *Annu Rev Pharmacol Toxicol* **39**:1–17.

Gurley BJ, Gardner SF, Hubbard MA, Williams DK, Gentry WB, Cui Y and Ang CY (2002) Cytochrome P450 phenotypic ratios for predicting herb-drug interactions in humans. *Clin Pharmacol Ther* **72**:276–287.

Guzey C, Norstrom A and Spigset O (2002) Change from the CYP2D6 extensive metabolizer to the poor metabolizer phenotype during treatment with bupropion. *Ther Drug Monit* **24**:436–437.

Hemeryck A and Belpaire FM (2002) Selective serotonin reuptake inhibitors and cytochrome P-450 mediated drug–drug interactions: an update. *Curr Drug Metab* **3**:13–37.

Hesse LM, Venkatakrishnan K, Court MH, von Moltke LL, Duan SX, Shader RI and Greenblatt DJ (2000) CYP2B6 mediates the in vitro hydroxylation of bupropion: potential drug interactions with other antidepressants. *Drug Metab Dispos* **28**:1176–1183.

Honig PK, Woosley RL, Zamani K, Conner DP and Cantilena LR, Jr. (1992) Changes in the pharmacokinetics and electrocardiographic pharmacodynamics of terfenadine with concomitant administration of erythromycin. *Clin Pharmacol Ther* **52**:231–238.

Honig PK, Wortham DC, Hull R, Zamani K, Smith JE and Cantilena LR (1993a) Itraconazole affects single-dose terfenadine pharmacokinetics and cardiac repolarization pharmacodynamics. *J Clin Pharmacol* **33**:1201–1206.

Honig PK, Wortham DC, Zamani K and Cantilena LR (1994) Comparison of the effect of the macrolide antibiotics erythromycin, clarithromycin, and azithromycin on terfenadine steady-state pharmacokinetics and electrocardiographic parameters. *Drug Invest* **7**:148–156.

Honig PK, Wortham DC, Zamani K, Conner DP, Mullin JC and Cantilena LR (1993b) Terfenadine-ketoconazole interaction. Pharmacokinetic and electrocardiographic consequences. *J Am Med Assoc* **269**:1513–1518.

Huang SM, Strong JM, Zhang L, Reynolds KS, Nallani S, Temple R, Abraham S, Habet SA, Baweja RK, Burckart GJ, Chung S, Colangelo P, Frucht D, Green MD, Hepp P, Karnaukhova E, Ko HS, Lee JI, Marroum PJ, Norden JM, Qiu W, Rahman A, Sobel S, Stifano T, Thummel K, Wei XX, Yasuda S, Zheng JH, Zhao H and Lesko LJ (2008) New era in drug interaction evaluation: US Food and Drug Administration update on CYP enzymes, transporters, and the guidance process. *J Clin Pharmacol* **48**:662–670.

Huang SM, Temple R, Throckmorton DC and Lesko LJ (2007) Drug interaction studies: study design, data analysis, and implications for dosing and labeling. *Clin Pharmacol Ther* **81**: 298–304.

Ito K, Brown HS and Houston JB (2004) Database analyses for the prediction of in vivo drug–drug interactions from in vitro data. *Br J Clin Pharmacol* **57**:473–486.

Ito K, Ogihara K, Kanamitsu S and Itoh T (2003) Prediction of the in vivo interaction between midazolam and macrolides based on in vitro studies using human liver microsomes. *Drug Metab Dispos* **31**:945–954.

Jeppesen U, Gram LF, Vistisen K, Loft S, Poulsen HE and Brøsen K (1996a) Dose-dependent inhibition of CYP1A2, CYP2C19 and CYP2D6 by citalopram, fluoxetine, fluvoxamine and paroxetine. *Eur J Clin Pharmacol* **51**:73–78.

Jeppesen U, Loft S, Poulsen HE and Brøsen K (1996b) A fluvoxamine-caffeine interaction study. *Pharmacogenetics* **6**:213–222.

Jones TE (1997) The use of other drugs to allow a lower dosage of cyclosporin to be used. Therapeutic and pharmacoeconomic considerations. *Clin Pharmacokinet* **32**:357–367.

Kanamitsu S, Ito K and Sugiyama Y (2000) Quantitative prediction of in vivo drug–drug interactions from in vitro data based on physiological pharmacokinetics: use of maximum unbound concentration of inhibitor at the inlet to the liver. *Pharm Res* **17**:336–343.

Kaplan GB, Cotreau MM and Greenblatt DJ (1992a) Effects of benzodiazepine administration on A1 adenosine receptor binding in-vivo and ex-vivo. *J Pharm Pharmacol* **44**:700–703.

Kaplan GB, Greenblatt DJ, Kent MA, Cotreau MM, Arcelin G and Shader RI (1992b) Caffeine-induced behavioral stimulation is dose-dependent and associated with A1 adenosine receptor occupancy. *Neuropsychopharmacol* **6**:145–153.

Kaplan GB, Greenblatt DJ, Kent MA and Cotreau-Bibbo MM (1993) Caffeine treatment and withdrawal in mice: relationships between dosage, concentrations, locomotor activity and A1 adenosine receptor binding. *J Pharmacol Exp Ther* **266**:1563–1572.

Knox TA, Oleson L, von Moltke LL, Kaufman RC, Wanke CA and Greenblatt DJ (2008) Ritonavir greatly impairs CYP3A activity in HIV infection with chronic viral hepatitis. *J Acquir Immune Defic Syndr* **49**:358–368.

Kotlyar M, Brauer LH, Tracy TS, Hatsukami DK, Harris J, Bronars CA and Adson DE (2005) Inhibition of CYP2D6 activity by bupropion. *J Clin Psychopharmacol* **25**:226–229.

Lacey LF, Keene ON, Pritchard JF and Bye A (1997) Common noncompartmental pharmacokinetic variables: are they normally or log-normally distributed? *J Biopharm Stat* **7**: 171–178.

Lee JI, Chaves-Gnecco D, Amico JA, Kroboth PD, Wilson JW and Frye RF (2002) Application of semisimultaneous midazolam administration for hepatic and intestinal cytochrome P450 3A phenotyping. *Clin Pharmacol Ther* **72**:718–728.

Lin JH (2006) CYP induction-mediated drug interactions: in vitro assessment and clinical implications. *Pharm Res* **23**:1089–1116.

Lin JH (2000) Sense and nonsense in the prediction of drug–drug interactions. *Curr Drug Metab* **1**:305–331.

Molden E, Garcia BH, Braathen P and Eggen AE (2005) Co-prescription of cytochrome P450 2D6/3A4 inhibitor-substrate pairs in clinical practice. A retrospective analysis of data from Norwegian primary pharmacies. *Eur J Clin Pharmacol* **61**:119–125.

Monahan BP, Ferguson CL, Killeavy ES, Lloyd BK, Troy J and Cantilena LR, Jr. (1990) Torsades de pointes occurring in association with terfenadine use. *J Am Med Assoc* **264**:2788–2790.

Obach RS, Walsky RL, Venkatakrishnan K, Gaman EA, Houston JB and Tremaine LM (2006) The utility of in vitro cytochrome P450 inhibition data in the prediction of drug–drug interactions. *J Pharmacol Exp Ther* **316**:336–348.

Obach RS, Walsky RL, Venkatakrishnan K, Houston JB and Tremaine LM (2005) In vitro cytochrome P450 inhibition data and the prediction of drug–drug interactions: qualitative relationships, quantitative predictions, and the rank-order approach. *Clin Pharmacol Ther* **78**:582–592.

Ochs HR, Greenblatt DJ, Arendt RM, Hübbel W and Shader RI (1984) Pharmacokinetic noninteraction of triazolam and ethanol. *J Clin Psychopharmacol* **4**:106–107.

Ohnhaus EE, Breckenridge AM and Park BK (1989) Urinary excretion of 6 beta-hydroxycortisol and the time course measurement of enzyme induction in man. *Eur J Clin Pharmacol* **36**:39–46.

Ohno Y, Hisaka A and Suzuki H (2007) General framework for the quantitative prediction of CYP3A4-mediated oral drug interactions based on the AUC increase by coadministration of standard drugs. *Clin Pharmacokinet* **46**:681–696.

Okudaira T, Kotegawa T, Imai H, Tsutsumi K, Nakano S and Ohashi K (2007) Effect of the treatment period with erythromycin on cytochrome P450 3A activity in humans. *J Clin Pharmacol* **47**:871–876.

Perucca E, Gatti G, Cipolla G, Spina E, Barel S, Soback S, Gips M and Bialer M (1994) Inhibition of diazepam metabolism by fluvoxamine: a pharmacokinetic study in normal volunteers. *Clin Pharmacol Ther* **56**:471–476.

Preskorn SH, Alderman J, Chung M, Harrison W, Messig M and Harris S (1994) Pharmacokinetics of desipramine coadministered with sertraline or fluoxetine. *J Clin Psychopharmacol* **14**:90–98.

Ragueneau-Majlessi I, Boulenc X, Rauch C, Hachad H and Levy RH (2007) Quantitative correlations among CYP3A sensitive substrates and inhibitors: literature analysis. *Curr Drug Metab* **8**:810–814.

Rampe D, Wible B, Brown AM and Dage RC (1993) Effects of terfenadine and its metabolites on a delayed rectifier K+ channel cloned from human heart. *Mol Pharmacol* **44**: 1240–1245.

Reese MJ, Wurm RM, Muir KT, Generaux GT, John-Williams L and McConn DJ (2008) An in vitro mechanistic study to elucidate the desipramine/bupropion clinical drug–drug interaction. *Drug Metab Dispos* **36**:1198–1201.

Rostami-Hodjegan A and Tucker GT (2007) Simulation and prediction of in vivo drug metabolism in human populations from in vitro data. *Nat Rev Drug Discov* **6**:140–148.

Sham HL, Kempf DJ, Molla A, Marsh KC, Kumar GN, Chen CM, Kati W, Stewart K, Lal R, Hsu A, Betebenner D, Korneyeva M, Vasavanonda S, McDonald E, Saldivar A, Wideburg N, Chen X, Niu P, Park C, Jayanti V, Grabowski B, Granneman GR, Sun E, Japour AJ, Leonard

JM, Plattner JJ and Norbeck DW (1998) ABT-378, a highly potent inhibitor of the human immunodeficiency virus protease. *Antimicrob Agents Chemother* **42**:3218–3224.

Smith DA, Abel SM, Hyland R and Jones BC (1998a) Human cytochrome P450s: selectivity and measurement in vivo. *Xenobiotica* **28**:1095–1128.

Smith G, Stubbins MJ, Harries LW and Wolf CR (1998b) Molecular genetics of the human cytochrome P450 monooxygenase superfamily. *Xenobiotica* **28**:1129–1165.

Stoch SA, Friedman E, Maes A, Yee K, Xu Y, Larson P, Fitzgerald M, Chodakewitz J and Wagner JA (2009) Effect of different durations of ketoconazole dosing on the single-dose pharmacokinetics of midazolam: shortening the paradigm. *J Clin Pharmacol* **49**:398–406.

Tanaka E, Kurata N and Yasuhara H (2003) How useful is the "cocktail approach" for evaluating human hepatic drug metabolizing capacity using cytochrome P450 phenotyping probes in vivo? *J Clin Pharm Ther* **28**:157–165.

Tsunoda SM, Velez RL, von Moltke LL and Greenblatt DJ (1999) Differentiation of intestinal and hepatic cytochrome P450 3A activity with use of midazolam as an in vivo probe: effect of ketoconazole. *Clin Pharmacol Ther* **66**:461–471.

Tucker GT, Houston JB and Huang SM (2001) Optimizing drug development: strategies to assess drug metabolism/transporter interaction potential – towards a consensus. *Br J Clin Pharmacol* **52**:107–117.

Venkatakrishnan K, von Moltke LL and Greenblatt DJ (2000) Effects of the antifungal agents on oxidative drug metabolism: clinical relevance. *Clin Pharmacokinet* **38**:111–180.

Venkatakrishnan K, von Moltke LL and Greenblatt DJ (2001) Human drug metabolism and the cytochromes P450: application and relevance of in vitro models. *J Clin Pharmacol* **41**: 1149–1179.

Venkatakrishnan K, von Moltke LL, Obach RS and Greenblatt DJ (2003) Drug metabolism and drug interactions: application and clinical value of in vitro models. *Curr Drug Metab* **4**: 423–459.

Volak LP, Greenblatt DJ and von Moltke LL (2007) In vitro approaches to anticipating clinical drug interactions, in *Drug–drug interactions in pharmaceutical development* (Li AP ed) pp 75–93, Wiley-interscience, Hoboken, N.J.

von Moltke LL, Greenblatt DJ, Cotreau-Bibbo MM, Duan SX, Harmatz JS and Shader RI (1994a) Inhibition of desipramine hydroxylation in vitro by serotonin-reuptake-inhibitor antidepressants, and by quinidine and ketoconazole: a model system to predict drug interactions in vivo. *J Pharmacol Exp Ther* **268**:1278–1283.

von Moltke LL, Greenblatt DJ, Court MH, Duan SX, Harmatz JS and Shader RI (1995) Inhibition of alprazolam and desipramine hydroxylation in vitro by paroxetine and fluvoxamine: comparison with other selective serotonin reuptake inhibitor antidepressants. *J Clin Psychopharmacol* **15**:125–131.

von Moltke LL, Greenblatt DJ, Duan SX, Harmatz JS and Shader RI (1994b) In vitro prediction of the terfenadine-ketoconazole pharmacokinetic interaction. *J Clin Pharmacol* **34**:1222–1227.

von Moltke LL, Greenblatt DJ, Harmatz JS, Duan SX, Harrel LM, Cotreau-Bibbo MM, Pritchard GA, Wright CE and Shader RI (1996) Triazolam biotransformation by human liver microsomes in vitro: effects of metabolic inhibitors and clinical confirmation of a predicted interaction with ketoconazole. *J Pharmacol Exp Ther* **276**:370–379.

von Moltke LL, Greenblatt DJ, Schmider J, Wright CE, Harmatz JS and Shader RI (1998) In vitro approaches to predicting drug interactions in vivo. *Biochem Pharmacol* **55**:113–122.

Williams JA, Hyland R, Jones BC, Smith DA, Hurst S, Goosen TC, Peterkin V, Koup JR and Ball SE (2004) Drug–drug interactions for UDP-glucuronosyltransferase substrates: a pharmacokinetic explanation for typically observed low exposure (AUCi/AUC) ratios. *Drug Metab Dispos* **32**:1201–1208.

Woosley RL, Chen Y, Freiman JP and Gillis RA (1993) Mechanism of the cardiotoxic actions of terfenadine. *J Am Med Assoc* **269**:1532–1536.

Yao C, Kunze KL, Kharasch ED, Wang Y, Trager WF, Ragueneau I and Levy RH (2001) Fluvoxamine-theophylline interaction: gap between in vitro and in vivo inhibition constants toward cytochrome P4501A2. *Clin Pharmacol Ther* **70**:415–424.

Yao C, Kunze KL, Trager WF, Kharasch ED and Levy RH (2003) Comparison of in vitro and in vivo inhibition potencies of fluvoxamine toward CYP2C19. *Drug Metab Dispos* **31**:565–571.

Youdim KA, Zayed A, Dickins M, Phipps A, Griffiths M, Darekar A, Hyland R, Fahmi O, Hurst S, Plowchalk DR, Cook J, Guo F and Obach RS (2008) Application of CYP3A4 in vitro data to predict clinical drug–drug interactions; predictions of compounds as objects of interaction. *Br J Clin Pharmacol* **65**:680–692.

Zhou H, Tong Z and McLeod JF (2004) "Cocktail" approaches and strategies in drug development: valuable tool or flawed science? *J Clin Pharmacol* **44**:120–134.

Zhou SF (2008) Drugs behave as substrates, inhibitors and inducers of human cytochrome P450 3A4. *Curr Drug Metab* **9**:310–322.

Zhu B, Ou-Yang DS, Chen XP, Huang SL, Tan ZR, He N and Zhou HH (2001) Assessment of cytochrome P450 activity by a five-drug cocktail approach. *Clin Pharmacol Ther* **70**:455–461.

Chapter 25
Toxicological Consequences of Drug–Drug Interactions

Rachel J. Walsh, Abhishek Srivastava, Daniel J. Antoine, Dominic P. Williams, and B. Kevin Park

Abstract Adverse drug reactions (ADRs) represent one of the major clinical challenges to patient's health and are a key reason for attrition in drug development. An understanding of how drug–drug interactions (DDIs) can influence and cause ADRs is critical in managing patients using several prescriptions at the same time. Whilst it may be possible to control on-target ADRs that occur from DDIs, due to predictable pharmacokinetic and pharmacodynamic interactions, the impact of DDIs in triggering off-target ADRs still remains difficult to understand. This chapter will examine clinical, in vivo and in vitro examples to look at how DDIs can potentially lead to both on- and off-target ADRs.

25.1 Introduction

25.1.1 Adverse Drug Reactions

Adverse drug reactions (ADRs) represent one of the major clinical challenges to patients' health and are a key reason for attrition in drug development (Lazarou et al., 1998; Pirmohamed et al., 2004; Kola and Landis, 2004). ADRs can be classified as "on-target" or "off-target" reactions. On-target reactions can be predicted from the known primary or secondary pharmacology of the drug and often represent an exaggeration of the pharmacological effect of the drug. They show simple dose–response relationships and, therefore, can usually be avoided by dose reduction and are only rarely life threatening. In contrast, off-target reactions cannot be predicted from knowledge of the basic pharmacology of the drug, show complex

R.J. Walsh (✉)
Department of Pharmacology and Therapeutics, MRC Centre for Drug Safety Science, University of Liverpool, Liverpool L69 3GE , UK
e-mail: rjwalsh@liv.ac.uk

K.S. Pang et al. (eds.), *Enzyme- and Transporter-Based Drug–Drug Interactions*,
DOI 10.1007/978-1-4419-0840-7_25,

dose relationships and can exhibit marked inter-individual susceptibility (idiosyncratic). Whilst not as common as on-target reactions, these can be serious and life threatening. Since many patients experiencing ADRs may be using several prescriptions simultaneously, it important to understand the role of drug–drug interactions (DDIs) in provoking both on- and off-target ADRs.

25.1.2 Off-Target Toxicity – Drug-Induced Liver Injury

ADRs are known to affect every organ system within the body. However, those involving the liver represent a significant proportion with around 52% of all cases of acute liver failure in the United States attributed to drug-induced liver injury (DILI) (Ostapowicz et al., 2002). DILI is the most frequent reason for the withdrawal of an approved drug from the market (Temple and Himmel, 2002), and it accounts for approximately 50% of all acute liver failure cases (Ostapowicz et al., 2002; Kaplowitz, 2001; Lee, 2003). Most drug-induced hepatotoxicities are unpredictable and poorly understood. The liver is the principal site of drug metabolism; therefore, it is often a site of off-target toxicity. The manifestations range from mild, asymptomatic changes in serum transaminases, which are relatively common, to fulminant hepatic failure, which although rare is potentially life threatening and may necessitate a liver transplant (Park et al., 1998).

The complete mechanisms of DILI remain largely unknown but appear to involve two pathways: direct hepatotoxicity and adverse immune reactions. In many instances, liver injury is thought to be initiated by the bioactivation of the drug to chemically reactive metabolites, which have the potential to modify the function of various critical cellular macromolecules and are therefore able to cause direct damage.

25.1.3 The Role of Metabolism in DILI

The biotransformation of lipophilic compounds into water-soluble derivatives that are more readily excreted is the physiological role of drug metabolism. The principal site of drug metabolism is the liver. The liver is exposed to drugs and other xenobiotics immediately after their absorption from the gastro-intestinal tract and has a high capacity for both Phase I and Phase II biotransformations. Usually this conversion from a lipid to water-soluble form results in loss of pharmacological/biological activity, but certain xenobiotics undergo biotransformation to toxic/reactive metabolites that can interfere with cellular functions and may have intrinsic chemical reactivity towards certain types of cellular macromolecules (Fig. 25.1). These toxic/reactive metabolites have the ability to interact with cellular proteins, lipids and nucleic acids, leading to protein dysfunction, lipid peroxidation, DNA damage and oxidative stress, potentially leading to an off-target ADR. This impairment of cellular function can result in cell death and possible liver failure.

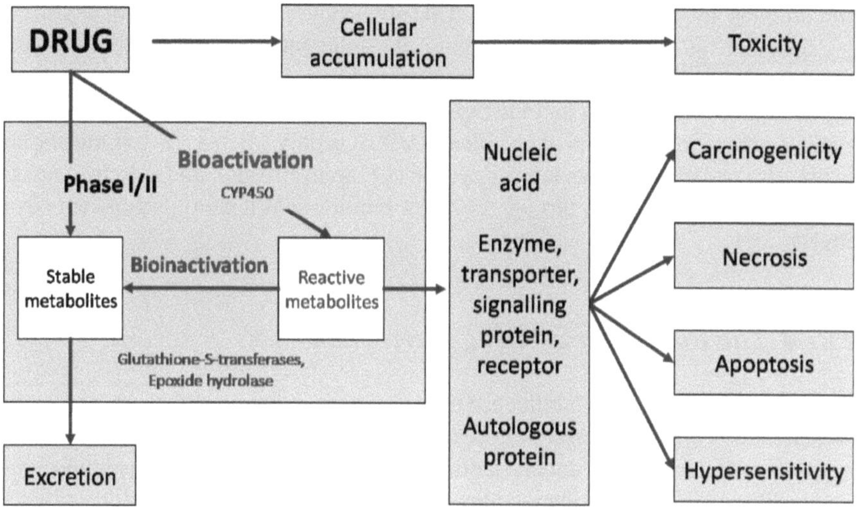

Fig. 25.1 The role of metabolism in drug-induced liver injury. Phase I oxidation can form reactive metabolites. These can interact with proteins to disrupt cellular processes and lead to toxicity

The propensity of a molecule to form either toxic and/or chemically reactive metabolites is simply a function of its chemistry. Such metabolites are typically short lived, with half-lives of generally less than 1 s, and are not usually detectable in plasma. Toxicophores are functional groups present in drugs which can be transformed into reactive species by normal biotransformations.

Formation of chemically reactive metabolites is mainly catalyzed by cytochromes P450, although products of Phase II metabolism can also lead to toxicity. Additionally, non-cytochrome P450 oxidative enzymes, such as myeloperoxidase and prostaglandin H synthetase, have been implicated in the bioactivation of drugs and other chemicals, the metabolites of which are thought to be responsible for observed toxicity, e.g. clozapine and agranulocytosis, benzene and aplastic anaemia (Smith et al., 1989; Ross et al., 1996; Mason and Fischer, 1992; Fischer et al., 1991). Cytochrome P450 isoforms are present in different proportions in many organs, though most abundantly in the liver, and thus bioactivate the chemicals to cause organ-specific toxicity (Uetrecht, 1992; Pelkonen and Raunio, 1997; Kao and Carver, 1990). The relationship between bioactivation and the occurrence of hepatic injury is not simple. For example, many chemicals undergo bioactivation in the liver but are not hepatotoxic. The best example is the absence of hepatotoxicity with therapeutic doses of acetaminophen. Tight coupling of bioactivation with bioinactivation may be one reason for this. Many enzymatic and nonenzymatic pathways of bioinactivation are present in the liver, which is perhaps the best equipped of all the organs in the body to deal with chemically reactive toxins. Typical examples of bioinactivation pathways include glutathione conjugation of quinones by glutathione *S*-transferases and hydration of arene oxides to dihydrodiols by epoxide hydrolases. The efficiency of a bioinactivation process is dependent on several factors including the inherent chemical reactivity of the substrate, substrate-selectivity of the enzymes, which is usually very broad, tissue expression of the enzymes and the rapid upregulation of enzyme(s) and co-factors mediated by cellular sensors of chemical stress. It is only when a reactive metabolite can bypass or saturate these defence systems of bioinactivation and thereby damage proteins and nucleic acids through covalent binding that it exerts significant toxic effects.

25.1.4 The Role of Drug–Drug Interactions

Drug–Drug interactions can influence both pharmacodynamic and pharmacokinetic properties of co-administered drugs. An understanding of the role of DDIs in on-target toxicity requires an understanding of these properties of all co-administered agents. Pharmacodynamic interactions can produce synergistic and additive effects, leading to toxicity. For example, consumption of alcohol whilst taking antihistamines causing drowsiness can lead to impaired psychomotor skills (Hindmarch and Bhatti, 1987). Pharmacokinetic interactions influence concentration of drug by alteration in absorption, distribution, metabolism and elimination of one agent by another. Co-administration of a drug with an agent which inhibits its clearance mechanisms leads to an increase in plasma levels, potentially exceeding the therapeutic window. The anticoagulant warfarin is metabolized by cytochrome P450 enzymes in the liver, specifically CYP2C9 and CYP3A4. Therefore agents which inhibit either of these two enzymes, such as fluconazole (Kunze et al., 1996), will lead to an increase in prothrombin time and possible bleeding.

The relationship between DDIs and off-target reactions is more complicated, with evidence tending to be anecdotal or confined to animal or in vitro models. The mechanisms of off-target reactions themselves are not yet fully understood, but what is known may help us to predict where and how DDIs could lead to toxicity.

25.2 Toxicological Consequences of DDI – On-Target Toxicity

25.2.1 Examples of On-Target Toxicity

As mentioned previously, on-target ADRs stem from a predictable exaggeration of the pharmacology of a drug. In multiple drug therapy this can follow from increased systemic drug concentrations due to impaired clearance (pharmacokinetic) or from synergistic and additive or antagonistic effects at receptor level (pharmacodynamic). In this chapter we will use interactions of anticoagulants to indicate how on-target toxicities can occur following these mechanisms.

25.2.2 Anticoagulants

Oral coumarin anticoagulants, such as warfarin, are hugely important in the management and prophylaxis of thrombosis and embolism. They act by inhibiting the conversion of vitamin K-epoxide to vitamin K. Vitamin K is a vital cofactor in the gamma-carboxylation of the N-terminus of coagulation factors (II, VII, IX and X). This carboxylation is inherent to the biological activity of these factors, therefore it is due to the resulting reduction in this process that an anticoagulant effect is produced (Choonara et al., 1988).

Over anticoagulation can lead to patients bleeding. This can present as ecchymoses, blood in the urine, uterine bleeding, nose bleeding, hematoma, gum bleeding and vomiting blood. In severe cases, hemorrhage can occur which is fatal in 1% of cases (Hanley, 2004). Patients are therefore carefully monitored by testing prothrombin time, i.e. the time it takes for plasma to clot.

Coumarin derivative combinations are associated with toxicological DDIs due to a combination of three properties: (i) high protein binding; (ii) cytochrome P450-dependent metabolism; and (iii) a narrow therapeutic range.

25.2.2.1 Enzyme Inhibition

Warfarin exists as a racemic mixture of two isomers. The S-isomer is five times more potent antagonist of vitamin K than the R-isomer (Breckenridge et al., 1974). Both isomers are metabolized to inactive hydroxyl metabolites, the S-isomer is metabolized by CYP2C9 to 7-hydroxywarfarin and the R-isomer by CYP1A2 to 6-hydroxywarfarin and 8-hydroxywarfarin and by CYP3A4 to 10-hydroxywarfarin (Kaminsky and Zhang, 1997). Due to the relative potencies of the isomers, it is inhibition of metabolism of the S-isomer which is toxicologically more important, and therefore co-administration of drugs inhibiting CYP2C9 can have toxicological consequences.

Fluconazole is a triazole antifungal. It is used to treat superficial or systemic fungal infections. In vitro studies have shown fluconazole to be an inhibitor of CYP2C9 and of CYP3A4 (Kunze et al., 1996). Several studies and case reports have indicated adverse reactions stemming from an interaction between fluconazole and coumarin derivatives. One case reports that a patient taking acenocoumarol suffered an intracranial hemorrhage 5 days after beginning fluconazole treatment (Isalska and Stanbridge, 1988). Several case reports have also indicated bleeding due to an interaction of warfarin and fluconazole (Baciewicz et al., 1994; Seaton et al., 1990; Kerr, 1993), including a report of intraocular hemorrhage (Mootha et al., 2002).

25.2.2.2 Protein Binding

Several drugs which potentiate the anticoagulant effects of warfarin are thought to do so via displacement of warfarin from plasma and tissue-binding sites, elevating the level of free drug. Warfarin has a high affinity for plasma with binding levels at 97% or more. Several drugs have been shown to or are thought to displace warfarin binding, but how much of a role this mechanism plays in the toxicology of DDIs is disputed (Rolan, 1994), with many drugs that work by this mechanism also thought to have secondary modes of potentiating warfarin's effects, such as enzyme inhibition. The interaction between warfarin and non-steroidal anti-inflammatory drugs is thought to occur via this mechanism. Azapropazone, phenylbutazone, naproxen, mefenamic acid and ibuprofen were all shown to increase free plasma warfarin, with azapropazone causing the biggest increase (39–46% free warfarin versus 2.5–6% free warfarin without competing drug) (Diana et al., 1989). This interaction has led to a range of toxicological effects, from bruising and bleeding to fatalities (Green et al., 1977; Powell-Jackson, 1977; Win et al., 1991).

25.2.2.3 Effects of Anti-platelet Drugs

Many patients receiving oral anticoagulation therapy also have coronary heart disease or peripheral arterial disease, hence requiring anti-platelet therapy. Anti-platelet drugs, such as aspirin and clopidogrel, inhibit platelet aggregation and therefore decrease the formation of thrombosis. The management of anticoagulant and antiplatelet combination therapy is important to prevent dangerous toxicological consequences. The risk of bleeding on this dual therapy is greatly increased. One study indicated a threefold increased risk of intracranial hemorrhage on combined warfarin antiplatelet therapy compared to warfarin alone (May et al., 2008). Bleeding, requiring hospitalization or blood transfusion, in artificial heart valve patients was three times more prevalent in patients receiving 500 mg daily aspirin plus warfarin compared to those on warfarin alone. Mainly gastrointestinal or cerebral bleeding was observed and all patients with intracranial bleeding died.

25.3 Toxicological Consequences of DDI – Off-Target Toxicity

25.3.1 Examples of Off-Target Toxicity

Whilst interactions causing on-target toxicity are well understood and well documented, there is little evidence for off-target toxicity caused by drug–drug interactions. As previously mentioned, drug-induced liver injury is the most frequent presentation of off-target toxicity, with formation of reactive metabolites thought to be involved (Park et al., 2005). It may therefore be the case that increased formation of reactive metabolites via enzyme induction may play a role in interactions leading to off-target toxicity, but other mechanisms involved are not presently understood.

25.3.1.1 Carbamazepine

The anti-convulsant carbamazepine, administered for the treatment of epilepsy, causes hypersensitivity reactions characterized by skin rash, hepatitis and eosinophilia (Leeder, 1998). Although CBZ undergoes bioactivation to toxic arene oxide and quinone metabolites (Ju and Uetrecht, 1999; Madden et al., 1996) definite evidence linking this to immunogenicity is lacking (Park et al., 1998). One of the adverse events that have been shown to occur with carbamazepine treatment is Stevens Johnson Syndrome (SJS). This is a rare but potentially fatal skin reaction where keratinocyte death causes separation of the dermis and the epidermis. Reports have linked the concurrent use of carbamazepine and antipsychotics with the development of this syndrome. In one report, three patients on antipsychotics (fluphenazine, haloperidol, trifluoperazine and chlorpromazine) developed SJS within 2 weeks of beginning carbamazepine (Wong, 1990). Another case report of development of SJS describes a patient taking lithium carbonate, trihexyphenidyl and haloperidol along with carbamazepine. Another case report describes development of SJS in a patient receiving carbamazepine and haloperidol (Huang and Tsai, 1995). It is not yet clear whether antipsychotic treatment does in fact increase the risk of developing SJS whilst taking carbamazepine or what the mechanisms behind this potential interaction may be. Further study is needed to clarify and understand this interaction.

25.3.1.2 Acetaminophen

One of the most well understood and most studied adverse drug reactions is DILI caused by acetaminophen (APAP) overdose. The toxicity is dose dependent and linked to formation of a reactive *N*-acetyl-*p*-benzoquinoneimine metabolite, which can bind to cellular proteins, disrupt mitochondria deplete cellular glutathione and ultimately lead to hepatocyte apoptosis and necrosis. At therapeutic doses, 70–90% of dose is glucuronidated or sulfated and excreted, with around 5–10% being converted by cytochrome P450 enzymes CYP1A2, CYP2E1 and CYP3A4 to the reactive metabolite (Raucy et al., 1989). This is then conjugated to glutathione and safely excreted as a mercapturic acid. However, in overdose or if the detoxification process is compromised, the levels of NAPQI accumulate and toxicity occurs. There is therefore a possibility that concurrent use of APAP and a drug-inducing enzyme responsible for formation of NAPQI may potentiate hepatotoxicity (Fig. 25.2). One report describes a female patient taking 2–4 g of APAP daily who began treatment with rifampicin, an antituberculosis treatment and inducer of CYP3A4. Her liver function deteriorated and rises in her serum transaminase levels were detected (Stephenson et al., 2001). Several other reports link another antituberculosis treatment and enzyme modulator, isoniazid, with increased APAP hepatotoxicity. A patient receiving isoniazid treatment for 6 months took approximately 3.5 g of APAP. Within 6 h, she had increased transaminase levels and hyperbilirubinaemia (Crippin, 1993). A patient who attempted suicide using APAP whilst on isoniazid developed near-fatal renal and hepatotoxicity despite serum APAP levels being within non-toxic range (Murphy et al., 1990). Both rifampicin and isoniazid have been shown to increase APAP toxicity in HepG2 cell line (Nicod et al., 1997).

Despite these two drugs causing enzyme induction, they have also both been associated with DILI (Menzies et al., 2008), and therefore other mechanisms of interaction may be involved instead of or as well as increased formation of NAPQI.

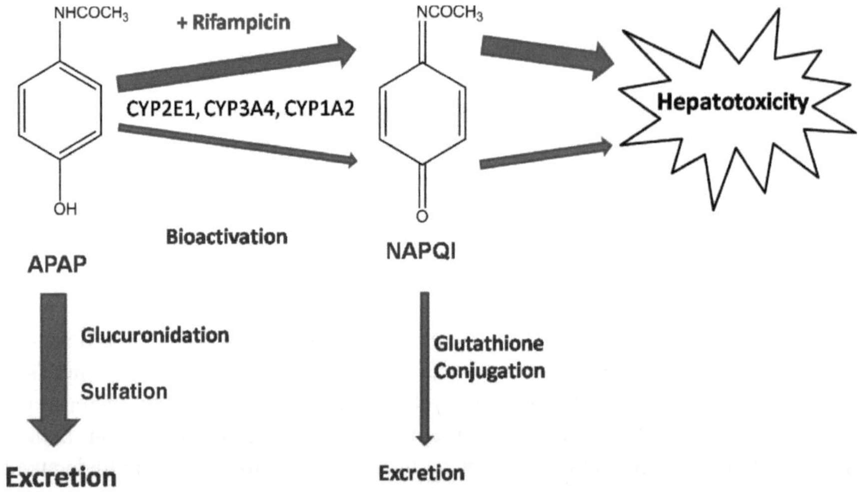

Fig. 25.2 Potentiation of the hepatotoxicity of APAP by the enzyme inducer, rifampicin. In normal therapeutic conditions, the majority of APAP dose is safely excreted following glucuronidation or sulfation. Rifampicin induces the formation of reactive metabolite, NAPQI, and may therefore increase hepatotoxicity

25.3.1.3 Nevirapine

Nevirapine (NVP), a non-nucleoside reverse transcriptase inhibitor, is widely used for the treatment of human immunodeficiency virus (HIV) infections. It is the main option for the first-line treatment of HIV-1, together with two nucleoside reverse transcriptase inhibitors, in countries with limited resources. NVP is associated with two serious clinically restrictive side effects: skin reactions and hepatotoxicity. Severe, life threatening and in some cases fatal hepatotoxicity, including fulminant and cholestatic hepatitis, hepatic necrosis and hepatic failure, has been reported in HIV-infected patients taking NVP. For this reason NVP is given a black box warning for hepatotoxicity, and concern has been raised over NVP-based treatment.

Although the role of an immune-mediated mechanism in NVP-induced skin rash and hepatotoxicity has been strongly advocated (Popovic et al., 2006), it is not yet clear whether immune induction in patients is due to a (reactive) metabolite or NVP itself. Highly circumstantial evidence for reactive metabolite involvement comes from a case reported by Claes et al. (2004), where a patient suffering from NVP-induced toxic epidermal necrolysis and toxic hepatitis was successfully treated with intravenous human immunoglobulins and high doses of *N*-acetylcysteine (300 mg/kg per day in a continuous infusion until recovery). As a precursor of glutathione, *N*-acetylcysteine may have helped to restore levels of glutathione, thus enhancing detoxification of toxic metabolites (De Rosa et al., 2000), and resulted in exceptionally fast clinical recovery in this case.

Hypothetically, there may be several pathways for NVP bioactivation. For example, the cyclopropylamine group has the potential to become bioactivated to an aminium cation radical (Shaffer et al., 2001). 12-Hydroxy NVP, which is the major Phase 1 metabolite in human liver microsomes, is a substrate for sulfotransferase, and it has been suggested that the sulfate ester dissociates to form a reactive quinonemethide intermediate (Chen et al., 2008; Uetrecht, 2006). However, in the case of the hepatotoxicity associated with NVP in humans, Chen et al. (Chen et al., 2008) have proposed that tissue injury is due to a quinonemethide produced by P450. NVP also has the potential to form an epoxide intermediate in either of the pyridine rings. Srivastava et al. (2009, in press) have found a pyridino-substituted mercapturate of NVP in human urine that could be derived from a glutathione conjugate of an epoxide. A reactive intermediate (aminium cation/epoxide/quinonemethide) might deplete hepatic glutathione (GSH) in certain patients because the synthesis of GSH is often reduced in HIV infection (Otis et al., 2008; Stehbens, 2004). Depletion of GSH, if it is uncompensated, may lead to oxidative stress and covalent binding of NVP to critical hepatic proteins, as a consequence triggering apoptosis and necrosis of liver cells. Adduct formation with proteins under these circumstances might also lead to the initiation of the immune response and possibly explain the hypersensitivity observed with NVP.

Due to the potential metabolic nature of NVP's toxicity, there is potential for similar interactions as seen with APAP. In one report of patients on concurrent nevirapine and rifampicin therapy over 6 months, whilst 80% of patients maintained normal liver function tests, one patient developed a grade 4 liver function test abnormality, and rifampicin treatment was suspended (Sathia et al., 2008). Concomitant use of rifampicin and nevirapine is contraindicated by the British HIV Association Guidelines (Pozniak et al., 2005).

25.3.1.4 Anti-tuberculosis Drugs

The previously mentioned anti-tuberculosis treatments, rifampicin and isoniazid, are often used concurrently as part of a regimen of anti-TB therapies that also includes pyrazinamide. All three of these drugs have been associated with DILI (Girling, 1978). Metabolic factors also play a role in the hepatotoxicity of these drugs via formation of reactive metabolites and enzyme induction. Isoniazid is metabolized by acetylation, leading to formation of acetylhydrazine which can be hydrolyzed to form hydrazine, which can then further be metabolized to hepatotoxic compounds causing irreversible cell damage (Nelson and Gordon, 1981). Rifampicin is a potent enzyme inducer and induces isoniazid hydrolase, increasing production of hydrazine. Rifampicin has been shown to increase the toxicity of isoniazid in vitro (Nicod et al., 1997), and there are several reports which suggest that combination use increases the risk of hepatotoxicity compared to isoniazid or rifampicin alone (Steele et al., 1991; Gangadharam, 1986; Askgaard et al., 1995; Pessayre et al., 1977). It is unknown if pyrazinamide toxicity is linked to metabolism or what other mechanisms might be involved. Pyrazinamide has been shown to be an inhibitor of certain rat CYP450s (Maffei and Carini, 1980), but has also shown to have no effect on the activity of human microsomes (Nishimura et al., 2004). However, the combination of rifampicin and pyrazinamide for 2 months caused more hepatotoxic events than a 6 month treatment of isoniazid alone (Jasmer et al., 2002; McNeill et al., 2003) and more than a 6 month cohort of all three drugs (van et al., 2004). Why this is the case is unclear but it has been suggested that isoniazid has an antagonistic effect, decreasing the potential for hepatotoxicity of pyrazinamide (Lee et al., 2002).

25.3.2 Other Interactions

Some reports exist in the literature of in vivo or in vitro interactions that may or may not have been described in the clinic. A study by Pontes et al. (2008) describes potentiation of the hepatotoxic effects of MDMA in mice following chronic ethanol exposure. Following 8 week exposure to ethanol and then 24 h exposure to 10 mg/kg MDMA (i.p.), they assessed liver histology, plasma transaminases and oxidative stress biomarkers. Animals that received both ethanol and MDMA showed significantly higher levels of liver injury than control animals or those that had received MDMA with no previous ethanol exposure. MDMA and ethanol hepatotoxicity have been linked to formation of metabolites shown to have higher toxicity than parent compound (Carmo et al., 2006; Gemma et al., 2007, 2006; Green et al., 2003). Ethanol has also been shown to induce the expression and activity of CYP450 (Klotz and Ammon, 1998).

An in vivo study has also been conducted on the effects of COX inhibitors on the autoimmunity of D-penicillamine (Seguin et al., 2003). Brown Norway rats received a dose of 20 mg/day or 100 mg/day of D-penicillamine in drinking water, following either a single dose or with continuous use of non-specific COX inhibitor, ketoprofen, the more selective diclofenac or COX-2 selective rofecoxib. It was found that single dose and continuous ketoprofen decreased the time of onset of immunity and continuous use ketoprofen also increased the incidence. Diclofenac had no effect, whilst single-dose rofecoxib appeared to lower the incidence of immunity. There are no reports, however, on the effects of ketoprofen on the immunity of penicillin in the clinic.

25.4 Conclusion

DDIs remain important for the clinician, as well as in drug discovery and development. Knowledge of the pharmacodynamics, pharmacokinetics and metabolism of drugs is essential in the prediction, prevention and management of toxicological interactions. On-target reactions can be predicted and controlled using this understanding and are a function of the pharmacology of the drug. Off-target toxicity proves more difficult to predict and is due to a biological factor present in a small number of susceptible patients. We need to more fully understand the biological basis of off-target ADRs in the susceptible patients in their own right prior to the additional complexity of DDIs.

References

Askgaard DS, Wilcke T and Dossing M (1995) Hepatotoxicity caused by the combined action of isoniazid and rifampicin. *Thorax* **50**:213–214.

Baciewicz AM, Menke JJ, Bokar JA and Baud EB (1994) Fluconazole-warfarin interaction. *Ann Pharmacother* **28**:1111.

Breckenridge A, Orme M, Wesseling H, Lewis RJ and Gibbons R (1974) Pharmacokinetics and pharmacodynamics of the enantiomers of warfarin in man. *Clin Pharmacol Ther* **15**: 424–430.

Carmo H, Brulport M, Hermes M, Oesch F, Silva R, Ferreira LM, Branco PS, Boer D, Remiao F, Carvalho F, Schon MR, Krebsfaenger N, Doehmer J, Bastos ML and Hengstler JG (2006) Influence of CYP2D6 polymorphism on 3,4-methylenedioxymethamphetamine ('Ecstasy') cytotoxicity. *Pharmacogenet Genom* **16**:789–799.

Chen J, Mannargudi BM, Xu L and Uetrecht J (2008) Demonstration of the metabolic pathway responsible for nevirapine-induced skin rash. *Chem Res Toxicol* **21**:1862–1870.

Choonara IA, Malia RG, Haynes BP, Hay CR, Cholerton S, Breckenridge AM, Preston FE and Park BK (1988) The relationship between inhibition of vitamin K1 2,3-epoxide reductase and reduction of clotting factor activity with warfarin. *Br J Clin Pharmacol* **25**:1–7.

Claes P, Wintzen M, Allard S, Simons P, De CA and Lacor P (2004) Nevirapine-induced toxic epidermal necrolysis and toxic hepatitis treated successfully with a combination of intravenous immunoglobulins and N-acetylcysteine. *Eur J Intern Med* **15**:255–258.

Crippin JS (1993) Acetaminophen hepatotoxicity: potentiation by isoniazid. *Am J Gastroenterol* **88**:590–592.

De Rosa SC, Zaretsky MD, Dubs JG, Roederer M, Anderson M, Green A, Mitra D, Watanabe N, Nakamura H, Tjioe I, Deresinski SC, Moore WA, Ela SW, Parks D, Herzenberg LA and Herzenberg LA (2000) N-acetylcysteine replenishes glutathione in HIV infection. *Eur J Clin Invest* **30**:915–929.

Diana FJ, Veronich K and Kapoor AL (1989) Binding of nonsteroidal anti-inflammatory agents and their effect on binding of racemic warfarin and its enantiomers to human serum albumin. *J Pharm Sci* **78**:195–199.

Fischer V, Haar JA, Greiner L, Lloyd RV and Mason RP (1991) Possible role of free radical formation in clozapine (clozaril)-induced agranulocytosis. *Mol Pharmacol* **40**:846–853.

Gangadharam PR (1986) Isoniazid, rifampin, and hepatotoxicity. *Am Rev Respir Dis* **133**:963–965.

Gemma S, Vichi S and Testai E (2006) Individual susceptibility and alcohol effects:biochemical and genetic aspects. *Ann Ist Super Sanita* **42**:8–16.

Gemma S, Vichi S and Testai E (2007) Metabolic and genetic factors contributing to alcohol induced effects and fetal alcohol syndrome. *Neurosci Biobehav Rev* **31**:221–229.

Girling DJ (1978) The hepatic toxicity of antituberculosis regimens containing isoniazid, rifampicin and pyrazinamide. *Tubercle* **59**:13–32.

Green AE, Hort JF, Korn HE and Leach H (1977) Potentiation of warfarin by azapropazone. *Br Med J* **1**:1532.

Green AR, Mechan AO, Elliott JM, O'Shea E and Colado MI (2003) The pharmacology and clinical pharmacology of 3,4-methylenedioxymethamphetamine (MDMA, "ecstasy"). *Pharmacol Rev* **55**:463–508.

Hanley JP (2004) Warfarin reversal. *J Clin Pathol* **57**:1132–1139.

Hindmarch I and Bhatti JZ (1987) Psychomotor effects of astemizole and chlorpheniramine, alone and in combination with alcohol. *Int Clin Psychopharmacol* **2**:117–119.

Huang SC and Tsai SJ (1995) Hyponatremia and Stevens-Johnson syndrome in a patient receiving carbamazepine. *Gen Hosp Psychiatry* **17**:458–460.

Isalska BJ and Stanbridge TN (1988) Fluconazole in the treatment of candidal prosthetic valve endocarditis. *BMJ* **297**:178–179.

Jasmer RM, Saukkonen JJ, Blumberg HM, Daley CL, Bernardo J, Vittinghoff E, King MD, Kawamura LM and Hopewell PC (2002) Short-course rifampin and pyrazinamide compared

with isoniazid for latent tuberculosis infection: a multicenter clinical trial. *Ann Intern Med* **137**:640–647.
Ju C and Uetrecht JP (1999) Detection of 2-hydroxyiminostilbene in the urine of patients taking carbamazepine and its oxidation to a reactive iminoquinone intermediate. *J Pharmacol Exp Ther* **288**:51–56.
Kaminsky LS and Zhang ZY (1997) Human P450 metabolism of warfarin. *Pharmacol Ther* **73**: 67–74.
Kao J and Carver MP (1990) Cutaneous metabolism of xenobiotics. *Drug Metab Rev* **22**:363–410.
Kaplowitz N (2001) Drug-induced liver disorders: implications for drug development and regulation. *Drug Saf* **24**:483–490.
Kerr HD (1993) Case report: potentiation of warfarin by fluconazole. *Am J Med Sci* **305**:164–165.
Klotz U and Ammon E (1998) Clinical and toxicological consequences of the inductive potential of ethanol. *Eur J Clin Pharmacol* **54**:7–12.
Kola I and Landis J (2004) Can the pharmaceutical industry reduce attrition rates?. *Nat Rev Drug Discov* **3**:711–715.
Kunze KL, Wienkers LC, Thummel KE and Trager WF (1996) Warfarin-fluconazole. I. Inhibition of the human cytochrome P450-dependent metabolism of warfarin by fluconazole: in vitro studies. *Drug Metab Dispos* **24**:414–421.
Lazarou J, Pomeranz BH and Corey PN (1998) Incidence of adverse drug reactions in hospitalized patients: a meta-analysis of prospective studies. *J Am Med Assoc* **279**:1200–1205.
Lee AM, Mennone JZ, Jones RC and Paul WS (2002) Risk factors for hepatotoxicity associated with rifampin and pyrazinamide for the treatment of latent tuberculosis infection: experience from three public health tuberculosis clinics. *Int J Tuberc Lung Dis* **6**:995–1000.
Lee WM (2003) Acute liver failure in the United States. *Semin Liver Dis* **23**:217–226.
Leeder JS (1998) Mechanisms of idiosyncratic hypersensitivity reactions to antiepileptic drugs. *Epilepsia* **39 Suppl 7**:S8–S16.
Madden S, Maggs JL and Park BK (1996) Bioactivation of carbamazepine in the rat in vivo. Evidence for the formation of reactive arene oxide(s). *Drug Metab Dispos* **24**:469–479.
Maffei FR and Carini M (1980) The inhibitory effect of pyrazinamide on microsomal monooxygenase activities is related to the binding to reduced cytochrome P-450. *Pharmacol Res Commun* **12**:523–537.
Mason RP and Fischer V (1992) Possible role of free radical formation in drug-induced agranulocytosis. *Drug Saf* **7 Suppl 1**:45–50.
May AE, Geisler T and Gawaz M (2008) Individualized antithrombotic therapy in high risk patients after coronary stenting. A double-edged sword between thrombosis and bleeding. *Thromb Haemost* **99**:487–493.
McNeill L, Allen M, Estrada C and Cook P (2003) Pyrazinamide and rifampin vs isoniazid for the treatment of latent tuberculosis: improved completion rates but more hepatotoxicity. *Chest* **123**:102–106.
Menzies D, Long R, Trajman A, Dion MJ, Yang J, Al JH, Memish Z, Khan K, Gardam M, Hoeppner V, Benedetti A and Schwartzman K (2008) Adverse events with 4 months of rifampin therapy or 9 months of isoniazid therapy for latent tuberculosis infection: a randomized trial. *Ann Intern Med* **149**:689–697.
Mootha VV, Schluter ML and Das A (2002) Intraocular hemorrhages due to warfarin fluconazole drug interaction in a patient with presumed Candida endophthalmitis. *Arch Ophthalmol* **120**:94–95.
Murphy R, Swartz R and Watkins PB (1990) Severe acetaminophen toxicity in a patient receiving isoniazid. *Ann Intern Med* **113**:799–800.
Nelson SD and Gordon WP (1981) Metabolic activation of hydrazines. *Adv Exp Med Biol* **136 Pt B**:971–981.
Nicod L, Viollon C, Regnier A, Jacqueson A and Richert L (1997) Rifampicin and isoniazid increase acetaminophen and isoniazid cytotoxicity in human HepG2 hepatoma cells. *Hum Exp Toxicol* **16**:28–34.

Nishimura Y, Kurata N, Sakurai E and Yasuhara H (2004) Inhibitory effect of antituberculosis drugs on human cytochrome P450-mediated activities. *J Pharmacol Sci* **96**:293–300.
Ostapowicz G, Fontana RJ, Schiodt FV, Larson A, Davern TJ, Han SH, McCashland TM, Shakil AO, Hay JE, Hynan L, Crippin JS, Blei AT, Samuel G, Reisch J and Lee WM (2002) Results of a prospective study of acute liver failure at 17 tertiary care centers in the United States. *Ann Intern Med* **137**:947–954.
Otis JS, Ashikhmin YI, Brown LA and Guidot DM (2008) Effect of HIV-1-related protein expression on cardiac and skeletal muscles from transgenic rats. *AIDS Res Ther* **5**:8.
Park BK, Kitteringham NR, Maggs JL, Pirmohamed M and Williams DP (2005) The role of metabolic activation in drug-induced hepatotoxicity. *Annu Rev Pharmacol Toxicol* **45**:177–202.
Park BK, Pirmohamed M and Kitteringham NR (1998) Role of drug disposition in drug hypersensitivity: a chemical, molecular, and clinical perspective. *Chem Res Toxicol* **11**:969–988.
Pelkonen O and Raunio H (1997) Metabolic activation of toxins: tissue-specific expression and metabolism in target organs. *Environ Health Perspect* **105 Suppl 4**:767–774.
Pessayre D, Bentata M, Degott C, Nouel O, Miguet JP, Rueff B and Benhamou JP (1977) Isoniazid-rifampin fulminant hepatitis. A possible consequence of the enhancement of isoniazid hepatotoxicity by enzyme induction. *Gastroenterology* **72**:284–289.
Pirmohamed M, James S, Meakin S, Green C, Scott AK, Walley TJ, Farrar K, Park BK and Breckenridge AM (2004) Adverse drug reactions as cause of admission to hospital: prospective analysis of 18 820 patients. *BMJ* **329**:15–19.
Pontes H, Duarte JA, de Pinho PG, Soares ME, Fernandes E, nis-Oliveira RJ, Sousa C, Silva R, Carmo H, Casal S, Remiao F, Carvalho F and Bastos ML (2008) Chronic exposure to ethanol exacerbates MDMA-induced hyperthermia and exposes liver to severe MDMA-induced toxicity in CD1 mice. *Toxicology* **252**:64–71.
Popovic M, Caswell JL, Mannargudi B, Shenton JM and Uetrecht JP (2006) Study of the sequence of events involved in nevirapine-induced skin rash in Brown Norway rats. *Chem Res Toxicol* **19**:1205–1214.
Powell-Jackson PR (1977) Interaction between azapropazone and warfarin. *Br Med J* **1**: 1193–1194.
Pozniak AL, Miller RF, Lipman MC, Freedman AR, Ormerod LP, Johnson MA, Collins S and Lucas SB (2005) BHIVA treatment guidelines for tuberculosis (TB)/HIV infection 2005. *HIV Med* **6 Suppl 2**:62–83.
Raucy JL, Lasker JM, Lieber CS and Black M (1989) Acetaminophen activation by human liver cytochromes P450IIE1 and P450IA2. *Arch Biochem Biophys* **271**:270–283.
Rolan PE (1994) Plasma protein binding displacement interactions – why are they still regarded as clinically important?. *Br J Clin Pharmacol* **37**:125–128.
Ross D, Siegel D, Schattenberg DG, Sun XM and Moran JL (1996) Cell-specific activation and detoxification of benzene metabolites in mouse and human bone marrow: identification of target cells and a potential role for modulation of apoptosis in benzene toxicity. *Environ Health Perspect* **104 Suppl 6**:1177–1182.
Sathia L, Obiorah I, Taylor G, Kon O, O'Donoghue M, Gibbins S, Walsh J and Winston A (2008) Concomitant use of nonnucleoside analogue reverse transcriptase inhibitors and rifampicin in TB/HIV type 1-coinfected patients. *AIDS Res Hum Retroviruses* **24**:897–901.
Seaton TL, Celum CL and Black DJ (1990) Possible potentiation of warfarin by fluconazole. *DICP* **24**:1177–1178.
Seguin B, Teranishi M and Uetrecht JP (2003) Modulation of D-penicillamine-induced autoimmunity in the Brown Norway rat using pharmacological agents that interfere with arachidonic acid metabolism or synthesis of inducible nitric oxide synthase. *Toxicology* **190**:267–278.
Shaffer CL, Morton MD and Hanzlik RP (2001) N-dealkylation of an N-cyclopropylamine by horseradish peroxidase. Fate of the cyclopropyl group. *J Am Chem Soc* **123**:8502–8508.
Smith MT, Yager JW, Steinmetz KL and Eastmond DA (1989) Peroxidase-dependent metabolism of benzene's phenolic metabolites and its potential role in benzene toxicity and carcinogenicity. *Environ Health Perspect* **82**:23–29.

Steele MA, Burk RF and DesPrez RM (1991) Toxic hepatitis with isoniazid and rifampin. A meta-analysis. *Chest* **99**:465–471.

Stehbens WE (2004) Oxidative stress in viral hepatitis and AIDS. *Exp Mol Pathol* **77**:121–132.

Stephenson I, Qualie M and Wiselka MJ (2001) Hepatic failure and encephalopathy attributed to an interaction between acetaminophen and rifampicin. *Am J Gastroenterol* **96**:1310–1311.

Temple RJ and Himmel MH (2002) Safety of newly approved drugs: implications for prescribing. *J Am Med Assoc* **287**:2273–2275.

Uetrecht J (2006) Evaluation of which reactive metabolite, if any, is responsible for a specific idiosyncratic reaction. *Drug Metab Rev* **38**:745–753.

Uetrecht JP (1992) The role of leukocyte-generated reactive metabolites in the pathogenesis of idiosyncratic drug reactions. *Drug Metab Rev* **24**:299–366.

van HR, Baars H, Kik S, van GP, Trompenaars MC, Kalisvaart N, Keizer S, Borgdorff M, Mensen M and Cobelens F (2004) Hepatotoxicity of rifampin-pyrazinamide and isoniazid preventive therapy and tuberculosis treatment. *Clin Infect Dis* **39**:488–496.

Win N, Mitchell DC, Jones PA and French EA (1991) Azapropazone and warfarin. *BMJ* **302**: 969–970.

Wong KE (1990) Stevens-Johnson syndrome in neuroleptic-carbamazepine combination. *Singapore Med J* **31**:432–433.

Part IV
Regulatory Aspects and Future Developments Involving DDI

Chapter 26
Complex Drug Interactions: Significance and Evaluation

Ping Zhao, Lei Zhang, and Shiew-Mei Huang

Abstract Complex drug interactions that result from a multiplicity of factors (e.g., concomitant medications, organ dysfunction, genetic polymorphism of enzyme/transporter) may be accompanied by important clinical relevance. However, it is difficult to properly design and evaluate all possible complex drug interactions during drug development. In this chapter, we review the types of complex drug interactions and recent advances in studying complex drug interactions using modeling and simulation approaches. Challenges in the quantitative evaluation of complex drug interactions include (1) the need to understand metabolism/transport pathways and their interplay, (2) accurate assessment of key parameters (e.g., fractional clearance) at the enzyme/transporter level, and (3) knowledge in how altered physiological conditions (e.g., by disease states) affect drug disposition and response. Additional research will provide confidence in the use of modeling and simulation to guide clinical study design and generate data for the informative labeling and effective use of medications.

26.1 Introduction

An individual's drug exposure can be altered by intrinsic factors (e.g., age, gender, race, organ dysfunction, genetic polymorphism of drug metabolizing enzymes and transporters) and extrinsic factors (e.g., food, juice, dietary supplements, concomitant medications). Metabolism- and transporter-based drug interactions have been under the spotlight after several drugs were withdrawn from the US market partly due to drug–drug interactions, and their evaluations have been recommended as

P. Zhao (✉)
Office of Clinical Pharmacology, Office of Translational Sciences, Center for Drug Evaluation and Research, Food and Drug Administration, Silver Spring, MD, USA
e-mail: ping.zhao@fda.hhs.gov

Disclaimer: The opinions contained in this chapter do not necessarily reflect the official views of the FDA.

K.S. Pang et al. (eds.), *Enzyme- and Transporter-Based Drug–Drug Interactions*, DOI 10.1007/978-1-4419-0840-7_26,

part of drug development (Huang et al., 2007; Huang and Temple, 2008). In 1997 and 1999, the US Food and Drug Administration (FDA) issued guidance documents for industry on studying metabolism-based drug interactions in vitro and in vivo, respectively. As science advances in this area, the FDA issued a revised guidance for industry in 2006 for public comments: "Drug Interaction Studies – Study Design, Data Analysis, and Implications for Dosing and Labeling" (draft DDI guidance, http://www.fda.gov/downloads/Drugs/GuidanceComplianceRegulatoryInformation/Guidances/ucm072101.pdf). The draft DDI guidance emphasizes the use of an integrated approach to evaluate drug interactions. Decision trees in determining the need to study drug interactions in vivo are presented for major cytochrome P450 enzymes (CYPs) as well as P-glycoprotein (P-gp) (draft DDI guidance; Zhang et al., 2008).

To date, the intrinsic and extrinsic factors affecting drug exposure are often separately evaluated. In addition, in vivo drug interaction studies have focused on "single-pair drug interaction" scenario. The results had been listed in the drug labeling with recommendations for dosing adjustment for a particular factor when appropriate. Table 26.1 lists an example of changes in systemic exposure resulted in corresponding dosing recommendation for rosuvastatin in patients with specific individual factors (Huang and Temple, 2008). However, the net exposure change and dose adjustment of a drug in patients with multiple factors are not always clear. For example, the systemic exposure change of a substrate by multiple inhibitors may be much greater than the product of the fold changes in area under the concentration – time profile (AUC) observed when the inhibitors are given individually (Huang et al., 2007, draft DDI guidance).

Table 26.1 Comparative systemic exposure and corresponding starting (and maintenance) dose recommendation in subgroups with various patient factors: young healthy male subjects (control); patients with various degrees of hepatic impairment based on the Child–Pugh scores, subgroups A or B (hepatic); patients with varying degrees of renal impairment (creatinine clearance of 50–80, 30–50, or <30 mL/min with hemodialysis) (renal); Asians compared with Caucasians (race); individuals taking concomitant medications such as cyclosporine, gemfibrozil, or lopinavir/ritonavir. (Data compiled from labeling for Crestor (rosuvastatin; AstraZeneca); labeling from http://www.accessdata.fda.gov/scripts/cder/drugsatfda) Reproduced with permission (Huang and Temple, 2008)

Group	Ethnic factor	Fold increase in exposure (AUC)	Initial dose (mg)	Daily dose (mg)
1	Control	1-fold	10–20	5–40
2	Hepatic impairment	1.1-fold (mild)	10–20	5–40
		1.2-fold (moderate)	10–20	5–40
3	Renal impairment	1-fold (mild)	10–20	5–40
		1-fold (moderate)	10–20	5–40
		3-fold (severe)	5	≤10
4	Race	2-fold (Asians)	5	5–20
5	Cyclosporine	7-fold		5
6	Gemfibrozil	1.9-fold		10
7	Lopinavir/ritonavir	5-fold		10

The clinical importance of complex drug interactions can not be overlooked. Poly-pharmacy can lead to complex drug interactions. A recent study indicated that 29% of the 3005 individuals (age 57–85) from households across the USA surveyed between 2005 and 2006 took more than five prescription drugs (Qato et al., 2008). The elderly population can be more sensitive to drug interactions since they can have additional age-related reduction in their liver and/or kidney functions.

A drug that is known to be eliminated by multiple metabolic pathways may also be subject to complex drug interactions. In January 2009, the FDA issued early communications regarding increased cardiovascular events of the antiplatelet agent clopidogrel in patients taking concomitant proton pump inhibitors (http://www.fda.gov/Drugs/DrugSafety/PostmarketDrugSafetyInformationforPatientsandProviders/DrugSafetyInformationforHeathcareProfessionals/ucm079520.htm). Clopidogrel is metabolized to its pharmacologically active moiety by several CYPs (Kurihara et al., 2005), of which CYP2C19 appears to be the most important. Polymorphism of CYP2C19 (Kim et al., 2008a), P-glycoprotein (Taubert et al., 2006), and concomitant use of proton pump inhibitors known to inhibit CYP2C19 and of the CYP3A inhibitor, ketoconazole (Farid et al., 2007; Kim et al., 2008b; Mega et al., 2009), have been associated with decreased formation of the active metabolite and/or clopidogrel resistance. Therefore, medical practitioners need to know all of the medications the individual patients are taking, in addition to critical genetic information, when appropriate.

The importance of evaluating complex drug interactions has been recognized by the FDA. The draft FDA DDI guidance recommends the use of multiple inhibitors to maximize the drug interaction potential for a substrate metabolized by multiple CYPs or eliminated by a combination of metabolizing enzymes and drug transporters when certain conditions are met (http://www.fda.gov/ohrms/dockets/ac/06/slides/2006-4248s1-6-FDAHuang_files/frame.htm; http://www.fda.gov/ohrms/dockets/ac/04/slides/2004-4079S1_06_Huang_files/frame.htm). Simulation results have indicated that the fold-change when two CYP pathways are simultaneously inhibited can be greater than the product of fold-changes when only one CYP pathway is inhibited, provided that there is a monir residual pathway that is not inhibited by inhibitors of either CYP (Thummel et al., 2007). The theoretical basis for quantitatively assessing multiple inhibitions has been proposed (Collins et al., 2006; Rostami-Hodjegan and Tucker, 2004; Zhang X et al., 2009).

The objectives of this chapter are 3-fold: (1) to review the types of complex drug interactions, (2) to review the advancement and challenges in studying complex drug interactions using modeling and simulation approaches, and (3) to provide scientific perspective on comprehensively evaluating complex drug interactions.

26.2 Intrinsic Clearance and Pharmacokinetic Outcome

The concept of clearance is at the heart of the quantitative evaluation of drug–drug interactions. The alteration of systemic clearance (CL) by an interacting drug at the level of drug metabolizing enzymes or drug transporters can lead to changes in the

Table 26.2 Mechanisms of major drug interactions at the level of drug metabolizing enzymes

	Major assumptions	Equations of R_{CLINT}	References
Reversible inhibition	Linear pharmacokinetics	$R_{\mathrm{CLint}} = 1 + \frac{I}{K_i}$	(Segel, 1975)
Time-dependent inhibition	Steady-state conditions and a simple enzyme homeostasis model: $E_{\mathrm{a}} = S_0/K_{\mathrm{deg}}$	$R_{\mathrm{CLint}} = \dfrac{K_{\mathrm{deg}} + \frac{k_{\mathrm{inact}} \times I}{K_I + I}}{K_{\mathrm{deg}}}$	(Ito et al., 1998; Kitz and Wilson, 1962)
Induction – increased protein synthesis	Steady-state conditions and a simple enzyme homeostasis model	$R_{\mathrm{CLint}} = \dfrac{S_{0,\mathrm{IND}}}{S_0}$	(Almond et al., 2009; Shou et al., 2008)
Induction – decreased protein degradation	Steady-state conditions and a simple enzyme homeostasis model	$R_{\mathrm{CLint}} = \dfrac{K_{\mathrm{deg}}}{K_{\mathrm{deg,IND}}}$	(Almond et al., 2009; Roberts et al., 1995)

I, inhibitor concentration; K_{i}, reversible inhibition constant; k_{inact}, maximum inactivation rate constant; K_I, inactivation constant; E_{a}, active enzyme level; S_0, apparent zero-order enzyme synthesis rates; K_{deg}, apparent first-order enzyme degradation rate constants; $S_{0,\mathrm{IND}}$, S_0 in the presence of an inducer that increases the protein synthesis; $K_{\mathrm{deg,IND}}$, K_{deg} in the presence of an inducer that stabilizes the enzyme

exposure of the substrate, such as AUC. Table 26.2 summarizes the mechanisms of major drug interactions with regard to drug metabolizing enzymes. Changes in the ratio of systemic clearance (CL) in vivo, without and with an interacting drug (R_{CLINT}, ∓I), dictate the ratio of substrate exposures, with and without an interacting drug (AUCR, ±I) (Fig. 26.1).

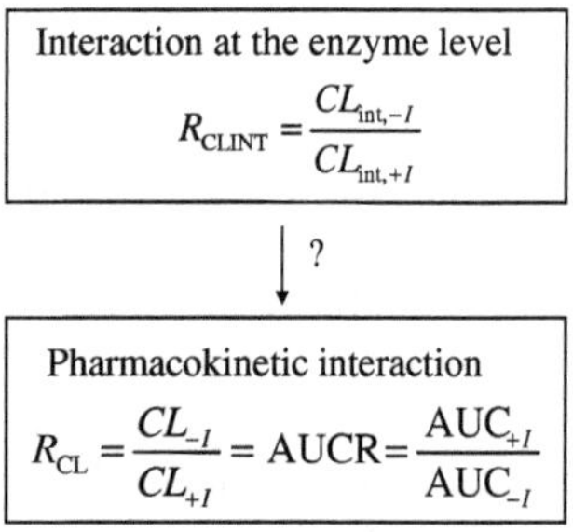

Fig. 26.1 Translating drug interaction at the level of drug metabolizing enzyme to the degree of drug interaction in vivo

The equations in Table 26.2 commonly serve as the basis for interpreting in vivo drug interaction using in vitro data, an exercise commonly known as in vitro–in vivo correlation or in vitro–in vivo extrapolation. However, a thorough understanding of the fundamental steps bridging R_{CLINT} and AUCR is critical to evaluate drug interactions. Generally, the value of $\mathrm{CL_{int}}$ at the enzyme level can be scaled up to organ intrinsic clearance, such as liver ($\mathrm{CL_{int,H}}$). The $\mathrm{CL_{int,H}}$ for a flow-limited substrate (absence of membrane barrier or transporters) can be translated into organ clearance ($\mathrm{CL_H}$) assuming a blood restriction model such as the well-stirred model (Pang and Rowland, 1977):

$$\mathrm{CL_H} = \frac{Q_\mathrm{H} f_\mathrm{u,b}\mathrm{CL_{int,H}}}{Q_\mathrm{H} + f_\mathrm{u,b}\mathrm{CL_{int,H}}} \tag{26.1}$$

Q_H and $f_\mathrm{u,b}$ are hepatic blood flow and free fraction of drug in the blood. On the in vivo side of the spectrum, CL is the sum of organ clearance values including $\mathrm{CL_H}$. Therefore, in order to understand pharmacokinetic drug interaction in light of the exposure change as expressed by AUCR (Fig. 26.1), one needs to understand the non-proportional relationship between the change in $\mathrm{CL_{int}}$ and the change in substrate exposure in vivo.

Below, we will focus our discussion on the non-proportional translation between R_CLINT and AUCR (Fig. 26.1). For in-depth discussion on drug interaction mechanisms summarized in Table 26.2, readers are directed to other chapters (e.g., Chapters 1, 7, 12, 13, 19 and 21) of this book. In addition, one needs to be aware of the negative impact on the interpretation of drug interaction by in vitro kinetic parameters obtained from poorly designed and poorly conducted experiments (Yang et al., 2007; 2008).

26.2.1 Non-proportionality Between Organ Intrinsic Clearance and Organ Clearance

Translation of interactions at the enzyme level into in vivo situation requires the estimation of the blood flow, free fraction of the drug, and the intrinsic clearance of the organ. The nonlinear relationship between $\mathrm{CL_{int,H}}$ and $\mathrm{CL_H}$ for liver metabolism is obvious from Equation (26.1). Assuming that $\mathrm{CL_H}$ is equal to CL and substrate is given intravenously, it may be shown that a proportional relationship exists between $R_\mathrm{CLINT,H}$ and AUCR only when $f_\mathrm{u,b}\mathrm{CL_{int,H}}$ is much smaller than Q_H (AUCR–$R_\mathrm{CLINT,H}$ panel when $f_\mathrm{u,b}\mathrm{CL_{int,H}}/Q_\mathrm{H} \Rightarrow 0$) (Fig. 26.2). As $f_\mathrm{u,b}\mathrm{CL_{int,H}}$ exceeds Q_H for drugs that exhibit flow-limited distribution (Pang and Rowland, 1977), AUCR becomes less sensitive toward inhibition of $\mathrm{CL_{int,H}}$ since $\mathrm{CL_{int,H}}$ is large and remains to be large. For example, when $f_\mathrm{u,b}\mathrm{CL_{int,H}}$ is five times greater than Q_H, a 20-fold decrease in $\mathrm{CL_{int,H}}$ translates into less than 5-fold increase in AUC.

26.2.2 Fractional Metabolic Clearance (f_m)

The value of the fractional metabolic clearance (f_m), namely the fraction of total clearance due to metabolism of the substrate, plays a critical role in determining the degree of drug interaction. An exponential increase in the inhibition potential as f_m increases to near unity has been theoretically demonstrated based on the additive clearance concept for drug inhibition (Bjornsson et al., 2003; Rowland and Matin, 1973). Equation (26.2) shows a general expression for the situation when one of the elimination pathways ($\mathrm{CL_1}$) is affected by an interacting drug:

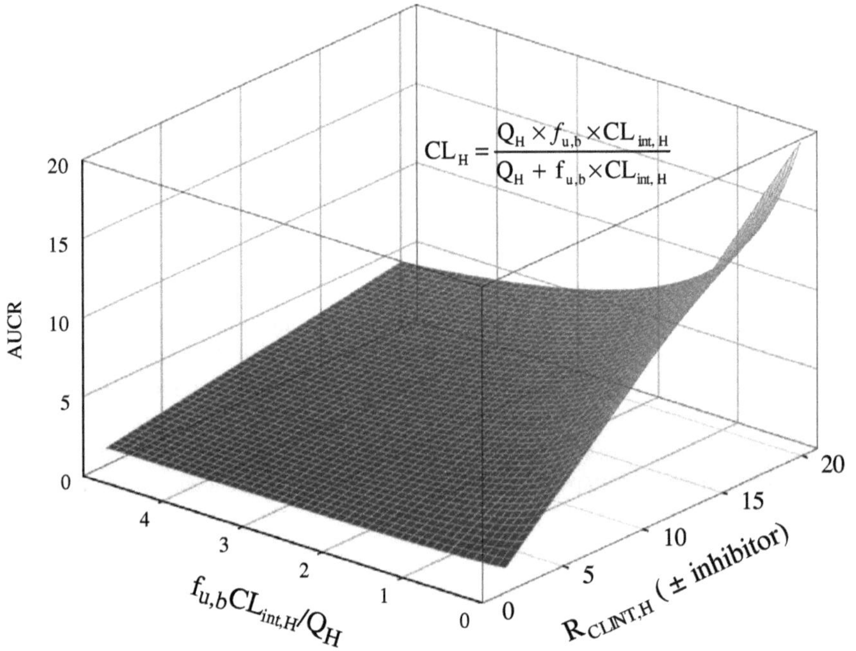

Fig. 26.2 Relationship between AUCR and the ratio of $CL_{int,H}(R_{CLINT,H})$ as a function of the ratio of $f_{u,b}CL_{int,H}/Q_H$. Assuming a well-stirred model for hepatic metabolism, $CL = CL_H$, and substrates are given intravenously

$$AUCR = CL_{-I}/CL_{+I} = \frac{1}{1 - f_{m,CL1} + \frac{f_{m,CL1}}{R_{CL1}}} \qquad (26.2)$$

where $f_{m,CL1}$ is the fraction of clearance that is subject to drug interaction and R_{CL1} is the ratio of this clearance pathway in the absence and presence of an interacting drug ($R_{CL1} > 1$ for inhibition and $R_{CL1} < 1$ for induction, respectively).

Figure 26.3 illustrates the relationship between AUCR and R_{CL1} as a function of $f_{m,CL1}$ for inhibition (upper panel) and induction (lower panel), respectively. The plots assume a maximum of 30-fold inhibition or induction on CL_1. Several features can be visualized from this figure: (1) only when $f_{m,CL1}$ equals 1, AUCR equals R_{CL1}, as depicted by the proportional relationship on the AUCR–R_{CL1} plane ($f_{m,CL1} = 1$); (2) the increase in AUCR by an inhibitor becomes steeper as $f_{m,CL1}$ approaches 1 (upper panel); on the other hand, the decrease in AUCR by an inducer becomes steeper as $f_{m,CL1}$ deviates from zero (lower panel). In addition, the steepness of these changes becomes more pronounced when interaction is stronger (greater R_{CL1} for inhibition and smaller R_{CL1} for induction).

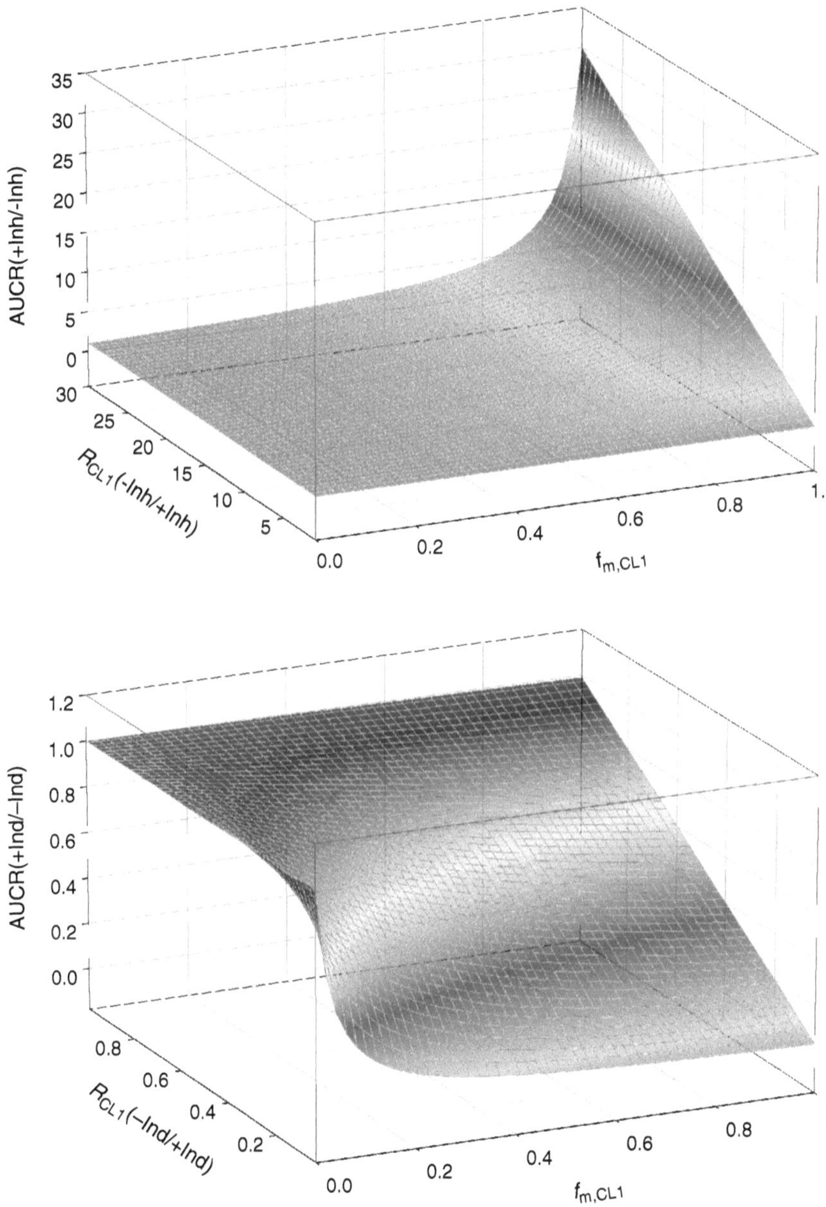

Fig. 26.3 Relationship between AUCR and R_{CL1} as a function of $f_{\mathrm{m,CL1}}$ for drug interactions (upper panel: inhibition; lower panel: induction). Inh: inhibitor; Ind: inducer

26.2.3 Drug Interaction Affecting First Pass Metabolism

Oral administration is the most commonly used route, and drug interaction at the level of first pass metabolism and transport will affect bioavailability (F_{oral}). In estimating F_{oral}, one needs to consider the sequential path of a drug molecule entering into systemic circulation from the absorption site (e.g., gut lumen), namely the fraction absorbed (f_{A}), fraction remained after the gut wall extraction (f_{G}), and fraction remained after the liver extraction (f_{H}):

$$F_{\text{oral}} = f_{\text{A}} f_{\text{G}} f_{\text{H}} \tag{26.3}$$

$$\text{AUCR} = \frac{\text{CL}_{-I}}{\text{CL}_{+I}} \times \frac{[f_{\text{A}} f_{\text{G}} f_{\text{H}}]_{+I}}{[f_{\text{A}} f_{\text{G}} f_{\text{H}}]_{-I}} \tag{26.4}$$

Drug exposure is thus dependent on both F_{oral} and CL (AUC = $F_{\text{oral}}\text{Dose}_{\text{oral}}$/CL). Common drug metabolizing enzymes and transporters such as CYP3A and P-gp are expressed in the small intestine. An approximation of oral clearance with organ intrinsic clearance (i.e., CL/F $\sim f_{\text{u,b}}\text{CL}_{\text{int,H}}$) may be valid for compounds with low intestinal extraction (Rowland and Tozer, 1995; Wilkinson, 1987). However, induction of CYP or P-gp may significantly decrease both f_{G} and f_{H}, causing a larger reduction of substrate AUC than expected if only f_{H} is considered (Rowland and Tozer, 1995; Wilkinson, 1987). In addition for compounds with significant intestinal extraction, inhibition may differentially affect f_{G} and f_{H} (Kirby and Unadkat, 2007; Malhotra et al., 2001). Further discussion of drug interaction at the level of intestinal metabolism and transport can be found in Chapters 5, 17, and 19.

26.2.4 Dynamic Nature of Drug Interaction

The draft DDI guidance recommends the use of I/K_i to determine the need for in vivo drug interaction study for reversible inhibition, where "I" is the mean maximum plasma concentration (C_{max}) value for total inhibitor (bound and unbound) at steady state of the highest clinically used dose. When I/K_i is >0.1 for a specific CYP, an in vivo interaction study with a probe substrate for that CYP in humans is recommended. Depending on the location of the transporters, concentration other than systemic exposure such as intestinal luminal or portal vein concentration may be more relevant to predict transporter-mediated interactions. For example, when evaluating P-gp inhibition by a new molecular entity (NME) following oral administration, inhibitor concentrations at the luminal side of intestine may be more relevant (Zhang et al., 2008). Therefore, in addition to the aforementioned inhibitor plasma concentrations (defined as I_1), I_2 was introduced as the ratio of the highest clinical dose to a volume of 250 mL (approximating intestinal volume). If an NME exhibits $I_1/\text{IC}_{50} > 0.1$ or $I_2/\text{IC}_{50} > 10$, in vivo study should be conducted to determine whether there is clinically relevant P-gp inhibition with digoxin, a P-gp substrate with a narrow therapeutic range. These empirically derived I/K_i criteria are used to determine the need for in vivo drug interaction studies and they are not intended to quantitatively project the extent of interaction. In addition, it has been debated with regard to which "*I*" should be used (e.g., total versus unbound concentration, systemic versus portal vein concentration) and what I/K_i cutoffs should be proposed to minimize "false negatives" and "false positives."

The outcome of an in vivo drug interaction study can be substantially influenced by its study design. The draft DDI guidance provides general guidelines on how to perform drug interaction studies. The key design feature is to maximize drug interactions to reveal the worst-case scenario. For example, pre-dosing of interacting drugs at their highest possible doses and shortest possible dose intervals is recommended. On the other hand, for substrate with a long $t_{1/2}$, multiple dosing of an inhibitor with a short $t_{1/2}$ subsequent to the coadministration with the substrate may be needed. In addition, timing of the administration of an inhibitor relative to that of the substrate is also critical if the substrate is extensively metabolized during absorption (Yang et al., 2003).

Using a strong reversible CYP3A inhibitor, ketoconazole, as an example, we investigated the effect of the dosing regimens of an inhibitor on CYP3A substrates with different pharmacokinetic characteristics (Zhao et al., 2009). Ketoconazole has a relatively short $t_{1/2}$ (3–5 h) and drug accumulation is expected to be minimal. We constructed 16 theoretical substrates whose elimination is predominantly by CYP3A metabolism (i.e., $f_{m,CYP3A} > 0.99$). The substrates differ in CL_{int} and volume of distribution (*V*), resulting in a wide range of $t_{1/2}$ ($t_{1/2} = 0.693$ V/CL) and *F*

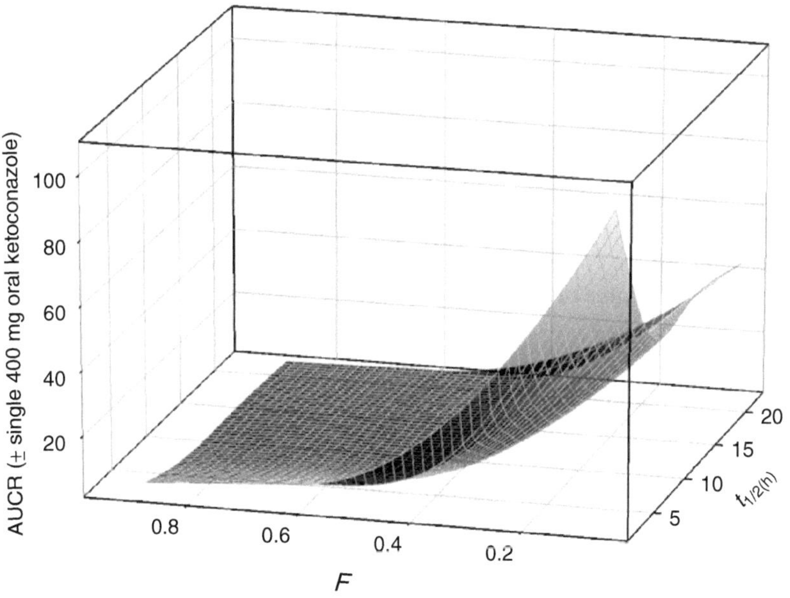

Fig. 26.4 Effect of substrate *F* and $t_{1/2}$ on single oral dose (400 mg) ketoconazole inhibition potential. Substrates are predominantly eliminated by CYP3A metabolism (f_m > 99%, reference Zhao et al. 2009)

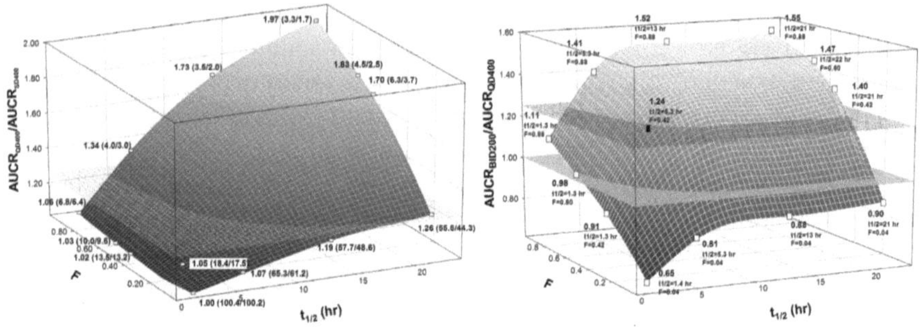

Fig. 26.5 *Left panel*. Effect of $t_{1/2}$ and F of CYP3A substrates on the inhibition potential of KTZ after QD400 mg versus SD 400 mg, expressed as $AUCR_{QD400}/AUCR_{SD400}$ (*left panel*) and here $AUCR = AUC_{KTZ}/AUC_{control}$. Values in parentheses are AUCR after QD and AUCR after SD. The plane represents $AUCR_{QD400}/AUCR_{SD400} = 1.25$. *Right panel*. Effect of $t_{1/2}$ and F of CYP3A substrates on the inhibition potential of KTZ after BID 200 mg versus QD 400 mg, expressed as $AUCR_{BID200}/AUCR_{QD400}$. Two planes represent $AUCR_{BID200}/AUCR_{QD400}$ of 1.25 (*upper*) and 1.00 (*lower*), respectively (Adapted from Zhao et al, 2009)

(assuming complete absorption and $F = f_G f_H$). The simulations performed using a physiologically based pharmacokinetic (PBPK) simulator, SimCYP (SimCYP Ltd, Sheffield, UK), assessed the inhibitory effect under different dosing schemes and utilized a time-based instead of a static inhibitor concentration as a forcing function to result in a dynamic change of CL_{int} of the substrates. Figure 26.4 shows the AUC ratios with and without coadministration of a single 400 mg ketoconazole. Although these test substrates are all predominantly eliminated by CYP3A metabolism ($f_{m,CYP3A} > 0.99$), the degree of inhibition upon single-dose (SD) coadministration of 400 mg ketoconazole ranges from less than 2-fold (for substrate with the longest $t_{1/2}$ and the highest F) to 100-fold (for substrate with the shortest $t_{1/2}$ and lowest F). For substrates with short $t_{1/2}$ and low F, 400 mg ketoconazole SD may result in maximal inhibition in vivo (Fig. 26.5, left panel). However, as the $t_{1/2}$ and F of the substrates increase, multiple dosing of ketoconazole is necessary to achieve maximal inhibition. For example, with 200 mg twice daily (BID) regimen appears to result in higher degree of inhibition than 400 mg once daily (Fig. 26.5, right panel).

26.3 Examples of Different Types of Complex Drug Interactions

The draft DDI guidance recommends the use of in vitro studies and associated decision trees in evaluating in vitro data to guide the need for in vivo drug interaction studies. Several measures to evaluate the drug interaction potential at the level of major CYPs and drug transporters have been implemented during contemporary drug discovery and development. As a result of optimizing absorption, distribution, metabolism, and excretion (ADME) properties to allow for convenient dosing such as once daily oral administration, NMEs may be developed with the following characteristics: relatively low clearance; substrate of multiple elimination pathways;

lack of CYP inhibition liability, etc (Grime et al., 2009; Obach et al., 2004). Such a strategy does not necessarily eliminate concerns for complex drug interactions, especially when multiple pathways are simultaneously affected. In this section, we will review several examples of complex drug interactions.

26.3.1 Two Interacting Drugs Affecting One Substrate

Repaglinide is metabolized by CYP3A4 and CYP2C8. It is also a substrate for an organic anion transporter protein OATP1B1. Gemfibrozil, an inhibitor of both CYP2C8 and OATP1B1, caused an 8.1-fold increase in repaglinide AUC. Itraconazole, a CYP3A4 and P-gp inhibitor, caused a 1.4-fold increase in repaglinide AUC. When these three drugs are coadministered, the AUC of repaglinide increased by 19-fold (Niemi et al., 2003), more than the multiple of individual fold changes. Similar synergistic effect has been observed when itraconazole and gemfibrozil were coadministered with loperamide (Niemi et al., 2006). In this study, the combined use of two interacting drugs caused a 12.6-fold increase in loperamide AUC, whereas itraconazole and gemfibrozil individually caused a 3.8- and a 2.2-fold increase in loperamide AUC, respectively.

These examples clearly represent poly-pharmacy scenarios, accentuating the need for medical professionals to "know ALL of a patient's medications – including over-the-counter products. . . – before prescribing an additional drug" (Huang et al., 2009a).

26.3.2 Drug Interaction in Patients with Organ Impairment

Hepatic and renal impairments constitute two major disease states requiring assessment of their effects on drug exposure and the safe and effective use of drugs cleared by the liver or kidneys. In 1998 and 2003, the FDA published "Guidance for industry: Pharmacokinetics in patients with impaired renal function – study design, data analysis, and impact on dosing and labeling" (http://www.fda.gov/downloads/Drugs/GuidanceComplainceRegulatoryInformation/Guidances/ucm072127.pdf, renal guidance) and "Guidance for industry: Pharmacokinetics in patients with impaired hepatic function: study design, data analysis, and impact on dosing and labeling" (http://www.fda.gov/downloads/Drugs/GuidanceComplianceRegulatoryInformation/Guidances/ucm072123.pdf, hepatic guidance), respectively. These documents describe the need to evaluate the influence of organ impairment on drug pharmacokinetics. More recently, the FDA published a preliminary concept paper updating its recommendation on when to conduct a pharmacokinetic study in subjects with renal impairment (www.fda.gov/ohrms/dockets/ac/08/briefing/2008-4351b1-00-index.htm; Zhang Y et al., 2009; Huang, 2009b).

When drug interaction and organ impairment are both identified to be critical for dose adjustment, their combined effects need to be considered. For example, one may expect some synergistic effect when an enzyme inhibitor is used in patients with renal impairment for drugs that are cleared by both metabolism and excretion. A recent limited observation showed that the use of ketoconazole in two severe renal impaired patients led to much higher AUC changes of telithromycin than those observed in healthy volunteers taking ketoconazole or in renal impaired subjects not taking ketoconazole (Shi et al., 2005). On the other hand, a progressive decrease in the degree of enzyme inhibition was observed in hepatic impaired subjects taking an enzyme inhibitor (Orlando et al., 2006), representing the less-than-additive situation.

The impairment of one eliminating organ may affect the function of the other eliminating organ or cause other physiological changes to alter drug disposition. A recent survey has shown clinically relevant effects of renal impairment on systemic exposure of substrates predominantly cleared via hepatic metabolism (Zhang Y et al., 2009). Orlando et al. (2009) recently discussed pronounced increase in free drug exposure of quinine by erythromycin via inhibition of CYP3A-mediated metabolism and displacement of quinine from its binding to plasma protein. The effect appeared to be further complicated by the decreased protein binding of quinine caused by liver cirrhosis. The pharmacokinetic parameter determining free quinine exposure (CL_u/F) decreased in healthy volunteers who were coadministered erythromycin and in patents with severe cirrhosis (Child–Pugh class C) by approximately 54 and 59%, respectively. When quinine was coadministered with erythromycin in patients with severe cirrhosis, the value of CL_u/F decreased by 73%. The authors suggested therapeutic monitoring on such occasions to prevent adverse events (Orlando et al., 2009).

26.3.3 Inhibition of an Enzyme or Transporter in Poor Metabolizers of Another Pathway

This type of complex interaction has been reviewed by others (Collins et al., 2006; Ito et al., 2005). The major concern is the unexpected, dramatic change in substrate exposure in poor metabolizers (PM) carrying the polymorphic gene of one enzyme and taking an inhibitor of another enzyme (Brynne et al., 1999). Collins and colleagues categorized such patient group as PMI (PM of one enzyme taking an inhibitor of another enzyme). A theoretical approach was proposed to predict the maximum exposure of the substrate in the PMI group. When f_m by polymorphic enzyme exceeds 0.75, PMI group may have the highest risk of experiencing elevated substrate exposure by more than 10-fold compared to exposure in extensive metabolizers (Collins, 2006).

26.3.4 Concurrent Inhibition and Induction

When inhibition and induction take place simultaneously, they tend to cancel the effect of each other. Ritonavir, the well-known potent CYP3A inhibitor, is also used at a subtherapeutic dose as a pharmacokinetic "booster" for other highly-metabolized CYP3A substrates, especially those from the same HIV protease inhibitor class, in order to maintain their systemic exposure (Hsu et al., 1998). Ritonavir also induces CYP enzymes and P-gp. The concurrent inhibition and induction by ritonavir has been proposed to be responsible for the various degree of drug interactions observed for different CYP3A substrates (Greenblatt et al., 1999; Hsu et al., 1998).

Rifampin is a well-known CYP enzyme and P-gp inducer (Combalbert et al., 1989). Recently, it has been demonstrated to be a potent inhibitor of OATP in vivo in humans (van Giersbergen et al., 2007; Zheng et al., 2009). The effect of different dosing regimen of rifampin on the pharmacokinetics of hypoglycemic agent, glyburide, was investigated. Coadministration of single intravenous dose of 600 mg rifampin and 1.25 mg oral dose of glyburide caused significant increase in AUC of glyburide and its pharmacologically active metabolite, along with decreased blood glucose level when compared to a control group. The "acute" effect was attributed to the inhibition of hepatic uptake of glyburide and its metabolite. When glyburide was dosed after multiple oral doses of rifampin (600 mg daily), glyburide AUC was markedly decreased, likely due to enzyme induction by rifampin. When multiple oral doses of rifampin were followed by coadministration with intravenous rifampin and oral glyburide, the inhibition of hepatic uptake transporters by rifampin appeared to cancel the induction effect (Zheng et al., 2009).

26.3.5 Inhibitory Metabolite(s)

Potent inhibitory metabolites formed significantly from parent inhibitors can contribute to the overall degree of drug interaction. For example, the metabolites of itraconazole appeared to have contributed to itraconazole's CYP3A inhibition as the extent of inhibition cannot be predicted solely based on the reversible inhibition K_i of the parent drug (Isoherranen et al., 2004; Templeton et al., 2008). In addition, time-dependent inactivators may form multiple species that are more potent reversible inhibitors or time-dependent inhibitors themselves. For example, the primary metabolite of diltiazem, *N*-desmethyl diltiazem, is a more potent time-dependent inactivator, which may be responsible for the observed apparent inactivation by the parent drug in vitro in both human liver microsomes and hepatocytes (Zhao et al., 2007). Furthermore, the sequential metabolite *N,N*-didesmethyl diltiazem has a reversible inhibition K_i that is much lower than those of its predecessors. However, the impact of each metabolite of diltiazem has not been evaluated in vivo. Another example of significant inhibitory metabolite is the gemfibrozil glucuronide, which appears to be more potent in the inhibition of both OATP and CYP2C8 than its parent drug. The IC_{50} values of gemfibrozil glucuronide toward the uptake and metabolism of cerivastatin by human OATP2 (OATP1B1)-expressing cells and CYP2C8 were 24 and 4 μM, as compared to 72 and 28 μM for gemfibrozil, respectively (Shitara et al., 2004). In addition to being a stronger inhibitor toward CYP2C8, gemfibrozil glucuronide is likely concentrated in the liver and has higher free fraction in plasma than its parent drug. Therefore, Shitara and colleagues (2004) concluded that the main mechanism for gemfibrozil–cerivastatin interaction was the inhibition of CYP2C8 by gemfibrozil glucuronide.

26.4 Assessing Complex Drug Interactions Using Modeling and Simulation

The draft DDI guidance indicates the conditions when there is a need to assess the exposure changes of an NME under multiple inhibitions: (1) the drug exhibits blood concentration-dependent safety concerns, (2) multiple CYP enzymes are responsible for the metabolic clearance of the drug, (3) the predicted residual or non-inhibitable drug clearance is low. Before investigating the impact of multiple CYP inhibitors on drug exposure, it is important to first characterize the individual effects of the CYP inhibitors and to estimate the combined effect of the inhibitors based on computer simulation. For safety concerns, lower doses of the investigational drug may be appropriate for evaluating the fold increase in systematic exposure when combined with multiple inhibitors (draft DDI guidance). These strategies are also applicable to the evaluation of other types of complex drug interactions.

Because not all complex drug interactions can be studied in vivo due to ethical concerns, using modeling and simulation approach to assess the combined effects of complex drug interactions appears to be a feasible alternative. In addition, a well-justified model may provide simulations to help clinical trial design and facilitate labeling recommendations in the safe and effective use of a drug.

26.4.1 Modeling and Simulation: Static Versus Dynamic Approaches

The static approach uses single concentration of the interacting drug to predict drug interaction potential. The concentrations evaluated include systemic plasma as well as portal venous concentrations. Although it is only semiquantitative, the static approach has been used to predict in vivo drug interaction potential. For example, FDA recommended using the threshold ratio (0.1) of total peak plasma inhibitor concentration (I) to the in vitro K_i value to determine the likelihood of an investigational drug as a reversible CYP inhibitor (draft DDI guidance, as discussed in Section 26.2). If complex drug interaction is of concern early in drug development, a static approach can also be used to project the degree of interactions (Collins et al., 2006; Ito et al., 2005; Rostami-Hodjegan and Tucker, 2004).

On the other hand, the dynamic approach considers the pharmacokinetic profiles of the substrate and interacting drugs using appropriate interaction mechanisms as illustrated in Table 26.2. The dynamic approach takes advantage of the sophisticated modeling and simulation tools. For example, the concentrations ("I" in Table 26.2) of a reversible CYP inhibitor changes according to the pharmacokinetics of the inhibitor. At different time points, different CL_{int} ratios can be calculated for the substrate. The simulation becomes more mechanistically meaningful in a physiologically based pharmacokinetic (PBPK) model, which incorporates homeostasis of the enzyme/transporter and uses tissue drug concentration rather than plasma concentration, along with other relevant physiological parameters.

A dynamic approach has several advantages over a static approach. First, the altered systemic exposure as a result of drug interaction can be expressed in a plasma concentration–time profile, in addition to global measures such as AUC and C_{max} (Einolf, 2007; Kanamitsu et al., 2000b; Kanamitsu et al., 2000; Kato et al., 2008; Rostami-Hodjegan and Tucker, 2007; Tsukamoto et al., 2001; Vossen et al., 2007). Second, the impact of different pharmacokinetic characteristics of both the interacting and the substrate drug on the degree of interaction can be assessed using a dynamic approach to guide optimal design of drug interaction studies (Chenel et al., 2008a, b; Chien et al., 2006; Johnson et al., 2009; Ozdemir et al., 2006; Yang et al., 2003; Zhao et al., 2009). Third, different mechanisms of drug interactions can be assessed individually or simultaneously once an appropriate and relevant model is constructed (Chien et al., 2006; Jamei et al., 2009a, b; Johnson et al., 2009; Watanabe et al., 2009).

26.4.2 Assessing Complex Drug Interactions Using Physiologically Based Pharmacokinetic (PBPK) Modeling and Simulation

Comprehensive knowledge of the physicochemical and ADME properties of a drug and/or metabolites and how their pharmacokinetics/pharmacodynamics change in the presence of various intrinsic and extrinsic factors when the drug is given under different dosing regimens is critical to fully understand both single-pair and complex drug interactions. The PBPK models have been used to assess the risk of environmental toxins (Clewell and Andersen, 1996) and more recently to study drug interactions of therapeutic agents in early stage of drug development and in academic research (Chenel et al., 2008a, b; Dickins and van de Waterbeemd 2004; Edginton et al., 2008; Edginton and Willmann, 2008; Jamei et al., 2009b; Rowland et al., 2004; Schmitt and Willmann, 2009; Tsukamoto et al., 2001; Watanabe et al., 2009). Emerging research has applied PBPK in drug discovery and development as a result of expanded knowledge base of human physiology related to drug disposition, enhanced characterization of the physicochemical and ADME properties of drugs, and the availability of user-friendly, highly sophisticated PBPK software tools.

A typical, simplified seven-compartment PBPK model is shown in Fig. 26.6, where each compartment is interconnected by systemic blood flow. In this example, substrate and interacting drugs can be dosed either via oral administration at the site of absorption (intestinal lumen) or via intravenous injection directly into the blood compartment. Drug distribution into different tissue is governed by organ blood flow, passive diffusion, and active transport. The elimination of

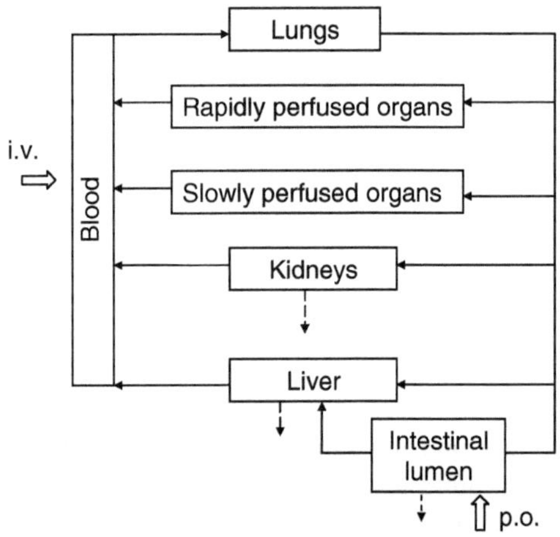

Fig. 26.6 A typical seven-compartmental PBPK model. Irreversible drug elimination (*dashed arrows*) may happen in the small intestine and liver during drug absorption, and liver and kidney after the drug is absorbed

the drugs can occur in intestine, liver, and kidney, as mediated by metabolism and transport. Such a system allows in vitro parameters (e.g., CL_{int} measured from liver microsomes to predict organ $CL_{int,H}$) to be evaluated together with system- or drug-related data (e.g., intrinsic and extrinsic factors) in an integrated, "system biology" manner (Dickins and van de Waterbeemd, 2004; Jamei et al., 2009b; Schmitt and Willmann, 2009). An important feature of such an integrated system is the incorporation of sources of variability in various parameters (Bouzom and Walther, 2008; Dickinson et al., 2007a, b; Jamei et al., 2009a, b; Kato et al., 2008). Consequently, different virtual populations in response to the system perturbation including drug interactions can be evaluated under a more realistic environment.

An in-depth review of the utility of PBPK modeling, the incorporation of parameter uncertainty and variability, the use of Bayesian population analysis, and the available tools oriented toward being user-friendly while preserving complexity and flexibility of PBPK modeling, is beyond the scope of this chapter. The readers are directed to several review and research articles (Clewell and Andersen, 1996; Covington et al., 2007; Dickins and van de Waterbeemd, 2004; Edginton et al., 2008; Hack et al., 2006; Jamei et al., 2009a, b; Rostami-Hodjegan and Tucker, 2007; Rowland et al., 2004; Schmitt and Willmann, 2009) as well as other chapters of this book (e.g., Chapters 12 and 13).

26.4.3 Applications of PBPK Approach in Studying Drug Interactions

A relatively simple PBPK model consisting of liver, portal vein, and systemic blood compartments was used to predict in vivo drug interaction between erythromycin and triazolam using in vitro data (Kanamitsu et al., 2000a). Inhibition of CYP3A by erythromycin was assumed to be time dependent. Sensitivity analyses were performed by varying the parameters with greater uncertainty such as K_{deg} of CYP3A4 (Table 26.2) and liver to blood concentration ratio of the inhibitor.

More recently, sophisticated PBPK models were applied to evaluate single-pair drug interactions and the impact of genetic polymorphism on drug exposure and response by incorporating interindividual variability of system- and drug-related parameters (Dickinson et al., 2007a, b; Hyland et al., 2008). Drug interaction prediction for maraviroc with a PBPK approach was recently reported (Hyland, et al., 2008). Maraviroc, a C-Chemokine receptor 5 (CCR5) inhibitor, is eliminated predominantly by CYP3A4-mediated metabolism. In a drug interaction prediction study, Hyland and colleagues compared the observed degree of drug interactions of maraviroc with ketoconazole and protease inhibitors in vivo with the predicted degree of drug interactions using a PBPK-based approach (Hyland et al., 2008). The predicted oral clearance of maraviroc using parameters from in vitro metabolism generally agreed with those in the placebo arm of the in vivo interaction studies. The CL_{int} values derived from recombinant CYP3A4 incubation and human liver microsomes appeared to be similar (35 μL/min/mg versus 40.8 μL/min/mg from recombinant CYP3A and liver microsomes, respectively). The authors also used the human mass balance data to obtain absorption rate constant (K_a) and renal clearance as input for maraviroc pharmacokinetic predictions. The predicted degrees of drug interaction (e.g., C_{max} and AUC changes) were generally in agreement with the clinical data.

The application of PBPK modeling and simulation to evaluate the impact of CYP polymorphism on the extent of drug interaction includes studies by Dickinson et al (Dickinson et al., 2007a, b). By including relative CL_{int} values of different genotypes of a polymorphic CYP (e.g., dextromethorphan data for various CYP2D6 genotypes and warfarin data for various CYP2C9 genotypes) and the population frequencies of various genotypes into a comprehensive PBPK model, the authors simulated in vivo CL as a function of genotype with relative accuracies when compared with clinical studies. With the established models, the authors also investigated the number of subjects required to detect specific AUC differences between different CYP genotypes. In addition to demonstration of the utility of PBPK simulation in optimizing study design, the authors also explored the possibility of combining preliminary a priori pharmacodynamic information to systematically evaluate the impact of different factors on drug response (Dickinson et al., 2007a, b).

Vossen and colleagues (2007) investigated CYP3A inhibition by itraconazole and its hydroxy metabolites using the PK-Sim® software. A midazolam base model was constructed using pharmacokinetic profiles of the parent drug, 1′-hydroxymidazolam (formed by CYP3A) and 1′-hydroxymidazolam glucuronide in male subjects whose CYP3A5 genotypes were determined. The metabolite formation kinetics of 1′-hydroxymidazolam (by CYP3A) was simulated to match the pharmacokinetic profiles of the observed clinical data. CYP3A activities after itraconazole dosing at any time point were predicted to be the product of the initial enzyme activity, the fraction of initial activity based on itraconazole concentration, and the fraction of initial activity based on hydroxyl-itraconazole concentration. The simulation appears to adequately predict the change in the pharmacokinetic profiles of midazolam and its metabolites upon inhibition by coadministration of itraconazole. In addition, the greater degree of overprediction in individuals with at least one CYP3A5*1 allele partially confirmed the specific inhibition of CYP3A4 rather than CYP3A5 by itraconazole based on in vitro data.

A PBPK system in human was constructed to describe pravastatin disposition mediated by hepatic uptake transporter (OATP1B1), passive diffusion between liver and blood, and biliary efflux transporter (MRP2) (Watanabe et al., 2009). The simulation indicated that changes in the OATP1B1 and MRP2 activities had larger impact on side effect and efficacy of pravastatin, respectively. The model appears to be applicable in the situation of complex drug interaction leading to the alteration of drug exposure by evaluating the interplay among sinusoidal transporters, biliary efflux transporter, and liver metabolism.

26.5 Challenges in Evaluating Complex Drug Interactions

Several challenges exist when studying both single-pair and complex drug interactions. Inconsistent in vitro parameters generated by different laboratories have been a major challenge faced by many ADME scientists (Yang et al., 2007; 2008). If a substrate is predominantly metabolized by one CYP isoenzyme, it may be sufficient to conclude the involvement of the CYP by a drug interaction study using a "specific" inhibitor. However, as science advances in the field of drug metabolism and transport, some historically deemed "specific" inhibitors or inducers are found to affect additional enzymes and transporters (Jones et al., 2008; Lu et al., 2007). For example, ketoconazole, an inhibitor used by many and recommended by the FDA guidance as a strong CYP3A inhibitor, has been shown to inhibit CYPs other than CYP3A, including CYP2C8, CYP2C9, CYP2C19 (University of Washington Metabolism and Transporter Drug Interaction Database, www.druginteractioninfo.org, accessed Dec 2008), and more recently CYP2J2 (Jones et al., 2008). When multiple pathways are involved in the metabolism or transport of the substrate, it becomes difficult to confirm a major clearance pathway when the clinical study used a "nonspecific" inhibitor that affects multiple pathways.

As discussed in 26.3.2, extents of drug interaction can be affected by concurrent dysfunction of drug eliminating organs, the liver and kidneys, in patients. For example, renal impairment can affect the metabolizing enzyme or transporter activities in the liver, kidneys, and small intestine. Similarly, besides its effect on CYP expression in the liver, hepatic impairment may also lead to physiological changes such as levels of plasma binding protein and decreased CYP expression in small intestine (McConn 2nd, et al., 2009; Orlando et al., 2009; Zhang Y et al., 2009). In a recent study by Edginton and Willmann, the authors constructed a PBPK model to systematically evaluate the pharmacokinetic changes of several drugs in patients with liver cirrhosis (Edginton and Willmann, 2008). The model incorporated hepatic as well as extrahepatic changes that may influence drug pharmacokinetics, including CYP activities, decreased kidney blood flow and glomerular filtration rate, and decreased levels of plasma binding proteins in cirrhotic patients. Therefore, when drug interaction in patients with organ impairment is investigated, multiple effects and other physiological changes need to be considered.

Although PBPK modeling and simulation appear promising in evaluating complex drug interactions, the approach has limitations. Often, a priori information for many parameters and/or their distributions in different populations is lacking and many assumptions are made when these values are used in the model. For example, volume of distribution for the glucuronide of 1′-hydroxymidazolam was estimated using the mean logarithmic difference of the lipophilicity of other compounds and their corresponding glucuronides (Vossen et al., 2007). In Watanabe's study on pravastatin, the initial *V*, scaling factor for the in vitro–in vivo correlation, and the liver free fraction of the drug were assumed to be the same across species (Watanabe et al., 2009). In addition, although methods to predict partition coefficient between organ and plasma using physicochemical properties of the drugs have been developed (Rodgers et al., 2005), these parameters require continuing refinement as knowledge on active processes (uptake and efflux) becomes available, which may significantly influence the estimated partitioning parameters (Edginton et al., 2008). Failure to include possible transporter processes in the model or inaccurate estimation of corresponding parameters may oversimplify the clearance process for both the substrate and the interacting drug. As stated by Rowland and colleagues, the tools of PBPK modeling and simulation were "diverse in both quality and scope of application" (Rowland et al., 2004). Standardization and the development of guidelines on good practice of PBPK approach may be needed.

26.6 Scientific Perspectives on Studying Complex Drug Interactions

Incorporation of in vivo human data when it becomes available is important when evaluating complex drug interactions. The K_a value of maraviroc collected from human mass balance studies provided more predictable drug interaction than K_a generated from the physicochemical property and in vitro permeability data using an in silico approach (Hyland et al., 2008). Vossen and colleagues used midazolam clearance obtained from in vivo human studies instead of in vitro CL_{int} to simulate CYP3A inhibition by itraconazole (Vossen et al., 2007). A similar approach is available in other PBPK simulation softwares to "scale down" in vivo data to obtain CL_{int} at enzyme level. One has to note that such approach utilizes the same scaling factor generally used in in vitro–in vivo extrapolation approaches (e.g., mg microsomal protein per gram liver, gram liver per kilogram bodyweight) to obtain a CL_{int} value, which can be different from that determined in vitro. However, utilizing parameters "extrapolated" from in vivo data seemingly added confidence when interpreting pharmacokinetics and drug interactions. The use of in vivo information in addition to in vitro data during the evaluation of drug interaction using PBPK model helps to fill the knowledge gap of the model and provides reality check of the in silico approach.

The construction of a comprehensive PBPK model to evaluate complex drug interactions relies on a few key parameters. Table 26.3 summarizes studies from which these critical parameters can be generated. For example, characterization of the f_m values for an investigational drug is perhaps the most important prerequisite for assessing its drug interaction potential, single pair or complex. Yet it is likely the most difficult datapoint to obtain with confidence (Also see Chapter 13). Generally, in vitro CYP phenotyping data and in vivo human mass balance data using radiolabeled material constitute key components for characterizing f_m. However, if a mass balance study was conducted using nonparental route of administration, the independent estimation of its absolute bioavailability (e.g., from Phase 1 pharmacokinetic studies) becomes important. To understand the impact of multiple drug interactions, dedicated studies such as single-pair drug interaction, pharmacokinetics/pharmacodynamics in organ impairment, and the study in PM subjects of particular enzymes are informative to characterize f_m. It is important to comprehensively evaluate data from these studies.

Table 26.3 Important studies for comprehensive evaluation of complex drug interaction

Type of study	Parameters evaluated
In vitro ADME and interaction	Enzyme/transporter involved in elimination and interaction Interaction mechanisms and parameters (e.g., K_i) Initial f_m estimation
Phase 1 dose escalation (oral administration)	CL/F_{oral} V/F_{oral} Likely CL_R and metabolite data
Absolute oral bioavailability	CL V/F_{oral} Likely CL_R and metabolite data
In vivo mass balance (using radiolabeled material)	Confirm f_m Confirm f_a

When the metabolite of an inhibitor drug is also an inhibitor for the same enzyme (e.g., itraconazole, gemfibrozil), or the substrate drug has an active metabolite (e.g., irinotecan, clopidogrel), characterization of the metabolite kinetics (systemic and presystemic formation and elimination) is critical (Pang et al., 2008; Sun and Pang, 2009). In both instances, the pharmacokinetics of metabolites needs to be characterized to quantitatively evaluate the degree of drug interactions (see Chapter 5). In many studies, the pharmacokinetic information of the metabolites was collected from studies of parent drug. However, the involvement of possible presystemic formation may make it difficult to elucidate the metabolite kinetics. It is therefore helpful when additional dedicated studies including administration of the metabolite are conducted.

When multiple factors take place simultaneously, the assumption that these factors are independent allows one to assess the contribution of individual mechanisms. However, such assumption may not be valid. Possible theoretical basis on multiple inhibition has been proposed and explored (Almond et al., 2009; Ito et al.,

1998; Jamei et al., 2009a; Rostami-Hodjegan and Tucker, 2004; Zhang X et al., 2009). Future studies are needed to provide mathematical elucidation of multiple interactions.

26.7 Conclusion

Unlike single-pair drug interaction, the effect of multiple drug interactions is difficult to design and evaluate properly in vivo. The use of modeling and simulation, especially with a PBPK approach, appears promising to understand complex drug interactions. Future studies need to focus on better defining key parameters required to quantitatively evaluate multiple factors and mechanistic understanding of the combined effects. Additional research will provide confidence in the use of modeling and simulation to guide clinical study design and generate data for the informative labeling and effective use of medications.

References

Almond LM, Yang J, Jamei M, Tucker GT and Rostami-Hodjegan A (2009) Towards a quantitative framework for the prediction of DDIs arising from cytochrome P450 induction. *Curr Drug Metab* **10**:420–432.

Bjornsson TD, Callaghan JT, Einolf HJ, Fischer V, Gan L, Grimm S, Kao J, King SP, Miwa G, Ni L, Kumar G, McLeod J, Obach RS, Roberts S, Roe A, Shah A, Snikeris F, Sullivan JT, Tweedie D, Vega JM, Walsh J and Wrighton SA (2003) The conduct of in vitro and in vivo drug-drug interaction studies: a Pharmaceutical Research and Manufacturers of America (PhRMA) perspective. *Drug Metab Dispos* **31**:815–832.

Bouzom F and Walther B (2008) Pharmacokinetic predictions in children by using the physiologically based pharmacokinetic modelling. *Funda Clin Pharmacol* **22**:579–587.

Brynne N, Forslund C, Hallen B, Gustafsson LL and Bertilsson L (1999) Ketoconazole inhibits the metabolism of tolterodine in subjects with deficient CYP2D6 activity. *Br J Clin Pharmacol* **48**:564–572.

Chenel M, Bouzom F, Aarons L and Ogungbenro K (2008a) Drug-drug interaction predictions with PBPK models and optimal multiresponse sampling time designs: application to midazolam and a phase I compound. Part 1: comparison of uniresponse and multiresponse designs using PopDes. *J Pharmacokinet Pharmacodyn* **35**:635–659.

Chenel M, Bouzom F, Cazade F, Ogungbenro K, Aarons L and Mentre F (2008b) Drug-drug interaction predictions with PBPK models and optimal multiresponse sampling time designs: application to midazolam and a phase I compound. Part 2: clinical trial results. *J Pharmacokinet Pharmacodyn* **35**:661–681.

Chien JY, Lucksiri A, Ernest CS 2nd, Gorski JC, Wrighton SA and Hall SD (2006) Stochastic prediction of CYP3A-mediated inhibition of midazolam clearance by ketoconazole. *Drug Metab Dispos* **34**:1208–1219.

Clewell H and Andersen M (1996) Use of physiologically based pharmacokinetic modeling to investigate individual versus population risk. *Toxicology* **111**:315–329.

Collins C, Levy R, Ragueneau-Majlessi I and Hachad H (2006) Prediction of maximum exposure in poor metabolizers following inhibition of nonpolymorphic pathways. *Curr Drug Metab* **7**:295–299.

Combalbert J, Fabre I, Fabre G, Dalet I, Derancourt J, Cano JP and Maurel P (1989) Metabolism of cyclosporin A. IV. Purification and identification of the rifampicin-inducible human liver cytochrome P-450 (cyclosporin A oxidase) as a product of P450IIIA gene subfamily. *Drug Metab Dispos* **17**:197–207.

Covington T, Gentry P, Van Landingham C, Andersen M, Kester J and Clewell H (2007) The use of Markov chain Monte Carlo uncertainty analysis to support a Public Health Goal for perchloroethylene. *Regul Toxicol Pharmacol* **47**:1–18.

Dickins M and van de Waterbeemd H (2004) Simulation models for drug disposition and drug interactions. *Drug Discov Today: Biosilico* **2**:38–45.

Dickinson GL, Lennard MS, Tucker GT and Rostami-Hodjegan A (2007a) The use of mechanistic DM-PK-PD modelling to assess the power of pharmacogenetic studies -CYP2C9 and warfarin as an example. *Br J Clin Pharmacol* **64**:14–26.

Dickinson GL, Rezaee S, Proctor NJ, Lennard MS, Tucker GT and Rostami-Hodjegan A (2007b) Incorporating in vitro information on drug metabolism into clinical trial simulations to assess the effect of CYP2D6 polymorphism on pharmacokinetics and pharmacodynamics: dextromethorphan as a model application. *J Clin Pharmacol* **47**:175–186.

Edginton A, Theil F, Schmitt W and Willmann S (2008) Whole body physiologically-based pharmacokinetic models: their use in clinical drug development. *Expert Opin Drug Metab Toxicol* **4**:1143–1152.

Edginton A and Willmann S (2008) Physiology-based simulations of a pathological condition prediction of pharmacokinetics in patients with liver cirrhosis. *Clinical Pharmacokinet* **47**:743–752.

Einolf HJ (2007) Comparison of different approaches to predict metabolic drug-drug interactions. *Xenobiotica* **37**:1257–1294.

Farid NA, Payne CD, Small DS, Winters KJ, Ernest CS 2nd, Brandt JT, Darstein C, Jakubowski JA and Salazar DE (2007) Cytochrome P450 3A inhibition by ketoconazole affects prasugrel and clopidogrel pharmacokinetics and pharmacodynamics differently. *Clin Pharmacol Ther* **81**:735–741.

Greenblatt DJ, von Moltke LL, Daily JP, Harmatz JS and Shader RI (1999) Extensive impairment of triazolam and alprazolam clearance by short-term low-dose ritonavir: the clinical dilemma of concurrent inhibition and induction. *J Clin Psychopharmacol* **19**:293–296.

Grime KH, Bird J, Ferguson D and Riley RJ (2009) Mechanism-based inhibition of cytochrome P450 enzymes: an evaluation of early decision making in vitro approaches and drug-drug interaction prediction methods. *Eur J Pharm Sci* **36**:175–191.

Hack C, Chiu W, Zhao Q and Clewell H (2006) Bayesian population analysis of a harmonized physiologically based pharmacokinetic model of trichloroethylene and its metabolites. *Regul Toxicol Pharmacol* **46**:63–83.

Hsu A, Granneman GR and Bertz RJ (1998) Ritonavir. Clinical pharmacokinetics and interactions with other anti-HIV agents. *Clin Pharmacokinet* **35**:275–291.

Huang, S-M, Temple R, Xiao S, Zhang L, Lesko, LJ (2009) When to conduct a renal impairment study during drug development- US Food and Drug Administration perspective, *Clin Pharmacol Ther* In press.

Huang SM, Lesko LJ and Temple R (2009a) Adverse drug reactions and pharmacokinetic drug interactions, in *Fundamental principles: Clinical pharmacology. "Pharmacology and therapeutics: Principles to practice"*, Elsevier, Amsterdam.

Huang SM and Temple R (2008) Is this the drug or dose for you? Impact and consideration of ethnic factors in global drug development, regulatory review, and clinical practice. *Clin Pharmacol Ther* **84**:287–294.

Huang SM, Temple R, Throckmorton DC and Lesko LJ (2007) Drug interaction studies: study design, data analysis, and implications for dosing and labeling. *Clin Pharmacol Ther* **81**: 298–304.

Hyland R, Dickins M, Collins C, Jones H and Jones B (2008) Maraviroc: in vitro assessment of drug-drug interaction potential. *Brit J Clin Pharmacol* **66**:498–507.

Isoherranen N, Kunze KL, Allen KE, Nelson WL and Thummel KE (2004) Role of itraconazole metabolites in CYP3A4 inhibition. *Drug Metab Dispos* **32**:1121–1131.

Ito K, Hallifax D, Obach RS and Houston JB (2005) Impact of parallel pathways of drug elimination and multiple cytochrome P450 involvement on drug-drug interactions: CYP2D6 paradigm. *Drug Metab Dispos* **33**:837–844.

Ito K, Iwatsubo T, Kanamitsu S, Ueda K, Suzuki H and Sugiyama Y (1998) Prediction of pharmacokinetic alterations caused by drug-drug interactions: metabolic interaction in the liver. *Pharmacol Rev* **50**:387–412.

Jamei M, Dickinson GL and Rostami-Hodjegan A (2009a) A framework for assessing inter-individual variability in pharmacokinetics using virtual human populations and integrating general knowledge of physical chemistry, biology, anatomy, physiology and genetics: A tale of 'bottom-up' vs 'top-down' recognition of covariates. *Drug Metab Pharmacokinet* **24**:53–75.

Jamei M, Marciniak S, Feng K, Barnett A, Tucker G and Rostami-Hodjegan A (2009b) The Simcyp((R)) population-based ADME simulator. *Expert Opin Drug Metab Toxicol* **3**:51–66.

Johnson TN, Kerbusch T, Jones B, Tucker GT, Rostami-Hodjegan A and Milligan PA (2009) Assessing the efficiency of mixed effects modelling in quantifying metabolism based drug-drug interactions: using in vitro data as an aid to assess study power. *Pharm Stat* Mar 16. [Epub ahead of print].

Jones JP, Katayama JH, Jiang Y, Lee, CA, and Totah RA (2008) Identification of danazol as a selective inhibitor of cytochrome P450 2J2. *Drug Metabo Rev.* **40** (Suppl 3):78.

Kanamitsu S, Ito K, Green CE, Tyson CA, Shimada N and Sugiyama Y (2000a) Prediction of in vivo interaction between triazolam and erythromycin based on in vitro studies using human liver microsomes and recombinant human CYP3A4. *Pharm Res* **17**:419–426.

Kanamitsu S, Ito K and Sugiyama Y (2000b) Quantitative prediction of in vivo drug-drug interactions from in vitro data based on physiological pharmacokinetics: use of maximum unbound concentration of inhibitor at the inlet to the liver. *Pharm Res* **17**:336–343.

Kanamitsu SI, Ito K, Okuda H, Ogura K, Watabe T, Muro K and Sugiyama Y (2000) Prediction of in vivo drug-drug interactions based on mechanism-based inhibition from in vitro data: inhibition of 5-fluorouracil metabolism by (E)-5-(2-Bromovinyl)uracil. *Drug Metab Dispos* **28**:467–474.

Kato M, Shitara Y, Sato H, Yoshisue K, Hirano M, Ikeda T and Sugiyama Y (2008) The quantitative prediction of CYP-mediated drug interaction by physiologically based pharmacokinetic modeling. *Pharm Res* **25**:1891–1901.

Kim KA, Park PW, Hong SJ and Park JY (2008a) The effect of CYP2C19 polymorphism on the pharmacokinetics and pharmacodynamics of clopidogrel: a possible mechanism for clopidogrel resistance. *Clin Pharmacol Ther* **84**:236–242.

Kim KA, Park PW and Park JY (2008b) Effect of CYP3A5*3 genotype on the pharmacokinetics and antiplatelet effect of clopidogrel in healthy subjects. *Eur J Clin Pharmacol* **64**:589–597.

Kirby BJ and Unadkat JD (2007) Grapefruit juice, a glass full of drug interactions?. *Clin Pharmacol Ther* **81**:631–633.

Kitz R and Wilson IB (1962) Esters of methanesulfonic acid as irreversible inhibitors of acetylcholinesterase. *J Biol Chem* **237**:3245–3249.

Kurihara A, Hagihara K, Kazui M, Ozeki T, Farid NA and Ikeda T (2005) In vitro metabolism of antiplatelet agent clopidogrel: cytochrome P450 isoforms responsible for two oxidation steps involved in the active metabolite formation. *Drug Metab Rev* **37**(Suppl 2):99.

Lu C, Miwa G, Prakash S, Gan L and Balani S (2007) A novel model for the prediction of drug-drug interactions in humans based on in vitro cytochrome P450 phenotypic data. *Drug Metab Disposition* **35**:79–85.

Malhotra S, Bailey DG, Paine MF and Watkins PB (2001) Seville orange juice-felodipine interaction: comparison with dilute grapefruit juice and involvement of furocoumarins. *Clin Pharmacol Ther* **69**:14–23.

McConn DJ 2nd, Lin YS, Mathisen TL, Blough DK, Xu Y, Hashizume T, Taylor SL, Thummel KE and Shuhart MC (2009) Reduced duodenal cytochrome P450 3A protein expression and catalytic activity in patients with cirrhosis. *Clin Pharmacol Ther* **85**:387–393.

Mega JL, Close SL, Wiviott SD, Shen L, Hockett RD, Brandt JT, Walker JR, Antman EM, Macias W, Braunwald E and Sabatine MS (2009) Cytochrome p-450 polymorphisms and response to clopidogrel. *N Engl J Med* **360**:354–362.

Niemi M, Backman JT, Neuvonen M and Neuvonen PJ (2003) Effects of gemfibrozil, itraconazole, and their combination on the pharmacokinetics and pharmacodynamics of repaglinide: potentially hazardous interaction between gemfibrozil and repaglinide. *Diabetologia* **46**:347–351.

Niemi M, Tornio A, Pasanen MK, Fredrikson H, Neuvonen PJ and Backman JT (2006) Itraconazole, gemfibrozil and their combination markedly raise the plasma concentrations of loperamide. *Eur J Clin Pharmacol* **62**:463–472.

Obach RS, Huynh P, Allen MC and Beedham C (2004) Human liver aldehyde oxidase: inhibition by 239 drugs. *J Clin Pharmacol* **44**:7–19.

Orlando R, De Martin S, Pegoraro P, Quintieri L and Palatini P (2009) Irreversible CYP3A inhibition accompanied by plasma protein-binding displacement: a comparative analysis in subjects with normal and impaired liver function. *Clin Pharmacol Ther* **85**:319–326.

Orlando R, Padrini R, Perazzi M, De Martin S, Piccoli P and Palatini P (2006) Liver dysfunction markedly decreases the inhibition of cytochrome P450 1A2-mediated theophylline metabolism by fluvoxamine. *Clin Pharmacol Ther* **79**:489–499.

Ozdemir M, Crewe HK, Tucker GT, and Rostami-Hodjegan A (2006) The impact of ketoconazole (KTZ) dosage regimen on midazolam clearance and its prediction using PBPK modelling. Ref Type: Conference Proceeding

Pang KS and Rowland M (1977) Hepatic clearance of drugs. I. Theoretical considerations of a "well-stirred" model and a "parallel tube" model. Influence of hepatic blood flow, plasma and blood cell binding, and the hepatocellular enzymatic activity on hepatic drug clearance. *J Pharmacokinet Biopharm* **5**:625–653.

Pang KS, Morris ME and Sun H (2008) Formed and preformed metabolites: facts and comparisons. *J Pharm Pharmacol* **60**:1247–1275.

Qato DM, Alexander GC, Conti RM, Johnson M, Schumm P and Lindau ST (2008) Use of prescription and over-the-counter medications and dietary supplements among older adults in the United States. *J Am Med Assoc* **300**:2867–2878.

Roberts BJ, Song BJ, Soh Y, Park SS and Shoaf SE (1995) Ethanol induces CYP2E1 by protein stabilization. Role of ubiquitin conjugation in the rapid degradation of CYP2E1. *J Biol Chem* **270**:29632–29635.

Rodgers T, Leahy D and Rowland M (2005) Physiologically based pharmacokinetic modeling 1: Predicting the tissue distribution of moderate-to-strong bases. *J Pharma Sci* **94**:1259–1276.

Rostami-Hodjegan A and Tucker GT (2004) *'In silico'* simulations to assess the *'in vivo'* consequences of *'in vitro'* metabolic drug–drug interactions. *Drug Discov Today: Technol* **1**:441–448.

Rostami-Hodjegan A and Tucker GT (2007) Simulation and prediction of in vivo drug metabolism in human populations from in vitro data. *Nat Rev Drug Discov* **6**:140–148.

Rowland M, Balant L and Peck C (2004) Physiologically based pharmacokinetics in drug development and regulatory science: a workshop report (Georgetown University, Washington, DC, May 29-30, 2002). *AAPS PharmSci* **6**:E6.

Rowland M and Matin SB (1973) Kinetics of drug-drug interactions. *J Pharmacokinet Biopharm* **1**:553–567.

Rowland M and Tozer TN (1995) *Clinical pharmacokinetics: concepts and applications*. Williams and Wilkins, New York.

Schmitt W and Willmann S (2009) Physiology-based pharmacokinetic modeling: ready to be used. *Drug Discov Today: Technol* 1:449–456.

Segel IH (1975) *Enzyme kinetics: behavior and analysis of rapid equilibrium and steady-state enzyme systems*. John Wiley and Sons, Inc, New York.

Shi J, Chapel S, Montay G, Hardy P, Barrett J, Sica D, Swan S, Noveck R, Leroy B and Bhargava V (2005) Effect of ketoconazole on the pharmacokinetics and safety of telithromycin and clarithromycin in older subjects with renal irnpairment. *Int J Clin Pharmacol Therap* **43**:123–133.

Shitara Y, Hirano M, Sato H and Sugiyama Y (2004) Gemfibrozil and its glucuronide inhibit the organic anion transporting polypeptide 2 (OATP2/OATP1B1:SLC21A6)-mediated hepatic uptake and CYP2C8-mediated metabolism of cerivastatin: analysis of the mechanism of the clinically relevant drug-drug interaction between cerivastatin and gemfibrozil. *J Pharmacol Exp Ther* **311**:228–236.

Shou M, Hayashi M, Pan Y, Xu Y, Morrissey K, Xu L and Skiles GL (2008) Modeling, prediction, and in vitro in vivo correlation of CYP3A4 induction. *Drug Metab Dispos* **36**:2355–2370.

Sun H and Pang KS (2009) Disparity in intestine disposition between formed and preformed metabolites and implications: a theoretical study. *Drug Metab Dispos* **37**:187–202.

Taubert D, von Beckerath N, Grimberg G, Lazar A, Jung N, Goeser T, Kastrati A, Schomig A and Schomig E (2006) Impact of P-glycoprotein on clopidogrel absorption. *Clin Pharmacol Ther* **80**:486–501.

Templeton IE, Thummel KE, Kharasch ED, Kunze KL, Hoffer C, Nelson WL and Isoherranen N (2008) Contribution of itraconazole metabolites to inhibition of CYP3A4 in vivo. *Clin Pharmacol Ther* **83**:77–85.

Thummel KE, Chung S, Nallani S, Reynolds KS, Strong JM, Yasuda S, Zhang L and Huang SM (2007) When is a multiple-inhibitor study necessary?. *Clin Pharmacol Ther* **81**(Suppl 1): PII–77.

Tsukamoto Y, Kato Y, Ura M, Horii I, Ishitsuka H, Kusuhara H and Sugiyama Y (2001) A physiologically based pharmacokinetic analysis of capecitabine, a triple prodrug of 5-FU, in humans: the mechanism for tumor-selective accumulation of 5-FU. *Pharm Res* **18**:1190–1202.

van Giersbergen PL, Treiber A, Schneiter R, Dietrich H and Dingemanse J (2007) Inhibitory and inductive effects of rifampin on the pharmacokinetics of bosentan in healthy subjects. *Clin Pharmacol Ther* **81**:414–419.

Vossen M, Sevestre M, Niederalt C, Jang IJ, Willmann S and Edginton AN (2007) Dynamically simulating the interaction of midazolam and the CYP3A4 inhibitor itraconazole using individual coupled whole-body physiologically-based pharmacokinetic (WB-PBPK) models. *Theor Biol Med Model* **4**:13.

Watanabe T, Kusuhara H, Maeda K, Shitara Y and Sugiyama Y (2009) Physiologically based pharmacokinetic modeling to predict transporter-mediated clearance and distribution of pravastatin in humans. *J Pharmacol Exp Ther* **328**:652–662.

Wilkinson GR (1987) Clearance approaches in pharmacology. *Pharmacol Rev* **39**:1–47.

Yang J, Jamei M, Yeo KR, Tucker GT and Rostami-Hodjegan A (2007) Theoretical assessment of a new experimental protocol for determining kinetic values describing mechanism (time)-based enzyme inhibition. *Eur J Pharm Sci* **31**:232–241.

Yang J, Kjellsson M, Rostami-Hodjegan A and Tucker GT (2003) The effects of dose staggering on metabolic drug-drug interactions. *Eur J Pharm Sci* **20**:223–232.

Yang J, Liao M, Shou M, Jamei M, Yeo KR, Tucker GT and Rostami-Hodjegan A (2008) Cytochrome p450 turnover: regulation of synthesis and degradation, methods for determining rates, and implications for the prediction of drug interactions. *Curr Drug Metab* **9**:384–394.

Zhang L, Zhang YD, Strong JM, Reynolds KS and Huang SM (2008) A regulatory viewpoint on transporter-based drug interactions. *Xenobiotica* **38**:709–724.

Zhang X, Jones DR and Hall SD (2009) Prediction of the effect of erythromycin, diltiazem, and their metabolites, alone and in combination, on CYP3A4 inhibition. *Drug Metab Dispos* **37**:150–160.

Zhang Y, Zhang L, Abraham S, Apparaju S, Wu TC, Strong JM, Xiao S, Atkinson AJ, Jr., Thummel KE, Leeder JS, Lee C, Burckart GJ, Lesko LJ and Huang SM (2009) Assessment of the impact of renal impairment on systemic exposure of new molecular entities: evaluation of recent new drug applications. *Clin Pharmacol Ther* **85**:305–311.

Zhao P, Lee CA and Kunze KL (2007) Sequential metabolism is responsible for diltiazem-induced time-dependent loss of CYP3A. *Drug Metab Dispos* **35**:704–712.

Zhao P, Ragueneau-Majlessi I, Zhang L, Strong JM, Reynolds KS, Levy RH, Thummel KE and Huang SM (2009) Quantitative evaluation of pharmacokinetic inhibition of CYP3A substrates by ketoconazole: a simulation study. *J Clin Pharmacol* **49**:351–359.

Zheng HX, Huang Y, Frassetto LA and Benet LZ (2009) Elucidating rifampin's inducing and inhibiting effects on glyburide pharmacokinetics and blood glucose in healthy volunteers: unmasking the differential effects of enzyme induction and transporter inhibition for a drug and its primary metabolite. *Clin Pharmacol Ther* **85**:78–85.

Useful Links

Huang SM (2006). Transporters and their role in drug interactions. Presented at Clinical Pharmacology Subcommittee of the Advisory Committee for Pharmaceutical Science. October 18, 2006. http://www.fda.gov/ohrms/dockets/ac/06/slides/2006-4248s1-6-FDAHuang_files/frame.htm

Early Communication about an Ongoing Safety Review of clopidogrel bisulfate (marketed as Plavix). http://www.fda.gov/Drugs/DrugSafety/PostmarketDrugSafety InformationforPatientsandProviders/DrugSafetyInformationforHeathcareProfessionals/ucm079520.htm

Pharmacokinetics in patients with impaired renal function-study design, data analysis, and impact on dosing and labeling. http://www.fda.gov/downloads/Drugs/GuidanceComplianceRegulatoryInformation/Guidances/ucm072127.pdf

Pharmacokinetics in patients with impaired hepatic function: study design, data analysis, and labeling. http://www.fda.gov/downloads/Drugs/GuidanceComplianceRegulatoryInformation/Guidances/ucm072123.pdf

Drug interaction studies - study design, data analysis, and implications for dosing and labeling. http://www.fda.gov/downloads/Drugs/GuidanceComplianceRegulatoryInformation/Guidances/ucm072101.pdf

The preliminary concept paper: Pharmacokinetics in Patients with Impaired Renal Function - Study Design, Data Analysis and Impact on Dosing and Labeling. http://www.fda.gov/ohrms/dockets/ac/08/briefing/2008-4351b1-00-index.htm

Chapter 27
Drug–Drug Interactions: Communicating Post-market Drug Safety Information in the USA

Soraya Madani and Helen Winter

Abstract It has been more than a decade since the withdrawal of a number of drugs from the market as a result of serious adverse events from drug–drug interactions (DDI). Since then, additional research has led to a much better understanding of the mechanisms behind many of these interactions. These advances have made DDI one of the best understood causes of observed adverse drug events. However, to reap the full benefit of all this research, much more could be done to translate this information into a form more useful in clinical care settings. Health-care providers rely on two principal sources – FDA approved product information (i.e., drug product label) and drug interaction databases – to help them prescribe medications safely. Greater standardization of definitions, classifications, and severity assessments of drug interactions would help prescribers to better understand the clinical significance of a particular DDI. Achieving greater standardization will require the continued input and coordination of efforts of important stake holders such as academic scientists, industry, and regulators.

27.1 Introduction

Drug–drug interactions (DDI) are one of the major causes of adverse drug reactions. The preceding chapters of this book highlight the progress that has been made in understanding the underlying mechanisms related to drug–drug interactions. This increased understanding should make adverse drug reactions more predictable, manageable, and preventable. Reducing adverse drug reactions has the potential to achieve considerable cost savings for the US health-care system

S. Madani (✉)
Novartis Pharmaceuticals Corporation (SM), East Hanover, NJ, USA
e-mail: soraya.madani@novartis.com

The opinions expressed in this chapter are solely those of the author and not necessarily those of Novartis Pharmaceuticals Corporation (NPC). NPC does not guarantee the accuracy or reliability of the information provided herein.

K.S. Pang et al. (eds.), *Enzyme- and Transporter-Based Drug–Drug Interactions*,
DOI 10.1007/978-1-4419-0840-7_27,

(Hillestad et al., 2005). One estimate by the RAND Corporation (Bigelow et al., 2005; http://www.rand.org/news/press.05/09.14.html) estimated cost savings of up to $3.5 billion annually. Achieving cost savings of this magnitude requires that the latest and most reliable scientific information is available to health-care providers to help them make the best therapeutic decisions for patients.

Improved prevention of serious drug–drug interactions would be a major step forward to enhancing patient safety. This requires effective communication of DDI information to patients as well as health-care providers. In this chapter, we discuss the presentation of drug interaction information in the FDA approved product information, which is a widely available tool for communicating information on a drug's efficacy and adverse events to health-care providers and patients. We also discuss the current state of DDI databases, and patient medication history which are additional important potential sources of drug interaction information for health-care providers.

27.1.1 FDA Approved Drug Product Information

A study performed by Smalley et al. (2000) indicates that, in multiple instances, prescribers made no change in the way they prescribed certain drugs, despite manufacturers and FDA efforts to include additional warnings regarding significant DDI to the product information of these agents.

Health-care professionals learn about drug interactions involving CYP450 induction and inhibition and transporter proteins from a variety of sources. However, there remains the challenge of having the most relevant information available exactly when it is needed – when a prescription is written. A review of the drug interaction or clinical pharmacology section of product information for many drugs metabolized by cytochrome P450s reveals that these sections contain a great deal of highly specialized scientific information that may or may not have clinical relevance to the prescribing of the medication.

A study on "Preventing Medication Errors" released in July 2006 by the Institute of Medicine (IOM, 2006) recommended that improving how product information is communicated could be one important tool used to reduce medication errors in the USA.

In an effort to improve the product information readability and usefulness, in January 2006 FDA issued a final rule requiring manufacturers to comply with a new format for package inserts (PI) (www.fda.gov/cber/rules/labelcf.pdf). The new format contains two main sections. The highlight section is a short half page of summary text, designed to be easy to read, which contains only the information considered the most important for the prescriber. The rest of the PI is the full package insert (FPI), which contains much more detailed scientific information. It is expected that a more standardized and simpler product label format will help physicians find the relevant information they need for making decisions about a patient's therapy. The ultimate goal is to decrease medication errors that have previously been attributed to the lack of readability and accessibility of information in the product information (i.e, product label). The real impact of the new format on medication errors in general and on drug–drug interactions specifically will need to be evaluated in future studies.

While the new product information format should be a major step forward, there is still much that could be done to improve the way in which drug interaction information is communicated. One issue that needs resolving is related to consistency of terminology and formatting of DDI information in the PI. As previously noted, there is a lack of clear guidelines to standardize the language that describe and classify DDI or indicate when a DDI should be considered clinically significant. Also, there is a lack of consistency as to where in the product label DDI information should appear. DDI information can appear in multiple subsections of the product information. To avoid confusing health-care providers, clearer criteria could be developed as to why information should be included in one subsection versus another. These inconsistencies are compounded by the fact that while newer drugs often have detailed DDI information in their product information, older drugs may have no DDI information at all. This is a reflection of the fact that the science related to DDI has matured and the inclusion of drug-specific DDI information has become more integrated into regulatory processes in recent years.

Direct patient communication is an additional strategy being used to reduce DDI. The drug product-specific medication guide is an example of such a communication tool. Medication guides (MedGuide) are a written means of safety communication that are formally reviewed and approved by FDA. Medication guides focus on the serious adverse effects of medications or medication classes and are intended for distribution to the patient at the pharmacy when specified products considered to pose "a serious and significant public health concern". MedGuides target patients, using simpler language than the PI and have easy to read format. However, MedGuides only exist for some drug products and not all FDA approved products (FDA review of use of medication guides. http://www.fda.gov/NewsEvents/Newsroom/PressAnnouncements/2007/ucm108886.htm).

Standardized and consistent language describing drug interactions across all FDA approved product information, although a challenging task, is an essential first step to improve methods for classifying drug interaction information so that it is much more obvious which types of drug interactions are considered clinically significant. Subsequently, depending on that classification, improved criteria can be established as to what type of information should enter which subsection of the product information. And, finally, all these clarifications and classifications should be presented such that the language is simple enough to communicate what the real clinical significance of a particular DDI is to health-care providers. The task of providing improved and more clinically useful FDA approved product information is one that can be accomplished by the various stake holders. Scientists, clinicians, pharmacists, regulators, industry, policy makers, and others need to work together to develop and establish standards that not only take into consideration the impact on each stake holder, but most importantly, the impact on the public health.

27.2 Patient Medication History Databases

A large under-utilized source of information relevant to preventing drug interactions is the patient's medication history. Currently, due to the fragmented state of the US health-care system, a patient's medical and medication history is scattered across various parts of the health-care system, such as pharmacies, physicians offices, hospitals, and managed care systems (HMOs, etc.) that patient has been participating in. There is no centralized repository of information that is readily available to the pharmacist or the physician to identify what current medications the patient is taking prior to dispensing the new medication. The setting up of a centralized, comprehensive up-to-date electronic health record database system is a hot topic that is currently receiving much attention at the national level (*Business Week*, April 24, 2009). Until such a system is in place, we will have to rely on a last voluntary dialogue between the pharmacist and the patient at the time of dispensing to obtain the relevant medication history. This is not optimal, given how busy pharmacists can be in the retail setting, and the fact that pharmacies often lack suitable amenities to facilitate a quality interaction between the pharmacist and the patient.

27.3 DDI Databases

Outside of FDA-approved product information there are other important sources of information that both physicians and pharmacists can use to obtain information about drug–drug interactions. However, these resources also have inconsistencies in the way DDI information is presented. In the case of DDI databases, each database has a different sponsor and their data are not interlinked. The size of these different databases varies and the interpretation of scientific information between databases is not always in agreement. As a result, there is much incongruity as to what is considered a clinically significant DDI. Many studies conducted in the USA and elsewhere have examined this issue as outlined below.

The Institute of Medicine (IOM) report on Adverse Events in 2007 discussed DDI database inconsistencies. In this report, reference was made to a study by Abarca et al., 2004. The Abarca study assessed the agreement of DDI severity ratings among four of the most commonly used US compendia: "Evaluations of Drug Interactions (2001)," "Drug Interaction Facts" (Mangini, 2001), "Drug Interactions: Analysis and Management" (Hansten and Horn, 2001), and the "Drug-REAX program" (Moore et al., 2001). These researchers found that of the total of 406 drug interactions that were considered clinically significant by at least one database, only 2% were listed in all four compendia.

Another comparative study published by Vitry in 2006 (Vitry, 2007) evaluated the consistency of inclusion and grading of major drug interactions for 50 drugs in four leading international drug interaction compendia: British National Formulary, the interaction supplement in the French drug compendium Vidal, and two US drug interaction compendia, "Drug Interaction Facts" and "The Micromedex (Drug-REAX) program." This assessment found that between 14 and 44% of the drug interactions classified as major in any one compendium were not listed as such in the other compendia.

The above studies and others published relatively recently (Abarca et al., 2004; Chao and Maibach, 2005; Vitry, 2007) all point out that the likely major cause of the discrepancies between databases is due to a lack of consensus on how to classify the potential severity of the drug interaction. The information sources used by the various databases, be it, published literature or unpublished manufacturer reports from post-marketing surveillance systems are subjected to a unique interpretation by each compendia or databasing system. As reported by the above publicat different systems are used by each database to describe important interactions identified from different sources of information. For example, Drug Interaction Facts classifies the severity of an interaction into three categories – major, moderate, and minor – and the degree of supporting documentation into five categories – established, probable, suspected, possible, and unlikely; it also assigns a significance of 1–5 to each drug interaction based on a combination of these two classification systems. The Micromedex Drug-REAX System classifies the severity of an interaction into three categories – major, moderate, and minor – and the degree of documentation into five categories – excellent, good, fair, poor, and unlikely.

An additional inherent complexity to the use of DDI databases is that data from the literature often characterize only a one- or two-way interaction between two potentially interacting drugs and does not address the potential for drug interactions between multiple concomitantly administered drugs. In reality, patients, and in particular the growing elderly population, are subject to increasing polypharmacy or being prescribed multiple drugs, which increase the likelihood of more than just two drugs interacting with each other. Furthermore, the information available from post-marketing surveillance systems often does not ascertain the degree of causality between the captured adverse event and the different drugs involved. Hence, unless the original source of information is the drug label itself, much of the information in these databases provides insufficient evidence for establishing the true clinical relevance of a potential interaction.

DDI databases also serve to alert practitioners about newly reported adverse events. In a recent report by the Institute of Medicine (2007, http://www.nap.edu/catalog/11897.html), concerns were raised about physicians and pharmacists becoming subject to "alert fatigue." In such databases alerts of dangerous drug interactions are diluted with alerts for other adverse events that may or may not be drug related. The IOM report indicates that the alerts are too sensitive and yet neither instructive nor specific, and as a result have become burdensome, while still not helping the health-care provider in making therapeutic decisions. The report noted that only one in nine alerts was deemed useful by health-care providers, according to a 2005 study (Spina et al., 2005). This can explain why up to 88% of all drug alerts end up being disregarded by community pharmacists (Chui and Rupp, 2000). Such a situation can be consequential for patients and can result in important drug interactions going undetected.

There have been various proposals to overcome the limitations of the currently available databases and compendia (Ferner and Aronson 2007). What these proposals have in common is the need for a standardized template that would communicate to the prescriber in relatively simple terms the nature of the risk of a particular drug interaction and the quality of the evidence on which that assessment is based. In order to achieve greater consistency in the information available across the various databases, there has also been a proposal to create one unified centralized database that integrates all the information from the other existing databases. This centralized database should be publicly accessible and easily updated as new information emerges. And, just as it has been highlighted for drug labels, there needs to be agreed and standardized definitions for which interactions are considered clinically significant. Only alerts for clinically significant interactions would be sent on to the health-care provider.

Once again, achieving such a system requires collaboration among the various stake holders such as regulators, manufacturers, scientists, clinicians, and pharmacists. Appropriately, the IOM report published in 2007, suggested the formation of a cross-disciplinary DDI working group to take on developing such a centralized database. The database could be informed by information from the drug product label, as well as from the literature and post-market surveillance data. This working group could establish commonly agreed terminology, create criteria to identify and classify DDIs, and create appropriate designations for the strength of evidence supporting the classification. To standardize this process, the working group could put in place a decision tree or other decision-making criteria that are based on and supported by scientific information and updated as new information becomes available.

In summary, safeguarding patients from serious adverse reactions and ensuring the public health requires the highest level of teamwork and co-operation between scientists, clinicians, and regulators. Releavant information from both the laboratory and patients must be sorted, harmonized, and communicated to the health-care provider before the patient leaves the pharmacy with their new prescription. Much progress has been made and yet it is the final step of communicating the real clinical significance of a potential DDI that is proving to be the most challenging aspect of all.

References

Abarca J, Malone DC, Armstrong EP, Grizzle AJ, Hansten PD, Van Bergen RC, and Lipton RB (2004) Concordance of severity ratings provided in four drug interaction compendia. *J Am Pharm Assoc (2003)* **44**:136–141.

Aspden P, Wolcott J, Bootman JL, Cronenwett LR, and Committee on Identifying and Preventing Medication Errors (2006) *Preventing Medication Errors: Quality Chasm Series*. National Academies Press, Washington, DC.

Bigelow JH, Fonkych K, Fung C, and Wang J (2005) *Analysis of Healthcare Interventions that Change Patient Trajectories*. RAND Health, RAND Corporation, Santa Monica, CA.

Chao SD and Maibach HI (2005) Lack of drug interaction conformity in commonly used drug compendia for selected at-risk dermatologic drugs. *Am J Clin Dermatol* **6**:105–111.

Chui MA and Rupp MT (2000) Evaluation of online prospective DUR (drug utilization review) programs in community pharmacy practice. *J Manag Care Pharm* **6**:27–32.

Drazen JM, Rainey J, Begg H, Stith Butler A, and Rapporteurs (2007) Forum on drug discovery, development, and translation, *Adverse Drug Event Reporting: The Roles of Consumers and Health-Care Professionals: Workshop Summary IOM of the National Academies*. National Academies Press, Washington, DC.

Ferner RE and Aronson JK (2007) Communicating drug safety. IOM report on adverse drug events. *Br Med J* **333**:143–145.

Hansten PD and Horn JR (2001) *Drug Interactions: Analysis and Management*. Facts and Comparisons, St. Louis, MO.

Hillestad R, Bigelow J, Bower A, Girosi F, Meili R, Scoville R, and Taylor R (2005) Can electronic medical record systems transform health care? Potential health benefits, savings, and costs. *Health Aff (Millwood)* **24**:1103–1117.

Mangini RJ (2001) *Drug Interaction Facts*. Facts and Comparisons, St. Louis, MO.

Moore LL, Minne K, and Moore KB (2001) DRUG-REAX System http:\\www.micromedex.com/products/drugreax/

Smalley W, Shatin D, Wysowski DK, Gurwitz J, Andrade SE, Goodman M, Chan KA, Platt R, Schech SD, and Ray WA (2000) Contraindicated use of cisapride. The impact of FDA regulatory action. *J Am Med Assoc* **284**:3036–3039.

Spina JR, Glassman PA, Belperio P, Cader R, Asch S. (2005) Clinical relevance of automated drug alerts from the perspective of medical providers. *American Journal of Medical Quality* **20**(1): 7–14.

Terhune C, Epstein K, and Arnst C (2009) The Dubious Promise of Digital Medicine. *Business Week*, April 23.

Vitry AI (2007) Comparative assessment of four drug interaction compendia. *Br J Clin Pharmacol* **63**:709–714.

Chapter 28
Drug–Drug Interactions: What Have We Learned and Where Are We Going?

K. Sandy Pang, Raimund M. Peter, and A. David Rodrigues

Abstract The study of drug–drug interactions (DDIs) has significantly progressed in recent years, sometimes at a considerable pace. The impetus has undoubtedly been due to the much publicized market withdrawal of a number of drugs (terfenadine, mibefradil, astemizole, and cisapride) due to unforeseen drug–drug interactions that placed the patients at considerable health risk or even resulted in fatalities. In response, subsequent actions by regulatory agencies like the FDA generated a more stringent DDI framework and guidance for drugs submitted for market approval. The pharmaceutical industry reacted with a more proactive approach in trying to screen out any undesirable drug–drug interaction liability from candidate drugs already at the early drug discovery stage. Looking forward, however, it is apparent that numerous challenges remain and opportunities still exist to develop an improved and more complete toolbox that can support the preclinical and clinical study of drug transporters and drug-metabolizing enzymes beyond the cytochrome P450s. At the same time, DDI models will have to become more comprehensive and enable the integration of enzyme and transporter data, taking into account the dynamic nature of both "perpetrator" and "victim" pharmacokinetics. This dynamic approach will have to consider the underlying mechanism(s) of the DDI in multiple elimination organs, transporter-mediated transport, as well as aspects like inter-subject variability, pharmacogenetics, or pharmacodynamics. In the future, the industry and regulatory agencies will need to integrate an even larger amount of preclinical and clinical information from different sources, deal with larger numbers of drug combinations, and consider complicated DDI scenarios involving a heterogeneous mix of small molecules and biologics, or when the interacting drug is a ligand of nuclear receptors that results in changes in transporter and enzyme function. Physiologically-based pharmacokinetic (PBPK) modeling approaches hold much promise in this regard and will be one of a number of steps en route to a more

K.S. Pang (✉)
Leslie Dan Faculty of Pharmacy, Department of Pharmaceutical Sciences, University of Toronto, Toronto, ON, Canada
e-mail: ks.pang@utoronto.ca

K.S. Pang et al. (eds.), *Enzyme- and Transporter-Based Drug–Drug Interactions*,
DOI 10.1007/978-1-4419-0840-7_28,

systems-based approach within a burgeoning model-based drug development environment. As the areas of molecular and cell biology and pharmacology progress toward such a goal, we can foresee that the study of DDIs, accompanied by the attendant data, will become more mechanistic and complete.

28.1 Drug Interactions Involving Metabolic Enzymes

Enzymatic reactions are quite ordered and are usually very specific toward select substrates. Their susceptibility to induction and inhibition, transcriptional regulation of nuclear receptors, and variation due to pharmacogenetics have been well appreciated (see Part I; Chapters 1, 3, and 4) over the recent years. Consequently, drug metabolism has become one of the most important factors in drug discovery and toxicological studies. In drug discovery, decisions focused on synthesis, testing, and further chemotype progression have to be made rapidly, requiring succinct information related to the metabolism of the compounds in question, the desired modifications based on available methodologies (Part II), in silico predictions (Chapter 6), and assessment of DDI potential (Chapter 23).

One can imagine a scenario where the inhibition of one enzyme will be compensated by another enzyme and thus lead to increased removal of the drug via alternate, metabolic, or excretion pathways (Morris and Pang, 1987; Sirianni and Pang, 1997). The compensation by a seemingly, unimportant pathway may even lead to an increase of a toxic pathway in some cases, and understanding of enzyme inhibition leading to an apparent induction of the alternate metabolic pathway is part and parcel of observations for competitive pathways. Enzyme inhibition can be observed in different types of kinetics: reversible inhibition as competitive, non-competitive, uncompetitive, or mixed type or as irreversible inhibition, which can show a time-dependent or even mechanism-based inactivation of the respective enzyme(s). The latter type involves either the formation of a metabolite intermediate complex or covalent binding (Part III, Chapters 19 and 20). Complex cases of multiple inhibition (Chapter 26) or inhibitory and inductive events (Chapter 21) are occasional observations that can complicate predictions of DDIs even further. Moreover, inhibitors can behave differently in various organ tissues, such that an inhibitor of an intestinal enzyme may not reach sufficiently high enough concentration levels to act as an inhibitor of the same enzyme in the liver. Clearly, there is a need for more comprehensive tools that link in vitro data with in vivo and biomarker data, and support decision making.

Due to the abundance of drug-metabolizing enzymes expressed in hepatic tissue, there has been an overwhelming emphasis on the liver as the metabolizing organ, and little and not enough attention given to extrahepatic organs or tissues. In addition to the liver, drug-metabolizing enzymes responsible for clearance processes are also expressed in significant levels in other organs/tissues, e.g., the intestine and the kidney (Chapters 1, 14, and 17). Organ-specific inhibition has been noted even for the same enzyme. Hence, prediction strategies for drug interactions should be expanded to include extrahepatic tissues, especially the intestine that is strategically placed as the gateway tissue to the liver in first-pass removal.

28.1.1 P450s Versus Phase II Enzymes

As can be surmised from these proceedings and other text references regarding DDIs, we currently possess ample knowledge on the cytochrome P450s (P450s), their multiplicity, and information related to the regulation, structure, and function of various subfamily members. The duplicity of CYP3A4 and CYP3A5 in the metabolism of common substrates, and a greater propensity of CYP3A4 toward inhibition by all inhibitors, render additional complexity in deciphering the degree of inhibition (McConn et al., 2004). This observation may suggest that one has now to consider *CYP3A5* genotype prior to conducting a CYP3A DDI study. Additionally, we can anticipate that further unknown P450s will be discovered and their functions be defined. The potential activities of orphan P450s have been examined repeatedly (Stark and Guengerich, 2007; Stark et al., 2008b), and some have been implicated in the metabolism of endogenous substrates, e.g., arachidonic acid (Stark et al., 2008a). As many of these P450s catabolize or metabolize pharmacologically active substrates like eicosanoids or steroids, DDIs involving their induction and inhibition should be increasingly taken into consideration.

Drug-metabolizing enzymes that catalyze conjugation reactions (phase II enzymes), such as UDP-glucuronosyltransferases (UGTs) and sulfotransferases (SULTs), share common substrates, albeit each proceeding with differing affinities. There is a general consensus that additional tools are required in order to enable more robust reaction phenotyping of these enzymes in vitro, improve the bridging of in vitro to in vivo, and support the conduct of more mechanistically meaningful clinical DDI studies (Zhang et al., 2007). Various UGTs and SULTs are expressed in the gut and liver, and contribute to first pass metabolism of various drugs, thus they cannot be dismissed in DDI risk assessment. Hence, although a particular drug may not inhibit the P450s, unexpected DDIs with UGT and SULT substrates can dominate in altering the clearance of a particular drug (Schwartz et al., 2009).

Among the three human UGT superfamilies (Tukey and Strassburg, 2000), multiple UGTs can be involved in the conjugation of the same substrate (Kostrubsky et al., 2005). Only a few inhibitors exist, though reaction phenotyping of glucuronidated substrates is currently feasible to some extent (Chapter 8) and online drug-metabolism systems, integrated into capillary electrophoresis that entail the encapsulation of microsomes in tetramethoxysilane (TMOS)-based silica matrices for the determination of UGT inhibitors in a single capillary, are available (Sakai-Kato et al., 2004). For example, retinoids are found to be inhibitors of UGT2B7 mRNA expression (Samokyszyn et al., 2000; Lu et al., 2008), whereas HIV protease inhibitors are UGT1A1 inhibitors (Zhang et al., 2005a).

In the past, the prediction of UGT-dependent clearance was problematic. However, Chapter 8 emphasized that the addition of bovine serum albumin (BSA) or HSAFAF (fatty acid free human serum albumin) to incubations of human liver microsomes (or recombinant UGTs) may permit the accurate prediction of in vivo clearance parameters and DDI potential. Recently, it has been proposed that cryopreserved hepatocytes may serve as an alternative model to assess UGT activity and DDI potential in vitro (Coughtrie et al., 2009). Irrespective of the model employed, however, prediction of UGT-mediated DDIs is difficult, because many substrates are metabolized by multiple UGTs, and inhibition constants (K_i) tend to be high. The implication of pumps at the endoplasmic reticulum membrane of glucuronides (Battaglia and Golan, 2001; Csala et al., 2004) further complicates the picture. Enterohepatic circulation of glucuronides to reappear as the unconjugated species renders an apparent observation of lessened exposure and therefore formation of the glucuronide metabolite. These complications add to the difficulties in addressing inhibition of the glucuronidation reaction.

In comparison, the sulfotransferase isoforms exert more stringent substrate specificities than the UGTs, and are subject to differential effects of inhibitors. Probes such as acetaminophen or 2-aminophenol (Riches et al., 2007) for phenosulfotransferase, SULT1A1, exist. Pentachloral phenol and mefenamic acid are potent and selective inhibitor of human liver SULT1A1 (SULT1A3 for mefenamic acid). Phthalates (used as plasticizers) inhibited estrogen sulfotransferase, SULT1E1, and hydroxysteroid sulfotransferase, SULT2A1 (Harris and Waring 2008). Hydroxylated polychlorinated biphenyls (PCBs), important persistent environmental contaminants, are substrates and inhibitors of human SULT2A1 (Liu et al., 2006). In addition, SULTs are inhibited by many dietary and environmental chemicals. SULT1A1 is strongly inhibited by flavonoids and a range of environmental chemicals and dietary components (see Chapter 22). Fruit and vegetable cytosols also inhibit SULT isoforms, as do long-chain alkylphenols and chlorinated phenols. Juices and green tea (Tamura and Matsui, 2000; Saruwatari et al. 2008) are known inhibitors of SULTS, and curcuinoids inhibit not only SULTs but UGTs and CYP3A4 (Volak et al., 2008). Quercetin, a flavonoid present in edible fruit, vegetable, and wine, was found to be a potent inhibitor of SULT1A1, and SULT1E1 activities and resveratrol sulfation (Pacifici, 2004).

The area of inhibitors of glutathione *S*-transferases (GST) is sparked by the notion that these GST enzymes are involved in the resistance to anticancer drugs, since elevated levels of GSTs are among the factors associated with an increased resistance of tumors to a variety of antineoplastic drugs. The inverse correlation between expression and prognosis in many tumors has provided a rationale for the design of GST inhibitors to enhance the therapeutic index. A major advancement to overcome GST-mediated detoxification of antineoplastic drugs is the development of GST inhibitors. Human GST inhibitors are multidrug resistance chemomodulators on human recombinant glutathione *S*-transferase (GSTs) activity, GST P1-1 by sulfinpyrazone, GST Al-1 by sulfasalazine, and camptothecin, GST M1-1 by sulfasalazine, camptothecin, and indomethacin, and progesterone as a potent inhibitor of GST P1-1 (Hayeshi et al., 2006). The α-glutamyl moiety plays an important role in modulating the affinity of the ligands to interact with GSH-dependent proteins. The glutathione *S*-conjugate,

L-γ-glutamyl-(S-9-fluorenylmethyl)-L-cysteinyl glycine, has been found to be a highly potent inhibitor of human GSTA1-1 in vitro, with lesser inhibitory activities toward GSTP1-1 and GSTM2-2 isoenzymes (Cacciatore et al., 2005). Thonningianin A (Th A), a novel antioxidant isolated from the medicinal herb, *Thonningia sanguinea*, inhibited rat liver GST and human GST P1-1 (Gyamfi et al., 2004).

28.1.2 Species Differences

The use of preclinical models to assess DDI in metabolism is commonplace but is complicated by species differences. What is encountered is that the expression level, functional activity, and/or tissue distribution differ. An appropriate animal model, when chosen and used properly, could be a valuable tool to provide the basis for extrapolating in vitro human data to clinical outcomes as well as mechanistic insight for the interpretation of interactions observed clinically (see Chapter 11). On occasion, relevance of the in vitro/in vivo animal models and gene-knockout animals is uncertain and needs to be questioned. The knockout animals that lack specific drug-metabolizing enzymes may exhibit altered morphology and flow dynamics in the liver (Schmidt et al., 1996; Lahvis et al., 2000), altered levels of ligands that affect receptors/transcription factors, and exhibit redundant or alternate pathways that are absent in humans (Kimura et al., 1999). Greater progress is needed, so that species differences are understood more fully and results of animal-based DDI models can be translated to man (Chapters 11 and 23).

28.1.3 Transgenic Animal Models

The recent development of transgenic animal models with humanized liver in mouse lines expressing specific drug transporters and/or metabolizing enzymes is of interest (see Chapter 11). Through the pioneering efforts of many investigators, humanized mice are now routinely used to rapidly advance research. Chimeric mice, constructed by transplanting human hepatocytes, are useful for predicting the human metabolism of drug candidates. Some success was seen with the metabolism of *S*-warfarin (Inoue et al., 2009) and induction of rifamycin by SXR (steroid X receptor) (Kim et al. 2008), and in the study of sex and developmental changes of the CYPs (Felmlee et al., 2008). In other instances, higher activities are found expressed in chimeric mice carrying humanized liver CYP1A2 (Uno et al., 2009). However, lots more need to be known about this model since the intra- and inter-organ characteristics as well as changes in hormonal and cytokine levels are unknown. Although quantitative assessments using these animal models are currently limited, it is conceivable that in the next decade, these models could become more valuable and validated in DDI assessments during drug discovery and early development processes.

28.2 Drug Interactions Involving Transporters

In contrast to the stringent specificities of the P450 CYPs and SULTs, a lot of redundancy exists among transporters that share the same substrates. Hence, it begs the question of whether the inhibition of certain transporters is important because of the sharing of substrates and the redundancy. The answer is yes, if transport is the rate-limiting step and especially when uptake or transport is mediated solely via the presence of the transporter. The situation is far more complex for compounds/inhibitors undergoing both metabolism and transport. We should also consider the drug and its metabolite as potential inhibitors (substrate and end product inhibition) as well as levels of the inhibitor and its metabolite, e.g., gemfibrozil glucuronide as inhibitor of OATP for cerivastatin transporter (Shitara et al., 2004).

28.2.1 Proteomics-Based Approach to Define Transporter Abundance in Tissues

Up until recently, we do not have a quantitative feel about how much of the transporter exists. There is a lack of the availability of absolute protein concentration levels in recombinant systems, isolated primary cells, and human tissues. For example, inhibition studies conducted in human hepatocytes ordinarily reflect the overall effect of inhibition (Shitara et al., 2003), and not on inhibition of any particular uptake transporter since the relative contribution of the sinusoidal transporters to uptake is unknown. In order to gain this insight, the relative amounts of the transporters must be known. The use of proteomics, with LC-MS/MS technologies, for example, will aid to decipher the problem. An absolute quantification method for membrane proteins in murine blood–brain barrier, liver, and kidney was determined by liquid chromatography–tandem mass spectrometer (LC/MS/MS). The method resulted in expression levels of 34 transporters in liver, kidney, and blood–brain barrier of mouse that showed an excellent correlation with the values obtained with existing methods using antibodies or binding molecules (Kamiie et al., 2008). The relative importance of transporter and function must now be correlated to the amount of protein present in the tissue or organ to those measured in in vitro systems to more quantitatively assess the relative contribution of transporters to drug transport in vivo, and define the significance of the DDIs.

28.2.2 Species Difference in Transport

An appropriate animal model, when chosen and used properly, could be a valuable tool to provide a basis for extrapolating in vitro human data to clinical outcomes as well as a mechanistic insight for the interpretation of interactions observed clinically. Although human orthologs may be identified among transporters, species difference can exist in drug transport and therefore in vitro data may not be that relevant in the prediction of DDI. Differing substrate specificities are often found among species and must be noted (Ho et al., 2006). Significant challenges still need to be overcome in terms of studying human drug transport.

28.2.3 Improved Tools to Examine Transporter Function

The identification of transporters involved in drug transport, as described in Chapters 2, 9, and 10, heavily relies on use of expression systems (Kopplow et al., 2005); this information is of vital importance for the determination of DDIs in transport. Otherwise, drug transport is assessed in hepatocytes or cells in culture. First, human (freshly prepared or cryopreserved) hepatocytes that contain the full complement of transporters are expensive and vary greatly from batch to batch for good viability measures. Hepatocytes are rarely prepared fresh and are reliant on the donor characteristics (age, sex, and disease), the source, and method of preparation, and handling of shipment during procurement. Hence, cryopreserved hepatocytes are used (Shitara et al., 2003). Recently, differences in transporter content assayed by LCMS were found between cryopreserved hepatocytes and liver but not between freshly prepared hepatocytes and whole liver, and more importantly, age dependency in MRP2 levels was suggested (Li et al., 2009).

Hepatocytes in culture undergo differentiation (Jigorel et al., 2005), and may not contain the complement of transporters or nuclear receptors in its native state (Nahmias et al., 2006; Ohno et al., 2008). Upon co-culture with sinusoidal endothelial cells, the transport function of low density lipoprotein is regained (Nahmias et al., 2006). Sandwich culture systems, even from human sources, contain reduced or elevated levels of transporters and enzymes, and may require use of dexamethasone, ligand of the glucocorticoid receptor (GR) and SXR (steroid X receptor) for culture. Transporter and enzyme genes in culture may further respond to induction by dexamethasone (1–100 μM) (Turncliff et al., 2004; Hoffmaster et al., 2004). Differential growth or loss of some transporters and enzymes exists in rat hepatocyte sandwich systems (Chandra et al., 2001); P-gp, the Mrp, the Oatps, Bsep, and the P450s are maintained, allowing for uptake, induction, and regulatory studies (Annaert et al., 2001; Zamek-Gliszczynski et al., 2003; Turncliff et al., 2004; Zhang et al., 2005b). Bile acids are secreted and retained within the bile canaliculus and may adversely present a cholestatic model; the accumulation of bile acids may affect enzymes and transporters via regulation by FXR (see Chapter 4). Although the micropatterned co-culture of human hepatocytes and murine fetal fibroblast show some promise (Khetani and Bhatia, 2008) (see Chapter 9), the flow circuitry, hepatocyte heterogeneity as well as acinar regions may not be maintained. For improved utilization of these systems, a thorough comparison of the levels of freshly prepared/whole liver vs. cultured hepatocytes and functional studies needs to be established for validation of the systems.

Other complementary tools for additional insights include knockout (KO) animals lacking specific drug transporters, and/or transgenic animal models with humanized mouse lines expressing specific drug transporters. Knockout animals for drug efflux: P-gp [*mdr1a*(–/–), *mdr1b*(–/–)] and the *mrp*(–/–) have provided some insight as to the involvement of transporters and alteration of ADME and drug and metabolite profiles. One of the drawbacks of the KO animals is the presence of redundant pathways. The *Abca1*(–/–) KOs that lacks the high density lipoprotein (HDL) transporter showed the same hepatic cholesterol, triglyceride, and phospholipid contents, and the same extent of biliary excretion rates of cholesterol, bile salts, and phospholipid, and unchanged uptake of cholesterol and cholesterol esters when fed high fat and cholesterol diet compared to the standard diet (Groen et al., 2001). The *mrp1(–/–)* KOs failed to protect aflatoxin B1 lung toxicity, and the same degree of tumor development was observed for the wildtype and KOs; the redundant pathways may be P-gp and other Mrps (Wijnholds et al., 1997; Lorico et al., 1997; Rappa et al., 1999; Allen et al., 2000; Lorico et al., 2002). In Bsep KOs, less than expected intrahepatic cholestasis was observed: there was 6% excretion and increased hydroxylation of bile salts (Wang et al., 2001). In Mrp2-KOs, increased P450s and Ugt1a were found to compensate for loss in excretory activity (Chu et al., 2006). Reduction in Cyp2b1/2 and Cyp3b1/2 was found in immunoblot analyses in TR^- that lack Mrp2 vs. control Wistar rats (Jäger et al., 1998). In Mrp deficient, Eisai hyperbilirubinemic rats (EHBR), a compensatory increase in Mrp3 was observed in comparison to Sprague Dawley rats, lending to the greater basolateral efflux of substrates (Akita et al., 2001). Many aspects of this have been covered in Chapter 9. Despite that the humanized liver is being used to examine drug transport (Okumura et al., 2007), many attendant changes are obscure and validation of this model as a useful tool remains unknown. It is envisioned that bioinformatics and in vitro/in vivo approaches are needed to assess the functional and regulatory differences between the human and mouse genes be characterized in these model to ensure a more complete picture

28.2.4 Improved Probes and Inhibitors to Examine Transporter Function

Due to the redundancy in transporters, there is a need for better/improved selective probes or inhibitors of exclusive or high selectivity to segregate the roles of each transporter in drug transport for the basolateral and apical membranes. Many inhibitors are not specific enough for single transporters, especially for the ATP-binding cassette proteins (Matsson et al., 2009). Rather, they serve as inhibitors of multiple transporters: for example, MK571 inhibits P-gp (Honda et al. 2004), MRP1 (Jedlitschky et al., 1996), MRP2 (Leier et al., 2000), MRP3 (Zelcer et al., 2003), MRP4 (Reid et al., 2003), and MRP4 (Chen et al., 2005). GF120918 is an inhibitor of both MDR1 (Tang et al., 2002; Taipalensuu et al. 2004) and BCRP (Maliepaard et al., 2001). Mitoxantrone inhibits P-gp (Polli et al., 2001), BCRP (Volk and Schneider 2003), and MRP1 (Morrow et al., 2006). Recently, the broad specificity of inhibitors has been summarized (Matsson et al., 2009). The lack of availability of specific inhibitors adds complexity to the identification of the relative contribution of transporters in drug uptake in hepatocytes and renders quantitative predictions difficult.

Among the transfection or expression systems available thus far, it is difficult to assess the relative contribution of uptake and efflux transporters in the net transepithelial flux of drugs. Normally, influx or uptake is studied within a short time frame within which drug efflux is negligible. However, drug efflux is seldom studied, except in indicator dilution (Schwab et al., 1990, 1992) or washout experiments (Akita et al., 2001). Inhibition of efflux would lend to increased accumulation of the victim substrate. In like fashion, estimation of apical efflux is reliant on initial permeation of substrates prior to utilization of available transporters. This does not pose as a problem for lipophilic substrates of P-gp. For polar substrates that utilize the MRP pumps, this presents more difficulty, and may require use of prodrugs (Nezasa et al., 2006 Reid et al., 2003; Dallas et al., 2004; Chen et al., 2005) that undergo intracellular cleavage to furnish the requisite substrate. Some high-throughput screens with facile fluorescent probes are needed.

28.3 Improved Methods for the Interpretation of Drug Interaction Data

28.3.1 PBPK Modeling

There has been a constant improvement in the way to interpret data. The equation on the ratio of area under the curve under inhibited and uninhibited conditions proposed by Rowland and Matin (1973) had served as the initial basis and underwent several modifications to correct for the presence of competing pathways and competing organs (Ito et al., 2005; Brown et al., 2005). An uncertain term that was debated over and over again is the appropriate estimate of the inhibitor concentration, $[I]$. The FDA suggests use of the steady state $C_{\max}$ value of the inhibitor, whereas the Japanese counterpart emphasizes use of the unbound value. In a comprehensive analysis, Obach and colleagues compared the appropriateness of total or unbound concentration of the estimated hepatic inlet concentrations, or systemic concentration as $[I]_{\max}$ in vivo and concluded that the unbound hepatic inlet $C_{\max}$ during the absorptive phase yielded the more accurate prediction of the magnitude of DDI (Obach et al., 2006). While this remains an outstanding issue, many recognized that these terms are fictitious correlates of $[I]$ since DDI occurs within the involved eliminating organ, and the corresponding concentration of the inhibitor at the locale of the enzyme or transporter is the only pertinent one.

There is common consensus for a push for improved predictions in an interactive and dynamic fashion. The physiologically based pharmacokinetic (PBPK) modeling and simulation approach is recognized to be the most appropriate method and ideal tool so far to more reliably integrate in vitro data to in vivo (see Chapters 5, 7, 13, 21, and 26). The PBPK model has been found to be superior compared to other methods in predicting drug kinetics and behavior in man (Parrott et al., 2006). Both transporters and enzymes and their inter-individual variations may be accommodated. The soundness of the PBPK model predicates on how the ADME parameters are fed into the model. In order to achieve good results, strategies of model representation, model parameterization, model simulation, and model evaluation must be used (Nestorov et al., 1998; Luttringer et al., 2003; Jones et al., 2006). Pre-existing animal PBPK models would lend insight to the development of a useful PBPK model for humans. The approach is to select the parts of the model that are pertinent; and the lumping of compartments of similar blood flow or partition (Nestorov et al., 1998). To tackle the distributional aspects, one needs to examine the tissue to plasma partition coefficients, and differences in protein binding need to be estimated (Lin et al., 1982; Poulin and Theil, 2000; Grime and Riley, 2006; Ito and Houston, 2005). Lastly, for input of human metabolic/elimination data, direct scaling factors or hepatocellularity per gram liver tissue need to be applied to in vitro human hepatocyte data ($CL_{int,in\ vitro}$ × cell density number × liver weight) to arrive at the intrinsic clearance in vivo, $CL_{int,in\ vivo}$ (Howgate et al., 2006; Barter et al., 2007). Then the blood to plasma ratio needs to be known to correlate blood clearance to blood flow, and binding to tissues at the target site should be considered. Some success was achieved in this kind of modeling and simulations on the appraisal of how inter-individual variations in metabolism can affect concentration–time profiles (Rostami-Hodjegan and Tucker, 2007).

In PBPK-DDI modeling, all physiological and ADME properties of both the "perpetrator" and "victim" drugs and their metabolites should be considered, together with information on study design. Mutual inhibition, if present, would be easily demonstrable. This type of modeling has been concentrated around the liver as the only eliminating organ. The efforts would reveal the temporal changes of the drug and inhibitor and their metabolites; the inhibiting species whether the inhibitor or its metabolite in transport or metabolism (modification of the K_m or V_{max}) may be included. Time-dependent destruction or induction of enzymes may further be modeled.

Various PBPK-DDI models have been used to examine in vivo kinetic consequences of mechanism-based inhibition (MBI) of CYP2D6 by 3,4-methylenedioxymethamphetamine (MDMA, ecstasy). A PBPK model with physiologically based components of drug metabolism, taking account of change in the hepatic content of active CYP2D6 due to enzyme inactivation with formation of a metabolic inhibitory complex that resulted in auto-inhibition, was used to explain dose and time dependence observed in the in vivo kinetics of MDMA (Yang et al. 2006). In other interactions comprising of midazolam/macrolides, triazolam/erythromycin, and 5-fluorouracil/sorivudine temporal changes of inhibitors were addressed and somewhat predicted using a semi-PBPK model, where the gut wall metabolism and interaction was accounted simplistically by incorporating a fixed F_g term to denote the fraction of the dose that is metabolized in gut wall (Kanamitsu et al., 2000; Ito et al., 2003). Further model development by Hall and colleagues demonstrated elegantly that excellent prediction of both reversible inhibition and MBI could be achieved with monitoring of the parent drug as well as the metabolite (Gorski et al., 1998; Pinto et al., 2005; Quinney et al., 2008a,b), including the case that intestinally formed metabolite may act as the inhibitor and exert downstream inhibitory effects on the liver. A semi-PBPK-DDI model that took into consideration the temporal changes in concentrations of the inhibitor, diltiazem (NTZ), gut wall interaction, and contribution from the inhibitory metabolite, *N*-desmethyl-DTZ (ND-NTZ), successfully predicted the nonlinear disposition of DTZ and the interaction between DTZ and midazolam (Zhang et al., 2009). These results considered the temporal disposition of the inhibitor and DDIs at the gut wall and changes in intestinal and hepatic enzymes into model development. Simulation results undertaken were demonstrative that both DTZ and ND-DTZ contributed to the overall inhibitory effect observed following the administration of DTZ. However, success of the method was highly dependent on precise determinations of k_{inact} and K_i. The improved predictions from this clinical study showed that DTZ treatment resulted in 4.1-fold and 1.6-fold increases in MDZ exposure following oral and intravenous MDZ administration, respectively, and divulged that the DDI in the gut wall played an important role in the DTZ/MDZ interaction. This improved model is superior compared to other models that examined midazolam/verapamil interaction with a single inhibitor concentration (the unbound average plasma concentration of the inhibitor at the steady state), when the intestinal intrinsic clearance of CYP3A4 was applied in an attempt to account for the gut wall metabolism (Wang et al., 2004).

However, the PBPK-DDI models so far have seldom included transporters into consideration. Pang and colleagues have made some progress in inclusion of transporter parameters for basolateral influx and efflux as well as excretion in PBPK models for organs to understand the role of transporters in areas under the curve of drug and metabolite (Pang et al., 2008; Sun and Pang, 2009b). Combined physiologically based models of different eliminating organs incorporating both uptake transporters and enzymes should provide a comprehensive prediction tool to explore and accommodate the range of possible outcomes and the added complexity that may result. There is the urgent need, therefore, to consider improved intestinal models for drug absorption and metabolism. The segregated flow model (SFM) consisting of reduced flow to the enterocyte region that was developed during the turn of the century to describe route-dependent intestinal removal (Cong et al., 2000; Pang, 2003, Sun and Pang, 2009a) has not been routinely adopted for improved PBPK modeling for orally administered drugs. The fact that greater inhibition was achieved resulted for oral over the intravenous dosing of the DTZ–midazolam interaction (Zhang et al., 2009) and the greater impact for the oral dose is suggestive that the SFM may better describe the DDI data over traditional, physiological models. Since segmental distribution of transporters and enzymes will further affect bioavailability (Tam et al., 2003), improved PBPK modeling for the intestine, achieved through recognition of segmental differences in enzymes and transporters, will further address the interplay of enzymes and transporters of the intestine. There is a recent development toward modeling segmental P-gp and CYP3A in the intestine in light of the segmental traditional PBPK model (Tam et al., 2003), and may have sparked resultant simulations on drugs of varying permeability to reveal the impact of competing metabolism and P-gp efflux (Badhan et al., 2009). The resultant PBPK-DDI model would reveal the interactions between transporters and enzymes, and show

how these would affect the rise and fall of the drug and metabolite profiles and those of the inhibitor and its metabolite. However, implementation would require a quantitative definition of the contribution of transporters and enzymes, as well as the flow rate to eliminating organs. It is highly recommended that metabolites be monitored and considered in the PBPK-DDI model.

28.3.2 In Vitro Estimates For In Vivo Extrapolation (IVIVE)

From the previous section (Section 28.3.1), it is recognized that modeling and simulation is the way to gain an improved understanding of DDI. The important task is how to implement this. Better success needs to be achieved with good in vitro estimates for in vivo extrapolation (IVIVE). Much more efforts have been devoted to IVIVE to address DDI potential. But many difficulties remain (see Chapters 7–10). First, recent investigations revealed that protein–protein interactions may further complicate IVIVE. Several examples were found. For the cytochromes in a purified, reconstituted enzyme system, CYP2D6 was found to decrease the substrate-binding affinity and rates of catalysis of CYP2C9 and inhibited the CYP2C9-mediated *S*-flurbiprofen metabolism in a protein concentration-dependent manner (Subramanian et al., 2009). UGT2B7 was reported to interact with UGT1A enzymes, affecting their kinetics in human liver microsomes and underscoring the complexities in glucuronidations in human liver (Fujiwara et al., 2009). Another example existed between CYP3A4 and UGT2B7, enzymes in close proximity to each other; the Leu331-to-Lys342 domain or the surrounding region of CYP3A4 played a role in the interaction with UGT2B7 and glucuronidation (Takeda et al., 2009). Second, it is difficult, and maybe impossible, to obtain the in vivo K_i value from DDI clinical studies. What is widely appreciated is that in vitro K_i values in PBPK model-based analyses do not necessarily reflect the result observed in vivo. In almost all cases, it is predicted that when the inhibitory potency is ≤ 1 μM, an in vivo drug–drug interaction would be observed; however, if the inhibitory potency is ≥ 10 μM, there is still the possibility that the drug would cause an interaction. Even with available programs such as Simcyp®, the program currently uses the method of predicting the in vivo K_i value based on correlations of clogP (calculated logP) with the ratio of the in vivo to in vitro K_i values. Improved accuracy in the estimation of in vivo K_i values must be achieved in the future. Simulation programs (e.g., Simcyp® or Simulation Plus®) must aim toward use of in vitro data for prediction purposes by allowing the incorporation of inter-individual variability and the easy exploration of various model options.

28.3.3 Software for Modeling

There are some attempts for Simcyp® to develop improved PBPK-DDI models that can account for the segregated flow effect, by employing only 40–59%

of intestinal flow for the assessment of intestinal clearance (Yang et al., 2007). Similar attempts have been made for Simulation Plus® (personal communication, Dr. Michael Bolger). It would be of interest to view how the above concepts on segregated flow of the intestine (Cong et al., 2000) and segmental distribution of transporters and enzymes (Tam et al., 2003), when incorporated into modeling and software, improve the level of prediction of drug absorption and bioavailability in clinical DDI studies. It was further stressed that the unbound fraction obtained in vitro in the incubation mixture, as well as the unbound and not total concentration of the inhibitor should be inputted for proper estimation of the intrinsic clearance of drug (see Chapter 7).

Present softwares that examine PBPK modeling or PBPK-DDI have yet to include transporters. It is hoped that some of the recently developed theoretical works on organ clearance concepts (Pang et al., 2008; Sun and Pang, 2009b) would fuel further development in this area. Such progress is sorely needed, because current models fail to fully rationalize long-standing and well-documented perpetrator–victim pairs such as gemfibrozil–repaglinide (involving CYP2C8 and/or organic anion transporting peptide 1B1 inhibition), 17α-ethinyl estradiol–selegiline (involving CYP2B6 inhibition), 17α-ethinyl estradiol–melatonin (involving CYP1A2 inhibition), and capecitabine–warfarin (involving CYP2C9 inhibition) (Chang et al., 2009; Hinton et al., 2008; Yildirim et al., 2006; Janney and Waterbury, 2005).

28.4 Difficulties Remaining

The inclusion of transporter activities (as influx or efflux intrinsic clearances) for the liver and kidney, and the combination of more physiologically relevant intestinal models would improve the IVIVE. Remaining challenges of accounting for variability in patient populations due to age, ethnicity, and genetic makeup on IVIVE need attention in simulation packages. The slowest process (or rate-limiting step) that affords the greatest change (or sensitivity) should be emphasized in the DDI reaction. The complex metabolic drug interactions such as inductive, mechanism-based, and allosteric DDIs would add more complexity and difficulty for prediction in vivo. Consideration should be given to the interactions between a drug ligand and nuclear receptors or cytokines that result in changes in transporters and enzymes (Le Vee et al., 2009; Fardel and Le Vee, 2009). These may be overcome by modeling and simulations, and development of softwares may be advanced to predict these kinds of occurrences and understand rate-determining steps. Metabolites as culprits should be seriously considered, and metabolite information is paramount (see Chapter 5). It is envisaged that proper validation of the simulation model with in vivo data would greatly strengthen the methodology and allow future classifications to be made.

28.4.1 Multiple Interactions

Unlike the DDIs that involve a single mechanism for pair of drugs, the effect of multiple drug interactions is difficult to design and evaluate properly in vivo. The use of modeling and simulation, especially with a PBPK approach, again appears promising to understand these complex drug interactions (see Chapter 26). Future studies need to focus on better defining key parameters required to quantitatively evaluate multiple factors and mechanistic understanding of the combined effects. Besides study design, the utility of modeling and simulation in labeling recommendations for the safe and effective use of a drug needs to be explored

28.4.2 Drug Interactions Involving Biologic Agents

In recent years, the pharmaceutical industry has employed increasing numbers of biologic agents, such as monoclonal antibodies (Mabs), domain antibodies (Dabs), inhibitory ribonucleic acid (RNAi), and protein or peptide-based derivatives (Mascelli et al., 2007; Zhou, 2007). Many of these agents are administered intravenously (or subcutaneously) and circulate at high concentrations, because of "pharmacokinetic enhancement" enabled by pegylation or coupling to proteins such as serum albumin. Unlike their small molecule counterparts, many biologics are characterized by a long plasma half-life of "days" rather than "hours," with clearance and distribution largely determined by the binding to, and turnover rate of, the biological target itself.

Although the manufacturing and safety of biologic agents has received considerable attention, the different aspects of DDIs have yet to be explored. For example, it is only recently that the possibility of DDIs involving therapeutic antibodies was considered and examined (Seitz and Zhou, 2007; Mahmood and Green, 2007). This is important, because many therapeutic antibodies interact with target proteins on the surface of cells and modulate circulating and intracellular cytokines. In turn, such cytokines can impact the expression of drug-metabolizing enzymes and transporters in tissues (Le Vee et al., 2009; Fardel and Le Vee, 2009). How would one go about assessing such DDI potential preclinically? What type of models and screening funnels would have to be in place? In the absence of validated in vivo (animal) and in vitro models, this would prove challenging. It is clear that the industry will have to adapt its current small molecule-oriented paradigms to encompass biologics. Moreover, regulatory agencies will have to more fully address this topic in future guidance documents.

28.4.3 Future Improvement of Prediction Strategies

From an industrial standpoint, a considerable amount of resources has been spent over the last decade building up databases (see Chapters 15 and 16) and developing fully automated (high-throughput) screening platforms. This has reached the point where most mid- to large-sized pharmaceutical companies are able to generate in vitro data for large numbers of compounds employing various (e.g., 384- and 1536-well) assay formats. For example, in vitro P450 inhibition and induction assays have become routine and the data are being used to screen out compounds, drive lead optimization campaigns, enable development of structure–activity relationships (SARs), and prioritize and guide the design of follow-up clinical DDI studies (Chapter 23). Although most groups can generate such data, the major challenge is what to do with it, how to integrate it, "bring it all together," and enable decision making.

Fortunately, the various members of industry organizations such as PhRMA (Pharmaceutical Research Manufacturers of America) have made the first steps in reaching a consensus on best practices for the *conduct of* preclinical and clinical DDI studies (Bjornsson et al., 2003; Chu et al., 2009; Grimm et al., 2009). In most cases, however, the focus has been on the implementation of empirical approaches (e.g., $[I]/K_i$ ratio guidance or the rank order approach) with little detail related to the actual modeling of DDIs. It is hoped that such consensus will pave the way for additional progress in the years to come, because DDI models in the future will need to be much more mechanistic and incorporate data from an ever increasing array of sources. For example, as microsomal-binding and protein-binding assays continue to become standard approaches, in silico methods for the prediction of free fractions and protein–protein interactions will become as common for DDI predictions as they are currently used for unbound clearance predictions. Similarly, as one P450 is inhibited, in silico models will eventually be able to predict the fraction metabolized (f_m) shift to another clearance mechanism, and the predicted impact on exposure in terms of AUC, C_{max}, and associated pharmacokinetic parameters. With greater emphasis being placed on pharmacokinetic–pharmacodynamics (PK-PD), it is also likely that DDI models will have to incorporate such information and enable assessment of likely impact on PD, not just PK. More broadly, the application of a systems biology approach to PK-ADME (absorption-distribution-metabolism-excretion) research will also enable improvements in DDI modeling and predictions (Ekins et al., 2007). Despite the obvious challenges, if the progress of the last two decades is anything to go by, the field of DDI research will continue to mature and develop in ways that cannot be imagined today.

Acknowledgments This work supported in part by CIHR grant (MOP89850 awarded to KSP).

References

Akita H, Suzuki H and Sugiyama Y (2001) Sinusoidal efflux of taurocholate is enhanced in Mrp2-deficient rat liver. *Pharm Res* **18**:1119–1125.

Allen JD, Brinkhuis RF, van Deemter L, Wijnholds J and Schinkel AH (2000) Extensive contribution of the multidrug transporters P-glycoprotein and Mrp1 to basal drug resistance. *Cancer Res* **60**:5761–5766.

Annaert P, Turncliff RZ, Booth CL, Thakker DR and Brouwer KLR (2001) P-Glycoprotein mediated in vitro biliary excretion in sandwich-cultured rat hepatocytes. *Drug Metab Dispos* **29**:1277–1283.

Badhan R, Penny J, Galetin A, and Houston JB (2009) Methodology for development of a physiological model incorporating CYP3A and P-glycoprotein for the prediction of intestinal drug absorption. *J Pharm Sci* **98**:2180–2197.

Barter ZE. Bayliss MK, Beaune PH, Boobis AR, Carlie DJ, Edwards RJ, Houston JB, Lake BG, Lipscomb JC, Pelkonen OR, Tucker GT and Rostami-Hodjegan A (2007) Scaling factors for the extrapolation of in vivo metabolic drug clearance from in vitro data: reaching a consensus on values on human microsomal protein and hepatocellurity per gram of liver. *Curr Drug Metab* **8**:33–45.

Battaglia E and Gollan J (2001) A unique multifunctional transporter translocates estradiol-17β-glucuronide in rat liver microsomal vesicles. *J Biol Chem* **276**:23492–23498.

Bjornsson TD, Callaghan JT, Einolf HJ, Fischer V, Gan L, Grimm S, Kao J, King SP, Miwa G, Ni L, Kumar G, McLeod J, Obach RS, Roberts S, Roe A, Shah A, Snikeris F, Sullivan JT, Tweedie D, Vega JM, Walsh J, and Wrighton SA Pharmaceutical Research and Manufacturers of America (PhRMA) Drug Metabolism/Clinical Pharmacology Technical Working Group; FDA Center for Drug Evaluation and Research (CDER) (2003) The conduct of in vitro and in vivo drug–drug interaction studies: a Pharmaceutical Research and Manufacturers of America (PhRMA) perspective. *Drug Metab Dispos* **31**:815–832.

Brown HS, Ito K, Galetin A and Houston JB (2005) Prediction of in vivo drug–drug interactions from in vitro data: impact of incorporating parallel pathways of drug elimination and inhibitor absorption rate constant. *Br J Clin Pharmacol* **60**:508–518.

Cacciatore I, Caccuri AM, Cocco A, De Maria F, Di Stefano A, Luisi G, Pinnen F, Ricci G, Sozio P and Turella P (2005) Potent isozyme-selective inhibition of human glutathione S-transferase A1-1 by a novel glutathione S-conjugate. *Amino Acids* **29**:255–261.

Chandra P, LeCluyse EL and Brouwer KLR (2001) Optimization of culture conditions for determining hepatobiliary disposition of taurocholate in sandwich-culture rat hepatocytes. In Vitro *Cell Dev Biol – Animal* **37**:380–385.

Chang SY, Chen C, Yang Z and Rodrigues AD (2009) Further assessment of 17α-ethinyl estradiol as an inhibitor of different human cytochrome P450 forms in vitro. *Drug Metab Dispos* **38**:1667–1675.

Chen ZS, Guo Y, Belinsky MG, Kotova E and Kruh GD (2005) Transport of bile acids, sulfated steroids, estradiol 17-beta-D-glucuronide, and leukotriene C4 by human multidrug resistance protein 8 (ABCC11). *Mol Pharmacol* **67**:545–557.

Chu XY, Strauss JR, Mariano MA, Li J, Newton DJ, Cai X, Wang RW, Yabut J, Hartley DP, Evans DC and Evers R (2006) Characterization of mice lacking the multidrug resistance protein MRP2 (ABCC2). *J Pharmacol Exp Ther* **317**:579–589.

Chu V, Einolf HJ, Evers R, Kumar GN, Moore D, Ripp SL, Silva J, Sinha V, Sinz M and Skerjanec A (2009) In vitro and in vivo induction of cytochrome P450: a survey of the current practices and recommendations: a Pharmaceutical Research and Manufacturers of America (PhRMA) perspective. *Drug Metab Dispos* **37**:1339–1354.

Cong D, Doherty M and Pang KS (2000) A new physiologically-based segregated flow model to explain route-intestinal metabolism. *Drug Metab Dispos* **28**:224–235.

Coughtrie M, Riches Z, Bloomer J, Patel A and Nolan A (2009) Assessment of cryopreserved human hepatocytes as a model system to investigate sulfation and glucuronidation and to evaluate inhibitors of drug conjugation. *Xenobiotica* **11**:1–8.

Csala M, Staines AG, Bánhegyi G, Mandl J, Coughtrie MW and Burchell B (2004) Evidence for multiple glucuronide transporters in rat liver microsomes. *Biochem Pharmacol* **68**:1353–1362.

Dallas S, Schlichter L and Bendayan R (2004) Multidrug resistance protein (MRP) 4- and MRP 5-mediated efflux of 9-(2-phosphonylmethoxyethyl)adenine by microglia. *J Pharmacol Exp Ther* **309**:1221–1229.

Ekins S, Ecker GF, Chiba P and Swaan PW (2007) Future directions for drug transporter modeling. *Xenobiotica* **37**:1152–1170.

Fardel O and Le Vee M (2009) Regulation of human hepatic drug transporter expression by pro-inflammatory cytokines. *Expert Opin Drug Metab Toxicol* September 28 [epub]

Felmlee MA, Lon HK, Gonzalez FJ and Yu AM (2008) Cytochrome P450 expression and regulation in CYP3A4/CYP2D6 double transgenic humanized mice. *Drug Metab Dispos* **36**:435–441.

Fujiwara R, Nakajima M, Oda S, Yamanaka H, Ikushiro SI, Sakaki T and Yokoi T (2009) Interactions between human UDP-glucuronosyltransferase (UGT) 2B7 and UGT1A enzymes. *J Pharm Sci* May 27. [Epub ahead of print].

Gorski JC, Jones DR, Haehner-Daniels BD, Hamman MA, O'Mara EM Jr and Hall SD (1998) The contribution of intestinal and hepatic CYP3A to the interaction between midazolam and clarithromycin. *Clin Pharmacol Ther* **64**:133–143.

Grime K, Riley RJ (2006) The impact of in vitro binding on in vitro-in vivo extrapolations, projection of metabolic clearance and clinical drug–drug interactions. *Curr Drug Metab* **7**:251–264.

Grimm SW, Einolf HJ, Hall SD, He K, Lim HK, Ling KH, Lu C, Nomeir AA, Seibert E, Skordos KW, Tonn GR, Van Horn R, Wang RW, Wong YN, Yang TJ and Obach RS The conduct of in vitro studies to address time-dependent inhibition of drug metabolizing enzymes: a perspective of the Pharmaceutical Research and Manufacturers of America (PhRMA). *Drug Metab Dispos* **37**:1355–1370.

Groen AK, Bloks VW, Bandsma RH, Ottenhoff R, Chimini G and Kuipers F (2001) Hepatobiliary cholesterol transport is not impaired in Abca1-null mice lacking HDL. *J Clin Invest* **108**: 843–850.

Gyamfi MA, Ohtani II, Shinno E and Aniya Y (2004) Inhibition of glutathione S-transferases by thonningianin A, isolated from the African medicinal herb, Thonningia sanguinea, in vitro. *Food Chem Toxicol* **42**:1401–1408.

Harris RM and Waring RH (2008) Sulfotransferase inhibition: potential impact of diet and environmental chemicals on steroid metabolism and drug detoxification. *Curr Drug Metab* **9**:269–275.

Hayeshi R, Chinyanga F, Chengedza S and Mukanganyama S (2006) Inhibition of human glutathione transferases by multidrug resistance chemomodulators in vitro. *J Enzyme Inhib Med Chem* **21**:581–587.

Hinton LK, Galetin A and Houston JB (2008) Multiple inhibition mechanisms and prediction of drug–drug interactions: status of metabolism and transporter models as exemplified by gemfibrozil-drug interactions. *Pharm Res* **25**:1063–1074.

Ho RH, Tirona RG, Leake BF, Glaeser H, Lee W, Lemke CJ, Wang Y and Kim RB (2006) Drug and bile acid transporters in rosuvastatin hepatic uptake: function, expression, and pharmacogenetics. *Gastroenterology* **130**:1793–1806.

Hoffmaster KA, Turncliff RZ, LeCluyse EL, Kim RB, Meier PJ and Brouwer KLR (2004) P-glycoprotein expression, localization, and function in sandwich-cultured primary rat and human hepatocytes: relevance to the hepatobiliary disposition of a opiod peptide. *Pharm Res* **21**: 1294–1302.

Honda Y, Ushigome F, Koyabu N, Morimoto S, Shoyama Y, Uchiumi T, Kuwano M, Ohtani H and Sawada Y (2004) Effects of grapefruit juice and orange juice components on P-glycoprotein- and MRP2-mediated drug efflux. *Br J Pharmcol* **143**:856–864.

Howgate EM, Rowland Yeo K, Proctor NJ, Tucker GT and Rostami-Hodjegan A (2006) Prediction of in vivo drug clearance from in vitro data. I: Impact of inter-individual variability. *Xenobiotica* **36**:473–497.

Inoue T, Sugihara K, Ohshita H, Horie T, Kitamura S and Ohta S (2009) Prediction of human disposition toward S-^{3}H-warfarin using chimeric mice with humanized liver. *Drug Metab Pharmacokinet* **24**:153–160.

Ito K and Houston JB (2005) Prediction of human drug clearance from in vitro and preclinical data using physiologically based and empirical approaches. *Pharm Res* **22**:103–112.

Ito K, Ogihara K, Kanamitsu S and Itoh T (2003) Prediction of the in vivo interaction between midazolam and macrolides based on in vitro studies using human liver microsomes. *Drug Metab Dispos* **31**:945–954.

Ito K, Hallifax D, Obach RS and Houston JB (2005) Impact of parallel pathways of drug elimination and multiple cytochrome p450 involvement on drug–drug interactions: CYP2D6 paradigm. *Drug Metab Dispos* **33**:837–844.

Jäger W, Sartori M, Herzog W and Thalhammer T (1998) Genistein metabolism in liver microsomes of Wistar and mutant TR(-)-rats. *Res Commun Mol Pathol Pharmacol* **100**: 105–116.

Janney LM and Waterbury NV (2005) Capecitabine-warfarin interaction. *Ann Pharmacother* **39**:1546–1551.

Jedlitschky G, Leier I, Buchholz U, Barnouin K, Kurz G and Keppler D (1996) Transport of glutathione, glucuronate, and sulfate conjugates by the MRP gene-encoded conjugate export pump. *Cancer Res* **56**:988–994.

Jigorel E, Le Vee M, Boursier-Neyret C, Bertrand M and Fardel O (2005) Functional expression of sinusoidal drug transporters in primary human and rat hepatocytes. *Drug Metab Dispos* **33**:1418–1422.

Jones HM, Parrott N, Jorga K and Lavé T (2006) A novel strategy for physiologicallly based predictions of human pharmacokinetics. *Clin Pharmacokinet* **45**:511–542.

Kamiie J, Ohtsuki S, Iwase R, Ohmine K, Katsukura Y, Yanai K, Sekine Y, Uchida Y, Ito S and Terasaki T (2008) Quantitative atlas of membrane transporter proteins: development and application of a highly sensitive simultaneous LC/MS/MS method combined with novel in-silico peptide selection criteria. *Pharm Res* **25**:1469–1483.

Kanamitsu SI, Ito K, Okuda H, Ogura K, Watabe T, Muro K and Sugiyama Y (2000) Prediction of in vivo drug–drug interactions based on mechanism-based inhibition from in vitro data: inhibition of 5-fluorouracil metabolism by (E)-5-(2-bromovinyl)uracil. *Drug Metab Dispos* **28**:467–474.

Kim S, Pray D, Zheng M, Morgan DG, Pizzano JG, Zoeckler ME, Chimalakonda A and Sinz MW (2008) Quantitative relationship between rifampicin exposure and induction of Cyp3a11 in SXR humanized mice: extrapolation to human CYP3A4 induction potential. *Drug Metab Lett* **2**:169–175.

Kimura S, Kawabe M, Ward JM, Morishima H, Kadlubar FF, Hammons GJ, Fernandez-Salguero P and Gonzalez FJ (1999) CYP1A2 is not the primary enzyme responsible for 4-aminobiphenyl-induced hepatocarcinogenesis in mice. *Carcinogenesis* **20**:1825–1830.

Khetani SR and Bhatia SN (2008) Microscale culture of human liver cells for drug development. *Nat Biotechnol* **26**:12012–12016.

Kopplow K, Letschert K, König J, Walter B and Keppler D (2005) Human hepatobiliary transport of organic anions analyzed by quadruple-transfected cells. *Mol Pharmacol* **68**:1031–1038.

Kostrubsky SE, Sinclair JF, Strom SC, Wood S, Urda E, Stolz DB, Wen YH, Kulkarni S and Mutlib A (2005) Phenobarbital and phenytoin increased acetaminophen hepatotoxicity due to inhibition of UDP-glucuronosyltransferases in cultured human hepatocytes. *Toxicol Sci* **87**:146–155.

Lahvis GP, Lindell SL, Thomas RS, McCuskey RS, Murphy C, Glover E, Bentz M, Southard J and Bradfield CA (2000) Portosystemic shunting and persistent fetal vascular structures in aryl hydrocarbon receptor-deficient mice. *Proc Natl Acad Sci USA* **97**:10442–10447.

Le Vee M, Lecureur B, Stieger B and Fardel O (2009) Regulation of transporter expression in human hepatocytes exposed to the proinflammatory cytokines tumor necresis factor-α or interleukin-6. *Drug Metab Dispos* **37**:685–593.

Leier I, Hummel-Eisenbeiss J, Cui Y and Keppler D (2000) ATP-dependent para-aminohippurate transport by apical multidrug resistance protein MRP2. *Kidney Int* **57**:1636–1642.

Lin JH, Sugiyama Y, Awazu S and Hanano M (1982) In vitro and in vivo evaluation of the tissue-to-blood partition coefficient for physiological pharmacokinetic models. *J Pharmacokinet Biopharm* **10**:637–647.

Li N, Zhang Y, Hua F and Lai Y (2009) Absolute difference of hepatobiliary transporter multidrug resistance-associated protein (MRP2/Mrp2) in liver tissues and isolated hepatocytes from rat, dog, monkey, and human. *Drug Metab Dispos* **37**:66–73.

Liu Y, Apak TI, Lehmler HJ, Robertson LW and Duffel MW (2006) Hydroxylated polychlorinated biphenyls are substrates and inhibitors of human hydroxysteroid sulfotransferase SULT2A1. *Chem Res Toxicol* **19**:1420–1425.

Lu Y, Bratton S, Heydel JM and Radominska-Pandya A (2008) Effect of retinoids on UDP-glucuronosyltransferase 2B7 mRNA expression in Caco-2 cells. *Drug Metab Pharmacookinet* **23**:364–372.

Lorico A, Nesland J, Emilsen E, Fodstad O and Rappa G (2002) Role of the multidrug resistance protein 1 gene in the carcinogenicity of aflatoxin B1: investigations using mrp1-null mice. *Toxicology* **171**:201–205.

Lorico A, Rappa G, Finch RA, Yang D, Flavell RA and Sartorelli AC (1997) Disruption of the murine MRP (multidrug resistance protein) gene leads to increased sensitivity to etoposide (VP-16) and increased levels of glutathione. *Cancer Res* **57**: 5238–5242.

Luttringer O, Theil F-P, Poulin P, Schmitt-Hoffmann AH, Guentert TW, Lavé T (2003) Physiologically based pharmacokinetic (PBPK) modelling of disposition of epiroprim in humans. *J Pharm Sci* **92**:1990–2007.

Mahmood I and Green MD (2007) Drug interaction studies of therapeutic proteins or monoclonal antibodies. *J Clin Pharmacol* **47**:1540–1554.

Mascelli MA, Zhou H, Sweet R, Getsy J, Davis HM, Graham M and Abernethy D (2007) Molecular, biologic, and pharmacokinetic properties of monoclonal antibodies: impact of these parameters on early clinical development. *J Clin Pharmacol* **47**:553–565.

Matsson P, Pedersen JM, Norinder U, Bergström CA and Artursson P (2009) Identification of novel specific and general inhibitors of the three major human ATP-binding cassette transporters P-gp, BCRP and MRP2 among registered drugs. *Pharm Res* **26**:1816–1831.

Maliepaard M, van Gastelen MA, Tohgo A, Hausheer FH, van Waardenburg RC, de Jong LA, Pluim D, Beijnen JH and Schellens JH (2001) Circumvention of breast cancer resistance protein (BCRP)-mediated resistance to camptothecins in vitro using non-substrate drugs or the BCRP inhibitor GF120918. *Clin Cancer Res* **7**:935–941.

McConn DJ 2nd, Lin YS, Allen K, Kunze KL and Thummel KE (2004) Differences in the inhibition of cytochromes P450 3A4 and 3A5 by metabolite-inhibitor complex-forming drugs. *Drug Metab Dispos* **32**:1083–1091.

Morris ME and Pang KS (1987) Competition between two enzymes for substrate removal in liver: modulating effects due to substrate recruitment of hepatocyte activity. *J Pharmacokinet Biopharm* **15**:473–496.

Morrow CS, Peklak-Scott C, Bishwokarma B, Kute TE, Smitherman PK and Townsend AJ (2006) Multidrug resistance protein 1 (MRP1, ABCC1) mediates resistance to mitoxantrone via glutathione-dependent drug efflux. *Mol Pharmacol* **69**:1499–1505.

Nahmias Y, Casali M, Barbe L, Berthiaume F and Yarmush ML (2006) Liver endothelial cells promote LDL-R expression and the uptake of HCV-like particles in primary rat and human hepatocytes. *Hepatology* **43**:257–265.

Nestorov IA, Aarons LJ, Arundel PA and Rowland M (1998) Lumping of whole-body physiologically-based pharmacokinetic models. *J Pharmcokinet Biopharm* **26**: 21–46.

Nezasa K, Tian X, Zamek-Gliszczynski MJ, Patel NJ, Raub TJ and Brouwer KL (2006) Drug Altered hepatobiliary disposition of 5 (and 6)-carboxy-2′,7′-dichlorofluorescein in Abcg2 (Bcrp1) and Abcc2 (Mrp2) knockout mice. *Drug Metab Dispos* **34**:718–723.

Obach RS, Walsky RL, Venkatakrishnan K, Gaman EA, Houston JB and Tremaine LM (2006) The utility of in vitro cytochrome p450 inhibition data in the prediction of drug–drug interactions. *J Pharmacol Exp Ther* **316**:336–348.

Ohno M, Motojima K, Okano T and Taniguchi A (2008) Up-regulation of drug-metabolizing enzyme genes in layered co-culture of a human liver cell line and endothelial cells. *Tissue Eng Part A* **14**:1861–1869.

Okumura H, Katoh M, Sawada T, Nakajima M, Soeno Y, Yabuuchi H, Ikeda T, Tateno C, Yoshizato K and Yokoi T (2007) Humanization of excretory pathway in chimeric mice with humanized liver. *Toxicol Sci* **97**:533–538.

Pacifici GM (2004) Inhibition of human liver and duodenum sulfotransferases by drugs and dietary chemicals: a review of the literature. *Int J Clin Pharmacol Ther* **42**:488–495.

Pang KS (2003) Modeling of intestinal drug absorption: roles of transporters and metabolic enzymes (For the Gillette Review Series). *Drug Metab Dispos* **31**:1507–1519.

Pang KS, Morris ME and Sun H (2008) Formed and preformed metabolite: facts and comparisons. *J Pharm Pharmacol* **60**:1247–1275.

Parrott N, Parquerau N, Coassolo P and Lavé T (2006) An evaluation of the utility of physiologically-based models of pharmacokinetics in early drug discovery. *J Pharm Sci* **94**:2327–2343.

Pinto AG, Wang Y-H, Chalasani N, Skaar T, Kolwankdar D, Gorski JC, Lngpunsakul S, Hamman MA, Arefayene M and Hall SD (2005) Inhibition of human intestinal wall metabolism by macrolide antibiotics: effect of clarithromycin on cytochrome P450 3A4/5 activity and expression. *ClinPharmacol Ther* **77**:178–188.

Polli JW, Wring SA, Humphreys JE, Huang L, Morgan JB, Webster LO and Serabjit-Singh CS (2001) Rational use of in vitro P-glycoprotein assays in drug discovery. *J Pharmacol Exp Ther* **299**:620–628.

Poulin P and Theil F-P (2000) A priori prediction of tissue:plasma partition coefficients of drugs to facilitate the use of physiologically-based pharmacokinetic models in drug discovery. *J Pharm Sci* **89**:16–35.

Quinney SK, Galinsky RE, Jiyamapa-Serna VA, Chen Y, Hamman MA, Hall SD and Kimura RE (2008a) Hydroxyitraconazole, formed during intestinal first-pass metabolism of itraconazole, controls the time course of hepatic CYP3A inhibition and the bioavailability of itraconazole in rats. *Drug Metab Dispos* **36**:1097–1101.

Quinney SK, Haehner BD, Rhoades MB, Lin Z, Gorski JC and Hall SD (2008b) Interaction between midazolam and clarithromycin in the elderly. *Br J Clin Pharmacol* **65**:98–109.

Rappa G, Finch RA, Sartorelli AC and Lorico A (1999) New insights into the biology and pharmacology of the multidrug resistance protein (MRP) from gene knockout models. *Biochem Pharmacol* **58**:557–562.

Reid G Wielinga P, Zelcer N, De Haas M, Van Deemter L, Wijnholds J, Balzarini J and Borst P (2003) Characterization of the transport of nucleoside analog drugs by the human multidrug resistance proteins MRP4 and MRP5. *Mol Phamracol* **63**:1094–1103.

Riches Z, Bloomer JC and Coughtrie MW (2007) Comparison of 2-aminophenol and 4-nitrophenol as in vitro probe substrates for the major human hepatic sulfotransferase, SULT1A1, demonstrates improved selectivity with 2-aminophenol. *Biochem Pharmacol* **74**:352–358.

Rostami-Hodjegan A and Tucker GT (2007) Simulation and prediction of in vivo drug metabolism in human populations from in vitro data. *Nature Rev* **6**:141–148.

Rowland M and Matin SB (1973) Kinetics of drug–drug interactions. *J Pharmacokinet Biopharm* **1**:553–567.

Sakai-Kato K, Kato M and Toyóoka T (2004) Screening of inhibitors of uridine diphosphate glucuronosyltransferase with a miniaturized on-line drug-metabolism system. *J Chromatgor A* **1051**:261–266.

Samokyszyn VM, Gall WE, Zawada G, Freyaldenhoven MA, Chen G, Mackenzie PI, Tephly TR and Radominska-Pandya A (2000) 4-hydroxyretinoic acid, a novel substrate for human liver microsomal UDP-glucuronosyltransferase(s) and recombinant UGT2B7. *J Biol Chem* **275**:6908–6914.

Saruwatari A, Okamura S, Nakajima Y, Narukawa Y, Takeda T and Tamura H (2008) Pomegranate juice inhibits sulfoconjugation in Caco-2 human colon carcinoma cells. *J Med Food* **11**:623–628.

Schmidt JV, Su GH, Reddy JK, Simon MC and Bradfield CA (1996) Characterization of a murine Ahr null allele: involvement of the Ah receptor in hepatic growth and development. *Proc Natl Acad Sci USA* **93**:6731–6736.

Schwab AJ, Barker III F, Goresky CA and Pang KS (1990) Transfer of enalaprilat across rat liver cell membranes is barrier-limited. *Am J Physiol* **258**:G461–G475.

Schwab AJ, de Lannoy IAM, Poon K, Goresky CA and Pang KS (1992) Enalaprilat handling by the rat kidney: barrier-limited cell entry. *Am J Physiol* **263**:F858–F869.

Schwartz J, Hunt T, Smith WB, Wong P, Larson P, Crumley T, Mehta A, Gottesdiener K and Agrawal N (2009) The Effect of etoricoxib on the pharmacokinetics of oral contraceptives in healthy participants. *J Clin Pharmacol* **49**:807–815.
Seitz K and Zhou H (2007) Pharmacokinetic drug–drug interaction potentials for therapeutic monoclonal antibodies: reality check. *J Clin Pharmacol* **47**:1104–1118.
Shitara Y, Hirano M, Sata H and Sugiyama Y (2004) Gemfibrozil and its glucuronide inhibit the organic anion transporting polypeptide 2 (OATP2/OATP1B1:SLC21A6)-mediated hepatic uptake and CYP2C8-mediated metabolism of cerivastatin: analysis of the mechanism of the clinically relevant drug–drug interaction between cerivastatin and gemfibrozil. *J Pharmacol Exp Ther* **311**:228–236.
Shitara Y, Itoh T, Sato H, Li AP and Sugiyama Y (2003) Inhibition of transporter-mediated hepatic uptake as a mechanism for drug–drug interaction between cerivastatin and cyclosporin A. *J Pharmacol Exp Ther* **304**:610–616.
Sirianni G and Pang KS (1997) Organ clearance concepts: new perspectives on old principles. *J Pharmacokinet Biopharm* **25**:449–470.
Stark K and Guengerich FP (2007) Characterization of orphan human cytochromes P450. *Drug Metab Rev* **39**:627–637.
Stark K, Dostalek M and Guengerich FP (2008a) Expression and purification of orphan cytochrome P450 4X1 and oxidation of anandamide. *FEBS J* **275**:3706–3717.
Stark K, Wu ZL, Bartleson CJ and Guengerich FP (2008b) mRNA distribution and heterologous expression of orphan cytochrome P450 20A1. *Drug Metab Dispos* **36**:1930–1937.
Subramanian M, Low M, Locuson CW and Tracy TS (2009) CYP2D6-CYP2C9 Protein-protein interactions and isoform-selective effects on substrate binding and catalysis. *Drug Metab Dispos* **37**:1682–1689.
Sun H and Pang KS (2009a) Disparity in intestine disposition between formed and preformed metabolites and implications: a theoretical study. *Drug Metab Dispos* **37**:187–202.
Sun H and Pang KS (2009b) Physiological modeling to understand the impact of enzymes and transporters on drug and metabolite data and bioavailability estimates. Submitted.
Taipalensuu J, Tavelin S, Lazorova L, Svensson AC and Artursson P (2004) Exploring the quantitative relationship between the level of MDR1 transcript, protein and function using digoxin as a marker of MDR1-dependent drug efflux activity. *Eur J Pharm Sci* **21**:69–75.
Takeda S, Ishii Y, Iwanaga M, Nurrochmad A, Ito Y, Mackenzie PI, Nagata K, Yamazoe Y, Oguri K and Yamada H (2009) Interaction of cytochrome P450 3A4 and UDP-glucuronosyl-transferase 2B7: evidence for protein-protein association and possible involvement of CYP3A4 J-helix in the interaction. *Mol Pharmacol* **75**:956–964.
Tam D, Tirona RG and Pang KS (2003) Segmental intestinal transporters and metabolic enzymes on intestinal drug absorption. *Drug Metab Dispos* **33**:373–383.
Tamura H and Matsui M (2000) Inhibitory effects of green tea and grape juice on the phenol sulfotransferase (SULT1A1) activity of human colon carcinoma cell line, Caco-2. *Biol Pharm Bull* **23**:695–699.
Tang F, Horie K and Borchardt RT (2002) Are MDCK cells transfected with the human MRP2 gene a good model of the human intestinal mucosa? *Pharm Res* **19**:773–779.
Tukey RH and Strassburg CP (2000) Human UDP-glucuronosyltransferases: metabolism, expression, and disease. *Annu Rev Pharmacol Toxicol* **40**:581–616.
Turncliff RZ, Meier PJ, Brouwer KLR (2004) Effect of dexamethasone treatment on the expression and function of transport proteins in sandwich-culture rat hepatocytes. *Drug Metab Dispos* **32**:834–839.
Uno S, Endo K, Ishida Y, Tateno C, Makishima M, Yoshizato K and Nebert DW (2009) CYP1A1 and CYP1A2 expression: comparing 'humanized' mouse lines and wild-type mice; comparing human and mouse hepatoma-derived cell lines. *Toxicol Appl Pharmacol* **237**:119–126.
Volak LP, Ghirmai S, Cashman JR and Court MH (2008) Curcuminoids inhibit multiple human cytochromes P450, UDP-glucuronosyltransferase, and sulfotransferase enzymes, whereas piperine is a relatively selective CYP3A4 inhibitor. *Drug Metab Dispos* **36**:1594–1605.

Volk EL and Schneider E (2003) Wild-type breast cancer resistance protein (BCRP/ABCG2) is a methotrexate polyglutamate transporter. *Cancer Res* **63**:5538–5543.

Wang R, Salem M, Yousef IM, Tuchweber B, Lam P, Childs SJ, Helgason CD, Ackerley C, Phillips MJ and Ling V (2001) Targeted inactivation of sister of P-glycoprotein gene (spgp) in mice results in nonprogressive but persistent intrahepatic cholestasis. *Proc Natl Acad Sci USA* **98**:2011–2016.

Wang YH, Jones DR and Hall SD (2004) Prediction of cytochrome P450 3A inhibition by verapamil enantiomers and their metabolites. *Drug MetabDispos* **32**:259–266.

Wijnholds J, Evers R, van Leusden MR, Mol CA, Zaman GJ, Mayer U, Beijnen JH, van der Valk M, Krimpenfort P and Borst P (1997) Increased sensitivity to anticancer drugs and decreased inflammatory response in mice lacking the multidrug resistance-associated protein. *Nature Med* **3**:1275–1279.

Yang J, Jamei M, Yeo KR, Tucker GT and Rostami-Hodjegan A (2007) Prediction of intestinal first-pass drug metabolism. *Curr Drug Metab* **8**:676–684.

Yang J, Jamei M, Yeo KR, de la Torre R, Farré M, Tucker GT and Rostami-Hodjegan A (2006) Implications of mechanism-based inhibition of CYP2D6 for the pharmacokinetics and toxicity of MDMA. *J Psychopharmacol* **20**:842–849.

Yildirim Y, Ozyilkan O, Akcali Z and Basturk B (2006) Drug interaction between capecitabine and warfarin: a case report and review of the literature. *Int J Clin Pharmacol Ther* **44**:80–82.

Zamek-Gliszczynski M, Xiong H, Patel NJ, Turncliff RZ, Pollack GM and Brouwer KLR (2003) Pharmacokinetics of (5 and 6)-carobyx-2'-7'-dichlorofluorescein and its diacetate promoiety in the liver. *J Pharmacol Ep Ther* **304**:801–809.

Zelcer N, Saeki T, Bot I, Kuil A and Borst P (2003) Transport of bile acids in multidrug-resistance-protein 3-overexpressing cells co-transfected with the ileal Na^+-dependent bile-acid transporter. *Biochem J* **369**(Pt 1):23–30.

Zhang D, Chando TJ, Everett DW, Patten CJ, Dehal SS and Humphreys WG (2005a) In vitro inhibition of UDP glucuronosyltransferases by atazanavir and other HIV protease inhibitors and the relationship of this property to in vivo bilirubin glucuronidation. *Drug Metab Dispos* **33**:1729–1739.

Zhang P, Tian X, Chandra P and Brouwer KLR (2005b) Role of glycosylation in trafficking of Mrp2 in sandwich culture hepatocytes. *Mol Pharmacol* **67**:1334–1341.

Zhang H, Davis CD, Sinz MW and Rodrigues AD (2007) Cytochrome P450 reaction-phenotyping: an industrial perspective. *Expert Opin Drug Metab Toxicol* **3**:667–687.

Zhang X, Jones DR and Hall SD (2009) Prediction of the effect of erythromycin, diltiazem, and their metabolites, alone and in combination, on CYP3A4 inhibition. *Drug Metab Dispos* **37**:150–160.

Zhou H (2007) Biologics in the pipeline: large molecules with high hopes or bigger risks? *J Clin Pharmacol* **47**:550–552.

Subject Index

K.S. Pang et al. (eds.), *Enzyme- and Transporter-Based Drug–Drug Interactions*,
DOI 10.1007/978-1-4419-0840-7,

B

C

E

F

K

L

N

Z

MIX
Papier aus verantwortungsvollen Quellen
Paper from responsible sources
FSC® C105338

If you have any concerns about our products,
you can contact us on
ProductSafety@springernature.com

In case Publisher is established outside the EU,
the EU authorized representative is:
Springer Nature Customer Service Center GmbH
Europaplatz 3, 69115 Heidelberg, Germany

Printed by Libri Plureos GmbH
in Hamburg, Germany